Progress in Mathematics
Volume 204

Shrawan Kumar

Kac–Moody Groups, their Flag Varieties and Representation Theory

Birkhäuser
Boston • Basel • Berlin

Shrawan Kumar
Department of Mathematics
University of North Carolina, Chapel Hill
Chapel Hill, NC 27599-3250
U.S.A.

Library of Congress Cataloging-in-Publication Data

Kumar, S. (Shrawan), 1953-
Kac–Moody groups, their flag varieties and representation theory / Shrawan Kumar.
p. cm. – (Progress in mathematics ; v. 204)
Includes bibliographical references and indexes.
ISBN 0-8176-4227-7 (acid-free paper) – ISBN 3-7643-4227-7 (acid-free paper)
1. Kac–Moody algebras. 2. Representation of groups. 3. Flag varieties. I. Title. II. Progress in mathematics (Boston, Mass) ; v. 204.

QA252.3.K86 2002
512'.55–dc21 2002018302
CIP

AMS Subject Classifications: Primary–22E46, 22E65, 14M15; Secondary–13D45, 14B05, 14C22, 14F05, 14F17, 14L30, 14M15, 14M17, 14N15, 19L47, 20E42, 20F55, 20G05, 20H15, 16E05, 17B10, 17B20, 17B35, 17B55, 17B56, 17B65, 17B67, 18G15, 18G40, 22E46, 22E65, 22E67, 35Q53, 55R20, 55R25, 55N10, 55N30, 55N33, 55N91, 55P35, 55P62, 55Q15, 55R20, 55R25, 57T10

Printed on acid-free paper

Birkhäuser ®

ISBN 0-8176-4227-7 SPIN 10798061
ISBN 3-7643-4227-7

Reformatted from author's files by TEXniques, Inc., Cambridge, MA.
Printed and bound by Hamilton Printing Company, Rensselaer, NY.
Printed in the United States of America.

9 8 7 6 5 4 3 2 1

Birkhäuser Boston • Basel • Berlin
A member of BertelsmannSpringer Science+Business Media GmbH

Dedicated to my parents

Contents

Preface

Kac–Moody Lie algebras $\mathfrak{g}$ were introduced in the mid-1960s independently by V. Kac and R. Moody, generalizing the finite-dimensional semisimple Lie algebras which we refer to as the finite case. The theory has undergone tremendous developments in various directions and connections with diverse areas abound, including mathematical physics, so much so that this theory has become a standard tool in mathematics. A detailed treatment of the Lie algebra aspect of the theory can be found in V. Kac's book [Kac–90]. This self-contained work treats the algebro-geometric and the topological aspects of Kac–Moody theory from scratch.

The emphasis is on the study of the Kac–Moody groups $\mathcal{G}$ and their flag varieties $\mathcal{X}^Y$, including their detailed construction, and their applications to the representation theory of $\mathfrak{g}$. In the finite case, $\mathcal{G}$ is nothing but a semisimple simply-connected algebraic group and $\mathcal{X}^Y$ is the flag variety $\mathcal{G}/\mathcal{P}_Y$ for a parabolic subgroup $\mathcal{P}_Y \subset \mathcal{G}$.

The main topics covered are the Weyl–Kac character formula; a result of Garland–Lepowsky on $\mathfrak{n}$-homology generalizing the celebrated result of Kostant; an introduction to the ind-varieties, pro-groups, pro-Lie algebras and Tits systems; a detailed construction of Kac–Moody groups and their flag varieties; the Demazure character formula; study of geometry of the Schubert varieties, including their normality and Cohen–Macaulay properties; the Borel–Weil–Bott theorem; the Bernstein–Gelfand–Gelfand (for short, BGG) resolution; the Kempf resolution; conjugacy theorem for the Cartan subalgebras and invariance of the generalized Cartan matrix; determination of the defining ideal of the flag varieties via the Plücker relations; the nil-Hecke ring; study of the T-equivariant and the singular cohomology of the flag varieties $\mathcal{X}^Y$ via the nil-Hecke ring, including some positivity result for the cup product; degeneracy of the Leray–Serre spectral sequence for the fibration $\mathcal{G} \to \mathcal{G}/T$ at E_3, and various criteria for the smoothness and rational smoothness of points on the Schubert varieties. Most of these topics are brought together here for the first time in book form.

We conclude with a chapter devoted to an explicit realization of the affine Kac–Moody algebras and the corresponding groups (as well as their flag varieties). These form the most important subclass of the Kac–Moody algebras and the groups beyond the finite case. For a detailed treatment of the loop groups we refer to the book [Pressley-Segal–86], although their approach is less algebro-geometric than ours.

Even though the book is devoted to the general Kac–Moody case, those who are primarily interested in the finite case will benefit as well. We particularly mention the following topics: the result on $\mathfrak{n}$-homology due to Kostant; the Demazure character formula and various geometric properties of the Schubert varieties; the Borel–Weil–Bott theorem; the BGG and the Kempf resolutions; determination of the defining ideal of the flag varieties; study of the T-equivariant and the singular cohomology of the flag varieties; the degeneracy of the Leray–Serre spectral sequence for the fibration $\mathcal{G} \to \mathcal{G}/T$; and various criteria of smoothness and rational smoothness for points on the Schubert varieties.

The book is devoted to the treatment of the theory over the complex numbers (or over an algebraically closed field of characteristic 0) and we do not use any characteristic p methods.

To keep the size of the book within reasonable limits several important related topics have not been treated here. These include, among others, the Kazhdan–Lusztig conjectures for the decomposition of Verma modules with positive and negative levels; the categorical equivalence between the negative level representations of affine Kac–Moody algebras under the fusion product and the representations of quantized enveloping algebras at roots of unity; construction of the Kac–Moody groups and their flag varieties over an arbitrary ring; the space of conformal blocks, moduli space of vector bundles and the Verlinde formula; study of the tensor product decomposition; Littelmann's LS-path model approach to the representation theory in the symmetrizable case; and the quantum cohomology of flag varieties. In addition, we have not treated the quantized enveloping algebras at all, but there are several books covering various aspects of these algebras.

Note: We do not require knowledge of Kac–Moody Lie algebras or of (finite-dimensional) algebraic groups although some basic understanding of finite-dimensional Lie algebras and algebraic groups will be helpful. We have included five appendices to recall some results we need from other areas including algebraic geometry, topology, homological algebra and spectral sequences. This book is suitable for an advanced graduate course.

I dedicate this book to my parents who showed me, by their example, the joy of work. I express my indebtedness to my brothers Pawan Kumar and Dipendra Prasad for innumerable discussions over dinner, providing the scientific stimulation I needed, absent otherwise in the sterile academic environment of my first college; to my brother Gopal Prasad who was the first in our family to choose mathematics in lieu of the traditional family business and show that it was a viable profession; to my wife Shyama and children Neeraj and Niketa from whom I stole numerous weekends to work on this book. I also want to express my gratitude to my high school and college teachers R.S. Yadav, K.N. Singh, S.B. Rao, and S.S. Shrikhande for all the encouragement they provided; and to H. Garland, B. Kostant, N. Mohan Kumar, M.S. Narasimhan, M.V. Nori, D. Peterson, H.V. Pittie, M.S. Raghunathan, S. Ramanan, M. Vergne, D.N. Verma, J. Wahl (to name only a few in the long list of mathematicians) for all that they taught me at various stages.

Many thanks are due to Lee Trimble, who did such a tremendous job of converting my hardly legible handwriting into a TEX document. The first few chapters were typed by Elaine Jackson.

I gave a course during the fall of 2001 covering many chapters of this book at the Tata Institute of Fundamental Research, Mumbai (India), hospitality of which is gratefully acknowledged. I thank many in the audience for their several useful comments. I particularly mention the names of S. Ilangovan, M.S. Raghunathan, R. Raghunathan, and D.N. Verma. S. Ilangovan looked at the whole book and pointed out various typos. Comments from several referees were very helpful. I specifically wish to acknowledge my gratitude to one referee who looked carefully at the chapters 4, 6, 7, 8, 10, 12, 13 and appendix A, and made numerous suggestions for improvement. It is my pleasure to thank Ann Kostant for the personal interest and care she took in preparing this book for publication. Finally I acknowledge the support of the NSF.

Shrawan Kumar
Chapel Hill

Convention

The subsections are numbered as (x, y, z) which means subsection z in section y in chapter x. The equations within a subsection are consecutively numbered as (1), (2), $\cdots$. They are referred to, within the same subsection, simply as (1), (2), $\cdots$, but the equation (t) in subsection (x, y, z) is referred in a different subsection as (x, y, z, t). The appendices are numbered as A,B, $\cdots$.

Unless otherwise explicitly stated, we take the base field to be the field of complex numbers $\mathbb{C}$. Thus the vector spaces are assumed to be over $\mathbb{C}$ and the linear maps as complex linear maps. Though bulk of the content of this book generalizes easily to any algebraically closed field of charactersitic 0. The identity map of a set X is denoted by I_X (or, when no confusion is likely, by I or I itself). We denote by $\mathbb{N}$ the set of positive integers $\{1, 2, \dots\}$, by $\mathbb{Z}_+$ the set of nonnegative integers $\{0, 1, 2, \dots\}$, and by $\mathbb{R}$ the set of real numbers. The cardinality of a finite set Y is denoted interchangeably by $|Y|$ or $\sharp Y$. The image $f(X)$ of a map $f : X \to Y$ is denoted by $\operatorname{Im} f$ and the kernel (when there is a preferred base point in Y) by $\operatorname{Ker} f$.

Kac–Moody Groups, their Flag Varieties and Representation Theory

I

Kac–Moody Algebras

Basic Theory

The aim of this chapter is to give the definition of Kac–Moody Lie algebras and establish their most basic properties, which we will need subsequently in the book. The basic reference for this chapter as well as Chapter 2 (barring the discussion of the Shapovalov form as in Subsections 2.3.1–2.3.8) is Kac's book [Kac–90] which exhaustively treats the algebraic theory of Kac–Moody Lie algebras and their representations.

In Section 1.1, we recall the definition of the Kac–Moody Lie algebra $\mathfrak{g} = \mathfrak{g}(A)$ associated to a generalized Cartan matrix A. Kac–Moody Lie algebras generalize the (finite-dimensional) semisimple Lie algebras, which we refer to as the *finite case*.

Section 1.2 is devoted to proving the triangular and root space decompositions of $\mathfrak{g}$ and establishing a presentation for the "positive part" $\mathfrak{n}$ of $\mathfrak{g}$ (cf. Theorem 1.2.1). This is achieved by constructing a family of representations (parametrized by $\mathfrak{h}^*$) of the Lie algebra $\tilde{\mathfrak{g}}$ in the tensor algebra $T(V)$ of a ℓ-dimensional vector space V, where $\tilde{\mathfrak{g}}$ is obtained by the same generators and relations as for $\mathfrak{g}$, except that we do not take the Serre relations. We also define various subalgebras of $\mathfrak{g}$, including the standard parabolic subalgebras. The center of $\mathfrak{g}$ is determined in Corollary 1.2.3.

In Section 1.3, we define the Weyl group W associated to $\mathfrak{g}$ and also define the notion of integrable representations of $\mathfrak{g}$. It is shown that the adjoint representation of $\mathfrak{g}$ is integrable and, moreover, that the set of weights of any integrable representation of $\mathfrak{g}$ is W-stable. Further, we define a general Coxeter group $\mathcal{W}$ and the Bruhat–Chevalley partial order $\leq$ in $\mathcal{W}$, and we establish various equivalent conditions for a group to be a Coxeter group. Using this we show that the Weyl group W associated to $\mathfrak{g}$ is a Coxeter group (in fact a crystallographic Coxeter group).

Section 1.4 is devoted to defining the dominant chamber $D_{\mathbb{R}}$ and the Tits cone C, and establishing their basic properties. In particular, we determine the isotropy group of any element $\lambda \in D_{\mathbb{R}}$ and show that $D_{\mathbb{R}}$ is a fundamental domain for the action of W on C. Moreover, C is convex.

In Section 1.5, we consider the subclass of *symmetrizable* Kac–Moody algebras $\mathfrak{g}$. We show the existence of an invariant bilinear form on $\mathfrak{g}$ and use this to define the Casimir–Kac element, which lives as a central element in a certain completion of the enveloping algebra $U(\mathfrak{g})$. The existence of the Casimir–Kac element plays an important role in the representation theory of $\mathfrak{g}$, and it is this element that distinguishes the symmetrizable Kac–Moody algebras from the non-symmetrizable ones.

We discuss an explicit realization of affine Kac–Moody Lie algebras, which is an important subclass of symmetrizable Kac–Moody algebras, as a central extension of "loop algebras" in Section 13.1. This section can be read without a knowledge of Chapters 2–12.

1.1. Definition of Kac–Moody Algebras

1.1.1 Definition. By a *generalized Cartan matrix* (GCM for short) of size ℓ, where ℓ is a positive integer, we mean a square matrix $A = (a_{i,j})_{1\le i,j\le \ell}$ with integral coefficients satisfying (for all i, j):

(a) $a_{i,i} = 2$,

(b) $a_{i,j} \le 0$ for $i \neq j$, and

(c) $a_{i,j} = 0$ if $a_{j,i} = 0$.

1.1.2 Definition. Let A be as above. Choose a triple $(\mathfrak{h}, \pi, \pi^\vee)$, where $\mathfrak{h}$ is a complex vector space of dimension $\ell +$ corank A,

$$\pi = \{\alpha_i\}_{1\le i\le \ell} \subset \mathfrak{h}^*, \text{ and } \pi^\vee = \{\alpha_i^\vee\}_{1\le i\le \ell} \subset \mathfrak{h}$$

are linearly independent indexed sets satisfying

$$\alpha_j(\alpha_i^\vee) = a_{i,j},$$

where corank $A := \ell -$ rank A. Such a triple exists and is unique up to isomorphism, in the sense that if $(\mathfrak{h}', \pi', \pi'^\vee)$ is another such triple, then there is an isomorphism of vector spaces $\theta : \mathfrak{h} \to \mathfrak{h}'$ such that $\theta(\alpha_i^\vee) = \alpha_i'^\vee$ and $\theta^*(\alpha_i') = \alpha_i$, where $\pi' = \{\alpha_i'\}_{1\le i\le \ell}$, $\pi'^\vee = \{\alpha_i'^\vee\}$ and θ^* is the induced isomorphism $\mathfrak{h}'^* \to \mathfrak{h}^*$ (cf. Exercises 1.1.E.1–2).

The triple $(\mathfrak{h}, \pi, \pi^\vee)$ above is called a *realization* of A. Clearly $(\mathfrak{h}^*, \pi^\vee, \pi)$ is a realization of the transposed matrix A^t.

The Kac–Moody algebra $\mathfrak{g} = \mathfrak{g}(A)$ is the Lie algebra over $\mathbb{C}$ generated by $\mathfrak{h}$ and the symbols e_i, f_i, $1 \le i \le \ell$, with the defining relations for all $1 \le i, j \le \ell$ and $h \in \mathfrak{h}$:

(R_1) $[\mathfrak{h}, \mathfrak{h}] = 0$,

(R$_2$) $[h, e_i] = \alpha_i(h)e_i$; $[h, f_i] = -\alpha_i(h)f_i$,
(R$_3$) $[e_i, f_j] = \delta_{i,j}\alpha_i^\vee$,
(R$_4$) $(\operatorname{ad} e_i)^{1-a_{i,j}}(e_j) = 0$, $i \neq j$, and
(R$_5$) $(\operatorname{ad} f_i)^{1-a_{i,j}}(f_j) = 0$, $i \neq j$.

This means that the Lie algebra $\mathfrak{g}$ is obtained by taking the quotient of the free Lie algebra (cf. Exercise 1.1.E.3) generated by $\mathfrak{h} \oplus \oplus_{i=1}^{\ell} (\mathbb{C}e_i \oplus \mathbb{C}f_i)$ by the ideal generated by the above elements: $\{[h, h'], [h, e_i] - \alpha_i(h)e_i, [h, f_i] + \alpha_i(h)f_i, [e_i, f_k] - \delta_{i,k}\alpha_i^\vee, (\operatorname{ad} e_i)^{1-a_{i,j}}(e_j), (\operatorname{ad} f_i)^{1-a_{i,j}}(f_j); h, h' \in \mathfrak{h}, 1 \leq i, j, k \leq \ell, i \neq j\}$. The relations (R$_4$)–(R$_5$) are called the *Serre relations*.

As we shall see in Theorem 1.2.1, the canonical map $\mathfrak{h} \to \mathfrak{g}$ is injective. We call $\mathfrak{h}$ the *Cartan subalgebra* of $\mathfrak{g}$. The elements e_i, f_i are called the *Chevalley generators*.

Let $\mathfrak{n}$, resp. $\mathfrak{n}^-$, be the Lie subalgebra of $\mathfrak{g}$ generated by $\{e_i;\ 1 \leq i \leq \ell\}$, resp. $\{f_i;\ 1 \leq i \leq \ell\}$. Then $\mathfrak{b} := \mathfrak{h} + \mathfrak{n}$ is a Lie subalgebra of $\mathfrak{g}$. It is called the *standard Borel subalgebra* of $\mathfrak{g}$. Similarly, we define the Lie subalgebra $\mathfrak{b}^- := \mathfrak{n}^- + \mathfrak{h}$.

We need to introduce an auxiliary Lie algebra $\tilde{\mathfrak{g}} = \tilde{\mathfrak{g}}(A)$, defined as the Lie algebra generated by $\mathfrak{h}$, and the symbols e_i, f_i, $1 \leq i \leq \ell$, subject only to the relations (R$_1$)–(R$_3$). Clearly, there is a surjective Lie algebra homomorphism $\gamma : \tilde{\mathfrak{g}} \to \mathfrak{g}$. Similar to $\mathfrak{n}$, resp. $\mathfrak{n}^-$, define the Lie subalgebra $\tilde{\mathfrak{n}}$, resp. $\tilde{\mathfrak{n}}^-$, of $\tilde{\mathfrak{g}}$ generated by $\{e_i;\ 1 \leq i \leq \ell\}$, resp. $\{f_i;\ 1 \leq i \leq \ell\}$.

Let $s\ell_2$ be the three-dimensional Lie algebra with basis $\{X, Y, H\}$, and the relations $[X, Y] = H$, $[H, X] = 2X$, and $[H, Y] = -2Y$. Of course, $s\ell_2$ is just the Lie algebra of 2×2 matrices of trace 0. For any $1 \leq i \leq \ell$, define the Lie subalgebra $\mathfrak{g}(i)$, resp. $\tilde{\mathfrak{g}}(i)$, of $\mathfrak{g}$, resp. $\tilde{\mathfrak{g}}$, spanned by $\{e_i, f_i, \alpha_i^\vee\}$. From the relations (R$_2$)–(R$_3$), it is clear that $\mathfrak{g}(i)$ as well as $\tilde{\mathfrak{g}}(i)$ is isomorphic with $s\ell_2$ (as Lie algebras) under the map $X \mapsto e_i$, $Y \mapsto f_i$, and $H \mapsto \alpha_i^\vee$.

From the defining relations, it is clear that the map $e_i \mapsto -f_i$, $f_i \mapsto -e_i$, and $h \mapsto -h$, for $1 \leq i \leq \ell, h \in \mathfrak{h}$, extends uniquely to a Lie algebra ($\mathbb{C}$-linear) automorphism ω of $\mathfrak{g}$ and also $\tilde{\omega}$ of $\tilde{\mathfrak{g}}$. The involution ω is called the *Cartan involution* of $\mathfrak{g}$. It is clear that $\omega(\mathfrak{n}) = \mathfrak{n}^-$ and $\omega(\mathfrak{n}^-) = \mathfrak{n}$.

Let A be a GCM as in 1.1.1. Then A is called *decomposable* if $\{1, \ldots, \ell\}$ is a disjoint union of nonempty subsets Y_1, Y_2 such that $a_{i,j} = 0$ for $i \in Y_1$, $j \in Y_2$; equivalently, $a_{j,i} = 0$ for $j \in Y_2$ and $i \in Y_1$. For any nonempty subset $Y \subset \{1, \ldots, \ell\}$, let $A_Y := (a_{i,j})_{i,j \in Y}$ be the sub GCM of A. Thus, A is the block sum up to a permutation of $\{1, \ldots, \ell\}$

$$A = \begin{pmatrix} A_{Y_1} & 0 \\ 0 & A_{Y_2} \end{pmatrix}.$$

Let $(\mathfrak{h}_1, \pi_1, \pi_1^\vee)$, resp. $(\mathfrak{h}_2, \pi_2, \pi_2^\vee)$, be a realization of A_{Y_1}, resp. A_{Y_2}. Then $(\mathfrak{h}, \pi, \pi^\vee)$ is a realization of A, where $\mathfrak{h} := \mathfrak{h}_1 \oplus \mathfrak{h}_2$, $\pi = \pi_1 \sqcup \pi_2$, $\pi^\vee = \pi_1^\vee \sqcup \pi_2^\vee$.

In this case it is easy to see that $\mathfrak{g}(A) \simeq \mathfrak{g}(A_{Y_1}) \oplus \mathfrak{g}(A_{Y_2})$ as Lie algebras. If A is not decomposable, it is called *indecomposable*.

1.1.E EXERCISES

(1) Let $A = (a_{i,j})$ be any square matrix over $\mathbb{C}$ of size ℓ. Assume that there is a vector space V with linearly independent elements $\pi = \{\alpha_i\}_{1 \leq i \leq \ell} \subset V^*$ and linearly independent elements $\pi^\vee = \{\alpha_i^\vee\}_{1 \leq i \leq \ell} \subset V$ satisfying: $\alpha_j(\alpha_i^\vee) = a_{i,j}$. Then show that $\dim V \geq \ell +$ corank A. Moreover, there exists a vector space $\mathfrak{h}$ of dimension $\ell +$ corank A satisfying the above.

(2) If $\mathfrak{h}$ is of dimension $\ell +$ corank A, prove that such a triple $(\mathfrak{h}, \pi, \pi^\vee)$ is unique, in the sense that if $(\mathfrak{h}', \pi', \pi'^\vee)$ is another such triple, then there is an isomorphism of vector spaces $\theta : \mathfrak{h} \to \mathfrak{h}'$ such that $\theta(\alpha_i^\vee) = \alpha_i'^\vee$ and $\theta^*(\alpha_i') = \alpha_i$, where $\pi' = \{\alpha_i'\}_{1 \leq i \leq \ell}$, $\pi'^\vee = \{\alpha_i'^\vee\}$.

Hint: Consider the map $\alpha = (\alpha_1, \ldots, \alpha_\ell) : \mathfrak{h} \to \mathbb{C}^\ell$. Since Ker α is of dimension equal to corank A and so is Ker $(\alpha_{|\oplus \mathbb{C}\alpha_i^\vee})$ (by definition of corank A), we get that $K := \cap_i (\operatorname{Ker} \alpha_i) \subset \oplus_i \mathbb{C}\alpha_i^\vee$. Define the map $\bar{\theta} : \oplus \mathbb{C}\alpha_i^\vee \to \oplus \mathbb{C}\alpha_i'^\vee$ by $\bar{\theta}(\alpha_i^\vee) = \alpha_i'^\vee$. Show that $\alpha_{|\oplus \mathbb{C}\alpha_i^\vee} = (\alpha'_{|\oplus \mathbb{C}\alpha_i'^\vee}) \circ \bar{\theta}$, where α' is defined similarly to α. Thus we conclude that $\theta(K) = K'$, where $K' := \cap_i (\operatorname{Ker} \alpha_i')$. Now extend $\bar{\theta}$ to an isomorphism $\theta : \mathfrak{h} \to \mathfrak{h}'$ such that $\alpha' \circ \theta = \alpha$. This θ does the job.

(3) Let V be a vector space and let $T(V)$ be its tensor algebra viewed as a Lie algebra under the bracket $[a, b] := ab - ba$. Let $F(V) \subset T(V)$ be the Lie subalgebra generated by the subspace $V = T^1(V)$. Then $F(V)$ is called the *free Lie algebra generated by* V.

Show that $F(V)$ has the following universal property:

For any Lie algebra $\mathfrak{s}$ and a linear map $\theta : V \to \mathfrak{s}$, there exists a unique Lie algebra homomorphism $\hat{\theta}$ making the following diagram commutative:

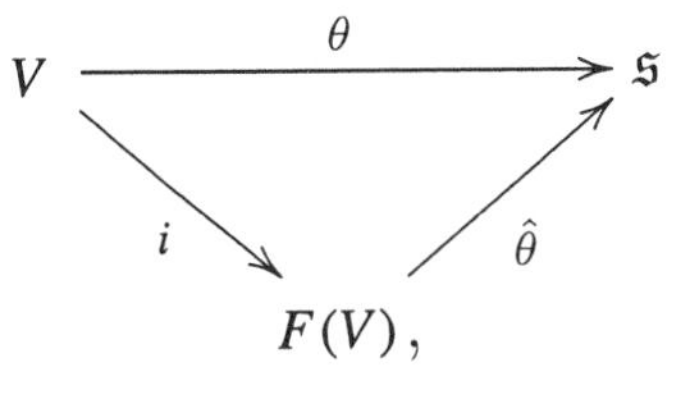

where $i : V \to F(V)$ is the standard inclusion.

Show further that the inclusion $V \subset T(V)$ induces an isomorphism $U(F(V)) \simeq T(V)$ of algebras.

1.2. Root Space Decomposition

Recall the definition of $\mathfrak{g}, \mathfrak{n}, \mathfrak{n}^-$ from 1.1.2. Define the $\mathbb{Z}$-submodule $Q := \oplus_{i=1}^{\ell}\mathbb{Z}\alpha_i \subset \mathfrak{h}^*$, to be called the *root lattice*. Set $Q^+ = \oplus_{i=1}^{\ell}\mathbb{Z}_+\alpha_i \subset Q$, where $\mathbb{Z}_+$ is the set of nonnegative integers. For any $\alpha \in \mathfrak{h}^*$, let $\mathfrak{g}_\alpha := \{x \in \mathfrak{g} : [h, x] = \alpha(h)X$, for all $h \in \mathfrak{h}\}$ be the eigenspace with respect to the $\mathfrak{h}$-action, corresponding to the eigenvalue α. Sometimes we denote $\mathfrak{n}$, resp. $\tilde{\mathfrak{n}}$, also by $\mathfrak{n}^+$, resp. $\tilde{\mathfrak{n}}^+$.

1.2.1 Theorem.

(a) *Triangular decomposition:* $\mathfrak{g} = \mathfrak{n}^- \oplus \mathfrak{h} \oplus \mathfrak{n}$ *(vector space direct sum).*
(b) *Root space decomposition:* $\mathfrak{n}^\pm = \bigoplus_{\alpha \in Q^+\setminus\{0\}} \mathfrak{g}_{\pm\alpha}.$
(c) $\dim \mathfrak{g}_\alpha < \infty$ *for any* α .
(d) *The Lie algebra* $\mathfrak{n}$, *resp.* $\mathfrak{n}^-$, *is generated by* $\{e_i;\ 1 \le i \le \ell\}$, *resp.* $\{f_i;\ 1 \le i \le \ell\}$, *subject to the only relations* R_4, *resp.* R_5, *of 1.1.2.*

Proof. We first prove the analogous statements for the Lie algebra $\tilde{\mathfrak{g}}$, where *(d)* is replaced by the assertion that $\tilde{\mathfrak{n}}$, resp. $\tilde{\mathfrak{n}}^-$, is freely generated by $\{e_i;\ 1 \le i \le \ell\}$, resp. $\{f_i;\ 1 \le i \le \ell\}$:

Let V be the ℓ-dimensional complex vector space with a basis $v_1, \dots, v_\ell$ and let us fix $\lambda \in \mathfrak{h}^*$. *From now on we often denote* $\lambda(h)$ *by* $\langle\lambda, h\rangle$ *for* $h \in \mathfrak{h}$. We define an action of the generators of $\tilde{\mathfrak{g}}(A)$ on the tensor algebra $T(V)$ over V by the following:

(α) $f_i(a) = v_i \otimes a$ for $a \in T(V)$;
(β) $h(1) = \langle\lambda, h\rangle 1$, and inductively on s,
$h(v_j \otimes a) = -\langle\alpha_j, h\rangle v_j \otimes a + v_j \otimes h(a)$ for $a \in T^{s-1}(V),\ h \in \mathfrak{h}$, and $j = 1, \dots, \ell$;
(γ) $e_i(1) = 0$, and inductively on s,
$e_i(v_j \otimes a) = \delta_{ij}\alpha_i^\vee(a) + v_j \otimes e_i(a)$ for $a \in T^{s-1}(V),\ j = 1, \dots, \ell.$

This defines a representation ρ_λ (depending on λ) of the Lie algebra $\tilde{\mathfrak{g}}(A)$ on the space $T(V)$. To see that, we have to check all the relations (R_1)–(R_3) of 1.1.2.

The relation (R_1) is obvious since $\mathfrak{h}$ operates diagonally. For the relation (R_3), we have

$$\begin{aligned}(e_i f_j - f_j e_i)(a) &= e_i(v_j \otimes a) - v_j \otimes e_i(a)\\ &= \delta_{ij}\alpha_i^\vee(a) + v_j \otimes e_i(a) - v_j \otimes e_i(a) = \delta_{ij}\alpha_i^\vee(a),\end{aligned}$$

by (α) and (γ).

For the second part of relation (R_2), we have

$$\begin{aligned}(hf_j - f_j h)(a) &= h(v_j \otimes a) - v_j \otimes h(a)\\ &= -\langle\alpha_j, h\rangle v_j \otimes a + v_j \otimes h(a) - v_j \otimes h(a)\\ &= -\langle\alpha_j, h\rangle f_j(a)\end{aligned}$$

by (α) and (β).

Finally, the first part of relation (R_2) is proved by induction on s. For $s = 0$, it clearly holds. For $s > 0$ take $a = v_k \otimes a_1$, where $a_1 \in T^{s-1}(V)$. We have

$$\begin{aligned}(he_j - e_jh)(v_k \otimes a_1) &= h(\delta_{jk}\alpha_j^\vee(a_1)) + h(v_k \otimes e_j(a_1)) \\ &\quad - e_j(-\langle \alpha_k, h\rangle(v_k \otimes a_1) + v_k \otimes h(a_1)) \\ &= \delta_{jk}\alpha_j^\vee(h(a_1)) - \langle \alpha_k, h\rangle v_k \otimes e_j(a_1) \\ &\quad + v_k \otimes h(e_j(a_1)) + \langle \alpha_k, h\rangle \delta_{jk}\alpha_j^\vee(a_1) \\ &\quad + \langle \alpha_k, h\rangle v_k \otimes e_j(a_1) - \delta_{jk}\alpha_j^\vee(h(a_1)) - v_k \otimes e_j(h(a_1)) \\ &= \langle \alpha_j, h\rangle \delta_{jk}\alpha_j^\vee(a_1) + v_k \otimes (he_j - e_jh)(a_1).\end{aligned}$$

To complete the proof, we apply the inductive assumption to the second summand.

Considering the above representation ρ_λ of $\tilde{\mathfrak{g}}(A)$ in $T(V)$ for all $\lambda \in \mathfrak{h}^*$, we easily see that the canonical map $\mathfrak{h} \to \tilde{\mathfrak{g}}(A)$ is injective.

Now we can deduce the analogues of all statements of the theorem for the algebra $\tilde{\mathfrak{g}}$. Using the relations (R_1)–(R_3) of 1.1.2, it is easy to show by induction on s that a bracket of s elements from the set $\{\mathfrak{h}, e_i, f_i; 1 \le i \le \ell\}$ lies in $\tilde{\mathfrak{n}}^- + \mathfrak{h} + \tilde{\mathfrak{n}}^+$. Let now $u = n_- + h + n_+ = 0$, where $n_\pm \in \tilde{\mathfrak{n}}^\pm$ and $h \in \mathfrak{h}$. Then, in the representation $T(V)$, we have $u(1) = n_-(1) + \langle \lambda, h\rangle 1 = 0$. It follows that $\langle \lambda, h\rangle = 0$ for every $\lambda \in \mathfrak{h}^*$, and hence $h = 0$. Furthermore, using the map $f_i \mapsto v_i$, we see that the tensor algebra $T(V)$ is a surjective image of the universal enveloping algebra of the Lie algebra $\tilde{\mathfrak{n}}^-$. Since $T(V)$ is a free associative algebra, we conclude that $T(V)$ is automatically the universal enveloping algebra $U(\tilde{\mathfrak{n}}^-)$ of $\tilde{\mathfrak{n}}^-$, the map $n_- \mapsto n_-(1)$ being the canonical embedding $\tilde{\mathfrak{n}}^- \hookrightarrow U(\tilde{\mathfrak{n}}^-) \simeq T(V)$. Hence $n_- = 0$ and (a) is proved. Moreover, by the Poincaré–Birkhoff–Witt theorem, $\tilde{\mathfrak{n}}^-$ is freely generated by $f_1, \dots, f_n$. Now applying $\tilde{\omega}$, we deduce that $\tilde{\mathfrak{n}}^+$ is freely generated by $e_1, \dots, e_n$, proving (d). The statement (c) is obvious in view of (a).

Using the relation (R_2) of 1.1.2 and the (a) part, we have

$$\tilde{\mathfrak{n}}^\pm = \oplus_{\alpha \in Q^+ \setminus \{0\}} \tilde{\mathfrak{g}}_{\pm\alpha}.$$

Now we come to the proof of the theorem for the Lie algebra $\mathfrak{g}$:

Let $\tilde{\mathfrak{r}} \subset \tilde{\mathfrak{g}}$ be the kernel of the homomorphism $\gamma : \tilde{\mathfrak{g}} \to \mathfrak{g}$ (cf. 1.1.2). Since $\tilde{\mathfrak{r}}$ is an ideal (in particular, $\mathfrak{h}$-stable under the adjoint action), by Exercise 1.2.E.2,

$$\tilde{\mathfrak{r}} = \left(\oplus_{\alpha \in Q^+ \setminus \{0\}} \tilde{\mathfrak{r}}_{-\alpha}\right) \oplus \tilde{\mathfrak{r}}_0 \oplus \left(\oplus_{\alpha \in Q^+ \setminus \{0\}} \tilde{\mathfrak{r}}_{\alpha}\right), \tag{1}$$

where $\tilde{\mathfrak{r}}_\beta := \tilde{\mathfrak{r}} \cap \tilde{\mathfrak{g}}_\beta$ (and $\tilde{\mathfrak{r}}_0 := \tilde{\mathfrak{r}} \cap \mathfrak{h}$).

We next prove that $\tilde{\mathfrak{r}}_0 = \{0\}$. Let $\tilde{\mathfrak{r}}^+$, resp. $\tilde{\mathfrak{r}}^-$, be the ideal of $\tilde{\mathfrak{n}}$, resp. $\tilde{\mathfrak{n}}^-$, generated by the elements $\{e_{i,j} := (\operatorname{ad} e_i)^{1-a_{ij}} e_j;\ i \neq j\}$, resp. $\{f_{i,j} := (\operatorname{ad} f_i)^{1-a_{ij}} f_j;\ i \neq j\}$. Then $\tilde{\mathfrak{r}}^+$, resp. $\tilde{\mathfrak{r}}^-$, is a $\mathfrak{h}$-weight submodule of $\tilde{\mathfrak{n}}$, resp. $\tilde{\mathfrak{n}}^-$. Further,

$$\text{(2)} \qquad \operatorname{ad} f_k(e_{i,j}) = 0, \text{ for all } k \text{ and } i \neq j :$$

For any $k \neq i$,

$$\begin{aligned} \operatorname{ad} f_k(e_{i,j}) &= (\operatorname{ad} e_i)^{1-a_{i,j}} (\operatorname{ad} f_k\, e_j) \\ &\begin{cases} = 0 & \text{if } k \neq j \\ = -(\operatorname{ad} e_i)^{1-a_{i,j}} (\alpha_j^\vee) & \text{if } k = j. \end{cases} \end{aligned}$$

But if $a_{i,j} < 0$, $(\operatorname{ad} e_i)^{1-a_{i,j}} (\alpha_j^\vee) = 0$, whereas for $a_{i,j} = 0$,

$$-(\operatorname{ad} e_i)(\alpha_j^\vee) = \alpha_i(\alpha_j^\vee) e_i = a_{j,i} e_i = 0 \text{ by (c) of 1.1.1.}$$

For $k = i$,

$$\begin{aligned} (\operatorname{ad} f_i)(e_{i,j}) &= \operatorname{ad} f_i (\operatorname{ad} e_i)^{1-a_{i,j}} e_j \\ &= (\operatorname{ad} e_i)^{1-a_{i,j}} ((\operatorname{ad} f_i)\, e_j) \\ &\quad + [(1 - a_{i,j}) a_{i,j} - (1 - a_{i,j}) a_{i,j}] (\operatorname{ad} e_i)^{-a_{i,j}} e_j = 0, \end{aligned}$$

by the commutation relation $[Y, X^k] = -k(k-1)X^{k-1} - kX^{k-1}H$ in the universal enveloping algebra $U(s\ell_2)$ (cf. Exercise 1.2.E.3) and the relations R_2–R_3 of 1.1.2. This proves (2).

By the analogue of (a) and (b) parts of the theorem for the algebra $\tilde{\mathfrak{g}}$ (already established), we get that

$$\text{(3)} \qquad \operatorname{ad} f_k(\tilde{\mathfrak{n}}) \subset \mathfrak{h} \oplus \tilde{\mathfrak{n}}.$$

Fix $1 \leq k \leq \ell$ and let $\mathfrak{s}_k := \{x \in \tilde{\mathfrak{r}}^+ : \operatorname{ad} f_k(x) \in \tilde{\mathfrak{r}}^+\}$. Then it is easy to see from (3) that $\mathfrak{s}_k$ is an ideal of $\tilde{\mathfrak{n}}$. Moreover, by (2), $e_{i,j} \in \mathfrak{s}_k$, for any $i \neq j$. Thus $\mathfrak{s}_k = \tilde{\mathfrak{r}}^+$, i.e., $\operatorname{ad} f_k(\tilde{\mathfrak{r}}^+) \subset \tilde{\mathfrak{r}}^+$, for all k. Hence $\tilde{\mathfrak{r}}^+$ is, in fact, an ideal of $\tilde{\mathfrak{g}}$. An identical argument shows that $\tilde{\mathfrak{r}}^-$ is an ideal of $\tilde{\mathfrak{g}}$.

Since, by definition, $\tilde{\mathfrak{r}}$ is the ideal of $\tilde{\mathfrak{g}}$ generated by the elements $\{e_{i,j}, f_{i,j};\ i \neq j\}$, we obtain

$$\text{(4)} \qquad \tilde{\mathfrak{r}} = \tilde{\mathfrak{r}}^+ \oplus \tilde{\mathfrak{r}}^-.$$

In particular, $\tilde{\mathfrak{r}}_0 = \{0\}$ and hence the canonical map $\mathfrak{h} \to \mathfrak{g}$ is injective. So $\mathfrak{g} \cong \tilde{\mathfrak{g}}/\tilde{\mathfrak{r}} = \tilde{\mathfrak{n}}^-/\tilde{\mathfrak{r}}^- \oplus \mathfrak{h} \oplus \tilde{\mathfrak{n}}/\tilde{\mathfrak{r}}^+$ (as vector spaces). But $\tilde{\mathfrak{r}}^- = \operatorname{Ker}(\gamma_{|\tilde{\mathfrak{n}}^-})$ (since

$\tilde{\mathfrak{r}} \cap \tilde{\mathfrak{n}}^- = \tilde{\mathfrak{r}}^-$ by (4)) and similarly $\tilde{\mathfrak{r}}^+ = \operatorname{Ker}(\gamma_{|\tilde{\mathfrak{n}}})$. This proves all the assertions (a)–(d) of the theorem. □

1.2.2 Definition. Define the *set of roots* $\Delta = \{\alpha \in Q\backslash\{0\} : \mathfrak{g}_\alpha \neq 0\}$. Any element of Δ is called a *root* and for any root $\alpha \in \Delta$, $\mathfrak{g}_\alpha$ is called the corresponding *root space*.

Then by the above theorem $\Delta = \Delta^+ \cup \Delta^-$ (disjoint union), where $\Delta^+ := \Delta \cap Q^+$ and $\Delta^- := \Delta \cap (-Q^+)$. The Cartan involution ω takes $\omega(\mathfrak{g}_\alpha) = \mathfrak{g}_{-\alpha}$ for any $\alpha \in \Delta$. In particular,

$$\Delta^- = -\Delta^+. \tag{1}$$

Any element of Δ^+, resp. Δ^-, is called a *positive*, resp. *negative*, root.

By Theorem 1.2.1, for any $1 \leq i \leq \ell$, $\alpha_i \in \Delta^+$ and, moreover, $\mathfrak{g}_{\alpha_i}$, resp. $\mathfrak{g}_{-\alpha_i}$, is the one-dimensional space spanned by e_i, resp. f_i. We call any α_i $(1 \leq i \leq \ell)$ a *simple root* and we call $\alpha_i^\vee$ a *simple coroot*. Moreover,

$$n\alpha_i \in \Delta \text{ if and only if } n = \pm 1. \tag{2}$$

Any $\alpha \in \Delta \cup \{0\}$ can be written as $\alpha = \sum_{i=1}^{\ell} n_i\alpha_i$, where either each n_i is nonnegative or else each n_i is nonpositive. We define the *principal gradation*, denoted $|\alpha|$, of α by

$$|\alpha| = \sum n_i. \tag{3}$$

We define the *height*, denoted $\operatorname{ht}\alpha$, of α by

$$\operatorname{ht}\alpha = |\sum n_i|. \tag{4}$$

We will see later in 13.1.4 that, unlike the finite-dimensional case, $\mathfrak{g}_\alpha$ is in general *not* one-dimensional. The dimension of $\mathfrak{g}_\alpha$ is called the *multiplicity of the root* α, denoted by $\operatorname{mult}\alpha$.

For any subset $Y \subset \{1, \ldots, \ell\}$ (including $Y = \emptyset$), define

$$\begin{aligned}
\mathfrak{g}_Y &= \mathfrak{h} \oplus \left(\oplus_{\alpha\in\Delta_Y} \mathfrak{g}_\alpha\right), \\
\mathfrak{u}_Y &= \oplus_{\alpha\in\Delta^+\backslash\Delta_Y^+} \mathfrak{g}_\alpha, \\
\mathfrak{u}_Y^- &= \oplus_{\alpha\in\Delta^-\backslash\Delta_Y^-} \mathfrak{g}_\alpha, \\
\mathfrak{p}_Y &= \mathfrak{g}_Y \oplus \mathfrak{u}_Y, \text{ and} \\
\mathfrak{p}_Y^- &= \mathfrak{g}_Y \oplus \mathfrak{u}_Y^-,
\end{aligned}$$

where

$$\Delta_Y = \Delta \cap (\oplus_{i\in Y}\mathbb{Z}\alpha_i)\,, \quad \text{and}$$
$$\Delta_Y^{\pm} = \Delta^{\pm} \cap \Delta_Y.$$

For any $\alpha, \beta \in \Delta \cup \{0\}$,

(5) $$[\mathfrak{g}_\alpha, \mathfrak{g}_\beta] \subset \mathfrak{g}_{\alpha+\beta}.$$

From this it is easy to see that $\mathfrak{g}_Y, \mathfrak{u}_Y, \mathfrak{u}_Y^-$ are subalgebras of $\mathfrak{g}$, and $\mathfrak{g}_Y$ normalizes $\mathfrak{u}_Y$ and $\mathfrak{u}_Y^-$. Hence $\mathfrak{p}_Y$ and $\mathfrak{p}_Y^-$ are subalgebras of $\mathfrak{g}$ and $\mathfrak{u}_Y \subset \mathfrak{p}_Y$ is an ideal; similarly $\mathfrak{u}_Y^- \subset \mathfrak{p}_Y^-$ is an ideal. We call $\mathfrak{p}_Y$ the *standard parabolic subalgebra* of $\mathfrak{g}$ corresponding to the subset Y, $\mathfrak{g}_Y$ the *standard Levi component* of $\mathfrak{p}_Y$ and $\mathfrak{u}_Y$ the *nil-radical* of $\mathfrak{p}_Y$. Of course, $\mathfrak{u}_\emptyset = \mathfrak{n}$, $\mathfrak{g}_\emptyset = \mathfrak{h}$, and $\mathfrak{p}_\emptyset = \mathfrak{b}$. If $Y = \{i\}$, for some $1 \leq i \leq \ell$, we denote $\mathfrak{p}_Y$ by $\mathfrak{p}_i$ and call it a *standard minimal parabolic subalgebra* of $\mathfrak{g}$.

Assume $Y \neq \emptyset$ and let $A_Y = (a_{i,j})_{i,j\in Y}$ be the sub GCM of A. We choose a subspace $\mathfrak{h}_Y \subset \mathfrak{h}$ of smallest dimension such that

(1) $\alpha_i^\vee \in \mathfrak{h}_Y$ for all $i \in Y$, and
(2) $\{\alpha_{i_{|\mathfrak{h}_Y}}\}_{i\in Y}$ are linearly independent.

Then automatically $\dim \mathfrak{h}_Y = |Y| +$ corank A_Y. The choice of $\mathfrak{h}_Y$ gives rise to a Lie algebra homomorphism $i_Y : \mathfrak{g}(A_Y) \to \mathfrak{g}_Y$; in fact it is an embedding by Exercise 1.2.E.1. We call a subset $Y \subset \{1, \ldots, \ell\}$ of *finite type* if $\mathfrak{g}_Y$ is finite-dimensional.

Define the *derived algebra* $\mathfrak{g}' = \mathfrak{g}'(A)$ as the commutator subalgebra $[\mathfrak{g}, \mathfrak{g}]$ of $\mathfrak{g}$.

1.2.3 Corollary. (a) $\mathfrak{g} = \mathfrak{g}' + \mathfrak{h}$ *and* $\mathfrak{h} \cap \mathfrak{g}' = \mathfrak{h}'$, *where* $\mathfrak{h}' := \oplus_{i=1}^{\ell}\mathbb{C}\alpha_i^\vee$.
(b) The center of $\mathfrak{g} = \{h \in \mathfrak{h} : \alpha_i(h) = 0$, for all $1 \leq i \leq \ell\}$. Moreover, the center of $\mathfrak{g} \subset \mathfrak{h}'$.

Proof. By the defining relations of $\mathfrak{g}$, we have $e_i, f_i \in \mathfrak{g}'$; in particular, by Theorem 1.2.1(d), $\mathfrak{n} \oplus \mathfrak{n}^- \subset \mathfrak{g}'$ and hence by Theorem 1.2.1(a), the assertion $\mathfrak{g} = \mathfrak{g}' + \mathfrak{h}$ follows.

Clearly, $\mathfrak{h} \cap \mathfrak{g}' \supset \mathfrak{h}'$ by the relation (R_3) of 1.1.2. For the reverse inclusion, it suffices, from the root space decomposition, to show that for any $\alpha \in \Delta^+$,

(1) $$[\mathfrak{g}_\alpha, \mathfrak{g}_{-\alpha}] \subset \mathfrak{h}'.$$

We prove (1) by induction on ht α. If ht $\alpha = 1$, (1) follows from the relation (R_3). Now let ht $\alpha > 1$ and take $x \in \mathfrak{g}_\alpha$, $y \in \mathfrak{g}_{-\alpha}$. By Theorem 1.2.1(d), we can assume that $x = [e_i, x']$, for some $1 \leq i \leq \ell$ and $x' \in \mathfrak{g}_{\alpha-\alpha_i}$. Thus $[x, y] = [e_i, [x', y]] - [x', [e_i, y]] \in \mathfrak{h}'$, by the induction hypothesis. This completes the induction and hence proves (1), thereby proving the (a) part of the corollary.

By the root space decomposition, the center $Z(\mathfrak{g})$ of $\mathfrak{g}$ satisfies

$$\begin{aligned} Z(\mathfrak{g}) &= \{h \in \mathfrak{h} : \alpha(h) = 0, \text{ for all } \alpha \in \Delta\} \\ &= \{h \in \mathfrak{h} : \alpha_i(h) = 0, \text{ for all } 1 \leq i \leq \ell\}, \end{aligned}$$

by Theorem 1.2.1(b). This proves the first assertion of the (b) part. In particular, from this, we get

$$\dim(Z(\mathfrak{g}) \cap \mathfrak{h}') = \text{ corank of } A.$$

But $\dim Z(\mathfrak{g}) = \dim \mathfrak{h} - \ell =$ corank of A. Hence $Z(\mathfrak{g}) \subset \mathfrak{h}'$. This proves the corollary completely. □

1.2.E EXERCISES

(1) For any nonempty subset $Y \subset \{1, \ldots, \ell\}$, recall the definition of the Lie algebra homomorphism $i_Y : \mathfrak{g}(A_Y) \to \mathfrak{g}_Y$ from 1.2.2. Show that it is an embedding.
Hint: From the triangular decomposition Theorem 1.2.1(a), it suffices to prove that $i_{Y|\mathfrak{n}_Y}$ is injective, where $\mathfrak{n}_Y$ is the sum of positive root spaces of $\mathfrak{g}(A_Y)$. Define the Lie algebra homomorphism $\delta : \mathfrak{n} \to \mathfrak{n}_Y$ by $e_i \mapsto e_i$ if $i \in Y$ and $e_i \mapsto 0$ otherwise.

(2) Let $\mathfrak{s}$ be a commutative Lie algebra and let M be a weight $\mathfrak{s}$-module, i.e., $M = \oplus_{\lambda \in \mathfrak{s}^*} M_\lambda$, where $M_\lambda := \{m \in M : hm = \lambda(h)m, \text{ for all } h \in \mathfrak{s}\}$. Then any $\mathfrak{s}$-submodule N of M is again a weight module, i.e., $N = \oplus_{\lambda \in \mathfrak{s}^*} (N \cap M_\lambda)$.
Hint: For $x = \sum_{i=1}^k x_i \in N$, where $x_i \in M_{\lambda_i}$ with $\{\lambda_1, \ldots, \lambda_k\}$ distinct, prove by induction on k that each x_i belongs to N.

(3) For any $m, n \geq 0$, prove the commutation relation in $U(s\ell_2)$:

$$X^{(m)} Y^{(n)} = \sum_{k=0}^{\min(m,n)} Y^{(n-k)} \binom{H - m - n + 2k}{k} X^{(m-k)},$$

where $X^{(m)} := X^m/m!$, similarly $Y^{(m)}$, and for any $a \in \mathbb{Z}$,

$$\binom{H+a}{k} := ((H+a)\cdots(H+a-k+1))/k!.$$

(4) Let V be a finite-dimensional $s\ell_2$-module and let $v \in V$ be a nonzero weight vector, i.e., $Hv = nv$ for some $n \in \mathbb{C}$. Then show that $n \in \mathbb{Z}$.

Assume further that $Xv = 0$ (and $Hv = nv$). Then show that $n \in \mathbb{Z}_+$ and, moreover, $Y^{n+1}v = 0$.

(This well-known result from $s\ell_2$-module theory can be found, e.g., in [Serre-87, Chap. 4, §3].)

1.3. Weyl Groups Associated to Kac–Moody Algebras

1.3.1 Definition. For any $1 \leq i \leq \ell$, define the reflection $s_i \in \operatorname{Aut} \mathfrak{h}^*$ by

$$s_i(\chi) = \chi - \langle \chi, \alpha_i^\vee \rangle \alpha_i, \qquad \text{for } \chi \in \mathfrak{h}^*.$$

Clearly s_i fixes the hyperplane $V_i = \{\chi \in \mathfrak{h}^* : \langle \chi, \alpha_i^\vee \rangle = 0\}$ pointwise and, moreover, has one remaining eigenvalue -1 (of multiplicity 1) with eigenvector α_i, and hence s_i is a reflection. In particular,

$$s_i^2 = 1\,. \tag{1}$$

Let $W \subset \operatorname{Aut}(\mathfrak{h}^*)$ be the subgroup generated by $\{s_i;\ 1 \leq i \leq \ell\}$. Then W is called the *Weyl group of the Kac–Moody algebra* $\mathfrak{g}$ and the s_i's are called *simple reflections*. The (faithful) representation of W in $\mathfrak{h}^*$, given by the inclusion $W \subset \operatorname{Aut} \mathfrak{h}^*$, is called the *standard representation* of W.

Dualizing this representation, we get the representation $W \subset \operatorname{Aut} \mathfrak{h}$, which is explicitly given for any $1 \leq i \leq \ell$ by

$$s_i(h) = h - \langle \alpha_i, h \rangle \alpha_i^\vee, \qquad \text{for } h \in \mathfrak{h}.$$

When we think of $w \in W$ acting on $\mathfrak{h}$, we denote it by $\bar{w}$, or if no confusion is likely by w itself. By the *length* of $w \in W$, denoted $\ell(w)$, we mean the smallest number $n \in \mathbb{Z}_+$ such that $w = s_{i_1} \ldots s_{i_n}$, $1 \leq i_j \leq \ell$. Further, such an expression with $n = \ell(w)$ is called a *reduced expression*. We set $\ell(e) = 0$.

Let $\rho \in \mathfrak{h}^*$ be an element such that $\langle \rho, \alpha_i^\vee \rangle = 1$ for all simple coroots $\alpha_i^\vee$. Now define the *shifted action* of W on $\mathfrak{h}^*$ by

$$w * \lambda = w(\lambda + \rho) - \rho, \text{ for } w \in W \text{ and } \lambda \in \mathfrak{h}^*.$$

It is easy to see that the action $*$ does *not* depend on the choice of $\rho \in \mathfrak{h}^*$.

1.3.2 Definition. Let $T : V \to V$ be a linear map of a complex vector space V. The map T is called *locally finite* if, for any $v \in V$, there exists a finite-dimensional subspace $v \in W \subset V$ such that $TW \subset W$. If, in addition, $T_{|W}$ is a nilpotent transformation, then we call T *locally nilpotent*.

For a locally finite $T : V \to V$, we can define an automorphism $\exp T : V \to V$ in the usual manner:

(1) $$\exp T = I + \sum_{n=1}^{\infty} \frac{T^n}{n!}.$$

Then

(2) $$\exp(kT) = (\exp T)^k, \qquad \text{for any } k \in \mathbb{Z}.$$

In an associative algebra R, we have the following identity for any $a, b \in R$ and $k \in \mathbb{N}$, where $\mathbb{N} := \mathbb{Z}_+ \setminus \{0\}$,

(3) $$(\operatorname{ad} a)^k b = \sum_{r=0}^{k} (-1)^r \binom{k}{r} a^{k-r} b a^r,$$

where $\operatorname{ad} a : R \to R$ is defined by

$$(\operatorname{ad} a)b = ab - ba.$$

To obtain (3), apply the Binomial Theorem to $(L_a - R_a)^k$ for the two commuting operators L_a and R_a, given respectively by $L_a b = ab,\ R_a b = ba$.

From (3) it is easy to see that for two linear maps $T, S : V \to V$ such that T is locally finite and $\{(\operatorname{ad} T)^n S;\ n \in \mathbb{N}\}$ spans a finite-dimensional subspace of $\operatorname{End} V$, we have that

(4) $$(\exp\ T)S\ \exp(-T) = \sum_{n \geq 0} \frac{(\operatorname{ad} T)^n}{n!}(S)$$

as operators on V, where $\operatorname{ad} T$ on the right side is to be thought of as an operator on the associative algebra $\operatorname{End} V$ (of all the linear operators of V).

A representation (V, π) of the Kac–Moody Lie algebra $\mathfrak{g}$ is called a *weight module* if V is the direct sum of its $\mathfrak{h}$-weight spaces, i.e., $V = \oplus_{\lambda \in \mathfrak{h}^*} V_\lambda$, where $V_\lambda := \{v \in V : h \cdot v = \lambda(h)v, \text{ for all } h \in \mathfrak{h}\}$. By the *multiplicity of a weight* λ, denoted $\operatorname{mult} \lambda$, in V, we mean the dimension of the weight space V_λ. This extends the notion of the multiplicity of a root defined in 1.2.2.

A weight module V of $\mathfrak{g}$ is called an *integrable representation* if all e_i and f_i $(1 \leq i \leq \ell)$ act locally nilpotently on V.

For an integrable representation (V, π) of $\mathfrak{g}$, and simple reflection s_i, set

(5) $$s_i(\pi) = (\exp f_i)(\exp -e_i)(\exp f_i) \in \operatorname{Aut} V,$$

where $\operatorname{Aut} V$ denotes the group of all the linear automorphisms of V, and $\exp f_i = \exp_\pi f_i$ denotes the automorphism $\exp(\pi(f_i))$ of V. Since $\pi(f_i)$ is locally nilpotent, $\exp(\pi(f_i))$ makes sense.

1.3.3 Lemma. (a) *Let $\mathfrak{s}$ be any Lie algebra and let $x \in \mathfrak{s}$. Define*

$$\mathfrak{s}_x := \{y \in \mathfrak{s} : (\operatorname{ad} x)^{n_y} y = 0, \text{ for some } n_y \in \mathbb{N}\}.$$

Then $\mathfrak{s}_x$ is a Lie subalgebra of $\mathfrak{s}$.

(b) *For any representation (V, π) of $\mathfrak{s}$ and $x \in \mathfrak{s}$ define $V_x = \{v \in V : \pi(x)^{n_v} v = 0$, for some $n_v \in \mathbb{N}\}$. Then V_x is a $\mathfrak{s}_x$-submodule of V.*

(c) *Let (V, π) be a representation of $\mathfrak{s}$ such that $\mathfrak{s}$ is generated (as a Lie algebra) by the set $F_V := \{x \in \mathfrak{s}$: $\operatorname{ad} x$ acting on $\mathfrak{s}$ is locally finite and $\pi(x)$ is locally finite$\}$. Then*

(c_1) *$\mathfrak{s}$ is spanned over $\mathbb{C}$ by F_V. In particular, if $\mathfrak{s}$ is generated by the set F of its* ad *locally finite vectors, then F spans $\mathfrak{s}$.*
(c_2) *If $\dim \mathfrak{s} < \infty$, then any $v \in V$ lies in a finite-dimensional $\mathfrak{s}$-submodule of V.*

Proof. (a) follows immediately from the Leibniz formula (i.e., $\operatorname{ad} x$ is a derivation)

$$(\operatorname{ad} x)^n [y, z] = \sum_{j=0}^{n} \binom{n}{j} [(\operatorname{ad} x)^j y, (\operatorname{ad} x)^{n-j} z].$$

Similar to the identity (1.3.2.3), considering the Binomial Theorem for the operator $L_x^n = (\operatorname{ad} x + R_x)^n$, we obtain in any associative algebra R and any elements $x, a \in R$,

$$x^n a = \sum_{j=0}^{n} \binom{n}{j} ((\operatorname{ad} x)^j a) x^{n-j}.$$

Applying the identity to v, the (b) part follows.

We first show that for $a, x \in F_V$ and $t \in \mathbb{C}$, $(\exp(t \operatorname{ad} a))\, x \in F_V$: since π is a Lie algebra representation, for any $y, z \in \mathfrak{s}$ and $n \in \mathbb{Z}_+$,

$$\pi((\operatorname{ad} y)^n z) = (\operatorname{ad} \pi(y))^n \pi(z), \tag{1}$$

as elements of $\operatorname{End}(V)$. In particular, for $a, x \in F_V$,

$$\begin{aligned} \pi((\exp(\operatorname{ad} a))x) &= (\exp(\operatorname{ad} \pi(a)))\, \pi(x) \\ &= (\exp(\pi a))\, \pi(x) \exp(-\pi a), \text{ by (1.3.2.4).} \end{aligned} \tag{2}$$

(Observe that since $a \in F_V$, $\pi(a)$ is locally finite and $\{(\operatorname{ad} \pi(a))^n \pi(x); n \in \mathbb{N}\}$ is finite dimensional by (1).) This shows that $\pi((\exp(t \operatorname{ad} a))x)$ is locally finite. Taking V to be the adjoint representation, we see that $(\exp(t \operatorname{ad} a))\, x \in F_V$.

Let $\mathfrak{s}_V \subset \mathfrak{s}$ be the $\mathbb{C}$-span of F_V. Since

$$\operatorname*{limit}_{t\to 0} \frac{(\exp(t \text{ ad } a))x - x}{t} = [a, x],$$

we see that $[a, x] \in \mathfrak{s}_V$ (for $a, x \in F_V$). In particular, $\mathfrak{s}_V$ is a Lie subalgebra of $\mathfrak{s}$. This proves (c_1). Now (c_2) follows from (c_1) by the PBW theorem. □

The following corollary follows immediately from the (a) part of the above lemma, Theorem 1.2.1(a)–(b), and the relations (R_2)–(R_5) of 1.1.2.

1.3.4 Corollary. ad *is an integrable representation of* $\mathfrak{g}$. □

1.3.5 Lemma. *For an integrable representation* (V, π) *of* $\mathfrak{g}$, *we have for any* $\lambda \in \mathfrak{h}^*$ *and simple reflection* s_i:

(a) $s_i(\pi)(V_\lambda) = V_{s_i\lambda}$.

In particular, for any $\lambda \in \mathfrak{h}^*$ *and* $w \in W$, $\text{mult}_V \lambda = \text{mult}_V w\lambda$.

(b) *For any* $v \in V$ *and* $x \in \mathfrak{g}$

$$s_i(\pi)(xv) = (s_i(\text{ad})\, x)(s_i(\pi)v).$$

In particular, $s_i(\text{ad})$ *is a Lie algebra automorphism of* $\mathfrak{g}$.

(c) *For* $v \in V_\lambda$, $s_i(\pi)^2 v = (-1)^{\langle\lambda,\alpha_i^\vee\rangle} v$. *(If* $v \neq 0$, *then* $\langle\lambda, \alpha_i^\vee\rangle \in \mathbb{Z}$.)

(d) *Let* $m_{i,j}$ *be the order of* $s_i s_j \in W$. *Then*

$$\underbrace{s_i(\pi)\, s_j(\pi)\, s_i(\pi) \ldots}_{m_{i,j} \text{ factors}} = \underbrace{s_j(\pi)\, s_i(\pi)\, s_j(\pi) \ldots}_{m_{i,j} \text{ factors}},$$

for $1 \le i \neq j \le \ell$ *such that* $m_{i,j} < \infty$.

Proof. Fix $v \in V_\lambda$. For any $h \in \mathfrak{h}$ such that $\langle\alpha_i, h\rangle = 0$, we have

$$h(s_i(\pi)v) = \langle\lambda, h\rangle s_i(\pi)v.$$

So, to prove the inclusion $s_i(\pi)(V_\lambda) \subset V_{s_i\lambda}$, it suffices to show that

$$\alpha_i^\vee(s_i(\pi)v) = -\langle\lambda, \alpha_i^\vee\rangle s_i(\pi)v, \text{ for } v \in V_\lambda.$$

Equivalently, we need to show that

(1) $$s_i(\pi)^{-1}\alpha_i^\vee s_i(\pi) = -\alpha_i^\vee \text{ as endomorphisms of } V.$$

By making repeated use of (1.3.2.4) and (1.3.3.1), the relations (R_2)–(R_3) give (1). Similarly, $s_i(\pi)^{-1}(V_\lambda) \subset V_{s_i\lambda}$. Thus the (a) part of the lemma follows since $s_i^2 = 1$.

We come to the proof of the (b) part:

$$\begin{aligned} s_i(\pi)(xv) &= s_i(\pi)(xs_i(\pi)^{-1}s_i(\pi)v) \\ &= ((\exp(\operatorname{ad} f_i)\exp(\operatorname{ad}(-e_i))\exp(\operatorname{ad} f_i))x)(s_i(\pi)v), \\ &\qquad\qquad \text{by successively using (1.3.2.4) and (1.3.3.1)} \\ &= (s_i(\operatorname{ad})\,x)(s_i(\pi)v). \end{aligned}$$

This proves the (b) part.

To prove (c), considering the subalgebra $\mathfrak{g}(i)$ of $\mathfrak{g}$ (cf. 1.1.2), we can assume that $\mathfrak{g} = s\ell(2)$. Further, by Lemma 1.3.3 (c_2), we can assume that V is a finite-dimensional $s\ell_2$-module. Moreover, if $v \in V_\lambda$ is nonzero, by the $s\ell_2$-module theory, $\langle \lambda, \alpha_i^\vee \rangle \in \mathbb{Z}$ (cf. Exercise 1.2.E.4). Considering the commutative diagram

(D) $$\begin{array}{ccc} s\ell_2 & \xrightarrow{\ \pi\ } & \operatorname{End} V \\ \big\downarrow{\scriptstyle \mathrm{Exp}} & & \big\downarrow{\scriptstyle \exp} \\ SL_2 & \longrightarrow & \operatorname{Aut} V, \end{array}$$

it suffices to prove that

(2) $$(\operatorname{Exp} Y \operatorname{Exp}(-X) \operatorname{Exp} Y)^2 = \operatorname{Exp}\begin{pmatrix} \pi\sqrt{-1} & 0 \\ 0 & -\pi\sqrt{-1} \end{pmatrix},$$

which can be easily checked by an explicit calculation, where π on the right side is the real number "π" (not to be confused with the representation π), and Exp is the exponential map from the Lie algebra to the group.

To prove (d), we can assume that $\mathfrak{g}$ is the Kac–Moody algebra corresponding to the 2×2 GCM $B = \begin{pmatrix} 2 & a_{i,j} \\ a_{j,i} & 2 \end{pmatrix}$. Further, by (a subsequent) Proposition 1.3.21, we can assume that $a = 0, 1, 2$ or 3 where $a := a_{i,j}a_{j,i}$, i.e., we can assume that $\mathfrak{g}$ is of type $A_1 \times A_1$, A_2, B_2 or G_2 (cf. [Serre–87, Theorem in the Appendix of Chap. 6]) and, using Lemma 1.3.3(c_2), V is finite dimensional. From an analogous diagram as (D) above, it suffices to show that, in the associated simply-connected group,

(3) $$C_1C_2C_1 \cdots = C_2C_1C_2 \ldots,$$

where $C_i := \operatorname{Exp}(f_i)\operatorname{Exp}(-e_i)\operatorname{Exp}(f_i)$, for $i = 1, 2$, and e_1, e_2, resp. f_1, f_2, are the simple positive, resp. negative, root vectors of $\mathfrak{g}$. In both sides of (3), there are 2, 3, 4 or 6 factors, depending on whether $\mathfrak{g}$ is of type $A_1 \times A_1$, A_2, B_2 or G_2 respectively (cf. Proposition 1.3.21). The identity (3) can be seen by case-by-case checking (cf. [Springer–98, Proposition 9.3.2]). This proves the (d) part. □

A root β is called *real* if $\beta = w\alpha_i$ for some $w \in W$ and some simple root α_i. By (a subsequent) Lemma 1.3.14 and Proposition 1.3.21, a positive root β is real iff $(W.\beta) \cap \Delta^- \neq \emptyset$. A root which is not real is called *imaginary*. Denote the set of real, resp. imaginary, roots by Δ_{re}, resp. Δ_{im}. The notation Δ^+_{re} stands for $\Delta_{\text{re}} \cap \Delta^+$, and Δ^+_{im} has a similar meaning. We record the following from [Kac–90, Proposition 5.5] together with [Moody–Pianzola–95, Corollary on page 329]:

(4) For $\alpha \in \Delta_{\text{im}}$, $n\alpha \in \Delta_{\text{im}}$ for any integer $n \neq 0$.

1.3.6 Corollary. (a) *The set of roots $\Delta \subset \mathfrak{h}^*$ is stable under the action of W. In fact, for any $\beta \in \Delta$ and $w \in W$,*

$$\text{mult}\ \beta = \text{mult}\ w\beta\,.$$

In particular, for a real root β, $\mathfrak{g}_\beta$ is one-dimensional and, moreover, $z\beta$ is not a root for $z \in \mathbb{C}$ unless $z = \pm 1$.

(b) *For any simple reflection s_i,*

$$s_i(\text{ad})_{|\mathfrak{h}} = \bar{s}_i \ \textit{as automorphisms of}\ \mathfrak{h}.$$

Proof. The (a) part follows immediately from Lemma 1.3.5(a) (applied to the adjoint representation) and Corollary 1.3.4. So we come to the proof of the (b) part:

Take an integrable representation (V, π) of $\mathfrak{g}$ and take a nonzero weight vector $v \in V$ (say of weight λ), and $x \in \mathfrak{h}$. Then, by Lemma 1.3.5(a)–(b),

$$\lambda(x)s_i(\pi)v = s_i(\pi)(xv) = (s_i(\text{ad})x)(s_i(\pi)v) = ((s_i\lambda)(s_i(\text{ad})x))s_i(\pi)v.$$

Hence $\lambda(x) = (s_i\lambda)(s_i(\text{ad})x)$. Replacing λ by $s_i\lambda$, we get

(1) $$\lambda(\bar{s}_i(x)) = (s_i\lambda)(x) = \lambda(s_i(\text{ad})x).$$

Since (1) is true for any λ which occurs as a weight in an integrable $\mathfrak{g}$-module, we get $\bar{s}_i = s_i(\text{ad})_{|\mathfrak{h}}$, proving the (b) part. By Lemma 2.1.7, any λ such that $\langle \lambda, \alpha_i^\vee \rangle \in \mathbb{Z}_+$, for all i, occurs as a weight of an integrable $\mathfrak{g}$-module and the span of such λ's is clearly $\mathfrak{h}^*$. □

1.3.7 Lemma. *Let $W \subset \mathrm{Aut}(\mathfrak{h}^*)$ be the Weyl group of any Kac–Moody algebra $\mathfrak{g}$ (cf. 1.3.1). Let $w \in W$ and $1 \leq i, j \leq \ell$ be such that $w\alpha_i = \alpha_j$. Then $\bar{w}(\alpha_i^\vee) = \alpha_j^\vee$.*

Proof. By Lemma 1.3.5(b) and Corollary 1.3.6(b), there exists a Lie algebra automorphism $\hat{w} : \mathfrak{g} \to \mathfrak{g}$ such that $\hat{w}_{|\mathfrak{h}} = \bar{w}$. So

$$[\hat{w}e_i, \hat{w}f_i] = \hat{w}[e_i, f_i] = \bar{w}(\alpha_i^\vee).$$

On the other hand, by Lemma 1.3.5(a), since $w\alpha_i = \alpha_j$,

$$[\hat{w}e_i, \hat{w}f_i] = z\alpha_j^\vee \qquad \text{for some } z \in \mathbb{C}.$$

Hence $\bar{w}(\alpha_i^\vee) = z\alpha_j^\vee$. Taking α_j of this identity, we get $z\alpha_j(\alpha_j^\vee) = \alpha_j(\bar{w}(\alpha_i^\vee)) = (w^{-1}\alpha_j)(\alpha_i^\vee) = \alpha_i(\alpha_i^\vee)$. Hence $z = 1$ and the lemma is proved. □

1.3.8 Definition. For any real root α, define the coroot $\alpha^\vee = \bar{w}\alpha_i^\vee \in \mathfrak{h}$ if $w\alpha_i = \alpha$. By the above lemma, this is well defined.

1.3.9 Lemma. *Let (V, π) be a locally finite $s\ell_2$-module, i.e., any $v \in V$ is contained in a finite-dimensional $s\ell_2$-submodule of V, and let $T \in \mathrm{End}\, V$ be such that*

(1) $$[\pi(X), T] = 0 \textit{ and } [\pi(H), T] = aT,$$

for some $a \in \mathbb{C}$. Then $a \in \mathbb{Z}_+$ and

$$(\mathrm{ad}\, \pi(Y))^{a+1}\, T = 0 \textit{ in } \mathrm{End}\, V.$$

Proof. Write $V = \oplus_{p \in I} V_p$ as a direct sum of finite-dimensional $s\ell_2$-submodules. For $p, q \in I$, let $T_{p,q} : V_q \to V_p$ be the map $T_{|V_q}$ followed by the projection on the p-th factor. Then $T_{p,q}$ again satisfies the relations (1) as an element of the $s\ell_2$-module $\mathrm{Hom}(V_q, V_p)$. Now, by (1), $T_{p,q}$ is a highest weight vector of $\mathrm{Hom}(V_q, V_p)$ with highest weight a, and hence, from the $s\ell_2$-module theory (cf. Exercise 1.2.E.4), $a \in \mathbb{Z}_+$ and $Y^{a+1} \cdot T_{p,q} = 0$. This proves the lemma. □

1.3.10 Corollary. *Let (V, π) be a representation of $\tilde{\mathfrak{g}}$ (cf. Definition 1.1.2) such that each e_i and f_i act locally finitely on V. Then V in fact is a module for $\mathfrak{g}$, i.e., the relations (R_4) and (R_5) of 1.1.2 are satisfied in* End V.

Proof. We prove the relations (R_5) in End V; the proof of (R_4) is similar. Fix $1 \leq i \neq j \leq \ell$ and consider V as a $\tilde{\mathfrak{g}}(i)$-module under restriction. Applying the above lemma to $T = \pi(f_j)$ (cf. Lemma 1.3.3 (c_2)) the corollary follows. □

Let $\mathcal{W}$ be a group generated by a fixed subset S of elements of order 2. For $w \in \mathcal{W}$, by *length* $\ell(w)$ of w (with respect to the subset S), we mean the smallest number ℓ such that w can be written as $w = s_1 \cdots s_\ell$, with $s_i \in S$. We take $\ell(1) = 0$. For $Y \subset S$, let $\mathcal{W}_Y$ be the subgroup of $\mathcal{W}$ generated by $\{s;\, s \in Y\}$.

1.3.11 Theorem. *Let $(\mathcal{W}, S)$ be as above. Then the following conditions are equivalent:*

(a) *Coxeter Condition: The group $\mathcal{W}$ is the quotient of the free group $\hat{\mathcal{W}}$ generated by the set S, modulo the following relations:*

(a_1) $s^2 = 1$, *for all* $s \in S$.
(a_2) *For all* $s \neq t \in S$, $(st)^{m_{s,t}} = 1$ *for some integers* $m_{t,s} = m_{s,t} \geq 2$ *or else* $m_{s,t} = \infty$.

The relation $(st)^\infty = 1$ *means that* st *is not of finite order; in particular, it does not contribute to a relation.*

(b) *Root-system Condition: There exists a representation V of $\mathcal{W}$ over $\mathbb{R}$ together with a subset $\Delta \subset V\backslash\{0\}$ such that*

(b_1) Δ *is symmetric, i.e.,* $-\Delta = \Delta$.
(b_2) Δ *is* $\mathcal{W}$*-invariant.*
(b_3) *There exists a subset* $\pi = \{\alpha_s\}_{s\in S} \subset \Delta$ *such that any* $\alpha \in \Delta$ *has the property that exactly one of* α *or* $-\alpha$ *belongs to* $\sum_{s\in S} \mathbb{R}_+\alpha_s$.
In the first case we say $\alpha > 0$ *and in the second* $\alpha < 0$. *The set* $\{\alpha \in \Delta; \alpha > 0\}$, *resp.* $\{\alpha \in \Delta; \alpha < 0\}$, *is denoted by* Δ^+, *resp.* Δ^-.
(b_4) $s\alpha_s < 0$ *and* $s\alpha > 0$ *for all* $s \in S$ *and* $\alpha \neq \alpha_s, \alpha > 0$.
(b_5) *For* $s, t \in S$ *and* $w \in \mathcal{W}$, $w\alpha_s = \alpha_t \Rightarrow wsw^{-1} = t$.

(c) *Strong Exchange Condition: Let* $s \in S$ *and* $v, w \in \mathcal{W}$ *be such that* $\ell(vsv^{-1}w) \leq \ell(w)$. *Then, for any expression (not necessarily reduced)*

$$w = s_1 \dots s_n \ (s_i \in S), \text{ we have } vsv^{-1}w = s_1 \dots \hat{s}_j \dots s_n \text{ for some } j.$$

(d) *Exchange Condition: Let* $s \in S$ *and* $w \in \mathcal{W}$ *be such that* $\ell(sw) \leq \ell(w)$. *Then for any reduced expression*

$$w = s_1 \dots s_n \ (s_i \in S), sw = s_1 \dots \hat{s}_j \dots s_n \text{ for some } j.$$

If S is finite, we can take V to be finite dimensional.

Proof. (a) $\Rightarrow$ (b). Let V be a vector space over $\mathbb{R}$ with $\{\alpha_s; s \in S\}$ as a basis. Define a bilinear form $(,)$ on V by setting

$$(\alpha_s, \alpha_s) = 1 \text{ for all } s \in S, \ (\alpha_{s_1}, \alpha_{s_2}) = (\alpha_{s_2}, \alpha_{s_1}) = -\cos\left(\frac{\pi}{m_{s_1,s_2}}\right)$$

for $s_1 \neq s_2 \in S$ and then extending bilinearly to $V \times V$.

For $s \in S$, $v \in V$, define $s(v) = v - 2(v, \alpha_s)\alpha_s$. It can be easily checked that $s(s(v)) = v$ and that $(s_1 s_2)^{m_{s_1,s_2}}(v) = v$ for all $v \in V$ if $s_1 \neq s_2$ and $m_{s_1,s_2} < \infty$. Hence this extends to an action of $\mathcal{W}$ on V. Note that $(s(v), s(v')) = (v, v')$ for all $v, v' \in V$, and hence $(w(v), w(v')) = (v, v')$ for all $v, v' \in V$, $w \in \mathcal{W}$. Let $\Delta := \bigcup_{s \in S} \mathcal{W}(\alpha_s)$. Then Δ is obviously $\mathcal{W}$-invariant. Note that $s(\alpha_s) = -\alpha_s$ and so $\Delta = -\Delta$ and $(\delta, \delta) = 1$ for all $\delta \in \Delta$.

We next prove by induction on $\ell(w)$ that for $s' \in S$,

$$(1) \qquad \ell(ws') \geq \ell(w) \Rightarrow w(\alpha_{s'}) = \sum_{s \in S} a_s \alpha_s \text{ with } a_s \geq 0 :$$

If $\ell(w) = 0$, then $w = 1$ and there is nothing to prove. So let $\ell(w) \geq 1$. Choose $s'' \in S$ such that $\ell(ws'') = \ell(w) - 1$. Since $\ell(ws') \geq \ell(w)$, $s' \neq s''$. Let $J = \{s', s''\}$ and let $\mathcal{W}_J$ be the subgroup of $\mathcal{W}$ generated by J. Let ℓ_J denote the length function in $\mathcal{W}_J$. Observe that $\ell \leq \ell_J$ on $\mathcal{W}_J$. Consider the set

$$A = \{z \in \mathcal{W} : z^{-1}w \in \mathcal{W}_J \text{ and } \ell(z) + \ell_J(z^{-1}w) = \ell(w)\}.$$

Clearly $w \in A$. Choose $x \in A$ such that $\ell(x)$ is minimum. Since $ws'' \in A$ (as can be easily checked), $\ell(x) \leq \ell(ws'') = \ell(w) - 1$. Next, if possible, let $\ell(xs') < \ell(x)$. Then $\ell(xs') = \ell(x) - 1$ and we have

$$\begin{aligned}\ell(w) &\leq \ell(xs') + \ell(s'x^{-1}w) \leq \ell(xs') + \ell_J(s'x^{-1}w) = \ell(x) - 1 + \ell_J(s'x^{-1}w)\\ &\leq \ell(x) - 1 + \ell_J(x^{-1}w) + 1 = \ell(x) + \ell_J(x^{-1}w) = \ell(w).\end{aligned}$$

Thus equality must hold at all places and so $\ell(w) = \ell(xs') + \ell_J(s'x^{-1}w)$. This means that $xs' \in A$, which is a contradiction since $\ell(xs') < \ell(x)$. Hence, $\ell(xs') \geq \ell(x)$. Similarly, we can prove that $\ell(xs'') \geq \ell(x)$. Since $\ell(x) < \ell(w)$, we can apply induction to the pairs (x, s') and (x, s'') to get $x(\alpha_{s'}) = \sum_{s \in S} c_s \alpha_s$ and $x(\alpha_{s''}) = \sum_{s \in S} d_s \alpha_s$ with $c_s, d_s \geq 0$ for all $s \in S$.

Let $y = x^{-1}w$. If possible, let $\ell_J(ys') < \ell_J(y)$. Then $\ell_J(ys') = \ell_J(y) - 1$ and

$$\begin{aligned}\ell(ws') &= \ell(xx^{-1}ws') \leq \ell(x) + \ell(x^{-1}ws') \leq \ell(x) + \ell_J(ys')\\ &= \ell(x) + \ell_J(y) - 1 = \ell(w) - 1,\end{aligned}$$

which is a contradiction since $\ell(ws') \geq \ell(w)$. Thus $\ell_J(ys') \geq \ell_J(y)$. Write a reduced expression for y in terms of generators s' and s''. It is clear that it ends with s''. Now either $m_{s',s''} = \infty$, in which case a direct computation shows that $y(\alpha_{s'}) = p\alpha_{s'} + q\alpha_{s''}$ with $p, q \geq 0$ (also, $|p - q| = 1$), or $m_{s',s''} < \infty$, in which case $\ell_J(y) < m_{s',s''}$. Note that $(s's'')^{m_{s',s''}} = 1$. Again, direct computation shows that $y(\alpha_{s'}) = p\alpha_{s'} + q\alpha_{s''}$ with $p, q \geq 0$. Thus, in either case, $y(\alpha_{s'}) = p\alpha_{s'} + q\alpha_{s''}$ with $p, q \geq 0$. Hence $w(\alpha_{s'}) = x(y(\alpha_{s'})) = x(p\alpha_{s'} + q\alpha_{s''}) = \sum_{s\in S}(pc_s + qd_s)\alpha_s$ with $a_s := pc_s + qd_s \geq 0$ for all $s \in S$. This verifies the induction hypothesis for w and so (1) is true.

Now given $\delta \in \Delta$, express $\delta = w(\alpha_{s'})$ for some $w \in \mathcal{W}$, $s' \in S$. If $\ell(ws') \geq \ell(w)$, then $\delta > 0$ by (1). If $\ell(ws') \leq \ell(w)$, then $ws'(\alpha_{s'}) > 0$ by (1). Hence $\delta < 0$ in this case. This proves (b_3). Note that we have proved a more precise statement than (b_3), viz.

$$\ell(ws') \geq \ell(w) \Rightarrow w(\alpha_{s'}) > 0. \tag{2}$$

We now come to the proof of (b_4). Obviously $s(\alpha_s) = -\alpha_s < 0$. Next, let $\delta > 0$ and $\delta \neq \alpha_s$. Since $(\delta, \delta) = 1$, it is clear that δ cannot be a multiple of α_s. Since $s(\delta) - \delta$ is a multiple of α_s, it is easy to see that $s(\delta) > 0$.

Next, let $w(\alpha_s) = \alpha_t$. Consider $y = wsw^{-1}t$. Then, for any $v \in V$,

$$\begin{aligned} y(v) &= wsw^{-1}(v - 2(v, \alpha_t)\alpha_t) = ws(w^{-1}(v) - 2(v, \alpha_t)w^{-1}(\alpha_t)) \\ &= w(w^{-1}(v) - 2(w^{-1}(v), \alpha_s)\alpha_s + 2(v, \alpha_t)\alpha_s) \\ &= w(w^{-1}(v) - 2(v, w(\alpha_s))\alpha_s + 2(v, \alpha_t)\alpha_s) = v. \end{aligned}$$

Now, if possible, let $y \neq 1$. Then there exists $s'' \in S$ such that $\ell(ys'') < \ell(y)$. By applying (2) to ys'', we get $ys''(\alpha_{s''}) > 0$, i.e., $y(-\alpha_{s''}) > 0$, i.e., $-\alpha_{s''} > 0$. This is a contradiction. Hence $y = 1$ and so $wsw^{-1} = t$. This proves (b_5).

The special representation constructed above is the so-called *geometric realization* of $\mathcal{W}$.

(b) $\Rightarrow$ (c). We first observe that $s(\alpha_s) = -\alpha_s$ since $-s(\alpha_s) > 0$ and $s(-s(\alpha_s)) = -\alpha_s < 0$ and so $-s(\alpha_s) = \alpha_s$ by (b_4).

Next we establish a one-to-one correspondence between T and the set $\{\delta > 0 : \delta = w(\alpha_s) \text{ for some } s \in S,\ w \in \mathcal{W}\}$, where $T = \{wsw^{-1} : w \in \mathcal{W}, s \in S\}$. For $\delta > 0$ such that $\delta = w(\alpha_s)$, define $t_\delta = wsw^{-1}$. Condition (b_5) then ensures that t_δ is independent of the choice of w and s. Conversely, let $t \in T$ be such that $t = wsw^{-1}$. Define $\delta_t = w(\alpha_s)$ or $-w(\alpha_s)$, whichever is > 0. We claim that δ_t is independent of the choice of w and s: So let $t = wsw^{-1} = w_1s_1w_1^{-1}$. Then $w^{-1}w_1s_1w_1^{-1}w = s$. Consider $\psi = w^{-1}w_1(\alpha_{s_1})$. Now

$$\begin{aligned} s(\psi) &= w^{-1}w_1s_1w_1^{-1}ww^{-1}w_1(\alpha_{s_1}) = w^{-1}w_1s_1(\alpha_{s_1}) \\ &= -w^{-1}w_1(\alpha_{s_1}) = -\psi. \end{aligned}$$

It is now clear from (b_4) that $\alpha_s = \psi$ or $-\psi$, whichever is positive. Our claim is now clear. It is easy to see that these two maps are inverses of each other. It is also easy to see that $t(\delta_t) = -\delta_t$.

We now prove the following:

(3) *Let $w = s_1 \dots s_p$ be any expression (not necessarily reduced) and let $t \in T$ be such that $w^{-1}(\delta_t) < 0$. Then $tw = s_1 \dots \hat{s}_i \dots s_p$ for some $1 \le i \le p$.*

To prove this, since $\delta_t > 0$ and $w^{-1}(\delta_t) = s_p \dots s_1(\delta_t) < 0$, there exists $1 \le i \le p$ such that

$$s_{i-1} \dots s_1(\delta_t) > 0 \text{ and } s_i \dots s_1(\delta_t) < 0.$$

By (b_4), $s_{i-1} \dots s_1(\delta_t) = \alpha_{s_i}$, i.e., $\delta_t = s_1 \dots s_{i-1}(\alpha_{s_i})$. Now, from the above correspondence, $t = s_1 \dots s_{i-1} s_i s_{i-1} \dots s_1$. Thus $tw = s_1 \dots \hat{s}_i \dots s_p$.

As a consequence of (3), for $w \in \mathcal{W}$, $t \in T$ such that $w^{-1}(\delta_t) < 0$, we have $\ell(tw) < \ell(w)$. Conversely, $\ell(tw) \le \ell(w) \Rightarrow w^{-1}(\delta_t) < 0$ (i.e., $w^{-1}(\delta_t) < 0$ iff $\ell(tw) < \ell(w)$ iff $\ell(tw) \le \ell(w)$). The Strong Exchange Condition is now clear. Hence (c) is proved.

(c) clearly implies (d).

(d) $\Rightarrow$ (a). Consider the canonical map $\eta : \tilde{\mathcal{W}} \to \mathcal{W}$, where $\tilde{\mathcal{W}}$ is the quotient of $\hat{\mathcal{W}}$, modulo the relations s^2, $s \in S$. For $s \in S$, let $\tilde{s}$ denote the same element in $\tilde{\mathcal{W}}$. For $s_1 \neq s_2 \in S$, let m_{s_1,s_2} denote the order of $s_1 s_2$ if it is finite. Let $\tilde{N}$ denote the normal subgroup of $\tilde{\mathcal{W}}$ generated by $\{(\tilde{s}_1\tilde{s}_2)^{m_{s_1,s_2}} : m_{s_1,s_2} < \infty\}$. It is then clear that $\tilde{N} \subseteq \operatorname{Ker} \eta$. We claim that $\tilde{N} = \operatorname{Ker} \eta$, which will, of course, prove (a):

If the claim is not true, choose $\tilde{z} = \tilde{s}_1 \dots \tilde{s}_k \in \operatorname{Ker} \eta$ such that $\tilde{z} \notin \tilde{N}$ and $\tilde{\ell}(\tilde{z}) = k$ is minimal with respect to this property ($\tilde{\ell}$ is the length function in $\tilde{\mathcal{W}}$). Now $1 = \eta(\tilde{z}) = s_1 \dots s_k$. Since $\ell(s_k) = 1$ and $\ell(s_1 \dots s_k) = 0$, there exists $i \le k-1$ such that $\ell(s_i \dots s_k) \le \ell(s_{i+1} \dots s_k)$ and $s_{i+1} \dots s_k$ is reduced. In fact, i satisfies $i \ge \frac{k}{2}$; otherwise $\ell(s_1 \dots s_k) = 0$ is not possible. Thus, by the Exchange Condition, there exists $i+1 \le j \le k$ such that $s_i \dots s_k = s_{i+1} \dots \hat{s}_j \dots s_k$, i.e., $s_i \dots s_j = s_{i+1} \dots s_{j-1}$. Now $\tilde{z}_o := \tilde{s}_i \dots \tilde{s}_j \tilde{s}_{j-1} \dots \tilde{s}_{i+1} \in \operatorname{Ker} \eta$ and $\tilde{\ell}(\tilde{z}_o) \le j - i + 1 + j - 1 - i = 2j - 2i \le 2k - k = k$. If the length is strictly smaller than k, then $\tilde{z}_o \in \tilde{N}$ by the minimality of k, and in that case

$$\begin{aligned}\tilde{z} = \tilde{s}_1 \dots \tilde{s}_k &= \tilde{s}_1 \dots \tilde{s}_{i-1} \cdot \tilde{z}_o \tilde{s}_{i+1} \dots \tilde{s}_{j-1} \cdot \tilde{s}_{j+1} \dots \tilde{s}_k \\ &= (\tilde{s}_1 \dots \tilde{s}_{i-1} \tilde{z}_o \tilde{s}_{i-1} \dots \tilde{s}_1) \tilde{s}_1 \dots \hat{\tilde{s}}_i \dots \hat{\tilde{s}}_j \dots \tilde{s}_k.\end{aligned}$$

So $\tilde{z} \in \tilde{N}$ as well, since $\tilde{s}_1 \dots \hat{\tilde{s}}_i \dots \hat{\tilde{s}}_j \dots \tilde{s}_k \in \operatorname{Ker} \eta$, of length $\le k-2$, and so $\in \tilde{N}$. This gives a contradiction. Hence $\tilde{\ell}(\tilde{z}_o) = k$ and $j = k = 2i$. Also, $s_1 \dots s_k = 1 = s_1 \dots \hat{s}_i \dots \hat{s}_k$ and so $\tilde{s}_1 \dots \hat{\tilde{s}}_i \dots \hat{\tilde{s}}_k \in \tilde{N}$. Thus, $\tilde{z} = \tilde{s}_1 \dots \tilde{s}_{i-1} \tilde{s}_i \tilde{s}_{i-1} \dots \tilde{s}_1 \tilde{s}_k \tilde{s}_k \tilde{s}_1 \dots \hat{\tilde{s}}_i \dots \hat{\tilde{s}}_k \tilde{s}_k \in \tilde{s}_1 \dots \tilde{s}_{i-1} \tilde{s}_i \tilde{s}_{i-1} \dots \tilde{s}_1 \cdot \tilde{s}_k \cdot \tilde{N}$.

Let $\tilde{z}_1 = \tilde{s}_k \cdot \tilde{s}_1 \dots \tilde{s}_{i-1} \cdot \tilde{s}_i \cdot \tilde{s}_{i-1} \dots \tilde{s}_1$. Then $\tilde{z}_1 \in \tilde{z}^{-1} \cdot \tilde{N}$ since $\tilde{N}$ is normal.

Now argue with $\tilde{z}_1$ instead of $\tilde{z}$. Note that $\tilde{\ell}(\tilde{z}_1) = k$ again, since $\tilde{z}_1 \in \operatorname{Ker}\eta$, $\tilde{\ell}(\tilde{z}_1) \leq k$, $\tilde{z}_1 \notin \tilde{N}$ and k is minimal with this property. Thus, we get

$$\tilde{z}_2 = \tilde{s}_1\tilde{s}_k\tilde{s}_1 \dots \tilde{s}_{i-2}\tilde{s}_{i-1}\tilde{s}_{i-2} \dots \tilde{s}_1 \cdot \tilde{s}_k \in (\tilde{z}_1)^{-1}\tilde{N} = \tilde{z}\tilde{N}, \text{ and so on.}$$

Finally, we get an element $\tilde{z}_r$ for a suitable r which is of the form $(\tilde{s}_1\tilde{s}_k)^i$ and such that $\tilde{z}_r \in (\tilde{z})^{\pm 1} \cdot \tilde{N}$. Since $\tilde{z}_r \in \operatorname{Ker}\eta$, it is clear that $m_{s_1,s_k} < \infty$, and it divides i, and so $\tilde{z}_r \in \tilde{N}$ by definition. Thus $\tilde{z} \in \tilde{N}$, which is a contradiction. This finally proves that $\tilde{N} = \operatorname{Ker}\eta$ and so (a) holds. This completes the proof of the theorem. □

1.3.12 Definition. Any group $(\mathcal{W}, S)$ satisfying the equivalent conditions of the above theorem is called a *Coxeter group.*

$\mathcal{W}$ admits a group homomorphism $\epsilon : \mathcal{W} \to \{\pm 1\}$ defined by $\epsilon(s) = -1$ for $s \in S$. It indeed extends to a group homomorphism, which follows from the Coxeter Condition (a) as above. The group homomorphism ϵ is called the *sign representation* of $\mathcal{W}$.

1.3.13 Lemma. *Let $(\mathcal{W}, S)$ be a Coxeter group and let V be any representation of $\mathcal{W}$ satisfying the root space Condition (b) of Theorem 1.3.11. Then, for $w \in \mathcal{W}$ and $s \in S$, $\ell(ws) > \ell(w)$ if and only if $w\alpha_s > 0$.*

More generally, for $t = vsv^{-1}$ with $v \in \mathcal{W}$, $\ell(wt) > \ell(w)$ iff $w(\delta_t) > 0$, where, as defined in the proof of Theorem 1.3.11, $\delta_t = v\alpha_s$ or $-v\alpha_s$, whichever is positive.

Proof. Follows easily from (1.3.11.3). □

1.3.14 Lemma. *Let the notation and assumptions be as in the above lemma. Then, for any $w \in \mathcal{W}$,*

$$\Phi_w := \{\phi \in \Delta^+ : w^{-1}\phi < 0\}$$

has cardinality exactly equal to $\ell(w)$. Further, if $w = s_1 \dots s_p$ is a reduced expression, then

$$\Phi_w = \{\alpha_{s_1}, s_1\alpha_{s_2}, \dots, s_1 \dots s_{p-1}\alpha_{s_p}\}. \tag{1}$$

In particular, the representation V of $\mathcal{W}$ is faithful.

Proof. Let $w = s_1 \dots s_p$ be a reduced expression. Let $\phi_i := s_1 \dots s_{i-1}\alpha_i$ (for $1 \leq i \leq p$), where $\alpha_i := \alpha_{s_i}$. Then we claim that $\phi_i \in \Phi_w$; $\phi_i \neq \phi_j$ for $i \neq j$; and $\Phi_w = \{\phi_1, \dots, \phi_p\}$:

By Lemma 1.3.13, $\phi_j > 0$ and $w^{-1}\phi_j < 0$, and thus $\phi_j \in \Phi_w$ for all j. If possible, let $\phi_i = \phi_j$ for some $i < j$. Then $\alpha_i = s_is_{i+1} \dots s_{j-1}\alpha_j$ and hence

$-\alpha_i = s_{i+1} \dots s_{j-1}\alpha_j$; in particular, $-\alpha_i > 0$ (a contradiction !). Finally, take $\phi \in \Phi_w$. Then there exists an $1 \le i \le p$ such that $s_i \dots s_1\phi > 0$ and $s_{i+1}s_i \dots s_1\phi < 0$. Hence by ($b_4$) of Theorem 1.3.11, $s_i \dots s_1\phi = \alpha_{i+1}$ and hence $\phi = s_1 \dots s_i\alpha_{i+1} = \phi_{i+1}$. This proves the lemma. □

1.3.15 Definition. In a Coxeter group $(\mathcal{W}, S)$, define the *Bruhat–Chevalley partial order* $\le$ as follows: For $v, w \in \mathcal{W}$, $v \le w$ iff there exist $t_1, \dots, t_p \in T := \{wsw^{-1};\ w \in \mathcal{W}, s \in S\}$ such that

(i) $v = t_p \cdots t_1 w$ and

(ii) $\ell(t_j \cdots t_1 w) \le \ell(t_{j-1} \cdots t_1 w)$ for all $1 \le j \le p$. (We set $w \le w$.)

Clearly $\le$ is a transitive relation, i.e., $u \le v$, $v \le w \Rightarrow u \le w$. The order $\le$ satisfies the following property:

For $v < w$, there exists a chain $v = v_0 < v_1 < \cdots < v_n = w$ such that $\ell(v_{i+1}) = \ell(v_i) + 1$, for all $0 \le i \le n-1$ (cf. [Deodhar–77, Theorem 1.1] or [Humphreys–90, Proposition 5.11]).

1.3.16 Lemma. *Let $(\mathcal{W}, S)$ be a Coxeter group and let $w \in \mathcal{W}$. Fix any reduced expression $w = s_1 \cdots s_n$. Then $v \le w \Leftrightarrow$ there exist $1 \le j_1 < j_2 < \cdots < j_p \le n$ such that*

$$(*) \qquad v = s_1 \cdots \hat{s}_{j_1} \cdots \hat{s}_{j_p} \cdots s_n.$$

In fact, for $v \le w$, we can choose $j := (j_1 < \cdots < j_p)$ such that $(*)$ is a reduced decomposition of v.

Proof. The implication '⇒' follows trivially from the Strong Exchange Property 1.3.11(c). For the converse, we take $v = s_1 \cdots \hat{s}_{j_1} \cdots \hat{s}_{j_p} \cdots s_n$ and prove that $v \le w$ by induction on $m(v, w) := (n+1)p - \sum_{i=1}^{p} j_i \ge 0$. If $m(v, w) = 0$, then $p = 0$, i.e., $v = w$. So assume $m(v, w) > 0$. Let $t = s_1 \cdots s_{j_1-1}s_{j_1}s_{j_1-1} \cdots s_1$. Then $v' := tv = s_1 \cdots \hat{s}_{j_2} \cdots \hat{s}_{j_p} \cdots s_n$. So, by induction, $v' \le w$. If $\ell(tv) \ge \ell(v)$, then $v \le v'$ and hence $v \le w$. So assume that $\ell(tv) < \ell(v)$. Using the expression $v = s_1 \cdots \hat{s}_{j_1} \cdots \hat{s}_{j_p} \cdots s_n$ in the Strong Exchange Property, there exists j ($j \ne j_r$, for any $1 \le r \le p$) such that tv has an expression obtained by deleting s_j from the above expression of v. We claim that $j > j_1$. If not, $tv = s_1 \cdots \hat{s}_j \cdots \hat{s}_{j_1} \cdots \hat{s}_{j_p} \cdots s_n$. It then follows that $t = s_1 \cdots s_{j_1}s_{j_1-1} \cdots s_1 = s_1 \cdots s_j s_{j-1} \cdots s_1$. This gives a contradiction to the fact that $w = s_1 \cdots s_n$ is a reduced expression. Hence $j > j_1$. Let $j_r < j < j_{r+1}$ for $r \ge 1$; we take $j_{p+1} = n$. Then we have $tv = s_1 \cdots \hat{s}_{j_1} \cdots \hat{s}_{j_r} \cdots \hat{s}_j \cdots \hat{s}_{j_{r+1}} \cdots \hat{s}_{j_p} \cdots s_n$. Hence, $v = t \cdot tv = s_1 \cdots \hat{s}_{j_2} \cdots \hat{s}_{j_r} \cdots \hat{s}_j \cdots \hat{s}_{j_{r+1}} \cdots \hat{s}_{j_p} \cdots s_n$. Now $m(v, w)$ associated with this expression is $(n+1)p - (j_2 + \cdots + j_r + j + j_{r+1} + \cdots + j_p)$. Since $j_1 < j$, it is clear that this number is smaller than $(n+1)p - (j_1 + \cdots + j_p)$. Hence the induction hypothesis applies and so $v \le w$.

To prove the "In fact" statement, take any decomposition $(*)$ of v such that p is the maximum. Then it is easy to see from the Exchange Property 1.3.11(d) that it is reduced. This proves the lemma. □

1.3.17 Definition. For any subset $Y \subset S$, recall that $\mathcal{W}_Y \subset \mathcal{W}$ is the subgroup generated by $\{s\}_{s\in Y}$. Define a subset of $\mathcal{W}$ by

$$(1) \qquad \mathcal{W}'_Y = \{w \in \mathcal{W} : \ell(wv) \geq \ell(w), \text{ for all } v \in \mathcal{W}_Y\},$$

i.e., $\mathcal{W}'_Y$ is the set of elements $w \in \mathcal{W}$ of minimal length in the cosets $w\mathcal{W}_Y$. In fact, each coset contains a unique element of minimal length:

Let w be an element of minimal length in its coset $w\mathcal{W}_Y$. Then, for any $v \in \mathcal{W}_Y$, we prove that

$$(2) \qquad \ell(wv) = \ell(w) + \ell(v) :$$

Consider the representation of $\mathcal{W}$ constructed in the proof of Theorem 1.3.11 ((a) $\Rightarrow$ (b) part). Take a reduced decomposition $v = s_1 \dots s_p$. It is easy to see that each $s_j \in Y$ by using Lemma 1.3.13. We claim that, for each $1 \leq j \leq p$,

$$(3)_j \qquad \ell(ws_1 \dots s_j) = j + \ell(w) :$$

Clearly $(3)_j$ is true for $j = 1$. Assume the validity of $(3)_j$ by induction on j and prove $(3)_{j+1}$: By Lemma 1.3.13, it suffices to show that

$$(4) \qquad ws_1 \dots s_j\alpha_{s_{j+1}} \in \Delta^+.$$

But since $s_1 \dots s_j s_{j+1}$ is reduced, by Lemma 1.3.13 again,

$$s_1 \dots s_j(\alpha_{s_{j+1}}) \in \Delta^+ \cap (\oplus_{s\in Y} \mathbb{R}\,\alpha_s) \subset \oplus_{s\in Y} \mathbb{R}_+\,\alpha_s.$$

But, by Lemma 1.3.13, $w\alpha_s \in \Delta^+$ for $s \in Y$, and hence (4) follows.

Now from (2) it immediately follows that each coset contains a unique element of minimal length.

1.3.18 Lemma. *Let $(\mathcal{W}, S)$ be a Coxeter group and let $Y \subset S$ be any subset. For any $w \in \mathcal{W}$, let w' denote the unique element of minimal length in the coset $w\mathcal{W}_Y$. Then, for any $v \in \mathcal{W}$, $v \leq w \Rightarrow v' \leq w'$.*

Proof. Follows easily from Lemma 1.3.16 used for $v' \leq w$ and (1.3.17.2). □

The following corollary follows easily from Lemma 1.3.16.

1.3.19 Corollary. *Let $(\mathcal{W}, S)$ be a Coxeter group and let $v \leq w \in \mathcal{W}$ and $s \in S$. Then*

(a) *either $sv \leq w$ or $sv \leq sw$,*

(b) *either $v \leq sw$ or $sv \leq sw$.* □

1.3.20 Lemma. *Let $(\mathcal{W}, S)$ be a Coxeter group and let $v, w \in \mathcal{W}$. Then there exists $u \in \mathcal{W}$ such that $v \leq u$ and $w \leq u$. Moreover, for any subset $Y \subset S$, if $v, w \in \mathcal{W}'_Y$, then we can take $u \in \mathcal{W}'_Y$ with $v, w \leq u$.*

Proof. We prove the lemma by induction on $n := \min(\ell(v), \ell(w))$. If $n = 0$, there is nothing to prove. So assume $n > 0$ and let $\ell(w) \leq \ell(v)$. Write $w = w's$, with $s \in S$ and $w' < w$. By induction, there exists $u' \in \mathcal{W}$ such that $v \leq u'$ and $w' \leq u'$. If $u' < u's$, taking $u = u's$ does the job. If $u' > u's$, then $u = u'$ itself does the job by Corollary 1.3.19.

To prove the "Moreover" statement, take any $u' \in \mathcal{W}$ with $v, w \leq u'$. Now write $u' = uu_1$, for $u_1 \in \mathcal{W}_Y$ and $u \in \mathcal{W}'_Y$. Then u does the job by Lemma 1.3.18. □

1.3.21 Proposition. *Let $W \subset \mathrm{Aut}(\mathfrak{h}^*)$ be the Weyl group of any Kac–Moody algebra $\mathfrak{g} = \mathfrak{g}(A)$ (cf. 1.3.1). Then $(W, \{s_i\}_{1 \leq i \leq \ell})$ is a Coxeter group. Moreover, the order $m_{i,j}$ of $s_i s_j$ $(1 \leq i \neq j \leq \ell)$ is given as follows:*

(1) $m_{i,j} = 2$ *if* $a = 0$
(2) $m_{i,j} = 3$ *if* $a = 1$
(3) $m_{i,j} = 4$ *if* $a = 2$
(4) $m_{i,j} = 6$ *if* $a = 3$, *and*
(5) $m_{i,j} = \infty$ *if* $a \geq 4$,

where $a := a_{ij} a_{ji}$.

Any Coxeter group with $m_{i,j} \in \{2, 3, 4, 6, \infty\}$ is called crystallographic. Hence the Weyl groups of Kac–Moody algebras are precisely the crystallographic Coxeter groups with finite S.

Proof. We prove that the condition (b) of Theorem 1.3.11 is satisfied by $(W, \{s_i\}_{1 \leq i \leq \ell})$ for the $\mathbb{R}$-subspace $V = \oplus_{i=1}^{\ell} \mathbb{R}\alpha_i \subset \mathfrak{h}^*$ and the set of roots $\Delta \subset V$. Note that the standard representation of W in $\mathfrak{h}^*$ keeps V stable. For the properties (b_1) and (b_3), see 1.2.2; the property (b_2) follows from Corollary 1.3.6(a); (b_4) follows since, for a simple root α_i, $n\alpha_i \in \Delta$ iff $n = \pm 1$ (cf. 1.2.2). Finally, we need to prove (b_5), i.e., if $w\alpha_i = \alpha_j$, for $1 \leq i, j \leq \ell$ and $w \in W$, then $ws_i w^{-1} = s_j$: For any $\chi \in \mathfrak{h}^*$ such that $\langle \chi, \alpha_j^\vee \rangle = 0$, we get $s_j \chi = \chi$. Also

$$\begin{aligned} ws_i w^{-1} \chi &= \chi - \langle w^{-1} \chi, \alpha_i^\vee \rangle w\alpha_i \\ &= \chi - \langle \chi, \bar{w}(\alpha_i^\vee) \rangle w\alpha_i = \chi, \end{aligned}$$

by Lemma 1.3.7. Similarly, $s_j \alpha_j = -\alpha_j$ and $ws_i w^{-1} \alpha_j = -w\alpha_i = -\alpha_j$. So the assertion $ws_i w^{-1} = s_j$ follows. This shows that $(W, \{s_i\})$ is a Coxeter group.

We now come to the proof of the assertion regarding $m_{i,j}$.

Fix $i \neq j$ and consider the subspace $L = V_{i,j} = \mathbb{C}\alpha_i \oplus \mathbb{C}\alpha_j \subset \mathfrak{h}^*$. Then L is stable under the subgroup $W_{i,j} \subset W$ generated by $\{s_i, s_j\}$. The action of s_i, resp.

s_j, in terms of the basis $\{\alpha_i, \alpha_j\}$ of L is given by the matrix

$$\begin{pmatrix} -1 & -a_{ij} \\ 0 & 1 \end{pmatrix} \left(\text{resp.} \begin{pmatrix} 1 & 0 \\ -a_{ji} & -1 \end{pmatrix}\right).$$

Hence $s_i s_{j|_L}$ is given by the matrix

$$M_{i,j} := \begin{pmatrix} -1+a & a_{ij} \\ -a_{ji} & -1 \end{pmatrix}.$$

The eigenvalues of this matrix are $\dfrac{a-2 \pm \sqrt{a(a-4)}}{2}$.

If $a \geq 5$, since one eigenvalue is > 1, $s_i s_{j|_L}$ (and hence $s_i s_j$) is of infinite order. If $a = 4$, both the eigenvalues of $s_i s_{j|_L}$ are 1, but $s_i s_{j|_L} \neq I$, and hence again $s_i s_{j|_L}$ is of infinite order. This takes care of the case (5). So assume $a < 4$. Then $\mathfrak{h}^* = L \oplus V_1$, where $V_1 := \{\chi \in \mathfrak{h}^* : \chi(\alpha_i^\vee) = \chi(\alpha_j^\vee) = 0\}$. Further, $s_{i|_{V_1}} = s_{j|_{V_1}} = I$, and moreover $M_{i,j}$ is of order 2, 3, 4, 6 resp. if $a = 0, 1, 2, 3$.

Conversely, any crystallographic Coxeter group is the Weyl group of a Kac-Moody algebra is easy to see by choosing a GCM which satisfies (1)–(5). This proves the proposition. □

Observe that the proof of the above proposition did not use Lemma 1.3.5(d), the proof of which used this proposition.

1.3.22 Corollary. *Let the notation and assumptions be as in Lemma 1.3.13. For $w \in \mathcal{W}$, fix a reduced decomposition $w = s_1 \cdots s_p$. Then, for any $\lambda \in V$,*

(1) $$\lambda - w\lambda = \sum_{j=1}^{p} 2(\lambda, \alpha_{s_j})\,(s_1 \cdots s_{j-1}\alpha_{s_j}),$$

where $(\,,\,)$ is any bilinear form on V satisfying $\lambda - s\lambda = 2(\lambda, \alpha_s)\alpha_s$ for all $\lambda \in V$ and $s \in S$. (Such a V with $(\,,\,)$ exists by the proof of Theorem 1.3.11(a) $\Rightarrow$ (b).)

In particular, for the Weyl group $W \subset \operatorname{Aut}(\mathfrak{h}^*)$ of any Kac–Moody algebra, we have, for $w \in W$ and $\lambda \in \mathfrak{h}^*$,

(2) $$\lambda - w\lambda = \sum_{j=1}^{p} \langle \lambda, \alpha_{s_j}^\vee \rangle (s_1 \cdots s_{j-1}\alpha_{s_j}).$$

Taking $\lambda = \rho$, where, as in 1.3.1, $\rho \in \mathfrak{h}^*$ is any element satisfying $\langle \rho, \alpha_i^\vee \rangle = 1$ for all the simple coroots $\alpha_i^\vee$, we get

(3) $$\rho - w\rho = \sum_{\phi \in \Phi_w} \phi,$$

where, as in Lemma 1.3.14, $\Phi_w := \Delta^+ \cap w\Delta^-$.

Proof. For $v \in W$ and $s \in S$, we have the identity

$$
\begin{aligned}
\lambda - vs\lambda &= \lambda - v\lambda + v(\lambda - s\lambda) \\
&= \lambda - v\lambda + 2(\lambda, \alpha_s) v\alpha_s. \qquad (4)
\end{aligned}
$$

Then (1) and (2) follow readily from (4) and the induction on $\ell(w)$. The identity (3) follows from (2) by using Lemma 1.3.14. □

1.3.E EXERCISE. Show that for the Weyl group W of a Kac–Moody Lie algebra $\mathfrak{g}$ and any subset $Y \subset \{1, \ldots, \ell\}$, one has

$$
\begin{aligned}
W'_Y &= \{w \in W : \Delta^+ \cap w^{-1}\Delta^- \subset \Delta^+ \backslash \Delta^+_Y\} \\
&= \{w \in W : w\Delta^+_Y \subset \Delta^+\},
\end{aligned}
$$

where W'_Y is defined in 1.3.17 and Δ^+_Y is defined in 1.2.2.

1.4. Dominant Chamber and Tits Cone

In this section W denotes the Weyl group associated to a Kac–Moody algebra $\mathfrak{g} = \mathfrak{g}(A)$ with Cartan subalgebra $\mathfrak{h}$.

1.4.1 Definition. Let us fix a real form $\mathfrak{h}_\mathbb{R}$ of $\mathfrak{h}$ (i.e., $\mathfrak{h}_\mathbb{R}$ is a real subspace of $\mathfrak{h}$ such that $\mathfrak{h}_\mathbb{R} \otimes_\mathbb{R} \mathbb{C} \approx \mathfrak{h}$) satisfying:

(a) $\{\alpha_1^\vee, \ldots, \alpha_\ell^\vee\} \subset \mathfrak{h}_\mathbb{R}$, and

(b) for any $1 \le i \le \ell$, $\alpha_i(\mathfrak{h}_\mathbb{R}) \subset \mathbb{R}$.

Clearly $\mathfrak{h}_\mathbb{R}$ is W-stable.

By the *dominant chamber* $D_\mathbb{R} \subset \mathfrak{h}^*_\mathbb{R} := \mathrm{Hom}_\mathbb{R}(\mathfrak{h}_\mathbb{R}, \mathbb{R})$ we mean

$$D_\mathbb{R} = \{\lambda \in \mathfrak{h}^*_\mathbb{R} : \lambda(\alpha_i^\vee) \in \mathbb{R}_{\ge 0} \text{ for all } i\}. \qquad (1)$$

The *Tits cone* C is defined as

$$C = \bigcup_{w \in W} w D_\mathbb{R}. \qquad (2)$$

1.4.2 Proposition. (a) *For* $\lambda \in D_\mathbb{R}$, *the isotropy group* $W_\lambda := \{w \in W : w(\lambda) = \lambda\}$ *is generated by the simple reflections which it contains.*

(b) *The dominant chamber* $D_\mathbb{R}$ *is a fundamental domain for the action of* W *on* C, *i.e., any orbit* $W \cdot \lambda$ *of* $\lambda \in C$ *intersects* $D_\mathbb{R}$ *at exactly one point.*

(c) $C = \{\lambda \in \mathfrak{h}^*_{\mathbb{R}} : \langle \lambda, h \rangle < 0$ *only for a finite number of* $h \in \Delta^{+\vee}\}$, *where* $\Delta^{+\vee} \subset \mathfrak{h}$ *is the set of positive roots for the Kac–Moody algebra* $\mathfrak{g}^\vee := \mathfrak{g}(A^t)$ *with respect to the realization* $(\mathfrak{h}^*, \pi^\vee, \pi)$ *of* A^t *(cf. 1.1.2). In particular, C is a convex cone.*

(d) $D_{\mathbb{R}} = \{\lambda \in \mathfrak{h}^*_{\mathbb{R}} :$ *for every* $w \in W$, $\lambda - w(\lambda) = \sum_i c_i \alpha_i$ *where* $c_i \geq 0\}$.

(e) *The following conditions are equivalent:*

(i) $|W| < \infty$

(ii) $C = \mathfrak{h}^*_{\mathbb{R}}$

(iii) $|\Delta^\vee| < \infty$

(iv) $|\Delta| < \infty$.

(f) *If* $\lambda \in C$, *then* $|W_\lambda| < \infty$ *if and only if* λ *lies in the interior of C under the Hausdorff topology on* $\mathfrak{h}^*_{\mathbb{R}}$.

Proof. Let $w \in W$ and let $w = s_{i_1} \dots s_{i_p}$ be a reduced decomposition of w. Take $\lambda \in D_{\mathbb{R}}$ and suppose that $\lambda' = w(\lambda) \in D_{\mathbb{R}}$. We have $\langle \alpha^\vee_{i_p}, \lambda \rangle \geq 0$ and therefore $\langle w(\alpha^\vee_{i_p}), \lambda' \rangle \geq 0$. But, by Lemma 1.3.13 (applied to the dual root system), $w(\alpha^\vee_{i_p}) < 0$, and hence $\langle w(\alpha^\vee_{i_p}), \lambda' \rangle \leq 0$. So, $\langle w(\alpha^\vee_{i_p}), \lambda' \rangle = 0$ and thus $\langle \alpha^\vee_{i_p}, \lambda \rangle = 0$. Hence $s_{i_p}(\lambda) = \lambda$ and both (a) and (b) follow by induction on $\ell(w)$.

Set $C' = \{\lambda \in \mathfrak{h}^*_{\mathbb{R}} : \langle \lambda, h \rangle < 0$ only for a finite number of $h \in \Delta^{+\vee}\}$. Clearly, $D_{\mathbb{R}} \subset C'$, and it follows from Proposition 1.3.21 and Theorem 1.3.11(b_4) that C' is s_i-invariant. Hence $C' \supset C$. To prove the reverse inclusion, let $\lambda \in C'$ and set $M_\lambda = \{h \in \Delta^\vee_+ : \langle \lambda, h \rangle < 0\}$. By definition, $|M_\lambda|$ is finite. If $M_\lambda \neq \emptyset$, then $\alpha^\vee_i \in M_\lambda$ for some i. But then it follows easily that $|M_{s_i(\lambda)}| < |M_\lambda|$. Induction on $|M_\lambda|$ completes the proof of (c).

The inclusion $\supset$ of (d) is obvious. We prove the reverse inclusion by induction on $p = \ell(w)$. For $\ell(w) = 1$, it follows from the definition of $D_{\mathbb{R}}$. If $\ell(w) = p$, let $w = s_{i_1} \dots s_{i_p}$. We have $\lambda - w(\lambda) = (\lambda - s_{i_1} \dots s_{i_{p-1}}(\lambda)) + s_{i_1} \dots s_{i_{p-1}}(\lambda - s_{i_p}(\lambda))$. Now we apply the inductive assumption to the first summand and Lemma 1.3.13 to the second summand, completing the proof.

Now we prove (e). For an element $\alpha = \sum r_i \alpha_i \in \sum_i \mathbb{R}\alpha_i$, define $|\alpha| = \sum r_i$. Take any $\lambda \in \mathfrak{h}^*_{\mathbb{R}}$. Since an element λ' of $W \cdot \lambda$ with maximal $|\lambda' - \lambda|$ lies in $D_{\mathbb{R}}$, (i) $\Rightarrow$ (ii). To show (ii) $\Rightarrow$ (iii), take λ in the interior of $D_{\mathbb{R}}$. Then $\langle -\lambda, h \rangle < 0$ for all $h \in \Delta^\vee_+$ and hence $|\Delta^\vee| < \infty$ by (c). Observe that (iii) $\Rightarrow$ (i), because the map $W \to P(\Delta^\vee)$ is injective by Lemma 1.3.13, where $P(\Delta^\vee)$ is the group of permutations of $\Delta^\vee$. The fact that (iv) is equivalent to (i) follows by using the dual root system. Note that the Weyl groups of a root system and the dual root system are isomorphic.

To prove (f), we may assume $\lambda \in D_{\mathbb{R}}$. Assume that $|W_\lambda| < \infty$. Let $Y = Y_\lambda = \{1 \leq i \leq \ell : \lambda(\alpha^\vee_i) = 0\}$. Then, by the (a) part of the proposition, W_λ is generated by $\{s_i\}_{i \in Y}$, i.e., $W_\lambda = W_Y$. For any $\theta \in \mathfrak{h}^*_{\mathbb{R}}$, by the (e) part, there exists a $w = w_\theta \in W_\lambda$ such that $w\theta(\alpha^\vee_i) \geq 0$, for all $i \in Y$. Let

$X_\lambda := \{\theta \in \mathfrak{h}_{\mathbb{R}}^* : \langle \lambda + w\theta, \alpha_j^\vee \rangle > 0$, for all $w \in W_\lambda$ and $j \notin Y\}$. Then $0 \in X_\lambda$ and X_λ is open in $\mathfrak{h}_{\mathbb{R}}^*$. Further, $\lambda + X_\lambda \subset C$, since $w_\theta(\lambda + \theta) \in D_{\mathbb{R}}$ for all $\theta \in X_\lambda$. Thus λ is an interior point of C.

Conversely, assume that $\lambda \in D_{\mathbb{R}}$ is an interior point of C. Let $Y = Y_\lambda$ be as above. Choose any $\theta \in \mathfrak{h}_{\mathbb{R}}^*$ such that $\theta(\alpha_i^\vee) = -1$ for all $i \in Y$. Since λ is an interior point of C, for all small $\epsilon > 0$, $\lambda + \epsilon\theta \in C$. Now $\langle \lambda + \epsilon\theta, h \rangle < 0$, for any $h \in \Delta_Y^{+\vee}$. Thus, by (c), $|\Delta_Y^{+\vee}| < \infty$ and hence, by Exercise 1.2.E.1 and (e), $|W_\lambda| < \infty$. This completes the proof of the proposition. □

1.4.E EXERCISES

(1) Prove that $C^o = \{\lambda \in \mathfrak{h}_{\mathbb{R}}^* : \langle \lambda, h \rangle \leq 0$ only for a finite number of $h \in \Delta^{+\vee}\}$, where C^o is the interior of C. In particular, C^o is convex. Use this and Proposition 1.4.2(f) to prove that C^o is the convex hull of $\cup_{w \in W} \, w D_{\mathbb{R}}^o$, where $D_{\mathbb{R}}^o := \{\lambda \in D_{\mathbb{R}} : \lambda(\alpha_i^\vee) > 0$, for all $i\}$.

(2) For $\lambda \in D$, show that $W \cdot \lambda \cap D = \{\lambda\}$ and, moreover, W_λ is generated by the simple reflections which it contains.

1.5. Invariant Bilinear Form and the Casimir Operator

In this section we consider a subclass of Kac–Moody Lie algebras, called symmetrizable Kac–Moody algebras (defined below).

1.5.1 Definition. Let $A = (a_{i,j})_{1 \leq i,j \leq \ell}$ be a GCM (cf. 1.1.1). The matrix A is called *symmetrizable* if there exists an invertible diagonal matrix $D = \text{diag}(\epsilon_1, \dots, \epsilon_\ell)$ with entries in $\mathbb{Q}$ such that $D^{-1}A$ is a symmetric matrix.

If A is symmetrizable, we can choose D such that all $\epsilon_i > 0$: Since $D^{-1}A$ is symmetric, we get

$$\epsilon_i^{-1} a_{i,j} = \epsilon_j^{-1} a_{j,i}. \tag{1}$$

In particular, whenever $a_{i,j} \neq 0$, $\frac{\epsilon_i}{\epsilon_j} > 0$, i.e., on each indecomposable component of A, ϵ takes the same sign. This proves the assertion that each ϵ_i can be chosen to be positive.

From (1) it is clear that for a symmetrizable GCM A, there exists a unique $D = \text{diag}(\epsilon_1, \dots, \epsilon_\ell)$ satisfying:

(a) $D^{-1}A$ is symmetric.
(b) Each ϵ_i is a positive integer.
(c) If $D' = \text{diag}(\epsilon_1', \dots, \epsilon_\ell')$ is another matrix satisfying (a) and (b), then $\epsilon_i \leq \epsilon_i'$ for all i.

We call D as above the *minimal* D.

The Kac–Moody Lie algebra $\mathfrak{g} = \mathfrak{g}(A)$ is called *symmetrizable* if A is symmetrizable. Clearly, for any subset $Y \subset \{1, \dots, \ell\}$, the sub GCM $A_Y := (a_{i,j})_{i,j \in Y}$ is again symmetrizable.

1.5.2 Proposition. *Let $\mathfrak{g} = \mathfrak{g}(A)$ be a symmetrizable Kac–Moody algebra. Then $\mathfrak{h}$ carries a nondegenerate symmetric W-invariant bilinear form.*

Proof. Let $\mathfrak{h}' := \bigoplus_{i=1}^{\ell} \mathbb{C}\alpha_i^\vee \subset \mathfrak{h}$ be as in 1.2.3. Pick a vector space complement $\mathfrak{h}''$ of $\mathfrak{h}'$ in $\mathfrak{h}$, i.e., $\mathfrak{h} = \mathfrak{h}' \oplus \mathfrak{h}''$. Set $\langle \mathfrak{h}'', \mathfrak{h}'' \rangle = 0$ and define, for any $1 \le i \le \ell$ and $h \in \mathfrak{h}$,

$$(1) \qquad \langle h, \alpha_i^\vee \rangle = \langle \alpha_i^\vee, h \rangle = \alpha_i(h)\epsilon_i,$$

where $D = \operatorname{diag}(\epsilon_1, \ldots, \epsilon_\ell)$ is minimal. Note that by (1.5.1.1), $\alpha_i(\alpha_j^\vee)\epsilon_i = \alpha_j(\alpha_i^\vee)\epsilon_j$ and hence (1) is well defined.

We first prove that $\langle\ ,\ \rangle$ is nondegenerate: Take $h_o \in \mathfrak{h}$ and assume that $\langle h_o, h \rangle = 0$, for all $h \in \mathfrak{h}$. In particular, $\alpha_i(h_o) = 0$ for all $1 \le i \le \ell$ and hence $h_o \in \mathfrak{h}'$ by Corollary 1.2.3(b). Write $h_o = \sum_{i=1}^{\ell} z_i \alpha_i^\vee$, with $z_i \in \mathbb{C}$. Hence $\langle h_o, h \rangle = \sum_{i=1}^{\ell} z_i \epsilon_i \alpha_i(h)$. But since $\{\alpha_1, \ldots, \alpha_\ell\}$ are linearly independent, we get $z_i = 0$ for all i. This proves that $h_o = 0$ and hence $\langle\ ,\ \rangle$ is nondegenerate.

Now we show that $\langle\ ,\ \rangle$ is W-invariant: For $h_1, h_2 \in \mathfrak{h}$ and simple reflection s_j:

$$\begin{aligned} \langle s_j h_1, s_j h_2 \rangle &= \langle h_1 - \alpha_j(h_1)\alpha_j^\vee, h_2 - \alpha_j(h_2)\alpha_j^\vee \rangle \\ &= \langle h_1, h_2 \rangle - \alpha_j(h_2)\alpha_j(h_1)\epsilon_j - \alpha_j(h_1)\alpha_j(h_2)\epsilon_j \\ &\quad + 2\alpha_j(h_1)\alpha_j(h_2)\epsilon_j = \langle h_1, h_2 \rangle. \end{aligned}$$

This shows that $\langle\ ,\ \rangle$ is W-invariant. ☐

1.5.3 Definition. Any nondegenerate symmetric W-invariant bilinear form $\langle\ ,\ \rangle$ on $\mathfrak{h}$ satisfying the condition (1.5.2.1) is called a *normalized invariant form on* $\mathfrak{h}$. Of course, a normalized invariant form is unique (not only up to scalar multiples) restricted to $\mathfrak{h}'$, but is *not* unique on the whole of $\mathfrak{h}$. It depends, e.g., on the choice of a vector space complement $\mathfrak{h}''$ of $\mathfrak{h}'$ in $\mathfrak{h}$. Also see Exercise 1.5.E.2.

Define the map $\nu : \mathfrak{h} \to \mathfrak{h}^*$ by

$$\nu(h)(h') = \langle h, h' \rangle, \qquad \text{for } h, h' \in \mathfrak{h}.$$

We transport the form $\langle\ ,\ \rangle$ via the W-equivariant isomorphism ν to get a W-equivariant nondegenerate symmetric bilinear form (again denoted by) $\langle\ ,\ \rangle$ on $\mathfrak{h}^*$ called a *normalized invariant form on* $\mathfrak{h}^*$.

1.5.4 Theorem. *Let $\mathfrak{g}$ be a symmetrizable Kac–Moody Lie algebra. Then there exists a bilinear form $\langle\ ,\ \rangle$ on $\mathfrak{g}$ satisfying the following:*

(a_1) *$\langle\ ,\ \rangle$ is invariant, i.e., $\langle [x, y], z \rangle + \langle y, [x, z] \rangle = 0$, for all $x, y, z \in \mathfrak{g}$,*

(a_2) *$\langle\ ,\ \rangle_{|\mathfrak{h}}$ is a normalized invariant form.*

Moreover, such a form is unique (if we fix its restriction $\langle\, ,\, \rangle_{|\mathfrak{h}}$ *) and automatically symmetric. In addition, this unique form satisfies:*

(b$_1$) $\langle \mathfrak{g}_\alpha, \mathfrak{g}_\beta\rangle = 0$ *unless* $\alpha + \beta = 0,\ \alpha, \beta \in \Delta \cup \{0\}$,

(b$_2$) $[x, y] = \langle x, y\rangle \nu^{-1}(\alpha)$ *for* $x \in \mathfrak{g}_\alpha,\ y \in \mathfrak{g}_{-\alpha}$ *and* $\alpha \in \Delta$.

Proof. For any $k \in \mathbb{Z}$, set $\mathfrak{g}_k = \bigoplus_{\alpha\in\Delta\cup\{0\}, |\alpha|=k} \mathfrak{g}_\alpha$ (cf. (1.2.2.3)), and set $\mathfrak{g}(N) = \bigoplus_{k=-N}^{N} \mathfrak{g}_k$ for $N \geq 0$. Extend the bilinear form of Proposition 1.5.2 to $\mathfrak{g}(1)$ by

(1)
$$\langle f_j, e_i\rangle = \langle e_i, f_j\rangle = \delta_{ij}\epsilon_i \ (i, j = 1, \dots, \ell);$$
$$\langle \mathfrak{g}_{k_1}, \mathfrak{g}_{k_2}\rangle = 0, \qquad \text{if } k_1 + k_2 \neq 0, \text{ and } |k_1|, |k_2| \leq 1.$$

Then the form $\langle\, ,\, \rangle$ on $\mathfrak{g}(1)$ satisfies condition (a$_1$) as long as all of x, y, z, $[x, y]$ and $[x, z]$ lie in $\mathfrak{g}(1)$. Indeed, it is sufficient to check that

(2)
$$\langle [e_i, f_j], h\rangle = -\langle f_j, [e_i, h]\rangle \quad \text{for } h \in \mathfrak{h},$$

or, equivalently,
$$\delta_{ij}\langle \alpha_i^\vee, h\rangle = \delta_{ij}\epsilon_i\langle \alpha_j, h\rangle,$$
which is true by the definition of $\langle\, ,\, \rangle$ on $\mathfrak{h}$.

Now extend $\langle\, ,\, \rangle$ to a bilinear form on the space $\mathfrak{g}(N)$ by induction on $N \geq 1$ so that $\langle \mathfrak{g}_{k_1}, \mathfrak{g}_{k_2}\rangle = 0$ if $|k_1|, |k_2| \leq N$ and $k_1 + k_2 \neq 0$, and also condition (a$_1$) is satisfied as long as all of x, y, z, $[x, y]$ and $[x, z]$ lie in $\mathfrak{g}(N)$. Suppose that this is already defined on $\mathfrak{g}(N-1)$; then we have only to define $\langle x, y\rangle$ for $x \in \mathfrak{g}_{\pm N},\ y \in \mathfrak{g}_{\mp N}$. We can write $y = \sum_i [u_i, v_i]$ where u_i and v_i are homogeneous elements of nonzero degree which lie in $\mathfrak{g}(N-1)$. Then $[x, u_i] \in \mathfrak{g}(N-1)$ and we set

(3)
$$\langle x, y\rangle = \sum_i \langle [x, u_i], v_i\rangle.$$

To show that this is well defined, we prove that if $i, j, s, t \in \mathbb{Z}$ are such that $|i + j| = |s + t| = N,\ i + j + s + t = 0,\ |i|, |j|, |s|, |t| < N$ and $x_i \in \mathfrak{g}_i, x_j \in \mathfrak{g}_j, x_s \in \mathfrak{g}_s, x_t \in \mathfrak{g}_t$, then we have (on $\mathfrak{g}(N-1)$)

(4)
$$\langle [[x_i, x_j], x_s], x_t\rangle = \langle x_i, [x_j, [x_s, x_t]]\rangle.$$

Indeed, using the invariance of $\langle\, ,\, \rangle$ on $\mathfrak{g}(N-1)$ and the Lie algebra axioms, we have

$$\begin{aligned}
\langle [[x_i, x_j], x_s], x_t\rangle &= \langle [[x_i, x_s], x_j], x_t\rangle - \langle [[x_j, x_s], x_i], x_t\rangle \\
&= \langle [x_i, x_s], [x_j, x_t]\rangle + \langle x_i, [[x_j, x_s], x_t]\rangle \\
&= \langle x_i, [x_s, [x_j, x_t]] + [[x_j, x_s], x_t]\rangle \\
&= \langle x_i, [x_j, [x_s, x_t]]\rangle.
\end{aligned}$$

If now $x = \sum_p [u'_p, v'_p]$, then by definition (3) and (4) we have

$$\langle x, y\rangle = \sum_i \langle [x, u_i], v_i\rangle = \sum_p \langle u'_p, [v'_p, y]\rangle.$$

Hence this is independent of the choice of the expressions for x and y.

We further show by induction on $N > 0$ that for $x \in \mathfrak{g}_N$, $y \in \mathfrak{g}_{-N}$ and $h \in \mathfrak{h}$,

$$\langle [x, y], h\rangle + \langle y, [x, h]\rangle = 0. \tag{5}$$

For $N = 1$, this is true because of (2). Now take $N > 1$ and write $x = \sum_i [u_i, v_i]$, for $u_i \in \mathfrak{g}_{k_i}$ and $v_i \in \mathfrak{g}_{m_i}$ with $0 < k_i, m_i < N$. Then

$$\begin{aligned}\langle y, [x, h]\rangle &= \sum_i \langle y, [[u_i, v_i], h]\rangle \\ &= \sum_i \langle y, [u_i, [v_i, h]]\rangle - \sum_i \langle y, [v_i, [u_i, h]]\rangle \\ &= -\sum \langle [u_i, y], [v_i, h]\rangle + \sum \langle [v_i, y], [u_i, h]\rangle \\ &= \sum \langle [v_i, [u_i, y]] - [u_i, [v_i, y]], h\rangle, \text{ by induction} \\ &= -\langle [x, y], h\rangle.\end{aligned}$$

This proves (5).

It is clear from the definition and (5) that (a_1) holds on $\mathfrak{g}(N)$ whenever $x, y, z, [x, y]$ and $[x, z]$ lie in $\mathfrak{g}(N)$. Hence we have constructed a bilinear form $\langle\, ,\, \rangle$ on $\mathfrak{g}$ such that (a_1) and (a_2) hold.

The form $\langle\, ,\, \rangle$ satisfies (b_1) since for $h \in \mathfrak{h}$, $x \in \mathfrak{g}_\alpha$ and $y \in \mathfrak{g}_\beta$ we have, by the invariance property,

$$0 = \langle [h, x], y\rangle + \langle x, [h, y]\rangle = \langle \alpha + \beta, h\rangle \langle x, y\rangle.$$

The verification of (b_2) is easy: For $x \in \mathfrak{g}_\alpha$, $y \in \mathfrak{g}_{-\alpha}$, where $\alpha \in \Delta$ and $h \in \mathfrak{h}$, we have

$$\langle [x, y] - \langle x, y\rangle \nu^{-1}(\alpha), h\rangle = \langle x, [y, h]\rangle - \langle x, y\rangle\langle \alpha, h\rangle = 0.$$

Now (b_2) follows from Proposition 1.5.2.

It follows from (a_2), (b_1), and (b_2) that the bilinear form $\langle\, ,\, \rangle$ is symmetric. The uniqueness of $\langle\, ,\, \rangle$ follows from (b_1)–(b_2). □

1.5.5 Remark. We call the invariant form $\langle , \rangle$ a *normalized invariant form on* $\mathfrak{g}$. By the above theorem this is unique once the choice of a normalized invariant form on $\mathfrak{h}$ is made.

As seen below, combined with Corollary 3.2.10, the form $\langle\, , \,\rangle$ is nondegenerate on $\mathfrak{g}$.

1.5.6 Definition. We take $\mathfrak{g} = \mathfrak{g}(A)$ to be any (not necessarily symmetrizable) Kac–Moody algebra. Let $\mathfrak{r} \subset \mathfrak{g} = \mathfrak{g}(A)$ be the maximal ideal such that $\mathfrak{r} \cap \mathfrak{h} = \{0\}$. Since any ideal $\mathfrak{s} \subset \mathfrak{g}$ is $\mathfrak{h}$-stable under the adjoint action; in particular, by 1.2.1,

$$\mathfrak{s} = (\oplus_{\alpha\in\Delta}\mathfrak{s} \cap \mathfrak{g}_\alpha) \oplus (\mathfrak{h} \cap \mathfrak{s}),$$

and hence the sum $\mathfrak{r}$ of all the ideals $\mathfrak{s}$ satisfying $\mathfrak{s} \cap \mathfrak{h} = \{0\}$ again satisfies $\mathfrak{r} \cap \mathfrak{h} = 0$. The ideal $\mathfrak{r}$ is called the *radical ideal* of $\mathfrak{g}$. Now define

$$\bar{\mathfrak{g}}(A) = \mathfrak{g}(A)/\mathfrak{r}.$$

(The Lie algebra $\bar{\mathfrak{g}}(A)$ should not be confused with $\tilde{\mathfrak{g}}(A)$ defined in 1.1.2.)

Observe that for any $\alpha \in \Delta$,

$$(1) \qquad [\mathfrak{r} \cap \mathfrak{g}_\alpha, \mathfrak{g}_{-\alpha}] \subset \mathfrak{r} \cap \mathfrak{h} = \{0\}.$$

Clearly, we have the triangular and root space decompositions, i.e., the analogue of Theorem 1.2.1(a)–(c) for $\bar{\mathfrak{g}}$. We denote the set of roots, resp. positive roots, resp. negative roots, of $\bar{\mathfrak{g}}$ by $\bar{\Delta}$, resp. $\bar{\Delta}^+$, resp. $\bar{\Delta}^-$. Similar to the Lie subalgebras $\mathfrak{b}$, $\mathfrak{b}^-$, $\mathfrak{n}$, $\mathfrak{n}^-$, $\mathfrak{p}_Y$, $\mathfrak{g}_Y$, $\mathfrak{u}_Y$, $\mathfrak{u}_Y^-$ of $\mathfrak{g}$, we define the Lie subalgebras $\bar{\mathfrak{b}}$, $\bar{\mathfrak{b}}^-$, $\bar{\mathfrak{n}}$, $\bar{\mathfrak{n}}^-$, $\bar{\mathfrak{p}}_Y$, $\bar{\mathfrak{g}}_Y$, $\bar{\mathfrak{u}}_Y$, $\bar{\mathfrak{u}}_Y^-$ respectively of $\bar{\mathfrak{g}}$.

It is easy to see that $\bar{\mathfrak{g}}_Y$ has no nonzero ideals $\mathfrak{r}_Y$ such that $\mathfrak{r}_Y \cap \mathfrak{h} = \{0\}$. Otherwise, take a nonzero root vector $x \in \mathfrak{r}_Y$ corresponding to a negative root α such that no negative $\gamma > \alpha$ is a root of $\mathfrak{r}_Y$ (cf. 2.1.1(a) for the definition of the relation $>$). Then $[\bar{\mathfrak{n}}, x] = 0$ since the set of elements $y \in \bar{\mathfrak{n}} : [y, x] = 0$ forms a Lie subalgebra, and hence the ideal $I_x \subset \bar{\mathfrak{g}}$ generated by x is nonzero and $I_x \cap \mathfrak{h} = \{0\}$. A contradiction to the definition of $\bar{\mathfrak{g}}$.

We will prove later (cf. Corollary 3.2.10) that, for symmetrizable A, $\mathfrak{r} = \{0\}$, hence $\bar{\mathfrak{g}}(A) = \mathfrak{g}(A)$.

The following result follows easily from Theorem 1.5.4 using (1) above.

1.5.7 Corollary. *With the same hypothesis and notation as in Theorem 1.5.4, the bilinear form $\langle\, , \,\rangle$ descends to a nondegenerate (invariant symmetric) bilinear form (denoted by $\langle\, , \,\rangle_{\bar{\mathfrak{g}}}$) on $\bar{\mathfrak{g}}$.* □

1.5.8 Definition. We define a certain completion $\hat{U}(\bar{\mathfrak{g}})$ of the universal enveloping algebra $U(\bar{\mathfrak{g}})$ as follows:

$$\hat{U}(\bar{\mathfrak{g}}) := \prod_{d\geq 0} U(\bar{\mathfrak{b}}^-) \otimes_{\mathbb{C}} U_d(\bar{\mathfrak{n}}),$$

where $U_d(\bar{\mathfrak{n}}) \subset U(\bar{\mathfrak{n}})$ is the set of homogeneous elements of (principal) degree d (cf. (1.2.2.3)).

The product in $U(\bar{\mathfrak{g}})$ extends to a product in $\hat{U}(\bar{\mathfrak{g}})$ defined as follows:

$$(1)\quad \left(\sum_{d\geq 0} x_d\right)\cdot\left(\sum_{m\geq 0} y_m\right) = \sum_k \sum_{d,m\geq o} (x_d y_m)_k, \text{ for } x_d, y_d \in U(\bar{\mathfrak{b}}^-)\otimes U_d(\bar{\mathfrak{n}}),$$

where $(x_d y_m)_k$ denotes the component of $x_d y_m$ in the factor $U(\bar{\mathfrak{b}}^-)\otimes U_k(\bar{\mathfrak{n}})$ under the decomposition

$$U(\bar{\mathfrak{g}}) = \oplus_{d\geq 0} U(\bar{\mathfrak{b}}^-)\otimes U_d(\bar{\mathfrak{n}}).$$

Observe that for any fixed k, there exist only finitely many $d, m \geq 0$ such that $(x_d y_m)_k \neq 0$ and hence the product given by (1) makes sense. The product defined above makes $\hat{U}(\bar{\mathfrak{g}})$ into an associative algebra.

We fix a normalized (nondegenerate) invariant form $\langle\, ,\, \rangle_{\bar{\mathfrak{g}}}$ on (symmetrizable) $\bar{\mathfrak{g}}$ (cf. Corollary 1.5.7). For any $\alpha \in \bar{\Delta}^+$ choose a basis $\{e^1_\alpha, \dots, e^p_\alpha\}$ of $\bar{\mathfrak{g}}_\alpha$ and a dual basis $\{e^1_{-\alpha}, \dots, e^p_{-\alpha}\}$ of $\bar{\mathfrak{g}}_{-\alpha}$, i.e., $\langle e^i_\alpha, e^j_{-\alpha}\rangle = \delta_{i,j}$. Now define the element

$$\Omega_\alpha = \sum_{i=1}^p e^i_{-\alpha} e^i_\alpha \in U(\bar{\mathfrak{g}}).$$

Similarly, define the element

$$\Omega_0 = \sum_k u_k u^k,$$

where $\{u_k\}$ and $\{u^k\}$ are dual bases of $\mathfrak{h}$.

Also choose (as earlier) an element $\rho \in \mathfrak{h}^*$ satisfying $\langle \rho, \alpha_i^\vee\rangle = 1$, for all simple coroots $\alpha_i^\vee$. Now define the *Casimir–Kac element*

$$\Omega := 2\nu^{-1}(\rho) + \Omega_0 + 2\sum_{\alpha\in\bar{\Delta}^+} \Omega_\alpha \in \hat{U}(\bar{\mathfrak{g}}),$$

where $\nu : \mathfrak{h} \to \mathfrak{h}^*$ is as defined in 1.5.3. Similarly, for any subset Y of $\{1, \dots, \ell\}$, define

$$\Omega^Y := 2\nu^{-1}(\rho) + \Omega_0 + 2\sum_{\alpha\in\bar{\Delta}^+_Y} \Omega_\alpha \in \hat{U}(\bar{\mathfrak{g}}_Y).$$

By the following lemma, Ω depends only on the choice of ρ.

1.5.9 Lemma. *For any* $\alpha \in \bar{\Delta}^+ \cup \{0\}$, *the element* Ω_α *is independent of the choice of dual bases as above, but, of course, depends on the choice of* $\langle\, ,\, \rangle_{\bar{\mathfrak{g}}}$.

Proof. Consider the element $x_\alpha := \sum_i e^i_{-\alpha} \otimes e^i_\alpha \in \bar{\mathfrak{g}}_{-\alpha} \otimes \bar{\mathfrak{g}}_\alpha \approx \bar{\mathfrak{g}}^*_\alpha \otimes \bar{\mathfrak{g}}_\alpha = \mathrm{End}(\bar{\mathfrak{g}}_\alpha)$, where $\bar{\mathfrak{g}}_{-\alpha} \approx \bar{\mathfrak{g}}^*_\alpha$ via the form $\langle\, ,\, \rangle$, and we set $\bar{\mathfrak{g}}_0 = \mathfrak{h}$. Now the element x_α is the unique element in $\bar{\mathfrak{g}}_{-\alpha} \otimes \bar{\mathfrak{g}}_\alpha$ which corresponds to the identity map of $\bar{\mathfrak{g}}_\alpha$ under the above identification. This shows that the element x_α does not depend on the choice of dual bases. In particular, this implies that the element Ω_α is independent of the choice of dual bases. □

1.5.10 Lemma. *If* $\alpha, \beta \in \bar{\Delta}$ *and* $z \in \bar{\mathfrak{g}}_{\beta-\alpha}$, *then, in* $\bar{\mathfrak{g}}(A) \otimes \bar{\mathfrak{g}}(A)$, *we have that*

$$\sum_s e^s_{-\alpha} \otimes [z, e^s_\alpha] = \sum_t [e^t_{-\beta}, z] \otimes e^t_\beta, \tag{1}$$

where $\{e^s_\alpha\}_s$, *resp.* $\{e^t_\beta\}_t$, *is a basis of* $\bar{\mathfrak{g}}_\alpha$, *resp.* $\bar{\mathfrak{g}}_\beta$. *In particular,*

$$\sum_s [e^s_{-\alpha}, [z, e^s_\alpha]] = -\sum_t [[z, e^t_{-\beta}], e^t_\beta] \ \text{ in } \bar{\mathfrak{g}}(A), \tag{2}$$

$$\sum_s e^s_{-\alpha}[z, e^s_\alpha] = -\sum_t [z, e^t_{-\beta}]e^t_\beta \ \text{ in } U(\bar{\mathfrak{g}}(A)). \tag{3}$$

Proof. We define the bilinear form $\langle\, ,\, \rangle$ on $\bar{\mathfrak{g}}(A) \otimes \bar{\mathfrak{g}}(A)$ by $\langle x \otimes y, x_1 \otimes y_1 \rangle = \langle x, x_1 \rangle \langle y, y_1 \rangle$. Take $e \in \bar{\mathfrak{g}}_\alpha$ and $f \in \bar{\mathfrak{g}}_{-\beta}$. It suffices to check that pairing both sides of (1) with $e \otimes f$ gives the same result. We have

$$\begin{aligned}\sum_s \langle e^s_{-\alpha} \otimes [z, e^s_\alpha], e \otimes f \rangle &= \sum_s \langle e^s_{-\alpha}, e \rangle \langle [z, e^s_\alpha], f \rangle \\ &= \sum \langle e^s_{-\alpha}, e \rangle \langle e^s_\alpha, [f, z] \rangle = \langle e, [f, z] \rangle\end{aligned}$$

by the invariance of $\langle\, ,\, \rangle$. Similarly,

$$\sum_t \langle [e^t_{-\beta}, z] \otimes e^t_\beta, e \otimes f \rangle = \sum_t \langle e^t_{-\beta}, [z, e] \rangle \langle e^t_\beta, f \rangle = \langle [z, e], f \rangle.$$

Applying again the invariance of $\langle\, ,\, \rangle$, we get (1). The identities (2) and (3) follow from (1) by applying the maps $\bar{\mathfrak{g}} \otimes \bar{\mathfrak{g}} \to \bar{\mathfrak{g}}$, $x \otimes y \mapsto [x, y]$ and $\bar{\mathfrak{g}} \otimes \bar{\mathfrak{g}} \to U(\bar{\mathfrak{g}})$, $x \otimes y \mapsto xy$. □

1.5.11 Theorem. *The element* Ω *defined in 1.5.8 lies in the center of* $\hat{U}(\bar{\mathfrak{g}})$.

Proof. Since the centralizer $Z(\Omega)$ of Ω in $\hat{U}(\bar{\mathfrak{g}})$ is a subalgebra, it suffices to show that e_i, f_i and any $h \in \mathfrak{h}$ commute with Ω. Since the weight of any Ω_α is 0, h

commutes with Ω_α and hence h commutes with Ω. Denote by $\bar{\Omega} = 2\sum_{\alpha\in\bar{\Delta}_+}\Omega_\alpha$. Then

$$[\bar{\Omega}, e_{\alpha_i}] = 2\sum_{\alpha\in\bar{\Delta}_+}\left(\sum_s [e^s_{-\alpha}, e_{\alpha_i}]e^s_\alpha + e^s_{-\alpha}[e^s_\alpha, e_{\alpha_i}]\right) = 2[e_{-\alpha_i}, e_{\alpha_i}]e_{\alpha_i}$$

$$+2\sum_{\alpha\in\bar{\Delta}_+\setminus\{\alpha_i\}}\left(\sum_s [e^s_{-\alpha}, e_{\alpha_i}]e^s_\alpha + \sum_t [e_{\alpha_i}, e^t_{-\alpha-\alpha_i}]e^t_{\alpha+\alpha_i}\right),$$

by (1.5.10.3)

$$= -2\nu^{-1}(\alpha_i)e_{\alpha_i}$$

(1) $$= -2\langle\alpha_i, \alpha_i\rangle e_{\alpha_i} - 2e_{\alpha_i}\nu^{-1}(\alpha_i).$$

For $x \in \bar{\mathfrak{g}}_\alpha$,

$$[\Omega_0, x] = \left[\sum u_k u^k, x\right]$$

$$= \sum_k \left(\alpha(u_k)xu^k + \alpha(u^k)u_k x\right)$$

$$= \sum_k \left(\alpha(u_k)xu^k + \alpha(u^k)xu_k + \alpha(u^k)\alpha(u_k)x\right)$$

(2) $$= 2x\nu^{-1}(\alpha) + \langle\alpha, \alpha\rangle x .$$

Further,

(3) $$\nu^{-1}(\alpha_i) = \alpha_i^\vee/\epsilon_i,$$

as can be seen by taking $\langle\nu^{-1}(\alpha_i), h\rangle$, $h \in \mathfrak{h}$. Thus

(4) $$2\langle\rho, \alpha_i\rangle = \langle\alpha_i, \alpha_i\rangle,$$

for all the simple roots α_i. Hence

$$[2\nu^{-1}(\rho) + \Omega_0, e_{\alpha_i}] = 2\langle\alpha_i, \nu^{-1}\rho\rangle e_{\alpha_i} + 2e_{\alpha_i}\nu^{-1}(\alpha_i) + \langle\alpha_i, \alpha_i\rangle e_{\alpha_i}$$

(5) $$= 2\langle\alpha_i, \alpha_i\rangle e_{\alpha_i} + 2e_{\alpha_i}\nu^{-1}(\alpha_i).$$

Combining (1) and (5) we get that $[\Omega, e_{\alpha_i}] = 0$. Similarly, it can be seen that $[\Omega, e_{-\alpha_i}] = 0$. This completes the proof of the theorem. □

1.5.12 Remark. The same proof as above shows that Ω^Y lies in the center of $\hat{U}(\bar{\mathfrak{g}}_Y)$.

1.5.E EXERCISES

(1) A GCM $A = (a_{i,j})_{1\le i,j\le \ell}$ is symmetrizable iff

$$a_{i_1,i_2}a_{i_2,i_3}\cdots a_{i_k,i_1} = a_{i_2,i_1}a_{i_3,i_2}\cdots a_{i_1,i_k}$$

for all $i_1, \dots, i_k$.

(2) Let $A = (a_{i,j})_{1\le i,j\le \ell}$ be an indecomposable GCM. Assume that there is a nonzero W-invariant bilinear form (,) on $\mathfrak{h}' := \oplus_{i=1}^{\ell} \mathbb{C}\alpha_i^\vee$. (We do not assume that (,) is nondegenerate nor do we assume that it is symmetric.) Then show the following:

(a) $\mathfrak{g}(A)$ is symmetrizable.

(b) (,) is unique up to scalar multiples, i.e., there exists a nonzero $c \in \mathbb{C}$ such that $(\, , \,) = c\langle \, , \, \rangle_{\mathfrak{h}'}$, where $\langle \, , \, \rangle$ is a normalized invariant symmetric bilinear form (cf. 1.5.3). In particular, (,) is automatically symmetric.

Hint: Denote $(\alpha_i^\vee, \alpha_i^\vee)/2$ by ϵ_i and show, by calculating $(s_j\alpha_i^\vee, s_j\alpha_i^\vee)$, that $a_{i,j}\epsilon_j = a_{j,i}\epsilon_i$ for all i, j. Thus, show that if one ϵ_i is zero, then so is each of ϵ_j. Assume now that one (and hence each) $\epsilon_j = 0$. In this case, calculate $(s_i\alpha_i^\vee, s_i\alpha_j^\vee)$ and show that $(\alpha_i^\vee, \alpha_j^\vee) = 0$ for all i, j. Thus $(\, , \,) \equiv 0$, a contradiction, and hence none of ϵ_i are zero. Taking D to be the diagonal matrix with entries $\epsilon_1, \dots, \epsilon_\ell$, we get that $D^{-1}A$ is symmetric. This proves (a). To prove (b), consider the form $(\, , \,) - c\langle \, , \, \rangle$, where $c := (\alpha_1^\vee, \alpha_1^\vee)/\langle \alpha_1^\vee, \alpha_1^\vee\rangle$. Then, by the above, it vanishes identically.

(3) Let $\mathfrak{g}(A)$ be finite dimensional. Show that, in this case, the Casimir element Ω takes the more traditional form

$$\Omega = \sum_k x_k y_k ,$$

where x_k is any basis of $\bar{\mathfrak{g}}(A)$ (not necessarily consisting of root vectors and elements of $\mathfrak{h}$) and y_k is the dual basis.

1.C Comments. Kac–Moody Lie algebras were introduced independently by [Kac–67] and [Moody–67]. The theory has grown tremendously in various directions ever since and has found numerous applications in diverse problems in mathematics and physics. The contents of this chapter are, by now, fairly standard, and, as mentioned in the beginning, a much more exhaustive treatment can be found in Kac's book [Kac–90]. Exercise 1.2.E.3 is taken from [Humphreys–72, Lemma 26.2]. Lemma 1.3.3 is taken from [Kac–85b]. Lemma 1.3.9 is taken from [Kac–Peterson–83]. The characterization of Coxeter groups as in Theorem 1.3.11 is taken from [Deodhar–86].

II

Representation Theory of Kac–Moody Algebras

The Bernstein–Gelfand–Gelfand category $\mathcal{O}$ is defined and its various basic properties are established (including the characterization of irreducible objects in $\mathcal{O}$ as irreducible quotients of Verma modules and characterization of integrable modules in $\mathcal{O}$) in Section 2.1. For any dominant integral weight λ, an integrable highest weight $\mathfrak{g}$-module $L^{\max}(\lambda)$ is explicitly constructed, and it is shown that any integrable highest weight $\mathfrak{g}$-module is a quotient of $L^{\max}(\lambda)$. An explicit expression for the action of the Casimir–Kac element (in the symmetrizable case) on any Verma module is obtained.

Section 2.2 is devoted to proving the Weyl–Kac character formula, which determines the character of an integrable highest weight $\mathfrak{g}$-module $V(\lambda)$ in terms of its highest weight λ, when $\mathfrak{g}$ is a symmetrizable Kac–Moody Lie algebra. This generalizes the classical Weyl character formula for finite-dimensional modules of (finite-dimensional) semisimple Lie algebras. The proof makes basic use of the Casimir–Kac operator.

As a consequence of the character formula, it is shown that any integrable highest weight $\mathfrak{g}$-module is irreducible and integrable modules in the category $\mathcal{O}$ are completely reducible for symmetrizable $\mathfrak{g}$. The Weyl–Kac character formula is the most central result in representation theory of Kac–Moody algebras and has led to various important applications in other branches of mathematics and physics, including Dedekind's η-function and Rogers–Ramanujan identities. A different proof of the Weyl–Kac character formula (still in the symmetrizable case) is given in the next chapter using the $\mathfrak{n}^-$-homology result of Garland–Lepowsky, extending the $\mathfrak{n}$-homology result of Kostant in the finite case. The Weyl–Kac character formula is extended for an arbitrary Kac–Moody algebra in Chapter 8 using geometric methods.

In Section 2.3, the Shapovalov bilinear form on the Verma modules $M(\lambda)$ is defined and its determinant is computed explicitly in the case of symmetrizable $\mathfrak{g}$. It turns out that the determinant breaks into a product of linear factors. This leads to the Jantzen character sum formula, which in turn determines the (irreducible) components of any Verma module $M(\lambda)$. A subset $K^{\text{w.g.}} \subset \mathfrak{h}^*$, which contains the Tits cone, is defined and it is shown that the Verma modules $M(\lambda)$, for $\lambda \in K^{\text{w.g.}}$, have a particularly simple description of their components.

For any $\lambda \in \mathfrak{h}^*_{\mathbb{R}}$, the Shapovalov form on $M(\lambda)$ gives rise to a contravariant nondegenerate Hermitian form on the irreducible $\mathfrak{g}$-module $L(\lambda)$. It is shown that this form on $L(\lambda)$ is positive definite iff $L(\lambda)$ is integrable. Further, an invariant symmetric bilinear form $\langle\cdot,\cdot\rangle$ on $\mathfrak{g}$, introduced in Chapter 1, gives rise to a contravariant Hermitian form $\{\ ,\ \}$ on $\mathfrak{g}$ by twisting $\langle\cdot,\cdot\rangle$ via a conjugate-linear antiautomorphism of order 2 of $\mathfrak{g}$. It is shown that $\{\cdot,\cdot\}$ restricted to $\mathfrak{n}^- \oplus \mathfrak{n}$ is positive definite.

Some results on the set $P(V)$ of weights of an integrable highest weight $\mathfrak{g}$-module V are contained in the Exercises 2.3.E.

Throughout this chapter $\mathfrak{g}$ *denotes an arbitrary Kac–Moody Lie algebra. We will impose the symmetrizability assumption on* $\mathfrak{g}$ *in Subsections 2.1.15–2.1.18 and Sections 2.2 and 2.3 (barring Subsections 2.3.1–2.3.3).*

2.1. Category $\mathcal{O}$

In this section $\mathfrak{g} = \mathfrak{g}(A)$ is an arbitrary Kac–Moody algebra. We impose the symmetrizability assumption on $\mathfrak{g}$ in Subsections 2.1.15–2.1.18.

2.1.1 Definition. (a) Define a partial order $\leq$ in $\mathfrak{h}^*$ by $\mu \leq \lambda$ if and only if $\lambda - \mu \in Q^+$, where Q^+ is as defined in the beginning of Section 1.2. Set $\mathfrak{h}^*_{\leq\lambda} = \{\mu \in \mathfrak{h}^* : \mu \leq \lambda\}$.

(b) For any $\lambda \in \mathfrak{h}^*$, let $\mathbb{C}_\lambda$ be the one-dimensional representation of $\mathfrak{h}$ in $\mathbb{C}$, defined by $x \cdot z = \lambda(x)z$, for $x \in \mathfrak{h}$ and $z \in \mathbb{C}$. We extend the $\mathfrak{h}$-module structure of $\mathbb{C}_\lambda$ to a $\mathfrak{b}$-module structure by demanding that $\mathfrak{n}$ act trivially on $\mathbb{C}_\lambda$.

We now define the *Verma module* $M(\lambda)$ as the induced module

(1) $$M(\lambda) := U(\mathfrak{g}) \otimes_{U(\mathfrak{b})} \mathbb{C}_\lambda,$$

where $U(\mathfrak{b})$ acts on $U(\mathfrak{g})$ via the right multiplication and the $\mathfrak{g}$-module structure on $M(\lambda)$ is obtained by the left multiplication on the $U(\mathfrak{g})$-factor.

By the PBW theorem, as a $\mathfrak{b}^-$-module,

(2) $$M(\lambda) \approx U(\mathfrak{n}^-) \otimes_{\mathbb{C}} \mathbb{C}_\lambda,$$

where $\mathfrak{n}^-$ acts on the right side via multiplication on the $U(\mathfrak{n}^-)$-factor and $\mathfrak{h}$ acts on the right side via the tensor product action ($\mathfrak{h}$ acting on $U(\mathfrak{n}^-)$ via the adjoint representation). In particular, $M(\lambda)$ is a weight module for the $\mathfrak{h}$-action and is free as a $U(\mathfrak{n}^-)$-module generated by $1 \otimes 1$.

Any $\mathfrak{g}$-module quotient $L \neq 0$ of a Verma module $M(\lambda)$ is called a *highest weight module with highest weight* λ. Observe that λ is a weight of L of multiplicity 1 and, moreover, any weight μ of L satisfies $\mu \leq \lambda$. From this it follows that $\mathrm{End}_{\mathfrak{g}}(L) = \mathbb{C}I_L$.

(c) Let $\mathcal{O}$ be the category of $\mathfrak{g}$-modules M satisfying the following:

(c_1) M is a weight module with finite-dimensional weight spaces,

(c_2) There exist finitely many $\lambda_1, \ldots, \lambda_k \in \mathfrak{h}^*$, depending on M, such that

$$P(M) \subset \cup_{j=1}^{k} \mathfrak{h}^*_{\leq \lambda_j},$$

where $P(M)$ is the set of weights of M.

The morphisms in $\mathcal{O}$ are all $\mathfrak{g}$-module morphisms.

Similarly, for any subset $Y \subset \{1, \ldots, \ell\}$, let $\mathcal{O}_Y$ be the category of $\mathfrak{g}_Y$-modules M satisfying (c_1) and (c_2) for the same $\mathfrak{h}$. The morphisms in $\mathcal{O}_Y$ are all the $\mathfrak{g}_Y$-module morphisms. Clearly, any submodule and quotient module of $M \in \mathcal{O}$ is again in the category $\mathcal{O}$. Further, for $M, N \in \mathcal{O}$, $M \oplus N$ and $M \otimes N \in \mathcal{O}$. It is easy to see that $M(\lambda) \in \mathcal{O}$ and hence any highest weight module $\in \mathcal{O}$.

The category $\mathcal{O}$ admits an involution σ defined as follows: Write $M \in \mathcal{O}$ as the sum of its weight spaces $M = \oplus_\mu M_\mu$, and define the *restricted dual* $M^\vee = \oplus_\mu M_\mu^*$. We put the $\mathfrak{g}$-module structure on $M^\vee$ by $(x \cdot f)v = -f(\omega(x)v)$, for $x \in \mathfrak{g}$, $f \in M^\vee$ and $v \in M$, where ω is the Cartan involution of Subsection 1.1.2. With this $\mathfrak{g}$-module structure we denote $M^\vee$ by M^σ.

(d) To $\lambda \in \mathfrak{h}^*$ associate the formal symbol e^λ, and let $\mathcal{A} = \mathcal{A}_\mathfrak{h}$ be the set of formal sums $a = \sum_{\lambda \in \mathfrak{h}^*} a_\lambda e^\lambda$ satisfying the following:

(d_1) $a_\lambda \in \mathbb{Z}$ for all λ;

(d_2) There exists a finite collection of elements $\lambda_1, \ldots, \lambda_k \in \mathfrak{h}^*$, depending on a, such that

$$a_\lambda = 0 \text{ for all } \lambda \notin \cup_{j=1}^{k} \mathfrak{h}^*_{\leq \lambda_j}.$$

Then $\mathcal{A}$ is an associative commutative algebra over $\mathbb{Z}$ under

$$\Big(\sum_\lambda a_\lambda e^\lambda\Big) \cdot \Big(\sum_\mu b_\mu e^\mu\Big) = \sum_{\theta \in \mathfrak{h}^*} \Big(\sum_{\lambda+\mu=\theta} a_\lambda b_\mu\Big) e^\theta.$$

Observe that for any θ, the sum $\sum_{\lambda+\mu=\theta} a_\lambda b_\mu$ is finite.

For any $M \in \mathcal{O}$, we define its *formal character*

$$\mathrm{ch}(M) = \sum_{\lambda \in \mathfrak{h}^*} (\dim M_\lambda)\, e^\lambda \in \mathcal{A}, \tag{3}$$

where $M = \oplus M_\lambda$ is the weight space decomposition. By the *multiplicity* of λ in M, denoted $\mathrm{mult}_\lambda(M)$, we mean the dimension of M_λ.

It is easy to see that, for $M, N \in \mathcal{O}$,

(4) $\operatorname{ch}(M \oplus N) = \operatorname{ch} M + \operatorname{ch} N$,
(5) $\operatorname{ch}(M \otimes N) = \operatorname{ch} M \cdot \operatorname{ch} N$,
(6) $\operatorname{ch}(M^\sigma) = \operatorname{ch} M$,
(7) $\operatorname{ch}(M/N) = \operatorname{ch} M - \operatorname{ch} N$, if N is a submodule of M,
(8) $(M^\sigma)^\sigma \simeq M$.

2.1.2 Lemma. *Any $M(\lambda)$ has a unique proper maximal $\mathfrak{g}$-submodule $M'(\lambda)$. Hence $M(\lambda)$ admits a unique irreducible quotient (denoted) $L(\lambda)$.*

Proof. Let M and N be two proper $\mathfrak{g}$-submodules of $M(\lambda)$. Then they are weight modules. In particular, the λ-th weight space $M_\lambda = N_\lambda = (0)$, since the λ-th weight space of $M(\lambda)$ generates $M(\lambda)$ as a $\mathfrak{g}$-module. So $M + N$ is again a proper $\mathfrak{g}$-submodule of $M(\lambda)$. The lemma follows by taking $M'(\lambda) = \sum_i M_i$, where the summation is taken over all proper $\mathfrak{g}$-submodules M_i of $M(\lambda)$. Since each $(M_i)_\lambda = (0)$, so is $(M'_\lambda)_\lambda = (0)$, and hence it is a proper submodule. □

2.1.3 Lemma. *For any irreducible module $L \in \mathcal{O}$, there exists a unique $\lambda \in \mathfrak{h}^*$ such that $L \simeq L(\lambda)$.*

Proof. Choose a maximal weight λ of L, i.e., if $\mu \geq \lambda$ and μ is a weight of L, then $\mu = \lambda$. By (c_2) of 2.1.1 such a λ always exists. Choose a nonzero vector $v_\lambda \in L$ of weight λ. Then since λ is a maximal weight, $\mathfrak{n} \cdot v_\lambda = 0$; in particular, there exists a $\mathfrak{b}$-module morphism $\mathbb{C}_\lambda \to L$ taking $1 \mapsto v_\lambda$, and hence a $\mathfrak{g}$-module morphism $\theta : M(\lambda) \to L$ taking $1 \otimes 1 \mapsto v_\lambda$. But since L is irreducible, θ is surjective and hence, by Lemma 2.1.2, $L \simeq L(\lambda)$.

To show uniqueness, let $L(\lambda) \simeq L(\mu)$. Any weight λ' of $L(\lambda)$ satisfies $\lambda' \leq \lambda$. In particular, $\lambda \leq \mu$ and $\mu \leq \lambda$. Hence $\lambda = \mu$. This proves the lemma. □

2.1.4 Lemma. *The module $L(\lambda)$, for any $\lambda \in \mathfrak{h}^*$, remains irreducible as a module for the commutator subalgebra $\mathfrak{g}' := [\mathfrak{g}, \mathfrak{g}]$.*

Proof. Let $V \subset L(\lambda)$ be a nonzero $\mathfrak{g}'$-submodule. Choose $0 \neq v = \sum v_k \in V$ such that $v_k \in L(\lambda)_{\lambda_k}$, $v_k \neq 0$, λ_k's are distinct and $\sum_k |\lambda - \lambda_k|$ is minimal. If $\lambda_k \neq \lambda$ for some k, then $e_j v_k \neq 0$ for some e_j. Otherwise, the $\mathfrak{g}$-submodule generated by v_k would be a proper $\mathfrak{g}$-submodule of $L(\lambda)$. Since $e_j v \in V$ and $e_j v \neq 0$, this contradicts the "minimal" choice of v. Thus we get that $v \in L(\lambda)_\lambda$. But then $V = L(\lambda)$, proving the lemma. □

2.1.5 Definition. Define the set of *dominant integral weights* $D \subset \mathfrak{h}^*$ by

(1) $$D = \{\lambda \in \mathfrak{h}^* : \langle \lambda, \alpha_i^\vee \rangle \in \mathbb{Z}_+, \text{ for all the simple coroots } \alpha_i^\vee\}.$$

For any $\lambda \in D$, define the $\mathfrak{g}$-submodule $M_1(\lambda) \subset M(\lambda)$ generated by the elements $\{f_i^{\lambda(\alpha_i^\vee)+1} \otimes 1\}_{1 \leq i \leq \ell}$. Now define the quotient module

$$L^{\max}(\lambda) := \frac{M(\lambda)}{M_1(\lambda)},$$

which we refer to as the *maximal integrable highest weight* $\mathfrak{g}$*-module with highest weight* λ (cf. Lemma 2.1.7).

2.1.6 Lemma. *For any* $1 \leq i, j \leq \ell$ *and* $\lambda \in D$, $e_j f_i^{\lambda(\alpha_i^\vee)+1} \otimes 1 = 0$ *as an element of* $M(\lambda)$. *In particular,* $M_1(\lambda)$ *is a proper submodule of* $M(\lambda)$.

Proof. For $j \neq i$, since e_j commutes with f_i, we have that

$$e_j f_i^{\lambda(\alpha_i^\vee)+1} \otimes 1 = f_i^{\lambda(\alpha_i^\vee)+1} e_j \otimes 1 = 0.$$

So take $j = i$. By the commutator identity in $U(s\ell_2)$ (cf. Exercise 1.2.E.3), for any $n \geq 1$,

$$(1) \qquad e_i f_i^n = f_i^n e_i + n f_i^{n-1}(\alpha_i^\vee - n + 1).$$

Applying the above identity for $n = \lambda(\alpha_i^\vee) + 1$, we get the lemma. □

2.1.7 Lemma. *For any* $\lambda \in D$, $L^{\max}(\lambda)$ *is an integrable* $\mathfrak{g}$*-module (cf. Subsection 1.3.2 for the definition of an integrable* $\mathfrak{g}$*-module). In particular, any quotient* $\mathfrak{g}$*-module of* $L^{\max}(\lambda)$ *is integrable.*

Conversely, any integrable highest weight $\mathfrak{g}$*-module* L *has highest weight* $\lambda \in D$*, and, moreover,* L *is a quotient of* $L^{\max}(\lambda)$.

Proof. Fix $1 \leq i \leq \ell$, and define $L_i = \{v \in L^{\max}(\lambda) : f_i^m v = 0$, for some $m = m(v)\}$. Then, by Lemma 1.3.3(b) and Corollary 1.3.4, L_i is a $\mathfrak{g}$-submodule of $L^{\max}(\lambda)$. But, by the definition of $L^{\max}(\lambda)$, $1 \otimes 1 \in L_i$. Hence $L_i = L^{\max}(\lambda)$ and thus f_i's act locally nilpotently on $L^{\max}(\lambda)$. Clearly e_i's act locally nilpotently on $L^{\max}(\lambda)$ since any weight μ of $L^{\max}(\lambda)$ satisfies $\mu \leq \lambda$. This proves the first part of the lemma.

To prove the converse part, let L be an integrable $\mathfrak{g}$-module which is a quotient of $M(\lambda)$. Then L is integrable as a $\mathfrak{g}(i) \approx s\ell(2)$-module (cf. Subsection 1.1.2); in particular, the $\mathfrak{g}(i)$-submodule of L generated by the weight space L_λ is finite-dimensional (by Lemma 1.3.3(c_2)) with highest weight λ. Hence, by the $s\ell_2$-module theory (cf. Exercise 1.2.E.4), $\lambda(\alpha_i^\vee) \in \mathbb{Z}_+$ and, moreover, $f_i^{\lambda(\alpha_i^\vee)+1} v_\lambda = 0$, where v_λ is a nonzero vector in L_λ. Thus L is a quotient of $L^{\max}(\lambda)$. □

2.1.8 Corollary. *The map* $\lambda \mapsto L(\lambda)$ *is a bijective correspondence between* D *and the set of isomorphism classes of integrable, irreducible highest weight* $\mathfrak{g}$*-modules.* □

2.1.9 Lemma. *Let* $V \in \mathcal{O}$ *and let* $\lambda \in \mathfrak{h}^*$. *Then there exists a filtration*

$$\{0\} = V_0 \subset V_1 \subset V_2 \subset \cdots \subset V_p = V$$

by $\mathfrak{g}$*-submodules* V_i *such that each* $W_i := V_{i+1}/V_i$ $(0 \leq i \leq p-1)$ *satisfies at least one of the following:*

(1) W_i *is irreducible with highest weight* $\lambda_i \geq \lambda$,
(2) $(W_i)_\mu = 0$ *for any* $\mu \geq \lambda$.

Proof. We prove the lemma by induction on $a(V, \lambda) := \sum_{\mu \geq \lambda} \dim V_\mu$. If $a(V, \lambda) = 0$, then the filtration $0 = V_0 \subset V_1 = V$ does the job. So assume that $a(V, \lambda) > 0$. Among the weights of V choose a maximal $\mu \geq \lambda$ and choose a nonzero vector $v_\mu \in V_\mu$. Then the $\mathfrak{g}$-submodule $W \subset V$ generated by v_μ is clearly a highest weight module with highest weight μ. By Lemma 2.1.2, W contains a maximal proper submodule W'. Then we have $0 \subset W' \subset W \subset V$. But $a(W', \lambda) < a(V, \lambda)$ and also $a(V/W, \lambda) < a(V, \lambda)$. Hence, by induction, W' and V/W admit filtrations satisfying the assumptions of the lemma. Combining these two filtrations we get a desired filtration for V. □

2.1.10 Lemma. *Let $M \in \mathcal{O}$. Then there exists a (possibly infinite) increasing filtration* $(0) = M_0 \subset M_1 \subset M_2 \subset \ldots$ *of $\mathfrak{g}$-submodules of M such that*

(i) $\cup_i M_i = M$,
(ii) M_i/M_{i-1} *is a highest weight module of highest weight λ_i, for all* $1 \leq i$,
(iii) $\lambda_i > \lambda_j \Rightarrow i < j$,
(iv) *for any weight λ of M, there exists r such that* $(M/M_r)_\lambda = 0$.

Proof. Since $M \in \mathcal{O}$, there exist $\mu_1, \ldots, \mu_k \in \mathfrak{h}^*$ such that the weights of M are contained in $\mathfrak{h}^*_{\leq \mu_1} \cup \cdots \cup \mathfrak{h}^*_{\leq \mu_k}$. It is easy to see that if $\mathfrak{h}^*_{\leq \mu_i} \cap \mathfrak{h}^*_{\leq \mu_j} \neq \emptyset$ for $i \neq j$, then there exists μ' such that $\mathfrak{h}^*_{\leq \mu_i} \cup \mathfrak{h}^*_{\leq \mu_j} \subseteq \mathfrak{h}^*_{\leq \mu'}$. Hence we may assume without loss of generality that $\mathfrak{h}^*_{\leq \mu_i} \cap \mathfrak{h}^*_{\leq \mu_j} = \emptyset$, for all $i \neq j$. Thus a weight μ of M is contained in a unique $\mathfrak{h}^*_{\leq \mu_i}$. Define $d(\mu) = \sum a_j$ where $\mu_i - \mu = \sum_{j=1}^{\ell} a_j \alpha_j$. Choose a weight λ_1 of M such that $d(\lambda_1)$ is minimal. Then clearly λ_1 is a maximal weight, i.e., if λ is a weight of M and $\lambda \geq \lambda_1$, then $\lambda = \lambda_1$. Choose a nonzero vector v_1 of weight λ_1 and let M_1 be the submodule of M generated by v_1. Then clearly M_1 is a highest weight module with v_1 as a generator. Consider M/M_1 and repeat the argument to get the required filtration. Since one chooses a weight λ with $d(\lambda)$ minimal at each stage, conditions (iii) and (iv) are automatically satisfied. Of course, (iv) implies (i). □

2.1.11 Definition. For $\mu \in \mathfrak{h}^*$ and $V \in \mathcal{O}$, we define the *multiplicity of $L(\mu)$ in V*, denoted $[V : L(\mu)]$ as follows:

Take any $\lambda \leq \mu$ and consider a filtration (denoted $\mathcal{F}_\lambda$) of V as in Lemma 2.1.9. Following the same notation as in Lemma 2.1.9, define

$$[V : L(\mu]_{\mathcal{F}_\lambda} = \#\{i : W_i \approx L(\mu)\}.$$

It is clear from Lemma 2.1.9 that

$$\text{(1)} \qquad \operatorname{ch} V = \sum_{\nu \geq \lambda} [V : L(\nu)]_{\mathcal{F}_\lambda} \operatorname{ch} L(\nu) + R_\lambda,$$

where $R_\lambda = \sum a_\theta(\lambda) e^\theta \in \mathcal{A}$ is such that $a_\theta(\lambda) = 0$ for $\theta \geq \lambda$; and there are only finitely many $\nu \geq \lambda$ such that $[V : L(\nu)]_{\mathcal{F}_\lambda} \neq 0$.

If we take a different filtration $\mathcal{F}'_{\lambda'}$ (corresponding to another $\lambda' \leq \mu$; we allow $\lambda' = \lambda$), then we claim that for any $\nu \geq \lambda$ and $\nu \geq \lambda'$,

$$[V : L(\nu)]_{\mathcal{F}_\lambda} = [V : L(\nu)]_{\mathcal{F}'_{\lambda'}}. \tag{2}$$

Otherwise, take a maximal $\nu_o \geq \lambda$ and $\nu_o \geq \lambda'$ such that $[V : L(\nu_o)]_{\mathcal{F}_\lambda} \neq [V : L(\nu_o)]_{\mathcal{F}'_{\lambda'}}$. But this would imply that the coefficient of e^{ν_o} on the right side of (1) is different for the two filtrations $\mathcal{F}_\lambda$ and $\mathcal{F}'_{\lambda'}$. This contradicts (1) and hence proves (2). In particular, $[V : L(\mu)]_{\mathcal{F}_\lambda}$ is independent of the choice of $\lambda \leq \mu$ and the filtration $\mathcal{F}_\lambda$. So we drop the subscript "$\mathcal{F}_\lambda$".

From the above, it is clear that $[V : L(\mu)] \neq 0$ if and only if $L(\mu)$ is a subquotient of V. In this case we say that $L(\mu)$ is a *component* of V.

A family of elements $\{a_i = \sum_\lambda a_i(\lambda)e^\lambda\}_{i\in I}$ of $\mathcal{A}$ indexed by a set I is called *locally finite* if for any $\lambda \in \mathfrak{h}^*$, the set $I_\lambda := \{i \in I : a_i(\lambda) \neq 0\}$ is finite. In this case, the sum $a = \sum_{i\in I} a_i$ clearly makes sense. (In general $a \notin \mathcal{A}$.)

2.1.12 Lemma. *For any $V \in \mathcal{O}$, the family $\{[V : L(\mu)] \operatorname{ch} L(\mu)\}_{\mu\in\mathfrak{h}^*}$ is locally finite. Further*

$$\operatorname{ch} V = \sum_{\mu\in\mathfrak{h}^*} [V : L(\mu)] \operatorname{ch} L(\mu). \tag{1}$$

Proof. Fix a $\lambda \in \mathfrak{h}^*$ and apply the decomposition (2.1.11.1) with respect to a filtration $\mathcal{F}_\lambda$. This gives that the coefficient of e^λ on both sides of (1) is the same. This proves the local finiteness of the family $\{[V : L(\mu)] \operatorname{ch} L(\mu)\}_{\mu\in\mathfrak{h}^*}$ as well as the decomposition (1). □

The following lemma follows easily from (2.1.1.2) and the PBW theorem.

2.1.13 Lemma. *For any $\lambda \in \mathfrak{h}^*$,*

$$\operatorname{ch} M(\lambda) = e^\lambda \prod_{\alpha\in\Delta^+} (1 - e^{-\alpha})^{-\operatorname{mult}\alpha},$$

where $(1 - e^{-\alpha})^{-\operatorname{mult}\alpha} := (1 + e^{-\alpha} + e^{-2\alpha} + \dots)^{\operatorname{mult}\alpha}$, and the (in general) infinite product $\prod_{\alpha\in\Delta^+}$ has the obvious meaning. □

2.1.14 Lemma. *A $\mathfrak{g}$-module $M \in \mathcal{O}$ is integrable if and only if all its components $L(\lambda)$ are integrable, i.e., $\lambda \in D$. In particular, for any integrable module $M \in \mathcal{O}$, M^σ is integrable.*

Proof. Of course the implication "⇒" is obvious. So we come to "⇐" and assume that all components of M are integrable. Since $M \in \mathcal{O}$, all e_i's act locally nilpotently on M. If M were not integrable, there exists a vector, which we can assume to be a weight vector of weight λ_o, $v \in M$ and some f_i, $1 \leq i \leq \ell$, such that no power of f_i kills v. Choose $q \in \mathbb{Z}_+$ large enough so that $\lambda_o + q'\alpha_i$ is not a weight of M for any $q' \geq q$. Now we claim that $\lambda_o - k\alpha_i$, for any $k \geq q + \langle \lambda_o, \alpha_i^\vee \rangle$, is not a weight of any component L of M: By Lemma 1.3.5(a), since L is integrable (by assumption)

$$\mathrm{mult}_{\lambda_o - k\alpha_i} L = \mathrm{mult}_{\lambda_o + [k - \langle \lambda_o, \alpha_i^\vee \rangle]\alpha_i}(L) = 0,$$

for $k \geq q + \langle \lambda_o, \alpha_i^\vee \rangle$. In particular, by Lemma 2.1.12, $\lambda_o - k\alpha_i$, for any $k \geq q + \langle \lambda_o, \alpha_i^\vee \rangle$, is not a weight of M. But this contradicts the assumption that no power of f_i kills v, proving that M is integrable.

Using (2.1.1.6) and Exercise 2.1.E, it is easy to see that the sets of components of M and M^σ are the same. Thus the "In particular" statement follows. □

2.1.15 Definition. Let $\mathfrak{g}$ be a symmetrizable Kac–Moody algebra. We fix a normalized invariant form $\langle\ \rangle$ on $\mathfrak{g}$. Recall the definition of the Casimir–Kac element $\Omega \in \widehat{U}(\bar{\mathfrak{g}})$ from Subsection 1.5.8. Similar to the definition of the category $\mathcal{O}$ for $\mathfrak{g}$-modules, we define the category $\bar{\mathcal{O}}$ of $\bar{\mathfrak{g}}$-modules. Also, for $\lambda \in \mathfrak{h}^*$, we define the Verma module $\bar{M}(\lambda)$ for $\bar{\mathfrak{g}}$. For any module $M \in \bar{\mathcal{O}}$, and $m \in M$, the element $\Omega(m)$ is well defined since it reduces to a finite sum. Moreover, Ω is a central element of $\widehat{U}(\bar{\mathfrak{g}})$ by Theorem 1.5.11. In particular, the map $\Omega_M : M \to M$, taking $m \mapsto \Omega m$, is a $\bar{\mathfrak{g}}$-module map. Moreover, for two $\bar{\mathfrak{g}}$-modules $M, N \in \bar{\mathcal{O}}$ and a $\bar{\mathfrak{g}}$-module homomorphism $\theta : M \to N$, the following diagram is commutative:

$$\begin{array}{ccc} M & \xrightarrow{\theta} & N \\ {\scriptstyle \Omega_M}\downarrow & & \downarrow{\scriptstyle \Omega_N} \\ M & \xrightarrow{\theta} & N. \end{array}$$

2.1.16 Lemma. *For any $\lambda \in \mathfrak{h}^*$,*

$$\Omega_{\bar{M}(\lambda)} = \langle \lambda, 2\rho + \lambda \rangle I,$$

where I is the identity map of $\bar{M}(\lambda)$ and $\langle\ \rangle$ is the normalized invariant form on $\mathfrak{h}^*$ induced from the restriction of $\langle\ \rangle$ to $\mathfrak{h}$ (cf. Subsection 1.5.3). In particular, for any subquotient M of $\bar{M}(\lambda)$, $\Omega_M = \langle \lambda, 2\rho + \lambda \rangle I_M$.

Proof. Since $\Omega_{\bar{M}(\lambda)}$ is a $\bar{\mathfrak{g}}$-module map, and $1 \otimes 1 \in \bar{M}(\lambda)$ generates $\bar{M}(\lambda)$ as a $\bar{\mathfrak{g}}$-module, it suffices to show that $\Omega_{\bar{M}(\lambda)}(1 \otimes 1) = \langle \lambda, \lambda + 2\rho \rangle (1 \otimes 1)$.

By the definition of Ω,

$$\begin{aligned} \Omega_{\bar{M}(\lambda)}(1 \otimes 1) &= (2\nu^{-1}(\rho) + \sum_k u_k u^k)(1 \otimes 1) \\ &= [\langle \lambda, 2\rho \rangle + \sum_k \lambda(u_k)\lambda(u^k)](1 \otimes 1) \\ &= \langle \lambda, 2\rho + \lambda \rangle (1 \otimes 1), \end{aligned}$$

where $\{u_k\}$ and $\{u^k\}$ are dual bases of $\mathfrak{h}$ as in Subsection 1.5.8. This proves the lemma. □

2.1.17 Definition. For any $M \in \bar{\mathcal{O}}$ and $z \in \mathbb{C}$, let M^z be the generalized z-eigenspace for the operator Ω_M, i.e., $M^z := \{v \in M : (\Omega_M - zI)^{n_v} v = 0$, for some $n_v > 0\}$. Clearly, M^z is a $\bar{\mathfrak{g}}$-submodule of M. Moreover, $\sum_{z\in\mathbb{C}} M^z \hookrightarrow M$ is a direct sum.

2.1.18 Lemma. *For any $M \in \bar{\mathcal{O}}$, $M = \oplus_{z\in\mathbb{C}} M^z$.*

Proof. By the definition of the category $\bar{\mathcal{O}}$, the weight space M_λ, for any $\lambda \in \mathfrak{h}^*$, is finite-dimensional. Moreover, $\Omega_M(M_\lambda) \subset M_\lambda$. In particular, $M_\lambda \subset \oplus_{z\in\mathbb{C}} M^z$. This proves the lemma. □

2.1.E EXERCISE. Let $a = \{a_\lambda\}_{\lambda\in\mathfrak{h}^*}$ be a family of integers such that $S_a := \{\lambda \in \mathfrak{h}^* : a_\lambda \neq 0\}$ is a subset of $\cup_{j=1}^k \mathfrak{h}^*_{\leq\lambda_j}$, for some $\lambda_j \in \mathfrak{h}^*$. Then show that the family $\{a_\lambda \operatorname{ch} L(\lambda)\}_{\lambda\in\mathfrak{h}^*}$ is locally finite. Moreover, assume that $\sum_\lambda a_\lambda \operatorname{ch} L(\lambda) = 0$. Then show that each $a_\lambda = 0$.

Hint: Choose a maximum λ_o such that $a_{\lambda_o} \neq 0$ and then show that the coefficient of e^{λ_o} in $\sum_\lambda a_\lambda \operatorname{ch} L(\lambda)$ is equal to a_{λ_o}.

2.2 Weyl–Kac Character Formula

In this section we assume that $\mathfrak{g}$ is symmetrizable and $\langle\ \rangle$ is any normalized invariant form on $\mathfrak{h}^*$ (cf. Subsection 1.5.3).

Similar to the definition of integrable highest weight $\mathfrak{g}$-modules, we can define integrable highest weight $\bar{\mathfrak{g}}$-modules. Of course, an integrable highest weight $\bar{\mathfrak{g}}$-module is an integrable highest weight $\mathfrak{g}$-module under the Lie algebra homomorphism $\mathfrak{g} \to \bar{\mathfrak{g}}$. Also any irreducible highest weight $\bar{\mathfrak{g}}$-module $\bar{L}(\lambda)$ (for any $\lambda \in \mathfrak{h}^*$) viewed as a $\mathfrak{g}$-module, is isomorphic to $L(\lambda)$.

2.2.1 Theorem. *Let L be an integrable highest weight $\bar{\mathfrak{g}}$-module with highest weight λ. Then*

$$\text{(1)} \qquad \operatorname{ch} L = \Big[\sum_{w\in W} \epsilon(w) e^{w*\lambda}\Big] \cdot \Big(\prod_{\alpha\in\bar{\Delta}^+} (1-e^{-\alpha})^{-\overline{\operatorname{mult}}\,\alpha}\Big), \text{ as elements of } \mathcal{A},$$

where $\bar{\Delta}^+$ is the set of positive roots for $\bar{\mathfrak{g}}$, $*$ is the shifted action of the Weyl group W (cf. Subsection 1.3.1), $\overline{\operatorname{mult}}\,\alpha$ denotes the dimension of the root space $\bar{\mathfrak{g}}_\alpha$ of $\bar{\mathfrak{g}}$, and ϵ is the sign representation of W (cf. Definition 1.3.12). In particular, we have the denominator formula:

$$\text{(2)} \qquad \prod_{\alpha\in\bar{\Delta}^+} (1-e^{-\alpha})^{\overline{\operatorname{mult}}\,\alpha} = \sum_{w\in W} \epsilon(w) e^{w*0}.$$

By Lemma 2.1.7, $\lambda \in D$. Observe further that $\sum_{w\in W} \epsilon(w)e^{w*\lambda} \in \mathcal{A}$, as can be easily seen from (1.3.22.2).

As preparation for the proof of the above theorem, we prove the following lemmas.

2.2.2 Lemma. *Let V be a highest weight $\bar{\mathfrak{g}}$-module with highest weight λ. Then*

(1)
$$\operatorname{ch} V = \sum_{\substack{\mu\le\lambda \\ \langle\mu+\rho,\mu+\rho\rangle=\langle\lambda+\rho,\lambda+\rho\rangle}} c_\mu \operatorname{ch} \bar{M}(\mu),$$

for $c_\mu \in \mathbb{Z}$ and $c_\lambda = 1$. Observe that the family $\{c_\mu \operatorname{ch} \bar{M}(\mu)\}_{\mu\le\lambda}$ for any $c_\mu \in \mathbb{Z}$ is locally finite.

Proof. By Lemmas 2.1.12 and 2.1.16, it suffices to prove (1) for $V = \bar{L}(\lambda)$. Let $S(\lambda) = \{\mu \in \mathfrak{h}^* : \mu \le \lambda \text{ and } \langle\mu+\rho, \mu+\rho\rangle = \langle\lambda+\rho, \lambda+\rho\rangle\}$. By Lemmas 2.1.12 and 2.1.16 again, we can write, for any $\mu \in S(\lambda)$,

$$\operatorname{ch} \bar{M}(\mu) = \sum_{\substack{\theta\le\mu \\ \langle\theta+\rho,\theta+\rho\rangle=\langle\mu+\rho,\mu+\rho\rangle}} a_{\mu,\theta} \operatorname{ch} \bar{L}(\theta),$$

with $a_{\mu,\theta} \in \mathbb{Z}_+$ and $a_{\mu,\mu} = 1$. In particular, the matrix $a = (a_{\mu,\theta})$ is lower triangular with diagonal entries 1, and hence is invertible. (The inverse is just $\sum_{n\ge 0} (I-a)^n$.) This proves the lemma. □

2.2.3 Lemma. *Denote the "Weyl denominator"*

$$\mathcal{D} = \prod_{\alpha\in\bar{\Delta}^+} (1-e^{-\alpha})^{\overline{\operatorname{mult}\alpha}}.$$

Then, for any $w \in W$,

(1)
$$w(e^\rho \cdot \mathcal{D}) = \epsilon(w)e^\rho \cdot \mathcal{D}.$$

Proof. It suffices to prove (1) for the simple reflections s_i. Now,

$$\begin{aligned} s_i(e^\rho \cdot \mathcal{D}) &= e^{\rho-\alpha_i}(1-e^{\alpha_i})\left[\prod_{\alpha\ne\alpha_i} (1-e^{-\alpha})^{\overline{\operatorname{mult}\alpha}}\right], \\ &= -e^\rho \mathcal{D}, \end{aligned}$$

where the first equality follows from Theorem 1.3.11(b_4) (cf. Proof of Proposition 1.3.21). □

2.2.4 Lemma. *Let $\lambda \in \mathfrak{h}^*$ be such that $\langle \lambda, \alpha_i^\vee \rangle \geq 0$ for all the simple coroots $\alpha_i^\vee$. Then, for any $\nu \in \mathfrak{h}^*$ satisfying*

(1) $\nu \leq \lambda + \rho$,
(2) $\langle \nu, \nu \rangle = \langle \lambda + \rho, \lambda + \rho \rangle$ *and*
(3) $\langle \nu, \alpha_i^\vee \rangle \geq 0$ *for all* i,

we have $\nu = \lambda + \rho$.

Proof. Write $\nu = \lambda + \rho - \sum n_i \alpha_i$, for some $n_i \in \mathbb{Z}_+$. Then

$$\begin{aligned}\langle \nu, \nu \rangle &= \langle \lambda + \rho, \lambda + \rho \rangle - \langle \lambda + \rho, \sum n_i \alpha_i \rangle - \langle \nu, \sum n_i \alpha_i \rangle \\ &= \langle \lambda + \rho, \lambda + \rho \rangle, \text{ by (2).}\end{aligned}$$

But then, by (3) and (1.5.11.3), we get $n_i = 0$, for each i. □

2.2.5 *Proof of Theorem 2.2.1.* By Lemma 2.2.2, write

$$(1)\ \operatorname{ch} L = \sum_{\substack{\mu \leq \lambda \\ \langle \mu+\rho, \mu+\rho \rangle = \langle \lambda+\rho, \lambda+\rho \rangle}} d_\mu \operatorname{ch} \bar{M}(\mu), \qquad \text{where } d_\mu \in \mathbb{Z} \text{ and } d_\lambda = 1.$$

So

$$(2)\, e^\rho \mathcal{D} \cdot (\operatorname{ch} L) = \sum d_\mu e^{\mu+\rho}, \text{ by the analogue of Lemma 2.1.13 for } \bar{M}(\lambda).$$

By Lemma 2.2.3 and Subsection 1.3.5(a),

$$(3) \qquad d_\mu = \epsilon(w) d_{w*\mu}, \qquad \text{for any } w \in W.$$

Fix μ such that $d_\mu \neq 0$. Then for any $w \in W$, $w * \mu \leq \lambda$. Choose $w_o = w(\mu) \in W$ such that $|\lambda - w_o * \mu|$ is minimum. Denote $\nu = w_o * \mu + \rho$. Then, by the minimality of $|\lambda - w_o * \mu|$, we get
(4) $\langle \nu, \alpha_i^\vee \rangle \geq 0$, for all simple coroots $\alpha_i^\vee$.
Of course, we also have
(5) $\nu \leq \lambda + \rho$ and
(6) $\langle \nu, \nu \rangle = \langle \lambda + \rho, \lambda + \rho \rangle$.

But then by Lemma 2.2.4, $\nu = \lambda + \rho$, i.e., $w_o * \mu = \lambda$.

So (2) reduces to (using (3)) $\mathcal{D} \cdot \operatorname{ch} L = \sum_{w \in W} \epsilon(w) e^{w*\lambda}$. This proves (1) of Theorem 2.2.1. The identity (2) of Theorem 2.2.1 follows by applying (1) to the trivial one-dimensional representation with highest weight $\lambda = 0$. □

2.2.6 Corollary. *Any integrable highest weight $\bar{\mathfrak{g}}$-module is irreducible.*

Proof. Let L be such a module and let L' be its irreducible quotient. By Theorem 2.2.1, $\operatorname{ch} L = \operatorname{ch} L'$. Hence $L \cong L'$. □

2.2.7 Corollary. *Any integrable $\bar{\mathfrak{g}}$-module $M \in \bar{\mathcal{O}}$ is a direct sum (as $\bar{\mathfrak{g}}$-modules)*

$$(1) \qquad M \cong \oplus_{\lambda \in D} \bar{L}(\lambda)^{\oplus [M:\bar{L}(\lambda)]}.$$

The corollary, in particular, applies to the tensor product $M = \bar{L}(\lambda) \otimes \bar{L}(\mu)$ for $\lambda, \mu \in D$.

Proof. Let $M_o \subset M$ be the subspace annihilated by $\bar{\mathfrak{n}}^+$. Then clearly M_o is $\mathfrak{h}$-stable. Take a basis $\{v_p\}$ of M_o consisting of weight vectors. Let v_p be of weight $\lambda_p \in D$.

By the previous corollary, there is an embedding $\delta_p : \bar{L}(\lambda_p) \hookrightarrow M$, taking $1 \otimes 1 \mapsto v_p$. These give rise to a $\bar{\mathfrak{g}}$-module map $\delta : \oplus_p \bar{L}(\lambda_p) \to M$, such that $\delta_{|_{\bar{L}(\lambda_p)}} = \delta_p$. We claim first that δ is injective:

Otherwise, take $0 \neq v \in \operatorname{Ker} \delta$ such that $\bar{\mathfrak{n}}^+ v = 0$, which is possible since $M \in \bar{\mathcal{O}}$. Then $v = \sum z_p 1_{\lambda_p}$, for some $z_p \in \mathbb{C}$, where 1_{λ_p} denotes $1 \otimes 1 \in \bar{L}(\lambda_p)$. So $\delta v = \sum z_p v_p = 0$, and hence each $z_p = 0$, i.e., $v = 0$, a contradiction. This proves that δ is injective.

We next claim that δ is surjective. If not, let $Q \neq 0$ be the cokernel of δ:

$$0 \to \oplus_p \bar{L}(\lambda_p) \xrightarrow{\delta} M \xrightarrow{\pi} Q \to 0.$$

Choose a submodule $\bar{L}(\mu) \subset Q$. This is possible by Corollary 2.2.6 since $Q \in \bar{\mathcal{O}}$. Replacing M by $\pi^{-1}(\bar{L}(\mu))$, we can assume that $Q = \bar{L}(\mu)$. By Lemmas 2.1.16 and 2.1.18, there exists a weight vector $v \in \pi^{-1}(1_\mu)$ such that $v \in M^z$, where $z = \langle \mu, 2\rho + \mu \rangle$.

Since $e_i v \in \operatorname{Ker} \pi$, write $e_i v = \sum_p w_p(i)$, with $w_p(i) \in \bar{L}(\lambda_p)$. Since $v \in M^z$, we can arrange that $w_p(i) = 0$ unless both of the following are satisfied:

(a) $\langle \lambda_p + \rho, \lambda_p + \rho \rangle = \langle \mu + \rho, \mu + \rho \rangle$; and

(b) $\mu + \alpha_i \leq \lambda_p$.

But then by Lemma 2.2.4, taking $\nu = \mu + \rho$ and $\lambda = \lambda_p$, we have that $\mu + \rho = \lambda_p + \rho$, which is a contradiction to (b). Hence $e_i v = 0$ for all i, i.e., $v \in M_o$. This is a contradiction and hence $Q = 0$. Now (1) follows by using Exercise 2.1.E. □

2.2.E EXERCISE. Let $\mathfrak{g}$ be an arbitrary Kac–Moody algebra and let $M \in \mathcal{O}$ be a module such that ch M is W-invariant. Then show that M is integrable.

Hint: Using $s\ell_2$-module theory show that, for any $\mu \in P(M)$, $\mu(\alpha_i^\vee) \in \mathbb{Z}$ for all i. Now write ch $M = \sum_\lambda [M : L(\lambda)]$ ch $L(\lambda)$. Since, for $\lambda \in D$, ch $L(\lambda)$ is W-invariant, $\sum_{\lambda \notin D} [M : L(\lambda)]$ ch $L(\lambda)$ is W-invariant. Taking maximum $\lambda_o \notin D$ such that $[M : L(\lambda_o)] \neq 0$, we get a contradiction.

2.3. Shapovalov Bilinear Form

From Subsection 2.3.4 until the end of this section we assume that $\mathfrak{g}$ is symmetrizable.

2.3.1 Definition. The triangular decomposition $\mathfrak{g} = \mathfrak{n}^- \oplus \mathfrak{h} \oplus \mathfrak{n}$ (cf. Theorem 1.2.1(a)) gives rise to the decomposition

(1) $$U(\mathfrak{g}) = (\mathfrak{n}^- U(\mathfrak{g}) + U(\mathfrak{g})\mathfrak{n}) \oplus S(\mathfrak{h}).$$

Observe that since $\mathfrak{h}$ is abelian, $U(\mathfrak{h})$ is the same as the symmetric algebra $S(\mathfrak{h})$.

The decomposition (1) gives rise to the projection (on the second factor) $\mathcal{H} : U(\mathfrak{g}) \to S(\mathfrak{h})$, where $\mathcal{H}$ stands for the Harish-Chandra map.

Recall the definition of the Cartan involution ω of $\mathfrak{g}$ (and hence of $U(\mathfrak{g})$) from 1.1.2. Let τ be the anti-automorphism of $U(\mathfrak{g})$ induced by $\tau(X) = -X$ for $X \in \mathfrak{g}$, and let σ be the anti-automorphism $\sigma := \tau\omega = \omega\tau$ of $U(\mathfrak{g})$ of order 2. Now define the *Shapovalov bilinear form*

$$S : U(\mathfrak{g}) \times U(\mathfrak{g}) \to S(\mathfrak{h})$$

by $S(a, b) = \mathcal{H}(\sigma(a)b)$.

Clearly S is symmetric (since $\sigma_{|S(\mathfrak{h})} = I$). Moreover, S satisfies the "contravariance" property

(2) $$S(ca, b) = S(a, \sigma(c)b).$$

Also,

$$S(U(\mathfrak{g})_{\gamma_1}, U(\mathfrak{g})_{\gamma_2}) = 0\,, \qquad \text{if } \gamma_1 \neq \gamma_2,$$

where $U(\mathfrak{g})_\gamma$ denotes the γ-weight space with respect to the adjoint action of $\mathfrak{h}$ on $U(\mathfrak{g})$.

For any $\lambda \in \mathfrak{h}^*$, the Shapovalov form S gives rise to the bilinear form $p_\lambda \circ S : U(\mathfrak{g}) \times U(\mathfrak{g}) \to \mathbb{C}$, where $p_\lambda : S(\mathfrak{h}) \to \mathbb{C}$ is the algebra homomorphism taking $h_1 \dots h_k \mapsto \lambda(h_1) \dots \lambda(h_k)$, for $h_i \in \mathfrak{h}$. It is easy to see that the bilinear form $p_\lambda \circ S$ descends to give a bilinear form $S_\lambda : M(\lambda) \times M(\lambda) \to \mathbb{C}$. Thus

(3) $$S_\lambda(a \otimes 1, b \otimes 1) = p_\lambda(S(a, b)), \qquad \text{for } a, b \in U(\mathfrak{g}).$$

Recall from Lemma 2.1.2 that $M'(\lambda)$ denotes the maximal proper submodule of $M(\lambda)$.

2.3.2 Proposition. *The symmetric bilinear form $S_\lambda : M(\lambda) \times M(\lambda) \to \mathbb{C}$, as defined above, is contravariant in the sense that*

(1) $S_\lambda(xv, w) = S_\lambda(v, \sigma(x)w)$, *for $v, w \in M(\lambda)$ and $x \in U(\mathfrak{g})$.*

In particular,

(2) $S_\lambda(M(\lambda)_\mu, M(\lambda)_\theta) = 0$, *if $\mu \neq \theta$.*

Further, $S_\lambda(M'(\lambda), M(\lambda)) = 0$ and hence S_λ induces a nondegenerate (symmetric contravariant) bilinear form, again denoted by S_λ on the irreducible quotient $L(\lambda)$. Moreover, any contravariant bilinear form on $L(\lambda)$ is a scalar multiple of S_λ, and hence is automatically symmetric.

Proof. Clearly (1) follows from (2.3.1.2) and (2.3.1.3), and (2) follows immediately from (1).

Take $a \otimes 1 \in M(\lambda)$ and $v \in M'(\lambda)$. Then

$$\begin{aligned} S_\lambda(v, a \otimes 1) &= S_\lambda(\sigma(a)v, 1 \otimes 1), \text{ by (1)}, \\ &= 0, \text{ by (2), since } \sigma(a)v \in M'(\lambda). \end{aligned}$$

Since Ker S_λ is a proper $\mathfrak{g}$-submodule, it coincides with $M'(\lambda)$.

Finally, let B be a contravariant bilinear form on $L(\lambda)$ and assume that $B_{|L(\lambda)_\lambda \times L(\lambda)_\lambda} = S_{\lambda_{|L(\lambda)_\lambda \times L(\lambda)_\lambda}}$. Then from the contravariance property (1) of S_λ and B, using (2), $B = S_\lambda$. □

2.3.3 Definition. For any $\lambda \in \mathfrak{h}^*$, the Verma module $M(\lambda)$ admits a filtration known as the *Jantzen filtration* defined below.

Choose an element $\rho \in \mathfrak{h}^*$ as in 1.3.1. We define a one-parameter family of (symmetric) bilinear forms $\hat{S}_\lambda$ on $M(\lambda)$ by

(1) $\hat{S}_\lambda(a \otimes 1, b \otimes 1) = p_{\lambda+t\rho}(S(a, b)) \in \mathbb{C}[t]$, for $a, b \in U(\mathfrak{n}^-)$,

where $p_{\lambda+t\rho} : S(\mathfrak{h}) \to \mathbb{C}[t]$ is the algebra homomorphism defined by $p_{\lambda+t\rho}(h_1 \dots h_k) = (\lambda(h_1) + t\rho(h_1)) \dots (\lambda(h_k) + t\rho(h_k))$. This defines a filtration of $M(\lambda)$ by $\mathfrak{g}$-submodules (cf. Exercise 2.3.E.5):

(2) $M(\lambda) = M^0(\lambda) \supset M^1(\lambda) \supset M^2(\lambda) \supset \dots,$

where

$$\begin{aligned} M^p(\lambda) := \{v_0 \in M(\lambda) :\ &\text{there exist } v_1, \dots, v_{p-1} \in M(\lambda) \text{ such that} \\ &\hat{S}_\lambda(v_0, w) + t\hat{S}_\lambda(v_1, w) + \dots + t^{p-1}\hat{S}_\lambda(v_{p-1}, w) \text{ is divisible by } t^p \\ &\text{for all } w \in M(\lambda)\}. \end{aligned}$$

Clearly

$$(3) \qquad M^1(\lambda) = M'(\lambda).$$

For $\eta \in Q^+$ (cf. the beginning of Section 1.2 for the notation Q^+), let $U(\mathfrak{n}^-)_\eta$ denote the $(-\eta)$-weight space of $U(\mathfrak{n}^-)$ under the adjoint action of $\mathfrak{h}$, and let

$$S_\eta^- : U(\mathfrak{n}^-)_\eta \times U(\mathfrak{n}^-)_\eta \to S(\mathfrak{h})$$

be the restriction of the Shapovalov form S (cf. 2.3.1). Further, define the Kostant partition function $P(\eta) \in \mathbb{Z}_+$ (for $\eta \in Q^+$) by

$$\operatorname{ch} M(0) = \sum_{\eta \in Q^+} P(\eta) e^{-\eta}.$$

Lemma 2.1.13 gives the standard expression for $P(\eta)$. We set $P(\eta) = 0$ if $\eta \notin Q^+$.

Choose a $\mathbb{C}$-basis $F = \{F_I\}$ of $U(\mathfrak{n}^-)_\eta$ and define

$$\det{}_F(S_\eta^-) := \det\,(S_\eta^-(F_I, F_{I'}))_{I,I'}\,,$$

which is an element of $S(\mathfrak{h})$. Of course, for a different basis G of $U(\mathfrak{n}^-)_\eta$, $\det_G(S_\eta^-)$ is a nonzero scalar multiple of $\det_F(S_\eta^-)$. We denote $\det_F(S_\eta^-)$ by $\det(S_\eta^-)$, but keep in mind that $\det(S_\eta^-)$ is defined only up to nonzero scalar multiples.

Recall the definition of the map $\nu : \mathfrak{h} \to \mathfrak{h}^*$ from Subsection 1.5.3. In the following result, we use the identification $\mathfrak{g} = \bar{\mathfrak{g}}$ (for symmetrizable $\mathfrak{g}$) to be proved in the next chapter. In particular, its proof does not rely on the following theorem.

As mentioned in the beginning of this section, from now until the end of this section, we assume that $\mathfrak{g}$ is symmetrizable.

2.3.4 Theorem. *Let $\mathfrak{g}$ be a symmetrizable Kac–Moody Lie algebra. Then for any $\eta \in Q^+$,*

$$\det(S_\eta^-) = \prod \left(\nu^{-1}(\alpha) + \langle \rho - \frac{n\alpha}{2}, \alpha\rangle\right)^{P(\eta - n\alpha)(\operatorname{mult}\alpha)},$$

where the product is taken over $(\alpha, n) \in \Delta^+ \times \mathbb{N}$ such that $\eta - n\alpha \in Q^+$ and $\langle\,,\,\rangle$ is any normalized invariant form on $\mathfrak{h}^*$ as in 1.5.3.

Proof. *Step 1: The Verma module $M(\lambda)$ is irreducible if $2\langle \lambda + \rho, \beta \rangle \neq \langle \beta, \beta \rangle$ for every $\beta \in Q^+ \backslash \{0\}$.*

Suppose $M(\lambda)$ is not irreducible. Then there exists a vector $v \in M(\lambda)$ of weight $\mu = \lambda - \beta_o$, for some $\beta_o \in Q^+ \backslash \{0\}$, such that $\mathfrak{n}^+ \cdot v = 0$. In particular, we get an injective $\mathfrak{g}$-module morphism $\varphi : M(\mu) \to M(\lambda),\ 1 \otimes 1 \mapsto v$. Injectivity of φ follows since, by the PBW theorem, $U(\mathfrak{n}^-)$ does not have zero-divisors. By Lemma 2.1.16, $\Omega_{M(\lambda)|M(\mu)} = \langle \lambda, 2\rho + \lambda \rangle I_{M(\mu)} = \Omega_{M(\mu)} = \langle \mu, 2\rho + \mu \rangle I_{M(\mu)}$. This gives that $2\langle \lambda + \rho, \beta_o \rangle = \langle \beta_o, \beta_o \rangle$, contradicting the assumption. So $M(\lambda)$ is irreducible.

Step 2: For any $\eta \in Q^+$, $\det(S_\eta^-) \neq 0$ and, moreover, the highest degree term of $\det(S_\eta^-)$ is equal to $\prod (\nu^{-1}\alpha)^{P(\eta - n\alpha)(\text{mult}\,\alpha)}$, where the product runs over $(\alpha, n) \in \Delta^+ \times \mathbb{N}$ such that $\eta - n\alpha \in Q^+$.

By Step 1, since Q^+ is countable, there exist irreducible Verma modules $M(\lambda)$. From this it follows that $\det(S_\eta^-) \neq 0$, for any $\eta \in Q^+$. Otherwise $M'(\lambda) \neq 0$ (cf. Proposition 2.3.2), contradicting the irreducibility of $M(\lambda)$.

Let $\tilde{\Delta}^+$ denote the set of positive roots of $\mathfrak{g}$ where each root occurs as many times as its multiplicity. Now choose a basis $\{F_\beta\}_{\beta \in \tilde{\Delta}^+}$ of $\mathfrak{n}^-$ and also a 'dual' basis $\{\bar{F}_\beta\}_{\beta \in \tilde{\Delta}^+}$ of $\mathfrak{n}^-$, such that F_β and $\bar{F}_\beta$ are root vectors corresponding to the root $-\beta$, and $S(\bar{F}_\beta, F_{\beta'}) = \delta_{\beta,\beta'}\, \nu^{-1}(\beta)$, which is possible by Theorem 1.5.4(b_2). Put a total order on $\tilde{\Delta}^+$. Let $\mathcal{P}(\eta) := \{\beta = (\beta_1 \leq \cdots \leq \beta_n) : \beta_i \in \tilde{\Delta}^+ \text{ and } \sum \beta_i = \eta\}$. By the PBW theorem, $\{F_\beta := F_{\beta_1} \ldots F_{\beta_n}\}_{\beta \in \mathcal{P}(\eta)}$ is a basis of $U(\mathfrak{n}^-)_\eta$ and so is $\{\bar{F}_\beta\}_\beta$. Define $\deg F_\beta = n$. Also, by deg of a nonzero element $x \in U(\mathfrak{h}) = S(\mathfrak{h})$, we mean its degree as a polynomial. The following three assertions are easy to verify from the commutation relations in $\mathfrak{g}$:

(1) $\deg S(\bar{F}_\beta, F_{\beta'}) \leq \min(\deg \bar{F}_\beta, \deg F_{\beta'})$.

(2) If $\deg \bar{F}_\beta = \deg F_{\beta'}$ and $\beta \neq \beta'$, then $\deg S(\bar{F}_\beta, F_{\beta'}) < \deg \bar{F}_\beta$.

(3) The highest degree homogeneous component of $S(\bar{F}_\beta, F_\beta)$ is equal to $c_\beta \nu^{-1}(\beta)$ for some $c_\beta \neq 0 \in \mathbb{C}$, where for $\beta = (\beta_1 \leq \cdots \leq \beta_n)$, $\nu^{-1}(\beta) := \prod_{i=1}^n \nu^{-1}(\beta_i)$.

From (1)–(3), it follows that the highest degree term of $\det(S_\eta^-)$ is equal to

$$\prod_{\beta \in \mathcal{P}(\eta)} \nu^{-1}(\beta).$$

Let $m(\beta, \alpha) := \#\{\beta_i : \beta_i = \alpha\}$. Now the assertion in Step 2 follows from the following identity for any $\alpha \in \tilde{\Delta}^+$:

$$\sum_{\beta \in \mathcal{P}(\eta)} m(\beta, \alpha) = \sum_{n \geq 1} n \,\#\{\beta \in \mathcal{P}(\eta) : m(\beta, \alpha) = n\}$$

$$= \sum_{n\geq 1} \#\{\beta \in \mathcal{P}(\eta) : m(\beta, \alpha) \geq n\}$$
$$= \sum_{n\geq 1} P(\eta - n\alpha).$$

Step 3. $\det(S_\eta^-)$ *breaks up as a product of linear factors of the form* $\nu^{-1}(\alpha) + \langle \rho - \frac{n\alpha}{2}, \alpha\rangle$ *for* $(\alpha, n) \in \Delta^+ \times \mathbb{N}$.

Let $Z_\eta \subset \mathfrak{h}^*$ be the hypersurface defined by the vanishing of the nonzero polynomial $\det(S_\eta^-)$. By Proposition 2.3.2,

$$Z_\eta \subseteq \{\lambda \in \mathfrak{h}^* : M(\lambda) \text{ is not irreducible}\}$$
$$\subseteq \bigcup_{\beta\in Q^+\backslash\{0\}} \{\lambda \in \mathfrak{h}^* : \langle \lambda + \rho - \frac{\beta}{2}, \beta\rangle = 0\}, \text{ by Step 1.}$$

In particular, there exist finitely many $\beta_1, \ldots, \beta_n \in Q^+\backslash\{0\}$ such that

$$Z_\eta = \bigcup_{\beta_k} H_{\beta_k},$$

where H_{β_k} is the hyperplane $\{\lambda \in \mathfrak{h}^* : \langle \lambda + \rho - \frac{\beta_k}{2}, \beta_k\rangle = 0\}$, i.e., $\det(S_\eta^-)$ is a product of linear factors (with some powers) of the form $\nu^{-1}(\beta_k) + \langle \rho - \frac{\beta_k}{2}, \beta_k\rangle$. Further, by Step 2, each $\beta_k = n_k\gamma_k$, for some $\gamma_k \in \Delta^+$ and $n_k \in \mathbb{Q}$. By [Kac–90, Proposition 5.5], we can, in fact, assume that $\beta_k = m_k\gamma_k'$, for some positive root γ_k' and $m_k \in \mathbb{N}$. This proves the assertion of Step 3.

Step 4. Take $\lambda \in \mathfrak{h}^*$. *Let* $\beta \in Q^+$ *be a quasiroot (i.e.,* $\beta = r\gamma$*, for some positive rational number* r *and* $\gamma \in \Delta^+$*) such that* $2\langle \lambda + \rho, \beta\rangle = \langle \beta, \beta\rangle$ *and suppose that* $2\langle \lambda + \rho, \gamma\rangle \neq \langle \gamma, \gamma\rangle$ *for any nonzero* $\gamma \in Q^+\backslash\{\beta\}$ *and* $2\langle \lambda + \rho - \beta, \gamma\rangle \neq \langle \gamma, \gamma\rangle$ *for any nonzero* $\gamma \in Q^+$. *Then the* $\mathfrak{g}$*-module* $M(\lambda)$ *has a submodule* N*, which is isomorphic to direct sum of a finite number (possibly zero) of modules isomorphic to* $M(\lambda - \beta)$*, and such that the* $\mathfrak{g}$*-module* $M(\lambda)/N$ *is irreducible.*

By Step 1, the $\mathfrak{g}$-module $M(\lambda - \beta)$ is irreducible. Therefore, the sum N of submodules in $M(\lambda)$, which are isomorphic to $M(\lambda - \beta)$, is direct sum of a finite number of submodules isomorphic to $M(\lambda - \beta)$. To prove that the $\mathfrak{g}$-module $L = M(\lambda)/N$ is irreducible, we consider the Casimir operator Ω on L. If $v \in L_{\lambda-\gamma}$ is a highest weight vector, then clearly $\gamma \neq \beta$. By Lemma 2.1.16, we have $\Omega(v) = \langle \lambda + 2\rho, \lambda\rangle v = \langle \lambda - \gamma + 2\rho, \lambda - \gamma\rangle v$; therefore $2\langle \lambda + \rho, \gamma\rangle = \langle \gamma, \gamma\rangle$ and hence $\gamma = 0$. This proves that L is irreducible.

Conclusion of the Proof of Theorem 2.3.4. By Step 3, $\det S_\eta^-$ is a product of linear factors of the form $\nu^{-1}\beta + \langle \rho - \frac{1}{2}\beta, \beta\rangle$, where $\beta \in \mathbb{N}\Delta^+ := \{n\alpha : n \in \mathbb{N}, \alpha \in$

$\Delta^+\}$. Step 2 implies that, in the case when $\langle\beta, \beta\rangle = 0$, the power of linear factor $\nu^{-1}\beta + \langle\rho, \beta\rangle$ in $\det S_\eta^-$ is exactly what we claim in the statement of the theorem.

So assume now that $\langle\beta, \beta\rangle \neq 0$. Consider $\lambda \in \mathfrak{h}^*$ such that $2\langle\lambda + \rho, \beta\rangle = \langle\beta, \beta\rangle$, $2\langle\lambda+\rho, \gamma\rangle \neq \langle\gamma, \gamma\rangle$ for any nonzero $\gamma \in Q^+\backslash\{\beta\}$ and $2\langle\lambda+\rho-\beta, \gamma\rangle \neq \langle\gamma, \gamma\rangle$ for any nonzero $\gamma \in Q^+$; such a λ clearly exists since $\langle\beta, \beta\rangle \neq 0$. Consider the Verma module

$$M(\lambda) = M^0(\lambda) \supset M^1(\lambda) \supset M^2(\lambda) \supset \dots,$$

with the Jantzen filtration. By Step 4, $M^1(\lambda)$ is a direct sum of submodules isomorphic to $M(\lambda - \beta)$. Let m_i be the multiplicity of $M(\lambda - \beta)$ in $M^i(\lambda)$; set $m_\beta = \sum_{i\geq 1} m_i$.

From the construction of the filtration, it follows that $\det(\hat{S}_{\lambda_{|M(\lambda)_{\lambda-\eta}}})$ is divisible exactly by the $m_\beta P(\eta - \beta)$-th power of t (cf. the proof of Corollary 2.3.5, particularly (2.3.5.4)) or, equivalently, $\det S_\eta^-$ is divisible exactly by the $m_\beta P(\eta - \beta)$-th power of $(\nu^{-1}(\beta) + \langle\rho - \frac{1}{2}\beta, \beta\rangle)$. In particular, the multiplicity of $\nu^{-1}(\beta)$ in the leading term of $\det S_\eta^-$ is equal to

$$\sum m_{\beta'} P(\eta - \beta'),$$

where the summation is taken over $\beta' \in \mathbb{N}\Delta^+$, which are (rationally) proportional to β. On the other hand, this multiplicity is, as given by Step 2, also equal to $\sum P(\eta - n\alpha)(\operatorname{mult}\alpha)$, where the summation runs over those $(\alpha, n) \in \Delta^+ \times \mathbb{N}$ such that α is proportional to β. Now the theorem follows from the simple fact that the functions $\{P(\eta - \beta)\}_{\beta\in Q^+\backslash\{0\}}$, considered as functions in the η-variable, are linearly independent over $\mathbb{Q}$. □

The following corollary generalizes Jantzen's character sum formula to the symmetrizable case.

2.3.5 Corollary. *Let $\mathfrak{g}$ be a symmetrizable Kac–Moody algebra. For any $\lambda, \mu \in \mathfrak{h}^*$, the weight space $M^p(\lambda)_\mu = 0$, for large enough p (depending on λ, μ). Moreover, .*

(1) $$\sum_{p\geq 1} \operatorname{ch} M^p(\lambda) = \sum_{(\alpha,n)\in D_\lambda} \operatorname{ch} M(\lambda - n\alpha),$$

where $D_\lambda := \{(\alpha, n) \in \tilde{\Delta}^+ \times \mathbb{N} : \langle\lambda + \rho - \frac{n}{2}\alpha, \alpha\rangle = 0\}$ and $M(\lambda) = M^0(\lambda) \supset M^1(\lambda) \supset M^2(\lambda) \supset \cdots$ is the Jantzen filtration (cf. 2.3.3).

Proof. Extend the bilinear form

$$\hat{S}_\lambda : M(\lambda) \otimes M(\lambda) \to \mathbb{C}[t] \qquad \text{(cf. (2.3.3.1))}$$

to a $\mathbb{C}[[t]]$-bilinear form

$$\hat{S}_\lambda(t) : (M(\lambda) \otimes_{\mathbb{C}} \mathbb{C}[[t]]) \otimes_{\mathbb{C}[[t]]} (M(\lambda) \otimes_{\mathbb{C}} \mathbb{C}[[t]]) \to \mathbb{C}[[t]],$$

where $\mathbb{C}[[t]]$ is the power series ring in one variable. Denote its restriction to $M(\lambda)_{\lambda-\eta}[[t]] := M(\lambda)_{\lambda-\eta} \otimes_{\mathbb{C}} \mathbb{C}[[t]]$, for any $\eta \in Q^+$, by $\hat{S}_{\lambda,\eta}(t)$.

By Step 1 of the proof of Theorem 2.3.4, there are some complex values of t such that $M(\lambda + t\rho)$ is irreducible. In particular, the (symmetric) bilinear form $\hat{S}_{\lambda,\eta}(t)$ is nondegenerate. By the structure theorem of modules over PID (cf. [Lang–65, Theorem 5, Section 2, Chapter 15]), there exist two bases over $\mathbb{C}[[t]]$, $(t_j)_{1\le j\le P(\eta)}$ and (s_j) of $M(\lambda)_{\lambda-\eta}[[t]]$ such that

(2) $\hat{S}_{\lambda,\eta}(t)(t_{j_1}, s_{j_2}) = t^{n_{j_1}} \delta_{j_1,j_2}$, for some integers $0 \le n_1 \le \cdots \le n_{P(\eta)}$.

By the definition of $M^p(\lambda)_{\lambda-\eta}$, it is easy to see that for any $p \ge 0$,

$$M^p(\lambda)_{\lambda-\eta} = \pi\{\sum \mathbb{C}[[t]]t_j : n_j \ge p\}, \tag{3}$$

where $\pi : M(\lambda) \otimes_{\mathbb{C}} \mathbb{C}[[t]] \to M(\lambda)$ is the canonical map induced by $t \mapsto 0$.

With this description of $M^p(\lambda)$, we come to the proof of the corollary. Of course, $M^p(\lambda)_\mu = 0$ unless μ is of the form $\lambda - \eta$ (for some $\eta \in Q^+$). By (3), $M^p(\lambda)_{\lambda-\eta} = 0$ for any $p > n_{P(\eta)}$. Now (for any $\eta \in Q^+$) by (2)–(3),

$$\begin{aligned} \operatorname{ord}_t(\det \hat{S}_{\lambda,\eta}(t)) &= \sum_{j=1}^{P(\eta)} n_j \\ &= \sum_{p\ge1} \dim M^p(\lambda)_{\lambda-\eta}. \end{aligned} \tag{4}$$

So

$$\begin{aligned} \sum_{p\ge1} \operatorname{ch} M^p(\lambda) &= e^\lambda \sum_{\eta\in Q^+} \operatorname{ord}_t(\det \hat{S}_{\lambda,\eta}(t))e^{-\eta} \\ &= \sum_{\eta\in Q^+} \sum_{(\alpha,n)\in D_\lambda} P(\eta - n\alpha)\, e^{\lambda-\eta}, \qquad \text{by Theorem 2.3.4} \\ &= \sum_{(\alpha,n)\in D_\lambda} \operatorname{ch} M(\lambda - n\alpha). \end{aligned}$$

This proves the corollary. □

We have the following corollary of Corollary 2.3.5.

2.3.6 Corollary. *For $\mu \leq \lambda \in \mathfrak{h}^*$, $L(\mu)$ is a component of $M(\lambda)$ if and only if there exists a sequence of positive roots $\beta_1, \dots, \beta_p$ $(p \geq 0)$ and positive integers $k_1, \dots, k_p$ such that*

(1) $\lambda - \mu = \sum_{i=1}^{p} k_i \beta_i$, and

(2) for all $1 \leq j \leq p$, we have $2\langle \lambda + \rho - \sum_{i=1}^{j-1} k_i \beta_i, \beta_j \rangle = k_j \langle \beta_j, \beta_j \rangle$.

Proof. We prove the corollary by induction on $|\lambda - \mu|$. If $|\lambda - \mu| = 0$, i.e., $\mu = \lambda$, take $p = 0$. So, assume $|\lambda - \mu| > 0$. By Corollary 2.3.5, $L(\mu)$ is a component of $M(\lambda)$ if and only if $L(\mu)$ is a component of $M(\lambda - k_1\beta_1)$, for some $(k_1, \beta_1) \in D_\lambda$. But $|\lambda - k_1\beta_1 - \mu| < |\lambda - \mu|$, so the corollary follows by induction. □

We define a subset $K^{\text{w.g.}} \subset \mathfrak{h}^*$ ("w.g." stands for *weakly good*) such that, for $\lambda \in K^{\text{w.g.}}$, the description of the components of $M(\lambda)$ takes a particularly simple form.

2.3.7 Definition. For any $\alpha \in \Delta^+_{\text{im}}$, where Δ^+_{im} is the set of positive imaginary roots as in Section 1.3, let us consider the set

$$C_\alpha := \{\lambda \in \mathfrak{h}^* : \langle \lambda + \rho, \alpha \rangle = \frac{\langle \alpha, \alpha \rangle}{2}\},$$

and define

$$C = \cup_{\alpha \in \Delta^+_{\text{im}}} C_\alpha ,$$
$$K^{\text{w.g.}} = \mathfrak{h}^* \backslash C.$$

Clearly C (and hence $K^{\text{w.g.}}$) is stable under the shifted W-action $*$ as in Subsection 1.3.1, defined by $w * \lambda = w(\lambda + \rho) - \rho$, for $w \in W$ and $\lambda \in \mathfrak{h}^*$.

For a real root $\beta \in \Delta^+$, define the reflection $s_\beta \in W$ corresponding to β by

$$s_\beta \lambda = \lambda - \langle \lambda, \beta^\vee \rangle \beta, \quad \text{where } \beta^\vee \text{ is the coroot defined in 1.3.8.}$$

Now, for any $\lambda \in \mathfrak{h}^*$, consider the subgroup $W(\lambda) \subset W$ generated by $\{s_\beta : \beta \in \Delta^+_{\text{re}} \text{ and } \langle \lambda + \rho, \beta^\vee \rangle \in \mathbb{Z}\}$.

2.3.8 Proposition. *Let $\lambda \in K^{\text{w.g.}}$. Then $[M(\lambda) : L(\mu)] > 0$ iff there exist reflections $s_{\beta_1}, \dots, s_{\beta_p} \in W(\lambda)$, for $\beta_1, \dots, \beta_p \in \Delta^+_{\text{re}}$, $p \geq 0$, such that*

$$\lambda > s_{\beta_1} * \lambda > (s_{\beta_2} s_{\beta_1}) * \lambda > \cdots > (s_{\beta_p} \dots s_{\beta_1}) * \lambda = \mu.$$

In particular, $\mu \in W(\lambda) * \lambda$ and hence $\mu \in K^{\text{w.g.}}$.

Proof. We first prove the "only if" part. Since $L(\mu)$ is a subquotient of $M(\lambda)$, of course $\lambda - \mu \in Q^+$. We prove the assertion by induction on $|\lambda - \mu|$. If $|\lambda - \mu| = 0$, then $p = 0$ and there is nothing to prove. So assume $|\lambda - \mu| > 0$. Since $M(\lambda)/M^1(\lambda)$ is the irreducible module $L(\lambda)$, we get from Corollary 2.3.5 that $L(\mu)$ is a subquotient of $M(\lambda - n\beta_1)$, for some $(\beta_1, n) \in D_\lambda$. But then $\beta_1 \in \Delta^+_{\text{re}}$ (otherwise, $n\beta_1 \in \Delta^+_{\text{im}}$ by (1.3.5.4), contradicting that $\lambda \in K^{\text{w.g.}}$) and, moreover, $n = \langle \lambda + \rho, \beta_1^\vee \rangle$ (in particular, $s_{\beta_1} \in W(\lambda)$). Hence $\lambda - n\beta_1 = s_{\beta_1} * \lambda$. Now, $K^{\text{w.g.}}$ being W-stable under $*$, $\lambda - n\beta_1 \in K^{\text{w.g.}}$ and of course $|s_{\beta_1} * \lambda - \mu| < |\lambda - \mu|$. So the "only if" part follows by the induction hypothesis and using the fact that $W(s_{\beta_1} * \lambda) = W(\lambda)$.

The "if" part follows immediately from Corollary 2.3.6. □

2.3.9 Definition. Fix a real form $\mathfrak{h}_\mathbb{R}$ of $\mathfrak{h}$ as in 1.4.1. Let $\mathfrak{g}_\mathbb{R}$ be the real Lie subalgebra of $\mathfrak{g}$ generated by $\{\mathfrak{h}_\mathbb{R}, e_i, f_i;\ 1 \le i \le \ell\}$. Then $\mathfrak{g}_\mathbb{R}$ is a real form of $\mathfrak{g}$, i.e., $\mathfrak{g}_\mathbb{R} \otimes_\mathbb{R} \mathbb{C} = \mathfrak{g}$. The choice of the real form $\mathfrak{g}_\mathbb{R}$ induces a conjugate-linear involution (i.e., a conjugate-linear algebra automorphism of order 2) of $U(\mathfrak{g})$, denoted by $a \mapsto \bar{a}$.

It is easy to see that, for the Cartan involution ω, $\omega(\mathfrak{g}_\mathbb{R}) \subset \mathfrak{g}_\mathbb{R}$. In particular, $\overline{\omega(a)} = \omega(\bar{a})$, for $a \in U(\mathfrak{g})$. Let $\sigma_o := \bar{\sigma}$ be the conjugate-linear antiautomorphism of order 2 of $U(\mathfrak{g})$, where σ is as in 2.3.1. We call σ_o the *compact involution of* $U(\mathfrak{g})$. Clearly, σ_o keeps $\mathfrak{g}$ stable. Consider the -1 eigenspace $\mathfrak{k}$ of $\sigma_o|_\mathfrak{g}$, i.e., $\mathfrak{k} := \{X \in \mathfrak{g} : \sigma_o(X) = -X\}$. Then $\mathfrak{k}$ is a real Lie subalgebra of $\mathfrak{g}$ such that $\mathfrak{k} \otimes_\mathbb{R} \mathbb{C} = \mathfrak{g}$. The Lie algebra $\mathfrak{k}$ is called the *standard unitary form* of $\mathfrak{g}$.

Let V be a complex vector space. Recall that a map $H : V \times V \to \mathbb{C}$ is called a *Hermitian form* on V if H is linear in the first variable, conjugate-linear in the second variable, and $H(v, w) = \overline{H(w, v)}$. A Hermitian form H on a $U(\mathfrak{g})$-module V is called *contravariant* if

(1) $$H(av, w) = H(v, \sigma_o(a)w),$$

for all $v, w \in V$ and $a \in U(\mathfrak{g})$.

Recall from Proposition 2.3.2 that there exists a contravariant nondegenerate symmetric bilinear form S_λ on $L(\lambda)$. Fix a highest weight vector $v_\lambda \in L(\lambda)$ such that

(2) $$S_\lambda(v_\lambda, v_\lambda) = 1.$$

For any $\lambda \in \mathfrak{h}^*_\mathbb{R}$, we get the real form $L(\lambda)_\mathbb{R} := U(\mathfrak{g}_\mathbb{R})v_\lambda \subset L(\lambda)$ (cf. Exercise 2.3.E.6), and hence a conjugate-linear involution of $L(\lambda)$, denoted by $v \mapsto \bar{v}$. Define the Hermitian form H_λ on $L(\lambda)$ by

(3) $$H_\lambda(v, w) = S_\lambda(v, \bar{w}), \qquad \text{for } v, w \in L(\lambda).$$

From the contravariance property of S_λ, it is easy to see that the Hermitian form H_λ on $L(\lambda)$ is contravariant. Moreover, H_λ is nondegenerate since S_λ is too.

It is easy to see that, once a choice of σ_o is made, any contravariant Hermitian form on $L(\lambda)$ is a scalar multiple of H_λ by using the corresponding property of S_λ.

Similarly, any nondegenerate symmetric invariant bilinear form $\langle\ ,\ \rangle$ on $\mathfrak{g}$ such that $\langle \mathfrak{g}_{\mathbb{R}}, \mathfrak{g}_{\mathbb{R}}\rangle \subset \mathbb{R}$ gives rise to a nondegenerate Hermitian form $\{\ ,\ \}$ on $\mathfrak{g}$ defined by

$$(4) \qquad \{x, y\} = \langle x, \sigma_o y\rangle.$$

Moreover, $\{\ ,\ \}$ is a contravariant Hermitian form on $\mathfrak{g}$. (Observe that by Theorem 1.5.4, the proof of Proposition 1.5.2, Corollary 1.5.7, and (1.5.11.3), there exists such a form $\langle\ ,\ \rangle$ on $\mathfrak{g}$.)

2.3.10 Definition. Define the operator $F : \mathfrak{n}^- \to \mathfrak{n}^-$ by

$$(1) \qquad F(y) = \sum_{\alpha\in\Delta^+}\sum_i [e^i_{-\alpha}, [e^i_\alpha, y]_-], \qquad y \in \mathfrak{n}^-,$$

where $\{e^i_\alpha\}_i$, $\{e^i_{-\alpha}\}_i$ are dual bases of $\mathfrak{g}_\alpha$ and $\mathfrak{g}_{-\alpha}$ respectively as in 1.5.8, and the notation $[e^i_\alpha, y]_-$ means the component of $[e^i_\alpha, y]$ in $\mathfrak{n}^-$ under the decomposition $\mathfrak{g} = \mathfrak{n}^- \oplus \mathfrak{h} \oplus \mathfrak{n}^+$. Observe that for any y, $[e^i_\alpha, y]_- = 0$ for all but finitely many α, and hence the sum (1) is finite for any $y \in \mathfrak{n}^-$.

2.3.11 Lemma. *For* $y \in \mathfrak{g}_{-\beta}$,

$$F(y) = (2\langle\rho, \beta\rangle - \langle\beta, \beta\rangle)y.$$

Proof. Let $v = 1 \otimes 1$ be the highest weight vector in $M(0)$. Then

$$\begin{aligned}\bar{\Omega}(yv) &= \Omega(yv) + 2\langle\rho, \beta\rangle yv - \Omega_0(yv)\\ &= 2\langle\rho, \beta\rangle yv - \sum_k \beta(u_k)\beta(u^k)yv, \qquad \text{by Lemma 2.1.16}\\ (1) \qquad &= [2\langle\rho, \beta\rangle - \langle\beta, \beta\rangle]yv,\end{aligned}$$

where $\bar{\Omega}$ is as defined in the proof of Theorem 1.5.11 and Ω is as in Subsection 1.5.8. On the other hand, by the definition of $\bar{\Omega}$, we have

$$\begin{aligned}\bar{\Omega}(yv) &= 2\sum_{\alpha\in\Delta^+}\sum_i e^i_{-\alpha}e^i_\alpha y(v)\\ &= 2\sum_{\alpha\in\Delta^+}\sum_i e^i_{-\alpha}[e^i_\alpha, y](v).\end{aligned}$$

Putting $S = S_\beta := \{\alpha \in \Delta^+ : \beta - \alpha \in \Delta^+\}$, we may write

$$\begin{aligned}\bar{\Omega}(yv) &= 2\sum_{\alpha\in S}\sum_i e^i_{-\alpha}[e^i_\alpha, y](v)\\ &= \sum_{\alpha\in S}\sum_i([e^i_{-\alpha}, [e^i_\alpha, y]] + [e^i_\alpha, y]e^i_{-\alpha} + e^i_{-\alpha}[e^i_\alpha, y])(v)\\ &= \sum_{\alpha\in S}\sum_i[e^i_{-\alpha}, [e^i_\alpha, y]](v), \qquad \text{by (1.5.10.3).}\end{aligned}$$

Comparing this with (1), we get

$$(2\langle\rho, \beta\rangle - \langle\beta, \beta\rangle)y(v) = \sum_{\alpha\in\Delta^+}\sum_i[e^i_{-\alpha}, [e^i_\alpha, y]_-](v).$$

As $M(0)$ is a free $U(\mathfrak{n}_-)$-module (cf. 2.1.1), the lemma follows. □

For $\lambda \in \mathfrak{h}^*$, let $P(\lambda)$ be the set of weights of $L(\lambda)$, that is, $P(\lambda) := \{\mu \in \mathfrak{h}^* : L(\lambda)_\mu \neq 0\}$.

2.3.12 Lemma. *Let $L(\lambda)$ be integrable, i.e., $\lambda \in D$. Then, for $\mu \in P(\lambda)$,*

$$(1) \qquad \langle\mu + \rho, \mu + \rho\rangle \leq \langle\lambda + \rho, \lambda + \rho\rangle.$$

Moreover, equality in (1) occurs if and only if $\lambda = \mu$.

Proof. First, $W \cdot \mu \cap D \neq \emptyset$ (in fact it is a singleton):

Let us choose a $w \in W$ such that $|\lambda - w\mu|$ is minimum. Denote $\mu_o = w\mu$. Then it is easy to see that $\mu_o \in D$. Further, by Exercise 1.4.E.2, $W \cdot \mu \cap D = \{\mu_o\}$.

Write $\lambda - \mu = \sum n_i\alpha_i$, $\lambda - \mu_o = \sum_i m_i\alpha_i$, for $n_i, m_i \in \mathbb{Z}_+$. Now

$$\begin{aligned}\langle\lambda + \rho, \lambda + \rho\rangle - \langle\mu + \rho, \mu + \rho\rangle &= \langle\lambda, \lambda\rangle - \langle\mu, \mu\rangle + 2\sum_i n_i\langle\alpha_i, \rho\rangle\\ &= \langle\lambda, \lambda\rangle - \langle\mu_o, \mu_o\rangle + 2\sum_i n_i\langle\alpha_i, \rho\rangle\\ &= \sum m_i\langle\alpha_i, \lambda + \mu_o\rangle + 2\sum_i n_i\langle\alpha_i, \rho\rangle.\end{aligned}$$

But since λ and $\mu_o \in D$, the lemma follows. □

2.3.13 Theorem.

(a) *Any contravariant Hermitian form $\{\ ,\ \}$ on $\mathfrak{g}$, as in 2.3.9, restricted to $\mathfrak{n}^- \oplus \mathfrak{n}$ is positive definite.*

(b) *Take $\lambda \in \mathfrak{h}^*_{\mathbb{R}}$. Then the Hermitian form H_λ on $L(\lambda)$ is positive definite if and only if $\lambda \in D$.*

Proof. We first prove (a). Using σ_o, it suffices to show that $\{\ ,\ \}$ is positive definite on $\mathfrak{g}_{-\beta}$ with $\beta \in \Delta^+$. We do it by induction on $|\beta|$. The case $|\beta| = 1$ is clear by (1.5.4.1). Otherwise, as earlier, define $S = S_\beta := \{\alpha \in \Delta^+ : \beta - \alpha \in \Delta^+\}$ and use the induction assumption to choose, for every $\alpha \in S$, an orthonormal basis $\{e^i_{-\alpha}\}$ of $\mathfrak{g}_{-\alpha}$ with respect to $\{\ ,\ \}$. Then, setting $e^i_\alpha = \sigma_o(e^i_{-\alpha})$, we have $\langle e^i_\alpha, e^j_{-\alpha}\rangle = \delta_{i,j}$. Now, we apply Lemma 2.3.11 with this choice of e^i_α and $e^i_{-\alpha}$ (for $\alpha \in S$). For $y \in \mathfrak{g}_{-\beta}$ we have

$$\begin{aligned}(2\langle\rho,\beta\rangle - \langle\beta,\beta\rangle)\{y,y\} &= \{F(y), y\}\\ &= \sum_{\alpha\in S}\sum_i \{[e^i_{-\alpha},[e^i_\alpha, y]], y\}\\ \text{(1)} \qquad &= \sum_{\alpha\in S}\sum_i \{[e^i_\alpha, y],[e^i_\alpha, y]\}.\end{aligned}$$

By the inductive assumption, the last sum is nonnegative. Further,

$$\text{(2)} \qquad 2\langle\rho,\alpha\rangle - \langle\alpha,\alpha\rangle > 0,$$

for any root $\alpha \in \Delta^+$ which is not simple: If $\alpha \in \Delta^+_{\text{im}}$, then $\langle\alpha,\alpha\rangle \le 0$ by Exercise 2.3.E.4. Of course, $\langle\rho,\alpha\rangle > 0$. So we can assume that α is real. Since α is not simple, $\langle\rho,\alpha^\vee\rangle \ge 2$, and hence $2\langle\rho,\alpha^\vee\rangle - \langle\alpha,\alpha^\vee\rangle > 0$. This proves (2). By (1)–(2) we get $\{y,y\} \ge 0$. Since $\{\ ,\ \}$ is nondegenerate on $\mathfrak{g}_{-\beta}$, we deduce that it is positive definite, proving (a).

We come to the proof of (b). Let $\lambda \in D$. Using Proposition 2.3.2, we need to show that the restriction of H_λ to $L(\lambda)_\mu$ is positive definite. We prove this by induction on $|\lambda-\mu|$. For $\lambda = \mu$ there is nothing to prove. Let $v \in L(\lambda)_\mu$, $\mu \ne \lambda$. By (a), for any basis $\{e^i_\alpha\}$ of $\mathfrak{g}_\alpha$, $\{\sigma_o(e^i_\alpha)\}$ is the dual (with respect to $\langle\ ,\ \rangle$) basis of $\mathfrak{g}_{-\alpha}$. Thus, we have

$$\text{(3)} \qquad \Omega(v) = \langle\mu+2\rho,\mu\rangle v + 2\sum_{\alpha\in\Delta^+}\sum_i \sigma_o(e^i_\alpha)e^i_\alpha(v).$$

Now, by Lemma 2.1.16,

$$\text{(4)} \qquad H_\lambda(\Omega v, v) = \langle\lambda, 2\rho+\lambda\rangle H_\lambda(v,v),$$

whereas by (3) and contragradience of H_λ we get

$$\text{(5)} \qquad H_\lambda(\Omega v, v) = \langle\mu, 2\rho+\mu\rangle H_\lambda(v,v) + 2\sum_{\alpha\in\Delta^+}\sum_i H_\lambda(e^i_\alpha v, e^i_\alpha v).$$

Equating (4) and (5), we get

$$\text{(6)} \quad (\langle\lambda+\rho,\lambda+\rho\rangle - \langle\mu+\rho,\mu+\rho\rangle)H_\lambda(v,v) = 2\sum_{\alpha\in\Delta^+}\sum_i H_\lambda(e^i_\alpha v, e^i_\alpha v).$$

By the inductive assumption, the right side of (6) is ≥ 0, and hence, by Lemma 2.3.12, $H_\lambda(v, v) \geq 0$. But since H_λ is nondegenerate, it is positive definite.

Conversely, let H_λ be positive definite. Then for any $k \in \mathbb{Z}_+$, by Exercise 1.2.E.3,

$$0 \leq H_\lambda(f_i^k v_\lambda, f_i^k v_\lambda) = H_\lambda(v_\lambda, e_i^k f_i^k v_\lambda) = \prod_{j=1}^{k} (j(\langle \lambda, \alpha_i^\vee \rangle + 1 - j)).$$

This forces $\langle \lambda, \alpha_i^\vee \rangle \in \mathbb{Z}_+$, which completes the proof of the theorem. □

2.3.E EXERCISES

(1) Let $\mathfrak{g}$ be symmetrizable and let $\lambda \in D$ and $w \in W$. Then show that $[M(w * \lambda) : L(\mu)] \neq 0$ for $\mu \in \mathfrak{h}^*$ iff μ is of the form $\mu = v * \lambda$ for some $v \geq w$. *Hint*: Use the Strong Exchange Condition Theorem 1.3.11(c), and Proposition 2.3.8.

(2) Let $\mathfrak{g}$ be an arbitrary Kac–Moody Lie algebra and $V = V(\lambda)$ an integrable highest weight $\mathfrak{g}$-module with highest weight λ. Let $P(V) \subset \mathfrak{h}^*$ be the set of weights of V. Fix a positive real root α and $\mu \in P(V)$. Prove the following:

(a) The set $S_{\mu,\alpha}(V) := (\mu + \mathbb{Z}\alpha) \cap P(V)$ is an unbroken string $\{\mu - p\alpha, \mu - (p-1)\alpha, \cdots, \mu + q\alpha\}$, for some $p, q \in \mathbb{Z}_+$.
The set $S_{\mu,\alpha}(V)$ is called the *α-string of V through μ*.

(b) $p - q = \langle \mu, \alpha^\vee \rangle$.

(c) $\text{mult}_{\mu - t\alpha}(V) = \text{mult}_{\mu - (\langle \mu, \alpha^\vee \rangle - t)\alpha}(V)$, for any $t \in \mathbb{Z}$.

(d) For $e_\alpha \in \mathfrak{g}_\alpha \backslash \{0\}$, the map $V_{\mu + t\alpha} \to V_{\mu + (t+1)\alpha}$, $v \mapsto e_\alpha(v)$, is injective for any $t < -\frac{1}{2}\langle \mu, \alpha^\vee \rangle$.

(e) $\mathfrak{g}_\alpha V_\mu \neq 0$ iff both $\mu, \mu + \alpha \in P(V)$.

(3) Let the notation and assumptions be as in the above Exercise 2. Show that $P(V) \subset [W \cdot \lambda]$, where $[W \cdot \lambda]$ denotes the convex hull of $\{w\lambda\}_{w \in W}$ in $\mathfrak{h}^*$.

Hint: Take $\mu \in P(V), \mu \neq \lambda$. Since $U(\mathfrak{n}^-) \cdot v_\lambda = V$, where v_λ is a highest weight vector of V, there exists a simple root α_i such that $\mu + \alpha_i \in P(V)$. Thus the α_i-string $S_{\mu,\alpha_i}(V)$ has $q > 0$ (in the notation of Exercise 2(a)). By induction on the height ht $(\lambda - \mu)$, conclude that $\mu + q\alpha_i \in [W \cdot \lambda]$. By Exercise 2(b), $s_i(\mu + q\alpha_i) = \mu - p\alpha_i$ and thus $\mu - p\alpha_i \in [W \cdot \lambda]$. Now use $\mu \in [\mu + q\alpha_i, \mu - p\alpha_i]$.

A stronger result for $L^{\max}(\lambda)$ is proved in [Kac–90, Proposition 11.3(a)].

(4) Let $\mathfrak{g}$ be symmetrizable and let $\alpha \in \Delta_{\text{im}}^+$. Then show that $\langle \alpha, \alpha \rangle \leq 0$.

Hint: Show that there exists $w \in W$ such that $w\alpha \in \Delta^+$ and $-w\alpha \in D$.

(5) With the notation as in 2.3.3, show that each $M^p(\lambda)$ is a $\mathfrak{g}$-submodule of $M(\lambda)$.

Hint: Extend the bilinear form $S : U(\mathfrak{g}) \times U(\mathfrak{g}) \to S(\mathfrak{h})$, of Subsection 2.3.1, $\mathbb{C}[t]$-bilinearly to $(U(\mathfrak{g}) \otimes_{\mathbb{C}} \mathbb{C}[t]) \times (U(\mathfrak{g}) \otimes_{\mathbb{C}} \mathbb{C}[t]) \to S(\mathfrak{h}) \otimes_{\mathbb{C}} \mathbb{C}[t]$. Show that it descends to a contravariant bilinear form on the quotient module $\big(U(\mathfrak{g}) \otimes_{\mathbb{C}} \mathbb{C}[t]\big)/\langle \mathfrak{n}^+, h - (\lambda + t\rho)h; h \in \mathfrak{h}\rangle$, where $\langle \mathfrak{n}^+, h - (\lambda + t\rho)h\rangle$ denotes the left ideal of $U(\mathfrak{g}) \otimes \mathbb{C}[t]$ generated by the elements under the bracket.

(6) For any $\lambda \in \mathfrak{h}^*_{\mathbb{R}}$, show that $L(\lambda)_{\mathbb{R}} := U(\mathfrak{g}_{\mathbb{R}})v_\lambda \subset L(\lambda)$ is a real form of $L(\lambda)$, i.e., the canonical map $\theta : L(\lambda)_{\mathbb{R}} \otimes_{\mathbb{R}} \mathbb{C} \to L(\lambda)$ is an isomorphism, where $\mathfrak{g}_{\mathbb{R}}$ is the real form of $\mathfrak{g}$ as in 2.3.9.

Hint: Since $L(\lambda)$ is an irreducible $\mathfrak{g}$-module, $\operatorname{Ker}\theta$ if nonzero contains a nonzero vector of $(L(\lambda)_{\mathbb{R}})_\lambda \otimes_{\mathbb{R}} \mathbb{C}$. This leads to a contradiction.

2.C Comments. The category $\mathcal{O}$ was introduced and its basic properties proved (in the finite case) by [Bernstein–Gelfand–Gelfand–75]. Its extension to Kac–Moody Lie algebras is straightforward. Lemma 2.1.4 appears in [Kac–90, Lemma 9.10]. Lemma 2.1.9 is due to [Deodhar–Gabber–Kac–82] and Lemma 2.1.10 is essentially due to [Garland–Lepowsky–76], except that properties (iii) and (iv) are not explicitly mentioned. The lemma in its present form is taken from [Deodhar–Gabber–Kac–82]. Lemma 2.1.16 is taken from [Kac–90, Section 9.8].

The Weyl–Kac character formula for symmetrizable $\mathfrak{g}$, Theorem 2.2.1, which generalizes the classical Weyl character formula in the finite case, and its Corollary 2.2.6 are due to [Kac–74]. The complete reducibility of integrable $\mathfrak{g}$-modules in the category $\mathcal{O}$, Corollary 2.2.7, is due to [Kac–Peterson–84a].

The Shapovalov bilinear form was introduced and its determinant calculated in the finite case by [Shapovalov–72]. [Jantzen–79] used this form to introduce his filtration and prove the character sum formula (in the finite case). Extension of these to symmetrizable Kac–Moody algebras (cf. Theorem 2.3.4 and Corollaries 2.3.5 and 2.3.6) was obtained by [Kac–Kazhdan–79].

The subset $K^{\text{w.g.}} \subset \mathfrak{h}^*$ (cf. Definition 2.3.7), which contains the Tits cone, was introduced and Proposition 2.3.8, determining the irreducible components of $M(\lambda)$ for $\lambda \in K^{\text{w.g.}}$, obtained by [Kumar–87b]. Kumar's result was influenced by the corresponding result for λ in the Tits Cone obtained by [Deodhar–Gabber–Kac-82]. Lemma 2.3.11 is independently due to [Kac–Peterson–84b] and [Kumar–84]. Theorem 2.3.13(a) is due to [Kac–Peterson–84b]. Theorem 2.3.13(b) in the affine case is due to [Garland–78] which was easily generalized by [Kac–Peterson–84b] to symmetrizable $\mathfrak{g}$ by using Theorem 2.3.13(a). Exercise 2.3.E.4 is taken from [Kac–90, Proposition 5.2(c)].

A precursor of the work [Kac–74] was the paper [Macdonald–72], which gave several Dedekind's η-function identities by considering the affine root systems. (See also the AMS review of the paper of Macdonald by D.N. Verma [Verma–75b], and [Kostant–76].) The work of Macdonald can be interpreted as the Denominator Formula (i.e., the Weyl–Kac character formula for the trivial representation) for affine Kac–Moody Lie algebras. The importance of the subject grew considerably after the paper of Macdonald and the paper of [Kac–74]. The Weyl–Kac character formula led to several results including the Rogers–Ramanujan identities. We refer the reader to two survey articles summarizing the developments in this direction: [Lepowsky–82] and [Misra–88].

III

Lie Algebra Homology and Cohomology

Let $\mathfrak{s}$ be a complex Lie algebra together with a subalgebra $\mathfrak{t}$ and V an $\mathfrak{s}$-module. In Section 3.1 we define the Lie algebra homology $H_*(\mathfrak{s}, \mathfrak{t}, V)$ and the cohomology $H^*(\mathfrak{s}, \mathfrak{t}, V)$ and derive their standard elementary properties including their connection with Tor and Ext groups introduced by Hochschild for the pair of algebras $(U(\mathfrak{s}), U(\mathfrak{t}))$. More specifically, it is shown that if $\mathfrak{s}$ is a finitely semisimple $\mathfrak{t}$-module under the adjoint action, then $H_*(\mathfrak{s}, \mathfrak{t}, V) \simeq \mathrm{Tor}_*^{(U(\mathfrak{s}),U(\mathfrak{t}))}(\mathbb{C}, V)$ and $H^*(\mathfrak{s}, \mathfrak{t}, V) \simeq \mathrm{Ext}^*_{(U(\mathfrak{s}),U(\mathfrak{t}))}(\mathbb{C}, V)$.

We use the "Hopf principle" (Proposition 3.1.10) to show that the tensor product of a free $U(\mathfrak{s})$-module with an arbitrary $\mathfrak{s}$-module is free. This is used to show that, for two $\mathfrak{s}$-modules V and W, $\mathrm{Ext}^*_{(U(\mathfrak{s}),U(\mathfrak{t}))}(V, W) \simeq H^*(\mathfrak{s}, \mathfrak{t}, \mathrm{Hom}_{\mathbb{C}}(V, W))$ and $\mathrm{Tor}_*^{(U(\mathfrak{s}),U(\mathfrak{t}))}(W^t, V) \simeq H_*(\mathfrak{s}, \mathfrak{t}, W \otimes V)$, where the right $\mathfrak{s}$-module W^t is as defined in Subsection 3.1.9.

Section 3.2 is devoted to proving the result of Garland–Lepowsky on homology of the nil-radical $\mathfrak{u}_Y^-$ of any (opposite) parabolic subalgebra $\mathfrak{p}_Y^-$ of a symmetrizable Kac–Moody Lie algebra $\mathfrak{g}$ with coefficients in any integrable highest weight $\mathfrak{g}$-module $L(\lambda)$ (cf. Theorem 3.2.7). This generalizes the fundamental $\mathfrak{n}$-homology result of Kostant in the finite case. The proof makes use of the standard resolution $U(\mathfrak{g}) \otimes_{U(\mathfrak{p}_Y)} \Lambda^*(\mathfrak{g}/\mathfrak{p}_Y)$ of the trivial $\mathfrak{g}$-module for the Lie algebra pair $(\mathfrak{g}, \mathfrak{p}_Y)$. On tensoring this resolution over $\mathbb{C}$ with $L(\lambda)$, we get a resolution of $L(\lambda)$. This resolution is decomposed as a direct sum of the generalized eigenspaces of the Casimir–Kac operator Ω. It follows from the basic theorem of homological algebra that only the generalized eigenspace corresponding to the eigenvalue by which Ω acts on $L(\lambda)$ contributes to the homology $H_*(\mathfrak{u}_Y^-, L(\lambda))$, from which the result of Garland–Lepowsky follows. As an immediate consequence of this result, one of course recovers the Weyl–Kac character formula for symmetrizable $\mathfrak{g}$ (proved by a different method in the last chapter).

As another consequence of this result (in fact, only the knowledge of $H_2(\mathfrak{n}^-, \mathbb{C})$ is needed), we obtain the vanishing of the radical in the symmetrizable case due to Gabber–Kac, although the proof given here is somewhat different from theirs. As yet another consequence of the Garland–Lepowsky result, we calculate the homology of the Lie algebra pair $(\mathfrak{g}, \mathfrak{g}_Y)$ for any subset Y of $\{1, \dots, \ell\}$ of finite type, where $\mathfrak{g}_Y$ is the standard Levi component of the parabolic subalgebra $\mathfrak{p}_Y$ (cf. Proposition 3.2.11).

Define a relation $\to$ in $\mathfrak{h}^*$ by $\lambda \to \mu$ if $L(\lambda)$ is a component of $M(\mu)$ and let $\sim$ be the equivalence relation in $\mathfrak{h}^*$ generated by $\to$. For any equivalence class $\Lambda \subset \mathfrak{h}^*$, let $\mathcal{O}_\Lambda$ be the full subcategory of $\mathcal{O}$ consisting of those $M \in \mathcal{O}$, such that all the components $L(\lambda)$ of M satisfy $\lambda \in \Lambda$. Section 3.3 is devoted to decomposing the BGG category $\mathcal{O}$ for symmetrizable $\mathfrak{g}$, as a direct sum of blocks $\{\mathcal{O}_\Lambda\}_\Lambda$ (a result due to Deodhar–Gabber–Kac), and also a certain full subcategory $\mathcal{O}^{\text{w.g.}}$ (due to Kumar) in terms of more refined blocks. The advantage to introducing the category $\mathcal{O}^{\text{w.g.}}$ lies in the fact that the description of blocks in $\mathcal{O}^{\text{w.g.}}$ is much more explicit. It is shown that for two modules $M, N \in \mathcal{O}$, which belong to distinct blocks, $\text{Ext}^*_{U(\mathfrak{g})}(M, N) = 0$. A similar result is obtained for modules in $\mathcal{O}^{\text{w.g.}}$.

This vanishing result gives the above mentioned decomposition of $\mathcal{O}$ and $\mathcal{O}^{\text{w.g.}}$. As another consequence of this vanishing, we obtain the vanishing (generalizing the two Whitehead lemmas in the finite case [Jacobson–62, Chap. III, Section 10]): $H_*(\mathfrak{g}, L(\lambda)) = H^*(\mathfrak{g}, L(\lambda)) = 0$, for the integrable highest weight modules $L(\lambda)$ with $\lambda \neq 0$. More generally, it is shown that $\text{Ext}^*_{U(\mathfrak{g})}(L(\lambda), L(\mu)) = 0$, for any two integrable highest weight $\mathfrak{g}$-modules $L(\lambda)$ and $L(\mu)$ with $\lambda \neq \mu$.

Section 3.4 is devoted to calculating the Hodge–Laplacian (due to Kumar) acting on the standard chain complex of $\mathfrak{u}_Y^-$ with coefficients in any integrable highest weight $\mathfrak{g}$-module, where $\mathfrak{g}$ is symmetrizable. This generalizes the corresponding result in the finite case due to Kostant. As an immediate consequence of the Laplacian computation, the homology result of Garland–Lepowsky (mentioned above) is recovered. This is the way Kostant originally obtained his $\mathfrak{n}$-homology result.

Yet another (geometric) proof, via the Kempf resolution, of the result of Garland–Lepowsky will be given in Section 9.3. In fact, this geometric proof does not require the symmetrizability assumption and hence we will have the theorem, as well as the Weyl–Kac character formula, for an arbitrary Kac–Moody Lie algebra $\mathfrak{g}$.

An alternative proof of Kostant's $\mathfrak{n}$-homology result (in the finite case), using the Hochschild–Serre spectral sequence and $s\ell_2$-representation theory, is outlined in the exercises. Also, Kostant's $\mathfrak{n}$-homology result is used to indicate a proof of the complete reducibility theorem in the finite case.

3.1. Basic Definitions and Elementary Properties

3.1.1 Lie algebra homology. For any Lie algebra $\mathfrak{s}$, and an $\mathfrak{s}$-module V, the

Lie algebra homology $H_*(\mathfrak{s}, V)$ of $\mathfrak{s}$ with coefficients in V is defined as follows: Consider the chain complex

(1) $$\to \Lambda_p \xrightarrow{\partial_p} \Lambda_{p-1} \dots \xrightarrow{\partial_1} \Lambda_0 \xrightarrow{\partial_0} 0,$$

where $\Lambda_p = \Lambda_p(\mathfrak{s}, V) := \Lambda^p(\mathfrak{s}) \otimes_{\mathbb{C}} V$ (in particular, $\Lambda_0 = V$), and

(2)
$$\partial_p(x_1 \wedge \dots \wedge x_p \otimes v) = \sum_{i<j} (-1)^{i+j} [x_i, x_j] \wedge x_1 \wedge \dots \wedge \hat{x}_i \wedge \dots \wedge \hat{x}_j \wedge \dots \wedge x_p \otimes v$$
$$+ \sum_i (-1)^i x_1 \wedge \dots \wedge \hat{x}_i \wedge \dots \wedge x_p \otimes (x_i \cdot v),$$

for $x_i \in \mathfrak{s}$ and $v \in V$.

For any $x \in \mathfrak{s}$, define the linear maps

$$\epsilon_x : \Lambda_p \to \Lambda_{p+1} \text{ by } \epsilon_x(y \otimes v) = x \wedge y \otimes v, \text{ for } y \in \Lambda^p(\mathfrak{s}), \ v \in V;$$

and

$$\theta_x : \Lambda_p \to \Lambda_p$$

as the map induced from the tensor product $\mathfrak{s}$-module structure on $\Lambda^p(\mathfrak{s}) \otimes V$ (with the adjoint action of $\mathfrak{s}$ on $\Lambda^p(\mathfrak{s})$).

It is easy to see that, for any $x \in \mathfrak{s}$,

(3) $$\partial_{p+1}\epsilon_x + \epsilon_x \partial_p = -\theta_x,$$

(4) $$\partial_p \circ \theta_x = \theta_x \circ \partial_p.$$

From (3) and (4) and induction on p, it is easy to see that

(5) $$\partial_p \partial_{p+1} = 0, \qquad \text{for all } p \geq 0.$$

Thus, (1) is indeed a chain complex. Define

(6) $$H_p(\mathfrak{s}, V) = \frac{\operatorname{Ker} \partial_p}{\operatorname{Im} \partial_{p+1}}.$$

By (4), θ_x is a chain map and hence $H_p(\mathfrak{s}, V)$ acquires a natural $\mathfrak{s}$-module structure; but by (3), $H_*(\mathfrak{s}, V)$ is a trivial $\mathfrak{s}$-module.

More generally, if $\mathfrak{s}$ is an ideal in another Lie algebra $\mathfrak{a}$ and the $\mathfrak{s}$-module structure on V is the restriction of an $\mathfrak{a}$-module structure on V, then the action θ_x $(x \in \mathfrak{a})$ of $\Lambda_p(\mathfrak{s}, V)$ gives rise to a natural $\mathfrak{a}/\mathfrak{s}$-module structure on $H_*(\mathfrak{s}, V)$.

3.1.2 Lie algebra cohomology. Analogously we define the Lie algebra cohomology $H^*(\mathfrak{s}, V)$ of $\mathfrak{s}$ with coefficients in a $\mathfrak{s}$-module V as follows:

Consider the cochain complex

$$(1) \qquad 0 \to C^0 \xrightarrow{d^0} \cdots \longrightarrow C^p \xrightarrow{d^p} C^{p+1} \longrightarrow \ldots,$$

where $C^p = C^p(\mathfrak{s}, V) := \operatorname{Hom}_{\mathbb{C}}(\Lambda^p(\mathfrak{s}), V)$ and

$$(2) \quad (d^p\omega)(x_1 \wedge \cdots \wedge x_{p+1}) = \sum_{i<j}(-1)^{i+j}\omega([x_i, x_j] \wedge x_1 \wedge \cdots \wedge \hat{x}_i \wedge \cdots \wedge \hat{x}_j \wedge \cdots \wedge x_{p+1}) + \sum_i (-1)^{i+1} x_i(\omega(x_1 \wedge \cdots \wedge \hat{x}_i \wedge \cdots \wedge x_{p+1})),$$

for $\omega \in C^p$ and $x_1, \ldots, x_{p+1} \in \mathfrak{s}$. Define, for $x \in \mathfrak{s}$,

$$i_x : C^p \to C^{p-1}$$

by

$$(3) \qquad (i_x\omega)(x_1 \wedge \cdots \wedge x_{p-1}) = \omega(x \wedge x_1 \wedge \cdots \wedge x_{p-1}),$$

and

$$L_x : C^p \to C^p$$

is induced from the canonical $\mathfrak{s}$-module structure on $\operatorname{Hom}(\Lambda^p(\mathfrak{s}), V)$, i.e.,

$$(4) \quad (L_x\omega)(x_1 \wedge \cdots \wedge x_p) = -\omega(\operatorname{ad} x(x_1 \wedge \cdots \wedge x_p)) + x \cdot (\omega(x_1 \wedge \cdots \wedge x_p)).$$

Then, as it is easy to see, for any $x \in \mathfrak{s}$,

$$(5) \qquad i_x d^p + d^{p-1} i_x = L_x \text{ on } C^p,$$

$$(6) \qquad L_x \circ d^p = d^p \circ L_x.$$

From (5), (6) and induction on p, we again get

$$(7) \qquad d^{p+1} \circ d^p = 0, \qquad \text{for all } p \geq 0.$$

Define

$$H^p(\mathfrak{s}, V) = \frac{\operatorname{Ker} d^p}{\operatorname{Im} d^{p-1}}.$$

The maps L_x induce a trivial $\mathfrak{s}$-module structure on $H^*(\mathfrak{s}, V)$. If $\mathfrak{s}$ is an ideal in a Lie algebra $\mathfrak{a}$ and the $\mathfrak{s}$-module structure on V is the restriction of an $\mathfrak{a}$-module structure on V, then $H^*(\mathfrak{s}, V)$ acquires a canonical $\mathfrak{a}/\mathfrak{s}$-module structure.

Let $V = \mathbb{C}$ be the trivial $\mathfrak{s}$-module. Then, under the usual wedge (also called Grassman) multiplication (cf., e.g., [Hochschild–Serre–53, §1]) $\wedge : C^p \times C^q \to C^{p+q}$, i_x is an antiderivation, i.e.,

(8) $$i_x(\omega \wedge \eta) = (i_x\omega) \wedge \eta + (-1)^p \omega \wedge i_x\eta\,,$$

for any $x \in \mathfrak{s}$, $\omega \in C^p$ and $\eta \in C^q$. From (8), by induction on $p+q$, it is easy to see that d is an antiderivation, i.e.,

(9) $$d^{p+q}(\omega \wedge \eta) = (d^p\omega) \wedge \eta + (-1)^p \omega \wedge d^q\eta\,,$$

for any $\omega \in C^p$ and $\eta \in C^q$. Thus the wedge product in C^* descends to make $H^*(\mathfrak{s}, \mathbb{C})$ into a graded commutative algebra over $\mathbb{C}$.

3.1.3 Lie algebra homology and cohomology of a pair. We also need Lie algebra homology (and cohomology) of a pair $(\mathfrak{s}, \mathfrak{t})$ (i.e., $\mathfrak{t}$ is a subalgebra of $\mathfrak{s}$) with coefficients in an $\mathfrak{s}$-module V.

Define the quotient $\Lambda_p(\mathfrak{s}, \mathfrak{t}, V)$ of $\Lambda_p(\mathfrak{s}, V)$ by

(1) $$\Lambda_p(\mathfrak{s}, \mathfrak{t}, V) := \frac{\Lambda^p(\mathfrak{s}/\mathfrak{t}) \otimes V}{\mathfrak{t} \cdot [\Lambda^p(\mathfrak{s}/\mathfrak{t}) \otimes V]}\,,$$

where the notation $\mathfrak{t} \cdot M$, for a $\mathfrak{t}$-module M, means the span of $x \cdot m$ for $x \in \mathfrak{t}$ and $m \in M$. It is easy to see (from (3.1.1.3)–(3.1.1.4)) that the map $\partial_p : \Lambda_p(\mathfrak{s}, V) \to \Lambda_{p-1}(\mathfrak{s}, V)$ descends to a map (again denoted by) $\partial_p : \Lambda_p(\mathfrak{s}, \mathfrak{t}, V) \to \Lambda_{p-1}(\mathfrak{s}, \mathfrak{t}, V)$. In particular, we get the quotient complex

(2) $$\cdots \to \Lambda_p(\mathfrak{s}, \mathfrak{t}, V) \xrightarrow{\partial_p} \Lambda_{p-1}(\mathfrak{s}, \mathfrak{t}, V) \to \cdots \to \Lambda_0(\mathfrak{s}, \mathfrak{t}, V) \to 0$$

of $\Lambda_*(\mathfrak{s}, V)$.

The homology of the complex (2) is denoted by $H_p(\mathfrak{s}, \mathfrak{t}, V)$ and is called the (Chevalley–Eilenberg) *Lie algebra homology of the pair* $(\mathfrak{s}, \mathfrak{t})$ *in the* $\mathfrak{s}$*-module* V.

Similarly, define the subspace

$$C^p(\mathfrak{s}, \mathfrak{t}, V) := \operatorname{Hom}_{\mathfrak{t}}(\Lambda^p(\mathfrak{s}/\mathfrak{t}), V) \subset C^p(\mathfrak{s}, V)$$

consisting of all $\mathfrak{t}$-module maps from $\Lambda^p(\mathfrak{s}/\mathfrak{t})$ to V. Again by (3.1.2.5)–(3.1.2.6) we get

(3) $$d^p(C^p(\mathfrak{s}, \mathfrak{t}, V)) \subset C^{p+1}(\mathfrak{s}, \mathfrak{t}, V);$$

in particular, we get a subcomplex

$$(4) \quad 0 \to C^0(\mathfrak{s}, \mathfrak{t}, V) \xrightarrow{d^0} \cdots \to C^p(\mathfrak{s}, \mathfrak{t}, V) \xrightarrow{d^p} C^{p+1}(\mathfrak{s}, \mathfrak{t}, V) \to \dots$$

of $C^*(\mathfrak{s}, V)$.

The cohomology of the complex (4) is denoted by $H^p(\mathfrak{s}, \mathfrak{t}, V)$, and is called the (Chevalley–Eilenberg) *Lie algebra cohomology of the pair* $(\mathfrak{s}, \mathfrak{t})$ *with coefficients in the* $\mathfrak{s}$*-module* V.

For the trivial $\mathfrak{s}$-module $V = \mathbb{C}$, the subspace $C^*(\mathfrak{s}, \mathfrak{t}, \mathbb{C}) \subset C^*(\mathfrak{s}, \mathbb{C})$ is, in fact, a subalgebra under wedge multiplication. Hence $H^*(\mathfrak{s}, \mathfrak{t}, \mathbb{C})$ acquires a graded commutative algebra structure.

3.1.4 Definition. Let $(\mathfrak{s}, \mathfrak{t})$ be any Lie algebra pair. Denote $U(\mathfrak{s})$, resp. $U(\mathfrak{t})$, by $\mathcal{S}$, resp. $\mathcal{T}$. Define the sequence

$$(\mathcal{D}) \qquad \cdots \to D_2 \xrightarrow{\partial_2} D_1 \xrightarrow{\partial_1} D_0 \xrightarrow{\epsilon} D_{-1} = \mathbb{C} \to 0$$

where, for $p \geq 0$,

$$D_p = D_p(\mathfrak{s}, \mathfrak{t}) = \mathcal{S} \otimes_{\mathcal{T}} \Lambda^p(\mathfrak{s}/\mathfrak{t}),$$

$\mathcal{S}$ is thought of as a $\mathcal{T}$-module under right multiplication and, for $p \geq 1$,

(1)

$$\partial_p(a \otimes \bar{y}_1 \wedge \cdots \wedge \bar{y}_p) = \sum_{i<j} (-1)^{i+j} a \otimes \overline{[y_i, y_j]} \wedge \bar{y}_1 \wedge \cdots \wedge \hat{\bar{y}}_i \wedge \cdots \wedge \hat{\bar{y}}_j \wedge \cdots \wedge \bar{y}_p$$
$$+ \sum_i (-1)^{i+1} (a y_i) \otimes \bar{y}_1 \wedge \cdots \wedge \hat{\bar{y}}_i \wedge \cdots \wedge \bar{y}_p,$$

for $a \in \mathcal{S}$, $\bar{y}_j = y_j \bmod \mathfrak{t} \in \mathfrak{s}/\mathfrak{t}$.

It is easy to see that ∂_p is well defined, i.e., it does not depend on the choice of the coset representatives y_j. (The map ϵ is induced from the standard augmentation map $\mathcal{S} \to \mathbb{C}$.) Further, $(\mathcal{D})$ is a chain complex. In fact, except for the last term D_{-1}, $(\mathcal{D})$ is a particular example of (3.1.3.2) where we take $V = \mathcal{S}$ considered as a $\mathfrak{s}$-module under $y \cdot a = -ay$, for $a \in \mathcal{S}$ and $y \in \mathfrak{s}$.

D_p is a $\mathcal{S}$-module under left multiplication on the $\mathcal{S}$-factor. Moreover, ∂_p is a $\mathcal{S}$-module map (and so is ϵ).

3.1.5 Proposition. *The above sequence* $(\mathcal{D})$ *is exact.*

Proof. Let $\{U_k(\mathfrak{s})\}_{k\geq 0}$ be the standard filtration of $U(\mathfrak{s})$, i.e., $U_k(\mathfrak{s})$ is the span of the elements $\{x_1 \dots x_\ell;\ \ell \leq k,\ x_i \in \mathfrak{s}\}$. Set $U_k(\mathfrak{s}) = 0$, if $k < 0$. Define an increasing filtration $\mathcal{F} = \{\mathcal{F}(s)\}_{s\geq 0}$ of the chain complex $(\mathcal{D})$ by setting $\mathcal{F}(s)_p$ $(p \geq 0)$ as the span of $a \otimes \Lambda^p(\mathfrak{s}/\mathfrak{t})$ with $a \in U_{s-p}(\mathfrak{s})$. Set $\mathcal{F}(s)_{-1} = \mathbb{C}$. It is easy to see that $\mathcal{F}(s)_*$ is a subcomplex of $(\mathcal{D})$. Let E^r be the corresponding homology spectral sequence converging to the homology $H_*(\mathcal{D})$ (cf. Appendix E.4). Then

$$(1)\qquad E^1_{s,t} = H_{s+t}(\mathcal{F}(s)/\mathcal{F}(s-1)).$$

By the PBW Theorem, for $p \geq 0$ and $s \geq 1$,

$$(2)\qquad [\mathcal{F}(s)/\mathcal{F}(s-1)]_p \approx S^{s-p}(\mathfrak{s}/\mathfrak{t}) \otimes \Lambda^p(\mathfrak{s}/\mathfrak{t}), \qquad \text{and}$$

$$(3)\qquad [\mathcal{F}(s)/\mathcal{F}(s-1)]_{-1} = 0\,,$$

where S^* is the symmetric algebra. Moreover, the induced differential (denoted)

$$\bar{\partial}_p : [\mathcal{F}(s)/\mathcal{F}(s-1)]_p \to [\mathcal{F}(s)/\mathcal{F}(s-1)]_{p-1}$$

is given by

$$(4)\qquad \bar{\partial}_p(a \otimes y_1 \wedge \cdots \wedge y_p) = \sum_i (-1)^{i+1}(ay_i) \otimes y_1 \wedge \cdots \wedge \hat{y}_i \wedge \cdots \wedge y_p,$$

for $a \in S^{s-p}(\mathfrak{s}/\mathfrak{t}),\ y_i \in \mathfrak{s}/\mathfrak{t}$. In particular, from the Koszul resolution (cf. Appendix D.13),

$$H_*(\mathcal{F}(s)/\mathcal{F}(s-1)) = 0, \ \text{for any } s \geq 1, \ \text{and also } H_*(\mathcal{F}(0)) = 0.$$

In particular, by (1), $E^1 \equiv 0$ and hence $E^\infty \equiv 0$, proving the proposition. □

Let R be a ring with identity, and S a subring containing the identity. For any (left) R-modules M, N, recall the definition of the relative Ext groups $\mathrm{Ext}^p_{(R,S)}(M,N)$ $(p \geq 0)$ and also the relative Tor groups $\mathrm{Tor}^{(R,S)}_p(M,N)$ for a right R-module M and a left R-module N from Appendix D.

3.1.6 Definition. Call a representation M of a (not necessarily finite-dimensional) Lie algebra $\mathfrak{t}$ *finitely semisimple* if M is the sum of its finite-dimensional irreducible $\mathfrak{t}$-submodules. Clearly, any quotient of a finitely semisimple module is finitely semisimple. Moreover, any submodule of a finitely semisimple module is a direct summand (cf. Exercise 3.1.E.3). Thus any submodule of a finitely semisimple module is again finitely semisimple. Further, the tensor product of two finitely semisimple modules is finitely semisimple (cf. [Hochschild–Serre–53, Proposition 1]).

In the following tensor product $R \otimes_S M$, R is thought of as a right S-module under right multiplication and $R \otimes_S M$ is a (left) R-module under left multiplication on the first factor. Recall the definition of an (R, S)-projective module from Appendix (D).

3.1.7 Lemma. *For any S-module M and R-module N, the restriction map*

$$\theta : \mathrm{Hom}_R(R \otimes_S M, N) \to \mathrm{Hom}_S(M, N), \tag{1}$$

given by $\chi \mapsto \chi_{|1\otimes M}$, *is an isomorphism.*

In particular, for any S-module M, $R \otimes_S M$ *is (R, S)-projective. Hence, any R-module is the quotient of a (R, S)-projective module. Similarly, for a left S-module M and right R-module N,*

$$N \otimes_R (R \otimes_S M) \simeq N \otimes_S M, \tag{2}$$

under the map $n \otimes (r \otimes m) \mapsto nr \otimes m$, *for* $n \in N,\ r \in R$ *and* $m \in M$.

Proof. To prove (1), we define the map $\beta : \mathrm{Hom}_S(M, N) \to \mathrm{Hom}_R(R\otimes_S M, N)$ as follows: For any $f \in \mathrm{Hom}_S(M, N)$, $\beta(f) : R \otimes_S M \to N$ is the map given by $\beta(f)(r \otimes m) = rf(m)$. It is easy to see that β is the inverse map of θ.

To prove (2), define the map $N \otimes_S M \to N \otimes_R (R \otimes_S M)$ by $n \otimes m \mapsto n \otimes (1 \otimes m)$. This provides the inverse to the map in (2). □

3.1.8 Corollary. *Let* $(\mathfrak{s}, \mathfrak{t})$ *be a Lie algebra pair such that* $\mathfrak{s}$ *is finitely semisimple under the adjoint action of* $\mathfrak{t}$. *Then the sequence* $(\mathcal{D})$ *of 3.1.4:*

$$\cdots \to D_2 \xrightarrow{\partial_2} D_1 \xrightarrow{\partial_1} D_0 \xrightarrow{\epsilon} D_{-1} = \mathbb{C} \to 0, \tag{$\mathcal{D}$}$$

where $D_p := \mathcal{S} \otimes_{\mathcal{T}} \wedge^p(\mathfrak{s}/\mathfrak{t})$, *is a* $(\mathcal{S}, \mathcal{T})$*-projective resolution of* $\mathbb{C}$. *Thus, for any* $\mathfrak{s}$*-module V, the following is a* $(\mathcal{S}, \mathcal{T})$*-projective resolution of V:*

$$\cdots \to D_2 \otimes_{\mathbb{C}} V \xrightarrow{\partial_2\otimes I_V} D_1 \otimes_{\mathbb{C}} V \xrightarrow{\partial_1\otimes I_V} D_0 \otimes_{\mathbb{C}} V \xrightarrow{\epsilon\otimes I_V} V \to 0. \tag{$\mathcal{D}_V$}$$

Proof. Each D_p is $(\mathcal{S}, \mathcal{T})$-projective by Lemma 3.1.7, and $(\mathcal{D})$ is an exact sequence by Proposition 3.1.5. Thus, it suffices to prove that $(\mathcal{D})$ is '$\mathfrak{t}$-split', i.e., each Ker ∂_p is a $\mathfrak{t}$-module direct summand in D_p and Ker ϵ is a $\mathfrak{t}$-module direct summand in D_0. It is easy to see that, as a $\mathfrak{t}$-module, D_p is a quotient of the tensor product $\mathfrak{t}$-module $\mathcal{S} \otimes_{\mathbb{C}} \wedge^p(\mathfrak{s}/\mathfrak{t})$, where $\mathfrak{t}$ acts on $\mathcal{S}$ via $X.a = Xa - aX$, for $X \in \mathfrak{t}$ and $a \in \mathcal{S}$. Since $\mathcal{S}$ is a $\mathfrak{t}$-module quotient of the tensor algebra $T(\mathfrak{s})$, where the $\mathfrak{t}$-module structure on $T(\mathfrak{s})$ is induced from the adjoint action, by Subsection 3.1.6, D_p is a finitely semisimple $\mathfrak{t}$-module. Thus, any $\mathfrak{t}$-submodule of D_p is a $\mathfrak{t}$-module direct summand by Exercise 3.1.E.3. This proves that $(\mathcal{D})$ is a $(\mathcal{S}, \mathcal{T})$-projective resolution.

Since $\bullet \otimes_{\mathbb{C}} V$ is an exact functor, the sequence $(\mathcal{D}_V)$ is exact and $\mathfrak{t}$-split. Finally, by Proposition 3.1.10, $D_p \otimes_{\mathbb{C}} V$ is $(\mathcal{S}, \mathcal{T})$-projective. Thus, $(\mathcal{D}_V)$ is a $(\mathcal{S}, \mathcal{T})$-projective resolution of V. □

3.1.9 Lemma. *For a Lie algebra pair* $(\mathfrak{s}, \mathfrak{t})$ *such that* $\mathfrak{s}$ *is finitely semisimple* $\mathfrak{t}$*-module under the adjoint action, and any* $\mathfrak{s}$*-module* V,

$$(1) \qquad H_*(\mathfrak{s}, \mathfrak{t}, V) \approx \mathrm{Tor}_*^{(\mathcal{S},\mathcal{T})}(V^t, \mathbb{C}) \approx \mathrm{Tor}_*^{(\mathcal{S},\mathcal{T})}(\mathbb{C}, V), \qquad \text{and}$$

$$(2) \qquad H^*(\mathfrak{s}, \mathfrak{t}, V) \approx \mathrm{Ext}^*_{(\mathcal{S},\mathcal{T})}(\mathbb{C}, V),$$

where $\mathcal{S} := U(\mathfrak{s})$ *(and* $\mathcal{T} := U(\mathfrak{t})$*) and* V^t *is the right* $\mathfrak{s}$*-module with the same underlying space as* V *and* $v \cdot X := -Xv$*, for* $v \in V$, $X \in \mathfrak{s}$*. In particular, the lemma is true for* $\mathfrak{t} = (0)$.

Proof. We prove (2); the proof of (1) is similar. Since $(\mathcal{D})$ is a $(\mathcal{S}, \mathcal{T})$-projective resolution of $\mathbb{C}$ (cf. Corollary 3.1.8), it suffices to prove that the complex obtained from the complex $(\mathcal{D})$:

$$0 \to \mathrm{Hom}_{\mathcal{S}}(D_0, V) \to \cdots \to \mathrm{Hom}_{\mathcal{S}}(D_p, V) \to \mathrm{Hom}_{\mathcal{S}}(D_{p+1}, V) \to \ldots$$

is isomorphic with the complex (3.1.3.4).

Define the map

$$\theta^p : \mathrm{Hom}_{\mathcal{S}}(D_p, V) = \mathrm{Hom}_{\mathcal{S}}(\mathcal{S} \otimes_{\mathcal{T}} \Lambda^p(\mathfrak{s}/\mathfrak{t}), V) \to C^p(\mathfrak{s}, \mathfrak{t}, V)$$
$$:= \mathrm{Hom}_{\mathfrak{t}}(\Lambda^p(\mathfrak{s}/\mathfrak{t}), V)$$

by $\theta^p(\chi) = \chi_{|_{1 \otimes \Lambda^p(\mathfrak{s}/\mathfrak{t})}}$, for $\chi \in \mathrm{Hom}_{\mathcal{S}}(\mathcal{S} \otimes_{\mathcal{T}} \Lambda^p(\mathfrak{s}/\mathfrak{t}), V)$. By Lemma 3.1.7, θ^p is an isomorphism. Moreover, it is easy to see that $\theta = (\theta^p)_{p \geq 0}$ is a cochain map. This proves (2). $\square$

3.1.10 Proposition. *Let* M *be a* $\mathfrak{t}$*-module and* N *a* $\mathfrak{s}$*-module. Then there is a natural isomorphism of* $\mathfrak{s}$*-modules, indicated in the proof below:*

$$(\mathcal{S} \otimes_{\mathcal{T}} M) \otimes_{\mathbb{C}} N \simeq \mathcal{S} \otimes_{\mathcal{T}} (M \otimes_{\mathbb{C}} N).$$

The left side is the tensor product of $\mathfrak{s}$*-modules,* $M \otimes_{\mathbb{C}} N$ *on the right is the tensor product of* $\mathfrak{t}$*-modules, with* N *regarded as a* $\mathfrak{t}$*-module by restriction, and* $\mathcal{S} := U(\mathfrak{s})$, $\mathcal{T} := U(\mathfrak{t})$.

Proof. In the following, the tensor product without a subscript means over $\mathbb{C}$. Let $\Delta : \mathcal{S} \to \mathcal{S} \otimes \mathcal{S}$ be the diagonal map, that is, the unique algebra homomorphism such that $\Delta(b) = b \otimes 1 + 1 \otimes b$, for all $b \in \mathfrak{s}$. We shall first construct a map from $\mathcal{S} \otimes_{\mathcal{T}} (M \otimes N)$ to $(\mathcal{S} \otimes_{\mathcal{T}} M) \otimes N$. Fix $b \in \mathcal{S}$, let $\Delta(b) = \sum_i b_{1i} \otimes b_{2i}$ $(b_{ji} \in \mathcal{S})$, and define

$$\varphi_b : M \otimes N \to (\mathcal{S} \otimes_{\mathcal{T}} M) \otimes N$$

by $\varphi_b(m \otimes n) = \sum_i (b_{1i} \otimes m) \otimes b_{2i} \cdot n$. Now define

$$\varphi : \mathcal{S} \otimes_{\mathcal{T}} (M \otimes N) \to (\mathcal{S} \otimes_{\mathcal{T}} M) \otimes N$$

by $\varphi(b \otimes x) = \varphi_b(x)$, for $b \in \mathcal{S},\ x \in M \otimes N$. This map is well defined because if $a \in \mathcal{T}$ and $\Delta(a) = \sum_j a_{1j} \otimes a_{2j}$, with $a_{ij} \in \mathcal{T}$, then

$$\Delta(ba) = \Delta(b)\Delta(a) = \sum_{i,j} b_{1i}a_{1j} \otimes b_{2i}a_{2j}, \text{ and so}$$

$$\begin{aligned}\varphi_{ba}(m \otimes n) &= \sum_{i,j} (b_{1i}a_{1j} \otimes m) \otimes b_{2i}a_{2j} \cdot n = \sum_{i,j} (b_{1i} \otimes a_{1j} \cdot m) \otimes b_{2i}a_{2j} \cdot n \\ &= \varphi_b(a \cdot (m \otimes n)).\end{aligned}$$

It is clear that φ is a $\mathfrak{s}$-module map.

Let $\tau : \mathcal{S} \to \mathcal{S}$ be the antiautomorphism as in 2.3.1. To define a map from $(\mathcal{S} \otimes_{\mathcal{T}} M) \otimes N$ to $\mathcal{S} \otimes_{\mathcal{T}} (M \otimes N)$, first fix $n \in N$ and let

$$\psi_n : \mathcal{S} \otimes_{\mathcal{T}} M \to \mathcal{S} \otimes_{\mathcal{T}} (M \otimes N)$$

be the map $\psi_n(b \otimes m) = \sum_i b_{1i} \otimes (m \otimes \tau(b_{2i}) \cdot n)$ with b and b_{ji} as above. This map is well defined. In fact, let $a \in \mathcal{T}$ with $\Delta(a) = \sum_j a_{1j} \otimes a_{2j}$ as above. Then

$$\begin{aligned}\psi_n(ba \otimes m) &= \sum_{i,j} b_{1i}a_{1j} \otimes (m \otimes \tau(b_{2i}a_{2j}) \cdot n) \\ &= \sum_{i,j} b_{1i} \otimes a_{1j} \cdot (m \otimes \tau(a_{2j})\tau(b_{2i}) \cdot n).\end{aligned}$$

Let $\Delta(a_{2j}) = \sum_s a_{2j1s} \otimes a_{2j2s}$, where the a's are in $\mathcal{T}$. Since

$$(\Delta \otimes 1) \circ \Delta = (1 \otimes \Delta) \circ \Delta : \mathcal{S} \to \mathcal{S} \otimes \mathcal{S} \otimes \mathcal{S},$$

we have

$$\psi_n(ba \otimes m) = \sum_{i,j,s} b_{1i} \otimes ((a_{1j} \cdot m) \otimes a_{2j1s}\tau(a_{2j2s})\tau(b_{2i}) \cdot n).$$

But

$$\sum_s a_{2j1s}\tau(a_{2j2s}) = \epsilon(a_{2j})1,$$

where $\epsilon : \mathcal{T} \to \mathbb{C}$ is the augmentation map, and 1 is the identity element of $\mathcal{T}$. Hence

$$\begin{aligned}\psi_n(ba \otimes m) &= \sum_{i,j} b_{1i} \otimes ((a_{1j} \cdot m) \otimes \epsilon(a_{2j})\tau(b_{2i}) \cdot n) \\ &= \sum_i b_{1i} \otimes ((a \cdot m) \otimes \tau(b_{2i}) \cdot n) = \psi_n(b \otimes a \cdot m).\end{aligned}$$

Thus ψ_n is well defined. Now define

$$\psi : (\mathcal{S} \otimes_{\mathcal{T}} M) \otimes N \to \mathcal{S} \otimes_{\mathcal{T}} (M \otimes N)$$

by $\psi(x \otimes n) = \psi_n(x)$, for $x \in \mathcal{S} \otimes_{\mathcal{T}} M$, $n \in N$. Computations similar to the above show that ψ is a $\mathfrak{s}$-module map. Using this it is easy to see that ψ is a left and right inverse of φ. □

3.1.11 Remarks. (a) Proposition 3.1.10 holds more generally (by the same proof) for an arbitrary Hopf algebra $\mathcal{S}$ and a Hopf subalgebra $\mathcal{T}$ in place of the universal enveloping algebras of $\mathfrak{s}$ and $\mathfrak{t}$.

(b) As mentioned by W. Soergel, one can alternatively prove that φ is an isomorphism by filtering both sides by the natural filtration on $\mathcal{S}$ and showing that φ induces an isomorphism on the associated gr.

As an immediate consequence of the above proposition, we get:

3.1.12 Corollary. *The tensor product of a free $\mathcal{S}$-module with an arbitrary $\mathcal{S}$-module is free.* □

3.1.13 Lemma. *For any Lie algebra pair* $(\mathfrak{s}, \mathfrak{t})$ *and $\mathfrak{s}$-modules V, W,*

(1) $\quad \mathrm{Tor}_*^{(\mathcal{S},\mathcal{T})}(W^t, V) \approx \mathrm{Tor}_*^{(\mathcal{S},\mathcal{T})}(\mathbb{C}, W \otimes V) \approx \mathrm{Tor}_*^{(\mathcal{S},\mathcal{T})}(V^t, W)$,

(2) $\quad \mathrm{Ext}^*_{(\mathcal{S},\mathcal{T})}(V, W) \approx \mathrm{Ext}^*_{(\mathcal{S},\mathcal{T})}(\mathbb{C}, \mathrm{Hom}_{\mathbb{C}}(V, W))$,

(3) $\quad \mathrm{Ext}^*_{(\mathcal{S},\mathcal{T})}(V, W^*) \approx (\mathrm{Tor}_*^{(\mathcal{S},\mathcal{T})}(V^t, W))^*$,

where the full vector space dual V^* of V is equipped with the dual representation, and V^t is the right $\mathfrak{s}$-module defined in Lemma 3.1.9.

Proof. Consider any $(\mathcal{S}, \mathcal{T})$-projective resolution of $\mathbb{C}$ (cf. Lemma D.4):

(4) $$\cdots \to P_p \to \quad \ldots \quad \to P_0 \to \mathbb{C} \to 0,$$

where each P_p is of the form $\mathcal{S} \otimes_{\mathcal{T}} M_p$, for a $\mathcal{T}$-module M_p. (Observe that the standard $(\mathcal{S}, \mathcal{T})$-projective resolution has this property, cf. D.5.) Tensoring (4) with V (over $\mathbb{C}$), we get the $(\mathcal{S}, \mathcal{T})$-exact sequence:

$$\cdots \to P_p \otimes_{\mathbb{C}} V \to \quad \ldots \quad \to P_0 \otimes_{\mathbb{C}} V \to V \to 0.$$

By Proposition 3.1.10, as $\mathcal{S}$-modules,

$$P_p \otimes_{\mathbb{C}} V \approx \mathcal{S} \otimes_{\mathcal{T}} (M_p \otimes_{\mathbb{C}} V)$$

and hence it is $(\mathcal{S}, \mathcal{T})$-projective. In particular, $\operatorname{Tor}_*^{(\mathcal{S},\mathcal{T})}(W^t, V)$ is the homology of the complex:

$$\text{(5)} \qquad \cdots \to W^t \otimes_{\mathcal{S}} (P_p \otimes_{\mathbb{C}} V) \to \quad \ldots \quad \to W^t \otimes_{\mathcal{S}} (P_0 \otimes_{\mathbb{C}} V) \to 0.$$

Now

$$\begin{aligned} W^t \otimes_{\mathcal{S}} (P_p \otimes_{\mathbb{C}} V) &\approx W^t \otimes_{\mathcal{S}} (\mathcal{S} \otimes_{\mathcal{T}} (M_p \otimes_{\mathbb{C}} V)) \\ \text{(6)} \qquad &\approx W^t \otimes_{\mathcal{T}} (M_p \otimes_{\mathbb{C}} V), \qquad \text{by (3.1.7.2).} \end{aligned}$$

Further, for any $\mathfrak{t}$-modules M, N, it is easy to see that

$$\text{(7)} \qquad M^t \otimes_{\mathcal{T}} N \approx \mathbb{C} \otimes_{\mathcal{T}} (M \otimes_{\mathbb{C}} N),$$

under the map $m \otimes n \mapsto 1 \otimes (m \otimes n)$. Hence, combining (6) and (7), we get

$$\text{(8)} \qquad W^t \otimes_{\mathcal{S}} (P_p \otimes_{\mathbb{C}} V) \approx \mathbb{C} \otimes_{\mathcal{T}} (W \otimes_{\mathbb{C}} M_p \otimes_{\mathbb{C}} V).$$

Moreover, under this identification, the differentials of the complex (5) correspond to the differentials of the same complex (5), where we replace W by $\mathbb{C}$ and V by $W \otimes_{\mathbb{C}} V$.

This proves the first isomorphism of (1). The second isomorphism of (1) of course follows from the first.

The proof of (2) is similar; we just need to use (3.1.7.1) (instead of 3.1.7.2) and the following isomorphism instead of (7).

For any $\mathfrak{t}$-modules M, N, P, we have a $\mathfrak{t}$-module isomorphism:

$$\text{(9)} \qquad \operatorname{Hom}_{\mathbb{C}}(M \otimes_{\mathbb{C}} N, P) \to \operatorname{Hom}_{\mathbb{C}}(M, \operatorname{Hom}_{\mathbb{C}}(N, P)),$$

taking $\chi \in \operatorname{Hom}_{\mathbb{C}}(M \otimes_{\mathbb{C}} N, P)$ to $\bar{\chi}$, where $(\bar{\chi}(m))(n) = \chi(m \otimes n)$, for $m \in M$, $n \in N$.

To prove (3), use the isomorphisms

$$\begin{aligned} \operatorname{Hom}_{\mathcal{S}}(\mathcal{S} \otimes_{\mathcal{T}} (M_p \otimes_{\mathbb{C}} V), W^*) &\approx \operatorname{Hom}_{\mathcal{T}}(M_p \otimes_{\mathbb{C}} V, W^*), \text{ by (3.1.7.1)} \\ &\approx \operatorname{Hom}_{\mathcal{T}}(M_p \otimes_{\mathbb{C}} V \otimes_{\mathbb{C}} W, \mathbb{C}), \text{by (9).} \end{aligned}$$

It is easy to see that, under the above identification, the cochain complex (with cohomology $\operatorname{Ext}^*_{(\mathcal{S},\mathcal{T})}(V, W^*)$):

$$\cdots \leftarrow \operatorname{Hom}_{\mathcal{S}}(\mathcal{S} \otimes_{\mathcal{T}} (M_p \otimes_{\mathbb{C}} V), W^*) \leftarrow \ldots \leftarrow \operatorname{Hom}_{\mathcal{S}}(\mathcal{S} \otimes_{\mathcal{T}} (M_0 \otimes_{\mathbb{C}} V), W^*) \leftarrow 0$$

is the dual of the chain complex (5) (under the identification (8)). This proves (3). $\square$

3.1.14 Lemma. *Let $\mathfrak{r} \supset \mathfrak{s} \supset \mathfrak{t}$ be Lie algebras such that $\mathfrak{r}$ is finitely semisimple under the adjoint action of $\mathfrak{t}$. Then, for any $\mathfrak{r}$-module M and $\mathfrak{s}$-module N such that N is finitely semisimple as a $\mathfrak{t}$-module, we have*

(1) $$Tor_*^{(\mathcal{R},\mathcal{T})}(M^t, \mathcal{R} \otimes_{\mathcal{S}} N) \simeq Tor_*^{(\mathcal{S},\mathcal{T})}(M^t, N), \text{ and}$$

(2) $$Ext^*_{(\mathcal{R},\mathcal{T})}(\mathcal{R} \otimes_{\mathcal{S}} N, M) \simeq Ext^*_{(\mathcal{S},\mathcal{T})}(N, M).$$

Proof. Consider the $(\mathcal{S}, \mathcal{T})$-projective resolution $(\mathcal{D}_N)$ of N (cf. Corollary 3.1.8), which we write, in view of Proposition 3.1.10, as

(3) $$\cdots \to N_1 \to N_0 \to N \to 0,$$

where $N_p := \mathcal{S} \otimes_{\mathcal{T}} (\wedge^p(\mathfrak{s}/\mathfrak{t}) \otimes_{\mathbb{C}} N)$. Since $\mathcal{R}$ is $\mathcal{S}$-free under the right multiplication (by the PBW theorem), (3) gives rise to the exact sequence of $\mathcal{R}$-modules:

(4) $$\cdots \to \mathcal{R} \otimes_{\mathcal{S}} N_1 \to \mathcal{R} \otimes_{\mathcal{S}} N_0 \to \mathcal{R} \otimes_{\mathcal{S}} N \to 0.$$

By (3.1.7.2),
$$\mathcal{R} \otimes_{\mathcal{S}} N_p \simeq \mathcal{R} \otimes_{\mathcal{T}} (\wedge^p(\mathfrak{s}/\mathfrak{t}) \otimes_{\mathbb{C}} N),$$
and thus it is $(\mathcal{R}, \mathcal{T})$-projective. But, as in the proof of Corollary 3.1.8, $\mathcal{R} \otimes_{\mathcal{T}} (\wedge^p(\mathfrak{s}/\mathfrak{t}) \otimes_{\mathbb{C}} N)$ is finitely semisimple $\mathfrak{t}$-module and thus (4) is a $(\mathcal{R}, \mathcal{T})$-projective resolution of $\mathcal{R} \otimes_{\mathcal{S}} N$. Now (1) follows from the commutativity of the following diagram, where the vertical isomorphisms are obtained from (3.1.7.2):

$$\begin{array}{ccccc} \cdots \to M^t \otimes_{\mathcal{R}} (\mathcal{R} \otimes_{\mathcal{S}} N_p) & \longrightarrow & \cdots & \longrightarrow & M^t \otimes_{\mathcal{R}} (\mathcal{R} \otimes_{\mathcal{S}} N_0) \to 0 \\ \downarrow \wr & & & & \downarrow \wr \\ \cdots \to M^t \otimes_{\mathcal{S}} N_p & \longrightarrow & \cdots & \longrightarrow & M^t \otimes_{\mathcal{S}} N_0 \to 0\cdot \end{array}$$

The proof of (2) is similar (use (3.1.7.1)). □

3.1.15 Remark. I learned from M. Duflo that Corollary 3.1.8 (and hence Lemma 3.1.9) remains true for any Lie algebra pair $(\mathfrak{s}, \mathfrak{t})$ such that $\mathfrak{t}$ is a $\mathfrak{t}$-module direct summand in $\mathfrak{s}$ (under the adjoint action). (No finite-dimensional restriction on $\mathfrak{t}$ is imposed.)

3.1.E EXERCISES

(1) Show that, for any Lie algebra pair $(\mathfrak{s}, \mathfrak{t})$ and any $\mathfrak{s}$-module V,

$$H^*(\mathfrak{s}, \mathfrak{t}, V^*) \approx H_*(\mathfrak{s}, \mathfrak{t}, V)^*.$$

(2) Give an example of a finite-dimensional Lie algebra $\mathfrak{s}$ and a subalgebra $\mathfrak{t}$ such that Lemma 3.1.9 is false (even) for the trivial module $V = \mathbb{C}$.

(3) Let $\mathfrak{t}$ be any Lie algebra. Show that any submodule N of a finitely semisimple $\mathfrak{t}$-module M is a direct summand.
Hint. Consider the family $\mathcal{F}$ of submodules Q of M such that $Q \cap N = (0)$. Under the inclusion, $\mathcal{F}$ is a partially ordered set. Use Zorn's lemma to show that $\mathcal{F}$ has a maximal element Q_o. Now show that $N \oplus Q_o = M$.

(4) Let $\mathfrak{t}$ be any ideal of $\mathfrak{s}$. Then show that Corollary 3.1.8 (and hence Lemma 3.1.9) is true for any $\mathfrak{s}$-module V.

(5) Let $\mathfrak{r} \supset \mathfrak{s} \supset \mathfrak{t}$ be Lie algebras. Assume that there exists a Lie subalgebra $\mathfrak{a}$ of $\mathfrak{r}$ such that $\mathfrak{r} = \mathfrak{s} \oplus \mathfrak{a}$ (just as a vector space) and $[\mathfrak{t}, \mathfrak{a}] \subset \mathfrak{a}$. Then, for any $\mathfrak{r}$-module M and $\mathfrak{s}$-module N, show that (3.1.14.1) and (3.1.14.2) are both true.
Hint. Use the fact that, for any $\mathfrak{s}$-module P, $\mathcal{R} \otimes_{\mathcal{S}} P \simeq U(\mathfrak{a}) \otimes_{\mathbb{C}} P$ as $\mathfrak{t}$-modules, where $\mathfrak{t}$ acts on the right side via the tensor product representation; $\mathfrak{t}$ acting on $U(\mathfrak{a})$ under the adjoint action.

3.2. Lie Algebra Homology of $\mathfrak{n}^-$: Results of Kostant–Garland–Lepowsky

In this section we assume that $\mathfrak{g}$ is a symmetrizable Kac–Moody algebra, though Lemmas 3.2.1, 3.2.3, 3.2.5, and 3.2.6 are true with the same proofs for an arbitrary, not necessarily symmetrizable, $\mathfrak{g}$.

For any $Y \subset \{1, \ldots, \ell\}$, let

$$D_Y := \{\lambda \in \mathfrak{h}^* : \langle \lambda, \alpha_i^\vee \rangle \in \mathbb{Z}_+, \text{ for all } i \in Y\}.$$

A $\mathfrak{g}_Y$-module V is called *integrable* if it is integrable as a $\mathfrak{g}(A_Y)$-module, where the sub GCM A_Y is defined in Subsection 1.1.2 and the subalgebra $\mathfrak{g}_Y \hookrightarrow \mathfrak{g}$, together with the Lie algebra homomorphism $i_Y : \mathfrak{g}(A_Y) \to \mathfrak{g}_Y$, is defined in Subsection 1.2.2. By Exercise 1.2.E.1, i_Y is an embedding. A $\mathfrak{g}_Y$-module V is called a *highest weight module* if it is a weight module with respect to the $\mathfrak{h}$-action and, moreover, is a highest weight $\mathfrak{g}(A_Y)$-module.

Recall the definition of $\bar{\mathfrak{g}}$, $\bar{\mathfrak{n}}$, $\bar{\mathfrak{n}}^-$, $\bar{\mathfrak{g}}_Y$, $\bar{\mathfrak{u}}_Y$, $\bar{\mathfrak{u}}_Y^-$, etc., from Subsection 1.5.6.

3.2.1 Lemma. *There is a bijective correspondence between irreducible, integrable, highest weight $\mathfrak{g}_Y$-modules and D_Y, taking a module V to its highest weight. Moreover, any such module remains irreducible as a module for the commutator subalgebra $\mathfrak{g}'_Y := [\mathfrak{g}_Y, \mathfrak{g}_Y]$.*

We denote by $L_Y(\lambda)$ the irreducible, highest weight, integrable $\mathfrak{g}_Y$-module with highest weight $\lambda \in D_Y$.

Proof. The proof of the bijective correspondence is the same as that of Corollary 2.1.8.

The same proof as that of Lemma 2.1.4 gives the irreducibility of $L_Y(\lambda)$ as a $\mathfrak{g}'_Y$-module. □

3.2.2 Remark. A $\bar{\mathfrak{g}}_Y$-module is called *integrable*, resp. *highest weight* if it is integrable, resp. highest weight, considered as a $\mathfrak{g}_Y$-module via the canonical homomorphism $\mathfrak{g}_Y \to \bar{\mathfrak{g}}_Y$. It is easy to see that, for $\lambda \in D_Y$, the $\mathfrak{g}_Y$-module structure on $L_Y(\lambda)$ descends to give a $\bar{\mathfrak{g}}_Y$-module structure. Then, $L_Y(\lambda)$, $\lambda \in D_Y$, are precisely the irreducible, integrable, highest weight $\bar{\mathfrak{g}}_Y$-modules. By the same proof as that of Corollary 2.2.7, any integrable $\bar{\mathfrak{g}}_Y$-module $M \in \bar{\mathcal{O}}_Y$ is a (possibly infinite) direct sum (as a $\bar{\mathfrak{g}}_Y$-module) of irreducible $\bar{\mathfrak{g}}_Y$-modules $L_Y(\lambda)$, for $\lambda \in D_Y$, where $\bar{\mathcal{O}}_Y$ is the BGG category $\mathcal{O}_Y$ (cf. 2.1.1) with $\mathfrak{g}_Y$ replaced by $\bar{\mathfrak{g}}_Y$.

With the notation $\bar{\Delta}^+$ and $\overline{\text{mult}}$ as in Theorem 2.2.1, we have the following:

3.2.3 Lemma. *Let $S \subset \mathfrak{h}^*$ be the set of weights of $\wedge^*(\bar{\mathfrak{n}})$, that is, $S = \{\lambda \in \mathfrak{h}^* : \lambda = \sum_{\alpha \in \bar{\Delta}^+} n(\alpha)\alpha$, where all but finitely many $n(\alpha)$ are zero and each $n(\alpha) \leq \overline{\text{mult}}\,\alpha\}$. Then $\rho - S$ is W-invariant.*

Proof. It suffices to show that $\rho - S$ is s_i-invariant for all simple reflections s_i. Since

$$s_i(\rho - \sum_{\alpha \in \bar{\Delta}^+} n(\alpha)\alpha) = \rho - \alpha_i - \sum_{\alpha \neq \alpha_i} n(\alpha) s_i\alpha + n(\alpha_i)\alpha_i,$$

the lemma follows from Theorem 1.3.11(b_4) and Lemma 1.3.5(a) applied to the integrable $\mathfrak{g}$-module $\bar{\mathfrak{g}}$. □

3.2.4 Lemma. *Let $\lambda \in D$ and let $P(\lambda) \subset \lambda - \sum_{i=1}^{\ell} \mathbb{Z}_+\alpha_i$ be any W-invariant subset containing λ. Then, for $\mu \in \mathfrak{h}^*$ of the form $\mu = -\nu + \theta$, $\nu \in S$ and $\theta \in P(\lambda)$,*

$$(1) \qquad \langle \mu + \rho, \mu + \rho \rangle \leq \langle \lambda + \rho, \lambda + \rho \rangle.$$

Moreover, equality in (1) *occurs if and only if there exists $w \in W$ such that*

$$(2) \qquad \nu = \rho - w\rho \text{ and } \theta = w\lambda.$$

Proof. Choose $v \in W$ such that $|\lambda + \rho - v(\mu + \rho)|$ is the minimum. Since $\rho - S + P(\lambda)$ is W-invariant, clearly $v(\mu + \rho) \in D$. Write

$$(3) \qquad v(\mu + \rho) = \lambda + \rho - \sum n_i\alpha_i, \text{ with } n_i \in \mathbb{Z}_+.$$

Then

$$\begin{aligned}\langle \mu+\rho, \mu+\rho\rangle &= \langle \lambda+\rho-\sum n_i\alpha_i\,, \lambda+\rho-\sum n_i\alpha_i\rangle \\ &= -\langle v(\mu+\rho), \sum n_i\alpha_i\rangle + \langle \lambda+\rho, \lambda+\rho\rangle - \langle \lambda+\rho\,, \sum n_i\alpha_i\rangle \\ &\leq \langle \lambda+\rho, \lambda+\rho\rangle, \qquad \text{since } \lambda+\rho,\ v(\mu+\rho)\in D.\end{aligned}$$

This proves (1). Further, equality in (1) occurs if and only if (a) $\langle \lambda+\rho, \sum n_i\alpha_i\rangle = 0$, and (b) $\langle v(\mu+\rho), \sum n_i\alpha_i\rangle = 0$. But since $\lambda \in D$, (a) is satisfied if and only if $\sum n_i\alpha_i = 0$, i.e., $v(\mu+\rho) = \lambda+\rho$; and then (b) is automatically satisfied. Now

(4) $v(\mu+\rho) = \lambda+\rho \Leftrightarrow v\theta + v(\rho-\nu) = \lambda+\rho \Leftrightarrow \rho - v(\rho-\nu) = v\theta - \lambda.$

But $-(v\theta-\lambda) \in \sum \mathbb{Z}_+\alpha_i$ and $\rho - v(\rho-\nu) \in \sum \mathbb{Z}_+\alpha_i$. Hence, by (4), $v(\mu+\rho) = \lambda+\rho \Leftrightarrow \rho = v(\rho-\nu)$ and $v\theta = \lambda$. Now, taking $w = v^{-1}$, the lemma follows. □

The following lemma follows immediately from (1.3.22.2) and Lemma 1.3.13.

3.2.5 Lemma. *For any $\mathfrak{g}$ (not necessarily symmetrizable), $\lambda \in D$ and $w \neq 1 \in W$, we have $w(\lambda+\rho) \neq (\lambda+\rho)$.* □

3.2.6 Lemma. *For any $w \in W$, the weight space of $\Lambda^*(\bar{\mathfrak{n}}^-)$ corresponding to the weight $w\rho-\rho$ is one-dimensional, and is spanned by*

(1) $$e_{-\phi_1}\wedge\cdots\wedge e_{-\phi_p}, \qquad p = \ell(w),$$

where $\Phi_w = \{\phi_1,\ldots,\phi_p\}$ is as in Lemma 1.3.14 and $e_{-\phi_i}$ is a nonzero root vector corresponding to the negative (real) root $-\phi_i$.

Proof. By (1.3.22.3), (1) is indeed a weight vector of weight $w\rho-\rho$. Now let $0 \neq e_{-\beta_1}\wedge\cdots\wedge e_{-\beta_q} \in \Lambda^q(\bar{\mathfrak{n}}^-)$ be such that $\sum \beta_j = \rho - w\rho$ (for some root vectors $e_{-\beta_j}$ corresponding to the negative roots $-\beta_j$; we allow $\beta_j = \beta_{j'}$ for imaginary roots). We prove the lemma by showing that $q = \ell(w)$ and $\{\beta_1,\ldots,\beta_q\} = \Phi_w$: Clearly, $w^{-1}(\rho - w\rho)$ is negative. In particular, $w^{-1}\beta_{j_1} \in \bar{\Delta}^-$ for some j_1, and hence $\beta_{j_1} \in \Phi_w$, say $\beta_{j_1} = \phi_{i_1}$. Hence $\beta^{(j_1)} := \sum_{j\neq j_1}\beta_j = \sum_{i\neq i_1}\phi_i$. Again $w^{-1}(\beta^{(j_1)}) < 0$. So there exists a $j_2 \neq j_1$ and $i_2 \neq i_1$ such that $\beta_{j_2} = \phi_{i_2}$. Continuing this way, we get the lemma. □

3.2.7 Theorem. *For any subset $Y \subset \{1,\ldots,\ell\}$ and any integrable highest weight $\bar{\mathfrak{g}}$-module L with highest weight λ, we have, for any $p \geq 0$,*

$$H_p(\bar{\mathfrak{u}}_Y^-, L) \approx \bigoplus_{\substack{w\in W'_Y \\ \ell(w)=p}} L_Y(w^{-1}*\lambda), \quad \textit{as } \bar{\mathfrak{g}}_Y\textit{-modules,}$$

where W_Y' is defined by (1.3.17.1) (also see Exercise 1.3.E). In fact, for $w \in W_Y'$ with $\ell(w) = p$, $L_Y(w^{-1} * \lambda)$ occurs with multiplicity 1 in the chain complex $\Lambda(\bar{\mathfrak{u}}_Y^-) \otimes L$ itself (and this occurs in $\Lambda^p(\bar{\mathfrak{u}}_Y^-) \otimes L$ with highest weight vector $\theta_{w^{-1}} := e_{-\phi_1} \wedge \cdots \wedge e_{-\phi_p} \otimes x_{w^{-1}\lambda}$, where $x_{w^{-1}\lambda}$ is a nonzero weight vector of L of weight $w^{-1}\lambda$ and $\Phi_{w^{-1}} = \{\phi_1, \ldots, \phi_p\}$).

Observe that $w^{-1} * \lambda \in D_Y$ for any $w \in W_Y'$ and $\lambda \in D$, and, since $\bar{\mathfrak{g}}_Y$ normalizes $\bar{\mathfrak{u}}_Y^-$, $H_*(\bar{\mathfrak{u}}_Y^-, L)$ has a natural $\bar{\mathfrak{g}}_Y$-module structure (cf. 3.1.1). Moreover, the $\bar{\mathfrak{g}}_Y$-module $\Lambda(\bar{\mathfrak{u}}_Y^-) \otimes L \in \bar{\mathcal{O}}_Y$, and it is integrable (cf. the proof below). Thus, by Remark 3.2.2, it is a direct sum of $\bar{\mathfrak{g}}_Y$-modules of the form $L_Y(\mu)$, $\mu \in D_Y$.

Proof. Consider the standard resolution of the trivial $\bar{\mathfrak{g}}$-module $\mathbb{C}$ for the pair $(\bar{\mathfrak{g}}, \bar{\mathfrak{p}}_Y)$ (cf. Proposition 3.1.5):

(1)
$$\cdots \to U(\bar{\mathfrak{g}}) \otimes_{U(\bar{\mathfrak{p}}_Y)} \Lambda^p(\bar{\mathfrak{g}}/\bar{\mathfrak{p}}_Y) \to \cdots \to U(\bar{\mathfrak{g}}) \otimes_{U(\bar{\mathfrak{p}}_Y)} \Lambda^0(\bar{\mathfrak{g}}/\bar{\mathfrak{p}}_Y) \to \mathbb{C} \to 0.$$

Tensoring the above sequence with L (over $\mathbb{C}$) and using the Hopf principle (cf. Proposition 3.1.10), we get the resolution:

$$\to X_p \to \cdots \to X_0 \to L \to 0, \tag{2}$$

where $X_p := U(\bar{\mathfrak{g}}) \otimes_{U(\bar{\mathfrak{p}}_Y)} (\Lambda^p(\bar{\mathfrak{g}}/\bar{\mathfrak{p}}_Y) \otimes_{\mathbb{C}} L)$. By the PBW theorem, as a $\bar{\mathfrak{p}}_Y^- := \bar{\mathfrak{u}}_Y^- \oplus \bar{\mathfrak{g}}_Y$-module,

$$X_p \approx U(\bar{\mathfrak{u}}^-) \otimes_{\mathbb{C}} \Lambda^p(\bar{\mathfrak{u}}^-) \otimes_{\mathbb{C}} L, \tag{3}$$

where $\bar{\mathfrak{u}}^- := \bar{\mathfrak{u}}_Y^-$ acts on the right side of (3) by (left) multiplication only on the first factor and $\bar{\mathfrak{g}}_Y$ acts via the adjoint action on the first two factors and it acts on the L-factor via the restriction of the $\bar{\mathfrak{g}}$-action. In particular, (2) provides a $U(\bar{\mathfrak{u}}^-)$-free resolution of L, and thus, by (3.1.9.1), $H_*(\bar{\mathfrak{u}}^-, L)$ is the homology of the complex

$$\cdots \to \mathbb{C} \otimes_{U(\bar{\mathfrak{u}}^-)} X_p \to \cdots \to \mathbb{C} \otimes_{U(\bar{\mathfrak{u}}^-)} X_0 \to 0. \tag{4}$$

But, by (3) and (3.1.7.2), as $\bar{\mathfrak{g}}_Y$-modules

$$\mathbb{C} \otimes_{U(\bar{\mathfrak{u}}^-)} X_p \approx \Lambda^p(\bar{\mathfrak{u}}^-) \otimes_{\mathbb{C}} L, \tag{5}$$

where $\bar{\mathfrak{g}}_Y$ acts on $\mathbb{C} \otimes_{U(\bar{\mathfrak{u}}^-)} X_p$ by $x \cdot (z \otimes v) = z \otimes x \cdot v$ for $x \in \bar{\mathfrak{g}}_Y$, $z \in \mathbb{C}$ and $v \in X_p$. In fact the complex (4), under the identification (5), is nothing but the standard complex to calculate the homology $H_*(\bar{\mathfrak{u}}^-, L)$ (cf. 3.1.1).

It is easy to see from (3) that $X_p \in \bar{\mathcal{O}}$. For $M \in \bar{\mathcal{O}}$, let $\Omega_M : M \to M$ be the Casimir operator (cf. 2.1.15). Abbreviating Ω_{X_p} by Ω_p, we have the commutative diagram:

$$\begin{array}{ccccccccc} \cdots \longrightarrow & X_p & \longrightarrow \cdots \longrightarrow & X_0 & \longrightarrow & L & \longrightarrow & 0 \\ & \downarrow{\scriptstyle \Omega_p} & & \downarrow{\scriptstyle \Omega_0} & & \downarrow{\scriptstyle \Omega_L} & & \\ \cdots \longrightarrow & X_p & \longrightarrow \cdots \longrightarrow & X_0 & \longrightarrow & L & \longrightarrow & 0. \end{array}$$

Since Ω_p are $\bar{\mathfrak{g}}$-module and hence $\bar{\mathfrak{u}}^-$-module maps, they induce the map

$$1 \otimes \Omega_p : \mathbb{C} \otimes_{U(\bar{\mathfrak{u}}^-)} X_p \approx Z_p \to Z_p, \text{ where } Z_p := \Lambda^p(\bar{\mathfrak{u}}^-) \otimes L$$

(cf. (5)). Further, by Lemma 2.1.16, $\Omega_L = \langle \lambda, 2p + \lambda \rangle I$, and hence by Theorem D.6 ((2) being a $U(\bar{\mathfrak{u}}^-)$-free resolution of L), we get that

(6) $\quad 1 \otimes \Omega_p - \langle \lambda, 2\rho + \lambda \rangle I$ induces the 0 map on the homology of the complex (4).

Analogous to the Casimir element $\Omega \in \hat{U}(\bar{\mathfrak{g}})$, recall the definition of the element $\Omega^Y \in \hat{U}(\bar{\mathfrak{g}}_Y)$ from 1.5.8. Then Ω^Y lies in the center of $\hat{U}(\bar{\mathfrak{g}}_Y)$ (cf. Remark 1.5.12). In particular, for any $\bar{\mathfrak{g}}_Y$-module $N \in \bar{\mathcal{O}}_Y$, Ω^Y induces a $\bar{\mathfrak{g}}_Y$-module map $\Omega^Y_N : N \to N$. Further, by the same proof as that of Lemma 2.1.16, for any highest weight $\bar{\mathfrak{g}}_Y$-module N with highest weight μ,

(7) $$\Omega^Y_N = \langle \mu, \mu + 2\rho \rangle I.$$

Observe that $Z_p \in \bar{\mathcal{O}}_Y$. By definition,

(8) $$\Omega = \Omega^Y + 2 \sum_{\alpha \in \bar{\Delta}^+ \setminus \bar{\Delta}^+_Y} \Omega_\alpha,$$

and hence

(9) $$1 \otimes \Omega_p = \Omega^Y_p, \qquad \text{where } \Omega^Y_p := \Omega^Y_{Z_p}.$$

In particular, $1 \otimes \Omega_p$ is a $\bar{\mathfrak{g}}_Y$-module map. By Corollary 1.3.4, $\bar{\mathfrak{g}}$ is an integrable $\bar{\mathfrak{g}}$-module (under the adjoint action), in particular, Z_p is an integrable $\bar{\mathfrak{g}}_Y$-module for any p. Hence, by Remark 3.2.2, Z_p is a direct sum of irreducible $\bar{\mathfrak{g}}_Y$-modules $L_Y(\mu)$, for $\mu \in D_Y$. For any $\mu \in D_Y$, let $Z_p(\mu)$ be the isotypical component of Z_p corresponding to the highest weight μ, i.e., $Z_p(\mu)$ is the sum of all the irreducible $\bar{\mathfrak{g}}_Y$-submodules of Z_p isomorphic to $L_Y(\mu)$. Since the differentials $\partial_p : Z_p \to Z_{p-1}$ are $\bar{\mathfrak{g}}_Y$-module maps, $Z_*(\mu) := \oplus_{p\geq 0} Z_p(\mu)$ is a subcomplex and, moreover, for any $p \geq 0$,

(10)
$$H_p(\bar{\mathfrak{u}}^-, L) \simeq H_p(Z_*) \simeq \oplus_{\mu \in D_Y} H_p(Z_*(\mu)) \simeq \bigoplus_{\substack{\mu\in D_Y \\ \langle \mu,\mu+2\rho\rangle = \langle \lambda,\lambda+2p\rangle}} H_p(Z_*(\mu)),$$

by (6), (7) and (9). Now fix $\mu \in D_Y$ such that $\langle \mu, \mu+2\rho\rangle = \langle \lambda, \lambda+2\rho\rangle$. Taking $P(\lambda)$ to be the set of weights of L in Lemma 3.2.4, we get (by Lemma 3.2.4) that for any decomposition

$$\mu = -\nu + \theta \qquad \text{for } \nu \in S \text{ and } \theta \in P(\lambda), \tag{11}$$

there exists $w \in W$ such that

$$\theta = w^{-1}\lambda \text{ and } \nu = \rho - w^{-1}\rho. \tag{12}$$

In particular,

$$\mu = w^{-1} * \lambda,$$

and hence such a $w \in W$ is unique (by Lemma 3.2.5). By Lemmas 3.2.6, 1.3.5(a), and (11)–(12), the μ-weight space of Z_* is of dim ≤ 1. Further, the μ-weight space of Z_*, for $\mu = w^{-1} * \lambda$, is one-dimensional, if and only if $w \in W_Y'$ (cf. Exercise 1.3.E). In this case (i.e., $w \in W_Y'$) it occurs in $\Lambda^{\ell(w)}(\bar{\mathfrak{u}}^-) \otimes L$, and is spanned by $\theta_{w^{-1}}$.

Finally, for any $i \in Y$ and $\mu = w^{-1} * \lambda$ for $w \in W_Y'$, $e_i\theta_{w^{-1}} = 0$. To prove this, we show that $\mu + \alpha_i$ is not a weight of $\Lambda(\bar{\mathfrak{u}}^-) \otimes L$:

$$\begin{aligned}\langle \mu+\rho+\alpha_i, \mu+\rho+\alpha_i\rangle &= \langle w^{-1}(\lambda+\rho)+\alpha_i, w^{-1}(\lambda+\rho)+\alpha_i\rangle \\ &= \langle \lambda+\rho, \lambda+\rho\rangle + \langle\alpha_i,\alpha_i\rangle + 2\langle \lambda+\rho, w\alpha_i\rangle .\end{aligned} \tag{13}$$

Now $\langle\alpha_i,\alpha_i\rangle > 0$ and, moreover, $\langle\lambda+\rho, w\alpha_i\rangle > 0$, since $w\alpha_i \in \Delta^+$ by Exercise 1.3.E, w being in W_Y' and $i \in Y$. But then (13) contradicts Lemma 3.2.4, proving $e_i\theta_{w^{-1}} = 0$.

Putting together the above, we obtain that any $\mu \in D_Y$ such that $\langle\mu, \mu+2\rho\rangle = \langle\lambda, \lambda+2\rho\rangle$ is of the form $\mu = w^{-1} * \lambda$. Moreover, for $\mu = w^{-1} * \lambda$, the μ-weight space in Z_* is nonzero if and only if $w \in W_Y'$ and, in this case, $Z_{\ell(w)}(w^{-1} * \lambda)$ is an irreducible $\bar{\mathfrak{g}}_Y$-module with highest weight vector $\theta_{w^{-1}}$ and $Z_p(w^{-1} * \lambda) = 0$ if $p \neq \ell(w)$. In particular, for any $w \in W_Y'$,

$$H_*(Z_*(w^{-1} * \lambda)) \cong Z_*(w^{-1} * \lambda),$$

and hence the theorem follows from (10). □

Observe that in the proof of the above theorem for $Y = \emptyset$, we have not used the Weyl–Kac character formula, i.e., Theorem 2.2.1. In fact, by the Euler–Poincaré principle, we can immediately deduce Theorem 2.2.1 from the above theorem.

3.2.8 Corollary. *With the same notation and assumptions as in Theorem 2.2.1,*

$$\text{(1)}\qquad \operatorname{ch} L = \left[\sum_{w\in W} \epsilon(w)e^{w*\lambda}\right]\cdot \prod_{\alpha\in\bar{\Delta}^+}(1-e^{-\alpha})^{-\overline{\operatorname{mult}\alpha}}.$$

In particular, any integrable highest weight $\bar{\mathfrak{g}}$-module L is irreducible.

Proof. By the Euler–Poincaré principle,

$$\begin{aligned}\sum_{p\geq 0}(-1)^p \operatorname{ch}(\Lambda^p(\bar{\mathfrak{n}}^-)\otimes L) &= \sum_{p\geq 0}(-1)^p \operatorname{ch} H_p(\bar{\mathfrak{n}}^-, L)\\ \text{(2)}\qquad &= \sum_{w\in W}\epsilon(w)e^{w*\lambda},\end{aligned}$$

by Theorem 3.2.7 for the case $Y = \emptyset$. But

$$\begin{aligned}\sum_{p\geq 0}(-1)^p \operatorname{ch}(\Lambda^p(\bar{\mathfrak{n}}^-)\otimes L) &= \left(\sum_{p\geq 0}(-1)^p \operatorname{ch}\Lambda^p(\bar{\mathfrak{n}}^-)\right)\cdot \operatorname{ch} L\\ \text{(3)}\qquad &= \left(\prod_{\alpha\in\bar{\Delta}^+}(1-e^{-\alpha})^{\overline{\operatorname{mult}\alpha}}\right)\cdot \operatorname{ch} L\,.\end{aligned}$$

Combining (2) and (3), we get (1). □

3.2.9 Lemma. *Let $\mathfrak{s}$ be a free Lie algebra and let $\mathfrak{a}\subset\mathfrak{s}$ be an ideal such that $\mathfrak{a}\subset[\mathfrak{s},\mathfrak{s}]$. Then*

$$H_2(\mathfrak{s}/\mathfrak{a}, \mathbb{C}) \approx \mathfrak{a}/[\mathfrak{s},\mathfrak{a}],$$

where $\mathbb{C}$ is the trivial one-dimensional module for $\mathfrak{s}/\mathfrak{a}$.

Proof. Let

$$\cdots \to \Lambda_p \xrightarrow{\partial_p} \Lambda_{p-1} \to \cdots \to \Lambda_0 \to 0$$

be the standard complex (cf. (3.1.1.1)) to calculate the homology $H_*(\mathfrak{s}/\mathfrak{a}, \mathbb{C})$, where $\Lambda_p := \Lambda^p(\mathfrak{s}/\mathfrak{a})$. Define the map $\varphi : \operatorname{Ker}\partial_2 \to \mathfrak{a}/[\mathfrak{s},\mathfrak{a}]$ by $\varphi(\sum_i \bar{x}_i\wedge\bar{y}_i) = \sum_i [x_i, y_i]$, for $x_i, y_i \in \mathfrak{s}$ where $\bar{x}$ denotes x mod $\mathfrak{a}$. Clearly φ is well defined, i.e., φ does not depend on the choice of the coset representatives of $\bar{x}_i, \bar{y}_i$. Further, the map φ is surjective:

Take $x\in\mathfrak{a}$; then $x = \sum_i [x_i, y_i]$, for some $x_i, y_i \in \mathfrak{s}$, since $\mathfrak{a}\subset[\mathfrak{s},\mathfrak{s}]$ by the assumption. Now clearly $\theta := \sum_i \bar{x}_i\wedge\bar{y}_i \in \operatorname{Ker}\partial_2$ and, moreover, $\varphi(\theta) = x$ mod $[\mathfrak{s},\mathfrak{a}]$.

Finally, $\operatorname{Ker}\varphi = \operatorname{Im}\partial_3$: Take $\theta = \sum_i \bar{x}_i\wedge\bar{y}_i \in \operatorname{Ker}\partial_2$ such that $\varphi(\theta) = 0$, i.e., $\sum_i [x_i, y_i] \in [\mathfrak{s},\mathfrak{a}]$. Write $\sum_i [x_i, y_i] = \sum_j [b_j, a_j]$, for $b_j \in \mathfrak{s}$, $a_j \in \mathfrak{a}$.

Then $\partial_2^{\mathfrak{s}}(\sum_i x_i \wedge y_i - \sum_j b_j \wedge a_j) = 0$, where $\partial_2^{\mathfrak{s}}$ denotes the differential for the Lie algebra $\mathfrak{s}$. But since $\mathfrak{s}$ is a free Lie algebra, $H_p(\mathfrak{s}, \mathbb{C}) = 0$ for all $p \geq 2$ (cf. Exercise 3.2.E.3); in particular, there exists a $\theta' \in \Lambda^3(\mathfrak{s})$ such that $\partial_3^{\mathfrak{s}}(\theta') = \sum_i x_i \wedge y_i - \sum_j b_j \wedge a_j$. But then $\partial_3(\bar{\theta}') = \theta$, where $\bar{\theta}'$ denotes the image of θ' in $\Lambda^3(\mathfrak{s}/\mathfrak{a})$. Conversely, it is easy to see that for any $\theta \in \operatorname{Im} \partial_3$, $\varphi(\theta) = 0$. This proves the lemma. □

3.2.10 Corollary. *The Lie algebra* $\mathfrak{g} = \bar{\mathfrak{g}}$, *i.e.,* $\mathfrak{r} = 0$ *for any symmetrizable* $\mathfrak{g}$.

Proof. Since $\mathfrak{r} = \mathfrak{r}^+ \oplus \mathfrak{r}^-$, where $\mathfrak{r}^\pm := \mathfrak{r} \cap \mathfrak{n}^\pm$ (cf. Definition 1.5.6), it suffices to show that $\mathfrak{n}^- = \bar{\mathfrak{n}}^-$ (as this will give $\mathfrak{n}^+ = \bar{\mathfrak{n}}^+$ by using the involution ω of Subsection 1.1.2).

Recall the definition of the Lie algebra $\tilde{\mathfrak{n}}^-$ from 1.1.2, which is free by the proof of Theorem 1.2.1. Then $\mathfrak{n}^- = \tilde{\mathfrak{n}}^-/\tilde{\mathfrak{r}}^-$ (cf. proof of Theorem 1.2.1). Let $\pi : \tilde{\mathfrak{n}}^- \to \mathfrak{n}^-$ be the canonical map, and consider the ideal $\hat{\mathfrak{r}}^- := \pi^{-1}(\mathfrak{r}^-)$ of $\tilde{\mathfrak{n}}^-$. Then, of course, $\tilde{\mathfrak{r}}^- \subset \hat{\mathfrak{r}}^-$, and $\tilde{\mathfrak{r}}^-, \hat{\mathfrak{r}}^-$ are both contained in $[\tilde{\mathfrak{n}}^-, \tilde{\mathfrak{n}}^-]$.

By Lemma 3.2.9,

(1) $$H_2(\bar{\mathfrak{n}}^-) \cong \hat{\mathfrak{r}}^-/[\hat{\mathfrak{r}}^-, \tilde{\mathfrak{n}}^-], \qquad \text{and}$$

(2) $$H_2(\mathfrak{n}^-) \cong \tilde{\mathfrak{r}}^-/[\tilde{\mathfrak{r}}^-, \tilde{\mathfrak{n}}^-].$$

From the proof of Lemma 3.2.9, it is easy to see that both of the above isomorphisms are $\mathfrak{h}$-module maps.

By Theorem 3.2.7, as $\mathfrak{h}$-modules,

(3) $$H_2(\bar{\mathfrak{n}}^-) \simeq \bigoplus_{\substack{w \in W \\ \ell(w)=2}} \mathbb{C}_{w\rho-\rho} = \bigoplus \mathbb{C}_{-(\alpha_j+(1-a_{i,j})\alpha_i)},$$

where the last summation is over $S :=$ {the ordered pairs (i, j), $1 \leq i \neq j \leq \ell$: $i < j$ if $a_{i,j} = 0$}. Consider the natural map

$$\gamma : H_2(\mathfrak{n}^-) \to H_2(\bar{\mathfrak{n}}^-),$$

which, under the identifications (1)–(2), is induced from the inclusion $\tilde{\mathfrak{r}}^- \subset \hat{\mathfrak{r}}^-$. We claim that γ is injective.

Clearly, by the definition of $\tilde{\mathfrak{r}}^-$, $\tilde{\mathfrak{r}}^-/[\tilde{\mathfrak{r}}^-, \tilde{\mathfrak{n}}^-]$ is spanned (over $\mathbb{C}$) by the elements $\{f_{i,j}\}_{(i,j)\in S}$ (cf. 1.2.1). Since the elements $\{f_{i,j}\}_{(i,j)}$ have distinct weights, Ker γ (if nonzero) contains some f_{i_o,j_o}, for $(i_o, j_o) \in S$, i.e., $f_{i_o,j_o} \in [\hat{\mathfrak{r}}^-, \tilde{\mathfrak{n}}^-]$. Pick a weight vector $x_{(i_o,j_o)} \in \hat{\mathfrak{r}}^-$ of weight μ such that the weight of $f_{i_o,j_o} < \mu$ and, moreover, no weight ν of $\hat{\mathfrak{r}}^-$ satisfies $\mu < \nu$. In particular, $x_{(i_o,j_o)} \notin [\hat{\mathfrak{r}}^-, \tilde{\mathfrak{n}}^-]$ and, moreover, $\mu = -(\alpha_{j_o} + k\alpha_{i_o})$ for some $0 < k < 1 - a_{i_o,j_o}$. This contradicts (3) and thereby proves that γ is injective; in particular, it proves that the elements $\{f_{i,j} \bmod [\tilde{\mathfrak{r}}^-, \tilde{\mathfrak{n}}^-]\}_{(i,j)\in S}$ form a basis of $H_2(\mathfrak{n}^-)$ under the identification (2).

Now since $\dim H_2(\tilde{\mathfrak{n}}^-) = \#S$ (by (3)), we obtain that γ, in fact, is an isomorphism. Thus, $\tilde{\mathfrak{r}}^- = \hat{\mathfrak{r}}^-$. Otherwise, pick a nonzero weight vector $x \in \hat{\mathfrak{r}}^-/\tilde{\mathfrak{r}}^-$ of weight μ such that any $\nu > \mu$ is not a weight of $\hat{\mathfrak{r}}^-/\tilde{\mathfrak{r}}^-$. Then $x \bmod [\hat{\mathfrak{r}}^-, \tilde{\mathfrak{n}}^-] \notin \operatorname{Im} \gamma$, a contradiction! This proves the corollary. □

So, from now on, we identify $\mathfrak{g}$ with $\bar{\mathfrak{g}}$, $\mathfrak{n}$ with $\bar{\mathfrak{n}}$, etc. in the symmetrizable case.

3.2.11 Proposition. *For any symmetrizable Kac–Moody algebra $\mathfrak{g}$ and any subset Y of finite type,*

$$H_i(\mathfrak{g}, \mathfrak{g}_Y) = H^i(\mathfrak{g}, \mathfrak{g}_Y) = 0, \quad \text{if } i \text{ is odd,}$$

and

$$\dim H_i(\mathfrak{g}, \mathfrak{g}_Y) = \dim H^i(\mathfrak{g}, \mathfrak{g}_Y) = \#\{w \in W'_Y : \ell(w) = i/2\}, \quad \text{if } i \text{ is even,}$$

where $\mathfrak{g}_Y$ is the standard Levi component of the parabolic subalgebra $\mathfrak{p}_Y$ as in 1.2.2.

Proof. By Exercise 3.1.E.1, it suffices to prove the result for $H^i(\mathfrak{g}, \mathfrak{g}_Y)$. The standard cochain complex C^* to compute $H^*(\mathfrak{g}, \mathfrak{g}_Y)$ is given by

$$C^n = \bigoplus_{p+q=n} \operatorname{Hom}_{\mathfrak{g}_Y}(\Lambda^p(\mathfrak{u}_Y) \otimes \Lambda^q(\mathfrak{u}_Y^-), \mathbb{C}).$$

Define a decreasing filtration $\{C^* = \mathcal{F}_0 \supset \mathcal{F}_1 \supset \mathcal{F}_2 \cdots\}$ of C^* by

$$(1)\quad \mathcal{F}_m := \bigoplus_{p,q} \operatorname{Hom}_{\mathfrak{g}_Y}\Big(\Lambda^p(\mathfrak{u}_Y) \otimes \Lambda^q(\mathfrak{u}_Y^-) / \sum_{s+t\le m-1} \Lambda^p_{(s)}(\mathfrak{u}_Y) \otimes \Lambda^q_{(t)}(\mathfrak{u}_Y^-), \mathbb{C}\Big),$$

where $\Lambda^p_{(s)}(\mathfrak{u}_Y)$, resp. $\Lambda^q_{(t)}(\mathfrak{u}_Y^-)$, denotes the subspace of $\Lambda^p(\mathfrak{u}_Y)$, resp. $\Lambda^q(\mathfrak{u}_Y^-)$, spanned by the weight vectors of weight β with *Y-relative height* $\operatorname{ht}_Y(\beta) = s$, resp. t, where, for $\beta = \sum_i n_i\alpha_i$ with $n_i \in \mathbb{Z}$, $\operatorname{ht}_Y(\beta) := |\sum_{i\notin Y} n_i|$.

It is easy to see that $\mathcal{F}_m$ is a subcomplex of C^* and, moreover, the corresponding spectral sequence has

$$(2)\quad E_1^{m,n} = \bigoplus_{s+t=m} \bigoplus_{p+q=m+n} \big[H^p(\operatorname{Hom}_{\mathbb{C}}(\Lambda^\bullet_{(s)}(\mathfrak{u}_Y), \mathbb{C})) \otimes H^q(\operatorname{Hom}_{\mathbb{C}}(\Lambda^\bullet_{(t)}(\mathfrak{u}_Y^-), \mathbb{C}))\big]^{\mathfrak{g}_Y}.$$

Now, there is a $\mathfrak{g}_Y$-module isomorphism

$$(3)\quad H^p(\operatorname{Hom}(\Lambda^\bullet_{(s)}(\mathfrak{u}_Y), \mathbb{C})) \simeq H_p(\Lambda^\bullet_{(s)}(\mathfrak{u}_Y^-)),$$

which follows from the nondegenerate invariant bilinear form on $\mathfrak{g}$. Using (3) and Theorem 3.2.7, we get that $E_1^{m,n} = 0$ unless $m+n$ is even. In particular, the above spectral sequence degenerates at the E_1-term itself. Thus, as a vector space,

$$H^i(\mathfrak{g}, \mathfrak{g}_Y) = \bigoplus_{m+n=i} E_1^{m,n}.$$

Combining (2) with Theorem 3.2.7, we get the proposition. □

3.2.E EXERCISES

(1) Let $\mathfrak{g}$ be of finite type. Here is an outline of an alternative proof of Kostant's $\mathfrak{n}$-cohomology result, by using the Hochschild–Serre spectral sequence and $s\ell_2$-representation theory.

(a) For any finite-dimensional $\mathfrak{g}$-module M, any weight $\mu \in \mathfrak{h}^*_{\mathbb{Z}}$, and any simple reflection s_i such that $\langle \mu+\rho, \alpha_i^\vee\rangle \geq 0$, prove that, for any $p \in \mathbb{Z}$,

$$H^p(\mathfrak{n}, M)_\mu \simeq H^{p+1}(\mathfrak{n}, M)_{s_i * \mu},$$

where the subscript μ denotes the μ-th weight space.
Hint: Use the Hochschild–Serre Lie algebra cohomology spectral sequence (cf. Theorem E.12) for the pair $(\mathfrak{n}, \mathfrak{u}_i)$ and the $s\ell_2$-representation theory, where, as in Subsection 1.2.2, $\mathfrak{u}_i$ is the nil-radical of the minimal parabolic subalgebra $\mathfrak{p}_i$.

(b) For any dominant integral weight μ and $w \in W$, iterating (a), prove that

$$H^p(\mathfrak{n}, M)_\mu \simeq H^{p+\ell(w)}(\mathfrak{n}, M)_{w*\mu}.$$

(c) From (b) obtain that, for any μ as in (b),

$$H^p(\mathfrak{n}, M)_\mu = 0 \text{ if } p \geq 1.$$

(d) Obtain Kostant's $\mathfrak{n}$-cohomology result, that for any irreducible finite-dimensional $\mathfrak{g}$-module $L(\lambda)$,

$$H^p(\mathfrak{n}, L(\lambda)) \simeq \bigoplus_{\substack{w\in W\\ \ell(w)=p}} \mathbb{C}_{w*\lambda}.$$

(2) Prove the complete reducibility of finite-dimensional $\mathfrak{g}$-modules by using (d), where $\mathfrak{g}$ is as in the above exercise (1).

(3) Let $\mathfrak{s}$ be the free Lie algebra generated by a vector space V (cf. Exercise 1.1.E.3). Then show that $H_p(\mathfrak{s}, \mathbb{C}) = 0$ for all $p \geq 2$. In fact, $H_p(\mathfrak{s}, M) = 0$ for all $p \geq 2$ and any $\mathfrak{s}$-module M.

Hint: Identify $H_p(\mathfrak{s}, M)$ as $\operatorname{Tor}_p^{U(\mathfrak{s})}(M^t, \mathbb{C})$ via (3.1.9.1). Consider the exact sequence induced from the standard augmentation map ϵ:

$$0 \to \operatorname{Ker}\epsilon \to U(\mathfrak{s}) \xrightarrow{\epsilon} \mathbb{C} \to 0.$$

Show that $\operatorname{Ker}\epsilon$ is a free $U(\mathfrak{s})$-module with a basis of V as its basis.

3.3. Decomposition of the Category $\mathcal{O}$ and some Ext Vanishing Results

We continue to make the assumption that $\mathfrak{g}$ is symmetrizable. Recall the definition of the subset $K^{\text{w.g.}} \subset \mathfrak{h}^*$ from Subsection 2.3.7.

3.3.1 Definition. Define a relation $\rightarrow$ in $\mathfrak{h}^*$ by $\lambda \rightarrow \mu$ if $L(\lambda)$ is a component of $M(\mu)$ (cf. 2.1.11). Let $\sim$ be the equivalence relation in $\mathfrak{h}^*$ generated by $\rightarrow$.

Define the full subcategory $\mathcal{O}^{\text{w.g.}}$ of $\mathcal{O}$ consisting of all those modules $M \in \mathcal{O}$ such that all the components $L(\mu)$ of M satisfy $\mu \in K^{\text{w.g.}}$.

Let $\rightarrow^o$ be the restriction of $\mathfrak{h}^*$ to the subset $K^{\text{w.g.}}$ and let $\sim^o$ be the equivalence relation in $K^{\text{w.g.}}$ thus generated. Of course, for $\lambda, \mu \in K^{\text{w.g.}}$, $\lambda \sim^o \mu \Rightarrow \lambda \sim \mu$. The converse, in general, is not true (cf. [Kumar–87b, Example 1.12]).

Let $\Lambda \subset \mathfrak{h}^*$, resp. $\Lambda^o \subset K^{\text{w.g.}}$, be an equivalence class under $\sim$, resp. $\sim^o$. A module $M \in \mathcal{O}$, resp. $\mathcal{O}^{\text{w.g.}}$, is said to be *of type* Λ, resp. Λ^o, if for all components $L(\mu)$ of M we have $\mu \in \Lambda$, resp. Λ^o.

As a corollary of Proposition 2.3.8, we get the following.

3.3.2 Lemma. *Let $\lambda \in K^{\text{w.g.}}$. Then for $\mu \in K^{\text{w.g.}}$, $\mu \sim^o \lambda$ if and only if $\mu \in W(\lambda) * \lambda$, where $W(\lambda)$ is as defined in 2.3.7.*

Proof. First observe that, for any $w \in W(\lambda)$,

$$W(w * \lambda) = W(\lambda). \tag{1}$$

Now the lemma follows easily from Proposition 2.3.8. □

3.3.3 Lemma. *$H_n(\mathfrak{g}, M) \approx H_n(\mathfrak{g}, M^\omega)$, for all $n \geq 0$ and any (left) $\mathfrak{g}$-module M. More generally, $H_n(\mathfrak{g}, \mathfrak{s}, M) \approx H_n(\mathfrak{g}, \mathfrak{s}, M^\omega)$ for any ω-stable subalgebra $\mathfrak{s}$ of $\mathfrak{g}$, and $H^n(\mathfrak{g}, \mathfrak{s}, M) \approx H^n(\mathfrak{g}, \mathfrak{s}, M^\omega)$, where M^ω denotes the twisted $\mathfrak{g}$-module with the same underlying space as M and with the new action*

$$a \odot v = \omega(a) \cdot v, \qquad \text{for } a \in \mathfrak{g} \text{ and } v \in M, \tag{1}$$

ω being the Cartan involution of $\mathfrak{g}$ as in 1.1.2.

Proof. We have the following commutative diagram:

$$\begin{array}{ccc} \Lambda^n(\mathfrak{g}) \otimes M & \xrightarrow{\partial} & \Lambda^{n-1}(\mathfrak{g}) \otimes M \\ \downarrow\wr & & \downarrow\wr \\ \Lambda^n(\mathfrak{g}) \otimes M^\omega & \xrightarrow{\partial} & \Lambda^{n-1}(\mathfrak{g}) \otimes M^\omega, \end{array}$$

where the horizontal maps are the standard chain maps corresponding to the Lie algebra $\mathfrak{g}$ with coefficients (resp.) in M and M^ω. The left vertical map is defined by $x_1 \wedge \cdots \wedge x_n \otimes m \to \omega(x_1) \wedge \cdots \wedge \omega(x_n) \otimes m$, for $x_1, \ldots, x_n \in \mathfrak{g}$ and $m \in M$; similarly, the other vertical map. This immediately proves the first part of the lemma. The proof for $H_n(\mathfrak{g}, \mathfrak{s}, M)$ and $H^n(\mathfrak{g}, \mathfrak{s}, M)$ is identical. □

3.3.4 Definition. A $\mathfrak{g}$-module M is called a $(\mathfrak{g}, \mathfrak{h})$-*module* if M is a weight module with respect to the $\mathfrak{h}$-action.

In the following, we abbreviate $\mathrm{Ext}^*_{U(\mathfrak{g})}$ by $\mathrm{Ext}^*_{\mathfrak{g}}$.

3.3.5 Lemma. *Let X, Y be $(\mathfrak{g}, \mathfrak{h})$-modules. Then*

(1) $$\mathrm{Ext}^n_{\mathfrak{g}}(X, Y^\vee) \simeq \mathrm{Ext}^n_{\mathfrak{g}}(X, Y^*), \qquad \textit{for all } n \geq 0\ ,$$

where $Y^\vee$ denotes the $\mathfrak{g}$-submodule of Y^* consisting of the $\mathfrak{h}$-semisimple part of Y^*. Hence, for any $n \geq 0$,

(2) $$\mathrm{Ext}^n_{\mathfrak{g}}(X, Y) \approx \mathrm{Tor}^{\mathfrak{g}}_n((X^\omega)^t, Y^\sigma)^*,$$

where Y^σ is as in 2.1.1.

Proof. By Corollary 3.1.8, Proposition 3.1.10 and Lemma 3.1.7, $\mathrm{Ext}_{\mathfrak{g}}(X, Y^*)$ is the cohomology of the cochain complex $C = \sum_{n\geq 0} C^n$, where

$$C^n := \mathrm{Hom}_{\mathbb{C}}(\Lambda^n(\mathfrak{g}) \otimes X, Y^*) \approx \mathrm{Hom}_{\mathbb{C}}(\Lambda^n(\mathfrak{g}) \otimes X \otimes Y, \mathbb{C}), \text{ by (3.1.13.9).}$$

Since $\mathfrak{g}$, X, and Y are all $(\mathfrak{g}, \mathfrak{h})$-modules, $\Lambda^n(\mathfrak{g}) \otimes X \otimes Y$ is again a $(\mathfrak{g}, \mathfrak{h})$-module, for any $n \geq 0$. Write $\Lambda^n(\mathfrak{g}) \otimes X \otimes Y = \sum_{\beta\in\mathfrak{h}^*} M^n_\beta$, where M^n_β is the weight space (with respect to the $\mathfrak{h}$-action) of $\Lambda^n(\mathfrak{g}) \otimes X \otimes Y$ corresponding to the weight $\beta \in \mathfrak{h}^*$. Clearly, $C^n = \prod_{\beta\in\mathfrak{h}^*}(M^n_\beta)^*$ and, for any $\beta \in \mathfrak{h}^*$, $\sum_n (M^n_\beta)^*$ is a subcomplex of C. For $f \in C^n$, denote by f_β its component in the $(M^n_\beta)^*$ factor.

Now $C_{\mathrm{res}} := \sum C^n_{\mathrm{res}}$ is a subcomplex of C, where

$$C^n_{\mathrm{res}} := \mathrm{Hom}_{\mathbb{C}}(\Lambda^n(\mathfrak{g}) \otimes X, Y^\vee),$$

and its cohomology is $\mathrm{Ext}_{\mathfrak{g}}(X, Y^\vee)$. Further, $f \in C^n$ belongs to C^n_{res} if and only if, for any $v \in \Lambda^n(\mathfrak{g}) \otimes X$, $f_\beta(v)$ is zero (as an element of Y^*) for all but finitely many β's depending on v. In particular, $\sum_n (M^n_0)^* \subset C_{\mathrm{res}}$.

Next, we prove that for any $f \in C^n$, resp. $f \in C^n_{\mathrm{res}}$, satisfying

(3) $$df = 0\ ,\ \ f_0 = 0,$$

there exists a $g \in C^{n-1}$, resp. C^{n-1}_{res}, satisfying

(4) $$dg = f\ ,\ \ g_0 = 0.$$

Of course, this will prove (1).

We now prove the existence of g. For any $\beta \neq 0 \in \mathfrak{h}^*$, fix a $h(\beta) \in \mathfrak{h}$ satisfying $\beta(h(\beta)) = 1$. Since $di_h + i_h d = L_h$, for any $h \in \mathfrak{h}$ (cf. (3.1.2.5)), we get $di_{h(\beta)}(f_\beta) + i_{h(\beta)}d(f_\beta) = f_\beta$. But, since $f = \sum_{\beta\neq 0} f_\beta$ and $df = \sum_{\beta\neq 0} df_\beta = 0$, we have, for all $\beta \neq 0 \in \mathfrak{h}^*$, $di_{h(\beta)}(f_\beta) = f_\beta$, that is

$$d(\sum_{\beta\neq 0} i_{h(\beta)}(f_\beta)) = f.$$

So the element $g := \sum_{\beta\neq 0} i_{h(\beta)}(f_\beta)$ does the job since $i_{h(\beta)}(f_\beta) \in (M_\beta^{n-1})^*$. Observe that if $f \in C^n_{\text{res}}$, then g again is in C^{n-1}_{res}.

To prove (2), since $(Y^\sigma)^\sigma \cong Y$,

$$\begin{aligned}
\operatorname{Ext}^n_{\mathfrak{g}}(X, Y) &\approx \operatorname{Ext}^n_{\mathfrak{g}}(X, (Y^\sigma)^\sigma)\\
&\approx \operatorname{Ext}^n_{\mathfrak{g}}(X, ((Y^\sigma)^\omega)^*), && \text{by (1)},\\
&\approx \operatorname{Tor}^{\mathfrak{g}}_n(X^t, (Y^\sigma)^\omega)^*, && \text{by (3.1.13.3)},\\
&\approx \operatorname{Tor}^{\mathfrak{g}}_n((X^\omega)^t, Y^\sigma)^*, && \text{by Lemma 3.3.3, (3.1.9.1) and (3.1.13.1).}
\end{aligned}$$

This proves (2). □

3.3.6 Remark. Observe that, for any ω-stable subalgebra $\mathfrak{s} \supset \mathfrak{h}$ of $\mathfrak{g}$ such that $\mathfrak{g}$ is finitely semisimple as a $\mathfrak{s}$-module under the adjoint action, (1) and (2) of the above lemma are true with $\mathfrak{g}$ replaced by the pair $(\mathfrak{g}, \mathfrak{s})$ by the same proof. In fact, in this case, the standard cochain complexes to calculate $\operatorname{Ext}_{(\mathfrak{g},\mathfrak{s})}(X, Y^*)$ and $\operatorname{Ext}_{(\mathfrak{g},\mathfrak{s})}(X, Y^\vee)$ coming from the $(U(\mathfrak{g}), U(\mathfrak{s}))$-projective resolution $(\mathcal{D}_X)$ of X (cf. 3.1.8) are themselves isomorphic.

3.3.7 Lemma. *For any* $(\mathfrak{g}, \mathfrak{h})$*-module* M *and any* $\lambda \in \mathfrak{h}^*$,

(1) $$\operatorname{Tor}^{(\mathfrak{g},\mathfrak{h})}_n((M(\lambda)^\omega)^t, M) \approx \operatorname{Tor}^{(\mathfrak{g},\mathfrak{h})}_n(M^t, M(\lambda)^\omega) \approx H_n(\mathfrak{n}^-, M)_\lambda,$$

where $H_n(\mathfrak{n}^-, M)_\lambda$ *denotes the* λ*-th weight space under the canonical action of* $\mathfrak{h}$ *on* $H_n(\mathfrak{n}^-, M)$.

Proof. It can easily be seen that, as a $\mathfrak{g}$-module,

(2) $$M(\lambda)^\omega \approx U(\mathfrak{g}) \otimes_{U(\mathfrak{b}^-)} \mathbb{C}_{-\lambda}.$$

By (3.1.14.1), we get

$$\begin{aligned}
\operatorname{Tor}^{(\mathfrak{g},\mathfrak{h})}_n(M^t, M(\lambda)^\omega) &\approx \operatorname{Tor}^{(\mathfrak{b}^-,\mathfrak{h})}_n(M^t, \mathbb{C}_{-\lambda})\\
&\approx H_n(\mathfrak{b}^-, \mathfrak{h}, M \otimes \mathbb{C}_{-\lambda}), \text{ by (3.1.13.1) and (3.1.9.1).}
\end{aligned}$$

But, since $\mathfrak{b}^-$ and M are $\mathfrak{h}$-weight modules, by definition of the standard complex to calculate the Lie algebra homology (cf. (3.1.3.2)),

$$H_n(\mathfrak{b}^-, \mathfrak{h}, M \otimes \mathbb{C}_{-\lambda}) \approx H_n(\mathfrak{n}^-, M \otimes \mathbb{C}_{-\lambda})^{\mathfrak{h}} \approx H_n(\mathfrak{n}^-, M)_\lambda.$$ □

3.3.8 Theorem. *(a) Let $\Lambda_1 \neq \Lambda_2$ be two equivalence classes in $\mathfrak{h}^*$ under $\sim$ and let $M \in \mathcal{O}$, resp. $N \in \mathcal{O}$, be of type Λ_1, resp. Λ_2. Then*

$$\operatorname{Ext}^n_{\mathfrak{g}}(M, N) = 0, \qquad \textit{for all } n \geq 0.$$

(b) Let $\Lambda_1^o \neq \Lambda_2^o$ be two equivalence classes in $K^{\text{w.g.}}$ under $\sim^o$ and let $M \in \mathcal{O}^{\text{w.g.}}$, resp. $N \in \mathcal{O}^{\text{w.g.}}$, be of type Λ_1^o, resp. Λ_2^o. Then

$$\operatorname{Ext}^n_{\mathfrak{g}}(M, N) = 0, \qquad \textit{for all } n \geq 0.$$

Proof. We first prove (a): From the definition of σ, it is easy to see that, for any irreducible $L(\mu)$,

$$\text{(1)} \qquad L(\mu)^\sigma \approx L(\mu), \qquad \text{as } \mathfrak{g}\text{-modules.}$$

In particular, N^σ is again of type Λ_2. By (3.3.5.2), we need to prove the vanishing of $\operatorname{Tor}^{\mathfrak{g}}_*((M^\omega)^t, N)$ for $M, N \in \mathcal{O}$ of types Λ_1, Λ_2 respectively.

By Lemma 2.1.10, for any $M \in \mathcal{O}$, there exists a (possibly infinite) increasing filtration $0 = M_0 \subset M_1 \subset M_2 \subset \dots$ of submodules of M such that $\cup M_i = M$ and M_i/M_{i-1} is a highest weight module for all $i \geq 1$.

Further, recall that, for any $\mathfrak{g}$-modules $M' \subset M''$, $N' \subset N''$, there are exact sequences (cf. Appendix D):

$$\text{(2)} \quad \operatorname{Tor}^{\mathfrak{g}}_n((M')^t, N'') \to \operatorname{Tor}^{\mathfrak{g}}_n((M'')^t, N'') \to \operatorname{Tor}^{\mathfrak{g}}_n((M''/M')^t, N'') \to \operatorname{Tor}^{\mathfrak{g}}_{n-1}((M')^t, N'') \to \dots,$$

and

$$\text{(3)} \quad \dots \to \operatorname{Tor}^{\mathfrak{g}}_n((M'')^t, N') \to \operatorname{Tor}^{\mathfrak{g}}_n((M'')^t, N'') \to \operatorname{Tor}^{\mathfrak{g}}_n((M'')^t, N''/N') \to \operatorname{Tor}^{\mathfrak{g}}_{n-1}((M'')^t, N') \to \dots .$$

Since Tor commutes with direct limits, from (2) and (3), we can assume that M, resp. N, is a quotient of a Verma module $M(\lambda)$, resp. $M(\mu)$, for $\lambda \in \Lambda_1$, resp. $\mu \in \Lambda_2$. Write $0 \to M' \to M(\lambda) \to M \to 0$, and $0 \to N' \to M(\mu) \to N \to 0$.

Assume, by induction, that for all $k < n$, $\operatorname{Tor}^{\mathfrak{g}}_k((M^\omega)^t, N) = 0$, for any $M \in \mathcal{O}$, resp. $N \in \mathcal{O}$, of type Λ_1, resp. Λ_2. Of course, induction starts at $n = 0$. Using the exact sequences (2) and (3) and the induction hypothesis, it suffices to show that $\operatorname{Tor}^{\mathfrak{g}}_n((M(\lambda)^\omega)^t, M(\mu)) = 0$ for $\lambda, \mu \in \mathfrak{h}^*$ with $\lambda \not\sim \mu$.

Now,

$$
\begin{aligned}
\mathrm{Tor}_*^{\mathfrak{g}}((M(\lambda)^\omega)^t, M(\mu)) &\approx \mathrm{Tor}_*^{\mathfrak{b}}((M(\lambda)^\omega)^t, \mathbb{C}_\mu), && \text{by Lemma 3.1.14}\\
&\approx \mathrm{Tor}_*^{\mathfrak{b}}(\mathbb{C}_\mu^t, M(\lambda)^\omega), && \text{by (3.1.13.1)}\\
&\approx \mathrm{Tor}_*^{\mathfrak{h}}(\mathbb{C}_\mu^t, \mathbb{C}_{-\lambda}), && \text{by Lemma 3.1.14 since, by (3.3.7.2),}\\
&\qquad M(\lambda)^\omega \simeq U(\mathfrak{b}) \otimes_{U(\mathfrak{h})} \mathbb{C}_{-\lambda} \text{ as } \mathfrak{b}\text{-modules}\\
&\approx H_*(\mathfrak{h}, \mathbb{C}_{\mu-\lambda}), && \text{by (3.1.9.1) and (3.1.13.1)}\\
&= 0,
\end{aligned}
$$

since $H_*(\mathfrak{h}, \mathbb{C}_{\mu-\lambda})$ is a trivial $\mathfrak{h}$-module by 3.1.1, whereas $\Lambda(\mathfrak{h}) \otimes \mathbb{C}_{\mu-\lambda}$ has no nonzero element of weight 0. Thus part (a) of the theorem is established.

Proof of part (b) is exactly the same. We just need to observe that for any $\lambda \in K^{\text{w.g.}}$ and $\mu \in \mathfrak{h}^*$ such that $L(\mu)$ is a component of $M(\lambda)$, we have $\mu \in K^{\text{w.g.}}$ by Proposition 2.3.8. □

3.3.9 Remark. The above proof can easily be adapted to prove that under the assumptions of Theorem 3.3.8 (a or b parts), $\mathrm{Ext}^*_{(\mathfrak{g},\mathfrak{h})}(M, N) = 0$.

3.3.10 Corollaries.
(a$_1$) *Let $M \in \mathcal{O}$. Then there exists a unique family $\{M_\Lambda\}$ of submodules of M parametrized by the set of equivalence classes $\{\Lambda\}$ of $\mathfrak{h}^*$ under $\sim$ such that*

(i) *M_Λ is of type Λ, and*
(ii) *$M = \oplus_\Lambda M_\Lambda$.*

(a$_2$) *For $M \in \mathcal{O}^{\text{w.g.}}$, there exists a unique family $\{M_{\Lambda^o}\}$ of submodules of M parametrized by the set of equivalence classes $\{\Lambda^o\}$ of $K^{\text{w.g.}}$ under $\sim^o$ such that*

(i) *M_{Λ^o} is of type Λ^o, and*
(ii) *$M = \oplus_{\Lambda^o} M_{\Lambda^o}$.*

(b) *Let $M \in \mathcal{O}$, resp. $\mathcal{O}^{\text{w.g.}}$, be such that no component $L(\lambda)$ of M satisfies $\lambda \sim 0$, resp. $\lambda \sim^o 0$. Then*

(b$_1$) *$H_*(\mathfrak{g}, M) = H_*(\mathfrak{g}, \mathfrak{h}, M) = 0$, and*
(b$_2$) *$H^*(\mathfrak{g}, M) = H^*(\mathfrak{g}, \mathfrak{h}, M) = 0$.*

In particular, for a highest weight module M of highest weight $\lambda \nsim 0$ (if $\lambda \in K^{\text{w.g.}}$, we just demand that $\lambda \notin W * 0$), we have the vanishing (b$_1$) and (b$_2$).

The following two special cases of (b) are of particular interest.

(b$_3$) *For any integrable highest weight module $L(\lambda)$, $\lambda \neq 0$,*

$$H_*(\mathfrak{g}, L(\lambda)) = H_*(\mathfrak{g}, \mathfrak{h}, L(\lambda)) = H^*(\mathfrak{g}, L(\lambda)) = H^*(\mathfrak{g}, \mathfrak{h}, L(\lambda)) = 0.$$

More generally, let $L(\lambda)$ and $L(\mu)$ be integrable highest weight modules with $\lambda \neq \mu$. Then

$$\mathrm{Ext}^*_{(\mathfrak{g},\mathfrak{h})}(L(\lambda), L(\mu)) = \mathrm{Ext}^*_{\mathfrak{g}}(L(\lambda), L(\mu)) = 0.$$

(b_4) *Let $M \in \mathcal{O}$ be such that the Casimir operator Ω_M acts as an automorphism on M. Then*

$$H_*(\mathfrak{g}, M) = H_*(\mathfrak{g}, \mathfrak{h}, M) = H^*(\mathfrak{g}, M) = H^*(\mathfrak{g}, \mathfrak{h}, M) = 0.$$

Proof. (a_1) For an equivalence class $\Lambda \subset \mathfrak{h}^*$ and an exact sequence

$$0 \to M_1 \to N \to M_2 \to 0,$$

observe that N is of type Λ if and only if M_1 and M_2 are both of type Λ.

Define $M_\Lambda \subset M$ as the sum of all submodules of M of type Λ. Then M_Λ itself is of type Λ and hence is the unique maximal submodule of M of type Λ. Further, for any distinct equivalence classes $\Lambda_0, \Lambda_1, \dots, \Lambda_p \subset \mathfrak{h}^*$,

$$M_{\Lambda_0} \cap (M_{\Lambda_1} + \cdots + M_{\Lambda_p}) = 0,$$

and hence the sum $\sum_\Lambda M_\Lambda$ is direct.

We next show that $M = \oplus_\Lambda M_\Lambda$. Otherwise, let $Q := M/\oplus_\Lambda M_\Lambda \neq 0$. There exists at least one Λ_1 such that $Q_{\Lambda_1} \neq 0$ by Lemma 2.1.10. Let M_1 be the module satisfying $\oplus_\Lambda M_\Lambda \subset M_1 \subset M$ such that $M_1/\oplus_\Lambda M_\Lambda = Q_{\Lambda_1}$. This gives rise to exact sequences:

(1) $$0 \to \oplus_{\Lambda \neq \Lambda_1} M_\Lambda \to M_1 \to Q_1 \to 0, \text{ and}$$

(2) $$0 \to M_{\Lambda_1} \to Q_1 \to Q_{\Lambda_1} \to 0,$$

where $Q_1 := M_1/\oplus_{\Lambda \neq \Lambda_1} M_\Lambda$. In particular, Q_1 is of type Λ_1. Now, by (3.3.5.2),

$$\begin{aligned}\mathrm{Ext}^n_{\mathfrak{g}}\left(Q_1, \oplus_{\Lambda\neq\Lambda_1} M_\Lambda\right) &\approx \mathrm{Tor}^{\mathfrak{g}}_n((Q_1^\omega)^t, \oplus_{\Lambda\neq\Lambda_1}(M_\Lambda^\sigma))^* \\ &\approx \prod_{\Lambda\neq\Lambda_1}\left(\mathrm{Tor}^{\mathfrak{g}}_n((Q_1^\omega)^t, M_\Lambda^\sigma)^*\right) \\ &\approx \prod_{\Lambda\neq\Lambda_1}\left(\mathrm{Ext}^n_{\mathfrak{g}}(Q_1, M_\Lambda)\right) \\ &= 0, \text{ by Theorem 3.3.8.}\end{aligned}$$

This gives that the sequence (1) splits as $\mathfrak{g}$-modules. Choose a splitting $\sigma : Q_1 \to M_1$ of (1). Then $\sigma(Q_1) \subset M_{\Lambda_1}$, M_{Λ_1} being the maximal submodule of M of type Λ_1, and hence $M_1 \subset \oplus_\Lambda M_\Lambda$, which is a contradiction since $Q_{\Lambda_1} \neq 0$. This completes the proof of (a_1).

The proof of (a_2) is similar.

(b) We first consider the case of the category $\mathcal{O}$. Write $M = \oplus_{\Lambda \neq O} M_\Lambda$, where $O \subset \mathfrak{h}^*$ is the equivalence class of 0. Then

$$H_*(\mathfrak{g}, M) = \oplus_{\Lambda \neq O} H_*(\mathfrak{g}, M_\Lambda)$$

and

$$H^*(\mathfrak{g}, M) = \prod_{\Lambda \neq O} H^*(\mathfrak{g}, M_\Lambda) \qquad \text{(cf. proof of the (a) part).}$$

Now

$$H_*(\mathfrak{g}, M_\Lambda)^* \approx \mathrm{Ext}^*_{\mathfrak{g}}(\mathbb{C}, M^\sigma_\Lambda), \quad \text{by (3.1.9.1) and (3.3.5.2),}$$

and

$$H^*(\mathfrak{g}, M_\Lambda) \approx \mathrm{Ext}^*_{\mathfrak{g}}(\mathbb{C}, M_\Lambda), \quad \text{by (3.1.9.2).}$$

So (b_1) and (b_2) for the category $\mathcal{O}$ follow from Theorem 3.3.8(a) and Remark 3.3.9. Proof of (b_1) and (b_2) for the category $\mathcal{O}^{\text{w.g.}}$ is similar.

The first part of (b_3) is a special case of (b_1) and (b_2), since any integrable $L(\lambda)$ belongs to $\mathcal{O}^{\text{w.g.}}$ by Exercise 2.3.E.4 and, moreover, if nonzero, $\lambda \notin W * 0$ by Exercise 1.4.E.2. For the second part, observe that for integrable $L(\lambda)$ and $L(\mu)$, $\lambda \in W * \mu$ if and only if $\lambda = \mu$. Now apply Theorem 3.3.8(b), Remark 3.3.9 and Lemma 3.3.2.

(b_4) Since Ω_M acts as an automorphism on M (by assumption), $\Omega_{L(\lambda)}$ is an automorphism for any component $L(\lambda)$ of M. We claim that $\lambda \not\sim 0$. Otherwise, $\Omega_{L(\lambda)} \equiv 0$, since $\Omega_{L(0)} \equiv 0$ (cf. Lemma 2.1.16). So (b_4) is a special case of (b_1) and (b_2). $\square$

3.3.E EXERCISES

(1) For any $M, N \in \mathcal{O}$, prove that $\mathrm{Ext}^*_{\mathcal{O}}(M, N) \simeq \mathrm{Ext}^*_{(\mathfrak{g},\mathfrak{h})}(M, N)$, where $\mathrm{Ext}_{\mathcal{O}}$ is the Ext functor for the category $\mathcal{O}$.

(2) For any Kac–Moody algebra $\mathfrak{g}$, prove that

$$H^1(\mathfrak{g}', \mathbb{C}) = H^2(\mathfrak{g}', \mathbb{C}) = 0,$$

where $\mathfrak{g}' := [\mathfrak{g}, \mathfrak{g}]$.

Hint: To prove the vanishing of $H^2(\mathfrak{g}', \mathbb{C})$, use the spectral sequence as in Theorem E.13 for the pair $(\mathfrak{g}', \mathfrak{h} \cap \mathfrak{g}')$.

3.4. Laplacian Calculation

In this section we continue to use the assumption that $\mathfrak{g} = \mathfrak{g}(A)$ is symmetrizable. Let $Y \subset \{1, \ldots, \ell\}$ be any subset and recall the definition of $\mathfrak{g}_Y$, $\mathfrak{p}_Y$, $\mathfrak{u}_Y$, and $\mathfrak{u}_Y^-$ from Subsection 1.2.2. Also recall from Definition 2.3.9 and Theorem 2.3.13 that $\mathfrak{u}_Y^-$ (and $\mathfrak{u}_Y$) admits a $\mathfrak{k}_Y := \mathfrak{k} \cap \mathfrak{g}_Y$-invariant positive definite Hermitian form $\{\ ,\ \}$ and also, for any $\lambda \in D \cap \mathfrak{h}_{\mathbb{R}}^*$, $L(\lambda)$ admits a $\mathfrak{k}$-invariant positive definite Hermitian form H_λ (unique up to positive scalar multiples). In particular, the components A_p of the following standard chain complex (cf. (3.1.1.1)) admit $\mathfrak{k}_Y$-invariant positive definite Hermitian forms (to be denoted by $\{\ ,\ \}$ itself), induced from $\{\ ,\ \}$ and H_λ,

$$\cdots \to A_p \xrightarrow{\partial_p} A_{p-1} \to \ldots \xrightarrow{\partial_1} A_0 \to 0,$$

where $A_p := \Lambda_p(\mathfrak{u}_Y^-, L(\lambda)) = \Lambda^p(\mathfrak{u}_Y^-) \otimes L(\lambda)$. We fix such a $\mathfrak{k}_Y$-invariant form $\{\ ,\ \}$ on A_p.

3.4.1 Definition. Let $\partial_p^* : A_{p-1} \to A_p$ be the adjoint of ∂_p, i.e.,

(1) $$\{\partial_p^* v, w\} = \{v, \partial_p w\}, \qquad \text{for } v \in A_{p-1} \text{ and } w \in A_p.$$

Observe that since the $\mathfrak{h}$-weight spaces of A_p are finite-dimensional, ∂_p are $\mathfrak{h}$-module maps and the weight spaces of A_p are mutually orthogonal with respect to $\{\ ,\ \}$, $\{\ ,\ \}$ being $\mathfrak{k}_Y$-invariant, the existence of the adjoint ∂_p^* is guaranteed.

Since ∂_p is a $\mathfrak{g}_Y$-linear map and $\{\ ,\ \}$ on A_p is $\mathfrak{k}_Y$-invariant, ∂_p^* is again $\mathfrak{g}_Y$-linear, $\mathfrak{k}_Y$ being a real form of $\mathfrak{g}_Y$.

Define the *Laplacian*

(2) $$\Delta_p : A_p \to A_p \text{ by } \Delta_p = \partial_p^* \partial_p + \partial_{p+1} \partial_{p+1}^*.$$

Then Δ_p is a $\mathfrak{g}_Y$-linear, selfadjoint, and, moreover, positive semidefinite operator on A_p, i.e.,

$$\{\Delta_p v, v\} \geq 0, \qquad \text{for all } v \in A_p.$$

Let $\mathcal{O}_Y$ denote the BGG category $\mathcal{O}$ for the algebra $\mathfrak{g}_Y$ (cf. Definition 2.1.1). Define

(3) $$Q_Y = \sum_{i \in Y} \mathbb{Z}\alpha_i \subset \mathfrak{h}^*.$$

For a coset $\bar{\mu} \in \mathfrak{h}^*/Q_Y$ and $\mathfrak{g}_Y$-module M such that M is a $\mathfrak{h}$-weight module, set

(4) $$M(\bar{\mu}) = \sum_{\mu \in \bar{\mu}} M_\mu.$$

Then, clearly, $M(\bar{\mu})$ is a $\mathfrak{g}_Y$-submodule of M and, moreover,

$$M = \oplus_{\bar{\mu}\in\mathfrak{h}^*/Q_Y} M(\bar{\mu}). \tag{5}$$

By Corollary 1.3.4, each A_p is an integrable $\mathfrak{g}_Y$-module; moreover, it is easy to see that $A_p \in \mathcal{O}_Y$. In particular, by Remark 3.2.2, each A_p is a direct sum of irreducible $\mathfrak{g}_Y$-modules of the form $L_Y(\theta)$, for $\theta \in D_Y$. For $\theta \in D_Y$, let $A_p(\theta)$ denote the isotypical component of A_p corresponding to the $\mathfrak{g}_Y$-module $L_Y(\theta)$.

Now we come to the main theorem of this subchapter.

3.4.2 Theorem. *Take $\lambda \in D \cap \mathfrak{h}^*_{\mathbb{R}}$. Then, with the notation as above, for any $p \geq 0$ and $\theta \in D_Y$,*

$$\Delta_{p\,|A_p(\theta)} = \frac{1}{2}[\langle\lambda+\rho,\lambda+\rho\rangle - \langle\theta+\rho,\theta+\rho\rangle]\, I_{A_p(\theta)}\,. \tag{1}$$

Proof. Choose an orthonormal basis $\{a_n\}_{n\in\mathbb{N}}$ of $\mathfrak{u}_Y^-$ consisting of root vectors and also choose an orthonormal basis $\{v_\varphi\}_{\varphi\in I}$ of $L(\lambda)$ consisting of weight vectors (for some indexing set I). Set

$$b_n = \sigma_o(a_n), \tag{2}$$

where σ_o is the compact involution as in 2.3.9. Then by (2.3.9.4),

$$\langle a_n, b_n\rangle = \{a_n, a_n\} = 1\,, \tag{3}$$

and, moreover, $\{b_n\}_{n\in\mathbb{N}}$ is an orthonormal basis of $\mathfrak{u}_Y$. Writing $\mathfrak{u}_Y^- = \mathfrak{g}/\mathfrak{p}_Y$, we see that $\mathfrak{u}_Y^-$ is a module for $\mathfrak{p}_Y$ under the adjoint action, denoted by $\bar{\mathrm{ad}}$. In the sequel, until the end of this subsection, we abbreviate $L(\lambda)$ by L and $\mathfrak{u}_Y^-$ by $\mathfrak{u}^-$.

3.4.3 Lemma. *For $X \in \Lambda^p(\mathfrak{u}^-)$ and $v \in L$,*

$$\partial^*(X\otimes v) = -\sum_{n\geq 1} a_n \wedge X \otimes b_n v - \frac{1}{2}\sum_{n\geq 1} a_n \wedge (\bar{\mathrm{ad}}\, b_n X)\otimes v. \tag{1}$$

Observe that the sum on the right is a finite sum.

Proof. Write $\partial = (\partial' \otimes I_L) + \partial''$, where $\partial' : \Lambda^p(\mathfrak{u}^-) \to \Lambda^{p-1}(\mathfrak{u}^-)$ is the differential of the Lie algebra $\mathfrak{u}^-$ with trivial coefficients, i.e.,

$$\partial'(x_1\wedge\cdots\wedge x_p) = \sum_{i<j}(-1)^{i+j}[x_i, x_j]\wedge x_1\wedge\cdots\wedge\hat{x}_i\wedge\cdots\wedge\hat{x}_j\wedge\cdots\wedge x_p, \tag{2}$$

and

$$\partial''(x_1\wedge\cdots\wedge x_p\otimes v) = \sum_i (-1)^i x_1\wedge\cdots\wedge\hat{x}_i\wedge\cdots\wedge x_p\otimes x_i v. \tag{3}$$

For $x \in \mathfrak{u}^-$,

$$\partial'^*(x) = -\sum_{i<j} \{x, [a_i, a_j]\}\, a_i \wedge a_j$$

$$= -\frac{1}{2}\sum_{i,j} \{x, [a_i, a_j]\}\, a_i \wedge a_j$$

$$= \frac{1}{2}\sum_{i,j} \{[b_j, x], a_i\}\, a_i \wedge a_j\,, \qquad \text{by the contravariance of } \{\ ,\ \}$$

(4) $$= -\frac{1}{2}\sum_{j} a_j \wedge (\bar{\mathrm{ad}}\, b_j\, x).$$

Consider the operator $\theta : \Lambda(\mathfrak{u}^-) \to \Lambda(\mathfrak{u}^-)$, defined by

(5) $$\theta(X) = \sum_{n \geq 1} a_n \wedge (\bar{\mathrm{ad}}\, b_n X).$$

Then it is easy to see that θ is an antiderivation (of degree 1). Also $\partial'^* : \Lambda(\mathfrak{u}^-) \to \Lambda(\mathfrak{u}^-)$ can be seen to be an antiderivation of degree 1 (cf. Exercise 3.4.E.1). Hence, by (4) and (5), for $X \in \Lambda(\mathfrak{u}^-)$,

(6) $$\partial'^*(X) = -\frac{1}{2}\sum_{n \geq 1} a_n \wedge (\bar{\mathrm{ad}}\, b_n X).$$

Now we seek an expression for ∂''^*: For $X \in \Lambda^p(\mathfrak{u}^-)$ and $v \in L$,

$$\partial''^*(X \otimes v)$$

$$= \sum_{\substack{j_1 < \cdots < j_{p+1} \\ \varphi \in I}} \{X \otimes v, \partial''(a_{j_1} \wedge \cdots \wedge a_{j_{p+1}} \otimes v_\varphi)\}\, a_{j_1} \wedge \cdots \wedge a_{j_{p+1}} \otimes v_\varphi$$

$$= \frac{1}{(p+1)!} \sum_{\substack{j_1, \ldots, j_{p+1} \in \mathbb{N} \\ \varphi \in I}} \{X \otimes v, \partial''(a_{j_1} \wedge \cdots \wedge a_{j_{p+1}} \otimes v_\varphi)\}\, a_{j_1} \wedge \cdots \wedge a_{j_{p+1}} \otimes v_\varphi$$

$$= \frac{1}{(p+1)!} \sum_{i=1}^{p+1} \sum_{\substack{j_1, \ldots, j_{p+1} \in \mathbb{N} \\ \varphi \in I}} (-1)^i \{X, a_{j_1} \wedge \cdots \wedge \hat{a}_{j_i} \wedge \cdots \wedge a_{j_{p+1}}\}$$

$$\cdot\, a_{j_1} \wedge \cdots \wedge a_{j_{p+1}} \otimes H_\lambda(v, a_{j_i} v_\varphi) v_\varphi$$

$$= \frac{1}{(p+1)!}\sum_{i=1}^{p+1}\sum_{\substack{j_1,\dots,j_{p+1}\in\mathbb{N}\\ \varphi\in I}}(-1)^i\{X, a_{j_1}\wedge\cdots\wedge\hat{a}_{j_i}\wedge\cdots\wedge a_{j_{p+1}}\}$$

$a_{j_1}\wedge\cdots\wedge a_{j_{p+1}}\otimes H_\lambda(b_{j_i}v, v_\varphi)v_\varphi$, from the contravariance of H_λ

$$(7)\qquad = -\frac{p!}{(p+1)!}\sum_{i=1}^{p+1}\sum_{j_i\in\mathbb{N}} a_{j_i}\wedge X\otimes b_{j_i}v = -\sum_{n\geq 1} a_n\wedge X\otimes b_n v.$$

Now the lemma follows by combining (6) and (7). □

3.4.4 Lemma. *For $X = x_1\wedge\cdots\wedge x_p\in\Lambda^p(\mathfrak{u}^-)$ and $v\in L$,*

$$2\Delta(X\otimes v) = \sum_n\Big[2(\operatorname{ad} a_n\, X)\otimes b_n v - \partial'(a_n\wedge \bar{\operatorname{ad}}\, b_n\, X)\otimes v$$

$$+ 2X\otimes a_n b_n v - a_n\wedge(\bar{\operatorname{ad}}\, b_n(\partial' X))\otimes v + 2\sum_{s=1}^{p}(-1)^s a_n\wedge X^{(s)}\otimes[x_s, b_n]v$$

$$+\sum_{s=1}^{p}(-1)^s a_n\wedge X^{(s)}\otimes(\bar{\operatorname{ad}}\, b_n\, x_s)v + (\bar{\operatorname{ad}}\, b_n\, X)\otimes a_n v\Big],$$

where $X^{(s)} := x_1\wedge\cdots\wedge\hat{x}_s\wedge\cdots\wedge x_p$.

Observe that the sum on the right is a finite sum.

Proof. By Lemma 3.4.3 and (3.4.3.3),

$$-2\Delta(X\otimes v) = \sum_{n\geq 1}\Big[2\partial'(a_n\wedge X)\otimes b_n v - 2X\otimes a_n b_n v$$

$$-2\sum_{s=1}^{p}(-1)^s a_n\wedge X^{(s)}\otimes x_s b_n v + \partial'(a_n\wedge\bar{\operatorname{ad}}\, b_n\, X)\otimes v$$

$$-\bar{\operatorname{ad}}\, b_n\, X\otimes a_n v + \sum_{s=1}^{p}(-1)^{s+1} a_n\wedge X^{(s)}\otimes(\bar{\operatorname{ad}}\, b_n x_s)v$$

$$+2a_n\wedge\partial' X\otimes b_n v + a_n\wedge(\bar{\operatorname{ad}}\, b_n(\partial' X))\otimes v + 2\sum_{s=1}^{p}(-1)^s a_n\wedge X^{(s)}\otimes b_n x_s v\Big].$$

Now the lemma follows by using (3.1.1.3). □

3.4.5 Lemma. *With the notation as in the above Lemma 3.4.4,*

$$(1)\qquad \sum_n\Big[\partial'(a_n\wedge\bar{\operatorname{ad}}\, b_n X) + a_n\wedge\bar{\operatorname{ad}}\, b_n(\partial' X)\Big]$$

$$= \sum_{s=1}^{p}(-1)^s F_Y(x_s)\wedge X^{(s)} + \sum_\gamma\sum_{s=1}^{p}(-1)^{s+1}(\operatorname{ad} h_\gamma\, x_s)\wedge \operatorname{ad} h_\gamma^{\#}(X^{(s)}),$$

where $\{h_\gamma\}_\gamma$ *is a basis of* $\mathfrak{g}_Y$ *consisting of* $\mathfrak{h}$*-weight vectors and* $\{h^\#_\gamma\}_\gamma$ *is the dual basis of* $\mathfrak{g}_Y$ *with respect to the Killing form* $\langle\ ,\ \rangle$ *on* $\mathfrak{g}$ *restricted to* $\mathfrak{g}_Y$ *(i.e.,* $\langle h_\gamma, h^\#_{\gamma'}\rangle = \delta_{\gamma,\gamma'}$*), and* $F_Y : \mathfrak{u}^- \to \mathfrak{u}^-$ *is defined by*

$$(2) \qquad F_Y(x) = \sum_n \operatorname{ad} a_n \bar{\operatorname{ad}}\, b_n(x), \qquad \text{for } x \in \mathfrak{u}^- \text{ (cf. Definition 2.3.10).}$$

Observe that, for a given X, there are only finitely many γ such that $\operatorname{ad} h_\gamma\, x_s \wedge \operatorname{ad} h^\#_\gamma(X^{(s)}) \neq 0$, and hence the right side of (1) makes sense.

Proof. By (3.1.1.3),

$$(3) \quad \partial'(a_n \wedge \bar{\operatorname{ad}}\, b_n X) + a_n \wedge \bar{\operatorname{ad}}\, b_n(\partial' X) \\ = -\operatorname{ad} a_n \bar{\operatorname{ad}}\, b_n\, X - a_n \wedge \partial'(\bar{\operatorname{ad}}\, b_n\, X) + a_n \wedge \bar{\operatorname{ad}}\, b_n(\partial' X).$$

Now

$$\begin{aligned}\sum_n \operatorname{ad} a_n \bar{\operatorname{ad}}\, b_n\, X &= \sum_n \sum_{s=1}^p (-1)^{s+1} \operatorname{ad} a_n((\bar{\operatorname{ad}}\, b_n\, x_s) \wedge X^{(s)}) \\ &= \sum_n \Bigg[\sum_{s=1}^p (-1)^{s+1} (\operatorname{ad} a_n(\bar{\operatorname{ad}}\, b_n\, x_s)) \wedge X^{(s)} \\ &\qquad + \sum_{s=1}^p (-1)^{s+1} (\bar{\operatorname{ad}}\, b_n x_s) \wedge \operatorname{ad} a_n X^{(s)}\Bigg] \\ &= \sum_{s=1}^p (-1)^{s+1} F_Y(x_s) \wedge X^{(s)} \\ (4) \qquad &\qquad + \sum_n \sum_{s=1}^p (-1)^{s+1} (\bar{\operatorname{ad}}\, b_n\, x_s) \wedge \operatorname{ad} a_n X^{(s)}.\end{aligned}$$

Recall the definition of the $U(\mathfrak{g})$-module map

$$\partial_p : U(\mathfrak{g}) \otimes_{U(\mathfrak{p}_Y)} \Lambda^p(\mathfrak{g}/\mathfrak{p}_Y) \to U(\mathfrak{g}) \otimes_{U(\mathfrak{p}_Y)} \Lambda^{p-1}(\mathfrak{g}/\mathfrak{p}_Y)$$

from (3.1.4.1) and consider the following commutative diagram:

$$\begin{array}{ccc} \Lambda^p(\mathfrak{u}^-) & \xrightarrow{\partial'} & \Lambda^{p-1}(\mathfrak{u}^-) \\ \downarrow{\scriptstyle \int} & & \downarrow{\scriptstyle \int} \\ \mathbb{C} \underset{U(\mathfrak{u}^-)}{\otimes} [U(\mathfrak{g}) \underset{U(\mathfrak{p}_Y)}{\otimes} \Lambda^p(\mathfrak{g}/\mathfrak{p}_Y)] & \xrightarrow{I \otimes \partial_p} & \mathbb{C} \underset{U(\mathfrak{u}^-)}{\otimes} [U(\mathfrak{g}) \otimes \Lambda^{p-1}(\mathfrak{g}/\mathfrak{p}_Y)], \end{array}$$

where the vertical maps are defined by $X \mapsto 1 \otimes (1 \otimes X)$ which, clearly, are isomorphisms by the PBW theorem. In particular,

$$
\begin{aligned}
1 \otimes (1 \otimes \partial'(\bar{\operatorname{ad}} b_n(X))) &= 1 \otimes \partial_p(b_n \otimes X) \\
&= 1 \otimes b_n(\partial_p(1 \otimes X)) \\
&= 1 \otimes (b_n \otimes \partial' X) + 1 \otimes \left(\sum_{s=1}^{p} (-1)^{s+1} b_n x_s \otimes X^{(s)}\right) \\
&= 1 \otimes (1 \otimes \bar{\operatorname{ad}} b_n(\partial' X)) \\
&\quad + 1 \otimes \left(\left(\sum_{s=1}^{p} (-1)^{s+1} [b_n, x_s] - (\bar{\operatorname{ad}} b_n\, x_s)\right) \otimes X^{(s)}\right).
\end{aligned}
$$

From this and the above commutative diagram, we obtain that

$$
\partial'(\bar{\operatorname{ad}} b_n(X)) = \sum_{s=1}^{p} (-1)^{s+1} \bar{\operatorname{ad}}\,([b_n, x_s] - \bar{\operatorname{ad}} b_n\, x_s) X^{(s)} + \bar{\operatorname{ad}} b_n(\partial' X). \tag{5}
$$

Further, for any $1 \leq s \leq p$,

$$
\begin{aligned}
&\sum_n a_n \wedge \bar{\operatorname{ad}}\,([b_n, x_s] - \bar{\operatorname{ad}} b_n\, x_s) X^{(s)} \\
&= \sum_n \sum_\gamma a_n \wedge \langle [b_n, x_s], h_\gamma \rangle (\operatorname{ad} h_\gamma^{\#}\, X^{(s)}) \\
&\quad + \sum_{n,m} a_n \wedge \langle [b_n, x_s], a_m \rangle (\bar{\operatorname{ad}} b_m\, X^{(s)}) \\
&= \sum_\gamma \sum_n \langle [x_s, h_\gamma], b_n \rangle a_n \wedge (\operatorname{ad} h_\gamma^{\#}\, X^{(s)}) \\
&\quad + \sum_{n,m} \langle [x_s, a_m], b_n \rangle\, a_n \wedge (\bar{\operatorname{ad}} b_m\, X^{(s)}) \\
&= -\sum_\gamma [h_\gamma, x_s] \wedge (\operatorname{ad} h_\gamma^{\#}\, X^{(s)}) - \sum_m [a_m, x_s] \wedge \bar{\operatorname{ad}} b_m\, X^{(s)}. \qquad (6)
\end{aligned}
$$

Finally, the following identity is easy to verify for any $a \in \mathfrak{u}^-$ and $b \in \mathfrak{u}$:

$$
\sum_{s=1}^{p} (-1)^s (\operatorname{ad} a\, x_s) \wedge \bar{\operatorname{ad}} b\, X^{(s)} = \sum_{s=1}^{p} (-1)^s (\bar{\operatorname{ad}} b\, x_s) \wedge \operatorname{ad} a\, X^{(s)}. \tag{7}
$$

Now the lemma follows by combining (3)–(7). □

3.4.6 Lemma. *With the notation as in Lemma 3.4.4,*

$$\sum_n \Big[\sum_{s=1}^p (-1)^s a_n \wedge X^{(s)} \otimes [x_s, b_n]v + (\operatorname{ad} a_n\, X) \otimes b_n v + (\bar{\operatorname{ad}}\, b_n\, X) \otimes a_n v\Big]$$

$$= -\sum_\gamma (\operatorname{ad} h_\gamma\, X) \otimes h_\gamma^\# v.$$

Proof.

$$\begin{aligned}
&\sum_n \sum_{s=1}^p (-1)^s a_n \wedge X^{(s)} \otimes [x_s, b_n]v \\
&\quad = \sum_n \sum_s \sum_m (-1)^s a_n \wedge X^{(s)} \otimes \langle [x_s, b_n], a_m\rangle\, b_m v \\
&\quad + \sum_n \sum_s \sum_m (-1)^s a_n \wedge X^{(s)} \otimes \langle [x_s, b_n], b_m\rangle\, a_m v \\
&\quad + \sum_n \sum_s \sum_\gamma (-1)^s a_n \wedge X^{(s)} \otimes \langle [x_s, b_n], h_\gamma\rangle\, h_\gamma^\# v \\
&\quad = \sum_m \sum_s (-1)^{s+1} [x_s, a_m] \wedge X^{(s)} \otimes b_m v \\
&\quad + \sum_m \sum_s (-1)^s (\bar{\operatorname{ad}}\, b_m\, x_s) \wedge X^{(s)} \otimes a_m v \\
&\quad + \sum_\gamma \sum_s (-1)^s (\operatorname{ad} h_\gamma\, x_s) \wedge X^{(s)} \otimes h_\gamma^\# v \\
&\quad = -\sum_n (\operatorname{ad} a_n\, X) \otimes b_n v - \sum_n (\bar{\operatorname{ad}}\, b_n\, X) \otimes a_n v - \sum_\gamma (\operatorname{ad} h_\gamma\, X) \otimes h_\gamma^\# v.
\end{aligned}$$

□

3.4.7 Lemma. *With the notation as in Lemma 3.4.4,*

$$\sum_n \sum_{s=1}^p (-1)^s a_n \wedge X^{(s)} \otimes (\bar{\operatorname{ad}}\, b_n\, x_s)v = \sum_n \bar{\operatorname{ad}}\, b_n\, X \otimes a_n v.$$

Proof.

$$\begin{aligned}
&\sum_n \sum_{s=1}^p (-1)^s a_n \wedge X^{(s)} \otimes (\bar{\operatorname{ad}}\, b_n\, x_s)v \\
&\quad = \sum_{n,m} \sum_s (-1)^s a_n \wedge X^{(s)} \otimes \langle [b_n, x_s], b_m\rangle\, a_m v \\
&\quad = \sum_m \sum_s (-1)^s \sum_n \langle [x_s, b_m], b_n\rangle\, a_n \wedge X^{(s)} \otimes a_m v \\
&\quad = \sum_m \sum_s (-1)^{s-1}\, (\bar{\operatorname{ad}}\, b_m\, x_s) \wedge X^{(s)} \otimes a_m v \\
&\quad = \sum_m (\bar{\operatorname{ad}}\, b_m\, X) \otimes a_m v.
\end{aligned}$$

This proves the lemma. □

3.4.8. Choose a basis (consisting of root vectors) $\{h_\phi^+\}_{\phi\in J}$ of $\mathfrak{g}_Y \cap \mathfrak{n}$, where J is an indexing set, and let $\{h_\phi^-\}_{\phi\in J}$ be the basis of $\mathfrak{g}_Y \cap \mathfrak{n}^-$, consisting again of root vectors, satisfying

$$(1) \qquad \langle h_\phi^+, h_{\phi'}^-\rangle = \delta_{\phi,\phi'}, \qquad \text{for } \phi, \phi' \in J.$$

Also take a basis $\{h_i\}$ of $\mathfrak{h}$. Then

$$(2) \qquad \{h_\phi^+, h_\phi^-, h_i;\ \phi \in J,\ 1 \le i \le \dim \mathfrak{h}\}$$

is a basis of $\mathfrak{g}_Y$ (consisting of $\mathfrak{h}$-weight vectors). Moreover, by (1),

$$(3) \qquad (h_\phi^+)^\# = h_\phi^-,\ (h_\phi^-)^\# = h_\phi^+.$$

3.4.9 Lemma. *For any $x \in \mathfrak{u}^-$,*

$$F_Y(x) = -\Omega^Y(x),$$

where Ω^Y is as defined in Definition 1.5.8.

In particular, F_Y does not depend on the particular choice of the basis $\{a_n\}$ of $\mathfrak{u}^-$.

Proof. By Remark 1.5.12, $\Omega^Y : \mathfrak{u}^- \to \mathfrak{u}^-$ is a $\mathfrak{g}_Y$-module map. Observe that on a given $x \in \mathfrak{u}^-$, all but finitely many of the h_ϕ^+ act by zero, and hence Ω^Y is well defined on $\mathfrak{u}^-$. Also, applying Lemmas 3.4.4 and 3.4.5 to the case when L is the trivial one-dimensional module, i.e., $L = L(0)$, and $x \in \mathfrak{u}^-$, we get $\Delta(x) = \frac{1}{2}F_Y(x)$. In particular, $F_Y : \mathfrak{u}^- \to \mathfrak{u}^-$ is a $\mathfrak{g}_Y$-module map as well. It can also be checked directly that F_Y is a $\mathfrak{g}_Y$-module map. But since $\mathfrak{u}^-$ is a direct sum of irreducible (highest weight) $\mathfrak{g}_Y$-modules of the form $L_Y(\theta)$ (cf. Subsection 3.4.1), it suffices to prove the lemma for $x \in \mathfrak{u}^-$ such that x is annihilated by $\mathfrak{g}_Y^+ := \mathfrak{g}_Y \cap \mathfrak{n}$.

For such an x, $F_Y(x) = F(x)$, where F is as defined in 2.3.10. So the lemma follows from Lemmas 2.3.11, 2.1.16 and Remark 1.5.12. □

3.4.10 *Proof of Theorem 3.4.2.* By Lemmas 3.4.4–3.4.7, we get, for $X = x_1 \wedge \cdots \wedge x_p \in \Lambda^p(\mathfrak{u}^-)$ and $v \in L$,

$$(1) \quad 2\Delta(X \otimes v) = \sum_{s=1}^{p}(-1)^{s-1} F_Y(x_s) \wedge X^{(s)} \otimes v - 2\sum_\gamma (\operatorname{ad} h_\gamma\, X) \otimes h_\gamma^\# v$$
$$+ \sum_\gamma \sum_{s=1}^{p} (-1)^s (\operatorname{ad} h_\gamma\, x_s) \wedge (\operatorname{ad} h_\gamma^\#\, X^{(s)}) \otimes v + 2X \otimes \sum_n a_n b_n v.$$

Taking for $\{h_\gamma\}$ the basis (3.4.8.2), the equation (1) reduces to the following:

$$
\begin{aligned}
2\Delta(X \otimes v) &= \sum_{s=1}^{p} (-1)^{s-1} F_Y(x_s) \wedge X^{(s)} \otimes v + 2X \otimes \sum_n a_n b_n v \\
&+ \sum_{\phi \in J} \sum_s (-1)^s \Big[(\mathrm{ad}\, h_\phi^+\, x_s) \wedge (\mathrm{ad}\, h_\phi^-\, X^{(s)}) \otimes v + (\mathrm{ad}\, h_\phi^-\, x_s) \wedge (\mathrm{ad}\, h_\phi^+\, X^{(s)}) \otimes v \Big] \\
&+ \sum_i \sum_s (-1)^s (\mathrm{ad}\, h_i\, x_s) \wedge (\mathrm{ad}\, h_i^\#\, X^{(s)}) \otimes v - 2 \sum_i (\mathrm{ad}\, h_i\, X) \otimes h_i^\# v \\
&- 2 \sum_{\phi \in J} \Big[(\mathrm{ad}\, h_\phi^+\, X) \otimes h_\phi^- v + (\mathrm{ad}\, h_\phi^-\, X) \otimes h_\phi^+ v \Big] \\
&= -2 \sum_\phi \sum_s (-1)^{s-1} (\mathrm{ad}\, h_\phi^-\, \mathrm{ad}\, h_\phi^+\, x_s) \wedge X^{(s)} \otimes v \\
&- 2 \sum_\phi \sum_s (-1)^{s-1} (\mathrm{ad}\, h_\phi^+\, x_s) \wedge (\mathrm{ad}\, h_\phi^-\, X^{(s)}) \otimes v \\
&- \sum_i \sum_s (-1)^{s-1} (\mathrm{ad}\, h_i\, x_s) \wedge (\mathrm{ad}\, h_i^\#\, X^{(s)}) \otimes v \\
&- \sum_i \sum_s (-1)^{s-1} (\mathrm{ad}\, h_i\, \mathrm{ad}\, h_i^\#\, x_s) \wedge X^{(s)} \otimes v \\
&- 2 \sum_s (-1)^{s-1} (\mathrm{ad}\, (\nu^{-1}\rho)\, x_s) \wedge X^{(s)} \otimes v + 2X \otimes \sum_n a_n b_n v \\
&-2 \sum_\phi \Big[(\mathrm{ad}\, h_\phi^+\, X) \otimes h_\phi^- v + (\mathrm{ad}\, h_\phi^-\, X) \otimes h_\phi^+ v \Big] - 2 \sum_i (\mathrm{ad}\, h_i\, X) \otimes h_i^\# v, \\
&\qquad \text{(by Lemma 3.4.9)} \\
&= -2 \sum_\phi h_\phi^- h_\phi^+ (X \otimes v) - \sum_i h_i h_i^\# (X \otimes v) + 2X \otimes \sum_\phi h_\phi^- h_\phi^+ v \\
&+ X \otimes \sum_i h_i h_i^\# v - 2(\nu^{-1}\rho)(X \otimes v) + 2X \otimes \sum_n a_n b_n v \\
&+ 2X \otimes (\nu^{-1}\rho) v \ \Big(\text{since} \sum_i (\mathrm{ad}\, h_i\, X) \otimes h_i^\# v = \sum_i (\mathrm{ad}\, h_i^\#\, X) \otimes h_i v\Big) \\
&= -\Omega^Y (X \otimes v) + X \otimes \Omega v \\
&\qquad \text{(by the definitions of } \Omega \text{ and } \Omega^Y \text{ as in Subsection 1.5.8).}
\end{aligned}
$$

Now the theorem follows from Lemma 2.1.16 and Remark 1.5.12. □

3.4.11 Corollary of Theorem 3.4.2. *Using the notation as in Theorem 3.4.2,*

$$\Delta_{p\,|A_p(\theta)} = 0 \text{ iff } \langle \lambda + \rho, \lambda + \rho \rangle = \langle \theta + \rho, \theta + \rho \rangle.$$

In particular (as $\mathfrak{g}_Y$-modules),

$$H_p(\mathfrak{u}^-, L(\lambda)) \simeq \oplus A_p(\theta), \tag{1}$$

where the summation runs over those $\theta \in D_Y$ such that $\langle \lambda + \rho, \lambda + \rho \rangle = \langle \theta + \rho, \theta + \rho \rangle$. □

From (1) and using the argument following (3.2.7.10) in the proof of Theorem 3.2.7 (assuming Corollary 3.2.10), we fully recover Theorem 3.2.7 for an arbitrary $Y \subset \{1, \ldots, \ell\}$.

3.4.E EXERCISES

(1) Let $\mathfrak{s}$ be a finite-dimensional Lie algebra equipped with any (not necessarily invariant) positive definite Hermitian form H. Then show that the adjoint $\partial^* : \Lambda(\mathfrak{s}) \to \Lambda(\mathfrak{s})$, of the differential ∂ as in (3.1.1.2) corresponding to the trivial $\mathfrak{s}$-module $V = \mathbb{C}$, is an antiderivation.

Hint: Using H, identify $\varphi : \Lambda^*(\mathfrak{s}) \simeq C^*(\mathfrak{s}) := \mathrm{Hom}_{\mathbb{C}}(\Lambda^*(\mathfrak{s}), \mathbb{C})$ under $\varphi(v)(w) = H(w, v)$. Show that $d \circ \varphi = \varphi \circ \partial^*$, where d is the differential of the cochain complex C^* as in (3.1.2.2). Show further that φ is a conjugate-linear ring homomorphism.

(2) Let $\mathfrak{k}$ be a finite-dimensional real Lie algebra with a $\mathfrak{k}$-invariant (positive definite) inner product. Such $\mathfrak{k}$ arise precisely as Lie K for compact Lie groups K (cf. [Milnor–63, Corollary 21.6]). This inner product induces an inner product on the cochain complex $C^* = \mathrm{Hom}_{\mathbb{C}}(\Lambda^*(\mathfrak{k}), \mathbb{C})$. Let d be the cochain differential (cf. 3.1.2), d^* its adjoint and $\Delta := dd^* + d^*d$ the Hodge Laplacian on C^*. Prove the following:

(a) $2\Delta = \Omega$, where Ω acting on C^* is the Casimir operator $-\sum_i e_i^2$, $\{e_i\}$ being an orthonormal basis of $\mathfrak{k}$ (cf. Exercise 1.5.E.3).

(b) Conclude thus that Ker $\Delta = (C^*)^{\mathfrak{k}}$.

(c) Use (b) to show that $H^*(\mathfrak{k}, \mathbb{C}) \simeq (C^*)^{\mathfrak{k}}$ (as algebras).

3.C Comments. The content of Section 3.1 is fairly standard. The basic reference for the homology and cohomology of a Lie algebra (but not for a pair) is [Cartan-Eilenberg–56]. The Lie algebra cohomology of a pair was introduced by [Chevalley–Eilenberg–48]. Relative Ext and Tor functors, for the pair (R, S) of a ring R and a subring S, were introduced by [Hochschild–56]. Proposition 3.1.5 is due to [Hochschild–56]. Proposition 3.1.10 and Corollary 3.1.12 are taken from [Garland–Lepowsky–76]. An introduction to Lie algebra homology and cohomology (including (3.1.13.1)) can also be found in [Lepowsky–79].

The main Theorem 3.2.7 of Section 3.2, in the case when Y is of finite type, is due to [Garland-Lepowsky–76], which generalizes to symmetrizable $\mathfrak{g}$ the corresponding celebrated result in the finite case due to [Kostant–61]. (A different proof of Kostant's theorem, in the finite case, can be found in [Wallach-88, §9.6.2].) The proof in the case of general Y is similar, and was done by [Liu–89]. The triviality of the radical for symmetrizable $\mathfrak{g}$, Corollary 3.2.10, was obtained by [Gabber–Kac–81]. However we have given a slightly different proof. The triviality of the radical for a general Kac–Moody Lie algebra $\mathfrak{g}$ is an important challenging problem. Proposition 3.2.11 is due to [Lepowsky–79], however his proof is different. The proof given here is due to [Kumar–84, Remark 3.3].

The decomposition of the category $\mathcal{O}$ in blocks given in Corollary 3.3.10(a_1) is due to [Deodhar–Gabber–Kac–82]. The decomposition of the full subcategory $\mathcal{O}^{\text{w.g.}}$ as in Corollary 3.3.10(a_2) is due to [Kumar–87b], which extends the corresponding decomposition of the subcategory $\mathcal{O}^{\text{g}} \subset \mathcal{O}^{\text{w.g.}}$ by [Deodhar–Gabber–Kac–82]. Lemma 3.3.5 is due to M. Duflo (unpublished). Theorem 3.3.8 on the vanishing of Ext is due to [Kumar–87b] and so is its Corollary 3.3.10(b). Corollary 3.3.10(b_3) is also due to Duflo (unpublished).

The main Theorem 3.4.2 of Section 3.4, giving an expression for the Hodge Laplacian acting on the chain complex of the Lie algebra $\mathfrak{u}_Y^-$ with coefficients in an integrable highest weight $\mathfrak{g}$-module L ($\mathfrak{g}$-symmetrizable), is due to [Kumar–84], which generalizes the corresponding well-known result in the finite case due to [Kostant–61]. A special case of Theorem 3.4.2 when $\mathfrak{g}$ is an affine Kac–Moody Lie algebra, $\mathfrak{u}_Y$ is the nil-radical of the standard maximal parabolic subalgebra and L is the trivial representation is due to [Garland–75].

IV

An Introduction to ind-Varieties and pro-Groups

Sections 4.1–4.3 are devoted to developing the basic definitions, examples and elementary properties of ind-varieties and ind-groups introduced by [Šafarevič–82], which is our basic reference.

More specifically, in Section 4.1, ind-varieties and morphisms between them are defined and various examples given. We also define the Zariski tangent space of an ind-variety.

An ind-algebraic group H (ind-group for short) is introduced in Section 4.2, and it is shown that its Zariski tangent space at the identity acquires a functorial Lie algebra structure denoted by Lie H. The notion of algebraic representations of an ind-group H is introduced, and it is shown that the derivative of an algebraic representation is a Lie algebra representation of Lie H. We introduce the notion of algebraic vector bundles on an ind-variety X; in particular, the notion of algebraic line bundles and Pic X, and also their equivariant analogue and cohomology. We give several examples of vector bundles.

An "algebraic" notion of smoothness of a point x in an ind-variety X is introduced in Section 4.3 and some examples are discussed. A "global" version of the Inverse Function Theorem for affine ind-varieties is obtained (cf. Proposition 4.3.6). It is shown that an ind-group is always algebraically smooth if the base field is of characteristic 0.

Section 4.4 develops the basic theory of pro-algebraic groups (following [Serre–60]) and pro-Lie algebras over an algebraically closed field of characteristic 0. We begin by defining pro-algebraic groups (pro-groups for short), the pro-topology on them, pro-subgroups and pro-group morphisms and we give some examples. A pro-subgroup acquires a natural pro-group structure and, moreover, the quotient of a pro-group by a normal pro-subgroup is given a canonical pro-group structure. It is shown that the image of a pro-group morphism is a pro-subgroup. In particular, a bijective pro-group morphism is an isomorphism.

A parallel theory of pro-Lie algebras is developed. In particular, we define the pro-topology on a pro-Lie algebra, and also define pro-Lie subalgebras, pro-Lie ideals, and pro-Lie algebra homomorphisms. It is shown that a pro-Lie subalgebra has a natural pro-Lie algebra structure and, moreover, for a pro-Lie ideal $\mathfrak{t}$ of $\mathfrak{s}$, $\mathfrak{s}/\mathfrak{t}$ acquires a natural pro-Lie algebra structure. Also, the image of a pro-Lie algebra homomorphism is a pro-Lie subalgebra. To any pro-group G, we associate a functorial pro-Lie algebra Lie G and the exponential map $\mathrm{Exp} : \mathrm{Lie}\, G \to G$. Further, it is shown that the category of pro-unipotent pro-groups is equivalent to the category of pro-nilpotent pro-Lie algebras, under the correspondence $G \rightsquigarrow \mathrm{Lie}\, G$. Finally, the notion of a pro-module of a pro-group (or a pro-Lie algebra) is introduced and the adjoint representation of G in Lie G is constructed. It is shown that the derivative of a pro-module π of a pro-group G, which is a pro-module of Lie G, determines π, provided G is connected. Furthermore, it is shown that, for a pro-unipotent pro-group G, a locally nilpotent pro-module of Lie G "integrates" to a pro-module of G.

In Sections 4.1–4.3, unless otherwise explicitly stated, we take k to be any algebraically closed field of an arbitrary characteristic. By a variety we mean a quasi-projective variety over k as in Appendix A. In particular, we do not put the irreducibility assumption on a variety.

4.1. Ind-Varieties: Basic Definitions

4.1.1 Definition. By an *ind-variety* over k we mean a set X together with a filtration

$$X_0 \subseteq X_1 \subseteq X_2 \subseteq \dots,$$

such that

(1) $\bigcup_{n\geq 0} X_n = X$, and

(2) each X_n is a finite-dimensional variety over k such that the inclusion $X_n \hookrightarrow X_{n+1}$ is a closed embedding.

An ind-variety X is said to be *projective*, resp. *affine*, if each X_n is projective, resp. affine.

For an ind-variety X, we define its *ring of regular functions* $k[X]$ by $k[X] = \mathrm{Inv.\,lt.}_n\, k[X_n]$, where $k[X_n]$ is the ring of regular functions on X_n. Putting the discrete topology on each $k[X_n]$ and taking the inverse limit topology on $k[X]$, i.e., the subspace topology induced from the product topology on $\prod_{n\geq 0} k[X_n]$, we obtain $k[X]$ as a topological k-algebra (with the discrete topology on k).

We define the *Zariski topology* on an ind-variety X by declaring a set $U \subseteq X$ open if and only if $U \cap X_n$ is Zariski-open in X_n for each n. It is easy to see that for an ind-variety X, a subset $Z \subseteq X$ is closed if and only if $Z \cap X_n$ is closed in X_n for each n.

Let X and Y be two ind-varieties with filtrations X_n and Y_n, respectively. A map $f : X \to Y$ is said to be a *morphism* if, for every $n \geq 0$, there exists a number $m(n) \geq 0$ such that $f(X_n) \subseteq Y_{m(n)}$ and, moreover, $f_{|X_n} : X_n \to Y_{m(n)}$ is a morphism. Clearly, a morphism $f : X \to Y$ is continuous and induces a continuous k-algebra homomorphism $f^* : k[Y] \to k[X]$. For morphisms $f : X \to Y$ and $g : Y \to Z$ between ind-varieties, their composite $g \circ f : X \to Z$ is a morphism. Observe that $k[X]$ is canonically isomorphic with the algebra of all the morphisms $X \to k$.

A morphism $f : X \to Y$ is said to be an *isomorphism* if f is bijective and $f^{-1} : Y \to X$ also is a morphism. Two ind-variety structures on the same set X are said to be *equivalent* if the identity map $I : X \to X$ is an isomorphism of ind-varieties. We shall not distinguish between two equivalent ind-variety structures on X.

It is easy to see that a morphism $f : X \to Y$ (where X and Y are affine ind-varieties) is an isomorphism iff the induced map $f^* : k[Y] \to k[X]$ is an isomorphism of topological k-algebras.

A map $f : X \to Y$ is called a *closed embedding*, also called a *closed immersion*, if, for every $n \geq 0$, there exists $m(n)$ such that $f(X_n) \subseteq Y_{m(n)}$ and $f_{|X_n} : X_n \to Y_{m(n)}$ is a closed embedding, $f(X)$ is closed in Y and, moreover, $f : X \to f(X)$ is a homeomorphism under the subspace topology on $f(X)$.

An open subset U of an ind-variety X is clearly an ind-variety itself, where we define the filtration U_n of U by $U_n := U \cap X_n$ and equip U_n with the (open) subvariety structure from X_n. With this structure U is called an *open ind-subvariety of* X.

Similarly, a closed subset Y of X acquires a canonical structure of an ind-variety defined by taking the filtration $Y_n := Y \cap X_n$ of Y and putting the unique closed (reduced) subvariety structure on Y_n from X_n. With this structure, Y is called a *closed ind-subvariety of* X. The inclusion $Y \hookrightarrow X$ is clearly a closed embedding, and, in fact, it is the unique ind-variety structure on Y making $Y \hookrightarrow X$ a closed embedding (use Lemma 4.1.2 given below). Conversely, for any closed embedding $f : Y \to X$, the ind-variety Y is isomorphic with the closed ind-subvariety $f(Y)$ of X (use Lemma 4.1.2 again). Thus, for a closed subset Y and an open subset U of X, $Y \cap U$ is canonically an ind-variety. With this ind-variety structure, we call $Y \cap U$ a *locally-closed ind-subvariety of* X.

An ind-variety X is called *irreducible* if the underlying topological space is irreducible, i.e., X is not the union of two proper closed subsets. Similarly, X is called *connected* if the underlying topological space is connected.

For an ind-variety X, define the presheaf of topological k-algebras on X by assigning the topological k-algebra $k[U]$ to any open subset $U \subset X$. It is easy to see that this presheaf is a sheaf, called the *structure sheaf* of X and is denoted by $\mathcal{O}_X$.

4.1.2 Lemma. *Let $g : Z \to X$ be a continuous map between ind-varieties. Then, for any $n \geq 0$, there exists $m(n) \geq 0$ such that*

$$g(Z_n) \subset X_{m(n)}. \tag{1}$$

As a consequence, for a closed embedding $f : X \to Y$, a map $g : Z \to X$ is a morphism, resp. closed embedding, iff $f \circ g : Z \to Y$ is a morphism, resp. closed embedding.

Proof. Fix $n \geq 0$ and assume that (if possible) there does not exist any m such that $g(Z_n) \subset X_m$. Then, there exists an infinite sequence of points $x_{m_i} \in g(Z_n) \cap (X_{m_i} \setminus X_{m_i-1})$ such that $1 \leq m_1 < m_2 < \cdots$. For any subset T of $S := \{x_{m_i}\}_{i\geq 1}$, since $T \cap X_m$ is finite for all m, T is closed in X and hence in S under the subspace topology on S. Thus the subspace topology on S is discrete. The map g being continuous by assumption, $g_n^{-1}(S)$ is closed in Z_n, where $g_n := g_{|Z_n}$; in particular, $g_n^{-1}(S)$ is a closed subvariety of Z_n and, moreover, $g_n^{-1}(x_{m_i})$ is open and nonempty in $g_n^{-1}(S)$, for all i. Hence $g_n^{-1}(S)$ has infinitely many connected components, which is a contradiction since $g_n^{-1}(S)$ is a variety. This proves (1).

The second statement follows from (1) and the corresponding result for closed embedding between (finite-dimensional) varieties. □

4.1.3 Examples. (1) Any (finite-dimensional) variety X is of course canonically an ind-variety, where we take each $X_n = X$.

An ind-variety X is isomorphic to a (finite-dimensional) variety Y, regarded as an ind-variety, iff $X_n = X$ for some n.

(2) If X and Y are ind-varieties, then $X \times Y$ is canonically an ind-variety, where we define the filtration by

$$(X \times Y)_n = X_n \times Y_n,$$

and we put the product variety structure on $X_n \times Y_n$ (cf. [Šafarevič – 94, Chapter I, Section 5]).

(3) $\mathbb{A}^\infty := \{(a_1, a_2, a_3, \cdots)$: each $a_i \in k$ and all but finitely many a_i's are zero$\}$ is an ind-variety under the filtration : $\mathbb{A}^1 \subset \mathbb{A}^2 \subset \mathbb{A}^3 \subset \cdots$, where $\mathbb{A}^n \subset \mathbb{A}^\infty$ is the set of all the sequences with $a_{n+1} = a_{n+2} = \cdots = 0$, which of course is the n-dimensional affine space.

(4) Any vector space V of countable dimension over k is canonically an affine ind-variety: Take a basis $\{e_i\}_{i\geq 1}$ of V. This gives rise to a k-linear isomorphism $\mathbb{A}^\infty \tilde{\to} V$, taking $(a_1, a_2, a_3, \cdots) \mapsto \sum a_i e_i$. By transporting the ind-variety structure from $\mathbb{A}^\infty$ via this isomorphism, we get an (affine) ind-variety structure

on V. It is easy to see that a different choice of basis of V gives an equivalent ind-variety structure on V. Similarly, the space $\mathbb{P}(V)$ of lines in V is canonically a projective ind-variety.

(5) Take any countable infinite set $S = \{x_0, x_1, \ldots\}$. Then S is an ind-variety under the filtration $S_n := \{x_0, x_1, \ldots, x_n\}$ with the only variety structure on S_n.

(6) Let V be the k-vector space $\mathbb{A}^\infty$ and let V_n be the subspace $\mathbb{A}^n$ for any $n \geq 1$. The inclusion $V_n \subset V_{n+1}$ induces a surjective k-algebra homomorphism $S(V_n^*) \leftarrow S(V_{n+1}^*)$ of the symmetric algebras. Let us consider the ring

$$\hat{S}(V^*) := \underset{n}{\text{Inv. lt.}}\, S(V_n^*).$$

(We view $\hat{S}(V^*)$ as a certain completion of $S(V^*)$.) Then, we have a canonical (surjective) ring homomorphism $\pi_n : \hat{S}(V^*) \to S(V_n^*)$, for any $n \geq 1$. Observe that $\hat{S}(V^*)$ is the ring $k[V]$ of regular functions on V.

For any set $P = \{P_\alpha\} \subset \hat{S}(V^*)$, let $Z(P) := \{a \in V : P_\alpha(a) = 0, \text{ for all } \alpha\}$ be the corresponding zero set. Then $Z(P) \subset \mathbb{A}^\infty$ is a closed subset, and thus acquires the structure of a closed ind-subvariety of $\mathbb{A}^\infty$. Hence $Z(P)$ is an affine ind-variety.

Similarly, for any $d \geq 0$, define

$$\hat{S}^d(V^*) = \underset{n}{\text{Inv. lt.}}\, S^d(V_n^*).$$

Then $\hat{S}_{\text{hom}}(V^*) := \bigoplus_{d \geq 0} \hat{S}^d(V^*)$ is a subring of $\hat{S}(V^*)$.

For any set $H = \{H_\alpha\}$, $H_\alpha \in \hat{S}^{d_\alpha}(V^*)$ (for some $d_\alpha \geq 0$), let

$$Z(H) := \{[a] : a \in V \text{ and } H_\alpha(a) = 0, \text{ for all } \alpha\},$$

where $[a]$ denotes the line in V passing through a. Then $Z(H) \subset \mathbb{P}^\infty$ is a closed subset (cf. Example 4), and thus acquires the structure of a closed ind-subvariety of $\mathbb{P}^\infty$.

4.1.4 Definition. Let X be an ind-variety with filtration (X_n). For any $x \in X$, define the *Zariski tangent space* $T_x(X)$ of X at x by

$$T_x(X) = \lim_{n\to\infty}\text{it}\; T_x(X_n),$$

where $T_x(X_n)$ is the Zariski tangent space of X_n at x (cf. Appendix A.3). Observe that $x \in X_n$ for all large enough n.

A morphism $f : X \to Y$ clearly induces a linear map $(df)_x : T_x(X) \to T_{f(x)}(Y)$, for any $x \in X$, called the *derivative* of f at x. Moreover, it satisfies the chain rule: $(d(g \circ f))_x = (dg)_{f(x)} \circ (df)_x$, for a morphism $g : Y \to Z$. Thus, an isomorphism $f : X \to Y$ of ind-varieties induces an isomorphism $(df)_x : T_x(X) \xrightarrow{\sim} T_{f(x)}(Y)$, for any $x \in X$. In particular, two equivalent ind-variety structures on X give rise to isomorphic Zariski tangent spaces $T_x(X)$.

4.1.E EXERCISE. Give an example of a closed ind-subvariety X of $V = \mathbb{A}^\infty$, which is *not* of the form $Z(P)$ for any $P \subset \hat{S}(V^*)$ (cf. Example 4.1.3 (6)).

4.2. Ind-Groups and their Lie Algebras

4.2.1 Definition. An ind-variety H is said to be an *ind-algebraic group* (for short an *ind-group*) if the underlying set H is a group such that the map $H \times H \to H$, taking $(x, y) \mapsto xy^{-1}$, is a morphism. A closed subgroup K of H, i.e., K is a subgroup of H and is a closed subset, is again an ind-group under the closed ind-subvariety structure on K (use Lemma 4.1.2).

We only have occasion to consider affine ind-groups, i.e., ind-algebraic groups H such that H is an affine ind-variety. *So this will be our tacit assumption on ind-groups.*

By a *group morphism* between two ind-groups H and K, we mean a group homomorphism $f : H \to K$ such that f is also a morphism of ind-varieties.

An abstract representation of an ind-group H in a countable dimensional k-vector space V is said to be *algebraic* if the map $H \times V \to V$, defined by $(h, v) \mapsto h.v$, is a morphism, where V is regarded as an ind-variety via Example 4.1.3(4).

For an ind-group H and ind-variety Y, we say that Y is an *H-ind-variety* if the group H acts on Y such that the action $H \times Y \to Y$ is a morphism of ind-varieties.

4.2.2 Proposition. *For an ind-group H, the Zariski tangent space $T_e(H)$ at the identity element e is endowed with a natural Lie algebra structure, described in the proof. We denote this Lie algebra by Lie H.*

Moreover, if $\alpha : H \to K$ is a group morphism between two ind-groups, then the derivative $(d\alpha)_e$: Lie $H \to$ Lie K is a Lie algebra homomorphism.

Proof. Denote $k[H]$ by A. The multiplication map $\mu = \mu_H : H \times H \to H$, taking $(h_1, h_2) \mapsto h_1 h_2$, induces a continuous homomorphism $\mu^* : A \to k[H \times H]$. There is a canonical inclusion $A \otimes_k A \hookrightarrow k[H \times H]$, and it is easy to see that the image is dense in $k[H \times H]$. So we denote $k[H \times H]$ by $A \hat{\otimes} A$, and view it as a certain completion of $A \otimes A$. Let $\epsilon : A \to k$ be the homomorphism, taking $f \mapsto f(e)$. Let $\mathfrak{m} = \operatorname{Ker} \epsilon$. Then for any $f \in \mathfrak{m}$

$$\mu^* f - f \otimes 1 - 1 \otimes f \in \mathfrak{m} \hat{\otimes} \mathfrak{m}, \tag{1}$$

where $\mathfrak{m} \hat{\otimes} \mathfrak{m}$ denotes the closure of $\mathfrak{m} \otimes \mathfrak{m}$ in $A \hat{\otimes} A$.

A continuous derivation $D : A \to A$ is said to be *invariant* if $L_h^* \circ D = D \circ L_h^*$, for all $h \in H$, where $L_h^* : A \to A$ is the algebra homomorphism induced from the left translation map $L_h : H \to H$ taking $g \mapsto hg$. The set $\mathrm{Der}_H A$ of continuous invariant derivations of A is a Lie algebra under

$$[D_1, D_2] := D_1 \circ D_2 - D_2 \circ D_1 , \qquad D_1, D_2 \in \mathrm{Der}_H\, A.$$

Define the map $\eta : T_e(H) \to \mathrm{Der}_H\, A$ as follows. Take $v \in T_e(H)$. Then $v \in T_e(H_n)$ for some n, where H_n is the filtration of H. By definition, $T_e(H_n) = \mathrm{Hom}_k(\mathfrak{m}_n/\mathfrak{m}_n^2, k)$, where $\mathfrak{m}_n := \{f \in k[H_n] : f(e) = 0\}$ is the maximal ideal of $k[H_n]$ corresponding to the point e. In particular, v gives rise to a k-linear map $\hat{v} : \mathfrak{m}_n \to k$. Let $\bar{v} : A \to k$ be the continuous linear map defined by $\bar{v}(1) = 0$, and $\bar{v}_{|\mathfrak{m}} = \hat{v} \circ \pi_n$, where $\pi_n : \mathfrak{m} \to \mathfrak{m}_n$ is the canonical restriction map.

Now the map $\eta(v) : A \to A$ is defined by

$$\eta(v) = (I \hat{\otimes} \bar{v}) \circ \mu^*,$$

where $I : A \to A$ is the identity map and $I \hat{\otimes} \bar{v} : A \hat{\otimes} A \to A \hat{\otimes} k = A$ is the continuous extension of the map $I \otimes \bar{v}$. By using (1), we get that $\eta(v)$ is a derivation. Further, it can be seen that $\eta(v)$ is invariant and hence $\eta(v) \in \mathrm{Der}_H\, A$.

Conversely, we define a map $\xi : \mathrm{Der}_H A \to T_e(H)$ as follows. Take $D \in \mathrm{Der}_H A$ and consider $\epsilon \circ D_{|\mathfrak{m}} : \mathfrak{m} \to k$. Since D is continuous, there exists some n such that $\epsilon \circ D_{|\mathfrak{m}}$ factors through $\mathfrak{m}_n$, giving rise to the map (denoted) $\beta_D : \mathfrak{m}_n \to k$. Since D is a derivation, $\beta_D(\mathfrak{m}_n^2) = 0$ and hence β_D gives rise to an element $\hat{\beta}_D \in T_e(H_n)$. Now set $\xi(D) = \hat{\beta}_D$.

It can easily be seen that $\xi \circ \eta$ and $\eta \circ \xi$ are both the identity maps; in particular, η and ξ are isomorphisms. We now transport the Lie algebra structure from $\mathrm{Der}_H A$ to $T_e(H)$ (via η).

Finally, we prove that, for any group morphism $\alpha : H \to K$, the derivative $\dot{\alpha} = (d\alpha)_e : \mathrm{Lie}\, H \to \mathrm{Lie}\, K$ is a Lie algebra homomorphism.

To prove this, it suffices to show that the following diagram is commutative for any $v \in \mathrm{Lie}\, H$.

(*)
$$\begin{array}{ccc} k[K] & \xrightarrow{\alpha^*} & k[H] \\ {\scriptstyle \eta(\dot{\alpha} v)}\downarrow & & \downarrow{\scriptstyle \eta(v)} \\ k[K] & \xrightarrow{\alpha^*} & k[H]. \end{array}$$

Take $f \in \mathfrak{m}_K$, where $\mathfrak{m}_K \subset k[K]$ is the maximal ideal corresponding to the point e. Then, by the definition of the map η,

(2) $$\eta(v)(\alpha^* f) = (I \hat{\otimes} \bar{v}) \mu_H^*(\alpha^* f), \quad \text{and}$$

(3) $$\eta(\dot{\alpha} v) f = (I \hat{\otimes} (\overline{\dot{\alpha} v})) \mu_K^*(f).$$

Further,

$$\alpha^*\eta(\dot{\alpha v})f = (\alpha^*\hat{\otimes}(\overline{\dot{\alpha v}}))\mu_K^*(f), \quad \text{whereas} \tag{4}$$

$$\overline{\dot{\alpha v}} = \bar{v} \circ \alpha^*, \quad \text{and} \tag{5}$$

$$(\alpha^*\hat{\otimes}\alpha^*) \circ \mu_K^* = \mu_H^* \circ \alpha^*. \tag{6}$$

Now combining (2)–(6), we get the commutativity of the diagram (∗). This proves the proposition. □

4.2.3 Definition. Let $\theta : H \times V \to V$ be an algebraic representation of an ind-group H in a (countable-dimensional) vector space V. This gives rise to the map $d\theta$: Lie $H \times V \to V$, called the *derivative* of θ, defined as follows:

Fix $v \in V$ and define the map $\theta_v : H \to V$ by $h \mapsto hv$. Consider the derivative $(d\theta_v)_e : T_e(H) = \text{Lie } H \to T_v(V) \approx V$ (cf. Exercise 4.2.E.4). Then the map $d\theta$: Lie $H \times V \to V$ is defined as $(x, v) \mapsto (d\theta_v)_e(x)$.

4.2.4 Lemma. *The map $d\theta$ is a representation of the Lie algebra Lie H (in V).*

Proof. We abbreviate $d\theta(x, v)$ by $x \cdot v$. For any $v \in V$, define the evaluation map $e(v) : k[V] \to k$ by $e(v)f = f(v)$, for $f \in k[V]$. Fix any $v_o \in V$. Then $v \in T_{v_o}(V) \approx V$ induces a k-linear map $\bar{v} : k[V] \to k$, such that $\bar{v}(1) = 0$ (cf. the proof of Proposition 4.2.2). Observe that $\bar{v}$ depends upon v as well as the choice of the base point v_o, so the notation $\bar{v}$ is to be understood with a fixed base point in mind. However, by (1), $\bar{v}_{|V^*}$ does not depend upon the base point. If $v, w \in T_{v_o}(V)$ are such that $\bar{v}_{|V^*} = \bar{w}_{|V^*}$, then $\bar{v} = \bar{w}$, where $V^* \subset k[V]$ denotes the full vector space dual of V. Moreover, it is easy to see that

$$\bar{v}_{|V^*} = e(v)_{|V^*} \,. \tag{1}$$

By definition (for any $v_o \in V$ and $x \in T_e(H)$), considering $x \cdot v_o \in T_{v_o}(V)$,

$$\overline{x \cdot v_o} = (\bar{x}\hat{\otimes}e(v_o)) \circ \theta^*, \tag{2}$$

where $\theta : H \times V \to V$ is the representation. Since θ is linear in the V-variable,

$$\theta^*(V^*) \subset k[H]\hat{\otimes}V^*, \tag{3}$$

where $k[H]\hat{\otimes}V^*$ is the closure of $k[H] \otimes V^*$ in $k[H]\hat{\otimes}k[V] := k[H \times V]$.

Consider the following commutative diagram, for any $x, y \in T_e(H)$ and $v \in V$, where $A = k[H]$:

$$
\begin{array}{ccc}
 & k[V] & \\
 & \Big\downarrow \theta^* & \\
 & A\hat{\otimes} k[V] & \\
\mu^*\hat{\otimes} I \swarrow & & \searrow I\hat{\otimes}\theta^* \\
(A\hat{\otimes} A)\hat{\otimes} k[V] & \approx & A\hat{\otimes}(A\hat{\otimes} k[V]) \\
(\bar{x}\hat{\otimes}\bar{y})\hat{\otimes} e(v) \searrow & & \swarrow \bar{x}\hat{\otimes}(\bar{y}\hat{\otimes} e(v)) \\
 & k\otimes k\otimes k = k. &
\end{array}
$$

The commutativity of the above diagram and (1)–(3) give the following for any $x, y \in T_e(H)$ and $v \in V$:

(4) $$e(x \cdot (y \cdot v))_{|V^*} = (((\bar{x}\hat{\otimes}\bar{y})\hat{\otimes} e(v)) \circ (\mu^*\hat{\otimes} I) \circ \theta^*)_{|V^*}.$$

By (4) we get

(5)
$$(e(x \cdot (y \cdot v)) - e(y \cdot (x \cdot v)))_{|V^*} = (((\bar{x}\hat{\otimes}\bar{y} - \bar{y}\hat{\otimes}\bar{x})\hat{\otimes} e(v)) \circ (\mu^*\hat{\otimes} I) \circ \theta^*)_{|V^*}.$$

But, as can be easily seen from the definition of the bracket in $T_e(H)$ (cf. proof of Proposition 4.2.2),

(6) $$(\bar{x}\hat{\otimes}\bar{y} - \bar{y}\hat{\otimes}\bar{x}) \circ \mu^* = \overline{[x, y]}.$$

Thus, by (5) and (6),

$$
\begin{aligned}
e(x \cdot (y \cdot v) - y \cdot (x \cdot v))_{|V^*} &= ((\overline{[x, y]}\hat{\otimes} e(v)) \circ \theta^*)_{|V^*}, \\
&= (\overline{[x, y] \cdot v})_{|V^*}, \text{ by (2)} \\
&= e([x, y] \cdot v)_{|V^*}, \text{ by (1)}.
\end{aligned}
$$

This gives that $x \cdot (y \cdot v) - y \cdot (x \cdot v) = [x, y] \cdot v$, proving the lemma. □

4.2.5 Lemma. *Let H be an ind-group. Then H is connected if and only if H is irreducible.*

Proof. Of course, if H is irreducible, then H is connected. So we need only show that H is irreducible if it is connected.

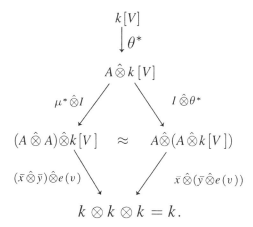

If possible, write $H = H' \cup H''$, where both H' and H'' are proper closed subsets. In particular, there exists a $g \in H' \backslash H''$. Translating by g^{-1}, we can assume that $e \in H' \backslash H''$. Let C_n be the union of all the irreducible components of H_n passing through e and let D_n be the union of all those irreducible components of H_n which do not pass through e. Hence

$$(1) \qquad C_n \subset H', \qquad \text{for all } n.$$

Moreover $C := \bigcup_n C_n$ is a subgroup of H: Let C_n^i, resp. C_m^j, be an irreducible component of C_n, resp. C_m. Choose N large enough such that $H_n \cdot H_m^{-1} \subset H_N$, where $H_n \cdot H_m^{-1}$ is, by definition, the image of $H_n \times H_m$ under $(g, h) \mapsto gh^{-1}$. Hence $C_n^i (C_m^j)^{-1}$ is an irreducible, but not necessarily closed, subset of H_N containing e, so $C_n^i (C_m^j)^{-1} \subset C_N$.

We further claim that C is an open subset of H: Since C is a subgroup, it suffices to show that C contains a neighborhood of e in H. To prove this, we first show that $D := \bigcup D_n$ is closed in H: For any fixed m, $D \cap H_m = \bigcup_n (D_n \cap H_m)$. For $n \geq m$, since $H_m = (C_n \cap H_m) \cup (D_n \cap H_m)$, $D_n \cap H_m$ is the union of certain irreducible components of H_m. But since $e \notin D_n \cap H_m$, $D_n \cap H_m \subset D_m$. This gives that $D \cap H_m = \bigcup_{n \leq m} D_n \cap H_m$. But $D_n \cap H_m$ is a closed subset of H_m for any n and hence $D \cap H_m$ is closed in H_m. Thus D is closed in H. Now clearly

$$e \in H \backslash D \subset C.$$

This concludes that C is open in H and hence closed, since C is a subgroup. But H being, by assumption, connected, we get $C = H$, thereby $H' = H$ by (1). But this is a contradiction since H' was chosen to be proper. This proves the lemma. □

4.2.6 Definition. For an ind-variety Y, by an *algebraic vector bundle of rank* r over Y, we mean an ind-variety E together with a morphism $\theta : E \to Y$ such that, for any n, $E_n \to Y_n$ is an algebraic vector bundle of rank r over the (finite-dimensional) variety Y_n, where $\{Y_n\}$ is the filtration of Y giving the ind-variety structure and $E_n := \theta^{-1}(Y_n)$ is equipped with the closed ind-subvariety (of E) structure. In particular, each ind-variety E_n, in fact, is a variety. If we equip E with a different ind-variety structure by taking the filtration: $E_0 \subseteq E_1 \subseteq \cdots$, then the ind-variety structure on E is equivalent to the original ind-variety structure, as is easy to see. If $r = 1$, we call E an *algebraic line bundle* over Y.

Let E and F be two algebraic vector bundles over Y. Then a morphism (of ind-varieties) $\varphi : E \to F$ is called a *bundle morphism* if the following diagram is commutative:

$$\begin{array}{ccc} E & \xrightarrow{\varphi} & F \\ & \searrow \quad \swarrow & \\ & Y & \end{array}$$

and, moreover, $\varphi_y : E(y) \to F(y)$ is a k-linear map for all $y \in Y$, where $\varphi_y := \varphi_{|E(y)}$ and $E(y)$ is the fiber over y. In particular, we have the notion of isomorphism, denoted by $\simeq$, of vector bundles over Y.

We define the *Picard group* Pic Y as the set of isomorphism classes of algebraic line bundles on Y. It is clearly an abelian group under the tensor product of line bundles (cf. Example 4.2.7(e) and Exercises 4.2.E.1 and 4.2.E.2 below).

We similarly define the notion of *principal H-bundles* on an ind-variety, for a finite-dimensional algebraic group H (cf. Definition A.41 for the notion of principal H-bundles on a variety as base).

For an ind-group H and H-ind-variety Y (cf. Definition 4.2.1), an algebraic vector bundle $\theta : E \to Y$ is said to be an *H-equivariant vector bundle* if the ind-variety E also is an H-ind-variety, such that the following diagram is commutative:

$$\begin{array}{ccc} H \times E & \longrightarrow & E \\ \big\downarrow {\scriptstyle I\times\theta} & & \big\downarrow {\scriptstyle \theta} \\ H \times Y & \longrightarrow & Y\,, \end{array}$$

and, moreover, for any $y \in Y$ and $h \in H$ the h-action $h : E(y) \to E(hy)$ is linear.

For an algebraic vector bundle $\theta : E \to Y$ over an ind-variety Y, we define

$$H^p(Y, E) := \operatorname*{Inv.\,lt.}_n H^p(Y_n, E_{|Y_n}). \tag{1}$$

It is easy to see that $H^0(Y, E)$ can canonically be identified with the space of all algebraic sections σ of the bundle E (i.e., the space of all the morphisms $\sigma : Y \to E$ such that $\theta \circ \sigma = I$).

If θ is an H-equivariant vector bundle for an ind-group H, then $H^p(Y, E)$ is canonically a module for the (abstract) group H.

More generally, let $\pi : E \to Y$ be an algebraic vector bundle with an ind-variety Y as base, and let an abstract group H (we do not assume that H is an ind-group) act on E as well as Y such that π is H-equivariant. Assume further that

(a) for all $h \in H$, the action of h on E and Y is algebraic, and

(b) for all $y \in Y$ and $h \in H$, the h-action $E(y) \to E(h \cdot y)$ is linear.

Then the cohomology $H^p(Y, E)$ has a natural structure of an H-module. Explicitly, the action of h on $H^0(Y, E)$ is given by

(2) $$(h \cdot \sigma)y = h \cdot \sigma(h^{-1}y), \quad \text{for } y \in Y \text{ and } \sigma \in H^0(Y, E).$$

4.2.7 Examples. (a) For any $n \geq 0$, and an ind-variety Y, consider the product ind-variety $Y \times k^n$ together with the projection $\theta : Y \times k^n \to Y$. The fibers of θ have canonical vector space structure, making θ an algebraic vector bundle ε_Y^n of rank n on Y. This is called the *trivial rank-n algebraic vector bundle on Y*.

(b) Let $\theta : E \to Y$ be an algebraic vector bundle and $f : Z \to Y$ a morphism of ind-varieties. Then we define the pullback algebraic vector bundle $f^*(E)$ on Z as follows: As a set

$$f^*(E) = \{(z, e) \in Z \times E : f(z) = \theta(e)\}.$$

Then $f^*(E)$ is a closed subset of the product ind-variety $Z \times E$. In particular, $f^*(E)$ is an ind-variety such that $f^*(E) \hookrightarrow Z \times E$ is a closed embedding (cf. 4.1.1). Define the projection $\theta_Z : f^*(E) \to Z$ by projecting on the first factor. Then, clearly θ_Z is a morphism and, moreover, each fiber of θ_Z inherits a vector space structure from that of E. It is easy to see that with these structures $\theta_Z : f^*(E) \to Z$ is an algebraic vector bundle called the *pullback of E via f*.

If Z' is another ind-variety with a morphism $g : Z' \to Z$, then

(1) $$g^*(f^*(E)) \simeq (f \circ g)^*(E).$$

(c) For a vector space V of countable dimension over k, recall the definition of the projective ind-variety $\mathbb{P}(V)$ from 4.1.3(4).

Let $\mathcal{L}_V := \{(x, v) \in \mathbb{P}(V) \times V : v \in x\} \subset \mathbb{P}(V) \times V$. Then $\mathcal{L}_V$ is a closed subset of the product ind-variety $\mathbb{P}(V) \times V$ and hence consider $\mathcal{L}_V$ as a closed ind-subvariety of $\mathbb{P}(V) \times V$. Now $\theta : \mathcal{L}_V \to \mathbb{P}(V)$, given by the projection on the first factor, is clearly a morphism, and the fibers of θ have a canonical 1-dimensional vector space structure. It is easy to see that, with these structures, $\theta : \mathcal{L}_V \to \mathbb{P}(V)$ is an algebraic line bundle, called the *tautological line bundle* on $\mathbb{P}(V)$.

(d) Let $\theta : E \to Z$ and $\theta' : E' \to Z'$ be two algebraic vector bundles of ranks n and n', respectively. Then, consider the product map $\theta \times \theta' : E \times E' \to Z \times Z'$ of the product ind-varieties. Equip the fibers of $\theta \times \theta'$ with the direct sum vector space structure. Then it is an algebraic vector bundle of rank $n + n'$, called the *Cartesian product* of θ and θ'.

If $Z' = Z$, consider the diagonal morphism $\Delta : Z \to Z \times Z$, $\Delta(z) = (z, z)$. Then the pullback bundle $\Delta^*(E \times E') \to Z$ is called the *Whitney sum* (also called the *direct sum*) of the bundles E and E', and is denoted $E \oplus E'$.

(e) Let $E \xrightarrow{\theta} Z$, $E' \xrightarrow{\theta'} Z$ be two algebraic vector bundles on the same ind-variety Z (with filtration Z_n) as base. Then, by definition, $E_n \to Z_n$ and $E'_n \to Z_n$ are algebraic vector bundles, where $E_n := \theta^{-1}(Z_n)$ and similarly for E'_n. Consider the tensor product (algebraic) vector bundle $E_n \otimes E'_n \to Z_n$ (cf. [Šafarevič–94, Chapter VI, Section 1.3]). Then, there is a canonical inclusion i_n making the following diagram commutative:

$$\begin{array}{ccc} E_n \otimes E'_n & \overset{i_n}{\hookrightarrow} & E_{n+1} \otimes E'_{n+1} \\ \downarrow & & \downarrow \\ Z_n & \hookrightarrow & Z_{n+1}. \end{array}$$

In fact, i_n is a closed embedding. Now define the ind-variety denoted by $E \otimes E' := \cup_n E_n \otimes E'_n$. Then, $E \otimes E' \to Z$ is an algebraic vector bundle called the *tensor product of* E *and* E'.

We similarly define the algebraic vector bundles $\Lambda^i(E) \to Z$ for any $i \geq 1$ and $E^* \to Z$ called the *i-th exterior-power bundle* and the *dual vector bundle* of E, respectively.

Let $E \to Z$ and $E' \to Z'$ be two algebraic vector bundles. Consider the product ind-variety $Z \times Z'$ with the projections $\pi_1 : Z \times Z' \to Z$ and $\pi_2 : Z \times Z' \to Z'$. Then the vector bundle $\pi_1^* E \otimes \pi_2^*(E') \to Z \times Z'$ is denoted by $E \boxtimes E'$ and is called the *external tensor product of* E *and* E'.

(f) Let Z be an ind-variety with filtration Z_n. Assume that, for each n, there is given an algebraic vector bundle $\pi_n : E_n \to Z_n$ together with a bundle isomorphism $i_n : E_n \xrightarrow{\sim} E_{n+1|Z_n}$. Then, there exists an algebraic vector bundle $\pi : E \to Z$ together with vector bundle isomorphisms

$$E_{|Z_n} \simeq E_n$$

for all n, constructed as follows:

The bundle isomorphism i_n gives rise to a closed embedding (again denoted by)

$$i_n : E_n \hookrightarrow E_{n+1}.$$

Consider the ind-variety $E := \cup_n E_n$ and let $\pi : E \to Z$ be the unique map such that $\pi_{|E_n} = \pi_n$ for all n. The verification that $\pi : E \to Z$ is an algebraic vector bundle is easy.

4.2.E EXERCISES

(1) For any algebraic line bundle $\mathcal{L}$ on an ind-variety Z,

$$\mathcal{L} \otimes \mathcal{L}^* \simeq \varepsilon_Z^1.$$

(2) For any algebraic vector bundle $E \to Z$,

$$E \otimes \varepsilon_Z^1 \simeq E.$$

(3) Let φ be a morphism of algebraic vector bundles

$$\begin{array}{ccc} E & \xrightarrow{\varphi} & F \\ & \searrow \quad \swarrow & \\ & Z\,. & \end{array}$$

Then φ is an isomorphism of algebraic vector bundles iff $\varphi_z : E(z) \to F(z)$ is an isomorphism for all $z \in Z$, where $E(z)$ is the fiber of E over z, similarly for $F(z)$ and $\varphi_z := \varphi_{|E(z)}$.

(4) For a finite-dimensional vector space V and any $v_o \in V$, show that the map $D_{v_o} : V \to T_{v_o}(V)$ is a k-linear isomorphism, where $D_{v_o}(v) : \mathfrak{m}_{v_o}/\mathfrak{m}_{v_o}^2 \to k$ is induced from the map $\bar{D}_{v_o}(v) : k[V] \to k$, $\bar{D}_{v_o}(v)(f) := \big[(f(tv + v_o) - f(v_o))/t\big]_{t=0}$, and $\mathfrak{m}_{v_o} := \{f \in k[V] : f(v_o) = 0\}$ is the maximal ideal at v_o.

Thus, for any vector space V of countable dimension and $v_o \in V$, $V \simeq T_{v_o}(V)$.

(5) Show that the total space E of an algebraic vector bundle $\pi : E \to Z$ is a (quasi-projective) variety if Z is a (quasi-projective) variety.

Hint (due to M. Brion): Embed Z in some $\mathbb{P}^n$ and take a large enough m such that $E^* \otimes \mathcal{O}_Z(m)$ is generated by global sections.

4.3. Smoothness of ind-Varieties

4.3.1 Definition. Let X be a variety and $x \in X$. We first recall the definition of the local ring $\mathcal{O}_{X,x}$ of X at x:

Let us consider the set S of all pairs (U, f), where $x \in U$ is an open subset of X and $f : U \to k$ is a regular map. We call $(U, f) \sim (V, g)$ if there exists an open set $x \in W \subset U \cap V$ such that $f_{|W} = g_{|W}$. This is an equivalence relation on the set S. We denote by $\mathcal{O}_{X,x}$ the set of equivalence classes. The set $\mathcal{O}_{X,x}$ is a ring under pointwise addition and multiplication of functions. Let $\mathfrak{m}_x := \{[U, f] : f(x) = 0\} \subset \mathcal{O}_{X,x}$, where $[U, f]$ denotes the equivalence class of (U, f). It is easy to see that $\mathfrak{m}_x$ is the unique maximal ideal of $\mathcal{O}_{X,x}$. The local ring $\mathcal{O}_{X,x}$ is called the *local ring of X at x*.

Now recall that a point x of a variety X is smooth if and only if, for all $p \geq 0$, the canonical map induced by multiplication

$$S^p(\mathfrak{m}_x/\mathfrak{m}_x^2) \to \mathfrak{m}_x^p/\mathfrak{m}_x^{p+1}$$

is an isomorphism. This algebraic notion of smoothness corresponds to the "geometric" notion of smoothness.

Now let X be an ind-variety with filtration (X_n) and let $x \in X$. Then, of course, $x \in X_n$ for all $n \geq n(x)$, for some $n(x)$. Let $\mathfrak{m}_x(n)$ denote the maximal ideal of the local ring $\mathcal{O}_{X_n,x}$ for any $n \geq n(x)$. We clearly have the restriction map $\mathcal{O}_{X_{n+1},x} \to \mathcal{O}_{X_n,x}$, which takes $\mathfrak{m}_x(n+1) \twoheadrightarrow \mathfrak{m}_x(n)$. This gives rise to the following commutative diagram for any $n \geq n(x)$ and $p \geq 0$:

$$\begin{array}{ccc} S^p(\mathfrak{m}_x(n+1)/\mathfrak{m}_x(n+1)^2) & \xrightarrow{\varphi_p(n+1)} & \mathfrak{m}_x(n+1)^p/\mathfrak{m}_x(n+1)^{p+1} \\ \downarrow{\scriptstyle \alpha_n} & & \downarrow{\scriptstyle \beta_n} \\ S^p(\mathfrak{m}_x(n)/\mathfrak{m}_x(n)^2) & \xrightarrow{\varphi_p(n)} & \mathfrak{m}_x(n)^p/\mathfrak{m}_x(n)^{p+1}. \end{array}$$

Define

$$\hat{S}^p(\mathfrak{m}_x/\mathfrak{m}_x^2) := \operatorname*{Inv.\,lt.}_n \; S^p(\mathfrak{m}_x(n)/\mathfrak{m}_x(n)^2),$$

and similarly define (for $0 \leq p < q$)

$$\widehat{\mathfrak{m}_x^p/\mathfrak{m}_x^q} := \operatorname*{Inv.\,lt.}_n \; \mathfrak{m}_x(n)^p/\mathfrak{m}_x(n)^q.$$

Thus the maps $\varphi_p(n)$ give rise to the map

$$\varphi_p : \hat{S}^p(\mathfrak{m}_x/\mathfrak{m}_x^2) \to \widehat{\mathfrak{m}_x^p/\mathfrak{m}_x^{p+1}}.$$

It is easy to see that an isomorphism $f : X \to Y$ of ind-varieties induces canonical isomorphisms $\hat{S}^p(\mathfrak{m}_{f(x)}/\mathfrak{m}_{f(x)}^2) \xrightarrow{\sim} \hat{S}^p(\mathfrak{m}_x/\mathfrak{m}_x^2)$ and $\widehat{\mathfrak{m}_{f(x)}^p/\mathfrak{m}_{f(x)}^q} \xrightarrow{\sim} \widehat{\mathfrak{m}_x^p/\mathfrak{m}_x^q}$.

We call a point $x \in X$ *algebraically smooth* if the k-linear maps

$$\varphi_p : \hat{S}^p(\mathfrak{m}_x/\mathfrak{m}_x^2) \to \widehat{\mathfrak{m}_x^p/\mathfrak{m}_x^{p+1}}$$

are isomorphisms for all $p \geq 0$. Observe that algebraic smoothness of a point is independent of equivalent ind-variety structures on X. It may be mentioned that [Šafarevič–82] calls such points *smooth*, but we prefer to call them *algebraically smooth* because of Example 4.3.8.

We call an ind-variety X *algebraically smooth* if every point of X is algebraically smooth.

4.3.2 Remark. I do not know if the set of algebraically smooth points in any ind-variety is always open.

4.3.3 Lemma. *For any ind-variety X and $x \in X$, the above map φ_p, for any $p \geq 0$, is surjective.*

Proof. We follow the same notation as above. For any $n \geq n(x)$, the canonical map

$$\varphi_p(n) : S^p(\mathfrak{m}_x(n)/\mathfrak{m}_x(n)^2) \to \mathfrak{m}_x(n)^p/\mathfrak{m}_x(n)^{p+1}$$

is of course surjective. Thus the lemma follows from Lemma 4.4.8. □

The following lemma is easy to verify.

4.3.4 Lemma. *Let $x \in X$ be such that, for any $n \geq 0$, there exists an integer $m \geq n$ with the property that x is a smooth point of X_m. Then x is an algebraically smooth point of X.*

In particular, $\mathbb{A}^\infty$ is algebraically smooth and so is $\mathbb{P}(V)$ (cf. Examples 3 and 4 of 4.1.3). □

Unlike varieties, ind-varieties may not contain a single algebraically smooth point.

4.3.5 Example. Let V be a nonsmooth variety and fix a singular point $v_o \in V$. Define $X_n = V^n$, the n-fold Cartesian product variety, and let $X_n \hookrightarrow X_{n+1}$ be the closed embedding $(x_1, \dots, x_n) \mapsto (x_1, \dots, x_n, v_o)$.

Now let X be the ind-variety $X = \cup X_n$. Then no point of X is algebraically smooth:

Take $x = (x_1, \dots, x_{n_o}) \in X_{n_o}$ for some n_o. For any $v \in V$, let $\mathfrak{m}_v$ denote the maximal ideal of the local ring $\mathcal{O}_{V,v}$. Then, for any $m \geq 0$,

(1)
$$\mathfrak{m}_x(n_o+m)/\mathfrak{m}_x(n_o+m)^2 \cong \bigoplus_{i=1}^{n_o+m} \big(1 \otimes \cdots \otimes 1 \otimes (\mathfrak{m}_{x_i}/\mathfrak{m}_{x_i}^2) \otimes 1 \otimes \cdots \otimes 1\big),$$

where $x_{n_o+1} = \cdots = x_{n_o+m} = v_o$ and, for $p < q$,

$$\mathfrak{m}_x(n_o+m)^p/\mathfrak{m}_x(n_o+m)^q$$
(2)
$$\cong \bigoplus_{i=1}^{n_o+m} \big(1 \otimes \cdots \otimes 1 \otimes (\mathfrak{m}_{x_i}^p/\mathfrak{m}_{x_i}^q) \otimes 1 \otimes \cdots \otimes 1\big) \oplus C\,,$$

for $C = C(m, p, q)$ a subquotient of $\sum \mathfrak{m}_{x_1}^{k_1} \otimes \cdots \otimes \mathfrak{m}_{x_{n_o+m}}^{k_{n_o+m}}$, where the summation runs over $k_1, \cdots, k_{n_o+m} \geq 0$ such that at least two of the k_i's are nonzero.

Now, the map

(3) $$S^p(\mathfrak{m}_x(n_o+m)/\mathfrak{m}_x(n_o+m)^2) \to \mathfrak{m}_x(n_o+m)^p/\mathfrak{m}_x(n_o+m)^{p+1}$$

has nonzero kernel for any $m \geq 1$ and any p such that

(4) $$S^p(\mathfrak{m}_{v_o}/\mathfrak{m}_{v_o}^2) \twoheadrightarrow \mathfrak{m}_{v_o}^p/\mathfrak{m}_{v_o}^{p+1}$$

has nonzero kernel. Observe that $v_o \in V$ being a singular point, (4) is not an isomorphism for some value of p. From this, it is easy to see that $x \in X$ is not an algebraically smooth point.

We prove the following "global" version of the Inverse Function Theorem for affine ind-varieties.

4.3.6 Proposition. *Let $f : X \to Y$ be a closed embedding between two affine ind-varieties. Let $x \in X$ be a point such that*

(1) $$(df)_x : T_x(X) \to T_{f(x)}(Y)$$

is an isomorphism. Assume further that:

(a) *X is algebraically smooth at x;*
(b) *for all $y \in Y$, there exists an irreducible component of $Y_{n(y)}$, for some $n(y)$, containing both $f(x)$ and y; and*
(c) *the map $f^* : k[Y] \to k[X]$ is surjective and is an open map.*

Then f is an isomorphism, i.e., f is surjective.

Proof. By the assumption (c), it suffices to show that f^* is injective (cf. 4.1.1).

We have the following commutative diagram for any $p \geq 0$:

$$\begin{array}{ccc}
\hat{S}^p(\mathfrak{m}_{x_o}/\mathfrak{m}_{x_o}^2) & \xrightarrow{\varphi_p(Y)} & \widehat{\mathfrak{m}_{x_o}^p/\mathfrak{m}_{x_o}^{p+1}} \\
\Big\downarrow{\scriptstyle \alpha(f^*)} & & \Big\downarrow{\scriptstyle \beta(f^*)} \\
\hat{S}^p(\mathfrak{m}_x/\mathfrak{m}_x^2) & \underset{\varphi_p(X)}{\xrightarrow{\sim}} & \widehat{\mathfrak{m}_x^p/\mathfrak{m}_x^{p+1}},
\end{array}$$

where $x_o := f(x)$, $\varphi_p(Y)$ (and $\varphi_p(X)$) is the map φ_p of Subsection 4.3.1, and the vertical maps $\alpha(f^*)$ and $\beta(f^*)$ are the canonical maps induced from f^*. By the assumption (a), $\varphi_p(X)$ is an isomorphism. Further, by (1) and Exercise 4.3.E.1, $\alpha(f^*)$ is an isomorphism. In particular, from the commutativity of the above diagram, $\varphi_p(Y)$ is injective. But, by Lemma 4.3.3, $\varphi_p(Y)$ is surjective, and hence $\varphi_p(Y)$ is an isomorphism. This forces $\beta(f^*)$ to be injective.

Now let $u \in \operatorname{Ker} f^*$. Then we claim that, for any n such that $x_o \in Y_n$,

(2) $$[u_{|Y_n}] \in \bigcap_{p\geq 0} \mathfrak{m}_{x_o}(n)^p,$$

where $[u_{|Y_n}]$ denotes the equivalence class of $u_{|Y_n}$ in $\mathcal{O}_{Y_n,x_o}$ (cf. 4.3.1). For, if not, choose $p_o > 0$ such that $[u_{|Y_n}] \in \mathfrak{m}_{x_o}(n)^{p_o}$ for all n such that $x_o \in Y_n$

and $[\widehat{u_{|Y_{n_o}}}] \notin \mathfrak{m}_{x_o}(n_o)^{p_o+1}$ for some n_o. Then u determines a nonzero element $\bar{u} \in \widehat{\mathfrak{m}_{x_o}^{p_o}/\mathfrak{m}_{x_o}^{p_o+1}}$, and hence $\beta(f^*)\bar{u} \neq 0$. But this is a contradiction since $u \in \operatorname{Ker} f^*$, establishing (2). This shows that, for any irreducible component C of any Y_n such that $x_o \in C$, $u_{|C} \equiv 0$ (use the Krull Intersection Theorem [Eisenbud–95, Corollary 5.4]). This implies, from the assumption (b), that $u \equiv 0$. So f^* is injective and the proposition is proved. □

4.3.7 Theorem. *Any affine ind-group H over k is algebraically smooth if characteristic $k = 0$.*

Proof. By homogeneity, it suffices to show that $e \in H$ is algebraically smooth.

We freely follow the notation as in the proof of Proposition 4.2.2. We prove that the surjective map (cf. Lemma 4.3.3)

$$\varphi_p : \hat{S}^p(\mathfrak{m}_e/\mathfrak{m}_e^2) \to \widehat{\mathfrak{m}_e^p/\mathfrak{m}_e^{p+1}}$$

is injective for all $p \geq 1$: Assume, by induction, that φ_p is injective and show the same for φ_{p+1}; of course, φ_1 is injective.

Recall the definition of the map $\eta : T_e(H) \to \operatorname{Der}_H A$ from the proof of Proposition 4.2.2, where $A := k[H]$. It is easy to see that, for any $v \in T_e(H)$ and $p \geq 1$,

(1) $$\eta(v)(\mathfrak{m}^p) \subset \mathfrak{m}^{p-1},$$

where, as in the proof of Proposition 4.2.2, $\mathfrak{m}$ is the (maximal) ideal of A consisting of functions vanishing at e. Also, $v \in T_e(H)$ gives rise to the map

$$\gamma(v) : \hat{S}^p(\mathfrak{m}_e/\mathfrak{m}_e^2) \to \hat{S}^{p-1}(\mathfrak{m}_e/\mathfrak{m}_e^2)$$

defined as follows: Fix m such that $v \in T_e(H_m) := (\mathfrak{m}_e(m)/\mathfrak{m}_e(m)^2)^*$, and define, for any $n \geq m$,

$$\gamma_n(v) : S^p(\mathfrak{m}_e(n)/\mathfrak{m}_e(n)^2) \to S^{p-1}(\mathfrak{m}_e(n)/\mathfrak{m}_e(n)^2)$$

by

(2) $$\chi_1 \cdots \chi_p \mapsto \sum_{i=1}^{p} \chi_1 \cdots \chi_{i-1}\langle \chi_i, v\rangle \chi_{i+1} \cdots \chi_p$$

for $\chi_i \in \mathfrak{m}_e(n)/\mathfrak{m}_e(n)^2$. Now $\gamma(v)$ is defined as the inverse limit of the maps $\{\gamma_n(v)\}_{n\geq m}$. Using (4.2.2.1), it is easy to see that the following diagram is commutative (for any $v \in T_e(H)$ and $p \geq 0$):

$$\begin{array}{ccc}
\hat{S}^{p+1}(\mathfrak{m}_e/\mathfrak{m}_e^2) & \xrightarrow{\varphi_{p+1}} & \widehat{\mathfrak{m}_e^{p+1}/\mathfrak{m}_e^{p+2}} \\
\big\downarrow{\scriptstyle \gamma(v)} & & \big\downarrow \\
\hat{S}^{p}(\mathfrak{m}_e/\mathfrak{m}_e^2) & \xrightarrow{\varphi_{p}} & \widehat{\mathfrak{m}_e^{p}/\mathfrak{m}_e^{p+1}},
\end{array}$$

where the right vertical map is induced by the derivation $\eta(v)$ (cf. (1)).

Now take $\alpha \in \operatorname{Ker}(\varphi_{p+1})$. Then $\gamma(v)\alpha = 0$ for all $v \in T_e(H)$ since, by induction, φ_p is injective. Represent $\alpha = (\alpha_n)_n$, where

$$\alpha_n \in S^{p+1}(\mathfrak{m}_e(n)/\mathfrak{m}_e(n)^2).$$

For $v \in T_e(H_m)$, $\gamma(v)\alpha$ is represented by the sequence $(\gamma_n(v)\alpha_n)_{n\geq m}$. Now take a k-basis $\{f_1, \ldots, f_q\}$ of $\mathfrak{m}_e(n)/\mathfrak{m}_e(n)^2$ and the dual basis $\{e_1, \ldots, e_q\}$ of $T_e(H_n)$. Then, we have the following Euler's identity, which follows by using (2),

$$(3) \qquad (p+1)\alpha_n = \sum_{j=1}^{q} (\gamma_n(e_j)\alpha_n) f_j.$$

Since $\gamma(v)\alpha = 0$ for all $v \in T_e(H)$, by (3), $(p+1)\alpha_n = 0$. But, by assumption, char. $k = 0$ and hence $\alpha_n = 0$ for all n. This gives that $\alpha = 0$, i.e., φ_{p+1} is injective, thereby completing the induction. $\square$

4.3.8 Example. We give below an example of an ind-variety X with defining filtration $(X_n)_{n\geq 0}$ and an algebraically smooth point $x \in X$ such that x is a singular point for all X_n containing x. In particular, the converse of Lemma 4.3.4 is false.

Let $X = \mathrm{SL}_2(k[t])$. Consider the filtration of X:

$$X_n = \left\{ \begin{pmatrix} X(t) & Y(t) \\ Z(t) & W(t) \end{pmatrix} : X(t), Y(t), Z(t), W(t) \in k[t]_{\leq n} \right.$$
$$\left. \text{and } X(t)W(t) - Y(t)Z(t) = 1 \right\},$$

where $k[t]_{\leq n}$ is the space of polynomials of degree $\leq n$. Writing $X(t) = 1 + \sum_{i=0}^{n} x_i t^i$, $Y(t) = \sum_{i=0}^{n} y_i t^i$, $Z(t) = \sum_{i=0}^{n} z_i t^i$ and $W(t) = 1 + \sum_{i=0}^{n} w_i t^i$, we realize X_n as the affine variety in the variables $\{x_i, y_i, z_i, w_i;\ 0 \leq i \leq n\}$ given by the common zero set of the polynomials R_m for $0 \leq m \leq 2n$, where

$$R_m := x_m + w_m + \sum_{i=0}^{m} (x_i w_{m-i} - y_i z_{m-i}), \qquad \text{for } 0 \leq m \leq n, \text{ and}$$

$$R_m := \sum_{i=m-n}^{n} (x_i w_{m-i} - y_i z_{m-i}), \qquad \text{for } n < m \leq 2n.$$

It is easy to see that X is an ind-group. We show below that the identity matrix $I = \begin{pmatrix} 1 & 0 \\ 0 & 1 \end{pmatrix} \in X_n$ is a singular point for all X_n. But, by Theorem 4.3.7, X, being an ind-group, is an algebraically smooth ind-variety when char. $k = 0$.

By [Šafarevič–94, Chap. II, §1.2], $T_I(X_n) \subseteq V := \{(x, y, z, w) \in \mathbb{A}^{4(n+1)} : x_m + w_m = 0, \text{ for all } 0 \leq m \leq n\}$, where x denotes $(x_0, \ldots, x_n)$ and similarly, for y, z, w. In fact,

$$(1) \qquad T_I(X_n) = V\,;$$

for let $P = L + M$ be a polynomial in the variables x, y, z and w with L linear and M having terms only of degree ≥ 2, such that

$$(2) \qquad P^q \in \langle R_m \rangle_{0 \leq m \leq 2n}$$

for some $q > 0$, where $\langle \bullet \rangle$ denotes the ideal generated by the elements $\bullet$. Then we claim that L is a linear combination of $\{x_m + w_m\}_{0 \leq m \leq n}$.

Setting all coordinates, except one fixed z_i, resp. y_i, 0, we see from (2) that the coefficient of z_i, resp. y_i, in L is 0. Using the relations $x_m + w_m$, we can further eliminate the w-variables from L, i.e., L is a linear combination of the x-variables only. Fix an $0 \leq i \leq n$ and set all $x_j,\ w_j,\ y_j,\ z_j = 0$ for $j \neq i$, and $x_i + w_i = 0$. From (2), we see that the coefficient of x_i in L is 0. So L itself is 0. This proves (1). In particular,

$$\dim T_I(X_n) = 3(n+1).$$

Now,

$$(3) \qquad R_{n+1}^q \notin \langle R_m \rangle_{0 \leq m \leq n} \quad \text{for any } q \geq 0,$$

as can be seen by setting all x and w variables to be 0 together with $z_0 = z_2 = z_3 = \cdots = z_n = y_0 = y_1 = \cdots = y_{n-1} = 0$. In fact, the same argument shows that (3) is true in the local ring $\mathcal{O}_{\mathbb{A}^{4(n+1)},0}$, thus

$$(4) \qquad \dim_I X_n < 3(n+1),$$

where $\dim_I X_n$ denotes the dimension of X_n at I, i.e., the maximum of the dimensions of the irreducible components of X_n through I. To prove (4), observe that the affine variety Y_n in the variables $\{x_i, y_i, z_i, w_i; 0 \leq i \leq n\}$ and the relations R_m, for $0 \leq m \leq n$, has $\dim_I(Y_n) = 3(n+1)$ because of the presence of the linear factors in each such R_m. Now, by (3), we see that $\dim_I(X_n) < \dim_I(Y_n)$, establishing (4). Hence I is a singular point of X_n.

4.3.E EXERCISES

(1) Let $\{V_n\}_{n \geq 0}$ be a direct system of vector spaces over a field k with direct limit V. Then show that the inverse limit of the inverse system $\{V_n^*\}_{n \geq 0}$ is canonically isomorphic with V^*.

(2) Let $\phi = (\phi_n)_{n\geq 0}$ be a morphism from a direct system $\{V_n\}_{n\geq 0}$ to another direct system $\{W_n\}_{n\geq 0}$, both consisting of finite-dimensional vector spaces over k, such that the induced map of the direct limits

$$\underset{n}{\text{limit}}\ V_n \to \underset{n}{\text{limit}}\ W_n$$

is an isomorphism. Then show that, for any $p \geq 1$, the induced morphism of inverse systems gives the isomorphism

$$S^p(\phi^*) : \underset{n}{\text{Inv. lt.}}\ S^p(W_n^*) \to \underset{n}{\text{Inv. lt.}}\ S^p(V_n^*).$$

Hint: Since there is a canonical isomorphism $\otimes^p(V_n^*) \simeq \left(\otimes^p V_n\right)^*$, prove, by using the above exercise (1), that

$$(*) \qquad \underset{n}{\text{Inv. lt.}}\ \left(\otimes^p(W_n^*)\right) \to \underset{n}{\text{Inv. lt.}}\ \left(\otimes^p(V_n^*)\right)$$

is an isomorphism.

Let $K_n := \ker S^p(W_n^*) \to S^p(V_n^*)$, and $\tilde{K}_n := \ker \otimes^p(W_n^*) \to \otimes^p(V_n^*)$. Then, by the Koszul resolution D.13, the canonical map $\otimes^p(W_n^*) \to S^p(W_n^*)$ maps $\tilde{K}_n$ surjectively onto K_n. Thus, by Lemma 4.4.8,

$$\underset{n}{\text{Inv. lt.}}\ \tilde{K}_n \twoheadrightarrow \underset{n}{\text{Inv. lt.}}\ K_n.$$

But, by $(*)$, $\underset{n}{\text{Inv. lt.}}\ \tilde{K}_n = (0)$ and thus so does $\underset{n}{\text{Inv. lt.}}\ K_n$. This gives the injectivity of $S^p(\phi^*)$. Prove the surjectivity of $S^p(\phi^*)$ from $(*)$ and Lemma 4.4.8.

4.4. An Introduction to pro-Groups and pro-Lie Algebras

Throughout this section k is an algebraically closed field of char. 0.

4.4.1 Definitions. Let G be a group and let $\mathcal{F}$ be a nonempty family of normal subgroups of G such that, for each $N \in \mathcal{F}$, G/N is given the structure of an affine k-algebraic group. Then the pair $(G, \mathcal{F})$ is a *pro-algebraic group* over k (to be abbreviated as *pro-group*) if the following axioms are satisfied:

(a_1) If $N_1, N_2 \in \mathcal{F}$, then $N_1 \cap N_2 \in \mathcal{F}$.

(a_2) If $N_1 \in \mathcal{F}$, then a normal subgroup $N_2 \supset N_1$ (of G) belongs to $\mathcal{F}$ iff N_2/N_1 is a closed (normal) subgroup of G/N_1.

(a_3) If $N_1, N_2 \in \mathcal{F}$ are such that $N_1 \subset N_2$, then the quotient map $\gamma_{N_2,N_1} : G/N_1 \to G/N_2$ is a morphism of k-algebraic groups.

(a_4) The natural homomorphism $\gamma : G \to \underset{N\in\mathcal{F}}{\text{Inv. lt.}}\ G/N$ is bijective, where $\mathcal{F}$ is a directed set under the partial order $N_1 \leq N_2$ iff $N_1 \supseteq N_2$.

The set $\mathcal{F}$ is called the *complete defining set* (*defining set* for short) of the pro-group G.

It is clear from (a_4) that

(1) $$\bigcap_{N\in\mathcal{F}} N = \{1\}.$$

Let G and G' be pro-groups with defining sets $\mathcal{F}$ and $\mathcal{F}'$, respectively. Then a group homomorphism $\phi : G \to G'$ is called a *pro-group morphism* if the following condition is satisfied:

($*$) For all $N' \in \mathcal{F}'$, $\phi^{-1}(N') \in \mathcal{F}$ and, moreover, the induced map $\phi_{N'}$: $G/\phi^{-1}(N') \to G'/N'$ is a morphism of algebraic groups.

Observe that from the existence of quotient groups of algebraic groups (cf. [Springer–98, Proposition 5.5.10]) and Exercise A.E.2, it suffices to check the condition (*) for N' in a cofinal subset of $\mathcal{F}'$. (We have used here the char. 0 assumption on k, since Exercise A.E.2 requires this assumption.) Recall that a subset S of a directed set X is called *cofinal* if, for any $x \in X$, there exists an $s \in S$ such that $s \geq x$.

It is easy to see that the composition of two pro-group morphisms is again a pro-group morphism. We define the *pro-topology* on a pro-group G by taking the inverse-limit topology on G via the identification γ of (a_4), where we endow each G/N (for $N \in \mathcal{F}$) with the Zariski topology. Recall that, as in Subsection 4.1.1, the *inverse limit topology* on the inverse limit of an inverse system of topological spaces $(X_\alpha)_{\alpha\in\Lambda}$ is, by definition, the subspace topology on $\operatorname{Inv.\,lt.}_{\alpha\in\Lambda} X_\alpha \subset \prod_{\alpha\in\Lambda} X_\alpha$, where $\prod_{\alpha\in\Lambda} X_\alpha$ is endowed with the product topology.

Alternatively, the pro-topology on G is the smallest topology such that each $\gamma_N : G \to G/N$ ($N \in \mathcal{F}$) is continuous, i.e., the sets $\{\gamma_N^{-1}(U);\ N \in \mathcal{F}$ and U open in $G/N\}$ form a base for the pro-topology on G. In particular, for any topological space Y, a map $f : Y \to G$ is continuous iff $\gamma_N \circ f$ is continuous for all $N \in \mathcal{F}$ (cf. [Spanier–66, Introduction, §2]). This immediately gives that a pro-group morphism $G \to G'$ is continuous. Of course, the converse is false in general, i.e., a continuous group homomorphism $G \to G'$ need not be a pro-group morphism. A pro-group G is called *connected* if G is connected under its pro-topology. Clearly, if G is connected, then so is each G/N, for $N \in \mathcal{F}$. The converse is true as well by Exercise 4.4.E.6.

For any subset $A \subset G$, its closure $\bar{A}$ is given by

(2) $$\bar{A} = \bigcap_{N\in\mathcal{F}} \gamma_N^{-1}(\overline{\gamma_N(A)}).$$

To prove this, observe that $\bar{A} \subset \bigcap_N \gamma_N^{-1}(\overline{\gamma_N(A)})$. Conversely, take $x \in G$ such that $\gamma_N(x) \in \overline{\gamma_N(A)}$ for all $N \in \mathcal{F}$. To prove (2), it suffices to show that, for all

$N \in \mathcal{F}$ and all open $U \subset G/N$ such that $\gamma_N(x) \in U$, $\gamma_N^{-1}(U) \cap A \neq \emptyset$, i.e., $\gamma_N(A) \cap U \neq \emptyset$. But this follows immediately since $\gamma_N(x) \in \overline{\gamma_N(A)}$.

By a *pro-algebraic subset*, for short *pro-subset*, (resp. *pro-algebraic subgroup*, for short *pro-subgroup*) $A \subset G$, we mean a closed subset, resp. closed subgroup, of G under pro-topology. Clearly, any $N \in \mathcal{F}$ is a pro-subgroup of G since $N = \gamma_N^{-1}\{1\}$.

4.4.2 Remark. In (a_4), the injectivity of γ is equivalent to the condition $\bigcap_{N \in \mathcal{F}} N = \{1\}$. The surjectivity of γ is equivalent to the following:

For any family of elements $\{g_N\}_{N \in \mathcal{F}}$ ($g_N \in G$) satisfying $g_N N \subset g_M M$ whenever $N \subset M$, the intersection $\bigcap_{N \in \mathcal{F}} g_N N \neq \emptyset$.

4.4.3 Examples. (1) Any algebraic k-group G is a pro-group, where we take $\mathcal{F}$ to be the family of all normal closed subgroups N of G where we equip G/N with the standard quotient algebraic group structure. Then, $\{e\} \in \mathcal{F}$. Conversely, any pro-group G such that $\{e\} \in \mathcal{F}$ is an algebraic group and, moreover, in this case the pro-group structure on G coincides with the above pro-group structure.

Clearly, the Zariski topology on G coincides with the pro-topology in this case. In particular, in this case, the pro-subsets, resp. pro-subgroups, are precisely the Zariski closed subsets, resp. closed subgroups.

(2) Let $\{G_i\}_{i \geq 1}$ be algebraic k-groups. Then the direct product $G = \prod_{i=1}^{\infty} G_i$ is a pro-group with the defining set $\mathcal{F}$ consisting of all normal subgroups N of G such that there exists some $j = j_N$ (depending on N) with $N \supset \prod_{i>j} G_i$, and $\pi_j(N) \hookrightarrow \prod_{i \leq j} G_i$ is a closed subgroup, where π_j is the projection on the first j-factors.

In fact, more generally, we have the following example:

(3) Let $\{G_i\}_{i \in \mathbb{N}}$ be a family of algebraic groups with surjective group morphisms $\phi_i : G_{i+1} \to G_i$ for all $i \in \mathbb{N}$. Then the inverse limit in the category of groups

$$G := \operatorname*{Inv.\ lt.}_{i \in \mathbb{N}} G_i$$

is a pro-group, under the family of normal subgroups

$$\mathcal{F} := \{\pi_i^{-1}(N_i) : N_i \text{ is a normal closed subgroup of } G_i,\ i \in \mathbb{N}\},$$

where $\pi_i : G \to G_i$ is the natural projection. Since each ϕ_i is surjective (by assumption), π_i is surjective as well. The verification that G is indeed a pro-group is easy.

(4) Let G and H be pro-groups over k. Then $G \times H$ is canonically a pro-group over k.

4.4.4 Lemma. *Let $A \subset G$ be a pro-subgroup of a pro-group G. Then, for any $N \in \mathcal{F}$, where $\mathcal{F}$ is the defining set of G, $\gamma_N(A)$ is closed in G/N. In particular, $A = \cap_{N\in\mathcal{F}} \gamma_N^{-1}(\gamma_N(A))$.*

Proof. By (4.4.1.2),

$$(1) \qquad A = \cap_{N\in\mathcal{F}} (\gamma_N^{-1}(\overline{\gamma_N(A)})).$$

Since any algebraic group morphism takes closed subgroups to closed subgroups (cf. A.40), for any $N' \subset N$ in $\mathcal{F}$,

$$(2) \qquad \gamma_{N,N'}(\overline{\gamma_{N'}(A)}) = \overline{\gamma_N(A)}$$

(cf. (a_3) of 4.4.1 for the notation $\gamma_{N,N'}$).

Let us define $B = \operatorname*{Inv.\,lt.}_{N\in\mathcal{F}} \overline{\gamma_N(A)}$. Then, by ($a_4$) of 4.4.1, $B \subset G$. Further, by (2) and Lemma 4.4.8, $\gamma_N(B) = \overline{\gamma_N(A)}$ for any $N \in \mathcal{F}$. We next prove that $B = A$: Take $g \in B$. Then $\gamma_N(g) \in \overline{\gamma_N(A)}$, i.e., $g \in \gamma_N^{-1}(\overline{\gamma_N(A)})$, for all $N \in \mathcal{F}$. Hence, by (1), $g \in A$. Of course, $A \subset B$ so the claim $A = B$ is established. In particular, $\gamma_N(B) = \gamma_N(A) = \overline{\gamma_N(A)}$, proving the lemma. □

4.4.5 Proposition. *Let H be a pro-subgroup of G. Then H is a pro-group with the defining set $\mathcal{F}' := \{$normal subgroups N' of H such that $N' \supset N \cap H$, for some $N \in \mathcal{F}$ and $N'/N\cap H \subset G/N$ is a closed subgroup$\}$. Further, the pro-topology on H coincides with the subspace topology.*

Proof. For any $N \in \mathcal{F}$, by Lemma 4.4.4, $H/H \cap N$ is a closed subgroup of G/N. In particular, $H/H \cap N$ acquires a natural structure of an algebraic group. Further, for any $N' \in \mathcal{F}$ such that $N' \cap H = N \cap H$, the algebraic structures on $H/H \cap N$ inherited from G/N and G/N' coincide, as can be easily seen by considering the morphisms $G/N \cap N' \to G/N$ and $G/N \cap N' \to G/N'$ and using the assumption that char. $k = 0$ (cf. Exercise A.E.2).

For any normal subgroup N' of H containing $N \cap H$, such that $N'/N \cap H$ is a closed subgroup of G/N, H/N' is a quotient group of $H/H \cap N$, and hence again is naturally an algebraic group. The axioms (a_1)–(a_3) of 4.4.1 are easy to verify for the family $\mathcal{F}'$.

Now we prove the axiom (a_4) for $\mathcal{F}'$. Let $\hat{H} := \operatorname*{Inv.\,lt.}_{N'\in\mathcal{F}'} H/N'$. Define the map $i : \hat{H} \to \operatorname*{Inv.\,lt.}_{N\in\mathcal{F}} G/N = G$ by $i((h_{N'}N')_{N'\in\mathcal{F}'}) = (g_N N)_{N\in\mathcal{F}}$, where $g_N := h_{N\cap H}$. Then i is clearly injective since the subset $\mathcal{F}'' = \{N \cap H : N \in \mathcal{F}\} \subset \mathcal{F}'$ is cofinal in $\mathcal{F}'$. Further, $i(\hat{H}) \supset H$. Conversely, take $x \in i(\hat{H})$ (with $x \in G$). Then $x \in \bigcap_{N\in\mathcal{F}} (HN)$. But, by Lemma 4.4.4, $\bigcap_{N\in\mathcal{F}} (HN) = H$, proving $\hat{H} \approx H$.

Finally, the assertion, that the pro-topology on H coincides with the subspace topology, follows easily from the characterization of a map $f : Y \to G$ (for any topological space Y) to be continuous as in 4.4.1. □

The following lemma is easy to verify (using Lemma 4.4.4), and hence is left to the reader.

4.4.6 Lemma. *Let H be a pro-subgroup of a pro-group G. Then H is a normal subgroup of G if and only if $\gamma_N(H)$ is normal in G/N for all $N \in \mathcal{F}$, where $\mathcal{F}$ is the defining set of G, and $\gamma_N : G \to G/N$ is the projection.* □

4.4.7 Definition. By an *inverse system of algebraic groups* we mean a directed set $\mathcal{F}$ together with an algebraic group K_N for each $N \in \mathcal{F}$ and an algebraic group morphism $f_{N_1,N_2} : K_{N_2} \to K_{N_1}$ for each $N_1 \leq N_2$, such that $f_{N,N} = I$ for all $N \in \mathcal{F}$ and the following "cocycle condition" is satisfied:

$$f_{N_1,N_2} \circ f_{N_2,N_3} = f_{N_1,N_3}, \qquad \text{for all } N_1 \leq N_2 \leq N_3.$$

We denote this inverse system of algebraic groups simply by $(K_N)_{N\in\mathcal{F}}$.

Given two inverse systems of algebraic groups $(K_N)_{N\in\mathcal{F}}$ and $(K'_N)_{N\in\mathcal{F}}$ parametrized by the same directed set $\mathcal{F}$, by a *morphism* from (K_N) to (K'_N) we mean an algebraic group morphism $\theta_N : K_N \to K'_N$, for each $N \in \mathcal{F}$, such that the following diagram is commutative (for all $N \leq M$):

$$\begin{array}{ccc} K_M & \xrightarrow{\theta_M} & K'_M \\ \big\downarrow{\scriptstyle f_{N,M}} & & \big\downarrow{\scriptstyle f'_{N,M}} \\ K_N & \xrightarrow{\theta_N} & K'_N\,. \end{array}$$

We denote this morphism by (θ_N). Clearly the morphism (θ_N) induces a natural group homomorphism (denoted) $\theta : \underset{N\in\mathcal{F}}{\text{Inv. lt.}}\ K_N \to \underset{N\in\mathcal{F}}{\text{Inv. lt.}}\ K'_N$, where the inverse limit is taken in the category of abstract groups.

4.4.8 Lemma. *Let (θ_N) be a morphism between two inverse systems of algebraic groups $\{G_N\}_{N\in\mathcal{F}}$ and $\{K_N\}_{N\in\mathcal{F}}$. Assume further that each θ_N is surjective. Then the induced map*

$$\theta : \underset{N\in\mathcal{F}}{\text{Inv. lt.}}\ G_N \to \underset{N\in\mathcal{F}}{\text{Inv. lt.}}\ K_N$$

is surjective.

Proof. Take an element $x = (x_N)_{N\in\mathcal{F}} \in \underset{N\in\mathcal{F}}{\text{Inv. lt.}}\ K_N$, with $x_N \in K_N$, and let $A_N = A_N(x) := \theta_N^{-1}(x_N) \subset G_N$. Since θ_N is surjective, $A_N \neq \emptyset$. Then, clearly, $f_{N,M}(A_M) \subset A_N$ for $M \geq N$, where $f_{N,M}$ is the morphism $G_M \to G_N$.

Now let $\mathfrak{S}$ be the set of families $(B_N)_{N\in\mathcal{F}}$ such that, for each $N \in \mathcal{F}$, B_N is a nonempty (closed) subset of A_N of the form $H_N g_N$ for some closed subgroup H_N of G_N and $g_N \in G_N$ and, moreover,

$$(1) \qquad f_{N,M}(B_M) \subset B_N \qquad \text{for each } N \leq M.$$

Since the family $(A_N)_{N\in\mathcal{F}}$ belongs to $\mathfrak{S}$, $\mathfrak{S}$ is nonempty. Further, $\mathfrak{S}$ is a partially ordered set under $(B_N) \leq (B'_N)$ if and only if $B_N \supset B'_N$ for each $N \in \mathcal{F}$. Since each G_N is a noetherian; in particular, quasicompact space, every chain in $\mathfrak{S}$ has a maximal element. Hence, by Zorn's Lemma, $\mathfrak{S}$ itself has a maximal element, say $(A^o_N)_N$. Now define $A'_N = \bigcap_{N\leq M} f_{N,M}(A^o_M)$. Then $(A'_N) \in \mathfrak{S}$: Since each $f_{N,M}(A^o_M)$ is closed (being the image of a closed subgroup up to a translate), and G_N is a noetherian space, there exist $N \leq M_1, \ldots, M_n$ such that $A'_N = \bigcap_{1\leq i\leq n} f_{N,M_i}(A^o_{M_i})$. Since $\mathcal{F}$ is directed, there exists $M' \in \mathcal{F}$ such that $M' \geq M_i$ for each $i = 1, \ldots, n$. Thus, $f_{N,M'}(A^o_{M'}) \subset \bigcap_{1\leq i\leq n} f_{N,M_i}(A^o_{M_i}) = A'_N$. But $A'_N \subset f_{N,M'}(A^o_{M'})$, and hence $A'_N = f_{N,M'}(A^o_{M'})$. This proves that $(A'_N) \in \mathfrak{S}$ and clearly $(A^o_N) \leq (A'_N)$. By maximality, this gives $A^o_N = A'_N$ for all N; in particular, $f_{N,M}(A^o_M) = A^o_N$ for all $N \leq M$. Finally, we claim that A^o_N is a singleton $\{y_N\}$ for all $N \in \mathcal{F}$:

Fix $N_o \in \mathcal{F}$ and choose a point $y_{N_o} \in A^o_{N_o}$ and define

$$\begin{aligned} A''_N &= (f_{N_o,N})^{-1}(y_{N_o}) \cap A^o_N, \qquad \text{if } N_o \leq N, \\ &= A^o_N, \text{ otherwise.} \end{aligned}$$

Then $(A''_N) \in \mathfrak{S}$ and $(A''_N) \geq (A^o_N)$ and hence, by maximality, $A''_N = A^o_N$ for all N; in particular, $A^o_{N_o} = \{y_{N_o}\}$. So $(A^o_N) = (y_N)$ and, by (1), $y = (y_N) \in \operatorname{Inv.lt.}_{N\in\mathcal{F}} G_N$. Further, by construction, $\theta(y) = x$. This proves the surjectivity of θ. □

4.4.9 Remark. From the above proof, we also obtain the following result.

Let $\{G_N\}_{N\in\mathcal{F}}$ and $\{K_N\}_{N\in\mathcal{F}}$ be two inverse systems of topological spaces whose points are closed and let (θ_N) be a morphism between them (in the category of inverse systems of topological spaces) such that each θ_N is surjective. Assume further that $f_{N,M} : G_M \to G_N$, for each $N \leq M$, is a closed map and each G_N is a noetherian topological space. Then the induced map

$$\theta : \operatorname{Inv.lt.}_{N\in\mathcal{F}} G_N \to \operatorname{Inv.lt.}_{N\in\mathcal{F}} K_N$$

is surjective.

4.4.10 Proposition. *Let H be a normal pro-subgroup of a pro-group G. Then G/H is a pro-group with the defining set $\mathcal{F}' := \{N/H,\ N \in \mathcal{F}$ and $N \supset H\}$,* *where $\mathcal{F}$ is the defining set of G. Further, the quotient map $\pi : G \to G/H$ is a pro-group morphism.*

Proof. For $H \subset N \in \mathcal{F}$, $(G/H)/(N/H) \approx G/N$, and hence $(G/H)/(N/H)$ is an algebraic group. The axioms (a_1)–(a_3) of 4.4.1 are trivial to verify. So, we come to the proof of (a_4).

It is clear that there is a canonical group homomorphism $\varphi : G/H \to \operatorname{Inv.\,lt.}_{H \subset N \in \mathcal{F}} G/N$. Also, φ is injective since $\varphi(gH) = 1 \Leftrightarrow gH \subset \bigcap_{H \subset N \in \mathcal{F}} N$. Moreover, $\bigcap_{H \subset N \in \mathcal{F}} N \subset \bigcap_{N \in \mathcal{F}} (HN)$ since, for any $N \in \mathcal{F}$, $HN \in \mathcal{F}$ by Lemma 4.4.4 and axiom (a_2) of 4.4.1. But, again by Lemma 4.4.4, $\bigcap_{N \in \mathcal{F}} (HN) = H$. This proves the injectivity of φ. To prove the surjectivity of φ, consider the canonical map $\theta : G = \operatorname{Inv.\,lt.}_{N \in \mathcal{F}} G/N \to \operatorname{Inv.\,lt.}_{N \in \mathcal{F}} G/NH$, induced from the projections $G/N \to G/NH$. By Lemma 4.4.8, θ is surjective. But it is easy to see that the canonical projection $p : \operatorname{Inv.\,lt.}_{N \in \mathcal{F}} G/NH \to \operatorname{Inv.\,lt.}_{H \subset N \in \mathcal{F}} G/N$ is an isomorphism. So $p \circ \theta : G \to \operatorname{Inv.\,lt.}_{H \subset N \in \mathcal{F}} G/N$ is surjective. From this the surjectivity of φ follows. This proves the first part of the proposition.

As is trivial to verify, the map $\pi : G \to G/H$ is a pro-group morphism. □

4.4.11 Corollary. *Let H be as in the above proposition. Then $H \in \mathcal{F}$ if and only if the pro-group G/H is, in fact, an algebraic group (cf. Example 4.4.3 (1)).*

Proof. The implication "$\Rightarrow$" follows from the definition. For the converse, assume that the pro-group G/H is, in fact, an algebraic group. Then $\{1\} \in \mathcal{F}'$, where $\mathcal{F}'$ is the defining set of G/H. But since $\pi : G \to G/H$ is a morphism of pro-groups, $\pi^{-1}\{1\} = H \in \mathcal{F}$. □

4.4.12 Proposition. *Let $\phi : G \to G'$ be a pro-group morphism between pro-groups. Then* $\operatorname{Im}\phi$ *is a pro-subgroup of G' and, moreover,*

$$G/\operatorname{Ker}\phi \approx \operatorname{Im}\phi \text{ as pro-groups.}$$

In particular, if ϕ is bijective, then ϕ is a pro-group isomorphism, i.e., ϕ^{-1} is again a pro-group morphism.

Proof. Let $\mathcal{F}$, resp. $\mathcal{F}'$, be the defining set of G, resp. G'. To prove that $\operatorname{Im}\phi$ is closed, we first show that $\gamma_{N'}(\operatorname{Im}\phi)$ is closed in G'/N', for all $N' \in \mathcal{F}'$: Since ϕ is a morphism, $\phi^{-1}(N') \in \mathcal{F}$ and we have

$$\text{(1)} \qquad \gamma_{N'} \circ \phi = \phi_{N'} \circ \gamma_{\phi^{-1}(N')},$$

where $\phi_{N'} : G/\phi^{-1}(N') \to G'/N'$ is the induced map. By (1), $\gamma_{N'}(\operatorname{Im}\phi) = \operatorname{Im}\phi_{N'}$ and, moreover, by A.40, $\operatorname{Im}\phi_{N'}$ is a closed subgroup of G'/N'.

Consider the subset of $\mathcal{F} \times \mathcal{F}'$:

$$\hat{\mathcal{F}} = \{(N, N') \in \mathcal{F} \times \mathcal{F}' : N \subset \phi^{-1}(N')\},$$

which is a directed set under the partial order $(N, N') \leq (N_1, N_1')$ if and only if $N \leq N_1$ and $N' \leq N_1'$. Define three inverse systems of algebraic groups $\{\hat{G}_{(N,N')}\}$, $\{\hat{G}'_{(N,N')}\}$, and $\{H'_{(N,N')}\}$ (all parametrized by $\hat{\mathcal{F}}$) as follows:

$\hat{G}_{(N,N')} := G/N$, $\hat{G}'_{(N,N')} := G'/N'$ and $H'_{(N,N')} :=$ Image of the canonical map $\phi_{N',N} : G/N \to G'/N'$. Then the maps $\phi_{N',N}$ induce the surjection (by Lemma 4.4.8):

$$\hat{\phi} : \underset{(N,N')\in\hat{\mathcal{F}}}{\text{Inv. lt.}} \hat{G}_{(N,N')} \twoheadrightarrow \underset{(N,N')\in\hat{\mathcal{F}}}{\text{Inv. lt.}} H'_{(N,N')}.$$

Of course, there is an injection:

$$i : \underset{(N,N')\in\hat{\mathcal{F}}}{\text{Inv. lt.}} H'_{(N,N')} \hookrightarrow \underset{(N,N')\in\hat{\mathcal{F}}}{\text{Inv. lt.}} \hat{G}'_{(N,N')}.$$

Moreover, under the inverse limit topology on the range, $\operatorname{Im} i$ is closed. To show this, observe that $\operatorname{Im} i = \left(\prod_{(N,N')\in\hat{\mathcal{F}}} H'_{(N,N')}\right) \cap \underset{(N,N')\in\hat{\mathcal{F}}}{\text{Inv. lt.}} \hat{G}'_{(N,N')}$. Now since $H'_{(N,N')} = \operatorname{Im} \phi_{N',N}$ is closed in $\hat{G}'_{(N,N')}$ for each $(N, N') \in \hat{\mathcal{F}}$, we get that $\operatorname{Im} i$ is closed.

Now, it is easy to see that the following canonical maps are isomorphisms of groups

$$p : G = \underset{N\in\mathcal{F}}{\text{Inv. lt.}}\, G/N \xrightarrow{\sim} \underset{(N,N')\in\hat{\mathcal{F}}}{\text{Inv. lt.}} \hat{G}_{(N,N')}, \qquad \text{and}$$

$$p' : G' = \underset{N'\in\mathcal{F}'}{\text{Inv. lt.}}\, G'/N' \xrightarrow{\sim} \underset{(N,N')\in\hat{\mathcal{F}}}{\text{Inv. lt.}} \hat{G}'_{(N,N')}.$$

Further, both p and p' are homeomorphisms under the inverse limit topologies on the spaces involved. Combining the maps p, $\hat{\phi}$, i, and p'^{-1}, we get

$$G \twoheadrightarrow \underset{(N,N')\in\hat{\mathcal{F}}}{\text{Inv. lt.}} H'_{(N,N')} \hookrightarrow G'.$$

Moreover, $\operatorname{Im} \phi$ coincides with the image of G in the above. Hence $\operatorname{Im} \phi$ is closed in G', proving the first part of the proposition.

To prove the second part, observe that $\operatorname{Ker} \phi$ is a normal pro-subgroup. The induced map $\bar{\phi} : G/\operatorname{Ker} \phi \to \operatorname{Im} \phi$ is clearly a morphism of pro-groups, and is an isomorphism of abstract groups. So, we need to show that a bijective morphism $\phi : G \to G'$ between pro-groups is an isomorphism. Take $N \in \mathcal{F}$. Then, from the first part of the proposition together with Proposition 4.4.5, $\phi(N)$ is closed in G' and, moreover, since ϕ is surjective, $\phi(N)$ is a normal subgroup of G'. In particular, the induced map $\bar{\phi}_N : G/N \to G'/\phi(N)$ is a bijective morphism of pro-groups. Hence, by Exercise 4.4.E.1, $G'/\phi(N)$ is an algebraic group. But then $\bar{\phi}_N$ is an isomorphism of algebraic groups since char. $k = 0$ by assumption (cf. Exercise A.E.2). Thus ϕ^{-1} is a morphism of pro-groups by Corollary 4.4.11. This completes the proof of the proposition. $\square$

Similar to the notion of pro-groups, there is the notion of pro-Lie algebras.

4.4.13 Definition. Let $\mathfrak{s}$ be a Lie algebra over k and let $\mathcal{F}$ be a nonempty family of ideals of $\mathfrak{s}$ of finite codimension. Then the pair $(\mathfrak{s}, \mathcal{F})$ is a *pro-Lie algebra* if the following axioms are satisfied:

(b_1) For $\mathfrak{a}_1, \mathfrak{a}_2 \in \mathcal{F}$, $\mathfrak{a}_1 \cap \mathfrak{a}_2 \in \mathcal{F}$.

(b_2) If $\mathfrak{a}_1 \in \mathcal{F}$ and $\mathfrak{a}_2 \supset \mathfrak{a}_1$ is an ideal, then $\mathfrak{a}_2 \in \mathcal{F}$.

(b_3) The canonical Lie algebra homomorphism $\gamma : \mathfrak{s} \to \underset{\mathfrak{a}\in\mathcal{F}}{\text{Inv. lt.}}\, \mathfrak{s}/\mathfrak{a}$ is an isomorphism, where $\mathcal{F}$ is a directed set under the partial order $\mathfrak{a}_1 \leq \mathfrak{a}_2$ if and only if $\mathfrak{a}_1 \supset \mathfrak{a}_2$.

The set $\mathcal{F}$ is called the *complete defining set* (for short, the *defining set*) of the pro-Lie algebra $\mathfrak{s}$.

As a simple consequence of (b_3), for a pro-Lie algebra, we obtain

$$(1) \qquad \bigcap_{\mathfrak{a}\in\mathcal{F}} \mathfrak{a} = \{0\}.$$

Putting the discrete topology on each $\mathfrak{s}/\mathfrak{a}$, for $\mathfrak{a} \in \mathcal{F}$, we can take the inverse limit topology on $\mathfrak{s}$ (via the isomorphism γ). This is called the *pro-topology* on $\mathfrak{s}$. Under the pro-topology, $\mathfrak{s}$ is a topological Lie algebra. Further, the ideals $\{\mathfrak{a}; \mathfrak{a} \in \mathcal{F}\}$ form a fundamental system of open neighborhoods of 0 in $\mathfrak{s}$.

Let $\mathfrak{s}$ and $\mathfrak{s}'$ be pro-Lie algebras with the defining sets $\mathcal{F}$ and $\mathcal{F}'$, respectively. Then a Lie algebra homomorphism $\phi : \mathfrak{s} \to \mathfrak{s}'$ is called a *pro-Lie algebra homomorphism* if the following condition is satisfied:

$$(*) \qquad \text{For all } \mathfrak{a}' \in \mathcal{F}',\ \phi^{-1}(\mathfrak{a}') \in \mathcal{F}.$$

As in the case of pro-groups, it suffices to check $(*)$ for $\mathfrak{a}'$ in a cofinal subset of $\mathcal{F}'$. Thus, one has the notion of a pro-Lie algebra isomorphism. Clearly, a pro-Lie algebra homomorphism is continuous. Conversely, a continuous Lie algebra homomorphism $\phi : \mathfrak{s} \to \mathfrak{s}'$ is a pro-Lie algebra homomorphism. Thus, an isomorphism of pro-Lie algebras is simply an isomorphism of Lie algebras which is bicontinuous.

By a *pro-Lie subalgebra*, resp. *pro-Lie ideal*, of $\mathfrak{s}$, we mean a Lie subalgebra, resp. ideal, $\mathfrak{t} \subset \mathfrak{s}$ such that $\mathfrak{t}$ is a closed subspace under the pro-topology on $\mathfrak{s}$. It is easy to see (by using (3) below) that a pro-Lie subalgebra $\mathfrak{t}$ is a pro-Lie ideal if and only if $\gamma_{\mathfrak{a}}(\mathfrak{t})$ is an ideal of $\mathfrak{s}/\mathfrak{a}$ for all $\mathfrak{a} \in \mathcal{F}$, where $\gamma_{\mathfrak{a}} : \mathfrak{s} \to \mathfrak{s}/\mathfrak{a}$ is the projection.

By the analogue of (4.4.1.2), obtained by the same proof, for any subset $A \subset \mathfrak{s}$,

(2) $$\overline{A} = \bigcap_{\mathfrak{a} \in \mathcal{F}} (A + \mathfrak{a}).$$

In particular, a subset $A \subset \mathfrak{s}$ is closed if and only if

(3) $$A = \bigcap_{\mathfrak{a} \in \mathcal{F}} (A + \mathfrak{a}).$$

4.4.14 Examples. We give some examples of pro-Lie algebras parallel to the examples of pro-groups given in 4.4.3:

(1) Any finite-dimensional Lie algebra $\mathfrak{s}$ is a pro-Lie algebra, where we take the family $\mathcal{F}$ to consist of all the ideals of $\mathfrak{s}$. In this case, the pro-topology on $\mathfrak{s}$ is the discrete topology. In fact, there is a unique pro-Lie algebra structure on a finite-dimensional Lie algebra.

(2) Let $\{\mathfrak{s}_i\}_{i \in \mathbb{N}}$ be a family of finite-dimensional Lie algebras with surjective Lie algebra homomorphisms $\varphi_i : \mathfrak{s}_{i+1} \to \mathfrak{s}_i$, for all $i \in \mathbb{N}$. Then the inverse limit

$$\mathfrak{s} = \underset{i \in \mathbb{N}}{\text{Inv. lt.}}\ \mathfrak{s}_i$$

is a pro-Lie algebra, under the family of ideals $\mathcal{F} := \{\pi_i^{-1}(\mathfrak{a}_i) : \mathfrak{a}_i$ is an ideal of $\mathfrak{s}_i,\ i \in \mathbb{N}\}$, where $\pi_i : \mathfrak{s} \to \mathfrak{s}_i$ is the canonical (surjective) projection.

(3) Let V be a vector space with a filtration $\mathcal{S}$ by finite-dimensional vector subspaces:

$$V_0 = \{0\} \subset V_1 \subset V_2 \subset \cdots,$$

such that $\bigcup V_i = V$. Let End V be the Lie algebra of all k-linear maps $f : V \to V$. Define the Lie subalgebra $\mathfrak{u}(\mathcal{S})$ of End V:

$$\mathfrak{u}(\mathcal{S}) = \{f \in \text{End}\ V : f(V_i) \subset V_{i-1}, \text{ for all } i \geq 1\}.$$

Then $\mathfrak{u}(\mathcal{S})$ is a pro-Lie algebra with the defining set $\{\mathfrak{a}\}$, where $\mathfrak{a}$ is any ideal of $\mathfrak{u}(\mathcal{S})$ containing $\mathfrak{a}_i := \{f \in \mathfrak{u}(\mathcal{S}) : f_{|V_i} \equiv 0\}$, for some $i \in \mathbb{N}$. Moreover, $\mathfrak{u}(\mathcal{S})$ is pro-nilpotent (cf. Definition 4.4.18).

Similarly, let Aut V be the group of all k-linear automorphisms of V. Then

$$U(\mathcal{S}) := \{f \in \text{Aut}\ V : (f - I)V_i \subset V_{i-1}, \text{ for all } i \geq 1\}$$

is a pro-group with the defining set $\{N\}$, where $N \subset U(\mathcal{S})$ is any normal subgroup containing

$$N_i := \{f \in U(\mathcal{S}) : f_{|V_i} = I\}, \text{ for some } i \in \mathbb{N},$$

and such that N/N_i is a closed subgroup of $\operatorname{Aut}(V_i)$. Moreover, $U(\mathcal{S})$ is pro-unipotent (cf. Definition 4.4.18).

It can be seen that Lie $U(\mathcal{S}) \approx \mathfrak{u}(\mathcal{S})$, as pro-Lie algebras, where the pro-Lie algebra Lie G of a pro-group G is defined in 4.4.16.

The following is a pro-Lie algebra analogue of Propositions 4.4.5, 4.4.10, and 4.4.12.

4.4.15 Proposition. (a) *A Lie subalgebra $\mathfrak{t}$ of a pro-Lie algebra $\mathfrak{s}$ is a pro-Lie subalgebra if and only if the canonical map*

$$\gamma_{\mathfrak{t}} : \mathfrak{t} \to \operatorname*{Inv.\ lt.}_{\mathfrak{a}' \in \mathcal{F}'} \mathfrak{t}/\mathfrak{a}'$$

is an isomorphism, where $\mathcal{F}' := \{$ideals $\mathfrak{a}'$ of $\mathfrak{t} : \mathfrak{a}' \supset \mathfrak{a} \cap \mathfrak{t}$ for some $\mathfrak{a} \in \mathcal{F}\}$. In particular, a pro-Lie subalgebra $\mathfrak{t}$ of $\mathfrak{s}$ is a pro-Lie algebra under the family $\mathcal{F}'$. Also, the pro-topology on a pro-Lie subalgebra $\mathfrak{t}$ is the subspace topology.

(b) *For a pro-Lie ideal $\mathfrak{t} \subset \mathfrak{s}$, $\mathfrak{s}/\mathfrak{t}$ is a pro-Lie algebra with the defining set $\mathcal{F}_{\mathfrak{t}} := \{\mathfrak{a}/\mathfrak{t} : \mathfrak{a} \in \mathcal{F}$ and $\mathfrak{a} \supset \mathfrak{t}\}$. The canonical map $\pi : \mathfrak{s} \to \mathfrak{s}/\mathfrak{t}$ is a pro-Lie algebra homomorphism.*

Given a pro-Lie algebra homomorphism $\varphi : \mathfrak{s} \to \mathfrak{s}'$ between two pro-Lie algebras, the image $\varphi(\mathfrak{s})$ is a pro-Lie subalgebra and, moreover, $\mathfrak{s}/\mathfrak{t} \approx \varphi(\mathfrak{s})$, as pro-Lie algebras, where $\mathfrak{t} :=$ Ker φ. In particular, a bijective pro-Lie algebra homomorphism is an isomorphism.

Proof. The proof of this proposition is similar (and, in fact, simpler) to the corresponding results for pro-groups, and hence we will give the proof only briefly.

Define $\hat{\mathfrak{t}} := \operatorname*{Inv.\ lt.}_{\mathfrak{a}' \in \mathcal{F}'} \mathfrak{t}/\mathfrak{a}'$. There is a canonical map $i : \hat{\mathfrak{t}} \to \operatorname*{Inv.\ lt.}_{\mathfrak{a} \in \mathcal{F}} \mathfrak{s}/\mathfrak{a} \cong \mathfrak{s}$ (analogous to the map i of the proof of Proposition 4.4.5), which is clearly injective. Further, it is easy to see that $i(\hat{\mathfrak{t}}) = \bigcap_{\mathfrak{a} \in \mathcal{F}} (\mathfrak{t} + \mathfrak{a})$. So, by (4.4.13.3), $i(\hat{\mathfrak{t}}) = \mathfrak{t}$ if $\mathfrak{t}$ is closed in $\mathfrak{s}$. Conversely, if $\gamma_{\mathfrak{t}}$ is an isomorphism, then it is easy to see that $i(\hat{\mathfrak{t}}) = \mathfrak{t}$ and hence $\bigcap_{\mathfrak{a} \in \mathcal{F}} (\mathfrak{t} + \mathfrak{a}) = \mathfrak{t}$. So, by (4.4.13.3) again, $\mathfrak{t} \subset \mathfrak{s}$ is closed. This proves (a). The assertion that the pro-topology on a pro-Lie subalgebra coincides with the subspace topology is easy to verify.

To prove that $\mathfrak{s}/\mathfrak{t}$ is a pro-Lie algebra, for $\mathfrak{t} \subset \mathfrak{s}$ a pro-Lie ideal, it suffices to show that the canonical map $\varphi : \mathfrak{s}/\mathfrak{t} \to \operatorname*{Inv.\ lt.}_{\substack{\mathfrak{a} \supset \mathfrak{t} \\ \mathfrak{a} \in \mathcal{F}}} \mathfrak{s}/\mathfrak{a}$ is an isomorphism. The injectivity of φ follows from (4.4.13.3). To prove the surjectivity of φ, we mimic the same argument as in the proof of Proposition 4.4.10. By Lemma 4.4.8, applied to inverse systems of finite-dimensional k-vector spaces (regarded as abelian algebraic groups), the canonical map

$$\mathfrak{s} \simeq \operatorname*{Inv.\,lt.}_{\mathfrak{a}\in\mathcal{F}} \mathfrak{s}/\mathfrak{a} \to \operatorname*{Inv.\,lt.}_{\mathfrak{a}\in\mathcal{F}} \mathfrak{s}/(\mathfrak{a}+\mathfrak{t})$$

is surjective. Further, the projection

$$\operatorname*{Inv.\,lt.}_{\mathfrak{a}\in\mathcal{F}} \mathfrak{s}/(\mathfrak{a}+\mathfrak{t}) \to \operatorname*{Inv.\,lt.}_{\substack{\mathfrak{a}\supset\mathfrak{t}\\ \mathfrak{a}\in\mathcal{F}}} \mathfrak{s}/\mathfrak{a}$$

is surjective (cf. proof of Proposition 4.4.10). This proves the surjectivity of φ and thus the first part of (b) is proved. To prove the second part, it suffices to show that $\varphi(\mathfrak{s})$ is closed in $\mathfrak{s}'$, which follows by the same argument as in the proof of Proposition 4.4.12. □

4.4.16 Definition. Let S be a pro-group with the defining set $\mathcal{F}$. For any $N \in \mathcal{F}$, let $\mathfrak{s}_N$ denote the Lie algebra of the algebraic group S/N. For $N_1 \subset N_2 \in \mathcal{F}$, the projection $\gamma_{N_2,N_1} : S/N_1 \to S/N_2$ induces the Lie algebra homomorphism

$$\dot{\gamma}_{N_2,N_1} : \mathfrak{s}_{N_1} \to \mathfrak{s}_{N_2}.$$

Thus, we get an inverse system of Lie algebras $\{\mathfrak{s}_N\}_{N\in\mathcal{F}}$. Let

$$\mathfrak{s} = \operatorname*{Inv.\,lt.}_{N\in\mathcal{F}} \mathfrak{s}_N.$$

Then, as it is easy to see (cf. Example 4.4.14 (2)), $\mathfrak{s}$ is a pro-Lie algebra with the defining set $\tilde{\mathcal{F}} := \{$ideals $\mathfrak{a}$ of $\mathfrak{s} : \mathfrak{a} \supset \operatorname{Ker}\pi_N,$ for some $N \in \mathcal{F}\}$, where $\pi_N : \mathfrak{s} \to \mathfrak{s}_N$ is the canonical projection. Observe that, to prove the axiom (b_3) of 4.4.13, we need to show that the canonical map $\gamma : \mathfrak{s} \to \operatorname*{Inv.\,lt.}_{N\in\mathcal{F}} \mathfrak{s}/\operatorname{Ker}\pi_N$ is an isomorphism. The map γ fits in the commutative diagram:

$$\begin{array}{ccc} & \mathfrak{s} & \\ {\scriptstyle\gamma}\swarrow & & \searrow{\scriptstyle\beta}\,\wr \\ \operatorname*{Inv.\,lt.}_{N\in\mathcal{F}} \mathfrak{s}/\operatorname{Ker}\pi_N & \underset{\alpha}{\hookrightarrow} & \operatorname*{Inv.\,lt.}_{N\in\mathcal{F}} \mathfrak{s}_N \end{array},$$

where α is induced from the maps π_N. Clearly, α is injective. Since β is an isomorphism (by definition), α is surjective and hence an isomorphism. Thus γ is an isomorphism.

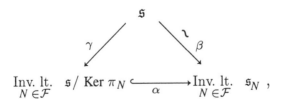

We call $\mathfrak{s}$ the *pro-Lie algebra of the pro-group* S and it is denoted by Lie S. Observe that the map $\pi_N : \mathfrak{s} \to \mathfrak{s}_N$ is surjective for any $N \in \mathcal{F}$. To prove this, consider the cofinal subset $\mathcal{F}_N := \{M \in \mathcal{F} : M \geq N\}$ of $\mathcal{F}$. Define the family $\{\mathfrak{t}_M^N\}_{M \in \mathcal{F}_N}$ of Lie algebras with $\mathfrak{t}_M^N := \mathfrak{s}_N$. For any $M \in \mathcal{F}_N$, the map $\dot{\gamma}_{N,M} : \mathfrak{s}_M \to \mathfrak{s}_N$ is surjective by [Springer–98, Theorem 4.3.7(iii)]. Now apply Lemma 4.4.8.

Let $\phi : S \to S'$ be a pro-group morphism between pro-groups S and S' with defining sets $\mathcal{F}$ and $\mathcal{F}'$. Thus, for any $N' \in \mathcal{F}'$ and $N \in \mathcal{F}$ such that $\phi(N) \subset N'$, the induced map $\phi_{N',N} : S/N \to S'/N'$ is a morphism of algebraic groups. This induces a Lie algebra homomorphism

$$\dot{\phi}_{N',N} : \mathfrak{s}_N \to \mathfrak{s}'_{N'}.$$

The Lie algebra homomorphisms $\{\dot{\phi}_{N',N}\}$ give rise to a pro-Lie algebra homomorphism (cf. proof of Proposition 4.4.12)

$$\dot{\phi} : \text{Lie } S \to \text{Lie } S',$$

satisfying $\pi'_{N'} \circ \dot{\phi} = \dot{\phi}_{N',N} \circ \pi_N$, for all N, N' as above.

This makes the assignment $S \rightsquigarrow$ Lie S into a covariant functor from the category of pro-groups and pro-group morphisms to the category of pro-Lie algebras and pro-Lie algebra homomorphisms.

If ϕ, ϕ' are two pro-group morphisms $S \to S'$ such that $\dot{\phi} = \dot{\phi}'$ and S is connected, then $\phi = \phi'$: For any $N' \in \mathcal{F}'$, choose $N \in \mathcal{F}$ such that $\phi(N) \subset N'$ and $\phi'(N) \subset N'$. Since $\dot{\phi}_{N',N} = \dot{\phi}'_{N',N}$, from the corresponding property of algebraic groups over a char. 0 field, $\phi_{N',N} = \phi'_{N',N}$. From this we conclude that $\phi = \phi'$.

Assume for the rest of this Subsection 4.4.16 that the base field $k = \mathbb{C}$.

For any $N \in \mathcal{F}$, let $\text{Exp}_N : \mathfrak{s}_N \to S/N$ be the exponential map. For $N_1 \subset N_2 \in \mathcal{F}$, we have the commutative diagram (cf. [Warner–71, Theorem 3.32]):

$$\begin{array}{ccc} \mathfrak{s}_{N_1} & \xrightarrow{\dot{\gamma}_{N_2,N_1}} & \mathfrak{s}_{N_2} \\ {\scriptstyle \text{Exp}_{N_1}}\downarrow & & \downarrow{\scriptstyle \text{Exp}_{N_2}} \\ S/N_1 & \xrightarrow[\gamma_{N_2,N_1}]{} & S/N_2 . \end{array}$$

Taking the inverse limit of the maps Exp_N $(N \in \mathcal{F})$, we get the continuous map (under the pro-topology)

$$\text{Exp} : \mathfrak{s} \to S.$$

It is easy to see that Exp : $\mathfrak{s} \to S$ is the unique map such that, for all $N \in \mathcal{F}$, the following diagram is commutative:

$$\begin{array}{ccc} \mathfrak{s} & \longrightarrow & \mathfrak{s}_N \\ {\scriptstyle \text{Exp}}\downarrow & & \downarrow{\scriptstyle \text{Exp}_N} \\ S & \longrightarrow & S/N\,, \end{array}$$

where the horizontal maps are the canonical projections.

Moreover, the following diagram is commutative for any pro-group morphism $\phi : S \to S'$:

$$\begin{array}{ccc} \text{Lie}\, S & \xrightarrow{\dot{\phi}} & \text{Lie}\, S' \\ {\scriptstyle \text{Exp}}\downarrow & & \downarrow{\scriptstyle \text{Exp}} \\ S & \xrightarrow[\phi]{} & S'\,. \end{array}$$

4.4.17 Remark. There are examples of finite-dimensional Lie algebras $\mathfrak{s}$, such that there is no algebraic group S with Lie $S \simeq \mathfrak{s}$ (cf. Exercise 4.4.E.5).

4.4.18 Definition. A pro-group S, resp. pro-Lie algebra $\mathfrak{s}$, is called *pro-unipotent*, resp. *pro-nilpotent*, if the algebraic group S/N is unipotent for all N in the defining set of S, resp. the Lie algebra $\mathfrak{s}/\mathfrak{a}$ is nilpotent for all $\mathfrak{a}$ in the defining set of $\mathfrak{s}$. Recall that an algebraic group is called *unipotent* if it is isomorphic to a closed subgroup of U_n for some n, where U_n is the group of upper-triangular $n \times n$ matrices over k with diagonal entries 1. A unipotent group is connected (cf. [Demazure–Gabriel–70, Chap. IV, §2, Proposition 4.1]).

Let $\mathcal{C}_{\text{uni}}$ be the category of pro-unipotent pro-groups over k and pro-group morphisms between them, where, following the assumption in this Section 4.4, k is an algebraically closed field of char. 0. By Exercise 4.4.E.6, any pro-unipotent pro-group is connected.

Similarly, let $\mathcal{C}_{\text{nil}}$ be the category of pro-nilpotent pro-Lie algebras over k and pro-Lie algebra homomorphisms between them.

4.4.19 Theorem. *The category $\mathcal{C}_{\text{uni}}$ is equivalent to the category $\mathcal{C}_{\text{nil}}$ under the functor $\mathfrak{L}$ taking $S \rightsquigarrow$ Lie S and a pro-group morphism $\phi \rightsquigarrow \dot{\phi}$ defined in 4.4.16. Moreover, the exponential map* Exp: Lie $S \to S$ *is bijective for any $S \in \mathcal{C}_{\text{uni}}$.*

Note that using [Demazure–Gabriel–70, Chap. II, §6, n⁰3], Exp : Lie $S \to S$ *is defined over k itself for any pro-unipotent pro-group S.*

Before we prove the above theorem, we need the following well-known result, the proof of which can be found, e.g., in [Demazure–Gabriel–70, Chap. IV, §2, $\text{n}^0 4$].

4.4.20 Lemma. *Let $\mathfrak{s}$ be a finite-dimensional nilpotent Lie algebra (over k). Then the Campbell–Hausdorff formula $H : \mathfrak{s} \times \mathfrak{s} \to \mathfrak{s}$ (cf. loc. cit.) defines a group structure on $\mathfrak{s}$, making it into a unipotent algebraic group (denoted by S). Moreover,* Lie $S = \mathfrak{s}$ *and the identity map $I : \mathfrak{s} \to S$ is the exponential map.*

Further, for any Lie algebra homomorphism $f : \mathfrak{s}_1 \to \mathfrak{s}_2$ between two finite-dimensional nilpotent Lie algebras, the same map $f : S_1 \to S_2$ is an algebraic group morphism. Moreover, $\dot{f} = f$.

We say that the (unipotent) algebraic group S as above is *associated* to the finite-dimensional nilpotent Lie algebra $\mathfrak{s}$.

4.4.21 *Proof of Theorem 4.4.19.* We define a functor $\mathfrak{I}$ from $\mathcal{C}_{\text{nil}}$ to $\mathcal{C}_{\text{uni}}$ as follows:

Let $\mathfrak{s}$ be a pro-nilpotent pro-Lie algebra with defining set $\mathcal{F}$. For any $\mathfrak{a} \in \mathcal{F}$, let $\mathfrak{s}_{\mathfrak{a}}$ be the nilpotent Lie algebra $\mathfrak{s}/\mathfrak{a}$ and let $S_{\mathfrak{a}}$ be the associated unipotent algebraic group. For any $\mathfrak{a}_1 \subset \mathfrak{a}_2 \in \mathcal{F}$, the homomorphism $\mathfrak{s}/\mathfrak{a}_1 \to \mathfrak{s}/\mathfrak{a}_2$ induces the algebraic group morphism $S_{\mathfrak{a}_1} \to S_{\mathfrak{a}_2}$. Thus, we have an inverse system of algebraic groups $\{S_{\mathfrak{a}}\}_{\mathfrak{a}\in\mathcal{F}}$. Let

$$S := \operatorname*{Inv.\,lt.}_{\mathfrak{a}\in\mathcal{F}} S_{\mathfrak{a}},$$

in the category of groups. We claim that S is a pro-group with the defining set $\tilde{\mathcal{F}}$ consisting of those subgroups $N \subset S$ such that $N = \pi_{\mathfrak{a}}^{-1}(N_{\mathfrak{a}})$, for some $\mathfrak{a} \in \mathcal{F}$ and some normal closed subgroup $N_{\mathfrak{a}} \subset S_{\mathfrak{a}}$, where $\pi_{\mathfrak{a}} : S \to S_{\mathfrak{a}}$ is the canonical projection. We first show that $\pi_{\mathfrak{a}} : S \to S_{\mathfrak{a}}$ is surjective for all $\mathfrak{a} \in \mathcal{F}$: Fix $\mathfrak{a} \in \mathcal{F}$ and define the (corresponding) cofinal subset $\mathcal{F}_{\mathfrak{a}} := \{\mathfrak{t} \in \mathcal{F} : \mathfrak{t} \geq \mathfrak{a}\}$. Define the inverse system of algebraic groups $\{H_{\mathfrak{t}}\}_{\mathfrak{t}\in\mathcal{F}_{\mathfrak{a}}}$ by $H_{\mathfrak{t}} = S_{\mathfrak{a}}$ and the identity maps between them, and define a morphism from $\{S_{\mathfrak{t}}\}_{\mathfrak{t}\in\mathcal{F}_{\mathfrak{a}}}$ to $\{H_{\mathfrak{t}}\}$ by the standard projection maps $S_{\mathfrak{t}} \to S_{\mathfrak{a}}$. Now, applying Lemma 4.4.8, the surjectivity of $\pi_{\mathfrak{a}}$ follows. From this it is easy to see that S is a pro-group. Moreover, Lie $S \cong \mathfrak{s}$.

As a set, since $S_{\mathfrak{a}} = \mathfrak{s}/\mathfrak{a}$ (for any $\mathfrak{a} \in \mathcal{F}$), we get (as sets) $S = \mathfrak{s}$. For any pro-Lie algebra homomorphism $\phi : \mathfrak{s} \to \mathfrak{s}'$ between pro-nilpotent pro-Lie algebras, the same set map, to be denoted by $\Phi : S \to S'$, can be seen to be a pro-group morphism and $\dot{\phi} = \phi$ (cf. Lemma 4.4.20).

Now a routine checking shows that the functors $\mathfrak{L}$ and $\mathfrak{I}$ are inverse to each other, if we use the following (cf. [Demazure–Gabriel–70, Chap. IV, §2, n^{0}4]): For a unipotent algebraic group H, the exponential map Exp : Lie $H \to H$ is a biregular isomorphism of varieties, where Lie H is thought of as an affine space. Moreover, for another unipotent algebraic group H', the map

$$\text{Mor}(H, H') \to \text{Mor}(\text{Lie } H, \text{Lie } H'),$$

taking $\phi \mapsto \dot{\phi}$, is a bijection, where Mor(H, H'), resp. Mor(Lie H, Lie H'), denotes the set of algebraic group morphisms from H to H', resp. Lie algebra homomorphisms. Further, for any $\phi \in \operatorname{Hom}(H, H')$, the following diagram is commutative:

$$\begin{array}{ccc} \operatorname{Lie} H & \xrightarrow{\dot{\phi}} & \operatorname{Lie} H' \\ {\scriptstyle \operatorname{Exp}}\big\downarrow & & \big\downarrow{\scriptstyle \operatorname{Exp}} \\ H & \xrightarrow{\phi} & H'. \end{array}$$

□

4.4.22 Definition. Let S be a pro-group with the defining set $\mathcal{F}$. An S-module V is called a *pro-module* (or *pro-representation*) of S if any $v \in V$ is contained in a finite-dimensional S-submodule $W \subset V$ satisfying the following:

(a_1) There exists $N = N_W \in \mathcal{F}$ such that N acts trivially on W, and

(a_2) the induced representation of the algebraic group S/N on W is algebraic.

It is easy to see that for any pro-module V of S, any finite-dimensional S-submodule W satisfies (a_1) and (a_2) as above.

Similarly, we define a pro-representation of a pro-Lie algebra:

4.4.23 Definition. Let $\mathfrak{s}$ be a pro-Lie algebra with defining set $\mathcal{F}$. Then, a $\mathfrak{s}$-module V is called a *pro-module* (or *pro-representation*) of $\mathfrak{s}$ if any $v \in V$ is contained in a finite-dimensional $\mathfrak{s}$-submodule $W \subset V$ such that there exists an ideal $\mathfrak{a} = \mathfrak{a}_W \in \mathcal{F}$ acting trivially on W.

Let S be a pro-group and let V_1, V_2 be two pro-representations of S. Then a linear map $f : V_1 \to V_2$ commuting with the S-actions is called a *S-module map* between pro-representations. We similarly define a $\mathfrak{s}$-module map between pro-representations of $\mathfrak{s}$.

Let $\mathfrak{M}(S)$, resp. $\mathfrak{M}(\mathfrak{s})$, be the category of pro-representations of S, resp. $\mathfrak{s}$, and S-module, resp. $\mathfrak{s}$-module, maps between them.

Let S be a pro-group with pro-Lie algebra $\mathfrak{s} = \operatorname{Lie} S$, and let V be a pro-representation of S under $\pi : S \to \operatorname{Aut} V$. Then V is canonically a pro-representation of the pro-Lie algebra $\mathfrak{s}$ denoted by $\dot{\pi}$, where $\dot{\pi}$ is obtained by "differentiating" π defined as follows:

Take a finite-dimensional S-submodule W of V. Then there exists $N = N_W \in \mathcal{F}$ such that π descends to an algebraic representation $\pi_W : S/N \to \operatorname{Aut}(W)$. On differentiating π_W, we get the Lie algebra representation $\dot{\pi}_W : \mathfrak{s}_N \to \operatorname{End} W$, where $\mathfrak{s}_N = \operatorname{Lie}(S/N)$ (cf. 4.4.16), and hence W becomes a $\mathfrak{s}$-module. It is easy to see that the $\mathfrak{s}$-module structure on W does not depend on the choice of N. Since finite-dimensional S-submodules W span V, we have an $\mathfrak{s}$-module structure on V denoted by $\dot{\pi}$. Clearly $\dot{\pi}$ is a pro-representation of $\mathfrak{s}$.

In the case $k = \mathbb{C}$, it is easy to see that the following diagram is commutative (cf. [Warner–71, Remark 3.36]):

$$\begin{array}{ccc}
\mathfrak{s} & \xrightarrow{\dot{\pi}} & \mathrm{End}_{\mathrm{fin}}\, V \\
{\scriptstyle \mathrm{Exp}}\downarrow & & \downarrow{\scriptstyle \exp} \\
S & \xrightarrow[\pi]{} & \mathrm{Aut}\, V\,,
\end{array}$$

where $\mathrm{End}_{\mathrm{fin}} V$ is the subset of End V consisting of all the locally finite endomorphisms of V and exp is the exponential map of 1.3.2.

4.4.24 Lemma. *Let π and ρ be two pro-representations of a connected pro-group S in the same vector space V. Then*

$$\pi = \rho \Leftrightarrow \dot{\pi} = \dot{\rho}.$$

Proof. We only need to prove the implication "$\Leftarrow$". So assume $\dot{\pi} = \dot{\rho}$ and we prove that $\pi = \rho$. We can assume without loss of generality, by base extension, that the base field $k = \mathbb{C}$.

Let W be a finite-dimensional $\mathfrak{s}$-submodule of V (under $\dot{\pi} = \dot{\rho}$), where $\mathfrak{s} = \mathrm{Lie}\, S$. Then there exists a finite-dimensional S-submodule $W_1 \supset W$ of V under π. Further, there exists $N = N_{W_1} \in \mathcal{F}$ such that π descends to an algebraic representation $\pi_N : S/N \to \mathrm{Aut}\, W_1$, and thus we get $\dot{\pi}_N : \mathfrak{s}_N \to \mathrm{End}\, W_1$, making the following diagram commutative:

$$\begin{array}{ccc}
\mathfrak{s}_N & \xrightarrow{\dot{\pi}_N} & \mathrm{End}\, W_1 \\
\downarrow{\scriptstyle \mathrm{Exp}} & & \downarrow{\scriptstyle \exp} \\
S/N & \xrightarrow[\pi_N]{} & \mathrm{Aut}\, W_1\,.
\end{array}$$

But since $\dot{\pi}_N$ keeps the subspace $W \subset W_1$ stable, from the commutativity of the above diagram, π_N (Exp $\mathfrak{s}_N$) as well keeps W stable. Further, since S is connected, Exp $\mathfrak{s}_N$ generates S/N as a group, and thus W is a S/N (and hence S)-submodule of W_1 under π. So, we can take $W_1 = W$. Commutativity of the above diagram, in addition, gives that π_N is determined on W by $\dot{\pi}_N$. Since finite-dimensional $\mathfrak{s}$-submodules W of V span V, the lemma follows. □

4.4.25 Definition (Adjoint representation). Let S be a pro-group with pro-Lie algebra $\mathfrak{s}$. We define a representation Ad of S in $\mathfrak{s}$ as follows.

For any $N \in \mathcal{F}$, where $\mathcal{F}$ is the defining set of S, let $\mathrm{Ad}_{S/N} : S/N \to \mathrm{Aut}\, \mathfrak{s}_N$ be the adjoint representation of the algebraic group S/N (cf. [Borel–91, 3.13]), and let $\mathrm{Ad}_N : S \to \mathrm{Aut}\, \mathfrak{s}_N$ be the composition $\mathrm{Ad}_{S/N} \circ \gamma_N$ where $\gamma_N : S \to S/N$

is the projection. For a fixed $g \in S$, we have the following commutative diagram for any $N_1 \subset N_2 \in \mathcal{F}$:

$$\begin{array}{ccc} \mathfrak{s}_{N_1} & \xrightarrow{\dot{\gamma}_{N_2,N_1}} & \mathfrak{s}_{N_2} \\ \Big\downarrow {\scriptstyle \mathrm{Ad}_{N_1}(g)} & & \Big\downarrow {\scriptstyle \mathrm{Ad}_{N_2}(g)} \\ \mathfrak{s}_{N_1} & \xrightarrow[\dot{\gamma}_{N_2,N_1}]{} & \mathfrak{s}_{N_2} \,. \end{array}$$

Thus, taking the inverse limit, we get a linear map denoted $\mathrm{Ad}(g) : \mathfrak{s} \to \mathfrak{s}$. Alternatively, $\mathrm{Ad}\, g = \dot{C}_g$, where C_g is the pro-group automorphism of S given by the conjugation $x \mapsto gxg^{-1}$. Thus, each $\mathrm{Ad}\, g$ is a pro-Lie algebra automorphism of $\mathfrak{s}$. The map $g \mapsto \mathrm{Ad}(g)$ gives a representation

$$\mathrm{Ad} : S \to \mathrm{Aut}(\mathfrak{s}),$$

called the *adjoint representation.* Over the base field $k = \mathbb{C}$, from the commutative diagram just above Remark 4.4.17, for $g \in S$ and $X \in \mathfrak{s}$,

$$\mathrm{Exp}(\mathrm{Ad}\, g\, X) = g\ \mathrm{Exp}\, X\, g^{-1}. \tag{1}$$

Moreover, for any $g \in S$, $X \in \mathfrak{s}$, and any pro-representation π of S, we have

$$\dot{\pi}(\mathrm{Ad}\, g\, X) = \pi(g)\, \dot{\pi}(X)\, \pi(g)^{-1}, \tag{2}$$

as can be seen from the commutative diagram of Subsection 4.4.23 and (1).

Observe that, in general, Ad is *not* a pro-representation of S (cf. Exercise 4.4.E.2).

Recall the definition of the categories $\mathfrak{M}(S)$ and $\mathfrak{M}(\mathfrak{s})$ from 4.4.23.

4.4.26 Proposition. *Let S be a pro-unipotent pro-group with pro-Lie algebra $\mathfrak{s}$. Then the category $\mathfrak{M}(S)$ is equivalent with the category $\mathfrak{M}_{\mathrm{nil}}(\mathfrak{s})$ under $(V, \pi) \rightsquigarrow (V, \dot{\pi})$ and $f \rightsquigarrow f$ for any S-module map $f : V \to V'$ between two pro-representations V and V' of S, where $\mathfrak{M}_{\mathrm{nil}}(\mathfrak{s})$ is the full subcategory of $\mathfrak{M}(\mathfrak{s})$ consisting of those pro-representations V of $\mathfrak{s}$ such that each element $x \in \mathfrak{s}$ acts locally nilpotently on V.*

Proof. We first show that if (V, π) is a pro-representation of $\mathfrak{s}$ in the category $\mathfrak{M}_{\mathrm{nil}}(\mathfrak{s})$, then V is a pro-module for S under ρ such that $\dot{\rho} = \pi$:

Let $W \subset V$ be a finite-dimensional subspace stable under the action of $\mathfrak{s}$. Then there exists $N = N_W \in \mathcal{F}$ (the defining set of S) such that the $\mathfrak{s}$-module structure on W descends to a $\mathfrak{s}_N$-module structure π_W, where $\mathfrak{s}_N := \mathrm{Lie}\, S/N$ (cf. 4.4.16). Since S/N is unipotent and each element of $\mathfrak{s}_N$ acts nilpotently on

W, W is an algebraic S/N-module under ρ_W such that $\dot{\rho}_W = \pi_W$. In particular, W is an S-module. Moreover, the S-module structure on W does not depend on the choice of N. Since such W's span V, we get an S-module structure ρ on V. Clearly ρ is a pro-representation of S and $\dot{\rho} = \pi$.

It is easy to see that a linear map $f : (V, \pi) \to (V', \pi')$ between two pro-representations of S is an S-module map iff it is a $\mathfrak{s}$-module map $(V, \dot{\pi}) \to (V', \dot{\pi}')$. In fact, this holds for an arbitrary connected pro-group S, not necessarily pro-unipotent.

Now the proposition follows from Lemma 4.4.24. □

4.4.E EXERCISES

(1) Let $\phi : G \to G'$ be a surjective morphism of pro-groups. Assume that G is an algebraic group. Then show that G' also is an algebraic group.

(2) In general Ad is *not* a pro-representation of a pro-group G. Take, e.g., $G = \overset{o}{G}(k[[t]])$, for a (finite-dimensional) non-abelian connected algebraic group $\overset{o}{G}$. Show that G has a canonical pro-group structure and Ad is not a pro-representation of G.

(3) Let G be a group and let $\mathcal{F}'$ be a directed nonempty family of normal subgroups satisfying the conditions (a_3) and (a_4) of 4.4.1. Then show that the family $\mathcal{F}'$ can uniquely be enlarged to a (complete) defining set of a pro-group structure on G.

(4) Let G be a pro-group with the defining set $\mathcal{F}$. For any $N \in \mathcal{F}$, let $H_N \subset G/N$ be a closed subgroup such that for $N_1 \subset N_2 \in \mathcal{F}$, $\gamma_{N_2,N_1}(H_{N_1}) = H_{N_2}$. Then, show that $H := \operatorname*{Inv.\,lt.}_{N \in \mathcal{F}} H_N$ is a pro-subgroup of G with $\gamma_N(H) = H_N$, for all $N \in \mathcal{F}$.

Show further that all the pro-subgroups of G are obtained this way.

(5) Give examples of finite-dimensional Lie algebras $\mathfrak{s}$ such that there is no algebraic group S with Lie $S \simeq \mathfrak{s}$.

(6) Let G be a pro-group such that G/N is connected for each $N \in \mathcal{F}$. Then show that G is connected. In particular, a pro-unipotent pro-group is connected.

4.C Comments. The notion of ind-varieties was introduced by [Šafarevič–82], which is our basic reference for Sections 4.1–4.3. In particular, Proposition 4.2.2, Lemma 4.2.5 and Theorem 4.3.7 are in [loc. cit.]. Also a stronger version of Proposition 4.3.6 (where the condition (b) is replaced by the irreducibility of Y) appears in [loc. cit., Theorem 2]. However, his argument for the stronger version is not complete. In particular, this also affects his Theorem 1. (See his corrigendum [Šafarevič–95] and also the *AMS Review* 96e:14054. The gap in [Šafarevič–82] is also pointed out in [Kambayashi–96], where a version of Proposition 4.3.6 is proved.)

The notion of pro-groups is due to [Serre–60] except that he only considers commutative groups, although the extension of his definition to noncommutative groups is straightforward; see [Kovacic–73] for the extension. For Subsections 4.4.1–4.4.12, the reader is referred to [Serre–60] and [Kovacic–73]. The definition and various properties of pro-Lie algebras are taken from [Slodowy–84a]. The classical version of Theorem 4.4.19 concerning unipotent algebraic groups and nilpotent Lie algebras in char. 0 (instead of pro-objects) can be found, e.g., in [Demazure–Gabriel–70, Chap. IV, §2, Corollary 4.5].

V

Tits Systems

Basic Theory

Most of the content of Section 5.1 is fairly standard. We recall the definition of a Tits system, also called a BN-pair, and then collect its various general properties in Theorem 5.1.3, which we will need subsequently, including the Bruhat decomposition. We also recall the definition of a topological Tits system. Its definition is modeled so that the associated "flag varieties" have a CW-complex structure with Bruhat cells as the cells (cf. Theorem 5.1.5). It is shown that for a Tits system (G, B, N, S), G is an amalgamated product of its minimal parabolic subgroups $\{P_s\}_{s\in S}$ and N. Conversely, given a finite family of groups B, $\{P_s\}_{s\in S}$ and N satisfying a number of properties, it is shown that (G, B, N, S) is a Tits system, where G is the amalgamated product of B, $\{P_s\}$ and N (cf. Theorem 5.1.8). This result, due to Tits, will be used in the construction of Kac–Moody groups in the next chapter.

The refined Tits system, introduced by Kac–Peterson, is defined in Section 5.2. It is shown that a refined Tits system canonically gives rise to a Tits system. We prove Theorem 5.2.3, collecting various general properties of a refined Tits system, including the Birkhoff decomposition.

5.1. An Introduction to Tits Systems

5.1.1 Definition. A *Tits system*, also called a *BN-pair*, (G, B, N, S) consists of a group G, subgroups B and N and a finite subset $S \subset N/(B \cap N)$ satisfying the following axioms:

(BN_1) $B \cap N$ is a normal subgroup of N and S generates $W := N/B \cap N$,
(BN_2) B and N generate G as a group,
(BN_3) for any $s \in S$, $sBs^{-1} \not\subseteq B$, and
(BN_4) for $w \in W$ and $s \in S$, $C(s)C(w) \subseteq C(w) \cup C(sw)$,

where, for $w \in W$, by $C(w)$ we mean the subset $B\bar{w}B \subset G$ for a coset representative $\bar{w}$ of w in N; of course, a different choice of the coset representative will give rise to the same subset.

The group W is called the *Weyl group* of the Tits system. Any subgroup P satisfying $B \subseteq P \subseteq G$ is called a *standard parabolic subgroup* of G. Any subgroup Q of G which is conjugate to a standard parabolic subgroup of G is called a *parabolic subgroup* of G.

The following is one of the most important examples of Tits systems:

5.1.2 Example. Let G be a connected reductive algebraic group over an algebraically closed field k (of an arbitrary char.), and let $B \subset G$ be a Borel subgroup and $T \subset B$ a maximal torus. Let N be the normalizer of T in G. Then, as is well known, $N \cap B = T$. Moreover, $W := N/(B \cap N)$ is, by definition, the Weyl group associated to G. Now, take for S the set of simple reflections of W. Then (G, B, N, S) is a Tits system (cf. [Humphreys–95, §29.1]).

Let (G, B, N, S) be a Tits system. For any subset $Y \subseteq S$, let W_Y be the subgroup of W generated by Y. We collect some of the main properties of Tits systems in the following theorem.

For $w \in W$, let $\ell(w)$ be the smallest number such that w can be written as $w = s_1 \ldots s_{\ell(w)}$ with each $s_i \in S$. (By the (a) part of the following theorem, each s_i is of order 2.) We set $\ell(e) = 0$. For a subset $V \subset G$, by $\langle V \rangle$ we mean the subgroup of G generated by V.

5.1.3 Theorem. *Let (G, B, N, S) be a Tits system. Then*

(a) *All the elements of S are of order 2.*

(b) *For any subset $Y \subset S$, $P_Y := BW_YB$ is a subgroup of G.*

(c) Bruhat decomposition: *We have the disjoint union*

(c_1) $G = \bigsqcup_{w \in W} C(w).$

More generally, for any $Y, Y' \subset S$, we have the disjoint union

(c_2) $G = \bigsqcup_{w \in W_Y \backslash W / W_{Y'}} P_Y w P_{Y'}.$

(d) $$C(s)C(w) = C(sw), \quad \textit{if } \ell(sw) \geq \ell(w)$$
$$= C(w) \cup C(sw), \quad \textit{if } \ell(sw) \leq \ell(w).$$

(e) *(W, S) is a Coxeter group.*

(f) *For any reduced decomposition $w = s_1 \ldots s_p$ with $s_i \in S$, $C(s_i) \subset \langle C(w) \rangle$ for $1 \leq i \leq p$. Moreover, $\langle B, wBw^{-1} \rangle = \langle C(w) \rangle$.*

(g) *For $Y, Y' \subset S$, $P_Y = P_{Y'}$ if and only if $Y = Y'$. Further, for any standard parabolic subgroup P of G, there exists $Y \subset S$ such that $P = P_Y$.*

In particular, there are exactly $2^{|S|}$ standard parabolic subgroups of G.

(h) *For any parabolic subgroup P of G, its normalizer $N(P) = P$. Moreover, two standard parabolic subgroups P and Q of G are conjugate if and only if $P = Q$.*

(i) *Let $Y \subseteq S$ and let $w \in W'_Y$* (cf. Definition 1.3.17). *For any decomposition $w = w_1 \dots w_k$ such that $\ell(w) = \sum_{i=1}^k \ell(w_i)$, the following holds:*
(i_1) *Let A_i be any subset of $C(w_i)$ such that the projection $A_i \to C(w_i)/B$ is bijective, resp. surjective, for all $1 \leq i \leq k$. Then $\varphi : A_1 \times \cdots \times A_k \to BwP_Y/P_Y$, taking $(a_1, \dots, a_k) \mapsto a_1 \dots a_k$* mod P_Y, *is bijective, resp. surjective.*
(i_2) *Assume, in addition, that each $w_i \in S$ and let $Z_i \subset P_{w_i}$ be any subset containing e such that the projection $Z_i \subset P_{w_i}/B$ is surjective, where $P_{w_i} := P_{\{w_i\}}$. Then the image of the product map $\varphi : Z_1 \times \cdots \times Z_k \to G/P_Y$ is precisely equal to $\bigcup_{v \leq w} BvP_Y/P_Y$, where "$\leq$" is the Bruhat–Chevalley order on the Coxeter group W (cf.* 1.3.15).

Proof. (a): Take $s \in S$. By (BN_4), taking $w = s^{-1}$,

(1) $$C(s)C(s^{-1}) \subseteq C(s^{-1}) \cup B\,.$$

Taking inverses in (1), we get

(2) $$C(s)C(s^{-1}) \subseteq C(s) \cup B.$$

By (BN_3), $C(s)C(s^{-1}) \neq B$, and hence, $C(s) \neq B$ and $C(s^{-1}) \neq B$. So, by (1) and (2),

(3) $$C(s)C(s^{-1}) = C(s^{-1}) \sqcup B = C(s) \sqcup B.$$

(Use the fact that all the sets involved in (3) are double B cosets.) By (3) we get

(4) $$C(s) = C(s^{-1}).$$

Similarly, by (BN_4), taking $w = s$, we get

(5) $$C(s)C(s) \subset C(s) \cup C(s^2).$$

By (3), (4) and (5), we get
$$C(s^2) = B.$$

This gives that $s^2 = 1$, proving (a).

(b): By (BN_4) and the (a)-part, clearly P_Y is a subgroup of G.

(c): By (BN_2) and the (b) part, $G = \bigcup_{w\in W} C(w)$. So, to prove ($c_1$), it suffices to show that $C(v) = C(w) \Rightarrow v = w$. We prove the above by induction on $d := \min\{\ell(v), \ell(w)\}$. If $d = 0$, i.e., (say) $w = e$, then $C(v) = C(w) \Rightarrow v = e$. So assume $d > 0$ and let $d = \ell(w) \leq \ell(v)$. Write $w = sw'$, with $\ell(w') < \ell(w)$ and $s \in S$. Then $sw'B \subseteq C(w) = C(v)$. Multiplying this by s and using (BN_4), we get

$$w'B \subset sC(v) \subset C(sv) \cup C(v).$$

Hence $C(w') = C(sv)$ or $C(w') = C(v) = C(w)$. In the first case, since $\ell(w') < \ell(w)$, by the induction hypothesis $w' = sv$, i.e., $v = w$. In the second case, by the induction hypothesis, $w' = w$, which is a contradiction. Hence this case is not possible. This completes the induction hypothesis and proves the (c_1) part.

We now prove (c_2). Since $\overline{G} := \bigcup_{w\in W_Y\backslash W/W_{Y'}} P_Y w P_{Y'}$ is bi B-stable and it contains $\bigcup_{w\in W_Y\backslash W/W_{Y'}} W_Y w W_{Y'} = W$, by ($c_1$), $\overline{G} = G$. Now, for any $w \in W$,

$$\begin{aligned} P_Y w P_{Y'} &= \bigcup_{v\in W_Y, v'\in W_{Y'}} C(v)C(w)C(v') \\ &\subset \bigcup_{v\in W_Y, v'\in W_{Y'}} C(s_1)\ldots C(s_p)C(w)C(s'_1)\ldots C(s'_{p'}) \\ (*) \qquad &\subset \bigcup_{\theta\in W_Y, \theta'\in W_{Y'}} C(\theta w \theta'), \text{ by } (BN_4), \end{aligned}$$

where $v = s_1 \ldots s_p$, $v' = s'_1 \ldots s'_{p'}$ are decompositions so that $s_i \in Y$ and $s'_j \in Y'$. The inclusion $C(v) \subset C(s_1)\ldots C(s_p)$ follows since the right side is bi B-stable and contains v. From $(*)$ and (c_1), it follows immediately that the union in (c_2) is disjoint, proving (c_2).

(d): We prove this by induction on $\ell(w)$. If $\ell(w) = 0$, it is trivially true. So assume $\ell(w) > 0$ and write $w = w't$ for $t \in S$ and $\ell(w') < \ell(w)$.

We first consider the case when $\ell(sw) \geq \ell(w)$. In this case, $\ell(sw') \geq \ell(w')$ since $\ell(w') + 1 = \ell(w) \leq \ell(sw) = \ell(sw't) \leq \ell(sw') + 1$. In particular, by the induction hypothesis,

$$(6) \qquad C(sw') = C(s)C(w').$$

Assume, if possible, that $C(s)C(w) \neq C(sw)$. Then, $sBw \bigcap C(w) \neq \emptyset$ by (BN_4), and hence $sBw' \bigcap C(w)t \neq \emptyset$. But, by taking the inverse of (BN_4), we get $C(w)t \subset C(w) \bigcup C(w')$, and hence $sBw' \bigcap (C(w) \bigcup C(w')) \neq \emptyset$. This gives (by (6) and the (c_1) part) $sw' = w$, since $sw' \neq w'$. But this is absurd since this would imply $\ell(w) - 1 = \ell(w') = \ell(sw) \geq \ell(w)$. This proves the (d) part in the case $\ell(sw) \geq \ell(w)$.

So consider now the case $\ell(sw) \leq \ell(w)$. Replacing w by sw, we get from the previous case, $C(s)C(sw) = C(w)$. So

$$\begin{aligned} C(s)C(w) &= C(s)C(s)C(sw) \\ &= (C(s) \cup B)C(sw), \qquad \text{by (3)–(4)}, \\ &= C(w) \cup C(sw). \end{aligned}$$

This completes the induction and hence proves (d).

(e): To prove (e), by Theorem 1.3.11, it suffices to check the exchange condition, i.e., for $s \in S$ and $w \in W$ such that $\ell(sw) \leq \ell(w)$ and any reduced expression $w = s_1 \dots s_n$ $(s_i \in S)$, we need to show that

$$sw = s_1 \dots \hat{s_j} \dots s_n, \quad \text{for some } j. \tag{7}$$

By the (d) part,

$$C(s)C(w) = C(sw) \cup C(w) \text{ and } C(w) = C(s_1) \dots C(s_n). \tag{8}$$

Choose the first $1 \leq j \leq n$ such that $\ell(ss_1 \dots s_j) \leq \ell(ss_1 \dots s_{j-1})$. Then, by using the (d) part and the identity obtained from it by taking inverses,

$$\begin{aligned} C(s)C(w) &= C(s)C(s_1) \dots C(s_{j-1})C(s_j) \dots C(s_n) \\ &= C(ss_1 \dots s_{j-1})C(s_j) \dots C(s_n) \\ &= [C(ss_1 \dots s_{j-1}s_j) \cup C(ss_1 \dots s_{j-1})] \cdot C(s_{j+1}) \dots C(s_n) \\ &\subset \bigcup \Big(C(ss_1 \dots s_j s_{j_1} \dots s_{j_p}) \cup C(ss_1 \dots s_{j-1} s_{j_1} \dots s_{j_p}) \Big), \end{aligned} \tag{9}$$

where the union is taken over $j + 1 \leq j_1 < \cdots < j_p \leq n$. Comparing (8) and (9), we see that, since $\ell(w) = n$, either $w = ss_1 \dots s_{j-1}s_{j+1} \dots s_n$ or $w = ss_1 \dots s_j s_{j+1} \dots \hat{s}_{j+k} \dots s_n$, for some $1 \leq k$. In either case (7) is established and hence the (e) part is proved.

(f): By abuse of notation, for any $v \in W$, we also denote by v its representative in N. Since $w \in C(w)$, and $wB \subset C(w)$, $\langle B \cup wBw^{-1} \rangle \subset \langle C(w) \rangle \subset \langle C(s_1) \cup \cdots \cup C(s_p) \rangle$. So, to prove both assertions of (f), it suffices to show that $\langle B \cup wBw^{-1} \rangle \supseteq \langle C(s_1) \cup \cdots \cup C(s_p) \rangle$. We prove this by induction on p. For $p = 0$, there is nothing to prove. So take $p \geq 1$. Since $\ell(s_1 w) < \ell(w)$, by the (d) part, $C(w) \subset C(s_1)C(w)$. In particular, $w = b_1 s_1 b_2 w b_3$, for some $b_i \in B$, and hence $s_1 \in \langle B \cup wBw^{-1} \rangle$ (use $s_1^{-1} = b_2 w b_3 w^{-1} b_1$). So $\langle B \cup wBw^{-1} \rangle \supset$

$\langle B \cup s_1 w B w^{-1} s_1^{-1}\rangle$. By induction, $\langle B \cup s_1 w B w^{-1} s_1^{-1}\rangle \supseteq \langle C(s_2) \cup \cdots \cup C(s_p)\rangle$, and hence $\langle B \cup w B w^{-1}\rangle \supset \langle C(s_1) \cup \cdots \cup C(s_p)\rangle$. This completes the induction and hence proves (f).

(g): By (c_1), $P_Y = P_{Y'}$ if and only if $W_Y = W_{Y'}$. Now, to prove that $W_Y = W_{Y'}$ if and only if $Y = Y'$, it suffices to show that for any $Y, Y' \subset S$, $Y \cap W_{Y'} \subset Y'$.

Write $s \in Y \cap W_{Y'}$ as $s = s_1 \ldots s_p$, with each $s_i \in Y'$. Since (W, S) is a Coxeter group (by the (e) part), by the Coxeter Condition (cf. Theorem 1.3.11(a)), unless $s \in Y'$, $s^{-1} s_1 \ldots s_p \neq 1$ since each relation in W involving s contains it at least twice. This also follows from the Root-system Condition (cf. Theorem 1.3.11(b)).This proves the assertion $Y \cap W_{Y'} \subset Y'$.

Now let $P \subset G$ be any standard parabolic subgroup. Write

$$P = \bigcup_{w \in W'} C(w),$$

where $W' \subset W$ is a subset. We claim that $W' = \langle W' \cap S\rangle$: By the (f) part, $\langle W' \cap S\rangle \supseteq W'$. But, P being a subgroup, $\langle W' \cap S\rangle \subset W'$. This proves (g).

(h): Let P and Q be two standard parabolic subgroups of G such that $gPg^{-1} = Q$ for some $g \in G$. Let $C(w)$ be the double coset containing g. Then clearly $wPw^{-1} = Q$, and hence Q contains $\langle B, wBw^{-1}\rangle = \langle C(w)\rangle$ (by (f)). In particular, $g \in Q$, showing that $P = Q$ and $N(P) = P$. This proves (h).

(i_1): We first show that the map φ is surjective, assuming that each $A_i \to C(w_i)/B$ is surjective. Im $\varphi = C(w_1) \ldots C(w_k) P_Y \pmod{P_Y}$. But, by the (d) part, $C(w_1) \ldots C(w_k) = C(w)$. Hence the surjectivity of φ follows.

We come to the injectivity of φ, assuming that all the projections $A_i \to C(w_i)/B$ are bijective. Let $\varphi(a_1, \ldots, a_k) = \varphi(a'_1, \ldots, a'_k)$, for $a_i, a'_i \in A_i$, i.e.,

$$a_1 \ldots a_k = a'_1 \ldots a'_k p, \tag{10}$$

for some $p \in C(v)$ and $v \in W_Y$. By (1.3.17.2), $\ell(wv) = \ell(w) + \ell(v)$ and hence, by the (d) part, the left side, resp. the right side, of (10) is contained in $C(w)$, resp. $C(wv)$. In particular, $v = e$ and $p \in B$. Now take a reduced decomposition $w_1 = s_1 \ldots s_n$ and choose a subset $D_j \subset C(s_j)$ such that $D_j \to C(s_j)/B$ is a bijection. Write (using the (d) part again)

$$a_1 = d_1 \ldots d_n b, \; a'_1 = d'_1 \ldots d'_n b', \tag{11}$$

for $d_i, d'_i \in D_i$ and $b, b' \in B$. Then, by (10), we get

$$d_1'^{-1} d_1 \ldots d_n b a_2 \ldots a_k = d'_2 \ldots d'_n b' a'_2 \ldots a'_k p. \tag{12}$$

By (BN_4), $d_1'^{-1}d_1 \in B \cup C(s_1)$. Assume, if possible, that $d_1'^{-1}d_1 \in C(s_1)$. Then the left side, resp. right side, of (12) would belong to $C(w)$, resp. $C(s_1 w)$, which is a contradiction. Hence $d_1'^{-1}d_1 \in B \Rightarrow d_1' = d_1$ since $D_1 \to C(s_1)/B$ is a bijection. Now, from the identity $d_2 \dots d_n b a_2 \dots a_k = d_2' \dots d_n' b' a_2' \dots a_k' p$, arguing the same way, we successively get $d_2 = d_2', \dots, d_n = d_n'$. So, by (11), $a_1 B = a_1' B \Rightarrow a_1 = a_1'$ since $A_1 \to C(w_1)/B$ is a bijection. Arguing the same way, $a_2 = a_2', \dots, a_{k-1} = a_{k-1}'$, and hence, by (10), $a_k = a_k' p \Rightarrow a_k = a_k'$ since $A_k \to C(w_k)/B$ is a bijection. This proves the injectivity of φ.

(i_2): We can assume that $Z_i = \{e\} \cup A_i$, where $A_i \subset C(w_i)$ is such that $A_i \to C(w_i)/B$ is surjective. Then $\operatorname{Im}\varphi = \bigcup C(w_{i_1}) \dots C(w_{i_p}) P_Y \pmod{P_Y}$, where the union is taken over all $1 \leq i_1 < \dots < i_p \leq k$. By the (d) part and Lemma 1.3.16,

$$\text{(13)} \qquad \bigcup C(w_{i_1} \dots w_{i_p}) \subseteq \bigcup C(w_{i_1}) \dots C(w_{i_p}) \subseteq \bigcup_{v \leq w} C(v).$$

Again by Lemma 1.3.16, for any $v \leq w$, we can write $v = w_{i_1} \dots w_{i_p}$, for some $1 \leq i_1 < \dots < i_p \leq k$, and hence both inclusions in (13) are actually equalities. This proves (i_2). So the proof of the theorem is completed. □

5.1.4 Definition. (a) A Tits system (G, B, N, S) is called a *topological Tits system* if the following additional axioms are satisfied:

(BN_5) G is a Hausdorff topological group, B and N are closed subgroups and $B \cap N$ is an open subgroup of N, i.e., W is discrete under the quotient topology.

(BN_6) For each $s \in S$, the space P_s/B, with the quotient topology, is homeomorphic to a sphere $S^{n(s)}$ of dimension $n(s) > 0$ and, moreover, the projection $\pi_s : P_s \to P_s/B$ has a local continuous cross-section and hence is a principal B-bundle (cf. [Steenrod–51, Part I, Corollary 7.4]).

For the reader's convenience, we recall the definition of a CW-complex from [Milnor–Stasheff–74, Definition 6.1].

(b) A *CW-complex* consists of a Hausdorff space X, called the *underlying space*, together with a partition of X into a collection $\{X_\alpha\}$ of disjoint subsets, called the *cells*, such that the following four properties are satisfied:

(C_1) For each α there exists a continuous map

$$f_\alpha : D^{n(\alpha)} \to X\,, \text{ for some } n(\alpha) \geq 0,$$

which carries the interior of the closed disc $D^{n(\alpha)} \subset \mathbb{R}^{n(\alpha)}$ homeomorphically onto X_α.

(C_2) For any α and any $x \in \bar{X}_\alpha \setminus X_\alpha$, x lies in a cell X_β of lower dimension $n(\beta)$.

(C_3) *Closure finiteness:* Each point of X is contained in a finite subcomplex, where, by a *finite subcomplex*, we mean a closed subset of X which is a finite union of cells.

(C_4) *Whitehead topology:* A subset of X is closed iff its intersection with each finite subcomplex is closed.

Recall that a subset of a topological space is called *locally closed* if it is open in its closure, or equivalently, it is the intersection of an open set with a closed set.

5.1.5 Theorem. *Let (G, B, N, S) be a topological Tits system and let $Y \subset S$ be any subset. Then:*

(a) *Assume that P_Y is closed and G/P_Y is Hausdorff under the quotient topology. Then, for any $w \in W'_Y$,*

$$\overline{BwP_Y} = \bigsqcup_{v \le w, v \in W'_Y} BvP_Y. \tag{1}$$

In particular, B-orbits in G/P_Y are locally closed.

(b) *Assume that G/B is Hausdorff under the quotient topology. Then $P_Y \subset G$ is closed if W_Y is finite.*

(c) *Let $P_Y \subset G$ be closed and G/P_Y Hausdorff. Define a new topology on G/P_Y, called the inductive limit topology, by declaring a set $A \subset G/P_Y$ closed if and only if $\pi_Y^{-1}(A) \cap \overline{BwP_Y}$ is closed in G for all $w \in W'_Y$, where $\pi_Y : G \to G/P_Y$ is the projection. With this topology, G/P_Y is a CW-complex with cells $\{BwP_Y/P_Y\}_{w \in W'_Y}$.*

Proof. (a): From the local triviality of π_s, it is easy to see that, for any $s \in S$, there exists a compact subset $A_s \subset P_s$ with $1 \in A_s$ and

$$A_s B = P_s \text{ and } \overline{(A_s \cap C(s))} = A(s). \tag{2}$$

Now let $w \in W'_Y$ and take a reduced decomposition $w = s_1 \dots s_n$. Then, by Theorem 5.1.3 (i_2),

$$A_{s_1} \dots A_{s_n} P_Y = \bigcup_{v \le w} BvP_Y = \bigsqcup_{\substack{v \le w, \\ v \in W'_Y}} BvP_Y. \tag{3}$$

But since each A_{s_i} is compact and G/P_Y is Hausdorff, $A_{s_1} \dots A_{s_n} P_Y$ is closed in G and, by (2),

$$\overline{A^\circ_{s_1} \dots A^\circ_{s_n} P_Y} = A_{s_1} \dots A_{s_n} P_Y, \tag{4}$$

where $A^\circ_{s_i} := A_{s_i} \bigcap C(s_i)$. But $A^\circ_{s_i} B = C(s_i)$ and hence, by Theorem 5.1.3 (i_1),

(5) $$A^\circ_{s_1} \dots A^\circ_{s_n} P_Y = BwP_Y.$$

Combining (3)–(5), (1) follows.

To prove the "In particular" statement, it suffices to show that BwP_Y is open in $\overline{BwP_Y}$ for $w \in W'_Y$. But, by (1),

$$\overline{BwP_Y} \backslash (BwP_Y) = \bigsqcup_{\substack{v<w \\ v \in W'_Y}} BvP_Y\,,$$

which is closed by (1). This proves (a).

(b) follows immediately from the (a) part, taking $P_Y = B$ in the (a) part and using the fact that, for any $w \in W_Y$, any reduced decomposition $w = s_1 \cdots s_n$ has $s_i \in Y$ (cf. 1.3.17).

(c): Recall the Bruhat decomposition of $X := G/P_Y$ from Theorem 5.1.3(c):

(6) $$X = \bigsqcup_{w \in W'_Y} BwP_Y/P_Y.$$

We show that X is a CW-complex under the inductive limit topology, with cells $\{BwP_Y/P_Y\}_{w \in W'_Y}$, BwP_Y/P_Y being a cell of dimension $d(w)$, where $d(w) := \sum_{i=1}^p n(s_i)$ for a reduced decomposition $w = s_1 \dots s_p$. Recall that, under the quotient topology, P_s/B is homeomorphic with the sphere $S^{n(s)}$. We will see later in the proof that $d(w)$ does not depend on the choice of the reduced decomposition of w. Of course, for $v < w$, $d(v) < d(w)$.

For each $s \in S$, by axiom (BN_6), $\pi_s : P_s \to P_s/B$ is a principal B-bundle over the sphere $S^{n(s)} \approx P_s/B$. Choose a continuous map $\alpha_s : (D^{n(s)}, S^{n(s)-1}) \to (P_s/B,\ B/B)$ such that, denoting $\overset{\circ}{D}{}^{n(s)} := D^{n(s)} \backslash S^{n(s)-1}$, $\alpha_{s\,|\overset{\circ}{D}{}^{n(s)}} : \overset{\circ}{D}{}^{n(s)} \to (P_s \backslash B)/B$ is a homeomorphism, where $D^{n(s)}$ is the closed unit disc in $\mathbb{R}^{n(s)}$ with boundary $S^{n(s)-1}$. The pullback bundle $\alpha_s^*(P_s)$, being a bundle over the contractible space $D^{n(s)}$, is trivial by [Steenrod–51, Part I, Corollary 11.6], i.e., it admits a continuous map $\sigma_s : D^{n(s)} \to P_s$ such that $\pi_s \circ \sigma_s = \alpha_s$. We can further assume that $1 \in \operatorname{Im} \sigma_s$.

Fix $w \in W'_Y$ and choose a reduced decomposition $w = s_1 \dots s_p$. Now define a continuous map (under the inductive limit topology on G/P_Y)

(7) $$\sigma_w : D^{n(s_1)} \times \cdots \times D^{n(s_p)} \to G/P_Y,$$

by $\sigma_w(t_1, \dots, t_p) = \sigma_{s_1}(t_1) \dots \sigma_{s_p}(t_p) \mod P_Y$. By Theorem 5.1.3 (i) and (1), (3), $\operatorname{Im} \sigma_w = (\overline{BwP_Y})/P_Y$ and $\overset{\circ}{\sigma}_w : \overset{\circ}{D} \to BwP_Y/P_Y$ is a continuous bijective map (where $\overset{\circ}{\sigma}_w$ is the restriction of σ_w to $\overset{\circ}{D} := \overset{\circ}{D}{}^{n(s_1)} \times \cdots \times \overset{\circ}{D}{}^{n(s_p)}$). Further, since $\sigma_s(S^{n(s)-1}) \subset B$, it is easy to see (cf. proof of Theorem 5.1.3 (i_2)) that

$$\sigma_w(D \backslash \overset{\circ}{D}) \subset \bigcup_{\substack{v<w \\ v \in W'_Y}} BvP_Y/P_Y, \tag{8}$$

where $D := D^{n(s_1)} \times \cdots \times D^{n(s_p)}$.

For any closed subset C of $\overset{\circ}{D}$, by (8),

$$\overset{\circ}{\sigma}_w(C) = \sigma_w(\bar{C}) \bigcap BwP_Y/P_Y,$$

and hence $\overset{\circ}{\sigma}_w$ is a homeomorphism, where $\bar{C}$ denotes the closure of C in D. In particular, $d(w)$ does not depend on the particular choice of the reduced decomposition of w.

By (1), each point x of $(\overline{BwP_Y/P_Y})\backslash(BwP_Y/P_Y)$ belongs to BvP_Y/P_Y, for some $v \in W'_Y$ and $v < w$; in particular, $d(v) < d(w)$, and hence x belongs to a cell of lower dimension.

The "Closure finiteness" axiom for a CW-complex follows again from (1). Finally, by definition, the topology on G/P_Y satisfies the Whitehead topology axiom. This completes the proof of (c). □

5.1.6 Definition. Let I be an indexing set and let $\{M_i\}_{i \in I}$ be a family of groups. For $i, j \in I$, let $M_{\{i,j\}} = M_{\{j,i\}}$ be a group together with a homomorphism $\varphi_{i,j} : M_{\{i,j\}} \to M_i$. The *amalgamated product* of the $\varphi_{i,j}$ is a pair $(M, (\varphi_i)_{i \in I})$, unique up to a unique isomorphism, satisfying the following:

(AP1) M is a group and $\varphi_i : M_i \to M$ are homomorphisms satisfying $\varphi_i \circ \varphi_{i,j} = \varphi_j \circ \varphi_{j,i}$, for all $i, j \in I$.

(AP2) If L is a group and $\psi_i : M_i \to L$ are homomorphisms satisfying $\psi_i \circ \varphi_{i,j} = \psi_j \circ \varphi_{j,i}$, for all $i, j \in I$, then there exists a unique homomorphism $\psi : M \to L$ satisfying $\psi_i = \psi \circ \varphi_i$ for all $i \in I$.

The existence and uniqueness of the amalgamated product is easy to see (cf. [Serre–80, Chap. I, §1, Proposition1]).

Let F be a set together with an indexed family of subsets $(M_i)_{i \in I}$ such that each M_i has a group structure. Then we say that $(M_i)_{i \in I}$ is a *system of groups* if $M_i \cap M_j$ is a subgroup of both M_i and M_j, for all $i, j \in I$; moreover, the group structure thus obtained on $M_i \cap M_j$ from M_i and M_j coincide. If we take $M_{\{i,j\}} := M_i \cap M_j$ and $\varphi_{i,j} : M_{\{i,j\}} \hookrightarrow M_i$ to be the inclusion, the amalgamated product of $\varphi_{i,j}$ in this case is just called the *amalgamated product of the system of groups* $(M_i)_{i \in I}$.

If F itself is a group and M_i are subgroups of F, then we say that F is the *amalgamated product of its indexed family of subgroups* $\{M_i\}_{i \in I}$ if $(F, (\varphi_i)_{i \in I})$ is the amalgamated product of $\varphi_{i,j}$, where $\varphi_i : M_i \to F$ is the inclusion.

5.1.7 Proposition. *Let (G, B, N, S) be a Tits system. Then G is the amalgamated product of its subgroups $\{N, P_s;\ s \in S\}$, where $P_s = P_{\{s\}}$ is as in Theorem 5.1.3 (b).*

Proof. Let G' be the amalgamated product of the groups $\{N, P_s;\ s \in S\}$ and let $\varphi_s : P_s \to G'$ and $\varphi = \varphi_N : N \to G'$ be the associated homomorphisms. Thus φ and φ_s coincide on $N \cap P_s$, and $\varphi_{s|B}$ does not depend on $s \in S$ since $P_S \cap P_t = B$ for $s \neq t$. Denote $\varphi_{s|B}$ by β. Thus G' is generated by $\beta(B)$ and $\varphi(N)$, and there exists a homomorphism $\psi : G' \to G$ such that $\psi \circ \varphi$ and $\psi \circ \beta$ are the canonical inclusions of N and B respectively. In particular, by the axiom (BN_2) (cf. 5.1.1), ψ is surjective. So we need to prove that ψ is injective.

We first show by induction on $\ell(\pi(n))$ that

$$(*) \qquad \text{for } b \in B, n \in N, \text{ and } s \in S \text{ such that } nbn^{-1} \in P_s, \\ \text{we have } \varphi_s(nbn^{-1}) = \varphi(n)\beta(b)\varphi(n^{-1}),$$

where $\pi : N \to W = N/T$ $(T := B \cap N)$ is the canonical projection.

For $\ell(\pi(n)) = 0$, $(*)$ is clearly true. Set $\pi(n) = w$ and choose $s' \in S$ such that $\ell(s'w) < \ell(w)$ (we choose $s' = s$ itself if $\ell(sw) < \ell(w)$). Set $n = mm'$ with $m \in \pi^{-1}(s')$. If $s \neq s'$, i.e., $\ell(sw) > \ell(w)$, then $nBn^{-1} \cap P_s \subset B$: This follows since $nBn^{-1} \cap BsB \subset (BnB \cap BsBn)n^{-1} \subset (BnB \cap BsnB)n^{-1} = \emptyset$, by Theorem 5.1.3 (c) and (d) parts. Thus, in all cases, $nbn^{-1} \in P_{s'}$, and we have (by the induction hypothesis)

$$\begin{aligned}
\varphi_s(nbn^{-1}) &= \varphi_{s'}(nbn^{-1}) \\
&= \varphi_{s'}(m)\varphi_{s'}(m'bm'^{-1})\varphi_{s'}(m^{-1}) \\
&= \varphi(m)\varphi(m')\beta(b)\varphi(m'^{-1})\varphi(m^{-1}) \\
&= \varphi(n)\beta(b)\varphi(n^{-1}),
\end{aligned}$$

which proves $(*)$.

We next show that

$$(1) \qquad G' = \beta(B)\varphi(N)\beta(B):$$

Since the group G' is generated by $\beta(B)$ and $\varphi(\pi^{-1}(S))$, to prove (1), it suffices to show that, for $s \in S$, $m \in \pi^{-1}(s)$, $n \in N$ and $b \in B$, one has

$$(2) \qquad \varphi(m)\beta(b)\varphi(n) \in \beta(B)\varphi(N)\beta(B).$$

Set $mbn = b'm'nb''$, with $b', b'' \in B$, $m' = 1$ or m. Then $nb''n^{-1} \in P_s$ and hence, by $(*)$,

$$\varphi(m)\beta(b) = \varphi_s(mb) = \varphi_s(b'm'nb''n^{-1}) = \varphi_s(b'm')\varphi_s(nb''n^{-1})$$

$$= \beta(b')\varphi(m')\varphi(n)\beta(b'')\varphi(n^{-1}),$$

which establishes (2) and hence (1).

Since, for any $n \in N$, $\psi(\beta(B)\varphi(n)\beta(B)) = BnB$, it follows from the Bruhat decomposition that $\operatorname{Ker}\psi \subset \beta(B)$. But, for $b \in B$, $\psi\beta(b) = b$, and hence $b = 1$ if $\beta(b) \in \operatorname{Ker}\psi$. This proves that ψ is injective, proving the proposition. $\square$

5.1.8 Theorem. *Let S be a finite indexing set and let $(B, N, P_s; s \in S)$ be a system of groups (cf. 5.1.6). Assume that this system satisfies the following properties:*

(P_1) *For $s \neq s' \in S$, $P_s \cap P_{s'} = B$.*

(P_2) *The subgroup $H := B \cap N$ is normal in N.*

(P_3) *For any $s \in S$, defining $N_s := N \cap P_s$, N_s/H is of order 2 denoted (by abuse of notation) $\{1, s\}$.*

(P_4) *$P_s = B \cup BsB$. Observe that, by (P_3), this is a disjoint union.*

(P_5) *The pair $(W := N/H, S)$ is a Coxeter group.*

(P_6) *Let $\pi : N \to W$ be the quotient map. For any $n \in N$ and any decomposition $n = n_1 \cdots n_p$ with $n_i \in N_{s_i} \setminus H$, for some $s_i \in S$, such that $\pi(n) = \pi(n_1) \cdots \pi(n_p)$ is a reduced decomposition, the subgroup $B(n_1, \ldots, n_p) \subset B$, defined below in the proof, depends only on $\pi(n)$ (to be denoted B_n or $B_{\pi(n)}$). Moreover, the homomorphism $\gamma(n_1, \ldots, n_p) : B_{\pi(n)} \to B$, again defined in the proof, depends only on n, to be denoted γ_n.*

(P_7) *For $w \in W$ and $s \in S$ such that $\ell(ws) = \ell(w) + 1$, we have*

$$B_w B_s = B.$$

(P_8) *Let $s, t \in S$ and $w \in W$ be such that $wtw^{-1} = s$ and $\ell(wt) = \ell(w) + 1$. Then, for any $m \in \pi^{-1}(s)$, $n \in \pi^{-1}(w)$ and $b \in B \backslash B_t$, there exist elements $y \in bB_t \cap B_n$ and $y', y'' \in B_n$ satisfying*
(a) *$m'^{-1}ym' = y'm'y''$ in P_t, and*
(b) *$m\gamma_n(y)m^{-1} = \gamma_n(y')m^{-1}\gamma_n(y'')$ in P_s, where $m' := n^{-1}m^{-1}n \in \pi^{-1}(t)$.*

(P_9) *B is not normal in any P_s.*

Then the canonical map of $Y := N \cup \bigcup_{s \in S} P_s$ *into the amalgamated product* G *of the system of groups* $\{B, N, P_s;\ s \in S\}$ *(cf. Definition 5.1.6) is injective. Moreover, denoting the images of* B *and* N *in* G *by the same symbols,* (G, B, N, S) *is a Tits system.*

Further, for any group $\bar{G}$ *with an injective map* $\bar{\varphi} : Y \to \bar{G}$ *such that* $\bar{\varphi}_{|N}$ *and* $\bar{\varphi}_{|P_s}$ *are group homomorphisms, and* $\bar{\varphi}(Y)$ *generates* $\bar{G}$ *as a group, the canonical homomorphism* $\bar{\psi} : G \to \bar{G}$ *is an isomorphism.*

Proof. Fix $n \in N$ and take a decomposition $n = n_1 \dots n_p$ with $n_i \in N_{s_i}$, for some $s_i \in S$, such that $\pi(n) = \pi(n_1) \dots \pi(n_p)$ is a reduced decomposition of $\pi(n)$. Define the subgroup $B(n_1, \dots, n_p) \subset B$ inductively by $B(n_1) = B \cap (n_1^{-1} B n_1)$ in P_{s_1}, and for $i > 1$,

(1) $$B(n_1, \dots, n_i) = B \cap (n_i^{-1} B(n_1, \dots, n_{i-1}) n_i) \text{ in } P_{s_i}.$$

(For $h \in H$, define $B(h) = B$.) Also, define the homomorphism

$$\gamma(n_1, \dots, n_p) : B(n_1, \dots, n_p) \to B,$$

by taking first the conjugation under n_p (in P_{s_p}), then conjugation under n_{p-1} (in $P_{s_{p-1}}$, and so on, and finally conjugation under n_1 (in P_{s_1}).

By (P_6), $B(n_1, \dots, n_p)$ depends only on $\pi(n)$ and $\gamma(n_1, \dots, n_p)$ depends only on n, to be denoted respectively by $B_{\pi(n)}$ (also by B_n) and γ_n. So $B_n = B_{n'}$ if $\pi(n) = \pi(n')$.

For any $n \in N$, $\gamma_n(B_n) \subset B_{n^{-1}}$ and, moreover,

(2) $$\gamma_n : B_n \to B_{n^{-1}} \text{ is a bijection with inverse } \gamma_{n^{-1}}.$$

Clearly, for $h \in H$ and $n \in N$,

(3) $$\gamma_{hn}(x) = h\gamma_n(x)h^{-1}, \qquad \text{for } x \in B_{\pi(n)}.$$

Also, each B_n contains H; in particular, H normalizes B_n. Further, for $n \in N$ and $m \in \pi^{-1}(s)$ for some $s \in S$ such that $\ell(\pi(nm)) = \ell(\pi(n)) + 1$,

(4) $$B_{nm} = B \cap m^{-1} B_n m \quad (\text{in } P_s).$$

In the product $\widetilde{X} = B \times N \times B$, introduce the relation $(b_1, n, b_2) \backsim (b_1', n', b_2')$ iff there exist $h \in H$ and $b \in B_n$ such that

$$n' = hn,\ b_2' = bb_2 \text{ and } b_1' = b_1 \gamma_n(b^{-1}) h^{-1}.$$

By using (3), it is easy to see that $\backsim$ is an equivalence relation.

Define $X = \widetilde{X}/\backsim$, and let $\theta : \widetilde{X} \to X$ be the canonical projection. The group B acts on $\widetilde{X}$ from the left as well as right via

$$b(b_1, n, b_2) = (bb_1, n, b_2), \text{ and}$$
$$(b_1, n, b_2)b = (b_1, n, b_2 b). \tag{5}$$

These actions descend to give left and right actions of B on X.

For any $s \in S$, we next define left and right actions of $P = P_s$ on $\widetilde{X}$.

Let $W' = W'_s := \{w \in W : \ell(ws) = \ell(w)+1\}$ and let $N' = N'_s := \pi^{-1}(W')$. By ($P_4$) and ($P_7$), for any $w \in W'$,

$$P = B_w N_s B. \tag{6}$$

For any $h \in H$, $n \in N$, and $b \in B_n$,

$$\gamma_n(bh) = \gamma_n(b)(nhn^{-1}). \tag{7}$$

Define a map $\alpha : B \times N' \times P \to X$ by $\alpha(b, n, p) = \theta(b\gamma_n(b_1), nn_1, b_2)$, where $p = b_1 n_1 b_2$ for $b_1 \in B_n$, $n_1 \in N_s$ and $b_2 \in B$ (cf. (6)). By using (4) and (7), it is easy to see that the map α does not depend on the choice of the decomposition of p as above. Further,

$$\alpha(b, n, p) = \alpha(b', n', p') \tag{8}$$

if and only if there exist $h \in H$ and $b'' \in B_n$ such that $n' = hn$, $p' = b''p$ and $b = b'h\gamma_n(b'')$. To prove this, observe that, for $n, n' \in N'$ and $n_1, n'_1 \in N_s$ such that $\pi(nn_1) = \pi(n'n'_1)$, we have $\pi(n) = \pi(n')$ and $\pi(n_1) = \pi(n'_1)$. Also, for $n \in N'$, $n_1 \in N_s$ and $b \in B_{nn_1}$, as an element of P_s, $n_1 b n_1^{-1} \in B_n$ and

$$\gamma_n(n_1 b n_1^{-1}) = \gamma_{nn_1}(b). \tag{9}$$

The group B, resp. P, operates on $B \times N' \times P$ from the left, resp. right, by the multiplication and, moreover, by (8), these actions descend via α to give a left B-action and right P-action on X. This left action of B on X coincides with the left action of B defined earlier. Moreover, the restriction of the right P-action on X to B coincides with the right B-action on X defined earlier.

For $x = \theta(b_1, n, b_2) \in X$, define

$$x^{-1} = \theta(b_2^{-1}, n^{-1}, b_1^{-1}) \in X. \tag{10}$$

Using (2) and (7), it is easy to see that x^{-1} is well defined, i.e., it does not depend on the choice of representatives.

Now we define a left action of P on X by setting

(11) $$p \cdot x = (x^{-1} \cdot p^{-1})^{-1}, \qquad \text{for } p \in P,\ x \in X.$$

It is easy to see that this left P-action on X restricted to B coincides with the earlier defined left B-action on X.

In the next lemma we shall prove that the left action of P_s on X commutes with the right action of P_t on X, for any $s, t \in S$. We assume its validity and continue with the proof of the theorem.

For any $s \in S$, both the (left and right) actions of P_s on X are effective. We prove this for the right action; the proof for the left action reduces to that of the right action. Let $p \in P_s$ be such that $xp = x$, for all $x \in X$. Taking $x = (1, 1, 1)$, we get $p \in B$. But then $(1, 1, 1) \backsim (1, 1, p)$ (in $\widetilde{X}$), which is possible only if $p = 1$.

Let G_ℓ, resp. G_r, be the group of bijections of X generated by the left, resp. right, action of $\{P_s;\ s \in S\}$ defined above. Both groups act transitively on X.

Define the map $\mathrm{e} : G_\ell \to X$ taking $g \mapsto g \cdot 1$, where 1 is the equivalence class of $(1, 1, 1)$. Since G_ℓ acts transitively on X, e is surjective. We next claim that e is injective: Let $g \in G_\ell$ be such that $g \cdot 1 = 1$. By the next lemma, $1 \cdot g' = (g \cdot 1) \cdot g' = g \cdot (1 \cdot g')$, for all $g' \in G_r$. In particular, $g = 1$, proving the injectivity of e. Now we put the group structure on X so that e is an isomorphism. For $p \in P_s$, we have

(12) $$p \cdot 1 = 1 \cdot p.$$

Next, define the map $\mu : N \to X$ by $n \mapsto \theta(1, n, 1)$. It is easy to see that μ is injective. We next prove that μ is a group homomorphism. It suffices to show that $\mu(n_1 n) = \mu(n_1)\,\mu(n)$, for $n_1 \in N_s$ and $n \in N$. But this follows from the following identities (13) and (14).

(13) $$\mu_{|N_s} = \mathrm{e}_{|N_s}, \quad \text{and}$$

(14) $$n_1 \cdot \theta(1, n, 1) = \theta(1, n_1 n, 1), \qquad \text{for } n_1 \in N_s,\ n \in N.$$

Let G be the amalgamated product of the following system of groups: $\{B, N, P_s;\ s \in S\}$. Then, by definition, there is a map $\varphi : Y (:= N \cup \cup_s P_s) \to G$, such that $\varphi_{|N}$ and $\varphi_{|P_s}$ are all group homomorphisms. By the defining property of the amalgamated products and (13), there is a unique group homomorphism $\psi : G \to X$ satisfying $\psi \circ \varphi_{|N} = \mu$ and $\psi \circ \varphi_{|P_s} = \mathrm{e}_{|P_s}$. We next claim that $\psi \circ \varphi$ is injective: Since e and μ are injective, it suffices to show that (a) $p \cdot 1 = p' \cdot 1$

(for $p \in P_s$, $p' \in P_{s'}$, $s \neq s'$) $\Rightarrow$ $p = p'$, and (b) $p \cdot 1 = \theta(1, n, 1)$ (for $p \in P_s$, $n \in N$) $\Rightarrow$ $p = n$. But both assertions are easy to verify. In particular, φ is injective. Since X is generated as a group by $e(B)$ and $\mu(N)$, ψ is surjective.

We next show that (G, B, N, S) is a Tits system. The axiom (BN_1) (cf. 5.1.1) follows from (P_2) and (P_5); (BN_2) follows from (P_4); (BN_3) follows from (P_4) and (P_9). So it remains to show (BN_4), i.e.,

$$n_1 Bn \subset BnB \cup Bn_1nB, \quad \text{for } n_1 \in N_s \text{ and } n \in N :$$

Assume first that $\ell(\pi(n_1 n)) > \ell(\pi(n))$. Then, by ($P_7$),

$$\begin{aligned} B &= B_{n_1} B_{n^{-1}}, \quad \text{and hence (in } \bar{G}) \\ n_1 Bn &= n_1 B_{n_1} n_1^{-1} n_1 nn^{-1} B_{n^{-1}} n \\ &= B_{n_1^{-1}} n_1 n B_n, \quad \text{by (2)} \\ &\subset Bn_1nB. \end{aligned}$$

So this case is taken care of. Next assume that $\ell(\pi(n_1 n)) < \ell(\pi(n))$. Then

$$\begin{aligned} n_1 Bn = n_1 Bn_1(n_1 n) &\subset (Bn_1 n) \cup Bn_1 Bn_1 n, \quad \text{by } (P_4) \\ &\subset (Bn_1 n) \cup BnB, \quad \text{by the earlier case.} \end{aligned}$$

This proves (BN_4).

Finally, we show that $\psi : G \to X$ is injective and hence an isomorphism. Let $bnb' \in \operatorname{Ker} \psi$ for $b, b' \in B$ and $n \in N$ (cf. Theorem 5.1.3(c_1)). Then $b'bn \in \operatorname{Ker} \psi$, so $\psi(b'b) = \psi(n^{-1})$ and hence $b'b = n^{-1}$ (since $\psi \circ \varphi$ is injective), proving the injectivity of φ. The assertion concerning $\bar{\psi}$ follows by the same argument as that for ψ. This completes the proof of the theorem modulo the following lemma. □

5.1.9 Lemma. *With the assumptions as in Theorem 5.1.8 and the notation as above, for any $s, t \in S$, the left action of P_s on X commutes with the right action of P_t on X, i.e., $(px)p' = p(xp')$, for $p \in P_s$, $p' \in P_t$ and $x \in X$.*

Proof. Let $X' := \{x \in X : (px)p' = p(xp'), \text{ for all } p \in P_s \text{ and } p' \in P_t\}$. We prove that

(1) $$X' = X :$$

Define $W' = \{w \in W : \ell(swt) = \ell(w) + 2\}$, and
$W'' = \{w \in W : w^{-1}sw = t \text{ and } \ell(sw) = \ell(w) + 1\}$.

Choose any set of representatives N' and N'' in N corresponding to W' and W'', respectively. Then

(2) $$N_s \cdot (N' \cup N'') \cdot N_t = N,$$

by the Strong Exchange Condition Theorem 1.3.11(c). Clearly, X' is stable under the left action of P_s and the right action of P_t.

To prove (1), it suffices to show that

(3) $$\mu(N' \cup N'') \subset X'.$$

Fix $n \in N' \cup N''$ and let $Q := \{p \in P_s : (p\mu(n))p' = p(\mu(n)p'),\ \text{for all } p' \in P_t\}$. It is easy to see that the left action of P_s on X commutes with the right action of B and, similarly, the left action of B on X commutes with the right action of P_t. Hence Q is stable under left B-action. Moreover, Q is stable under right $B_{n^{-1}}$-action, since, for $b \in B_{n^{-1}}$, $p \in Q$ and $p' \in P_t$,

$$\begin{aligned}(pb\mu(n))p' &= (p\mu(n)\,\gamma_{n^{-1}}(b))p' = (p\mu(n))(\gamma_{n^{-1}}(b)p') = p(\mu(n)\,\gamma_{n^{-1}}(b)p') \\ &= p(b\mu(n)p') = (pb)(\mu(n)p').\end{aligned}$$

Choose any $m \in \pi^{-1}(s)$ and $m' \in \pi^{-1}(t)$. Since $\ell(\pi(m)\,\pi(n)) = \ell(\pi(n)) + 1$, by the properties ($P_4$) and ($P_7$), $P_s = B \cup BmB = B \cup (BmB_m B_{n^{-1}}) = B \cup BmB_{n^{-1}}$, i.e.,

(4) $$P_s = B\{1, m\}B_{n^{-1}}.$$

Further, for any $Y \subset B$ such that $YB_t = B$, we get

(5) $$((Ym') \cup \{1\})B = P_t.$$

Hence, to prove (3), it suffices to show that

(6) $$(m\mu(n))(ym') = m(\mu(n)\,ym'), \qquad \text{for all } y \in Y.$$

Suppose first that $n \in N'$. Consider $n_1 := mn$ and choose $Y \subset B_{n_1}$ (which is possible by (P_7)) and hence $Y \subset B_n$, since, for $n', n \in N$ such that $\ell(\pi(n'n)) = \ell(\pi(n')) + \ell(\pi(n))$, we have

(7) $$B_{n'n} \subset B_n\,.$$

Now (6) follows easily for such a n $(\in N')$ and $y \in B_{n_1}$, since both sides of (6) are equal to $\gamma_{n_1}(y) \cdot \mu(n_1 m')$.

Next take $n \in N''$ and choose $m' \in \pi^{-1}(t)$ satisfying $m' = n^{-1}m^{-1}n$. If $y \in Y \cap B_t$, we can take $y = 1$ and then, in this case, (6) is easy to verify. So assume that $y \in Y \backslash B_t$. By (P_8), we can make a choice of Y so that $Y \subset B_n$ (and of course $YB_t = B$) and such that for all $y \in Y$ there exist elements $y', y'' \in B_n$ satisfying

(8) $$m'^{-1} y m' = y' m' y'' \text{ in } P_t , \quad \text{and}$$

(9) $$m \gamma_n(y) m^{-1} = \gamma_n(y') m^{-1} \gamma_n(y''), \quad \text{in } P_s .$$

With these choices, the left side of (6) is equal to (by (8))

(10) $$\mu(mn)(m'y'm'y'') = \mu(n)(y'm'y'') = \gamma_n(y')\, \mu(nm')y'',$$

and the right side of (6) is equal to (by (9))

(11) $$(m\gamma_n(y))\, \mu(nm') = (\gamma_n(y')m^{-1}\gamma_n(y'')m)\mu(nm') = \gamma_n(y')\mu(nm')y''.$$

Combining (10) and (11), we get (6) in this case as well. This completes the proof of the lemma and thus finishes the proof of Theorem 5.1.8. □

5.1.E EXERCISE. Show that, for a Tits system (G, B, N, S), it is not true in general that N is the normalizer of $N \cap B$ in G.
Hint: Take any Tits system (G, B, N, S) and any group H and consider the Tits system $(G \times H, B \times H, N, S)$.

5.2. Refined Tits Systems

Following Kac–Peterson, a refined Tits system is defined as follows:

5.2.1 Definition. A 6-tuple (G, N, U_+, U_-, H, S) consisting of a group G and subgroups $H \subset N, U_+, U_-$ together with a subset $S \subset N/H$ is called a *refined Tits system* if the following three axioms are satisfied:

(RT1) The group G is generated by N and U_+; H is a normal subgroup of N; H normalizes U_+ and U_-; S generates the group $W := N/H$ and $s^2 = 1$ for all $s \in S$.

(RT2) For $s = \bar{s}H \in S$ and $w = \bar{w}H \in W$ we have:

- (a) $U_s^s \backslash \{1\} \subset U_s \bar{s} H\, U_s$,
- (b) $U_s \neq \{1\}$,
- (c) $U_s^w \subset U_+$ or $U_s^w \subset U_-$,
- (d) $U_+ = U_s\, (U_+ \cap U_+^s)$,

where, for any subgroup $M \subset G$ normalized by H, $M^w := \bar{w}^{-1} M \bar{w}$ and $U_s := U_+ \cap U_-^s$.

(RT3) For $u_\pm \in U_\pm$ and $n \in N$ such that $u_- n u_+ = 1$, we have $u_+ = u_- = n = 1$.

Define the subgroup $B_\pm := U_\pm \cdot H$ of G. We abbreviate B_+ by B. In the following Lemma, (G, N, U_+, U_-, H, S) is an arbitrary refined Tits system.

5.2.2 Lemma.

(a) *For* $s = \bar{s}H \in S$ *and* $w = \bar{w}H \in W$*, we have:*

(a_1) $sBw \subset (BswU_s^w) \cap ((Bsw) \cup (BwU_s^{sw}))$*, where* $sBw := \bar{s}B\bar{w}$ *etc.*

(a_2) *Exactly one of the following holds:*

(i) $U_s^w \subset U_+$ *and* $U_s^{sw} \subset U_-$*, or*
(ii) $U_s^{sw} \subset U_+$ *and* $U_s^w \subset U_-$.

(b) (G, B, N, S) *is a Tits system. In particular,* (W, S) *is a Coxeter group.*

(c) *The following three assertions for any* $s \in S$ *and* $w \in W$ *are equivalent:*

(c_1) $U_s^w \subset U_+$
(c_2) $U_s^{sw} \subset U_-$
(c_3) $\ell(sw) > \ell(w)$.

Proof.

$$(a_1): \qquad sBw = s(U_+ \cap U_+^s)HU_s w, \qquad \text{by RT2(d)}$$
$$= (U_+ \cap U_+^s)HswU_s^w.$$

Similarly,

$$\begin{aligned} sBw &= (U_+ \cap U_+^s)U_s^s Hsw \\ &\subset (U_+ \cap U_+^s)(\{1\} \cup (U_s HsU_s))\, Hsw\,, \quad \text{by RT2(a)} \\ &\subset (Bsw) \cup (BwU_s^{sw}). \end{aligned}$$

(a_2): By RT2(c), U_s^w is contained in exactly one of U_+ and U_- since, by (RT3), $U_+ \cap U_- = \{1\}$. Similarly, U_s^{sw} also is contained exactly in one of U_+ and U_-. But, by RT2(a), $(U_s^w U_s^{sw} U_s^w) \cap N \neq \{1\}$. In particular, U_s^w and U_s^{sw} can not both be contained in U_+ or in U_- since, by (RT3), $U_- \cap N = U_+ \cap N = \{1\}$.

(b): From (RT3), we get $N \cap B = H$. The axiom (BN_3) (cf. 5.1.1) follows from RT2(b) and (RT3). The axiom (BN_4) follows readily from the (a) part, and the axioms (BN_1)–(BN_2) are consequences of (RT1). This proves the (b) part.

(c): The equivalence of (c_1) and (c_2) follows from the (a_2) part. We now show (c_1)⇒(c_3). Assume $U_s^w \subset U_+$. Then $sBw \subset BswB$ (by the (a_1) part). By Theorem 5.1.3(d) and the (b) part, this gives $\ell(sw) > \ell(w)$. Conversely, assume $\ell(sw) > \ell(w)$. If possible, let $U_s^{sw} \subset U_+$. Then, by the (a_1) part, $sBsw \subset BwB$.

Again, using Theorem 5.1.3(d), we get $\ell(w) > \ell(sw)$, a contradiction to the assumption. So $U_s^{sw} \subset U_-$, proving (c). $\square$

For a group K and subsets $K', K_1, \ldots, K_n$, we write $K' = K_1 \odot K_2 \odot \ldots \odot K_n$ to mean that every element k of K' can be uniquely written as $k = k_1 \ldots k_n$ with $k_i \in K_i$ and $K_1 \ldots K_n = K'$.

5.2.3 Theorem. *Let (G, N, U_+, U_-, H, S) be a refined Tits system. For $v, w \in W$ such that $\ell(vw) = \ell(v) + \ell(w)$, we have the following:*

(a) $U_- \cap U_+^{vw} = (U_- \cap U_+^v)^w \odot (U_- \cap U_+^w)$.
(b) $U_+ \cap U_-^{vw} = (U_+ \cap U_-^v)^w \odot (U_+ \cap U_-^w)$.
(c) $U_+ \cap U_+^w = (U_+ \cap U_-^v)^w \odot (U_+ \cap U_+^{vw})$.
(d) $U_+ vB = U_+ \odot (vH) \odot (U_+ \cap U_-^v)$.
(e) $U_- vB = U_- \odot (vH) \odot (U_+ \cap U_+^v)$.
(f) $G = U_+ U_- N = U_- U_+ N$.
(g) *Birkhoff decomposition:* $G = \bigsqcup_{n \in N} U_- n U_+$.
More generally, for any subset $Y \subset S$,
(g_1) $G = \bigsqcup_{w \in W'_Y} U_- w P_Y$, *where the notation P_Y is as in Theorem 5.1.3(b).*
(h) $\bigcap_{w \in W} U_-^w = \{1\}$.
(i) U_- *is generated by its subgroups $\{U_s^w;\ s \in S, w \in W$ such that $\ell(sw) < \ell(w)\}$.*

Proof. By RT2(d), for any $s \in S$,

(1) $$B^s B = U_s^s \odot B.$$

Take a reduced decomposition $v = s_1 \ldots s_p$ ($s_i \in S$). Then, by Theorem 5.1.3(i_1) and (1),

(2) $$B^v B = U_{s_1}^{s_1 \ldots s_p} \odot U_{s_2}^{s_2 \ldots s_p} \odot \ldots \odot U_{s_p}^{s_p} \odot B.$$

By Lemma 5.2.2(c),

(3) $$U_{s_1}^{s_1 \ldots s_p} \odot U_{s_2}^{s_2 \ldots s_p} \odot \ldots \odot U_{s_p}^{s_p} \subset U_- \cap U_+^v.$$

Further,

(4) $$(U_- \cap U_+^v) B \subset B^v B.$$

By (2)–(4) we get (since $U_- \cap B = \{1\}$):

(5) $$U_{s_1}^{s_1 \ldots s_p} \odot U_{s_2}^{s_2 \ldots s_p} \odot \ldots \odot U_{s_p}^{s_p} = U_- \cap U_+^v,$$

and

$$B^v B = (U_- \cap U_+^v) \odot B. \tag{6}$$

Now (a) follows from (5); (d) follows from (6) applied to v^{-1} and taking inverses; and (b) follows by using (a) for $U_- \cap U_+^{w^{-1}v^{-1}}$ and conjugating by $(vw)^{-1}$ and then taking inverses.

By induction on $\ell(w)$, we next prove that

$$U_+^w = (U_+^w \cap U_-) \odot (U_+^w \cap U_+) \ : \tag{7}$$

For $w = e$, the above is obviously true. Choose $s \in S$ such that $\ell(sw) < \ell(w)$. By RT2(d),

$$U_+ \subset U_s U_+^s, \text{ and hence}$$
$$U_+^w \subset U_s^w U_+^{sw}. \tag{8}$$

By Lemma 5.2.2(c),

$$U_s^w \subset U_-, \tag{9}$$

and hence by the induction hypothesis and (8),

$$U_+^w \subset U_- U_+.$$

Therefore, by (RT3),

$$U_+^w \cap B = U_+^w \cap (U_- U_+ \cap B) = U_+^w \cap U_+. \tag{10}$$

Since $U_+^w \subset B^w B$, (7) follows from (6) and (10).

We now prove (c): By (7) applied to w^{-1} and v (and conjugating by w^{-1}) we get:

$$U_+ = (U_+ \cap U_-^w) \odot (U_+ \cap U_+^w), \text{ and} \tag{11}$$
$$U_+^{vw} = (U_+^{vw} \cap U_-^w) \odot (U_+^{vw} \cap U_+^w). \tag{12}$$

Taking the intersection of (11) and (12), we get (since $U_-^w \cap U_+^w = \{1\}$):

$$U_+ \cap U_+^{vw} = (U_+ \cap U_+^{vw} \cap U_-^w) \odot (U_+ \cap U_+^{vw} \cap U_+^w). \tag{13}$$

By (a), $(U_-^w \cap U_+^{vw}) \odot (U_- \cap U_+^w) \subset U_-$; in particular, $U_-^w \cap U_+^{vw} \subset U_-$. Therefore (13) reduces to $U_+ \cap U_+^{vw} = U_+ \cap U_+^{vw} \cap U_+^w$; in particular,

$$(14) \qquad U_+ \cap U_+^{vw} \subset U_+^w.$$

By (7) applied to $(vw)^{-1}$ and conjugating by $(vw)^{-1}$ we obtain:

$$\begin{aligned} U_+ &= (U_+ \cap U_-^{vw}) \odot (U_+ \cap U_+^{vw}) \\ (15) \qquad &= (U_+ \cap U_-^v)^w \odot (U_+ \cap U_-^w) \odot (U_+ \cap U_+^{vw}), \quad \text{by (b)}. \end{aligned}$$

Since the first and the third factors of the right side of the above equation are contained in U_+^w (using (14)), we get $U_+ \cap U_+^w = (U_+ \cap U_-^v)^w \odot (U_+ \cap U_+^{vw})$. This proves (c).

To prove the (e) part, write

$$\begin{aligned} U_- vB &= U_- H\, U_+^{v^{-1}} v \\ &= U_- H(U_+^{v^{-1}} \cap U_-)(U_+^{v^{-1}} \cap U_+)v, \quad \text{by (7)} \\ &= U_- \odot Hv \odot (U_+ \cap U_+^v), \quad \text{by (RT3)}. \end{aligned}$$

This proves (e).

For any $n \in N$, we get by (7) applied to $w = nH$,

$$(16) \qquad U_+ n \subset nU_- U_+.$$

In particular, by Lemma 5.2.2(b) and the Bruhat decomposition (cf. Theorem 5.1.3(c_1)), $G = \bigcup_{n \in N} nU_- U_+$. On taking the inverses, we get the first equality in (f).

By Lemma 5.2.2(a), for any $w \in W$ and $s \in S$,

$$(17) \qquad sBw \subset BswU_- \cup BwU_-.$$

In particular, $U_+ NU_-$ is stable under the left multiplication by N (and of course by U_+). So, by the Bruhat decomposition again, $U_+ NU_- = G$. Taking inverses we get,

$$(18) \qquad U_- NU_+ = G.$$

If $n, n' \in N$ are such that $U_- nU_+ \cap U_- n' U_+ \neq \emptyset$, then $n' \in U_- nU_+$. Moreover, by (7),

$$(19) \qquad U_- nU_+ \subset U_- U_+ n.$$

In particular, there exists $u_\pm \in U_\pm$ such that $u_-^{-1} n' n^{-1} u_+^{-1} = 1$. By (RT3) we get $n' = n$. This, together with (18), proves (g); and (18)–(19) prove the second equality of (f).

To prove (g_1), in view of (g), it suffices to show that for any $w \in W'_Y$, $U_- w P_Y = \bigsqcup_{v \in W_Y} U_- wvB$. Since $U_- w P_Y$ is left U_- and right P_Y (in particular B)-stable, clearly $\bigsqcup_{v \in W_Y} U_- wvB \subset U_- w P_Y$. To prove the reverse inclusion, it suffices to show that, for any $s \in Y$ and $v_o \in W_Y$, $wv_oBs \subset \bigsqcup_{v \in W_Y} U_- wvB$. But this follows easily from Lemma 5.2.2(a). This proves (g_1).

(h): Take $u \in \bigcap_{w \in W} U_-^w$. By the Bruhat decomposition and (d), write $u = u_+ u_- n$, for some $u_+ \in U_+$, $u_- \in U_+^{n^{-1}} \cap U_-$ and $n \in N$. Then $u_-(nun^{-1})^{-1} n u_+ = 1$, and $nun^{-1} \in U_-$ (by assumption). This gives (from (RT3)), $n = u_+ = u_-(nun^{-1})^{-1} = 1$. But $u_- \in U_+^{n^{-1}} \cap U_-$ and $u_- = u \in U_-^{n^{-1}}$ (by assumption). Hence $u = 1$. This proves (h).

(i): Let U' be the subgroup of U_- generated by these U_s^w. (Observe that $U_s^w \subset U_-$ for $\ell(sw) < \ell(w)$, by Lemma 5.2.2(c).) Then, by the same argument as given in the proof of the (g) part, we get $G = U' N U_+$. Hence $U' \subset U_- \subset U' N U_+$, which gives $U' = U_-$ by (RT3). This completes the proof of the theorem. □

5.C Comments. For an exhaustive treatment of Tits system, including Theorem 5.1.3, the reader is referred to, apart from the original papers of Tits, the books [Brown–89], [Ronan–89]. The definition of topological Tits system is taken from [Mitchell–88], though with some modifications, and Theorem 5.1.5 is taken from [loc. cit.]. The proof of Theorem 5.1.5 is similar to the one given in [Kac–85b] for the Kac–Moody groups. Proposition 5.1.7 is due to [Tits–74] and Theorem 5.1.8 is due to [Tits–81a]. The notion of a refined Tits system and all the results of Section 5.2 are due to [Kac–Peterson–85a].

VI

Kac–Moody Groups

Basic Theory

This chapter is devoted to the construction of the "maximal" Kac–Moody group $\mathcal{G}$ associated to a Kac–Moody Lie algebra $\mathfrak{g} = \mathfrak{g}(A)$, for A any $\ell \times \ell$ GCM, due to Tits. There are other versions of groups associated to $\mathfrak{g}(A)$, e.g., we will discuss the "minimal" Kac–Moody group $\mathcal{G}^{\min}$ defined by Kac–Peterson in Section 7.4. We first construct certain groups $\mathcal{B}$, N and $\{\mathcal{P}_i\}_{1 \le i \le \ell}$, and then $\mathcal{G}$ is constructed as the amalgamated product of these groups. We now give an outline of the construction of these groups. Let $\mathfrak{n} \subset \mathfrak{g}$ be the positive part of $\mathfrak{g}$, i.e., the direct sum of positive root spaces of $\mathfrak{g}$. Consider the completion $\hat{\mathfrak{n}}$ of $\mathfrak{n}$ got by taking the direct *product* of the positive root spaces. Then $\hat{\mathfrak{n}}$ is canonically a pro-nilpotent pro-Lie algebra. Let $\mathcal{U}$ be the pro-unipotent pro-group with Lie $\mathcal{U} = \hat{\mathfrak{n}}$ (guaranteed by Theorem 4.4.19). Similarly, let $\mathcal{U}_i$, $1 \le i \le \ell$, be the pro-unipotent pro-group with Lie algebra Lie $\mathcal{U}_i = \hat{\mathfrak{u}}_i$, where $\hat{\mathfrak{u}}_i \subset \hat{\mathfrak{n}}$ is the direct product of all the positive root spaces except the one corresponding to the simple root α_i.

Next, we fix an integral form $\mathfrak{h}_{\mathbb{Z}}$ of the Cartan subalgebra $\mathfrak{h} \subset \mathfrak{g}$ satisfying certain properties given in 6.1.6 and consider the torus $T := \mathrm{Hom}_{\mathbb{Z}}(\mathfrak{h}_{\mathbb{Z}}^*, \mathbb{C}^*)$, where $\mathfrak{h}_{\mathbb{Z}}^* := \mathrm{Hom}_{\mathbb{Z}}(\mathfrak{h}_{\mathbb{Z}}, \mathbb{Z})$. The adjoint action of $\mathfrak{h}$ on $\mathfrak{g}$ exponentiates to give an action of T on $\mathfrak{g}$, thereby giving an action of T on $\mathcal{U}$ and also on $\mathcal{U}_i$. Define $\mathcal{B}$ as the semidirect product $\mathcal{U} \ltimes T$, which acquires a natural pro-group structure. Moreover, define N as the group generated by T and $\{\tilde{s}_i\}_{1 \le i \le \ell}$, subject to certain relations given in 6.1.6. Finally we define the pro-group $\mathcal{P}_i$: Let $\mathcal{G}_i$ be the connected reductive group with Lie algebra $\mathbb{C}f_i \oplus \mathfrak{h} \oplus \mathbb{C}e_i$ and satisfying certain "root datum" conditions (cf. 6.1.12–6.1.13). Then it is shown that $\mathcal{G}_i$ acts on $\mathcal{U}_i$. Define the semidirect product $\mathcal{P}_i := \mathcal{U}_i \ltimes \mathcal{G}_i$. It is further shown that the family of groups $\{\mathcal{B}, N, \mathcal{P}_i; 1 \le i \le \ell\}$ satisfies the hypotheses of Theorem 5.1.8, thus giving rise to the Kac–Moody group $\mathcal{G}$, as their amalgamated product.

Further, $(\mathcal{G}, \mathcal{B}, N, S)$ is a Tits system, where $S = \{s_i\}_{1 \le i \le \ell}$ is the set of simple reflections in the Weyl group of $\mathfrak{g}$. In fact, it is shown that $(\mathcal{G}, N, \mathcal{U}, \mathcal{U}^-, T, S)$ is a refined Tits system, where $\mathcal{U}^-$ is the subgroup of $\mathcal{G}$ generated by the one-parameter subgroups U_β corresponding to the negative real roots β. In particular, from the general property of Tits systems, we get the Bruhat decomposition: $\mathcal{G} = \bigsqcup_{\bar{n} \in N/T} \mathcal{U}\bar{n}\mathcal{B}$, and also the Birkhoff decomposition $\mathcal{G} = \bigsqcup_{n \in N} \mathcal{U}^- n \mathcal{U}$. There is a more general version of both of these decompositions for parabolic subgroups $\mathcal{P}_Y$.

In Section 6.2, we study representations of the Kac–Moody groups $\mathcal{G}$. A representation π of $\mathcal{G}$ is called a pro-representation if the restriction of π to any minimal parabolic subgroup $\mathcal{P}_i$ is a pro-representation. It is shown that (suitably defined, cf. 6.2.1) the category of pro-representations $\mathfrak{M}(\mathcal{G})$ of $\mathcal{G}$ is equivalent to the category of $\mathfrak{g}$-modules π such that π is an integrable $\mathfrak{g}$-module, $\pi_{|\mathfrak{n}}$ extends to a pro-representation of $\hat{\mathfrak{n}}$, and $\pi_{|\mathfrak{h}}$ integrates to a locally finite algebraic T-module. In particular, any integrable highest weight $\mathfrak{g}$-module $V(\lambda)$, with highest weight $\lambda \in \mathfrak{h}^*_{\mathbb{Z}}$, acquires a functorial $\mathcal{G}$-module structure. Further, for any $g \neq 1$ in $\mathcal{G}$, there exists a $V \in \mathfrak{M}(\mathcal{G})$ such that g acts nontrivially on V.

We show that the normalizer of T in $\mathcal{G}$ equals N, and we also determine the center of $\mathcal{G}$ in Lemma 6.2.9. Finally, the adjoint representation Ad of $\mathcal{G}$ in the completion $\hat{\mathfrak{g}} := \mathfrak{n}^- \oplus \mathfrak{h} \oplus \hat{\mathfrak{n}}$ of $\mathfrak{g}$ is defined. Also the exponential map $\mathrm{Exp} : \mathfrak{g}_{\text{fin}} \to \mathcal{G}$, satisfying the standard properties, is defined on the "finite" part $\mathfrak{g}_{\text{fin}}$ of $\mathfrak{g}$ consisting of all the conjugates of completed parabolic subalgebras of finite type (cf. Proposition 6.2.11).

6.1. Definition of Kac–Moody Groups and Parabolic Subgroups

In this section we consider an arbitrary (not necessarily symmetrizable) Kac–Moody Lie algebra $\mathfrak{g}$ (cf. Section 1.1). Also recall the definition of its subalgebras $\mathfrak{h}, \mathfrak{n}, \mathfrak{n}^-, \mathfrak{g}_Y, \mathfrak{u}_Y, \mathfrak{u}_Y^-, \mathfrak{p}_Y$ from Sections 1.1–1.2.

6.1.1 The group $\mathcal{U}$. From the root space decomposition (cf. Theorem 1.2.1) we get $\mathfrak{n} = \oplus_{\alpha \in \Delta^+} \mathfrak{g}_\alpha$ where Δ^+ is the set of positive roots (cf. 1.2.2). Now define its "completion"

$$(1) \qquad \hat{\mathfrak{n}} = \prod_{\alpha \in \Delta^+} \mathfrak{g}_\alpha \,.$$

This is a Lie algebra under the bracket

$$(2) \qquad \left[\sum_{\alpha \in \Delta^+} x_\alpha, \sum_{\beta \in \Delta^+} y_\beta \right] = \sum_{\gamma \in \Delta^+} \sum_{\substack{\alpha, \beta \in \Delta^+ \\ \alpha + \beta = \gamma}} [x_\alpha, y_\beta],$$

for $x_\alpha, y_\alpha \in \mathfrak{g}_\alpha$. Since, for any $\gamma \in \Delta^+$, there are only finitely many $\alpha, \beta \in \Delta^+$ such that $\alpha + \beta = \gamma$, (2) is well defined. Similarly, define the completion $\hat{\mathfrak{g}}$ of $\mathfrak{g}$ by

$$\hat{\mathfrak{g}} = \mathfrak{n}^- \oplus \mathfrak{h} \oplus \hat{\mathfrak{n}}.$$

By the same bracket as (2), $\hat{\mathfrak{g}}$ is a Lie algebra extending the Lie algebra structure of $\mathfrak{g}$. In fact $\hat{\mathfrak{g}}$ should be viewed as a completion of $\mathfrak{g}$ in the "positive" direction.

Any $\mathfrak{g}$-module $M \in \mathcal{O}$ (cf. Definition 2.1.1(c)) canonically becomes a $\hat{\mathfrak{g}}$-module, extending the $\mathfrak{g}$-module structure.

For any $k \geq 1$, define the ideal of $\hat{\mathfrak{n}}$:

(3) $$\hat{\mathfrak{n}}(k) = \prod_{\substack{\alpha \in \Delta^+ \\ |\alpha| \geq k}} \mathfrak{g}_\alpha,$$

where $|\alpha|$ is the principal gradation of α (cf. 1.2.2). The Lie algebra $\hat{\mathfrak{n}}$ is a pro-Lie algebra under the family $\mathcal{F}$ of ideals defined by

$$\mathcal{F} = \{ \text{ ideals } \mathfrak{a} \text{ of } \hat{\mathfrak{n}} \text{ such that } \mathfrak{a} \supset \hat{\mathfrak{n}}(k) \text{ for some } k \geq 1\}.$$

The verification that $\mathcal{F}$ indeed defines a pro-Lie algebra structure on $\hat{\mathfrak{n}}$ is easy. Clearly $\hat{\mathfrak{n}}$ is pro-nilpotent (cf. Definition 4.4.18).

Let $\mathcal{U}$ be the pro-unipotent pro-group with Lie $\mathcal{U} \approx \hat{\mathfrak{n}}$ (cf. Theorem 4.4.19). Set-theoretically, we can take $\mathcal{U} = \hat{\mathfrak{n}}$ and then the map Exp: $\hat{\mathfrak{n}} \to \mathcal{U}$ is nothing but the identity map (cf. Definition 4.4.16, Lemma 4.4.20 and Subsection 4.4.21).

Observe that, as a vector space,

$$\hat{\mathfrak{g}} = \operatorname*{Inv.}_{k} \text{ lt. } \hat{\mathfrak{g}}/\hat{\mathfrak{n}}(k).$$

In particular, taking the discrete topology on each $\hat{\mathfrak{g}}/\hat{\mathfrak{n}}(k)$, we endow $\hat{\mathfrak{g}}$ with the inverse limit topology. Of course, $\hat{\mathfrak{g}}/\hat{\mathfrak{n}}(k) = \mathfrak{g}/\mathfrak{n}(k)$, where $\mathfrak{n}(k) = \oplus_{\alpha \in \Delta^+, |\alpha| \geq k} \mathfrak{g}_\alpha$.

For any $x \in \hat{\mathfrak{n}}$, consider the one-dimensional pro-Lie subalgebra $\mathbb{C}x$ of $\hat{\mathfrak{n}}$ and let

(4) $$\mathcal{U}_x := \text{Exp}\,(\mathbb{C}x)$$

be the corresponding pro-subgroup of $\mathcal{U}$ (cf. Theorem 4.4.19 and Proposition 4.4.12). Then $\mathcal{U}_x$ is a one-dimensional algebraic group isomorphic with the additive group $\mathbb{C}$, called the *one-parameter subgroup of* $\mathcal{U}$ corresponding to x.

Let $\Theta \subset \Delta^+$ be a subset such that for $\alpha, \beta \in \Theta$, if $\alpha + \beta \in \Delta^+$, then $\alpha + \beta \in \Theta$. We call such a subset Θ *bracket closed.* Define

(5) $$\hat{\mathfrak{n}}_\Theta := \prod_{\alpha \in \Theta} \mathfrak{g}_\alpha.$$

Then $\hat{\mathfrak{n}}_\Theta$ is a pro-Lie subalgebra of $\hat{\mathfrak{n}}$ (use (4.4.13.3)). Set

(6) $$\mathcal{U}_\Theta = \text{Exp}\,(\hat{\mathfrak{n}}_\Theta).$$

With the notation as above, we have:

6.1.2 Lemma. *$\mathcal{U}_\Theta$ is a pro-subgroup of $\mathcal{U}$. Suppose we further assume that Θ satisfies the following condition:*

$$(*) \qquad \textit{For } \alpha \in \Delta^+,\ \beta \in \Theta \textit{ with } \alpha + \beta \in \Delta^+, \textit{ we have } \alpha + \beta \in \Theta,$$

then $\mathcal{U}_\Theta$ is a normal subgroup of $\mathcal{U}$.

Proof. Since $\hat{\mathfrak{n}}_\Theta$ is a pro-nilpotent pro-Lie algebra, by Theorem 4.4.19, there is a pro-unipotent pro-group $\mathcal{U}'_\Theta$ with Lie $\mathcal{U}'_\Theta = \hat{\mathfrak{n}}_\Theta$ and a pro-group morphism $\phi : \mathcal{U}'_\Theta \to \mathcal{U}$ such that $\dot{\phi}$ is the canonical inclusion $\hat{\mathfrak{n}}_\Theta \hookrightarrow \hat{\mathfrak{n}}$. In particular, ϕ is injective since Exp is bijective. Moreover, we can take $\mathcal{U}'_\Theta = \hat{\mathfrak{n}}_\Theta$ as sets with identity as the exponential map. Hence $\phi(\mathcal{U}'_\Theta) = \hat{\mathfrak{n}}_\Theta = \text{Exp}\,(\hat{\mathfrak{n}}_\Theta) = \mathcal{U}_\Theta$. But then the first part of the lemma follows from Proposition 4.4.12.

For the second part, observe that (because of $(*)$) $\hat{\mathfrak{n}}_\Theta$ is an ideal of $\hat{\mathfrak{n}}$. In particular, the image $\hat{\mathfrak{n}}_\Theta(k)$ of $\hat{\mathfrak{n}}_\Theta$ in $\hat{\mathfrak{n}}/\hat{\mathfrak{n}}(k)$ is an ideal of $\hat{\mathfrak{n}}/\hat{\mathfrak{n}}(k)$, for any $k \geq 1$. Hence the subgroup $\text{Exp}(\hat{\mathfrak{n}}_\Theta(k)) \subset \mathcal{U}/\text{Exp}\,\hat{\mathfrak{n}}(k)$ is normal, for all $k \geq 1$. This shows that $\mathcal{U}_\Theta$ is normal in $\mathcal{U}$ by Lemma 4.4.6, proving the lemma. □

Let $\Theta \subset \Delta^+$ be a bracket closed subset such that Θ is *bracket coclosed* as well, i.e., $\Delta^+\backslash\Theta$ is bracket closed.

6.1.3 Lemma. *For a bracket closed and bracket coclosed subset $\Theta \subset \Delta^+$, the multiplication map $m : \mathcal{U}_\Theta \times \mathcal{U}_{\Delta^+\backslash\Theta} \to \mathcal{U}$ is a bijection.*

Proof. Since $\hat{\mathfrak{n}}_\Theta \cap \hat{\mathfrak{n}}_{\Delta^+\backslash\Theta} = \{0\}$, taking Exp, we get $\mathcal{U}_\Theta \cap \mathcal{U}_{\Delta^+\backslash\Theta} = \{1\}$. This gives the injectivity of m. Let H be the unique map, making the following diagram commutative:

$$\begin{array}{ccc}
\hat{\mathfrak{n}}_\Theta \times \hat{\mathfrak{n}}_{\Delta^+\backslash\Theta} & \xrightarrow{\ H\ } & \hat{\mathfrak{n}} \\
\wr \downarrow \text{Exp}\times\text{Exp} & & \downarrow \wr\, \text{Exp} \\
\mathcal{U}_\Theta \times \mathcal{U}_{\Delta^+\backslash\Theta} & \xrightarrow{\ m\ } & \mathcal{U}.
\end{array}$$

To prove the surjectivity of m, it suffices to show that H is surjective. Fix $x \in \hat{\mathfrak{n}}$. We denote the "β-th component" of x by x_β for any $\beta \in \Delta^+$, i.e., $x \in \sum_{\beta\in\Delta^+} x_\beta$. Assume, by induction on $p \geq 0$, that there are elements $y^p \in \hat{\mathfrak{n}}_\Theta$ and $z^p \in \hat{\mathfrak{n}}_{\Delta^+\backslash\Theta}$ such that

(1) $y^p_\beta = z^p_\beta = 0$ for all $|\beta| > p$,
(2) $H(y^p, z^p)_\beta = x_\beta$ for all $|\beta| \leq p$, and
(3) $y^p_\beta = y^{p-1}_\beta$ and $z^p_\beta = z^{p-1}_\beta$ for all $|\beta| \leq p - 1$.

The induction starts at $p = 0$ by taking $y^0 = z^0 = 0$.

Define

$$y^{p+1} = y^p + \sum_{\substack{|\beta|=p+1\\ \beta\in\Theta}} \big(x_\beta - H(y^p, z^p)_\beta\big), \text{ and}$$

$$z^{p+1} = z^p + \sum_{\substack{|\beta|=p+1\\ \beta\in\Delta^+\setminus\Theta}} \big(x_\beta - H(y^p, z^p)_\beta\big).$$

Using the formula for H as given in [Bourbaki–75b, Ch. II, §6, n° 4], we see that the elements y^{p+1}, z^{p+1} again satisfy (1)–(3), completing the induction. Now

$$H\left(\sum_{p\geq 1}(y^p - y^{p-1}), \sum_{p\geq 1}(z^p - z^{p-1})\right) = x.$$

This proves the surjectivity of H and hence the lemma is proved. □

We state the following lemma (cf. [Bourbaki–75b, Chap. III, §9, n° 5, Proposition 17]) for its application later on.

6.1.4 Lemma. *Let $\Theta \subset \Delta^+$ be a bracket closed finite subset. Assume further that Θ is a disjoint union $\Theta_1 \sqcup \cdots \sqcup \Theta_p$ of bracket closed subsets Θ_i. Then the multiplication map*

$$m : \mathcal{U}_{\Theta_1} \times \cdots \times \mathcal{U}_{\Theta_p} \to \mathcal{U}_\Theta, \;\; m(g_1, \dots, g_p) = g_1 \cdots g_p,$$

is a biregular isomorphism of affine varieties. Observe that, since Θ is finite, $\mathcal{U}_\Theta$ and each $\mathcal{U}_{\Theta_i}$ is a unipotent algebraic group. □

6.1.5 Examples. (a) Let $\alpha \in \Delta^+$ be a real root. Then, by Corollary 1.3.6(a), $\Theta = \{\alpha\}$ is bracket closed, and $\mathfrak{g}_\alpha$ is one-dimensional. So, for any $x \neq 0 \in \mathfrak{g}_\alpha$, $\mathbb{C}x = \mathfrak{g}_\alpha$. In this case, the one-parameter subgroup $\mathcal{U}_x$ (cf. 6.1.1) is denoted by $\mathcal{U}_\alpha$.

If α is a simple root, then $\Theta = \{\alpha\}$ is bracket closed as well as bracket coclosed. Moreover, $\Delta^+\setminus\{\alpha\}$ satisfies the condition $(*)$ of Lemma 6.1.2. In particular, $\mathcal{U}_{\Delta^+\setminus\{\alpha\}}$ is a normal subgroup of $\mathcal{U}$, and $\mathcal{U}$ is a semidirect product of $\mathcal{U}_\alpha$ and $\mathcal{U}_{\Delta^+\setminus\{\alpha\}}$ by Lemma 6.1.3.

(b) For any $w \in W$ (the Weyl group associated to $\mathfrak{g}$), let $\Phi_w = \{\alpha \in \Delta^+ : w^{-1}\alpha \in \Delta^-\}$ be as in Lemma 1.3.14. Then Φ_w is bracket closed as well as bracket coclosed. By Lemma 1.3.14, $|\Phi_w| = \ell(w)$ and, moreover, all the roots in Φ_w are real. In particular, $\hat{\mathfrak{n}}_{\Phi_w}$ is a (nilpotent) Lie algebra of dimension $\ell(w)$ and $\mathcal{U}_{\Phi_w}$ is a unipotent algebraic group (of dim $\ell(w)$).

(c) For any subset $Y \subset \{1, \dots, \ell\}$, recall the definition of the subset $\Delta_Y^+ \subset \Delta^+$ from 1.2.2. Then Δ_Y^+ is bracket closed and bracket coclosed. Moreover, $\Delta^+\setminus\Delta_Y^+$

satisfies the condition $(*)$ of Lemma 6.1.2. In particular, $\mathcal{U}$ is a semidirect product of its subgroup $\mathcal{U}_{\Delta_Y^+}$ and the normal subgroup $\mathcal{U}_{\Delta^+\setminus\Delta_Y^+}$.

6.1.6 The groups T and N. Recall the definition of the Cartan subalgebra $\mathfrak{h} \subset \mathfrak{g}$ from Definition 1.1.2. We choose a finitely generated $\mathbb{Z}$-submodule $\mathfrak{h}_{\mathbb{Z}} \subset \mathfrak{h}$ satisfying the following conditions:

(c_1) $\mathfrak{h}_{\mathbb{Z}}$ is an integral form of $\mathfrak{h}$, i.e., the natural map $\mathfrak{h}_{\mathbb{Z}} \otimes_{\mathbb{Z}} \mathbb{C} \to \mathfrak{h}$ is an isomorphism,

(c_2) all the simple coroots $\alpha_i^\vee \in \mathfrak{h}_{\mathbb{Z}}$,

(c_3) $\mathfrak{h}_{\mathbb{Z}}^* := \operatorname{Hom}_{\mathbb{Z}}(\mathfrak{h}_{\mathbb{Z}}, \mathbb{Z}) \subset \mathfrak{h}^*$ contains all the simple roots α_i, and

(c_4) $\mathfrak{h}_{\mathbb{Z}}/\sum_{i=1}^n \mathbb{Z}\alpha_i^\vee$ is torsion free.

In particular, $\mathfrak{h}_{\mathbb{Z}}$, resp. $\mathfrak{h}_{\mathbb{Z}}^*$, is W-stable under the canonical W-action on $\mathfrak{h}$, resp. $\mathfrak{h}^*$, (cf. Definition 1.3.1). We call $\mathfrak{h}_{\mathbb{Z}}$ as above an *integral Cartan subalgebra* of $\mathfrak{g}$.

In the sequel we fix an integral Cartan subalgebra $\mathfrak{h}_{\mathbb{Z}}$ of $\mathfrak{g}$.

Set

$$D_{\mathbb{Z}} := D \cap \mathfrak{h}_{\mathbb{Z}}^*,$$

where D is as in (2.1.5.1). Any element of $D_{\mathbb{Z}}$ is called a *dominant totally integral weight* (with respect to the choice of $\mathfrak{h}_{\mathbb{Z}}$).

Define the algebraic group T by

$$T := \operatorname{Hom}_{\mathbb{Z}}(\mathfrak{h}_{\mathbb{Z}}^*, \mathbb{C}^*). \tag{1}$$

Then T is a torus of dimension the same as $\dim \mathfrak{h} = \ell +$ corank A. The action of W on $\mathfrak{h}_{\mathbb{Z}}^*$ induces a natural action of W on T.

The Lie algebra Lie T of T can be identified with $\operatorname{Hom}_{\mathbb{Z}}(\mathfrak{h}_{\mathbb{Z}}^*, \mathbb{C}) \approx \mathfrak{h}$ and the exponential map Exp: Lie $T \to T$ is given by $f \mapsto e^f$, for $f \in \operatorname{Hom}_{\mathbb{Z}}(\mathfrak{h}_{\mathbb{Z}}^*, \mathbb{C})$, where $e^f(\lambda) = e^{\langle f,\lambda\rangle}$.

For any $\lambda \in \mathfrak{h}_{\mathbb{Z}}^*$, define the character $e^\lambda : T \to \mathbb{C}^*$ by $e^\lambda(t) = t(\lambda)$. Let $X(T)$ denote the group of all characters, i.e., algebraic group homomorphisms $T \to \mathbb{C}^*$. Then it is easy to see that the map $\lambda \mapsto e^\lambda$ is an isomorphism of groups $\mathfrak{h}_{\mathbb{Z}}^* \to X(T)$. We call $X(T)$ the *character group* of T.

Any $\mathfrak{h}$-module (V, π) which is a weight module, i.e., V is the sum of its weight spaces V_λ, such that all the weights of V lie in $\mathfrak{h}_{\mathbb{Z}}^*$, integrates to a T-module under

$$t \cdot v_\lambda = t(\lambda)v_\lambda, \quad \text{for } t \in T \text{ and } v_\lambda \in V_\lambda. \tag{2}$$

For $h \in \mathfrak{h}_{\mathbb{Z}}$ and $z \in \mathbb{C}^*$, define the element $z^h \in T$ by

$$z^h(\lambda) = z^{\langle\lambda,h\rangle}, \quad \text{for } \lambda \in \mathfrak{h}_{\mathbb{Z}}^*. \tag{3}$$

Let N be the group generated by the set $T \cup \{\tilde{s}_i\}_{1 \le i \le \ell}$ (for certain symbols $\tilde{s}_i$), and subject to the following relations:

(d_1) system of relations defining the group T,

(d_2) $\tilde{s}_i t \tilde{s}_i^{-1} = s_i(t)$, for $1 \le i \le \ell$,

(d_3) $\tilde{s}_i^2 = (-1)^{\alpha_i^\vee}$, for $1 \le i \le \ell$, and

(d_4) $\underbrace{\tilde{s}_i \tilde{s}_j \tilde{s}_i \dots}_{m_{i,j} \text{ factors}} = \underbrace{\tilde{s}_j \tilde{s}_i \tilde{s}_j \dots}_{m_{i,j} \text{ factors}}$, for $1 \le i \ne j \le \ell$ such that $m_{i,j} < \infty$,

where $m_{i,j}$ is the order of $s_i s_j \in W$ (cf. Proposition 1.3.21).

For an integrable representation (V, π) of $\mathfrak{g}$, and simple reflection s_i, recall the definition of $s_i(\pi) := (\exp f_i)(\exp -e_i)(\exp f_i) \in \operatorname{Aut} V$ from (1.3.2.5).

6.1.7 Lemma. *Any integrable representation (V, π) of $\mathfrak{g}$, such that all the weights of V lie in $\mathfrak{h}_{\mathbb{Z}}^*$, is a module for the group N under the action of T given by (6.1.6.2) and $\tilde{s}_i$ acts on V via $s_i(\pi)$.*

Proof. We need to check the relations (d_1)–(d_4) with $\tilde{s}_i$ replaced by $s_i(\pi)$: Since V is a T-module, (d_1) is clearly satisfied. The relations (d_2)–(d_4) follow from Lemma 1.3.5. □

6.1.8 Corollary. *The canonical map $\theta : T \cup \{\tilde{s}_i\}_{1 \le i \le \ell} \to N$ is injective. Further, there is an exact sequence of groups:*

(1)
$$1 \to T \xrightarrow{\theta_{|T}} N \xrightarrow{\pi} W \to 1\,,$$

(2)
$$\textit{with } \pi(\tilde{s}_i) = s_i, \textit{ for all } 1 \le i \le \ell\,.$$

Proof. Using the N-module structure, given in the previous lemma, on the integrable $\mathfrak{g}$-modules $L^{\max}(\lambda)$ for any $\lambda \in D_{\mathbb{Z}}$ (cf. Lemma 2.1.7), the injectivity of θ follows easily. (Use the fact that any $\lambda \in \mathfrak{h}_{\mathbb{Z}}^*$ can be written as $\lambda = \lambda_1 - \lambda_2$ for $\lambda_1, \lambda_2 \in D_{\mathbb{Z}}$.)

From the defining relations of N and (1.3.1.1), the map π defined by (2) extends to a (surjective) group homomorphism $\pi : N \to W$. By the relation (d_2) of 6.1.6, $T \subset N$ is a normal subgroup. Let $\overline{\pi} : N/T \to W$ be the induced (surjective) homomorphism. The elements $\tilde{s}_i T \in N/T$ satisfy the following relations: (a) $(\tilde{s}_i T)^2 = 1$ and (b) $[(\tilde{s}_i T)(\tilde{s}_j T)]^{m_{i,j}} = 1$ for $i \ne j$ such that $m_{i,j} < \infty$. But, by Proposition 1.3.21, W is a Coxeter group (with precisely these two as relations) with $\pi(\tilde{s}_i T) = s_i$ as the Coxeter generators. Hence $\overline{\pi}$ is an isomorphism. This proves the corollary. □

6.1.9 Remark. Since the group T is abelian, the conjugation action of N on T descends to an action of N/T on T. By virtue of the relation (d_2) of 6.1.6, this action of $N/T \approx W$ (under $\overline{\pi}$) on T coincides with the W-action defined in 6.1.6.

6.1.10 Completed parabolic subalgebras. Recall, for any subset $Y \subset \{1, \dots, \ell\}$, the definition of the standard parabolic subalgebra $\mathfrak{p}_Y$, its nilradical $\mathfrak{u}_Y$, and the Levi component $\mathfrak{g}_Y$ from 1.2.2.

We impose the restriction that Y is of finite type, i.e., $\mathfrak{g}_Y$ is finite dimensional (cf. 1.2.2). Similar to the completion $\hat{\mathfrak{n}}$ of $\mathfrak{n}$, define the "completion" $\hat{\mathfrak{p}}_Y$ of $\mathfrak{p}_Y$ by

$$\hat{\mathfrak{p}}_Y = \mathfrak{g}_Y \oplus \hat{\mathfrak{u}}_Y,$$

where $\hat{\mathfrak{u}}_Y := \hat{\mathfrak{n}}_{\Delta^+ \setminus \Delta_Y^+}$ (cf. (6.1.1.5) and Example 6.1.5 (c)). We define the Lie algebra bracket on $\hat{\mathfrak{p}}_Y$ by the same definition as (6.1.1.2). We endow $\hat{\mathfrak{p}}_Y$ with a pro-Lie algebra structure as follows:

For any $\beta = \sum n_i \alpha_i \in \Delta$, recall its Y-relative height from 3.2.11:

(1) $$\operatorname{ht}_Y \beta = |\sum_{i \notin Y} n_i|.$$

In particular, for any $\beta \in \Delta_Y$,

(2) $$\operatorname{ht}_Y \beta = 0.$$

Now define the family $\hat{\mathcal{F}}_Y$ of ideals of $\hat{\mathfrak{p}}_Y$ by

$$\hat{\mathcal{F}}_Y = \{\text{ ideals } \mathfrak{a} \subset \hat{\mathfrak{p}}_Y : \mathfrak{a} \supset \hat{\mathfrak{u}}_Y(k), \text{ for some } k > 0\},$$

where

$$\hat{\mathfrak{u}}_Y(k) := \prod_{\substack{\beta \in \Delta^+, \\ \operatorname{ht}_Y \beta \geq k}} \mathfrak{g}_\beta.$$

Each $\hat{\mathfrak{u}}_Y(k)$, being an ideal of $\hat{\mathfrak{p}}_Y$, is a $\mathfrak{g}_Y$-module under the adjoint action.

6.1.11 Lemma. *For any $k \geq 1$, $\hat{\mathfrak{u}}_Y(k)/\hat{\mathfrak{u}}_Y(k+1)$ is a finite-dimensional vector space. In particular, $\hat{\mathcal{F}}_Y$ is the defining set of a pro-Lie algebra structure on $\hat{\mathfrak{p}}_Y$.*

Clearly, $\hat{\mathfrak{n}}$ is a pro-Lie subalgebra of $\hat{\mathfrak{p}}_Y$, and $\hat{\mathfrak{u}}_Y$ is a pro-Lie ideal of $\hat{\mathfrak{p}}_Y$.

Proof. For $k = 1$, $M_k := \hat{\mathfrak{u}}_Y(k)/\hat{\mathfrak{u}}_Y(k+1)$ is generated as a $\mathfrak{g}_Y$-module by the simple root vectors $\{e_i\}_{i \in \{1,\dots,\ell\} \setminus Y}$. Hence, for $k = 1$, M_k is finite dimensional by Lemma 1.3.3(c) and Corollary 1.3.4.

Assume now by induction that M_k is finite dimensional. Take a basis $\{e_\beta\}_{\beta \in I}$ of M_k consisting of root vectors. Then M_{k+1} is generated by $\{[e_\beta, e_i]; \beta \in I, i \in$

$\{1, \ldots, \ell\} \backslash Y\}$ as a $\mathfrak{g}_Y$-module; so again it is finite dimensional. This completes the induction. □

We first recall the definition of a root datum (cf. [Springer–98, §7.4]).

6.1.12 Definition. A *root datum* is a quadruple $\Psi = (X, R, X^\vee, R^\vee)$ consisting of two finitely generated free abelian groups $X, X^\vee$ and finite subsets $R \subset X$, $R^\vee \subset X^\vee$, together with a pairing $\langle \cdot, \cdot \rangle : X \times X^\vee \to \mathbb{Z}$, and a bijection $\alpha \mapsto \alpha^\vee$ of R onto $R^\vee$, satisfying the following properties for any $\alpha \in R$:

(RD0) $\langle \cdot, \cdot \rangle$ identifies X with $\mathrm{Hom}_{\mathbb{Z}}(X^\vee, \mathbb{Z})$,
(RD1) $\langle \alpha, \alpha^\vee \rangle = 2$,
(RD2) $s_\alpha R \subset R$, $\bar{s}_\alpha(R^\vee) \subset R^\vee$,

where s_α, resp. $\bar{s}_\alpha$, is the endomorphism of X, resp. $X^\vee$, defined by

$$s_\alpha x = x - \langle x, \alpha^\vee \rangle \alpha \quad (\text{resp. } \bar{s}_\alpha \bar{x} = \bar{x} - \langle \alpha, \bar{x} \rangle \alpha^\vee) \text{ for } x \in X, \text{ resp. } \bar{x} \in X^\vee.$$

By (RD1), s_α and $\bar{s}_\alpha$ are involutive automorphisms of X and $X^\vee$, respectively.

6.1.13 Parabolic (sub)groups associated to Kac–Moody algebra $\mathfrak{g}$. Fix a subset $Y \subset \{1, \ldots, \ell\}$ of finite type and an integral Cartan subalgebra $\mathfrak{h}_{\mathbb{Z}}$ of $\mathfrak{g}$ (cf. 6.1.6). Define the root datum

$$\Psi_Y = (\mathfrak{h}_{\mathbb{Z}}^*, \Delta_Y, \mathfrak{h}_{\mathbb{Z}}, \Delta_Y^\vee), \tag{1}$$

where $\Delta_Y^\vee := \{\alpha^\vee : \alpha \in \Delta_Y\}$ (cf. Definition 1.3.8). Observe that, since Y is of finite type, all the roots of Δ_Y are real (use (1.3.5.4)). We take the standard pairing $\mathfrak{h}_{\mathbb{Z}}^* \times \mathfrak{h}_{\mathbb{Z}} \to \mathbb{Z}$. The map $\alpha \mapsto \alpha^\vee$ of Δ_Y onto $\Delta_Y^\vee$ is a bijection by Lemma 1.3.7.

Let $\mathcal{G}_Y$ be the unique (up to isomorphism) connected reductive (finite dimensional) linear algebraic group associated to the root datum Ψ_Y (cf. [Springer–98, Theorem 10.1.1]). Then Lie $\mathcal{G}_Y = \mathfrak{g}_Y$ and $T := \mathrm{Hom}_{\mathbb{Z}}(\mathfrak{h}_{\mathbb{Z}}^*, \mathbb{C}^*)$ is a maximal torus of $\mathcal{G}_Y$ with Lie $T = \mathfrak{h}$. Further, any finite-dimensional $(\mathfrak{g}_Y, T)$-module V is a $\mathcal{G}_Y$-module under $\pi : \mathcal{G}_Y \to \mathrm{Aut}\,(V)$ extending the T-module structure and such that the derivative $\dot{\pi}$ is the original $\mathfrak{g}_Y$-module structure on V. Recall that a finite-dimensional $\mathfrak{g}_Y$-module V is called a $(\mathfrak{g}_Y, T)$-module if the $\mathfrak{h}$-module structure on V, obtained by the restriction, integrates to an algebraic T-module structure.

For any $k \geq 1$, the finite-dimensional space

$$M_k := \bigoplus_{\substack{\beta \in \Delta^+ \\ \mathrm{ht}_Y \beta = k}} \mathfrak{g}_\beta \tag{2}$$

is a $(\mathfrak{g}_Y, T)$-module under the adjoint action (cf. (6.1.6.2)). In particular, M_k is a $\mathcal{G}_Y$-module, and hence so is $\hat{\mathfrak{u}}_Y$. Further, for any $g \in \mathcal{G}_Y$, g acts on $\hat{\mathfrak{u}}_Y$ as a Lie algebra automorphism. To see this, fix $x \in \mathfrak{g}_Y$, $v \in M_{k_1}$ and $w \in M_{k_2}$, for some $k_1, k_2 > 0$, and define the function $f : \mathbb{R} \to M_{k_1+k_2}$ by

$$f(t) = [\text{Exp}\,(tx)v,\ \text{Exp}\,(tx)w] - \text{Exp}\,(tx)[v, w].$$

The function f satisfies the first order differential equation $\frac{d}{dt} f(t) = x \cdot f(t)$, with the boundary condition $f(0) = 0$. By the uniqueness of the solution, we get $f \equiv 0$. This shows that $\text{Exp}\, x$ acts on $\hat{\mathfrak{u}}_Y$ as a Lie algebra automorphism for any $x \in \mathfrak{g}_Y$. Since $\text{Exp}\, \mathfrak{g}_Y$ generates $\mathcal{G}_Y$ as a group, we obtain that any $g \in \mathcal{G}_Y$ acts as a Lie algebra automorphism on $\hat{\mathfrak{u}}_Y$. The action of g is clearly continuous (under the pro-topology on $\hat{\mathfrak{u}}_Y$) and hence g acts as a pro-Lie algebra automorphism. In particular, by Theorem 4.4.19, we get a group homomorphism

$$\phi : \mathcal{G}_Y \to \text{Aut}\, \mathcal{U}_Y, \tag{3}$$

where $\mathcal{U}_Y$ is the pro-unipotent pro-group with Lie algebra $\hat{\mathfrak{u}}_Y$ (cf. Theorem 4.4.19) and $\text{Aut}\,\mathcal{U}_Y$ is the group of all the pro-group automorphisms of $\mathcal{U}_Y$.

Now we define the *standard parabolic group* $\mathcal{P}_Y$, associated to the subset Y of finite type and the Kac–Moody algebra $\mathfrak{g}$, as the semidirect product

$$\mathcal{P}_Y = \mathcal{U}_Y \ltimes \mathcal{G}_Y\,. \tag{4}$$

Recall that as a set $\mathcal{U}_Y \ltimes \mathcal{G}_Y$ is nothing but $\mathcal{U}_Y \times \mathcal{G}_Y$ and the product is given by

$$(u_1, g_1) \cdot (u_2, g_2) = (u_1\phi(g_1)u_2, g_1g_2), \tag{5}$$

for $u_i \in \mathcal{U}_Y$ and $g_i \in \mathcal{G}_Y$.

The following lemma is easy to verify from the definition of pro-groups.

6.1.14 Lemma. *$\mathcal{P}_Y$ is a pro-group with the defining set*

$$\mathcal{F}_Y = \{\textit{normal subgroups } N \subset \mathcal{P}_Y : N \supset \mathcal{U}_Y(k), \textit{for some } k = k(N) \geq 1, \\ \textit{and } N/\mathcal{U}_Y(k) \textit{ is Zariski closed in } (\mathcal{U}_Y/\mathcal{U}_Y(k)) \ltimes \mathcal{G}_Y\},$$

where $\mathcal{U}_Y(k)$ is the pro-subgroup of $\mathcal{U}_Y$ corresponding to the pro-Lie subalgebra $\hat{\mathfrak{u}}_Y(k) \subset \hat{\mathfrak{u}}_Y$. Observe that $\mathcal{U}_Y(k)$ is normal in $\mathcal{P}_Y$ since $\mathcal{G}_Y$ keeps $\hat{\mathfrak{u}}_Y(k)$, and hence $\mathcal{U}_Y(k)$, stable under ϕ.

Moreover, $\mathcal{U}_Y$ is a normal pro-subgroup of $\mathcal{P}_Y$, and the induced pro-structure on $\mathcal{U}_Y$ via Proposition 4.4.5 coincides with its original pro-structure. Further, Lie $\mathcal{P}_Y \approx \hat{\mathfrak{p}}_Y$ *as pro-Lie algebras.* □

For any subsets $Y_1 \subset Y_2 \subset \{1, \dots, \ell\}$ such that Y_2 (and hence Y_1) is of finite type, there is an inclusion of algebraic groups $\mathcal{G}_{Y_1} \hookrightarrow \mathcal{G}_{Y_2}$ corresponding to the inclusion $\mathfrak{g}_{Y_1} \subset \mathfrak{g}_{Y_2}$.

Consider the nilpotent Lie subalgebra $\mathfrak{s} = \oplus_{\alpha \in \Delta^+_{Y_2} \setminus \Delta^+_{Y_1}} \mathfrak{g}_\alpha$ of $\mathfrak{g}_{Y_2}$ and let $S \subset \mathcal{G}_{Y_2}$ be the corresponding algebraic subgroup. It is easy to see that S exists. Then the pro-subgroup $\mathcal{U}_{Y_2} \ltimes S$ of $\mathcal{P}_{Y_2}$ can be identified with the pro-subgroup $\mathcal{U}_{Y_1}$ of $\mathcal{P}_{Y_1}$ (since they have the same pro-Lie algebras). Since $\mathcal{G}_{Y_1}$ normalizes S inside $\mathcal{G}_{Y_2}$,

$$\mathcal{P}_{Y_1} := \mathcal{U}_{Y_1} \ltimes \mathcal{G}_{Y_1} = (\mathcal{U}_{Y_2} \ltimes S) \ltimes \mathcal{G}_{Y_1} = \mathcal{U}_{Y_2} \ltimes (S \ltimes \mathcal{G}_{Y_1}) \subset \mathcal{P}_{Y_2}.$$

The last inclusion follows since $S \cap \mathcal{G}_{Y_1} = \{1\}$, as can easily be seen.

So we have shown the following:

6.1.15 Lemma. *For subsets $Y_1 \subset Y_2 \subset \{1, \dots, \ell\}$ such that Y_2 (and hence Y_1) is of finite type, there is an injective (unique) pro-group morphism $\gamma = \gamma_{Y_2,Y_1}$: $\mathcal{P}_{Y_1} \hookrightarrow \mathcal{P}_{Y_2}$ such that $\dot{\gamma}$ is the inclusion $\hat{\mathfrak{p}}_{Y_1} \subset \hat{\mathfrak{p}}_{Y_2}$.* □

6.1.16 Kac–Moody group. For any $i \in \{1, \dots, \ell\}$, let $\mathcal{P}_i$ be the standard parabolic group $\mathcal{P}_{\{i\}}$ and let $\mathcal{B} = \mathcal{P}_\phi$. By Lemma 6.1.15,

$$\mathcal{B} \subset \mathcal{P}_i, \qquad \text{for any } i.$$

Also let $N_i \subset N$ be the subgroup (cf. 6.1.6 for the definition of N) generated by $T \cup \{\tilde{s}_i\}$. Then by (6.1.8.1),

$$N_i = T \cup T\tilde{s}_i \qquad \text{(disjoint union)}. \tag{1}$$

Set $\mathcal{G}_i = \mathcal{G}_{\{i\}}$ and define the embedding $\theta_i : N_i \hookrightarrow \mathcal{G}_i \subset \mathcal{P}_i$ as follows:

$$\theta_{i|T} = I, \text{ and } \theta_i(\tilde{s}_i) = \operatorname{Exp}(f_i)\operatorname{Exp}(-e_i)\operatorname{Exp}(f_i) \in \mathcal{G}_i\,, \tag{2}$$

where "Exp" is the exponential map $\mathfrak{g}_i \to \mathcal{G}_i$ and $\mathfrak{g}_i := \mathfrak{g}_{\{i\}}$. From Lemma 1.3.5(c), it follows that $\theta_i(\tilde{s}_i)^2 = (-1)^{\alpha_i^\vee}$, and hence θ_i defined by (2) indeed extends to a group homomorphism $\theta_i : N_i \to \mathcal{P}_i$. Moreover, θ_i is injective, as follows from Lemma 1.3.5(a).

Define $Z = ((\sqcup_{i\in\{1,\dots,\ell\}}\mathcal{P}_i) \sqcup N)/\sim$, where we identify the following elements:

(a) $\gamma_i(b) \sim \gamma_j(b)$, for any i, j and $b \in \mathcal{B}$;
(b) $n \sim \theta_i(n)$, for $n \in N_i \subset N$,

where $\gamma_i : \mathcal{B} \hookrightarrow \mathcal{P}_i$ is the inclusion as in Lemma 6.1.15.

Each $\mathcal{P}_i$ and N inject in Z, and $\mathcal{B} \cap N = T$. To show this, observe that

$$\theta_i(N_i \setminus T) \subset \mathcal{P}_i \setminus \mathcal{B}, \tag{3}$$

which follows from Lemma 1.3.5(a).

We come to the crucial definition: Let $\mathcal{G}$ be the amalgamated product of the system of groups $(N, \mathcal{P}_i;\ 1 \le i \le \ell)$ (cf. Definition 5.1.6), where we think of each of N and $\mathcal{P}_i$ as a subset of Z.

6.1.17 Theorem. *The canonical map $Z \to \mathcal{G}$ is injective. In particular, the canonical group homomorphisms $\mathcal{P}_i \to \mathcal{G}$ (for any $i \in \{1, \dots, \ell\}$) and $N \to \mathcal{G}$ are all injective. Hence, each of $\mathcal{P}_i$ and N (in particular $\mathcal{B}$) can (and will) be considered as a subgroup of $\mathcal{G}$. Further, $(\mathcal{G}, \mathcal{B}, N, S)$ is a Tits system, where $S := \{\tilde{s}_i T \in N/T\}_{1 \le i \le \ell}$.*

The group $\mathcal{G}$ is called the *"maximal" Kac–Moody group* associated to the Kac–Moody Lie algebra $\mathfrak{g}$, although we will drop the adjective "maximal" in the sequel. The subgroup $\mathcal{B}$, resp. T, is called the *standard Borel subgroup*, resp. *standard maximal torus* of $\mathcal{G}$.

In the finite case, i.e., when $\mathfrak{g}$ is a finite-dimensional semisimple Lie algebra, $\mathcal{G}$ coincides with the simply-connected (semisimple) complex algebraic group with Lie algebra $\mathfrak{g}$ (cf. Exercise 6.1.E.2).

Proof. We apply Theorem 5.1.8 to prove the above theorem, so we need to check all the properties (P_1)–(P_9) of Theorem 5.1.8 for the system of groups $(N, \mathcal{P}_i;\ 1 \le i \le \ell)$. (An alternative proof of the theorem, not using Theorem 5.1.8, is outlined in Exercise 6.2.E.3.)

(P_1) and (P_2) follow from (6.1.16.3); (P_3) follows from (6.1.16.1). From the Bruhat decomposition for rank-1 groups, we get

(1) $$\mathcal{G}_i = (T \operatorname{Exp}(\mathbb{C} e_i)) \cup (\operatorname{Exp}(\mathbb{C} e_i) \cdot \tilde{s}_i T \operatorname{Exp}(\mathbb{C} e_i)),$$

where we identified N_i with a subgroup of $\mathcal{P}_i$ under θ_i. From this (P_4) follows. (P_5) follows from Corollary 6.1.8 and Proposition 1.3.21.

For any $1 \le i \le \ell$, $n \in T\,\tilde{s}_i$, and a bracket closed subset $\Theta \subset \Delta^+$ (cf. 6.1.1), we have by Corollary 1.3.6:

(2) $$\mathcal{U} \cap (n^{-1} \mathcal{U}_\Theta n) = \mathcal{U}_{(s_i \Theta) \cap \Delta^+} \text{in } \mathcal{P}_i.$$

Let $\pi : N \to W$ be the homomorphism defined in 6.1.8. Fix $n = n_{i_1} \cdots n_{i_p} \in N$, where $n_j \in N_j \setminus T$, such that $\pi(n) = \pi(n_{i_1}) \cdots \pi(n_{i_p})$ is a reduced decomposition. We call such a decomposition of n a *reduced decomposition*. Then, following the notation of Theorem 5.1.8, by making successive use of (2), we get

(3) $$\mathcal{B}(n_{i_1}, \dots, n_{i_p}) = T \cdot \mathcal{U}_{\pi(n)^{-1}\Delta^+ \cap \Delta^+}.$$

Observe that, in obtaining the above, we used the following: For $w \in W$ and $1 \le j \le \ell$, such that $ws_j > w$, we have $ws_j\, \Delta^+ \cap \Delta^+ \subset w\, \Delta^+ \cap \Delta^+$, as follows from Lemma 1.3.14.

Clearly (3) proves the first part of (P_6). To prove the second part of(P_6), observe that, by Lemma 6.1.7 and Corollary 1.3.4, $\mathfrak{g}$ is a module for N induced by the adjoint representation.

For any $n \in T\tilde{s}_i$ and a bracket closed subset $\Theta \subset \Delta^+$ such that $s_i\Theta \subset \Delta^+$, we have the following commutative diagram:

$$(4)\qquad \begin{array}{ccc} \hat{\mathfrak{n}}_\Theta & \xrightarrow{\;n\;} & \hat{\mathfrak{n}}_{s_i\Theta} \\ \wr\downarrow \text{Exp} & & \wr\downarrow \text{Exp} \\ \mathcal{U}_\Theta & \xrightarrow[\gamma_n]{} & \mathcal{U}_{s_i\Theta}\,, \end{array}$$

where $\hat{\mathfrak{n}}_\Theta$ is as in (6.1.1.5), the top horizontal map n is induced from the action of n on $\mathfrak{g}$ and γ_n is the conjugation $g \mapsto ngn^{-1}$ in $\mathcal{P}_i$ (cf. (2)). Since $\mathfrak{g}$ is an N-module, by (4), we get the validity of the second condition of (P_6). (P_7) follows from (3), Lemma 1.3.14 and Lemma 6.1.3. We come to (P_8) (following the same notation, except that we take $s = s_i$, $t = s_j$). By (3),

$$\mathcal{B}_t = \mathcal{U}_{\Delta^+\setminus\{\alpha_j\}}T.$$

Take $y = \text{Exp}(ze_j) \in \mathcal{G}_j$, for any $z \neq 0 \in \mathbb{C}$. Then $y \in \mathcal{B}_n$ by (3), since $\ell(wt) = \ell(w) + 1$ by assumption. By the Bruhat decomposition for rank -1 groups, we can choose $y', y'' \in \mathcal{U}_{\{\alpha_j\}}T \subset \mathcal{B}_n$ such that

$$(5)\qquad m'^{-1}ym' = y'm'y'' \text{ in } \mathcal{G}_j \subset \mathcal{P}_j,$$

proving the (a) part of (P_8).

Consider the isomorphism $\gamma_n : \mathcal{B}_n \to \mathcal{B}_{n^{-1}}$. Then, by (4),

$$(6)\qquad \gamma_n(\mathcal{U}_{\{\alpha_j\}}T) \subset \mathcal{U}_{\{\alpha_i\}}T,$$

since $w\alpha_j = \alpha_i$. Now by [Humphreys–95, Proposition 32.3] or [Jantzen–87, Part II, Proposition 1.14], the homomorphism $\gamma_{n|(\mathcal{U}_{\{\alpha_j\}}T)}$ can be (uniquely) extended to an isomorphism $\tilde{\gamma}_n : \mathcal{G}_j \to \mathcal{G}_i$ such that

$$\tilde{\gamma}_n(x) = nxn^{-1}, \text{ for } x \in N_j.$$

Now applying the homomorphism $\tilde{\gamma}_n$ to (5) we get the (b) part of (P_8).

For any $1 \le i \le \ell$ and any $n_i \in N_i \setminus T$,

$$n_i\mathcal{U}_{\alpha_i}n_i^{-1} = \mathcal{U}_{-\alpha_i} \qquad \text{in } \mathcal{G}_i,$$

where $\mathcal{U}_{\alpha_i} = \mathcal{U}_{\{\alpha_i\}} = \text{Exp}\,\mathfrak{g}_{\alpha_i}$ and $\mathcal{U}_{-\alpha_i} = \text{Exp}\,\mathfrak{g}_{-\alpha_i}$. Hence $\mathcal{B}$ is not normal in any $\mathcal{P}_i$, proving (P_9).

Finally, the theorem follows from Theorem 5.1.8. □

6.1.18 Definition. For any subset $Y \subset \{1, \ldots, \ell\}$ (not necessarily of finite type), define the *standard parabolic subgroup* $\mathcal{P}_Y$ of $\mathcal{G}$ by

$$(1) \qquad \mathcal{P}_Y = \mathcal{B}\, W_Y\, \mathcal{B},$$

where $W_Y \subset W$ is the subgroup generated by $\{s_i\}_{i \in Y}$. By Theorems 5.1.3(b) and 6.1.17, $\mathcal{P}_Y$ is indeed a subgroup of $\mathcal{G}$ and, moreover, by Theorem 5.1.3(g), for any subgroup $\mathcal{P}$ of $\mathcal{G}$ containing $\mathcal{B}$ there exists a unique Y such that $\mathcal{P} = \mathcal{P}_Y$. For $Y = \{i\}$, $1 \le i \le \ell$, we denote $\mathcal{P}_{\{i\}}$ by $\mathcal{P}_i$ itself and call it a *minimal parabolic subgroup*.

Also recall the definition of the group $\mathcal{P}_Y$, for any subset $Y \subset \{1, \ldots, \ell\}$ of finite type, from 6.1.13. To distinguish, let us temporarily denote this $\mathcal{P}_Y$ (i.e., that of 6.1.13) by $\mathcal{Q}_Y$. The following lemma justifies the same notation $\mathcal{P}_Y$.

6.1.19 Lemma. *For any subset of finite type $Y \subset \{1, \ldots, \ell\}$, there exists a unique isomorphism of groups $\theta : \mathcal{Q}_Y \approx \mathcal{P}_Y$, such that $\theta_{|\mathcal{Q}_i} = I$, for all $i \in Y$ (cf. Lemma 6.1.15).*

Proof. By (6.1.13.4), $\mathcal{Q}_Y = \mathcal{U}_Y \ltimes \mathcal{G}_Y$. By Example 5.1.2, $(\mathcal{G}_Y, \mathcal{B}_Y, N_Y, S_Y)$ is a Tits system, where $\mathcal{B}_Y \subset \mathcal{G}_Y$ is the Borel subgroup corresponding to the subalgebra $\mathfrak{b}_Y := \mathfrak{h} \oplus \bigoplus_{\alpha \in \Delta_Y^+} \mathfrak{g}_\alpha$ of $\mathfrak{g}_Y$, N_Y is the normalizer of T in $\mathcal{G}_Y$, and $S_Y = \{\theta_i(\tilde{s}_i) \bmod T\}_{i \in Y}$ (cf. (6.1.16.2)). Hence, $(\mathcal{Q}_Y, \mathcal{U}_Y \ltimes \mathcal{B}_Y, N_Y, S_Y)$ is a Tits system. So, by Proposition 5.1.7, $\mathcal{Q}_Y$ is the amalgamated product of its subgroups $\{N_Y, \mathcal{Q}_i;\ i \in Y\}$. Further, N_Y is isomorphic with the subgroup of N generated by $\{T, \tilde{s}_i;\ i \in Y\}$ where $\theta_i(\tilde{s}_i) \mapsto \tilde{s}_i$ and $t \mapsto t$ for $t \in T$ (cf. [Tits–66]). Hence, there exists a unique group homomorphism $\theta : \mathcal{Q}_Y \to \mathcal{G}$ such that $\theta_{|\mathcal{Q}_i} = \mathrm{I}$, and $\theta(\theta_i(\tilde{s}_i)) = \theta_i(\tilde{s}_i)$, for all $i \in Y$. Since $\mathcal{Q}_Y$, resp. $\mathcal{P}_Y$, is generated (as a group) by $\{\mathcal{Q}_i;\ i \in Y\}$, resp. $\{\mathcal{P}_i;\ i \in Y\}$, we get that $\operatorname{Im} \theta = \mathcal{P}_Y$.

Finally we show that θ is injective: Let $g \in \operatorname{Ker} \theta$. By the Bruhat decomposition (cf. Theorem 5.1.3(c)), write $g = bnb'$, for some $b, b' \in \mathcal{U}_Y \ltimes \mathcal{B}_Y$ and $n \in N_Y$. Then $\theta(g) = 1$ implies $\theta(n) = \theta(b'b)^{-1}$. But since $N \cap \mathcal{B} = T$ (in the Kac–Moody group $\mathcal{G}$), we get $n \in T$. (We are using the fact that the canonical map $N_Y \to N$ is injective.) Hence $g \in \mathcal{U}_Y \ltimes \mathcal{B}_Y$; in particular, $g = 1$ since θ is injective on $\mathcal{U}_Y \ltimes \mathcal{B}_Y$. This proves the lemma. □

Combining Theorems 5.1.3(c) and 6.1.17, we get the following Bruhat decomposition.

6.1.20 Corollary. *For any subsets $Y, Y' \subset \{1, \ldots, \ell\}$, we have*

$$\mathcal{G} = \bigsqcup_{w \in W_Y \backslash W / W_{Y'}} \mathcal{P}_Y w \mathcal{P}_{Y'}.$$

In particular, taking $Y = Y' = \emptyset$, we get

$$\mathcal{G} = \bigsqcup_{w \in W} \mathcal{B} w \mathcal{B}.$$

□

6.1.E EXERCISES

(1) Let M be a $\mathfrak{g}$-module as well as a pro-representation of $\hat{\mathfrak{n}}$ (cf. 6.1.1), such that the two $\mathfrak{n}$-module structures on M (coming from $\mathfrak{g}$ and $\hat{\mathfrak{n}}$) coincide. Then show that there is a (unique) $\hat{\mathfrak{g}}$-module structure on M extending the two structures.

(2) Show that in the finite case, i.e., when $\mathfrak{g}$ is a finite-dimensional semisimple Lie algebra, $\mathcal{G}$ coincides with the simply-connected (semisimple) complex algebraic group with Lie algebra $\mathfrak{g}$.
Hint. Use Proposition 5.1.7.

6.2. Representations of Kac–Moody Groups

6.2.1 Pro-representations of $\mathcal{G}$ and $\hat{\mathfrak{g}}$. A representation (V, π) of $\mathcal{G}$, resp. $\hat{\mathfrak{g}}$, is called a *pro-representation* of $\mathcal{G}$, resp. of $\hat{\mathfrak{g}}$, if, for all $1 \le i \le \ell$, the representation $\pi_{|\mathcal{P}_i}$, resp. $\pi_{|\hat{\mathfrak{p}}_i}$, is a pro-representation of the pro-group $\mathcal{P}_i$, resp. of the pro-Lie algebra $\hat{\mathfrak{p}}_i$ (cf. Definitions 4.4.22 and 4.4.23).

Let $\mathfrak{M}(\mathcal{G})$, resp. $\mathfrak{M}(\hat{\mathfrak{g}})$, be the category of pro-representations of $\mathcal{G}$, resp. pro-representations of $\hat{\mathfrak{g}}$, and the $\mathcal{G}$-module, resp. $\hat{\mathfrak{g}}$-module, maps between them.

A $\hat{\mathfrak{g}}$-module V is called a *$(\hat{\mathfrak{g}}, T)$-module* if the $\mathfrak{h}$-module structure on V, obtained by the restriction of the $\hat{\mathfrak{g}}$ action, integrates to a locally finite algebraic T-module structure on V. In particular, any $(\hat{\mathfrak{g}}, T)$-module is a weight module for the $\mathfrak{h}$-action. Let $\mathfrak{M}_T(\hat{\mathfrak{g}})$ be the full subcategory of $\mathfrak{M}(\hat{\mathfrak{g}})$ consisting of $(\hat{\mathfrak{g}}, T)$-modules $V \in \mathfrak{M}(\hat{\mathfrak{g}})$. All the three categories $\mathfrak{M}(\mathcal{G})$, $\mathfrak{M}(\hat{\mathfrak{g}})$, and $\mathfrak{M}_T(\hat{\mathfrak{g}})$ are closed under taking direct sums, tensor products, submodules, and quotient modules.

6.2.2 Lemma. *Let (V, π) be a pro-representation of $\mathcal{G}$. Then there exists a unique pro-representation $(V, \dot{\pi})$ (in the same vector space) of $\hat{\mathfrak{g}}$ such that, for all $1 \le i \le \ell$, $\dot{\pi}_{|\hat{\mathfrak{p}}_i} = \dot{\pi}_i$, where $\pi_i := \pi_{|\mathcal{P}_i}$ (cf. Lemma 6.1.14 and Definition 4.4.23).*

The representation $\dot{\pi}$ is called the *derivative* of π, and π is called the *integral* of $\dot{\pi}$. (By Lemma 4.4.24, π is indeed uniquely determined by $\dot{\pi}$.)

Proof. Clearly $\dot{\pi}_{i|\hat{\mathfrak{b}}} = \dot{\pi}_{j|\hat{\mathfrak{b}}}$, for any $1 \le i, j \le \ell$. For any i, define

$$\dot{\pi}(f_i) = \dot{\pi}_i(f_i).$$

Also define $\dot{\pi}_{|\hat{\mathfrak{b}}} = \dot{\pi}_{i|\hat{\mathfrak{b}}}$.

To prove the lemma, in view of Exercise 6.1.E.1, it suffices to show that $\dot{\pi}$ as defined above extends to a representation of $\mathfrak{g}$, i.e., $\dot{\pi}(f_i)$'s satisfy the relation (R_5) of Definition 1.1.2. But this follows from Corollary 1.3.10. □

6.2.3 Theorem. *The category $\mathfrak{M}(\mathcal{G})$ is equivalent to the category $\mathfrak{M}_T(\hat{\mathfrak{g}})$ under $(V, \pi) \rightsquigarrow (V, \dot{\pi})$ and $f \rightsquigarrow f$ for any $\mathcal{G}$-module map $f : V \to V'$ between two pro-representations of $\mathcal{G}$.*

Further, the category $\mathfrak{M}_T(\hat{\mathfrak{g}})$ *consists precisely of those* $(\hat{\mathfrak{g}}, T)$*-modules* (V, π) *such that* V *is integrable as a* $\mathfrak{g}$*-module and* $(V, \pi_{|\hat{\mathfrak{n}}})$ *is a pro-representation of* $\hat{\mathfrak{n}}$*. In particular, any integrable highest weight* $\mathfrak{g}$*-module* V *with highest weight* $\lambda \in D_{\mathbb{Z}}$ *belongs to* $\mathfrak{M}_T(\hat{\mathfrak{g}})$ *and hence acquires a* $\mathcal{G}$*-module structure.*

Proof. Since $\{\mathcal{P}_i\}_{1\leq i\leq \ell}$ generate $\mathcal{G}$ as a group and $\{\hat{\mathfrak{p}}_i\}_{1\leq i\leq \ell}$ generate $\hat{\mathfrak{g}}$ as a Lie algebra, from Lemmas 6.2.2 and 4.4.24 and the proof of Proposition 4.4.26, to prove the first assertion of the above theorem, it suffices to show that for any $(V, \sigma) \in \mathfrak{M}_T(\hat{\mathfrak{g}})$ there exists a $(V, \pi) \in \mathfrak{M}(\mathcal{G})$ such that $\sigma = \dot{\pi}$.

So take $(V, \sigma) \in \mathfrak{M}_T(\hat{\mathfrak{g}})$. Fix $1 \leq i \leq \ell$. Since $\sigma_{|\hat{\mathfrak{p}}_i}$ is a pro-representation of $\hat{\mathfrak{p}}_i$, for any finite-dimensional $\hat{\mathfrak{p}}_i$-submodule $W \subset V$, there exists an ideal $\hat{\mathfrak{u}}_i(k)$ for some $k = k(W)$ of $\hat{\mathfrak{p}}_i$ which annihilates W, i.e., W is a module for $\hat{\mathfrak{p}}_i/\hat{\mathfrak{u}}_i(k) = (\hat{\mathfrak{u}}_i/\hat{\mathfrak{u}}_i(k)) \oplus \mathfrak{g}_i$ (semidirect product), where $\hat{\mathfrak{u}}_i(k)$ is as in 6.1.10. Thus, W is a module for the associated simply-connected group. But, by 6.1.13, W is a $\mathcal{G}_i$-module and also W is a pro-representation of $\mathcal{U}_i$ (cf. Proposition 4.4.26). (Observe that each element of $\hat{\mathfrak{u}}_i$ acts locally nilpotently on W since W is a weight module for the $\mathfrak{h}$-action.) From this it is easy to see that W is a pro-representation of $\mathcal{P}_i$. From the uniqueness as in Lemma 4.4.24, and since finite-dimensional $\hat{\mathfrak{p}}_i$-submodules $W \subset V$ span V, V is a pro-representation of $\mathcal{P}_i$. Since V is a locally finite $(\mathfrak{g}_i, T)$-module; in particular, it is an integrable $\mathfrak{g}$-module and hence is a module for N (cf. Lemma 6.1.7). So V becomes a pro-representation for $\mathcal{G}$ (say under π). By construction, $\dot{\pi} = \sigma$. This proves the first part of the theorem.

The second assertion follows easily from Lemma 1.3.3(c_2). The "In particular" part of the theorem follows readily from the second assertion. □

6.2.4 Remark. I do not know any example of a $(\hat{\mathfrak{g}}, T)$-module (V, π) such that V is integrable as a $\mathfrak{g}$-module but $(V, \pi_{|\hat{\mathfrak{n}}})$ is *not* a pro-representation of $\hat{\mathfrak{n}}$.

6.2.5 Lemma. *For a positive root* β *and nonzero* $X_\beta \in \mathfrak{g}_\beta$*, there exists an integrable* $\mathfrak{g}$*-module* $M \in \mathcal{O}$ *such that* X_β *acts nontrivially on* M*. In fact, we can choose* M *to be the* $(\mathfrak{g}, T)$*-module* $L^{\max}(\lambda)^\sigma$ *for some* $\lambda \in D_{\mathbb{Z}}$*, where* σ *is the involution of the category* $\mathcal{O}$ *defined in 2.1.1(c).*

Proof. Write $\beta = \sum n_i\alpha_i$ and let $n := \max_i n_i$. Take any $\lambda \in D_{\mathbb{Z}}$ such that $\langle \lambda, \alpha_i^\vee \rangle \geq n$, for all i. Set $M = L^{\max}(\lambda)^\sigma$. Then $M \in \mathcal{O}$, and is a $(\mathfrak{g}, T)$-module. Moreover, M is integrable (cf. Lemmas 2.1.7 and 2.1.14). We next show that X_β acts nontrivially on M.

Let v_λ be a nonzero highest weight vector in $L^{\max}(\lambda)$. Then, for any nonzero $Y_\beta \in \mathfrak{g}_{-\beta}$, from the definition of $L^{\max}(\lambda)$ (cf. 2.1.5) and Lemma 2.1.6,

$$Y_\beta v_\lambda \neq 0. \tag{1}$$

Let $f \in M$ be any element such that $f(Y_\beta v_\lambda) \neq 0$ for $Y_\beta := \omega(X_\beta)$, where

ω is the Cartan involution (cf. 1.1.2). Then $(X_\beta \cdot f)v_\lambda = -f(Y_\beta v_\lambda) \neq 0$. This proves the lemma. □

6.2.6 Corollary. *For any $g \neq 1 \in \mathcal{U}$, there exists $\hat{M} \in \mathfrak{M}(\mathcal{G})$ such that g acts nontrivially on $\hat{M}$.*

Proof. Express $g = \operatorname{Exp}(\sum_{\beta \in \Delta^+} X_\beta)$, with $X_\beta \in \mathfrak{g}_\beta$. Choose a $\beta = \beta_o$ such that $X_{\beta_o} \neq 0$ and, moreover, $|\beta_o|$ is minimum with this property. Now take any M, as guaranteed by the previous lemma, such that X_{β_o} acts nontrivially on M. By Theorem 6.2.3, M integrates to a $\mathcal{G}$-module $\hat{M} \in \mathfrak{M}(\mathcal{G})$. Let $m \in M$ be a weight vector such that $X_{\beta_o} m \neq 0$. Since $|\beta_o|$ is minimum (by choice), it is easy to see that $gm \neq m$. □

6.2.7 Definition. For any positive real root α, recall the definition of the one-parameter subgroup $\mathcal{U}_\alpha \subset \mathcal{U}$ from 6.1.5(a). Now we want to define, for any real root α, the one-parameter subgroup $\mathcal{U}_\alpha \subset \mathcal{G}$: Choose $n \in N$ and a simple root α_i such that $w\alpha_i = \alpha$, where $w := \pi(n)$ under the canonical map $\pi : N \to W$ (cf. 6.1.8). Set

(1) $$\mathcal{U}_\alpha := n\mathcal{U}_{\alpha_i} n^{-1} \subset \mathcal{G}.$$

Clearly $\mathcal{U}_\alpha$ does not depend on the choice of the representative $n \in N$ of w. Further, by the proof of Theorem 6.1.17 particularly (6.1.17.6), together with Theorem 1.3.11(b_5) and Lemma 1.3.13, we see that $\mathcal{U}_\alpha$ does not depend on the choice $w\alpha_i = \alpha$. If α is a positive real root, then $\mathcal{U}_\alpha$ as defined by (1) coincides with the subgroup $\mathcal{U}_\alpha$ defined in 6.1.5(a). This can be easily seen by making successive use of the diagram (4) in the proof of Theorem 6.1.17. The subgroup $\mathcal{U}_\alpha \subset \mathcal{G}$ is called the *one-parameter subgroup of $\mathcal{G}$ corresponding to the real root* α.

For a real root α and $n \in N$, we have, from (1),

(2) $$\mathcal{U}_{\pi(n)\alpha} = n\mathcal{U}_\alpha n^{-1}.$$

Now let $\mathcal{U}^-$ be the subgroup of $\mathcal{G}$ generated by $\{\mathcal{U}_\alpha : \alpha \in \Delta_{\text{re}}^-\}$. Clearly the torus T normalizes each $\mathcal{U}_\alpha$ and hence T normalizes $\mathcal{U}^-$. Let $\mathcal{B}^- \subset \mathcal{G}$ be the subgroup $T \cdot \mathcal{U}^-$. The subgroup $\mathcal{B}^-$ is called the *standard negative Borel subgroup* of $\mathcal{G}$, and $\mathcal{U}^- \subset \mathcal{B}^-$ is called its *unipotent radical.*

6.2.8 Theorem. *The 6-tuple $(\mathcal{G}, N, \mathcal{U}, \mathcal{U}^-, T, S)$ is a refined Tits system (cf. Definition 5.2.1) for any Kac–Moody group $\mathcal{G}$. In particular, we have the Birkhoff decomposition $\mathcal{G} = \bigsqcup_{n \in N} \mathcal{U}^- n \mathcal{U}$ by Theorem 5.2.3 (g). More generally, for any $Y \subset \{1, \dots, \ell\}$,*

$$\mathcal{G} = \bigsqcup_{w \in W'_Y} \mathcal{U}^- w \mathcal{P}_Y .$$

Proof. The axiom (RT1) of Definition 5.2.1 is easy to verify (cf. Theorem 6.1.17 and Theorem 5.1.3(a)).

We next show that

$$(1) \qquad \mathcal{U} \cap \mathcal{U}^- = \{1\}.$$

Otherwise let $g \neq 1 \in \mathcal{U} \cap \mathcal{U}^-$. By Corollary 6.2.6, there exists $\hat{M} \in \mathfrak{M}(\mathcal{G})$ and a weight vector $v \in \hat{M}$ (say of weight λ) such that $gv \neq v$. Since $g \in \mathcal{U}$,

$$(2) \qquad gv = \sum_{\mu \geq \lambda} v_\mu, \qquad v_\lambda = v,$$

where v_μ is of weight μ. Similarly, since $g \in \mathcal{U}^-$,

$$(3) \qquad gv = \sum_{\theta \leq \lambda} w_\theta, \qquad w_\lambda = v.$$

From (2) and (3) we see that $gv = v$, a contradiction. This proves (1).

Fix a simple reflection s_i, $1 \leq i \leq \ell$. Following the notation as in 5.2.1, we prove that

$$(4) \qquad \mathcal{U}_{s_i} = \mathcal{U}_{\alpha_i}:$$

Clearly $\mathcal{U}_{\alpha_i} \subset \mathcal{U}_{s_i} := \mathcal{U} \cap \bar{s}_i \mathcal{U}^- \bar{s}_i^{-1}$, where $\bar{s}_i \in \pi^{-1}(s_i)$. Moreover, by Example 6.1.5(a),

$$(5) \qquad \mathcal{U} = \mathcal{U}_{\alpha_i} \cdot \mathcal{U}_{\Delta^+ \setminus \{\alpha_i\}}.$$

Further,

$$\begin{aligned} \mathcal{U}_{\Delta^+\setminus\{\alpha_i\}} \cap \bar{s}_i \mathcal{U}^- \bar{s}_i^{-1} &= \bar{s}_i \left((\bar{s}_i^{-1} \mathcal{U}_{\Delta^+\setminus\{\alpha_i\}} \bar{s}_i) \cap \mathcal{U}^- \right) \bar{s}_i^{-1} \\ &= \bar{s}_i \left(\mathcal{U}_{\Delta^+\setminus\{\alpha_i\}} \cap \mathcal{U}^- \right) \bar{s}_i^{-1} \qquad \text{(since } \mathcal{U}_{\Delta^+\setminus\{\alpha_i\}} \text{ is a normal subgroup of } \mathcal{P}_i\text{)} \\ &= \{1\}, \qquad \text{by (1).} \end{aligned}$$

This proves (4) (in view of (5)).

Now we prove the axiom (RT2) of Definition 5.2.1, following the same notation, except that we replace s by s_i and H by T.

(a) By (6.2.7.2) and (4),

$$(6) \qquad \mathcal{U}_{s_i}^{s_i} = \mathcal{U}_{-\alpha_i}.$$

In the rank-1 group $\mathcal{G}_i \subset \mathcal{P}_i$, we clearly have

$$\mathcal{U}_{-\alpha_i} \cap \mathcal{U}_{\alpha_i} T = \{1\}.$$

Hence $(\mathcal{U}_{-\alpha_i} \backslash \{1\}) \cap \mathcal{U}_{\alpha_i} T = \emptyset$.

In particular, from the Bruhat decomposition of $\mathcal{G}_i$, and (4), (6), (a) of (RT2) follows. Clearly (b) follows from (6), (c) follows from (4) and (6.2.7.2), and (d) follows easily from (5).

Finally, we prove (RT3). Take $u_\pm \in \mathcal{U}^\pm$ and $n \in N$ such that $u_- n u_+ = 1$. Let $V(\lambda)$ be an integrable highest weight $\mathfrak{g}$-module with highest weight $\lambda \in D_\mathbb{Z}$ (and hence a $\mathcal{G}$-module) and let v_λ be a nonzero highest weight vector. Then

(7) $$v_\lambda = u_- n u_+ v_\lambda = u_- n v_\lambda.$$

For $w := \pi(n) \in W$, by Lemma 1.3.5(a), nv_λ is a nonzero weight vector of weight $w\lambda$. Further, $u_- n v_\lambda$ can be written as (cf. (3)):

(8) $$u_- n v_\lambda = \sum_{\theta \le w\lambda} v_\theta\,, \qquad \text{for } v_\theta \text{ of weight } \theta.$$

But $w\lambda \le \lambda$ since $w\lambda$ is a weight of $V(\lambda)$.

Now, if $w \neq 1$, choose $\lambda \in D_\mathbb{Z}$ such that $w\lambda \neq \lambda$. Then (7) and (8) contradict each other and hence $w = 1$, i.e., $n \in T$. Thus, by (3) and (7), we get

$$v_\lambda = \sum_{\theta < \lambda} v_\theta + n(\lambda) v_\lambda.$$

This gives $n(\lambda) = 1$ for all $\lambda \in D_\mathbb{Z}$. This forces $n = 1$, so we are reduced to $u_- u_+ = 1$. By (1), this gives $u_- = u_+ = 1$, proving (RT3). So the theorem is proved. □

6.2.9 Lemma. (a) *For any $g \neq 1 \in \mathcal{G}$, there exists a module $V \in \mathfrak{M}(\mathcal{G})$ such that g acts nontrivially on V.*

(b) *The normalizer $N_\mathcal{G}(T)$ of T in $\mathcal{G}$ equals N.*

(c) *The center Z of $\mathcal{G}$ is given by*

$$Z = \{t \in T : t(\alpha_i) = 1, \textit{ for all the simple roots } \alpha_i\}.$$

Proof. (a) Write $g = u_1 n u_2$, for $u_1, u_2 \in \mathcal{U}$ and $n \in N$. For any integrable highest weight module $V(\lambda)$ with $\lambda \in D_\mathbb{Z}$ and highest weight vector $v_\lambda \neq 0$,

$$g v_\lambda = u_1 n v_\lambda = n v_\lambda + \sum_{\theta > w\lambda} v_\theta,$$

where $w := n \mod T \in W$. If

$$g v_\lambda = v_\lambda, \text{ then } g v_\lambda = n v_\lambda.$$

But for any $n \in N$, $n \neq 1$, we can choose $\lambda \in D_{\mathbb{Z}}$ such that $n v_\lambda \neq v_\lambda$. So we can assume that $n = 1$. Thus $g \in \mathcal{U}$ and hence there exists $V \in \mathfrak{M}(\mathcal{G})$ such that g acts nontrivially on V by Corollary 6.2.6. This proves the (a) part.

(b) Take $g \in N_{\mathcal{G}}(T)$ and write (cf. Theorem 5.2.3(d))

$$g = g_1 n g_2, \text{ for } g_1 \in \mathcal{U},\ n \in N \text{ and } g_2 \in \mathcal{U} \cap (\mathcal{U}^-)^n.$$

For any $t \in T$, there exists $s = s(t) \in T$ such that

$$tg = gs, \text{ i.e., } (tg_1t^{-1})(tn)g_2 = g_1(ns)(s^{-1}g_2s).$$

By the uniqueness of the decomposition in Theorem 5.2.3(d), we get

(1) $$tg_1t^{-1} = g_1\,, \quad \text{for all } t \in T\,, \text{ and}$$

(2) $$g_2 = s^{-1}g_2s\,, \quad \text{for all } s \in T\,.$$

From (1) and (2), we get $g_1 = g_2 = 1$, i.e., $g \in N$. Conversely, N normalizes T, thus (b) follows.

(c) Of course $Z \subset N_{\mathcal{G}}(T) = N$. Since the action of W on $\mathfrak{h}^*_{\mathbb{Z}}$ is faithful, the action of W on T is faithful as well (cf. Remark 6.1.9). In particular, $Z \subset T$. Let $Z' = \{t \in T : t(\alpha_i) = 1, \text{ for all the simple roots } \alpha_i\}$. By ($d_2$) of 6.1.6, $t \in Z \Rightarrow s_i(t) = t$ and hence $t \in Z'$. Conversely, for $t \in Z'$, t commutes with $\tilde{s}_i$ and hence t commutes with N. Further, for $t \in Z'$, t commutes with $\mathcal{U}$ and hence t commutes with $\mathcal{G}$, i.e., $Z' = Z$, proving (c). □

6.2.10 Definition (Adjoint Representation). We define the *adjoint representation* Ad of $\mathcal{G}$ in $\hat{\mathfrak{g}}$: First, $\hat{\mathfrak{g}}$ is a T-module under

(1) $$t \cdot \Big(\sum_{\beta \in \Delta \cup \{0\}} x_\beta \Big) = \sum t(\beta) x_\beta, \qquad \text{for } x_\beta \in \mathfrak{g}_\beta.$$

Since $\mathfrak{g}$ is an integrable $\mathfrak{g}$-module under the adjoint action (cf. Corollary 1.3.4), $\mathfrak{g}$ is a module for N by Lemma 6.1.7. We next show that

$$s_i(\mathrm{ad}) := \exp(\mathrm{ad}\ f_i)\exp(-\,\mathrm{ad}\ e_i)\exp(\mathrm{ad}\ f_i)$$

extends to an automorphism (again denoted by $s_i(\mathrm{ad})$) of $\hat{\mathfrak{g}}$: In view of Lemma 1.3.5 (a), it suffices to observe that $s_i\Delta^+ \cap \Delta^-$ is finite (in fact, it is a singleton).

Now we fix $1 \le i \le \ell$ and show that $\hat{\mathfrak{g}}$ is a module for $\mathcal{P}_i$: Recall the filtration $\{\hat{\mathfrak{u}}_i(k)\}_{k>0}$ by ideals of the pro-Lie algebra $\hat{\mathfrak{p}}_i$ from 6.1.10. Then, for any $k > 0$, $\hat{\mathfrak{g}}_i(k) := \hat{\mathfrak{g}}/\hat{\mathfrak{u}}_i(k)$ is a pro-representation of $\hat{\mathfrak{p}}_i$ under the adjoint action. Moreover, $\hat{\mathfrak{g}}_i(k)$ is a $(\hat{\mathfrak{p}}_i, T)$-module. In particular, $\hat{\mathfrak{g}}_i(k)$ is a module for $\mathcal{G}_i$ by 6.1.13 and also a pro-module for $\mathcal{U}_i$ by Proposition 4.4.26. By the same argument as that of the proof of Theorem 6.2.3, we get that $\hat{\mathfrak{g}}_i(k)$ is a pro-representation of $\mathcal{P}_i$ denoted by Ad_i^k. Since

$$(2) \qquad \hat{\mathfrak{g}} \simeq \operatorname*{Inv.}_{k} \text{ lt. } \hat{\mathfrak{g}}_i(k),$$

taking the inverse limit of the representations Ad_i^k, we get the representation Ad_i of $\mathcal{P}_i$ in the space $\hat{\mathfrak{g}}$. It is easy to see that $\mathrm{Ad}_{i|\mathcal{B}} = \mathrm{Ad}_{j|\mathcal{B}}$ for any $1 \le i, j \le \ell$, by using Lemma 4.4.24.

Finally, it is easy to see that, for any $1 \le i \le \ell$,

$$s_i(\mathrm{ad}) = \mathrm{Ad}_i(\theta_i(\tilde{s}_i)),$$

where the notation $\theta_i(\tilde{s}_i)$ is as in 6.1.16. Thus, $\hat{\mathfrak{g}}$ becomes a representation of $\mathcal{G}$ denoted by Ad such that $\mathrm{Ad}_{|\mathcal{P}_i} = \mathrm{Ad}_i$ and $\mathrm{Ad}(\theta_i(\tilde{s}_i)) = s_i(\mathrm{ad})$, for all $1 \le i \le \ell$.

By an argument similar to the one used in 6.1.13, it follows that $\mathrm{Ad}\,(g)$ for any $g \in \mathcal{P}_i$ is a Lie algebra automorphism of $\hat{\mathfrak{g}}$. But since $\{\mathcal{P}_i;\ 1 \le i \le \ell\}$ generate $\mathcal{G}$ as a group, $\mathrm{Ad}(g)$ is a Lie algebra automorphism of $\hat{\mathfrak{g}}$ for any $g \in \mathcal{G}$.

By Lemma 6.1.11, it is easy to see that the inverse limit topology on $\hat{\mathfrak{g}}$ defined by (2) (putting the discrete topology on each $\hat{\mathfrak{g}}_i(k)$) is the same as defined in 6.1.1. Hence, for any $g \in \mathcal{P}_i$ $(1 \le i \le \ell)$, $\mathrm{Ad}\,(g) : \hat{\mathfrak{g}} \to \hat{\mathfrak{g}}$ is continuous. Since $\mathcal{P}_i$'s generate $\mathcal{G}$ as a group, we get that, for any $g \in \mathcal{G}$,

$$\mathrm{Ad}\,(g) : \hat{\mathfrak{g}} \to \hat{\mathfrak{g}} \text{ is continuous.}$$

We next show that for any $(V, \pi) \in \mathfrak{M}(\mathcal{G})$, $x \in \hat{\mathfrak{g}}$ and $g \in \mathcal{G}$:

$$(3) \qquad \dot{\pi}(\mathrm{Ad}\ g\,x) = \pi(g)\,\dot{\pi}(x)\,\pi(g)^{-1}.$$

In particular, $\dot{\pi}(\mathrm{Ad}\ g\,x)$ is also locally finite if $\dot{\pi}(x)$ is locally finite.

To prove (3), we first take $x \in \mathfrak{g}$ and $g = \mathrm{Exp}\ y$, for $y \in \mathfrak{g}_i$. Then

$$(4) \qquad \mathrm{Ad}\ g\,x = [\exp(\mathrm{ad}\ y)](x),$$

and hence, $\dot{\pi}$ being a Lie algebra homomorphism,

$$\begin{aligned} \dot{\pi}(\mathrm{Ad}\ g\,x) &= [\exp(\mathrm{ad}(\dot{\pi}\ y))]\,\dot{\pi}(x) \\ &= \exp(\dot{\pi}\ y)\,\dot{\pi}(x)\,\exp(-\dot{\pi}\ y), \text{ by (1.3.2.4)} \\ (5) \qquad &= \pi(g)\,\dot{\pi}(x)\,\pi(g^{-1}), \end{aligned}$$

since $\exp(\dot{\pi}(y)) = \pi(g)$, and $\dot{\pi}(y)$, resp. ad y, is locally finite on V, resp. $\mathfrak{g}$. This proves (3) for $g = \operatorname{Exp} y$ and $x \in \mathfrak{g}$ (for $y \in \mathfrak{g}_i$). Further, we have by (4.4.25.2) and Lemma 6.1.14,

$$\text{(6)} \qquad \dot{\pi}(\operatorname{Ad} g\, x) = \pi(g)\,\dot{\pi}(x)\,\pi(g^{-1}), \text{ for } g \in \mathcal{P}_i \text{ and } x \in \hat{\mathfrak{p}}_i.$$

Combining (5) and (6), we get the validity of (3) for $g = \operatorname{Exp} y$ (for $y \in \mathfrak{g}_i$) and $x \in \hat{\mathfrak{g}}$. Fix $g \in \mathcal{G}$ and consider the set

$$\mathfrak{s}_g = \{x \in \hat{\mathfrak{g}} : \dot{\pi}(\operatorname{Ad} g\, x) = \pi(g)\,\dot{\pi}(x)\,\pi(g^{-1})\}.$$

Then $\mathfrak{s}_g$ is a Lie subalgebra of $\hat{\mathfrak{g}}$ since Ad g acts as a Lie algebra automorphism. For $g \in \mathcal{B}$, by (6), $\mathfrak{s}_g$ contains $\bigoplus_{i=1}^{\ell} \hat{\mathfrak{p}}_i$ and hence $\mathfrak{s}_g = \hat{\mathfrak{g}}$. But since the elements $\mathcal{B} \cup \{\operatorname{Exp} y : y \in \bigcup_{i=1}^{\ell} \mathfrak{g}_i\}$ generate $\mathcal{G}$ as a group, we get the validity of (3) for any $g \in \mathcal{G}$ and $x \in \hat{\mathfrak{g}}$.

For any vector space V, let $\operatorname{End}_{\text{fin}}(V)$ denote the set of all the locally finite endomorphisms of V. Also, set

$$\mathfrak{g}_{\text{fin}} := \cup_Y \cup_{g \in \mathcal{G}} \operatorname{Ad} g(\hat{\mathfrak{p}}_Y) \subset \hat{\mathfrak{g}},$$

where $\cup_Y$ runs over subsets $Y \subset \{1, \ldots, \ell\}$ of finite type.

6.2.11 Proposition. *There exists a unique map*

$$\operatorname{Exp} : \mathfrak{g}_{\text{fin}} \to \mathcal{G},$$

such that, for all $(V, \pi) \in \mathfrak{M}(\mathcal{G})$, the following diagram is commutative:

$$(*) \qquad \begin{array}{ccc} \mathfrak{g}_{\text{fin}} & \xrightarrow{\dot{\pi}} & \operatorname{End}_{\text{fin}}(V) \\ {\scriptstyle \operatorname{Exp}}\downarrow & & \downarrow{\scriptstyle \exp} \\ \mathcal{G} & \xrightarrow{\pi} & \operatorname{Aut} V, \end{array}$$

where exp is the map defined in 1.3.2.

Moreover, for any $g \in \mathcal{G}$ and $X \in \mathfrak{g}_{\text{fin}}$,

$$\text{(1)} \qquad g(\operatorname{Exp} X)g^{-1} = \operatorname{Exp}(\operatorname{Ad} g(X)).$$

(It is seen in the proof that $\dot{\pi}(\mathfrak{g}_{\text{fin}})$ indeed lies in $\operatorname{End}_{\text{fin}}(V)$.*)*

Proof. We first observe that $\dot{\pi}(\hat{\mathfrak{p}}_Y) \subset \operatorname{End}_{\text{fin}}(V)$ for Y of finite type (use Theorem 6.2.3 and Lemma 1.3.3 (c)).

Take any $X \in \mathfrak{g}_{\text{fin}}$ and write $X = \text{Ad}\, g(x)$ for some $g \in \mathcal{G}$ and $x \in \hat{\mathfrak{p}}_Y$ (Y of finite type). Define

$$\text{(2)} \qquad \text{Exp}(X) := g(\text{Exp}\, x)g^{-1},$$

where Exp x denotes the image of x under

$$\text{Exp} : \hat{\mathfrak{p}}_Y \to \mathcal{P}_Y \qquad \text{(cf. 4.4.16).}$$

If $X = \text{Ad}\, g(x) = \text{Ad}\, g'(x')$, for some $g' \in \mathcal{G}$ and $x' \in \hat{\mathfrak{p}}_{Y'}$ (Y' again of finite type), we get by (6.2.10.3)

$$\text{(3)} \qquad \exp\left(\pi(g)\,\dot{\pi}(x)\,\pi(g^{-1})\right) = \exp\left(\pi(g')\,\dot{\pi}(x')\,\pi(g'^{-1})\right), \text{ i.e.,}$$

$$\text{(4)} \qquad \pi(g)\,\pi(\text{Exp}\, x)\,\pi(g^{-1}) = \pi(g')\,\pi(\text{Exp}\, x')\,\pi(g'^{-1}), \text{ by 4.4.23,}$$

for any $(V, \pi) \in \mathfrak{M}(\mathcal{G})$. Observe that any $(V, \pi) \in \mathfrak{M}(\mathcal{G})$ is a pro-representation of $\mathcal{P}_Y$, for any finite type Y.

From (4) and Lemma 6.2.9(a), we get

$$g(\text{Exp}\, x)g^{-1} = g'(\text{Exp}\, x')g'^{-1},$$

which proves that Exp(X) is well defined (by (2)). Commutativity of the diagram (∗) follows from (6.2.10.3), and (1) follows from the definition (2). The uniqueness of Exp : $\mathfrak{g}_{\text{fin}} \to \mathcal{G}$, making the diagram (∗) commutative for all $(V, \pi) \in \mathfrak{M}(\mathcal{G})$, follows from Lemma 6.2.9(a). This proves the proposition. □

6.2.E EXERCISES

(1) For any $w \in W$, define the map $\theta : \mathcal{U}_{\Phi_w} \to \mathcal{B}w\mathcal{B}/\mathcal{B}$ by $\theta(g) = g\bar{w}\mathcal{B}$, where $\bar{w}$ is a representative of w in N, $\mathcal{U}_{\Phi_w}$ is as in Example 6.1.5(b), and $\mathcal{B}w\mathcal{B}$ denotes $\mathcal{B}\bar{w}\mathcal{B}$. Then prove that θ is bijective.
Hint. Use Theorem 5.2.3(b), (d) parts, and Lemma 6.1.4.

(2) For any real root α and $g \in \mathcal{U}_\alpha$, show that the adjoint representation Ad g keeps $\mathfrak{g} \subset \hat{\mathfrak{g}}$ stable.
Hint. Use (6.2.10.4).

(3) Prove Theorem 6.1.17 without using Theorem 5.1.8; instead, take the following steps:

(a) Any module $V \in \mathfrak{M}_T(\hat{\mathfrak{g}})$ is a module for each of the standard minimal parabolic subgroups $\mathcal{P}_i$ (cf. Proof of Theorem 6.2.3) and also a module for N (cf. Lemma 6.1.7). Thus V is a module for the amalgamated product $\mathcal{G}$ of the system of groups $(N, \mathcal{P}_i;\ 1 \leq i \leq \ell)$. In particular, for any $\lambda \in D_{\mathbb{Z}}$, $L^{\max}(\lambda)$ and $L^{\max}(\lambda)^{\sigma}$ are $\mathcal{G}$-modules.

(b) Use the above $\mathcal{G}$-modules and Corollary 6.2.6 to prove that the canonical map $Z \to \mathcal{G}$ is injective, where Z is defined in 6.1.16.

(c) Finally, prove that $(\mathcal{G}, \mathcal{B}, N, S)$ is a Tits system. The axioms (BN_1)–(BN_3) of 5.1.1 are trivial to verify. To prove (BN_4), take a simple reflection s_i and any $w \in W$. Observe that $\mathcal{B}s_i\mathcal{B} = \mathcal{B}s_i\mathcal{U}_{\alpha_i}$. Now if $w^{-1}\alpha_i > 0$, $s_i\mathcal{U}_{\alpha_i}w = s_iw\mathcal{U}_{w^{-1}\alpha_i} \subset s_iw\mathcal{B}$. If $w^{-1}\alpha_i < 0$, use the $s\ell_2$-result $s_i\mathcal{U}_{\alpha_i} \subset (T\mathcal{U}_{\alpha_i}\mathcal{U}_{-\alpha_i}) \cup (T\mathcal{U}_{\alpha_i}s_i)$ to show that $s_i\mathcal{U}_{\alpha_i}w \subset \mathcal{B}w\mathcal{B} \cup \mathcal{B}s_iw\mathcal{B}$.

6.C Comments. [Moody–Teo–72] were the first to associate a group $\mathcal{G}_{\text{ad}}$ over a field of char. 0 (more generally over any field of char. $p > \max\{|a_{ij}|\}$) to any Kac–Moody Lie algebra $\mathfrak{g}$. They "integrated" the adjoint representation of $\mathfrak{g}$, and thus obtained the "minimal" version of only the adjoint group. Their construction was modified by [Marcuson–75] who replaced the adjoint representation by an integrable highest weight irreducible $\mathfrak{g}$-module $L(\lambda)$. This way he got a family of "formal" groups $\mathcal{G}(\lambda)$ (depending on λ) over a field of char. 0 each having a "minimal" subgroup covering $\mathcal{G}_{\text{ad}}$. [Garland–80] considered the (untwisted) affine Kac–Moody Lie algebras and constructed the corresponding groups essentially over $\mathbb{Z}$ (still using the integrable representations $L(\lambda)$ and a specific $\mathbb{Z}$-lattice of the enveloping algebra $U(\mathfrak{g})$), which enabled him to define the corresponding group $\mathcal{G}(\lambda)$ over arbitrary fields via base change. The question of dependence of $\mathcal{G}(\lambda)$ (on λ) was not addressed by Marcuson but Garland answered it in the affine case by showing that the group he constructs is a central extension of the loop group and by computing the cocycle which describes the extension (in terms of λ). All these works also constructed the underlying Tits system. Subsequently [Kac–Peterson–83], [Peterson–Kac–83] (also see [Kac–85b]) constructed the minimal group $\mathcal{G}^{\min}$ over a field of char. 0, still using the integrable $\mathfrak{g}$-modules, but they considered all the integrable $\mathfrak{g}$-modules simultaneously; thereby the group they obtained was independent of the choice of the representation. We discuss theirconstruction in Section 7.4.

[Tits–81a,b] (also see [Slodowy–84a] for a more detailed exposition), on the other hand, gave an "intrinsic" construction of the formal as well as the minimal group associated to any Kac–Moody Lie algebra over any field (of an arbitrary char.). This is the construction we have followed in Section 6.1 although we take the base field to be $\mathbb{C}$.

For constructions of the group(s) over an arbitrary commutative ring associated to any Kac–Moody Lie algebra $\mathfrak{g}$, the reader is referred to [Tits–82, 85, 87], [Mathieu–88(b), 89], [Kashiwara–89]. Whether the group of $\mathbb{C}$-points of the ind-group scheme constructed by [Mathieu–88(b), 89] coincides with the abstract group defined in this book does not seem to be known (even in the affine case) (cf.

[Tits–89]). A unitary form of the group $\mathcal{G}^{\min}$ was studied by [Kac–Peterson–83, 84b, 85a] and [Peterson–Kac–83] (cf. Exercise 7.4.E.10). [Goodman–Wallach–84] constructed the affine Kac–Moody group at various analytic levels (also see [Pressley-Segal–86, Chapter 6]). We refer the reader to the article [Tits–89] which surveys various constructions of the groups associated to a Kac–Moody Lie algebra.

Most of the content of Section 6.1 is taken from [Tits–81a, b] and [Slodowy–84a]. (The definition of N as in 6.1.6 appears in [Tits–66].) Theorem 6.2.3 is essentially contained in [Tits–81b]. Theorem 6.2.8 is due to [Kac–Peterson–85a]. The alternative way to prove Theorem 6.1.17 outlined in Exercise 6.2.E.3 was suggested by M.S. Raghunathan.

VII

Generalized Flag Varieties of Kac–Moody Groups

Let $\mathcal{G}$ be a Kac–Moody group and $\mathcal{P}_Y \subset \mathcal{G}$ any parabolic subgroup. The aim of Section 7.1 is to realize the homogeneous space $\mathcal{X}^Y := \mathcal{G}/\mathcal{P}_Y$ as a projective ind-variety so that the Schubert subvarieties $X_w^Y \subset \mathcal{G}/\mathcal{P}_Y$ are indeed closed finite-dimensional (projective) irreducible subvarieties. Fix a (dominant integral) weight $\lambda \in D_{\mathbb{Z}}$ such that, for $1 \leq i \leq \ell$, $\lambda(\alpha_i^\vee) = 0$ iff $i \in Y$. Such a λ is called Y-regular. Let $V(\lambda)$ be an integrable highest weight $\mathfrak{g}$-module with highest weight λ. From the last chapter, $V(\lambda)$ acquires a $\mathcal{G}$-module structure. Then the orbit through the highest weight vector in the projective space (of lines) $\mathbb{P}(V(\lambda))$ determines an injective map $i_{V(\lambda)} : \mathcal{X}^Y \hookrightarrow \mathbb{P}(V_\lambda)$. Consider the increasing filtration of $\mathcal{X}^Y$ given by $X_n^Y := \bigcup_{\substack{v \in W_Y' \\ \ell(v) \leq n}} \mathcal{B}v\mathcal{P}_Y/\mathcal{P}_Y$. For any sequence of simple reflections $\mathfrak{w} = (s_{i_1}, \ldots, s_{i_n})$, let $Z_{\mathfrak{w}}$ be the Bott–Samelson–Demazure–Hansen variety, which is a smooth irreducible projective variety of dimension n; it is built from successive locally trivial $\mathbb{P}^1$-fibrations. There exists a morphism of ind-varieties $m_{\mathfrak{w}}^\lambda : Z_{\mathfrak{w}} \to \mathbb{P}(V(\lambda))$, and hence its image is closed. This is used to show that $i_{V(\lambda)}(X_n^Y)$ is closed in $\mathbb{P}(V(\lambda))$ for any $n \geq 0$ and so is $i_{V(\lambda)}(X_w^Y)$ for any $w \in W_Y'$.

Thus, we can endow $\mathcal{X}^Y$ with a (unique) projective ind-variety structure with filtration $\{X_n^Y\}_{n \geq 0}$ so that $i_{V(\lambda)}$ is a closed embedding. Now take $V(\lambda)$ to be the maximal integrable module $L^{\max}(\lambda)$; the ind-variety, resp. variety, structure thus obtained on $\mathcal{X}^Y$, resp. X_n^Y, is denoted by $\mathcal{X}^Y(\lambda)$, resp. $X_n^Y(\lambda)$. If $\mathfrak{w}$ is reduced, i.e., $w = s_{i_1} \cdots s_{i_n}$ is a reduced decomposition such that $w \in W_Y'$, then $m_{\mathfrak{w}}^\lambda(Z_{\mathfrak{w}}) = X_w^Y$ and $m_{\mathfrak{w}}^\lambda : Z_{\mathfrak{w}} \to X_w^Y(\lambda)$ is a (surjective) birational morphism, where $X_w^Y(\lambda)$ refers to the closed subvariety structure on X_w^Y coming from $X_n^Y(\lambda)$. It is shown that the set-theoretic identity map $\mathcal{X}^Y(\lambda + \mu) \to \mathcal{X}^Y(\lambda)$, for any $\mu \in D_{\mathbb{Z}}$ such that $\mu(\alpha_i^\vee) = 0$ for all $i \in Y$, is a morphism. To prove this, we use the Segre embedding $S : \mathbb{P}(V) \times \mathbb{P}(W) \to \mathbb{P}(V \otimes W)$ given by $[v] \times [w] \mapsto [v \otimes w]$, for any two countable-dimensional vector spaces V and W.

Thus there exists a "large" enough Y-regular $\lambda_n \in D_{\mathbb{Z}}$ (depending on n) such that the identity map $X_n^Y(\lambda_n+\mu) \to X_n^Y(\lambda_n)$ is a biregular isomorphism for any μ as above. The variety structure $X_n^Y(\lambda_n)$ is referred to as the "stable" variety structure on X_n^Y (and denoted just by X_n^Y). Now put the (projective) ind-variety structure on $\mathcal{X}^Y$ so that each X_n^Y is endowed with the stable variety structure. This is called the stable ind-variety structure on $\mathcal{X}^Y$ and denoted simply by $\mathcal{X}^Y$. Then, for any $w \in W_Y'$, $X_w^Y \subset \mathcal{X}^Y$ is a closed irreducible subvariety of dimension $\ell(w)$. (For symmetrizable $\mathfrak{g}$, using the irreducibility of $L^{\max}(\lambda)$, it is shown that the identity map $\mathcal{X}^Y(\lambda+\mu) \to \mathcal{X}^Y(\lambda)$ is a biregular isomorphism for any Y-regular $\lambda \in D_{\mathbb{Z}}$ and any μ as above; in particular, the ind-variety $\mathcal{X}^Y(\lambda)$ is equivalent to the stable ind-variety $\mathcal{X}^Y$.)

As defined in 6.2.7, let $\mathcal{U}^- \subset \mathcal{G}$ be the subgroup generated by one-parameter subgroups $\mathcal{U}_\beta$ corresponding to the negative real roots β. We determine the Zariski closure of $\mathcal{U}^-$-orbits in $\mathcal{X}^Y$ by interpreting the Bruhat–Chevalley order in W in terms of the $U(\mathfrak{n}^-)$-orbits of extremal weight vectors in an integrable highest weight module. The T-fixed points of $\mathcal{X}^Y$, with respect to the left multiplication of T, are shown to be precisely $\{\dot{w}\}_{w\in W_Y'}$, where $\dot{w} := w\mathcal{P}_Y \in \mathcal{X}^Y$.

For any Y-regular $\lambda \in D_{\mathbb{Z}}$, the pullback of the tautological line bundle on $\mathbb{P}(L^{\max}(\lambda))$ via $i_{L^{\max}(\lambda)}$ is a $\mathcal{G}$-equivariant algebraic line bundle denoted $\mathcal{L}^Y(-\lambda)$ on $\mathcal{X}^Y$. This is used to define any $\mathcal{G}$-equivariant algebraic line bundle on $\mathcal{X}^Y$ in Section 7.2.

Section 7.3 is devoted to the study of the subgroup $\mathcal{U}^- \subset \mathcal{G}$. Actually, we think of $\mathcal{U}^-$ as a subset of $\mathcal{X}$ via its orbit through the base point 1. As shown in this section, $\mathcal{U}^-$ is open in $\mathcal{X}$, and thus acquires an ind-variety structure from $\mathcal{X}$. We show that, with this structure, $\mathcal{U}^-$ is an affine ind-group and any integrable highest weight $\mathfrak{g}$-module $V(\lambda)$, for $\lambda \in D_{\mathbb{Z}}$, is an algebraic representation of $\mathcal{U}^-$. To prove this, a crucial use is made of the embedding $\mathcal{U}^- \hookrightarrow \hat{\mathcal{U}}^-$, where $\hat{\mathcal{U}}^-$ is the pro-unipotent pro-group with the completion $\hat{\mathfrak{n}}^-$ as its Lie algebra.

In Section 7.4, we study the subgroup $\mathcal{G}^{\min} \subset \mathcal{G}$ generated by the one-parameter subgroups $\{\mathcal{U}_\alpha\}_{\alpha\in\Delta_{\text{re}}}$ and T. This was originally defined and studied by Kac–Peterson (though they defined it differently). It follows easily from the corresponding property of $\mathcal{G}$ that $(\mathcal{G}^{\min}, \mathcal{B}^{\min}, N, S)$ is a Tits system, where $\mathcal{B}^{\min} := \mathcal{B} \cap \mathcal{G}^{\min}$. We endow $\mathcal{G}^{\min}$ with an affine ind-variety structure by embedding it as a closed subset of the direct sum of certain maximal integrable highest weight $\mathfrak{g}$-modules $L^{\max}(\lambda_i)$ and maximal integrable lowest weight $\mathfrak{g}$-modules $L^{\max}(\lambda_i)^-$ and then "stabilizing" the structure thus obtained in a way that is similar to that of $\mathcal{X}^Y$. With this ind-variety structure, $\mathcal{G}^{\min}$ is an (affine) ind-group such that, for any $\lambda \in D_{\mathbb{Z}}$, any integrable highest weight $\mathfrak{g}$-module $V(\lambda)$ as well as any integrable lowest weight $\mathfrak{g}$-module $V(\lambda)^-$ is an algebraic representation of $\mathcal{G}^{\min}$. It is shown that the quotient map $\mathcal{G}^{\min} \to \mathcal{X}$ is a locally trivial principal $\mathcal{B}^{\min}$-bundle. (A similar result can be obtained for $\mathcal{X}$ replaced by $\mathcal{X}^Y$.) Further, $(\mathcal{G}^{\min}, \mathcal{B}^{\min}, N, S)$ is a topological Tits system, where $\mathcal{G}^{\min}$ is endowed with the analytic topology. In particular, under the analytic topology, for any subset $Y \subset \{1, \ldots, \ell\}$, $\mathcal{X}^Y$ is a CW-complex with cells $\{\mathcal{B}w\mathcal{P}_Y/\mathcal{P}_Y\}_{w\in W'_Y}$.

Finally, we outline the construction of the unitary form of $\mathcal{G}^{\min}$ (given by Kac–Peterson) in one of the exercises.

7.1. Generalized Flag Varieties—Ind-Variety Structure

Let $\mathcal{G}$ be the Kac–Moody group associated to an arbitrary Kac–Moody Lie algebra $\mathfrak{g}$ as in 6.1.16. For any subset $Y \subset \{1, \ldots, \ell\}$, let $\mathcal{P}_Y$ be the standard parabolic subgroup (cf. 6.1.18). The aim of this section is to put a projective ind-variety structure on the coset space $\mathcal{G}/\mathcal{P}_Y$.

7.1.1 Definition. For any integrable highest weight $\mathfrak{g}$-module $V = V(\lambda)$ with highest weight $\lambda \in D_{\mathbb{Z}}$ (and hence $\mathcal{G}$-module by Theorem 6.2.3), define the map

$$\bar{i}_V : \mathcal{G} \to \mathbb{P}(V), \quad \bar{i}_V(g) = [gv_\lambda],$$

where v_λ is any nonzero highest weight vector of V, $\mathbb{P}(V)$ is the space of lines in V, and $[gv_\lambda]$ denotes the line through gv_λ.

For any $Y \subset \{1, \ldots, \ell\}$, define the set D_Y^o of *Y-regular dominant totally integral weights* (for short *Y-regular weights)* by

$$D_Y^o := \{\lambda \in D_{\mathbb{Z}} : \langle\lambda, \alpha_i^\vee\rangle \; = 0 \text{ iff } i \in Y\}.$$

For $Y = \emptyset$, we denote D_Y^o simply by D^o.

7.1.2 Lemma. *For $\lambda \in D_Y^o$, the map $\bar{i}_V$ factors through the coset space $\mathcal{G}/\mathcal{P}_Y$ to give the injective map*

$$i_V : \mathcal{G}/\mathcal{P}_Y \hookrightarrow \mathbb{P}(V).$$

Proof. Of course, it suffices to prove that the stabilizer S of $[v_\lambda]$ is $\mathcal{P}_Y$. Clearly, $S \supset \mathcal{B}$ and also $\tilde{s}_i \in S$ for $i \in Y$ by Lemma 1.3.5(a), and hence $S \supset \mathcal{P}_Y$. By Theorem 5.1.3(g), the lemma follows since $s_i\lambda \neq \lambda$ for $i \notin Y$. □

7.1.3 Bott–Samelson–Demazure–Hansen variety. Let $\mathfrak{W}$ be the set of all the ordered sequences $\mathfrak{w} = (s_{i_1}, \ldots, s_{i_n})$ of simple reflections, for any $n \geq 0$. The sequence corresponding to $n = 0$ is called the *empty sequence* and is denoted by ϕ. For any $\mathfrak{w} = (s_{i_1}, \ldots, s_{i_n})$, define the *Bott–Samelson–Demazure–Hansen variety*

$$Z_{\mathfrak{w}} = \mathcal{P}_{i_1} \times \cdots \times \mathcal{P}_{i_n} \,/\, \mathcal{B}^n, \tag{1}$$

where the product group $\mathcal{B}^n$ acts on $\mathcal{P}_{i_1} \times \cdots \times \mathcal{P}_{i_n}$ from the right via

$$(p_1, \ldots, p_n)(b_1, \ldots, b_n) = (p_1 b_1, b_1^{-1} p_2 b_2, \ldots, b_{n-1}^{-1} p_n b_n),$$

for $p_j \in \mathcal{P}_{i_j}$ and $b_j \in \mathcal{B}$. We denote the $\mathcal{B}^n$-orbit of $(p_1, \ldots, p_n)$ by $[p_1, \ldots, p_n]$. For $n = 0$, we set $Z_{\mathfrak{w}}$ as the one-point space.

We will endow $Z_{\mathfrak{w}}$ with a projective variety structure later in this section.

Define a partial order $\leq$ in $\mathfrak{W}$ by declaring $\mathfrak{v} \leq \mathfrak{w}$ iff $\mathfrak{v}$ can be obtained from $\mathfrak{w}$ by deleting some entries, i.e., $\mathfrak{v} = (s_{i_{j_1}}, \ldots, s_{i_{j_m}})$ for some $1 \leq j_1 < \cdots < j_m \leq n$. Thus $\phi \leq \mathfrak{w}$ for any $\mathfrak{w}$. Note that there may be a different sequence $1 \leq j_1' < \cdots < j_m' \leq n$ as well, such that $\mathfrak{v} = (s_{i_{j_1'}}, \ldots, s_{i_{j_m'}})$. Clearly $\mathfrak{W}$ is a directed set under $\leq$.

Any element $\mathfrak{w} = (s_{i_1}, \ldots, s_{i_n}) \in \mathfrak{W}$ is called a *word of length n*. Consider the map $\pi : \mathfrak{W} \to W$, $\mathfrak{w} \mapsto s_{i_1} \cdots s_{i_n}$. A word $\mathfrak{w}$ is called a *reduced word* if $\pi(\mathfrak{w}) = s_{i_1} \cdots s_{i_n}$ is a reduced decomposition. For two words $\mathfrak{v} = (s_{j_1}, \ldots, s_{j_m})$, and $\mathfrak{w} = (s_{i_1}, \ldots, s_{i_n})$, their *concoction* $\mathfrak{v} * \mathfrak{w}$ is, by definition, the word $(s_{j_1}, \ldots, s_{j_m}, s_{i_1}, \ldots, s_{i_n})$.

For $\mathfrak{w} \in \mathfrak{W}$ as above and for any sequence $J = (j_1, \ldots, j_m)$ with $1 \leq j_1 < \cdots < j_m \leq n$, by $\hat{\mathfrak{w}}_J$ we mean the word $\mathfrak{w}_J := (s_{i_{j_1}}, \ldots, s_{i_{j_m}})$ together with the specific realization as $\mathfrak{w}_J \leq \mathfrak{w}$ where $s_{i_{j_1}}, \ldots, s_{i_{j_m}}$ go in the $j_1, \ldots, j_m$ slots, respectively. For any $\mathfrak{w}$ and J as above, $\hat{\mathfrak{w}}_J$ is called a *subword* of $\mathfrak{w}$. For example, take $\mathfrak{w} = (s_1, s_2, s_1)$, $J = (1)$, and $J' = (3)$. Then the word $\mathfrak{w}_J = \mathfrak{w}_{J'} = (s_1)$, but, as subwords, $\hat{\mathfrak{w}}_J$ and $\hat{\mathfrak{w}}_{J'}$ are distinct. When no confusion is likely, we will denote $\hat{\mathfrak{w}}_J$ by $\mathfrak{w}_J$ itself.

For a subword $\mathfrak{v} = \mathfrak{w}_J$ of $\mathfrak{w}$, there is a natural inclusion

$$i_{\mathfrak{v},\mathfrak{w}} : Z_{\mathfrak{v}} \hookrightarrow Z_{\mathfrak{w}} \tag{2}$$

defined by $i_{\mathfrak{v},\mathfrak{w}}\left([p_{j_1}, \ldots, p_{j_m}]\right) = [p_1, \ldots, p_n]$, where, if $i \notin \{j_1, \ldots, j_m\}$, then $p_i = 1$.

Also define

$$Z_{\mathfrak{w}}^o := \{[p_1, \ldots, p_n] \in Z_{\mathfrak{w}} : p_j \in \mathcal{B} s_{i_j} \mathcal{B}\}. \tag{3}$$

Then

$$Z_{\mathfrak{w}} = \bigsqcup i_{\mathfrak{w}_J,\mathfrak{w}}(Z_{\mathfrak{w}_J}^o), \tag{4}$$

where the union runs over all the subwords $\mathfrak{w}_J$ of $\mathfrak{w}$ including $J = \emptyset$. (The image $i_{\mathfrak{w}_\emptyset,\mathfrak{w}}(Z_{\mathfrak{w}_\emptyset}^o)$ is by definition $[1, \ldots, 1]$.) More generally, for any $\mathfrak{a} = (a_1, \ldots, a_n) \in \mathcal{P}_{i_1} \times \cdots \times \mathcal{P}_{i_n}$, define the subset

$$Z_{\mathfrak{w}}(\mathfrak{a}) = \{[p_1, \ldots, p_n] \in Z_{\mathfrak{w}} : p_j \in a_j \mathcal{U}_{-\alpha_{i_j}} \text{ for all } 1 \leq j \leq n\}. \tag{5}$$

Then, by Theorem 5.2.3(d) (actually taking the inverse of the identity 5.2.3(d)) and (6.2.8.4),

$$(6) \qquad Z_{\mathfrak{w}}(\tilde{s}_{i_1}, \dots, \tilde{s}_{i_n}) = Z^o_{\mathfrak{w}}.$$

Fix $\mathfrak{w} = (s_{i_1}, \dots, s_{i_n}) \in \mathfrak{W}$. Recall the definition of the normal subgroups $\mathcal{U}_j(k) \subset \mathcal{P}_j$ from Lemma 6.1.14. For any sequence of positive integers $\mathfrak{k} = (k_1, \dots, k_n)$ such that

$$(7) \qquad \mathcal{U}_{i_1}(k_1) \subset \mathcal{U}_{i_2}(k_2) \subset \cdots \subset \mathcal{U}_{i_n}(k_n),$$

set

$$\mathcal{P}_{\mathfrak{w}}/\mathcal{U}_{\mathfrak{w}}(\mathfrak{k}) := \mathcal{P}_{i_1}/\mathcal{U}_{i_1}(k_1) \times \cdots \times \mathcal{P}_{i_n}/\mathcal{U}_{i_n}(k_n),$$

and define the map

$$(8) \qquad \theta = \theta_{\mathfrak{w},\mathfrak{k}} : \mathcal{P}_{\mathfrak{w}}/\mathcal{U}_{\mathfrak{w}}(\mathfrak{k}) \to Z_{\mathfrak{w}}$$

by $\theta(\overline{p}_1, \dots, \overline{p}_n) = [p_1, \dots, p_n]$, for $p_j \in \mathcal{P}_{i_j}$, where $\overline{p}_j := p_j \bmod \mathcal{U}_{i_j}(k_j)$. From the assumption (7), it is easy to see that the map θ is well defined. Observe that by Lemma 6.1.11, we can always choose $\mathfrak{k}$ satisfying (7). Further (under the assumption (7)), the group $\mathcal{B}^n/\mathcal{U}_{\mathfrak{w}}(\mathfrak{k}) := \mathcal{B}/\mathcal{U}_{i_1}(k_1) \times \cdots \times \mathcal{B}/\mathcal{U}_{i_n}(k_n)$ acts on $\mathcal{P}_{\mathfrak{w}}/\mathcal{U}_{\mathfrak{w}}(\mathfrak{k})$ from the right via

$$(9) \qquad (\overline{p}_1, \dots, \overline{p}_n)(\bar{b}_1, \dots, \bar{b}_n) = (\overline{p_1 b_1}, \overline{b_1^{-1} p_2 b_2}, \dots, \overline{b_{n-1}^{-1} p_n b_n}),$$

for $p_j \in \mathcal{P}_{i_j}$ and $b_j \in \mathcal{B}$, where $\bar{b}_j := b_j \mod \mathcal{U}_{i_j}(k_j)$. It is easy to see that (9) indeed gives a well-defined action under the assumption (7). The following lemma is easy toverify.

7.1.4 Lemma. *Under the assumption (7) as above, any fiber of the map* $\theta : \mathcal{P}_{\mathfrak{w}}/\mathcal{U}_{\mathfrak{w}}(\mathfrak{k}) \to Z_{\mathfrak{w}}$ *is a single* $\mathcal{B}^n/\mathcal{U}_{\mathfrak{w}}(\mathfrak{k})$*-orbit. Further, the action of* $\mathcal{B}^n/\mathcal{U}_{\mathfrak{w}}(\mathfrak{k})$ *on* $\mathcal{P}_{\mathfrak{w}}/\mathcal{U}_{\mathfrak{w}}(\mathfrak{k})$ *is free.* □

7.1.5 Definition. Let H be a pro-group (with the defining set $\mathcal{F}$) acting on an ind-variety X. We say that the action is *regular* if, for any $n \in \mathbb{Z}_+$, there exists $m = m(n) \in \mathbb{Z}_+$ and $N = N(n) \in \mathcal{F}$ such that $H \cdot X_n \subset X_m$ and, moreover, the H-action $H \times X_n \to X_m$ factorizes over the quotient group H/N to give an algebraic map $H/N \times X_n \to X_m$. In this case, we call X as an *H–ind-variety*.

An algebraic vector bundle $\pi : E \to X$ is said to be an *H-equivariant algebraic vector bundle* if E and X are H–ind-varieties such that π is H-equivariant and, for any $h \in H$ and $x \in X$, the map $E_x \to E_{hx}$, $v \mapsto h \cdot v$, is linear, where $E_x := \pi^{-1}(x)$.

7.1.6 Lemma. *Let H be a pro-group and $H' \subset H$ a pro-subgroup such that the following assumption $(*)$ is satisfied:*

$(*)$ For all $N' \in \mathcal{F}'$, there exists $N \in \mathcal{F}$, $N \subset N'$, where $\mathcal{F}, \mathcal{F}'$ are the defining sets of H and H', respectively.

Then, for any regular action of H' on a (quasi-projective) variety X, the set $\tilde{X} := H \times_{H'} X$ carries a natural structure of a variety such that the left action of H on $\tilde{X}$ is regular. In particular, H/H' is a variety such that the left action of H is regular. Moreover, the canonical map $q : \tilde{X} \to H/H'$ is an (H-equivariant) isotrivial fibration with fiber X. Thus by [Altman–Kleiman–70], if X is smooth, then so is $\tilde{X}$.

Recall that an isotrivial fibration means a locally trivial fibration in the étale topology.

Proof. Choose $N' \in \mathcal{F}'$ such that the action of H' on X factors through a regular action of H'/N' on X. Now choose $N \in \mathcal{F}$ with $N \subset N'$. Then $\tilde{X} = H/N \times_{H'/N} X$. Hence the lemma follows from [Serre–58]. It is easy to see that a different choice of $N \in \mathcal{F}$ will give rise to the same variety structure on $\tilde{X}$. $\square$

7.1.7 Example. Let $\mathcal{P}_Y \subset \mathcal{G}$ be a standard parabolic subgroup of finite type (i.e., Y is of finite type) and let $\mathcal{P}_{Y'} \subset \mathcal{P}_Y$ be another standard parabolic subgroup which is automatically of finite type. Then $H' := \mathcal{P}_{Y'} \subset H := \mathcal{P}_Y$ satisfies the assumption $(*)$ of the above lemma, as follows from Lemma 6.1.11. In particular, $\mathcal{P}_Y/\mathcal{B}$ is a variety. In fact, $\mathcal{P}_Y/\mathcal{B}$ is a projective variety since

$$\mathcal{P}_Y/\mathcal{B} \cong (\mathcal{P}_Y/\mathcal{U}_Y)/(\mathcal{B}/\mathcal{U}_Y) \cong \mathcal{G}_Y/\mathcal{B}_Y,$$

where $\mathcal{B}_Y := \mathcal{B}/\mathcal{U}_Y$ is a Borel subgroup of $\mathcal{G}_Y$.

7.1.8 Proposition.

(a) *For any $\mathfrak{w} = (s_{i_1}, \ldots, s_{i_n})$, $Z_{\mathfrak{w}}$, defined in 7.1.3, acquires a natural irreducible smooth variety structure. Moreover, the left action of $\mathcal{P}_{i_1}$ on $Z_{\mathfrak{w}}$, given by the left multiplication on the first factor, is regular.*

(b) *For any sequence of positive integers $\mathfrak{k} = (k_1, \ldots, k_n)$ satisfying the condition (7.1.3.7), the map $\theta = \theta_{\mathfrak{w},\mathfrak{k}} : \mathcal{P}_{\mathfrak{w}}/\mathcal{U}_{\mathfrak{w}}(\mathfrak{k}) \to Z_{\mathfrak{w}}$ defined in 7.1.3 is a (Zariski) locally trivial principal $\mathcal{B}^n/\mathcal{U}_{\mathfrak{w}}(\mathfrak{k})$-bundle with respect to the right action of $\mathcal{B}^n/\mathcal{U}_{\mathfrak{w}}(\mathfrak{k})$ on $\mathcal{P}_{\mathfrak{w}}/\mathcal{U}_{\mathfrak{w}}(\mathfrak{k})$ defined by (7.1.3.9).*

Proof. We first prove the (a) part by induction on n. For $n = 0$, Z_ϕ is a point and there is nothing to prove inthis case. Now, we come to the general n assuming the result for $n - 1$. We can canonically write

$$Z_{\mathfrak{w}} \cong \mathcal{P}_{i_1} \underset{\mathcal{B}}{\times} Z_{\mathfrak{w}'},$$

where $\mathfrak{w}' := (s_{i_2}, \ldots, s_{i_n})$, $\mathcal{B}$ acts on $\mathcal{P}_{i_1}$ via the right multiplication and $\mathcal{P}_{i_2}$, and hence $\mathcal{B}$, acts on $Z_{\mathfrak{w}'}$ via left multiplication on the first factor.

By induction, $Z_{\mathfrak{w}'}$ has an irreducible, smooth variety structure such that the left action of $\mathcal{P}_{i_2}$ (and hence of $\mathcal{B}$) on $Z_{\mathfrak{w}'}$ is regular. Hence, by Lemma 7.1.6 and Example 7.1.7, $\mathcal{P}_{i_1} \underset{\mathcal{B}}{\times} Z_{\mathfrak{w}'}$ acquires a natural irreducible smooth variety structure such that the left action of $\mathcal{P}_{i_1}$ is regular. This proves (a).

(b): We first prove that θ is an algebraic map by induction on the length $\ell(\mathfrak{w})$ of $\mathfrak{w}$: So assume that $\theta_{\mathfrak{w}',\mathfrak{k}'} : \mathcal{P}_{\mathfrak{w}'}/\mathcal{U}_{\mathfrak{w}'}(\mathfrak{k}') \to Z_{\mathfrak{w}'}$ is an algebraic map, where $\mathfrak{k}' := (k_2, \ldots, k_n)$, and hence

$$\mathrm{I} \times \theta_{\mathfrak{w}',\mathfrak{k}'} : \mathcal{P}_{\mathfrak{w}}/\mathcal{U}_{\mathfrak{w}}(\mathfrak{k}) \to (\mathcal{P}_{i_1}/\mathcal{U}_{i_1}(k_1)) \times Z_{\mathfrak{w}'}$$

is an algebraic map. But since $\mathcal{P}_{i_1}$ acts regularly on $Z_{\mathfrak{w}}$ by the (a) part, there exists some $k_1' \geq k_1 > 0$ such that $\mathcal{U}_{i_1}(k_1')$ acts trivially on $Z_{\mathfrak{w}}$ and, moreover, the induced map $\theta_1' : (\mathcal{P}_{i_1}/\mathcal{U}_{i_1}(k_1')) \times Z_{\mathfrak{w}} \to Z_{\mathfrak{w}}$ is algebraic. Next, observe that the action of $\mathcal{U}_{i_1}(k_1)$ on $Z_{\mathfrak{w}}$ itself is trivial because of the condition (7.1.3.7). So we have a commutative diagram:

$$\begin{array}{ccc} (\mathcal{P}_{i_1}/\mathcal{U}_{i_1}(k_1')) \times Z_{\mathfrak{w}} & \xrightarrow{\theta_1'} & Z_{\mathfrak{w}} \\ & \searrow \qquad \nearrow{\scriptstyle \theta_1} & \\ & (\mathcal{P}_{i_1}/\mathcal{U}_{i_1}(k_1)) \times Z_{\mathfrak{w}}, & \end{array}$$

where the downward map is induced from the inclusion $\mathcal{U}_{i_1}(k_1') \subset \mathcal{U}_{i_1}(k_1)$. It is easy to see that $\theta = \theta_1 \circ (\mathrm{I} \times \theta')$, where θ' denotes $\theta_{\mathfrak{w}',\mathfrak{k}'}$ followed by the inclusion $Z_{\mathfrak{w}'} \subset Z_{\mathfrak{w}}$. Now, by [Borel–91, §6.1 and Theorem 6.8], θ_1 is an algebraic map and hence so is θ. Also, it is easy to see that the action of $\mathcal{B}^n/\mathcal{U}_{\mathfrak{w}}(\mathfrak{k})$ on $\mathcal{P}_{\mathfrak{w}}/\mathcal{U}_{\mathfrak{w}}(\mathfrak{k})$ is regular.

Further, we show that the map θ admits algebraic sections on an open cover of $Z_{\mathfrak{w}}$: Fix $\mathfrak{a} = (a_1, \ldots, a_n) \in \mathcal{P}_{i_1} \times \cdots \times \mathcal{P}_{i_n}$. Then, we claim that $Z_{\mathfrak{w}}(\mathfrak{a}) \subset Z_{\mathfrak{w}}$ defined by (7.1.3.5) is open.

Assume, by induction, that $Z_{\mathfrak{w}'}(\mathfrak{a}')$ is open in $Z_{\mathfrak{w}'}$, where $\mathfrak{a}' := (a_2, \ldots, a_n)$. Consider the map $\beta : ((a_1\mathcal{U}_{-\alpha_{i_1}}\mathcal{B})/\mathcal{U}_{i_1}(k_1)) \times \mathcal{P}_{i_2}/\mathcal{U}_{i_2}(k_2) \times \cdots \times \mathcal{P}_{i_n}/\mathcal{U}_{i_n}(k_n) \to \mathcal{P}_{\mathfrak{w}'}/\mathcal{U}_{\mathfrak{w}'}(\mathfrak{k}')$ given by

$$\beta(\overline{a_1 z b}, \overline{p}_2, \ldots, \overline{p}_n) = (\overline{bp_2}, \overline{p}_3, \ldots, \overline{p}_n),$$

for $p_j \in \mathcal{P}_{i_j}$ $(2 \leq j \leq n)$ and $z \in \mathcal{U}_{-\alpha_{i_1}}$, $b \in \mathcal{B}$. Then, it is easy to see that $(\theta_{\mathfrak{w}',\mathfrak{k}'} \circ \beta)^{-1}(Z_{\mathfrak{w}'}(\mathfrak{a}')) = \theta^{-1}(Z_{\mathfrak{w}}(\mathfrak{a}))$. In particular, $\theta^{-1}(Z_{\mathfrak{w}}(\mathfrak{a}))$ is open in $\mathcal{P}_{\mathfrak{w}}/\mathcal{U}_{\mathfrak{w}}(\mathfrak{k})$. But θ is a smooth morphism, as can be seen by induction on $\ell(\mathfrak{w})$

together with A.14 and hence is an open map by A.14. This proves that $Z_{\mathfrak{w}}(\mathfrak{a})$ is open in $Z_{\mathfrak{w}}$.

It is easy to see that

$$\theta_{\mathfrak{a}} := \theta_{|(a_1\mathcal{U}_{-\alpha_{i_1}})\times\cdots\times(a_n\mathcal{U}_{-\alpha_{i_n}})} : (a_1\mathcal{U}_{-\alpha_{i_1}}) \times \cdots \times (a_n\mathcal{U}_{-\alpha_{i_n}}) \to Z_{\mathfrak{w}}$$

is injective with image, of course, equal to $Z_{\mathfrak{w}}(\mathfrak{a})$, where $a_j\mathcal{U}_{-\alpha_{i_j}}$ is to be thought of as a closed subvariety of $\mathcal{P}_{i_j}/\mathcal{U}_{i_j}(k_j)$. In particular, by Theorem A.11, $\theta_{\mathfrak{a}}$ is a biregular isomorphism onto $Z_{\mathfrak{w}}(\mathfrak{a})$. So $\theta_{\mathfrak{a}}^{-1}$ gives rise to the desired section on the open set $Z_{\mathfrak{w}}(\mathfrak{a})$.

Finally, we have the following commutative diagram:

$$\begin{array}{ccc} Z_{\mathfrak{w}}(\mathfrak{a}) \times H & \xrightarrow{\sigma} & \theta^{-1}(Z_{\mathfrak{w}}(\mathfrak{a})) \\ & \searrow \quad \swarrow \theta & \\ & Z_{\mathfrak{w}}(\mathfrak{a}) & \end{array}$$

where $H := \mathcal{B}^n/\mathcal{U}_{\mathfrak{w}}(\mathfrak{k})$ and σ is the algebraic map $\sigma(z,h) = \theta_{\mathfrak{a}}^{-1}(z) \cdot h$. By Lemma 7.1.4, σ is a bijection and hence, by Theorem A.11, it is a biregular isomorphism. This proves the local triviality of θ and hence the proposition is proved. □

7.1.9 Remark. It is easy to see that, for any $1 \le i \le \ell$, the projective variety $\mathcal{P}_i/\mathcal{B}$ is isomorphic with $\mathbb{P}^1$ (cf. Example 7.1.7).

7.1.10 Corollary. *Fix any* $\mathfrak{w} = (s_{i_1}, \dots, s_{i_n}) \in \mathfrak{W}$.

(a) *The canonical projection map* $Z_{\mathfrak{w}} \to \mathcal{P}_{i_1}/\mathcal{B} \cong \mathbb{P}^1$ *is a (Zariski) locally trivial fibration with fiber* $Z_{\mathfrak{w}'}$*, where* $\mathfrak{w}' := (s_{i_2}, \dots, s_{i_n})$.

(b) *The map* $\psi = \psi_{\mathfrak{w},n-1} : Z_{\mathfrak{w}} \to Z_{\mathfrak{w}_1}$ *is a locally trivial* $\mathbb{P}^1$*-fibration, where* $\mathfrak{w}_1 = \mathfrak{w}[n-1] := (s_{i_1}, \dots, s_{i_{n-1}})$ *and* $\psi[p_1, \dots, p_n] = [p_1, \dots, p_{n-1}]$. *Moreover, it admits a regular section* $\sigma = \sigma_{\mathfrak{w}} : Z_{\mathfrak{w}_1} \to Z_{\mathfrak{w}}$ *defined by* $\sigma[p_1, \dots, p_{n-1}] = [p_1, \dots, p_{n-1}, 1]$.

(c) $Z_{\mathfrak{w}}$ *is a projective variety.*

(d) *For any subword* $\mathfrak{v} = \mathfrak{w}_J$ *of* $\mathfrak{w}$*, the inclusion* $i_{\mathfrak{v},\mathfrak{w}} : Z_{\mathfrak{v}} \to Z_{\mathfrak{w}}$ *given by (7.1.3.2) is a closed embedding.*

Proof. (a) follows easily from the proof of Proposition 7.1.8(a) together with Proposition 7.1.8(b).

(b) The maps ψ and σ are indeed morphisms, as follows from Proposition 7.1.8(b) and the following commutative diagram (following the notation of Proposition 7.1.8):

$$\begin{array}{ccc} \mathcal{P}_{\mathfrak{w}}/\mathcal{U}_{\mathfrak{w}}(\mathfrak{k}) & \xrightarrow{\theta_{\mathfrak{w},\mathfrak{k}}} & Z_{\mathfrak{w}} \\ \bar{\psi}\downarrow & & \downarrow\psi \\ \mathcal{P}_{\mathfrak{w}_1}/\mathcal{U}_{\mathfrak{w}_1}(\mathfrak{k}_1) & \xrightarrow{\theta_{\mathfrak{w}_1,\mathfrak{k}_1}} & Z_{\mathfrak{w}_1}, \end{array}$$

where $\bar{\psi}$ is the projection on the first $(n-1)$-factors and $\mathfrak{k}_1 := (k_1, \ldots, k_{n-1})$. (In more detail, since $\bar{\psi}$ is a morphism, so is the composite $\psi \circ \theta_{\mathfrak{w},\mathfrak{k}}$. But since $\theta_{\mathfrak{w},\mathfrak{k}}$ is a locally trivial fibration, ψ is a morphism by using local sections of $\theta_{\mathfrak{w},\mathfrak{k}}$.) Clearly, the map $\bar{\psi}$ is a trivial principal $\mathcal{P}_{i_n}/\mathcal{U}_{i_n}(k_n)$-bundle, where $\mathcal{P}_{i_n}/\mathcal{U}_{i_n}(k_n)$ acts on $\mathcal{P}_{\mathfrak{w}}/\mathcal{U}_{\mathfrak{w}}(\mathfrak{k})$ via the right multiplication on the last factor. Moreover, by Proposition 7.1.8(b), $\theta_{\mathfrak{w},\mathfrak{k}}$, resp. $\theta_{\mathfrak{w}_1,\mathfrak{k}_1}$, is a locally trivial principal $\mathcal{B}^n/\mathcal{U}_{\mathfrak{w}}(\mathfrak{k})$-bundle, resp. $\mathcal{B}^{n-1}/\mathcal{U}_{\mathfrak{w}_1}(\mathfrak{k}_1)$-bundle. In particular, $\psi \circ \theta_{\mathfrak{w},\mathfrak{k}}$ is a locally trivial principal $H := \big((\mathcal{B}^{n-1}/\mathcal{U}_{\mathfrak{w}_1}(\mathfrak{k}_1)) \times \mathcal{P}_{i_n}/\mathcal{U}_{i_n}(k_n)\big)$-bundle. Hence, ψ is a locally trivial fiber bundle with fiber $H/(\mathcal{B}^n/\mathcal{U}_{\mathfrak{w}}(\mathfrak{k})) \approx \mathcal{P}_{i_n}/\mathcal{B} \approx \mathbb{P}^1$. This proves the (b) part.

(c) Assume, by induction, that $Z_{\mathfrak{w}'}$ is a projective variety and consider the locally trivial fibration $p : Z_{\mathfrak{w}} \to \mathbb{P}^1$ with fiber $Z_{\mathfrak{w}'}$, as in the (a) part. By Proposition 7.1.8(a), $Z_{\mathfrak{w}}$ is smooth and by [Šafarevič–94, Chap.VI, Exercises to §1, no. 4], $Z_{\mathfrak{w}}$ is complete. Further, by the proof of Proposition 7.1.8(b), p is a trivial fibration restricted to$\mathbb{P}^1\backslash\{x_o\}$, for any $x_o \in \mathbb{P}^1$. Hence, the projectivity of $Z_{\mathfrak{w}}$ follows from the Chevalley–Kleiman criterion asserting that a smooth complete variety is projective if and only if any finite set of its points is contained in an affine open subset [Šafarevič–94, Chap.VI, §2.4].

(d) The map $i_{\mathfrak{v},\mathfrak{w}}$ is a morphism, as follows from Proposition 7.1.8(b). Further, since $Z_{\mathfrak{v}}$ is projective (by (c)), $Z := \operatorname{Im} i_{\mathfrak{v},\mathfrak{w}}$ is closed. So it suffices to show that $i_{\mathfrak{v},\mathfrak{w}}^{-1} : Z \to Z_{\mathfrak{v}}$ is a morphism under the (reduced) subvariety structure on Z.

Let $\theta = \theta_{\mathfrak{w},\mathfrak{k}} : \mathcal{P}_{\mathfrak{w}}/\mathcal{U}_{\mathfrak{w}}(\mathfrak{k}) \to Z_{\mathfrak{w}}$ be the principal $\mathcal{B}^n/\mathcal{U}_{\mathfrak{w}}(\mathfrak{k})$-bundle, as in Proposition 7.1.8(b), and denote $\tilde{Z} := \theta^{-1}Z$. To prove that $i_{\mathfrak{v},\mathfrak{w}}^{-1}$ is a morphism, it suffices to show that $\gamma := i_{\mathfrak{v},\mathfrak{w}}^{-1} \circ \theta_{|\tilde{Z}} : \tilde{Z} \to Z_{\mathfrak{v}}$ is a morphism. Following the same notation as in (7.1.3.2), $\theta^{-1}Z = (Q_1/\mathcal{U}_{i_1}(k_1)) \times \cdots \times (Q_n/\mathcal{U}_{i_n}(k_n))$, where $Q_j := \mathcal{P}_{i_j}$ if $j \in \{j_1, \ldots, j_m\}$ and $Q_j = \mathcal{B}$ otherwise. Now, it is easy to see that, for $(\bar{p}_1, \ldots, \bar{p}_n) \in \theta^{-1}Z$,

$$\gamma(\bar{p}_1, \ldots, \bar{p}_n) \\ = \big[(\bar{p}_1 \cdots \bar{p}_{j_1-1}\bar{p}_{j_1}), (\bar{p}_{j_1+1} \cdots \bar{p}_{j_2-1}\bar{p}_{j_2}), \ldots, (\bar{p}_{j_{m-1}+1} \cdots \bar{p}_{j_m})\big];$$

which is clearly a morphism. □

7.1.11 Definition. For any $\mathfrak{w} = (s_{i_1}, \ldots, s_{i_n}) \in \mathfrak{W}$, define the map $m_{\mathfrak{w}} : Z_{\mathfrak{w}} \to \mathcal{G}/\mathcal{B}$ by $m_{\mathfrak{w}}[p_1, \ldots, p_n] = p_1 \cdots p_n\, \mathcal{B}$, for $p_j \in \mathcal{P}_{i_j}$.

If $\mathfrak{w}$ is reduced, by Theorem 5.1.3(i_2) and (c),

$$\text{(1)} \qquad \operatorname{Im} m_{\mathfrak{w}} = \bigsqcup_{v \le w} \mathcal{B}\, v\, \mathcal{B}/\mathcal{B}, \qquad \text{where } w := \pi(\mathfrak{w}).$$

Similarly, for any $Y \subset \{1, \dots, \ell\}$, define the map $m^Y_{\mathfrak{w}} : Z_{\mathfrak{w}} \to \mathcal{G}/\mathcal{P}_Y$ to be the map $m_{\mathfrak{w}}$ followed by the projection $\mathcal{G}/\mathcal{B} \to \mathcal{G}/\mathcal{P}_Y$. If $\mathfrak{w}$ is reduced and $w \in W'_Y$ (cf. Definition 1.3.17), we have by (1) and 1.3.17:

$$\text{(2)} \qquad \operatorname{Im} m^Y_{\mathfrak{w}} = \bigsqcup_{\substack{v \le w \\ v \in W'_Y}} \mathcal{B}\, v\, \mathcal{P}_Y/\mathcal{P}_Y.$$

7.1.12 Proposition. *For any countable-dimensional $V \in \mathfrak{M}(\mathcal{G})$ (cf. 6.2.1), $\mathfrak{w} \in \mathfrak{W}$, and $\mathcal{B}$-fixed line $[v_o]$ in V, the map*

$$m_{\mathfrak{w}}(v_o) : Z_{\mathfrak{w}} \to \mathbb{P}(V), \ \textit{taking } x \mapsto m_{\mathfrak{w}}(x)[v_o],$$

is a morphism of ind-*varieties (cf. Example 4.1.3(4)).*

Proof. Since $V \in \mathfrak{M}(\mathcal{G})$, for any minimal parabolic subgroup $\mathcal{P}_j \subset \mathcal{G}$ and a finite-dimensional subspace $M \subset V$, there exists a $\mathcal{P}_j$-stable subspace $M' \subset V$ containing M and a positive integer k such that the action of $\mathcal{P}_j$ on M' descends to an algebraic action $(\mathcal{P}_j/\mathcal{U}_j(k)) \times M' \to M'$. Making an iterated use of this, we obtain the following:

For any finite-dimensional subspace $M \subset V$, there exists a finite-dimensional subspace $M' \subset V$ such that the map

$$\theta : \mathcal{P}_{i_1} \times \cdots \times \mathcal{P}_{i_n} \times M \to V, \ (p_1, \dots, p_n; v) \mapsto p_1 \cdots p_n v,$$

for $p_j \in \mathcal{P}_{i_j}$ and $v \in M$, has its image in M' and, moreover, θ descends to a morphism

$$\bar{\theta} : \mathcal{P}_{i_1}/\mathcal{U}_{i_1}(k_1) \times \cdots \times \mathcal{P}_{i_n}/\mathcal{U}_{i_n}(k_n) \times M \to M'$$

for some choice of $\mathfrak{k} = (k_1, \dots, k_n)$ satisfying (7.1.3.7). Now the proposition follows from Proposition 7.1.8(b). □

7.1.13 Definition. For any $Y \subset \{1, \dots, \ell\}$ and $w \in W'_Y$, define the *Schubert variety* (cf. Corollary 6.1.20):

$$\text{(1)} \qquad X^Y_w := \bigsqcup_{\substack{v \le w \\ v \in W'_Y}} \mathcal{B}\, v\, \mathcal{P}_Y/\mathcal{P}_Y \subset \mathcal{G}/\mathcal{P}_Y.$$

(For the moment X_w^Y is only a set. The variety structure on X_w^Y will be defined in 7.1.19.) For an arbitrary $w \in W$, we set $X_w^Y = X_{w'}^Y$, where $w' \in W'_Y$ is the unique element such that $w'w'' = w$ for some $w'' \in W_Y$. Then $X_w^Y = \bigcup_{\substack{v \leq w \\ v \in W}} \mathcal{B}\, v\, \mathcal{P}_Y/\mathcal{P}_Y$ (by Lemma 1.3.18) and, for $v \leq w \in W$, $X_v^Y \subset X_w^Y$.

If $Y = \emptyset$, we abbreviate $X_w^\emptyset$ simply by X_w. Also denote

(2)
$$\mathcal{X}^Y := \mathcal{G}/\mathcal{P}_Y,$$

and set, for $n \geq 0$,

(3)
$$X_n^Y = \bigcup_{\substack{w \in W'_Y \\ \ell(w) \leq n}} X_w^Y = \bigsqcup_{\substack{v \in W'_Y \\ \ell(v) \leq n}} \mathcal{B}\, v\, \mathcal{P}_Y/\mathcal{P}_Y.$$

Again, for $Y = \emptyset$, we abbreviate X_n^Y by X_n.

For any $\lambda \in D_Y^o$ (cf. Definition 7.1.1), consider the integrable highest weight $\mathfrak{g}$-module $V = L^{\max}(\lambda)$ (cf. Definition 2.1.5) and the injective (set) map $i_V : \mathcal{G}/\mathcal{P}_Y \hookrightarrow \mathbb{P}(V)$ (cf. Lemma 7.1.2). We denote i_V by i_λ.

For any $w \in W'_Y$ and a reduced word $\mathfrak{w}$ with $\pi(\mathfrak{w}) = w$, consider the morphism $m_{\mathfrak{w}}(v_\lambda) : Z_{\mathfrak{w}} \to \mathbb{P}(V)$ as in 7.1.12, denoted here simply by $m_{\mathfrak{w}}^\lambda$. Observe that $m_{\mathfrak{w}}^\lambda = i_\lambda \circ m_{\mathfrak{w}}^Y$ and thus has the image (by (7.1.11.2)):

(4)
$$\operatorname{Im} m_{\mathfrak{w}}^\lambda = i_\lambda(X_w^Y).$$

In particular, by Proposition 7.1.12, $i_\lambda(X_w^Y)$ (and hence any X_n^Y) is a Zariski closed subset of $\mathbb{P}(M)$, for some finite-dimensional subspace $M = M_w \subset V$; use Theorem A.2.

We equip X_w^Y, resp. X_n^Y, with the (reduced) variety structure so that $i_{\lambda|X_w^Y}$, resp. $i_{\lambda|X_n^Y}$, is a closed embedding. Equipped with this variety structure, we denote X_w^Y by $X_w^Y(\lambda)$ and X_n^Y by $X_n^Y(\lambda)$. In particular, for any $w \in W'_Y$, resp. $n \geq 0$, $X_w^Y(\lambda)$, resp. $X_n^Y(\lambda)$, is a projective variety. Further, $X_w^Y(\lambda)$ is irreducible by (4), since so is $Z_{\mathfrak{w}}$ by Proposition 7.1.8(a). Moreover, $\mathcal{X}^Y$ together with the filtration

(5)
$$X_0^Y \subset X_1^Y \subset \cdots$$

and the projective variety structure $X_n^Y(\lambda)$ on X_n^Y becomes a projective ind-variety denoted by $\mathcal{X}^Y(\lambda)$. It is easy to see that $i_\lambda : \mathcal{X}^Y(\lambda) \hookrightarrow \mathbb{P}(V)$ is a closed embedding of ind-varieties. (Use the Bruhat decomposition (3) and the fact that $W'_Y \to \mathfrak{h}_{\mathbb{Z}}^*$, $w \mapsto w\lambda$, is injective, to show that for any finite-dimensional $\mathcal{B}$-stable subspace $M \subset V$, $i_\lambda^{-1}(\mathbb{P}(M)) \subset X_n^Y$ for some large enough $n = n_M$. Now, any finite-dimensional subspace of V is contained in a finite-dimensional $\mathcal{B}$-stable subspace of V, as is clear from (6.2.8.2). Thus the assertion, that i_λ is a closed embedding, follows.)

We now analyze the dependence of the variety structure $X_w^Y(\lambda)$ on $\lambda \in D_Y^o$. But before this, we need the following preparatory lemma.

7.1.14 Lemma. *Let V and W be two countable-dimensional vector spaces. Regard $\mathbb{P}(V)$ and $\mathbb{P}(W)$ as ind-varieties via Example 4.1.3(4). Then the Segre map*

$$S : \mathbb{P}(V) \times \mathbb{P}(W) \to \mathbb{P}(V \otimes W), \quad [v] \times [w] \mapsto [v \otimes w],$$

is a closed embedding, where $[v]$ denotes the line through the vector $v \in V$.

Proof. For any finite-dimensional subspaces $V_1 \subset V$ and $W_1 \subset W$, the map $\mathbb{P}(V_1) \times \mathbb{P}(W_1) \to \mathbb{P}(V_1 \otimes W_1)$ is a closed embedding (cf. [Šafarevič–94, Chap. I, §5.1]). So, to prove that S is a closed embedding, it suffices to show that for any closed subset $Z \subset \mathbb{P}(V) \times \mathbb{P}(W)$, $S(Z) \cap \mathbb{P}(V_1 \otimes W_1)$ is closed in $\mathbb{P}(V_1 \otimes W_1)$. Observe that

(1) $$S(Z) \cap \mathbb{P}(V_1 \otimes W_1) = S(Z \cap (\mathbb{P}(V_1) \times \mathbb{P}(W_1))).$$

Since $Z \cap (\mathbb{P}(V_1) \times \mathbb{P}(W_1))$ is closed in $\mathbb{P}(V_1) \times \mathbb{P}(W_1)$, by (1), $S(Z) \cap \mathbb{P}(V_1 \otimes W_1)$ is closed in $\mathbb{P}(V_1 \otimes W_1)$. This proves the lemma. □

For any $Y \subset Y'$, let $\sigma_{Y',Y} : \mathcal{X}^Y \to \mathcal{X}^{Y'}$ be the canonical projection map.

7.1.15 Proposition. *Fix any reduced $\mathfrak{w}$ such that $w := \pi(\mathfrak{w}) \in W'_Y$.*

(a) *The morphism $m^Y_{\mathfrak{w}} : Z_{\mathfrak{w}} \to X^Y_w(\lambda)$ is a surjective birational map for any $\lambda \in D^o_Y$. In fact, $Z^o_{\mathfrak{w}}$ and $m^Y_{\mathfrak{w}}(Z^o_{\mathfrak{w}})$ are open in $Z_{\mathfrak{w}}$ and $X^Y_w(\lambda)$, respectively, and $m^Y_{\mathfrak{w}|Z^o_{\mathfrak{w}}}$ is a biregular isomorphism onto its image $\mathcal{B}w\mathcal{P}_Y/\mathcal{P}_Y$, where $Z^o_{\mathfrak{w}}$ is defined in (7.1.3.3). In particular, $X^Y_w(\lambda) = \overline{\mathcal{B}w\mathcal{P}_Y/\mathcal{P}_Y}$, where the closure is taken with respect to the Zariski topology.*

(b) *The Zariski topology on $X^Y_n(\lambda)$, for any $n \geq 0$, (and hence on $X^Y_w(\lambda)$) does not depend on the choice of $\lambda \in D^o_Y$.*

Further, for $\lambda \in D^o_Y$ and $\mu \in D_{\mathbb{Z}}$ such that $\mu(\alpha_i^\vee) = 0$ for all $i \in Y$, the identity map $I_n = I_n(\lambda, \lambda+\mu) : X^Y_n(\lambda+\mu) \to X^Y_n(\lambda)$ is a morphism. In particular, the identity map $I_w : X^Y_w(\lambda+\mu) \to X^Y_w(\lambda)$ is a morphism.

Also $i_\mu \circ \sigma_{Y_\mu,Y} : \mathcal{X}^Y(\lambda+\mu) \to \mathbb{P}(L^{\max}(\mu))$ is a morphism, where $Y_\mu := \{1 \leq i \leq \ell : \mu(\alpha_i^\vee) = 0\}$.

(c) *In the case of symmetrizable $\mathfrak{g}$, I_n (and hence I_w) is a biregular isomorphism.*

Proof. (a) For any subword $\mathfrak{v} = \mathfrak{w}_J$ of $\mathfrak{w}$, the map $i_{\mathfrak{v},\mathfrak{w}} : Z_{\mathfrak{v}} \to Z_{\mathfrak{w}}$ defined in (7.1.3.2) is a morphism (in fact, it is a closed embedding by Corollary 7.1.10(d)). In particular, $Z^o_{\mathfrak{w}}$ is open in $Z_{\mathfrak{w}}$ since, by (7.1.3.4), $Z_{\mathfrak{w}} \backslash Z^o_{\mathfrak{w}} = \bigsqcup_{\mathfrak{w}_J \neq \mathfrak{w}} i_{\mathfrak{w}_J,\mathfrak{w}}(Z^o_{\mathfrak{w}_J}) = \bigcup_{\mathfrak{w}_J \neq \mathfrak{w}} i_{\mathfrak{w}_J,\mathfrak{w}}(Z_{\mathfrak{w}_J})$. Further,

(1) $$m^Y_{\mathfrak{w}}(Z_{\mathfrak{w}} \backslash Z^o_{\mathfrak{w}}) \cap m^Y_{\mathfrak{w}}(Z^o_{\mathfrak{w}}) = \emptyset,$$

as follows from the axiom (BN_4) of 5.1.1. From (1), we obtain that $m^Y_{\mathfrak{w}}(Z^o_{\mathfrak{w}})$ is open in $X^Y_w(\lambda)$.

The map $m^Y_{\mathfrak{w}|_{Z^o_{\mathfrak{w}}}} : Z^o_{\mathfrak{w}} \to \mathcal{B}w\,\mathcal{P}_Y/\mathcal{P}_Y$ is a bijection, as follows from Theorem 5.1.3(i_1). To complete the proof of (a), it suffices to show that $m^Y_{\mathfrak{w}|_{Z^o_{\mathfrak{w}}}}$ is biregular isomorphic with the open subvariety $(\mathcal{B}w\,\mathcal{P}_Y/\mathcal{P}_Y)(\lambda)$ of $X^Y_w(\lambda)$. Since, for any $b \in \mathcal{B}$, the left action of b on $(\mathcal{B}w\,\mathcal{P}_Y/\mathcal{P}_Y)(\lambda)$ is a morphism, the variety $(\mathcal{B}w\,\mathcal{P}_Y/\mathcal{P}_Y)(\lambda)$ is smooth. Thus, $m^Y_{\mathfrak{w}|_{Z^o_{\mathfrak{w}}}}$ is a biregular isomorphism by Theorem A.11. This proves (a).

(b) Abbreviate $V_1 := L^{\max}(\lambda)$, $V_2 := L^{\max}(\mu)$ and $V_3 := L^{\max}(\lambda + \mu)$. Then there exists a unique $\mathfrak{g}$-module map $\theta : V_3 \to V_1 \otimes V_2$, which takes $v_{\lambda+\mu} \mapsto v_\lambda \otimes v_\mu$, where v_λ is a fixed nonzero highest weight vector of V_1, etc.. By Theorem 6.2.3, θ is a $\mathcal{G}$-module map as well. Observe that $V_1 \otimes V_2 \in \mathfrak{M}(\mathcal{G})$ under the tensor product $\mathcal{G}$-module structure. Let $K := \operatorname{Ker}\theta$. Then θ gives rise to a morphism of ind-varieties $\hat{\theta} : \mathbb{P}(V_3)\backslash\mathbb{P}(K) \to \mathbb{P}(V_1 \otimes V_2)$. Further, by Lemma 7.1.14, we have a closed embedding $S : \mathbb{P}(V_1) \times \mathbb{P}(V_2) \hookrightarrow \mathbb{P}(V_1 \otimes V_2)$. The map $i_{\lambda+\mu} : \mathcal{X}^Y \hookrightarrow \mathbb{P}(V_3)$ has its image in $\mathbb{P}(V_3)\backslash\mathbb{P}(K)$. Indeed, since θ is a $\mathcal{G}$-module map, K is $\mathcal{G}$-stable and hence so is $\mathbb{P}(V_3)\backslash\mathbb{P}(K)$. Since the latter contains $[v_{\lambda+\mu}]$, it contains the orbit $\mathcal{G} \cdot [v_{\lambda+\mu}] = \operatorname{Im} i_{\lambda+\mu}$. Thus, we have the following commutative diagram:

$$
\text{(D)} \qquad
\begin{array}{ccc}
\mathcal{X}^Y(\lambda+\mu) & \xrightarrow{\;i_{\lambda+\mu}\;} & \mathbb{P}(V_3)\backslash\mathbb{P}(K) \\
\Big\downarrow{\scriptstyle (i_\lambda,\, i_\mu \circ \sigma)} & & \Big\downarrow{\scriptstyle \hat{\theta}} \\
\mathbb{P}(V_1) \times \mathbb{P}(V_2) & \xhookrightarrow{\;S\;} & \mathbb{P}(V_1 \otimes V_2),
\end{array}
$$

where $\sigma := \sigma_{Y_\mu, Y}$. Since $i_{\lambda+\mu}$ and $\hat{\theta}$ are morphisms and S is a closed embedding, we get that $(i_\lambda, i_\mu \circ \sigma)$ is a morphism as well (by Lemma 4.1.2). In particular, $i_\lambda : \mathcal{X}^Y(\lambda+\mu) \to \mathbb{P}(V_1)$ is a morphism (and so is $i_\mu \circ \sigma$). But $i_\lambda : \mathcal{X}^Y(\lambda) \to \mathbb{P}(V_1)$ is a closed embedding (cf. 7.1.13), and hence the identity map $I : \mathcal{X}^Y(\lambda+\mu) \to \mathcal{X}^Y(\lambda)$ is a morphism. In particular, $I_n : X^Y_n(\lambda+\mu) \to X^Y_n(\lambda)$ is a morphism. Since $X^Y_n(\lambda+\mu)$ is a projective variety, I_n is a homeomorphism by Theorem A.2.

(c) For symmetrizable $\mathfrak{g}$, since V_3 is an irreducible $\mathfrak{g}$-module (by Corollaries 2.2.6 and 3.2.10), $K = \{0\}$. In particular, $\hat{\theta} : \mathbb{P}(V_3) \to \mathbb{P}(V_1 \otimes V_2)$ is a closed embedding. First take $\mu = \lambda$ and consider the following commutative diagram (analagous to (D)), with $V_1 = V_2 = L(\lambda)$ and $V_3 = L(2\lambda)$,

$$\begin{array}{ccc} \mathcal{X}^Y(\lambda) & \xrightarrow{\hat{i}_{2\lambda}} & \mathbb{P}(V_3) \\ \Big\downarrow {\scriptstyle (i_\lambda, i_\lambda)} & & \Big\downarrow {\scriptstyle \hat{\theta}} \\ \mathbb{P}(V_1) \times \mathbb{P}(V_2) & \xhookrightarrow{S} & \mathbb{P}(V_1 \otimes V_2), \end{array}$$

where the top horizontal map $\hat{i}_{2\lambda}$ is defined set theoretically to be the morphism $i_{2\lambda} : \mathcal{X}^Y(2\lambda) \hookrightarrow \mathbb{P}(V_3)$. Since $\hat{\theta}$ is a closed embedding, we get that $\hat{i}_{2\lambda}$ is a morphism. Thus, for any $n \geq 0$, the identity map $I_n : X_n^Y(\lambda) \to X_n^Y(2\lambda)$ is a morphism and hence a biregular isomorphism (by the (b) part). Finally, the assertion that the identity map $I_n : X_n^Y(\lambda+\mu) \to X_n^Y(\lambda)$ is a biregular isomorphism, for any $\lambda \in D_Y^o$ and $\mu \in D_{\mathbb{Z}}$ such that $\mu(\alpha_i{}^\vee) = 0$ for all $i \in Y$, follows from the above by taking a large enough k such that $(2^k - 1)\lambda - \mu \in D_{\mathbb{Z}}$ and considering the identity maps $X_n^Y(2^k\lambda) \overset{I_n}{\to} X_n^Y(\lambda+\mu) \overset{I_n}{\to} X_n^Y(\lambda)$. This proves the (c) part. $\square$

7.1.16 Remark. Even though we do not prove this in the book, it is known that, for any Kac–Moody Lie algebra $\mathfrak{g}$ and any λ, μ as in 7.1.15(b) for some Y, the map $\theta : L^{\max}(\lambda+\mu) \to L^{\max}(\lambda) \otimes L^{\max}(\mu)$, defined above, is injective (cf. [Kumar–89a, Theorem 2.7] for the case when $Y = \emptyset$; the same proof works for an arbitrary Y). (This was also proved by [Mathieu–89] using char. p methods.) Thus, from the above proof, $I : \mathcal{X}^Y(\lambda+\mu) \to \mathcal{X}^Y(\lambda)$ is a biregular isomorphism for any $\mathfrak{g}$.

7.1.17 Definition. Define a partial order $\preceq$ in $D_{\mathbb{Z}}$ by $\lambda \preceq \mu$ iff $\mu - \lambda \in D_{\mathbb{Z}}$. (This relation should not be confused with the partial order $\leq$ in $\mathfrak{h}^*$ defined in 2.1.1(a).)

7.1.18 Corollary (of Proposition 7.1.15). *With the notation as in Proposition 7.1.15, for any fixed $w \in W_Y'$, there exists a large enough (with respect to the partial order $\preceq$) $\lambda = \lambda_w \in D_Y^o$ such that, for any $\lambda \preceq \mu$, the identity map $X_w^Y(\mu) \to X_w^Y(\lambda)$ is a biregular isomorphism. Observe that automatically $\mu \in D_Y^o$.*

Thus, for any $n \geq 0$, there exists $\lambda = \lambda_n \in D_Y^o$ such that, for any $\lambda \preceq \mu$, the identity map $X_n^Y(\mu) \to X_n^Y(\lambda)$ is a biregular isomorphism.

Proof. Fix any $\rho_Y \in D_Y^o$. Choose also a reduced $\mathfrak{w} \in \mathfrak{W}$ with $\pi(\mathfrak{w}) = w$. Then, by Proposition 7.1.15, we have the following commutative diagram, where all the maps are birational morphisms.

$$
\begin{array}{ccccccc}
 & & & Z_{\mathfrak{w}} & & & \\
 & \swarrow m_{\mathfrak{w}}^{Y} & & \downarrow m_{\mathfrak{w}}^{Y} & & \searrow m_{\mathfrak{w}}^{Y} & \\
\cdots \xrightarrow[I]{} X_w^Y(m\rho_Y) & \xrightarrow[I]{} \cdots & \xrightarrow[I]{} & X_w^Y(2\rho_Y) & \xrightarrow[I]{} & X_w^Y(\rho_Y). &
\end{array}
$$

By Exercise 7.1.E.6, $X_w^Y(m\rho_Y)$ have the same normalization for all $m \geq 1$. Now, by Exercise 7.1.E.7, there exists a $m_o = m_o(w) \geq 1$ such that $X_w^Y(m\rho_Y) \to X_w^Y(m_o\rho_Y)$ is a biregular isomorphism for all $m \geq m_o$. From this, the first part of the corollary follows easily.

The "Thus" part follows from the first part since, for any $n \geq 0$, there exists $w = w_n \in W'_Y$ such that $X_n^Y \subset X_w^Y$ (by Lemma 1.3.20). □

7.1.19 Definition (Stable variety structure on X_w^Y). For any $w \in W'_Y$, choose $\lambda = \lambda_w \in D_Y^o$ large enough such that the identity map $X_w^Y(\mu) \to X_w^Y(\lambda)$ is a biregular isomorphism for all $\lambda \preceq \mu$ (cf. Corollary 7.1.18). Unless otherwise stated, from now on, we will endow X_w^Y with the irreducible projective variety structure $X_w^Y(\lambda)$ (for such a λ) and refer to this as the *stable variety structure on* X_w^Y and denote it simply by X_w^Y. Similarly, we define the stable variety structure on X_n^Y, for any $n \geq 0$, and with this variety structure denote it by X_n^Y itself.

The inclusion $X_n^Y \hookrightarrow X_{n+1}^Y$ is a closed embedding and hence $\mathcal{X}^Y$, together with the filtration $X_0^Y \subset X_1^Y \subset \cdots$ and the stable variety structure on each X_n^Y, becomes a projective ind-variety, which we refer to as the *stable ind-variety* $\mathcal{X}^Y$. By Proposition 7.1.15(b), it is easy to see that, for any subsets $Y_1 \subset Y_2$, the canonical map $\mathcal{X}^{Y_1} \to \mathcal{X}^{Y_2}$ is a morphism of ind-varieties. The Kac–Moody group $\mathcal{G}$ of course acts on $\mathcal{X}^Y$ via left multiplication.

Let Z be an ind-variety on which $\mathcal{G}$ acts. We say that the action is *regular* if, for all $1 \leq i \leq \ell$, the action of $\mathcal{P}_i$ on Z, obtained by the restriction, is regular (cf. Definition 7.1.5). Since $\mathcal{G}$ is generated as an abstract group by $\{\mathcal{P}_i : 1 \leq i \leq \ell\}$, for any fixed $g \in \mathcal{G}$, the induced map $g : Z \to Z$ is a biregular isomorphism if the action of $\mathcal{G}$ on Z is regular. In particular, g acts as a homeomorphism on Z. It is easy to see that the action of $\mathcal{G}$ on $\mathcal{X}^Y$ is regular.

7.1.20 Proposition. *Let $\lambda \in D^o$ and let $V(\lambda)$ be any integrable highest weight $\mathfrak{g}$-module with highest weight λ. For any $w \in W$, let $x_{w\lambda} \in V(\lambda)_{w\lambda}$ be a nonzero vector. Then, for $v, w \in W$,*

$$w \leq v \quad \textit{iff}\; x_{v\lambda} \in U(\mathfrak{n}^-)\, x_{w\lambda}.$$

Proof. We first prove the implication "⇒": It suffices to show that, for $v = s_\beta w$ with $\beta \in \Delta_{\text{re}}^+$ and $w < v$,

(1) $$x_{v\lambda} \in U(\mathfrak{n}^-)x_{w\lambda},$$

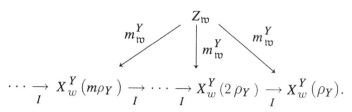

where $s_\beta \in W$ is defined by $s_\beta \chi = \chi - \langle \chi, \beta^\vee \rangle \beta$ for $\chi \in \mathfrak{h}^*$. Consider the Lie subalgebra $s\ell_2(\beta) \subset \mathfrak{g}$ generated by $\mathfrak{g}_\beta$ and $\mathfrak{g}_{-\beta}$. Then $s\ell_2(\beta)$ is isomorphic with the Lie algebra $s\ell_2$. Choose a nonzero vector $e_\beta \in \mathfrak{g}_\beta$ and also $e_{-\beta} \in \mathfrak{g}_{-\beta}$. We claim that

(2) $$e_\beta \cdot x_{w\lambda} = 0.$$

Clearly (2) is equivalent to $e_{w^{-1}\beta} x_\lambda = 0$. But, by Lemma 1.3.13, $w^{-1}\beta \in \Delta^+$ and hence (2) is established. Let $V_1 \subset V(\lambda)$ be the $s\ell_2(\beta)$-submodule generated by $x_{w\lambda}$. Then, by $s\ell_2$-representation theory (cf. Exercise 1.2.E.3),

$$e_{-\beta}^{\langle w\lambda, \beta^\vee \rangle} x_{w\lambda} \neq 0\,.$$

Since $w^{-1}\beta \in \Delta^+$, $\langle w\lambda, \beta^\vee \rangle \in \mathbb{Z}_+$. But since $w\lambda - \langle w\lambda, \beta^\vee \rangle \beta = v\lambda$,

$$e_{-\beta}^{\langle w\lambda, \beta^\vee \rangle} x_{w\lambda} = c\, x_{v\lambda}, \qquad \text{for some nonzero } c\,,$$

since the weight space $V(\lambda)_{v\lambda}$ is one-dimensional by Lemma 1.3.5(a).

We now come to the implication "$\Leftarrow$": Take any $v, w \in W$ such that $x_{v\lambda} \in U(\mathfrak{n}^-)\, x_{w\lambda}$. Then, for any simple reflection s_i such that $s_i v < v$, we prove that

(3) $$x_{s_i v\lambda} \in U(\mathfrak{n}^-)\, x_{s_i w\lambda} \cup U(\mathfrak{n}^-)\, x_{w\lambda} :$$

Express

(4) $$x_{v\lambda} = P x_{w\lambda}, \qquad \text{for some } P \in U(\mathfrak{n}^-).$$

Since $s_i v < v$, by an argument as above, up to a nonzero scalar multiple,

(5) $$x_{s_i v\lambda} = e_i^m x_{v\lambda}, \qquad \text{for some } m \in \mathbb{Z}_+\,.$$

We first consider the case $s_i w > w$. Then, by the same proof as that of (2),

(6) $$e_i\, x_{w\lambda} = 0\,.$$

Combining (4)–(5), we get

$$x_{s_i v\lambda} \in U(\mathfrak{p}_i^-)\, x_{w\lambda}, \text{ where } \mathfrak{p}_i^- \text{ is as in 1.2.2.}$$

In particular, we can write

$$\begin{aligned} x_{s_i v\lambda} &= \left(\sum P_j P_j' + Q e_i\right) x_{w\lambda}, \quad \text{for some } P_j \in U(\mathfrak{n}^-),\ P_j' \in U(\mathfrak{h}) \\ &\qquad\qquad\qquad\qquad\qquad\qquad\qquad \text{and } Q \in U(\mathfrak{p}_i^-) \\ &= \sum P_j P_j'\, x_{w\lambda}, \text{ by (6).} \end{aligned}$$

Thus,

$$x_{s_i v\lambda} \in U(\mathfrak{n}^-)\, x_{w\lambda} \ \text{ (in the case } s_i w > w). \tag{7}$$

This proves (3) in this case, i.e., if $s_i w > w$.

If $s_i w < w$, similar to (5), we get (up to nonzero scalar multiples)

$$x_{w\lambda} = f_i^{m'}\, x_{s_i w\lambda}, \quad \text{for some } m' \in \mathbb{Z}_+ . \tag{8}$$

Hence,

$$x_{v\lambda} \in U(\mathfrak{n}^-) f_i^{m'}\, x_{s_i w\lambda} \subset U(\mathfrak{n}^-) x_{s_i w\lambda}.$$

Applying (7) for $w' := s_i w$, we get

$$x_{s_i v\lambda} \in U(\mathfrak{n}^-)\, x_{s_i w\lambda}, \quad \text{in the case } s_i w < w.$$

This proves (3) in this case as well, completing the proof of (3).

Finally, we prove the implication "$\Leftarrow$" in the proposition by induction on $\ell(v)$. For $\ell(v) = 0$, i.e., $v = e$, $x_{v\lambda} \in U(\mathfrak{n}^-)\, x_{w\lambda}$ only if $w = e$, since $\lambda \in D^o$ by assumption (cf. Lemma 3.2.5). So take $\ell(v) > 0$ and choose a simple reflection s_i such that $s_i v < v$. By (3) and the induction hypothesis,

$$w \leq s_i v < v \quad \text{or} \quad s_i w \leq s_i v.$$

In either case, using Corollary 1.3.19, we get $w \leq v$. This completes the induction and finishes the proof of the proposition. □

For any $w \in W'_Y$, define the following subsets of $\mathcal{G}/\mathcal{P}_Y$ (cf. Theorem 6.2.8):

$$\mathcal{B}^w_Y := \mathcal{U}^- w\, \mathcal{P}_Y/\mathcal{P}_Y$$
$$X^w_Y := \bigsqcup_{\substack{v \geq w \\ v \in W'_Y}} \mathcal{B}^v_Y,$$

where $\mathcal{U}^- \subset \mathcal{G}$ is the subgroup as in 6.2.7. We abbreviate $X^w_\emptyset$ by X^w and $\mathcal{B}^w_\emptyset$ by $\mathcal{B}^w$.

7.1.21 Proposition. *For any $w \in W$,*

$$\overline{\mathcal{B}^w} = X^w, \tag{1}$$

where the closure is taken with respect to the Zariski topology on $\mathcal{G}/\mathcal{B}$ (cf. 7.1.19).

More generally, for any $w \in W'_Y$,

$$\overline{\mathcal{B}^w_Y} = X^w_Y. \tag{2}$$

Proof. Fix any $\lambda \in D^o$ and recall the definition of the closed embedding i_λ: $\mathcal{X}(\lambda) \hookrightarrow \mathbb{P}(L^{\max}(\lambda))$ from 7.1.13, where $\mathcal{X}(\lambda) := \mathcal{G}/\mathcal{B}$ as a set. The Zariski topology on $\mathcal{X}(\lambda)$ does not depend on λ by Proposition 7.1.15(b). Consider the closed subspace $\mathbb{P}(U(\mathfrak{n}^-)\, x_{w\lambda}) \subset \mathbb{P}(L^{\max}(\lambda))$, where $x_{w\lambda}$ is a nonzero weight vector of $L^{\max}(\lambda)$ of weight $w\lambda$. Since i_λ is a $\mathcal{G}$ (in particular $\mathcal{U}^-$)-equivariant map and $U(\mathfrak{n}^-)\, x_{w\lambda}$ is $\mathcal{U}^-$-stable (because, by definition, $\mathcal{U}^-$ is generated by one-parameter subgroups $\mathcal{U}_\alpha$ corresponding to the negative real roots α, and $\mathcal{U}_\alpha$ keeps $U(\mathfrak{n}^-)\, x_{w\lambda}$ stable by Proposition 6.2.11), we get, by using Proposition 7.1.20 and the Birkhoff decomposition (cf. Theorem 6.2.8),

$$i_\lambda^{-1}\left(\mathbb{P}(U(\mathfrak{n}^-)\, x_{w\lambda})\right) = X^w. \tag{3}$$

From (3) we get

$$\overline{\mathcal{B}^w} \subset X^w. \tag{4}$$

Conversely, since each $g \in \mathcal{G}$ acts as a homeomorphism on $\mathcal{G}/\mathcal{B}$ (cf. 7.1.19), to prove the reverse inclusion in (4), it suffices to show that for $v \geq w$, $v \in \overline{\mathcal{B}^w}$. We can further assume that v is of the form $v = s_\beta w$, for $\beta \in \Delta^+_{\text{re}}$ (and of course $w \leq v$). By Lemma 1.3.13, $w^{-1}\beta \in \Delta^+_{\text{re}}$. Now, by (6.2.7.2),

$$(\mathcal{U}_{-\beta}\mathcal{U}_\beta w\mathcal{B})/\mathcal{B} = (\mathcal{U}_{-\beta} w \mathcal{U}_{w^{-1}\beta}\mathcal{B})/\mathcal{B} \subset \mathcal{B}^w.$$

In particular,

$$\overline{(\mathcal{U}_{-\beta}\mathcal{U}_\beta w\mathcal{B}/\mathcal{B})} \subset \overline{\mathcal{B}^w}. \tag{5}$$

By Exercise 7.1.E.2, $s_\beta w \in \overline{\mathcal{B}^w}$. This completes the proof of (1).

To prove (2), observe that by the same proof as that of (4) (using Exercise 7.1.E.4 instead of Proposition 7.1.20), we obtain

$$\overline{\mathcal{B}^w_Y} \subset X^w_Y. \tag{6}$$

The reverse inclusion follows easily from (1) by using the continuous map $\mathcal{G}/\mathcal{B} \to \mathcal{G}/\mathcal{P}_Y$ (cf. 7.1.19). This proves (2). □

7.1.22 Lemma.

(a) *The T-fixed points $(\mathcal{G}/\mathcal{P}_Y)^T$, under the action of T on $\mathcal{G}/\mathcal{P}_Y$ via left multiplication, are precisely $\{\dot{w}\}_{w \in W'_Y}$, where $\dot{w} := w\mathcal{P}_Y \in \mathcal{G}/\mathcal{P}_Y$.*

(b) *For $v, w \in W'_Y$,*

$$X_w^Y \cap X_Y^v \neq \emptyset \quad \textit{iff} \quad v \leq w.$$

Proof. (a) From the Bruhat decomposition Corollary 6.1.20, it suffices to show that for any $w \in W'_Y$, $(\mathcal{B}w\mathcal{P}_Y/\mathcal{P}_Y)^T = \{\dot{w}\}$. By Theorem 5.1.3($\mathrm{i}_1$), we can assume that $Y = \emptyset$. In this case, the assertion $(\mathcal{B}w\mathcal{B}/\mathcal{B})^T = \{\dot{w}\}$ follows from Exercise 6.2.E.1.

(b) If nonempty, the projective variety $X^Y_w \cap X^v_Y \subset \mathcal{G}/\mathcal{P}_Y$ (cf. 7.1.19) is T-stable. In particular, by Exercise 7.1.E.5, $(X^Y_w \cap X^v_Y)^T \neq \emptyset$ iff $X^Y_w \cap X^v_Y \neq \emptyset$. Now the (b) part follows from (a). □

7.1.E EXERCISES

(1) With the notation and assumptions as in Proposition 7.1.20, for $v, w \in W$, prove that $w \geq v$ iff $x_{v\lambda} \in U(\mathfrak{n})\, x_{w\lambda}$.

(2) With the notation as in (7.1.21.5), prove that

$$s_\beta w \in \overline{(\mathcal{U}_{-\beta}\mathcal{U}_\beta w\mathcal{B}/\mathcal{B})}.$$

Hint. By (6.2.7.1), we can assume that $\beta = \alpha_i$ for a simple root α_i. This reduces the problem to SL_2.

(3) Show that the canonical morphism $\mathcal{G}/\mathcal{B} \to \mathcal{G}/\mathcal{P}_Y$ of ind-varieties (cf. 7.1.19) is a closed map (under the Zariski topologies) iff $Y \subset \{1, \dots, \ell\}$ is of finite type.

(4) Let $Y \subset \{1, \dots, \ell\}$ be any subset and let $V(\lambda)$ be an integrable highest weight $\mathfrak{g}$-module with highest weight $\lambda \in D^o_Y$. For any $v, w \in W'_Y$ prove that $x_{v\lambda} \in U(\mathfrak{n}^-)\, x_{w\lambda} \Leftrightarrow w \leq v$, where $x_{v\lambda}$ is a nonzero vector of $V(\lambda)$ of weight $v\lambda$.

(5) Let T be an algebraic torus acting algebraically on a (nonempty) projective variety X. Then $X^T \neq \emptyset$.
Hint. Prove that any T-orbit of minimum dimension in X is a point.

(6) Let $f : X \to Y$ be a birational proper morphism with finite fibers between irreducible varieties over an algebraically closed field. Then show that the induced map at the normalization level (cf. Proposition A.7) $\tilde{f} : \tilde{X} \to \tilde{Y}$ is a biregular isomorphism.
Hint. Use a form of Zariski's theorem [Mumford–88, III.9, page 288, Original form].

(7) Let $\{X_n\}_{n\geq 1}$ be irreducible varieties over an algebraically closed field with birational proper morphisms f_n with finite fibers:

$$\cdots \to X_3 \overset{f_2}{\to} X_2 \overset{f_1}{\to} X_1.$$

Then show that there exists an $i \geq 1$ such that $f_j : X_{j+1} \to X_j$ is a biregular isomorphism for all $j \geq i$.

Hint. By the previous exercise, all the X_n's have isomorphic normalizations. Take an affine open subset U in X_n. Then $f_n^{-1}(U)$ is again affine (since, by a result of Chevalley, a proper morphism with finite fibers is a finite morphism; in particular, an affine morphism, cf. [Grothendieck-61, Proposition 4.4.2]). Thus, for any affine open subset $U_1 \subset X_1$, $U_n \subset X_n$ is affine, where $U_n := (f_1 \circ \cdots \circ f_{n-1})^{-1} U_1$. Now, we get a chain of k-algebras

$$k[U_1] \subset k[U_2] \subset \cdots \subset k[\tilde{U}_1],$$

where $\tilde{U}_1$ is the normalization of U_1 (thus of any U_n). From the noetherian property, conclude that there exists an i such that $k[U_i] = k[U_{i+1}] = \cdots$. Thus $f_{j|_{U_{j+1}}}$ is an isomorphism for all $j \geq i$.

The following exercise provides an affine variant of the above, where we do not require f_n's to be proper.

(8) Let $\{X_n\}_{n\geq 1}$ be a sequence of affine varieties with bijective morphisms f_n between them:

$$\cdots \to X_3 \xrightarrow{f_2} X_2 \xrightarrow{f_1} X_1.$$

Assume that f_n's are homeomorphisms under the Zariski topology and, moreover, there exist open dense sets $U_n \subset X_n$ such that $f_{n|_{U_{n+1}}} : U_{n+1} \to X_n$ is a biregular isomorphism onto U_n. Then prove that there exists a large enough n_o such that, for all $n \geq n_o$, the maps $X_{n+1} \xrightarrow{f_n} X_n$ are biregular isomorphisms.
Hint. Assume first that X_1 (and hence each X_n) is irreducible. For any n, let $\tilde{f}_n : \tilde{X}_{n+1} \to \tilde{X}_n$ be the induced map of the corresponding normalizations $\pi_n : \tilde{X}_n \to X_n$. By a form of Zariski's theorem (cf. [Mumford–88, III.9, page 288, Original form]), $\tilde{f}_n$ is an open embedding. Thus we can identify $\tilde{X}_{n+1}$ as an open subvariety of $\tilde{X}_n$. Let Σ_n be the closed subvariety of X_n consisting of the nonnormal points. Then $f_n(\Sigma_{n+1}) \subset \Sigma_n$. Conclude from this that

$$\tilde{X}_1 \backslash \tilde{X}_2 \subset \tilde{X}_1 \backslash \tilde{X}_3 \subset \cdots \subset \pi_1^{-1}(\Sigma_1).$$

Conclude further from the above that there exists a large enough k such that

$$(\tilde{X}_1 \backslash \tilde{X}_k)' = (\tilde{X}_1 \backslash \tilde{X}_{k+1})' = \cdots,$$

where $(\tilde{X}_1 \backslash \tilde{X}_k)'$ denotes the union of all the irreducible components of $\tilde{X}_1 \backslash \tilde{X}_k$ of codimension 1. Thus all the irreducible components of $\tilde{X}_n \backslash \tilde{X}_{n+1}$ are of codimension ≥ 2 for any $n \geq k$. Conclude that $k[\tilde{X}_n] = k[\tilde{X}_{n+1}]$, thus $\tilde{X}_n$ is isomorphic with $\tilde{X}_{n+1}$ for all $n \geq k$. From this conclude that X_n is isomorphic with X_{n+1} for all large n.

The general case can be handled by taking the irreducible decomposition $X_1 = \bigcup_{i=1}^p X_1(i)$. This gives rise to the irreducible decomposition $X_n = \bigcup_{i=1}^p X_n(i)$, where $X_n(i) := (f_1 \circ \cdots \circ f_{n-1})^{-1} X_1(i)$. By the above, $X_n(i)$ is isomorphic with $X_{n+1}(i)$ for all $1 \le i \le p$ and all large enough n. Finally, use the finite morphism $\bigsqcup_{i=1}^p X_n(i) \to X_n$, where $\bigsqcup_{i=1}^p X_n(i)$ denotes the disjoint union of the varieties $X_n(i)$.

7.2. Line Bundles on $\mathcal{X}^Y$

We follow the notation from Section 7.1.

7.2.1 Line bundles on $\mathcal{X}^Y$. For any $\lambda \in D_Y^o$, define the algebraic line bundle $\mathcal{L}^Y(-\lambda)$ on $\mathcal{X}^Y$ to be the pullback of the tautological line bundle on $\mathbb{P}(L^{\max}(\lambda))$ via the morphism $i_\lambda : \mathcal{X}^Y \hookrightarrow \mathbb{P}(L^{\max}(\lambda))$ (cf. 7.1.13 and Example 4.2.7(c)). Observe that the identity map $\mathcal{X}^Y \to \mathcal{X}^Y(\lambda)$ is a morphism.

Let

$$\mathfrak{h}^*_{\mathbb{Z},Y} := \{\lambda \in \mathfrak{h}^*_{\mathbb{Z}} : \langle \lambda, \alpha_i^\vee \rangle = 0, \text{ for all } i \in Y\}.$$

Now we define $\mathcal{L}^Y(\lambda)$, for any $\lambda \in \mathfrak{h}^*_{\mathbb{Z},Y}$, as follows: Write $\lambda = \lambda_1 - \lambda_2$ for some $\lambda_1, \lambda_2 \in D_Y^o$ and set

(1) $$\mathcal{L}^Y(\lambda) = \mathcal{L}^Y(-\lambda_2) \otimes \mathcal{L}^Y(-\lambda_1)^*$$

(cf. Example 4.2.7(e)). Since Pic X is an abelian group for any ind-variety X (cf. 4.2.6), from the following lemma, the algebraic line bundle $\mathcal{L}^Y(\lambda)$ on $\mathcal{X}^Y$ does not depend (up to isomorphism) on the choice of the decomposition of λ as above. For any $w \in W$, the restriction of $\mathcal{L}^Y(\lambda)$ to $X_w^Y \subset \mathcal{X}^Y$ is denoted by $\mathcal{L}_w^Y(\lambda)$. When the reference to Y is clear, we drop the superscript Y from $\mathcal{L}_w^Y(\lambda)$.

7.2.2 Lemma. *For $\lambda, \mu \in D_Y^o$, the line bundle $\mathcal{L}(-\lambda) \otimes \mathcal{L}(-\mu)$ is isomorphic with the line bundle $\mathcal{L}(-(\lambda + \mu))$.*

Proof. Recall the diagram (D) from the proof of Proposition 7.1.15. Following the same notation as in (D), it is easy to see that

(1) $$S^*(\mathcal{L}_{V_1 \otimes V_2}) \simeq \mathcal{L}_{V_1} \boxtimes \mathcal{L}_{V_2}, \qquad \text{and}$$

(2) $$\hat{\theta}^*(\mathcal{L}_{V_1 \otimes V_2}) \simeq \mathcal{L}_{V_3|(\mathbb{P}(V_3)\setminus\mathbb{P}(K))},$$

where $\mathcal{L}_V$ denotes the tautological line bundle on $\mathbb{P}(V)$, for a vector space V of countable dimension (cf. Example 4.2.7(c)). Hence, by (4.2.7.1) and (1), (2),

(3) $$(i_\lambda, i_\mu)^*(\mathcal{L}_{V_1} \boxtimes \mathcal{L}_{V_2}) \simeq i^*_{\lambda+\mu}(\mathcal{L}_{V_3}) = \mathcal{L}(-(\lambda + \mu)).$$

But

(4) $$(i_\lambda, i_\mu)^*(\mathcal{L}_{V_1} \boxtimes \mathcal{L}_{V_2}) \simeq i_\lambda^*(\mathcal{L}_{V_1}) \otimes i_\mu^*(\mathcal{L}_{V_2}) = \mathcal{L}(-\lambda) \otimes \mathcal{L}(-\mu),$$

so (3) and (4) together give the lemma. □

7.2.3 Definition (Action of $\mathcal{G}$ on the line bundles $\mathcal{L}(\lambda)$). As in 7.1.19, the left action of $\mathcal{G}$ on $\mathcal{X}^Y$ is regular. In fact, for any $\lambda \in D_Y^o$, the left action of $\mathcal{G}$ on $\mathcal{X}^Y(\lambda)$ is regular, since $L^{\max}(\lambda)$ is a pro-representation of $\mathcal{G}$ and $i_\lambda : \mathcal{X}^Y(\lambda) \hookrightarrow \mathbb{P}(L^{\max}(\lambda))$ is a closed embedding (cf. 7.1.13).

Further, the actions of $\mathcal{G}$ on $L^{\max}(\lambda)$ and $\mathcal{X}^Y$ induce a regular action of $\mathcal{G}$ on the ind-variety $\mathcal{L}(-\lambda)$, making the following diagram commutative.

$$\begin{array}{ccc} \mathcal{G} \times \mathcal{L}(-\lambda) & \longrightarrow & \mathcal{L}(-\lambda) \\ \downarrow & & \downarrow \\ \mathcal{G} \times \mathcal{X}^Y & \longrightarrow & \mathcal{X}^Y. \end{array}$$

Moreover, the action of $\mathcal{G}$ is linear on the fibers. We denote the action of $g \in \mathcal{G}$ on $\mathcal{L}(-\lambda)$ by $\hat{L}_g(-\lambda)$. Now, for an arbitrary $\lambda \in \mathfrak{h}^*_{\mathbb{Z},Y}$, we define a regular action of $\mathcal{G}$ on the ind-variety $\mathcal{L}(\lambda)$ by

(1) $$\hat{L}_g(\lambda) = \hat{L}_g(-\lambda_2) \otimes (\hat{L}_{g^{-1}}(-\lambda_1))^*, \quad \text{for } g \in \mathcal{G},$$

where we write $\lambda = \lambda_1 - \lambda_2$ for $\lambda_1, \lambda_2 \in D_Y^o$. Then again we have a commutative diagram:

$$\begin{array}{ccc} \mathcal{L}(\lambda) & \xrightarrow{\hat{L}_g(\lambda)} & \mathcal{L}(\lambda) \\ \downarrow & & \downarrow \\ \mathcal{X}^Y & \xrightarrow{L_g} & \mathcal{X}^Y, \end{array}$$

where L_g is the action of g on $\mathcal{X}^Y$.

By an argument similar to the proof of Lemma 7.2.2, it is easy to see that the definition (1) of $\hat{L}_g(\lambda)$ does not depend on the choice of the decomposition $\lambda = \lambda_1 - \lambda_2$.

7.2.4 Definition (Certain line bundles on $Z_{\mathfrak{w}}$). For any $\mathfrak{w} \in \mathfrak{W}$, recall the definition of the map $m_{\mathfrak{w}} : Z_{\mathfrak{w}} \to \mathcal{G}/\mathcal{B}$ from 7.1.11. By Proposition 7.1.12, $m_{\mathfrak{w}}$ is a morphism. Thus, for any $\lambda \in \mathfrak{h}^*_{\mathbb{Z}}$, we can pull back the algebraic line bundle $\mathcal{L}^{\emptyset}(\lambda)$ on $\mathcal{G}/\mathcal{B}$ via $m_{\mathfrak{w}}$ to get the line bundle $\mathcal{L}_{\mathfrak{w}}(\lambda)$ on $Z_{\mathfrak{w}}$. These line bundles will play an important role in the next chapter.

7.2.E EXERCISES

(1) Let $Y \subset \{1, \dots, \ell\}$ be any subset. For any $\lambda \in D_{\mathbb{Z}} \cap \mathfrak{h}^*_{\mathbb{Z},Y}$, prove that

$$i_\lambda^*(\mathcal{L}_{L^{\max}(\lambda)}) \cong \mathcal{L}^Y(-\lambda),$$

where $i_\lambda : \mathcal{X}^Y \to \mathbb{P}(L^{\max}(\lambda))$ is the morphism $g \mod \mathcal{P}_Y \mapsto [gv_\lambda]$. Observe that i_λ is indeed a morphism by Proposition 7.1.15(b).

(2) Let $Y \subset \{1, \dots, \ell\}$ be any subset and let $\lambda \in \mathfrak{h}^*_{\mathbb{Z},Y}$. Then show that the line bundle $\mathcal{L}^Y(\lambda)_{|X_w^Y}$ is ample for all $w \in W'_Y$ iff $\lambda \in D_Y^o$.

7.3. Study of the Group $\mathcal{U}^-$

7.3.1 Definition. Similar to the definition of the pro-Lie algebra $\hat{\mathfrak{n}}$ (cf. 6.1.1), define the pro-Lie algebra

$$\hat{\mathfrak{n}}^- := \prod_{\alpha \in \Delta^-} \mathfrak{g}_\alpha \tag{1}$$

under the defining family of ideals $\{\hat{\mathfrak{n}}^-(k)\}_{k \geq 1}$, where

$$\hat{\mathfrak{n}}^-(k) := \prod_{\substack{\alpha \in \Delta^- \\ \operatorname{ht}\alpha \geq k}} \mathfrak{g}_\alpha \tag{2}$$

(cf. (1.2.2.4) for the definition of $\operatorname{ht}\alpha$).

Let $\hat{\mathcal{U}}^-$ be the pro-unipotent pro-group with Lie $\hat{\mathcal{U}}^- \approx \hat{\mathfrak{n}}^-$ (cf. Theorem 4.4.19). For any bracket closed subset $\Theta \subset \Delta^+$, define (analogously to the definition of $\mathcal{U}_\Theta$ in 6.1.1)

$$\hat{\mathcal{U}}_\Theta^- := \operatorname{Exp}(\hat{\mathfrak{n}}_\Theta^-), \tag{3}$$

where $\hat{\mathfrak{n}}_\Theta^-$ is the pro-Lie subalgebra of $\hat{\mathfrak{n}}^-$ defined by

$$\hat{\mathfrak{n}}_\Theta^- := \prod_{\alpha \in \Theta} \mathfrak{g}_{-\alpha}. \tag{4}$$

For any $k \geq 1$, let $\hat{\mathcal{U}}^-(k) := \operatorname{Exp}\hat{\mathfrak{n}}^-(k)$ be the corresponding normal pro-subgroup of $\hat{\mathcal{U}}^-$ (cf. Lemma 6.1.2). For any integrable highest weight $\mathfrak{g}$-module $V = V(\lambda)$ with highest weight $\lambda \in D_{\mathbb{Z}}$, and any $k \geq 1$, define

$$V^{(k)} = V/\Big(\oplus_{\substack{\beta \in Q^+ \\ |\beta| \geq k}} V_{\lambda-\beta}\Big),$$

where Q^+ is as in the beginning of Section 1.2 and $|\beta|$ denotes the principal gradation as in (1.2.2.3).

Then $V^{(k)}$ is a finite-dimensional T-module and, moreover, it is canonically a pro-representation of $\hat{\mathfrak{n}}^-$ such that each element of $\hat{\mathfrak{n}}^-$ acts locally nilpotently on $V^{(k)}$. Hence $V^{(k)}$ is a pro-representation of $\hat{\mathcal{U}}^-$ (cf. Proposition 4.4.26). In fact, $\hat{\mathfrak{n}}^-(k)$ acts trivially on $V^{(k)}$ and hence so is $\hat{\mathcal{U}}^-(k)$, so that the unipotent algebraic group $\hat{\mathcal{U}}^{-(k)} := \hat{\mathcal{U}}^-/\hat{\mathcal{U}}^-(k)$ acts algebraically on $V^{(k)}$.

7.3.2 Lemma. *Fix $k \geq 1$. Then, for any $\lambda \in D_{\mathbb{Z}}$ such that $\langle \lambda, \alpha_i^\vee \rangle \geq k-1$ for all simple coroots $\alpha_i^\vee$, the map*

$$\theta_k(\lambda) : \hat{\mathcal{U}}^{-(k)} \to V^{(k)}, \quad g \mapsto g v_\lambda,$$

is a closed embedding, where $V := L^{\max}(\lambda)$.

Proof. Since the orbits of any unipotent algebraic group in any affine variety are closed (cf. [Borel–91, Proposition 4.10] or the original source [Rosenlicht–61, Theorem 2]), $\operatorname{Im}\theta_k(\lambda)$ is closed in $V^{(k)}$. We next show that $\theta_k(\lambda)$ is injective, i.e., the isotropy subgroup S of $\hat{\mathcal{U}}^{-(k)}$ corresponding to the vector v_λ is trivial:

Let $\mathfrak{s} \subset \hat{\mathfrak{n}}^-/\hat{\mathfrak{n}}^-(k)$ be the annihilator of $v_\lambda \in V^{(k)}$. Then, clearly, $\mathfrak{s}$ is $\mathfrak{h}$-stable under the adjoint action. But, by Lemma 2.1.6 and the assumption on λ (i.e., $\langle \lambda, \alpha_i^\vee \rangle \geq k-1$), the root space $\mathfrak{s}_\beta = 0$ for any $\beta \in \Delta^-$ with $\operatorname{ht}\beta < k$. Hence $\mathfrak{s} = 0$. Since S is connected (S being a unipotent group), it follows that $S = (e)$.

Now the lemma follows from Theorem A.11. Observe that $\operatorname{Im}\theta_k(\lambda)$ is smooth; in particular, normal since $\hat{\mathcal{U}}^{-(k)}$ acts transitively on $\operatorname{Im}\theta_k(\lambda)$. □

Recall from Theorem 6.2.3 that any integrable highest weight $\mathfrak{g}$-module $V = V(\lambda)$, with highest weight $\lambda \in D_{\mathbb{Z}}$, is a $\mathcal{G}$-module; in particular, is a $\mathcal{U}^-$-module. Moreover, for any $k \geq 1$, this induces a $\mathcal{U}^-$-module structure on $V^{(k)}$ (cf. (6.2.8.3)).

7.3.3 Lemma. *There exists an embedding of groups $i : \mathcal{U}^- \hookrightarrow \hat{\mathcal{U}}^-$ such that $i(\mathcal{U}_\beta) = \operatorname{Exp}(\mathfrak{g}_\beta)$, for any $\beta \in \Delta^-_{\text{re}}$.*

Moreover, for any integrable highest weight $\mathfrak{g}$-module $V = V(\lambda)$ with $\lambda \in D_{\mathbb{Z}}$ and $k \geq 1$, the following diagram is commutative:

$$\begin{array}{ccc} \mathcal{U}^- & \xrightarrow{\;i\;} & \hat{\mathcal{U}}^- \\ & \searrow \quad \swarrow & \\ & \operatorname{Aut} V^{(k)} & . \end{array}$$

Proof. Let F be the free product of the additive groups $\{\mathfrak{g}_\beta\}_{\beta \in \Delta^-_{\text{re}}}$. Then there is a unique group homomorphism $\hat{i} : F \to \hat{\mathcal{U}}^-$ such that $\hat{i}_{|\mathfrak{g}_\beta} = \operatorname{Exp}_{|\mathfrak{g}_\beta}$, for all $\beta \in \Delta^-_{\text{re}}$, where $\operatorname{Exp} : \hat{\mathfrak{n}}^- \to \hat{\mathcal{U}}^-$ is the exponential map of the pro-group $\hat{\mathcal{U}}^-$ (cf. 4.4.16). Similarly, there is a unique group homomorphism $\pi : F \to \mathcal{U}^-$ such that $\pi_{|\mathfrak{g}_\beta} = \operatorname{Exp}_{|\mathfrak{g}_\beta}$, where Exp is the exponential map defined in Proposition 6.2.11. Then clearly π is surjective. By 4.4.23 and Proposition 6.2.11, we have the following commutative diagram:

$$\begin{array}{ccccc} & & \mathcal{U}^- & & \\ & \overset{\pi}{\nearrow} & & \overset{\rho}{\searrow} & \\ F & & & & \operatorname{Aut} V^{(k)} \\ & \underset{\hat{i}}{\searrow} & & \underset{\hat{\rho}}{\nearrow} & \\ & & \hat{\mathcal{U}}^- & & \end{array}$$

Thus, for $g \in \operatorname{Ker} \pi$, we have $\hat{\rho}\,\hat{i}(g) = I$, and hence by Lemma 7.3.2 (taking λ and k sufficiently large), we see that $\hat{i}(g) = 1$. Hence, the group homomorphism $\hat{i}$ factors through $\mathcal{U}^-$ to give the group homomorphism $i : \mathcal{U}^- \to \hat{\mathcal{U}}^-$. Further, by Theorems 5.2.3(e), 6.2.8 and Lemma 7.1.2, for any $g \neq 1 \in \mathcal{U}^-$ and any $\lambda \in D^o$, $gv_\lambda \neq v_\lambda$ as elements of $V(\lambda)$. Thus, taking k sufficiently large (depending on g and $V(\lambda)$), we see that $\rho(g) \neq I$. Hence $i(g) \neq 1$. This proves the lemma. □

Recall that, for any $v \in W'_Y$, the subsets $\mathcal{B}^v_Y$ and X^v_Y of $\mathcal{G}/\mathcal{P}_Y$ are defined just before 7.1.21 and X^Y_v is defined in 7.1.13.

7.3.4 Definition. For any $v \in W'_Y$, set

$$\partial X^v_Y := X^v_Y \backslash \mathcal{B}^v_Y.$$

Then, for $w \in W'_Y$,

$$\begin{aligned} X^Y_w \cap \partial X^v_Y &= \bigcup_{\substack{v<\theta \\ \theta \in W'_Y}} (X^Y_w \cap \mathcal{B}^\theta_Y) \\ &= \bigcup_{\substack{v<\theta \\ \theta \in W'_Y}} (X^Y_w \cap X^\theta_Y) \\ &= \bigcup_{\substack{v<\theta\leq w \\ \theta \in W'_Y}} (X^Y_w \cap X^\theta_Y), \quad \text{by Lemma 7.1.22(b).} \end{aligned}$$

So, by Proposition 7.1.21, for any $w \in W'_Y$, $X^Y_w \cap \partial X^v_Y$ is (Zariski) closed in X^Y_w (being a finite union of closed sets) and hence ∂X^v_Y is closed in $\mathcal{G}/\mathcal{P}_Y$. In particular, $X^Y_w \cap \mathcal{B}^v_Y = (X^Y_w \cap X^v_Y) \backslash (X^Y_w \cap \partial X^v_Y)$ is a locally closed subset of X^Y_w. Hence, it acquires a subvariety structure from X^Y_w; and $\mathcal{B}^e_Y = (\mathcal{G}/\mathcal{P}_Y) \backslash \partial X^e_Y$ is an open subset of $\mathcal{G}/\mathcal{P}_Y$. Thus, $X^Y_w \cap v\mathcal{B}^e_Y$ is an open subvariety of X^Y_w.

7.3.5 Lemma. *For any $v, w \in W'_Y$ with $v \leq w$, the varieties $X^Y_w \cap \mathcal{B}^v_Y$ and $X^Y_w \cap v\mathcal{B}^e_Y$ are both affine.*

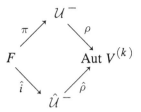

Proof. We first prove that $X^Y_w \cap v\mathcal{B}^e_Y \simeq (v^{-1}X^Y_w) \cap \mathcal{B}^e_Y$ is affine: Choose a large enough $\lambda \in D^o_Y$ such that, as a variety, $X^Y_w = X^Y_w(\lambda)$ (cf. Definition 7.1.19). Fix a nonzero highest weight vector $e_\lambda \in L^{\max}(\lambda)$, and consider the linear function $e_\lambda^* : L^{\max}(\lambda) \to \mathbb{C}$, taking $e_\lambda \mapsto 1$ and all other weight vectors to 0. This gives rise to a section σ of the line bundle $\mathcal{L}^Y(\lambda)_{|v^{-1}X_w^Y}$ (cf. 7.2.1). By the Birkhoff decomposition (Theorem 6.2.8) together with (6.2.8.3), it is easy to see that $(v^{-1}X_w^Y)\backslash Z(\sigma) = (v^{-1}X_w^Y) \cap \mathcal{B}_Y^e$, where $Z(\sigma)$ is the zero set of σ. In particular, $(v^{-1}X_w^Y) \cap \mathcal{B}_Y^e$ is affine by Lemma A.20.

The proof for $X_w^Y \cap \mathcal{B}_Y^v$ is similar. Consider the projective variety $D := X_w^Y \cap \overline{\mathcal{B}_Y^v}$, a closed subvariety of X_w^Y, and let $e_{v\lambda}^* : L^{\max}(\lambda) \to \mathbb{C}$ be the linear function taking $e_{v\lambda} \mapsto 1$ and all other weight vectors to 0 (where $e_{v\lambda}$ is a fixed weight vector of weight $v\lambda$). By (7.1.21.2) and Exercise 7.1.E.5, it is easy to see that the section σ_v of the line bundle $\mathcal{L}^Y(\lambda)_{|D}$, determined by $e_{v\lambda}^*$, satisfies

$$D\backslash Z(\sigma_v) = X_w^Y \cap \mathcal{B}_Y^v. \qquad \square$$

7.3.6 Definition. Recall from Theorems 5.2.3(e) and 6.2.8 that $\mathcal{U}^- \hookrightarrow \mathcal{G}/\mathcal{B}$ under $g \mapsto g\mathcal{B}$. We will freely identify $\mathcal{U}^-$ as a subset of $\mathcal{G}/\mathcal{B}$ under this embedding. Also recall the filtration $\{X_n\}$ of $\mathcal{X} = \mathcal{G}/\mathcal{B}$ from 7.1.13. This gives rise to the filtration $\{\mathcal{U}_n^-\}_{n\in\mathbb{Z}_+}$ of $\mathcal{U}^-$ by

$$\mathcal{U}_n^- := \mathcal{U}^- \cap X_n. \tag{1}$$

Since $\mathcal{X}$ is an ind-variety (cf. 7.1.19) and $\mathcal{U}^-$ is an open subset of $\mathcal{X}$ (cf. 7.3.4), $\mathcal{U}^-$ is canonically an ind-variety under the filtration $\{\mathcal{U}_n^-\}$. Moreover, by Lemma 7.3.5, it is an affine ind-variety.

7.3.7 Proposition. *For any $n \in \mathbb{Z}_+$ and $k \geq 1$, the map*

$$i(n,k) : \mathcal{U}_n^- \to \hat{\mathcal{U}}^{-(k)}$$

(induced by the inclusion i of Lemma 7.3.3 followed by the projection) is a morphism. Moreover, for any fixed n there exists a large enough $k(n)$ such that for all $k \geq k(n)$, $i(n,k)$ is a closed embedding.

Proof. Choose a large enough $\lambda \in D^o$ such that $X_n = X_n(\lambda)$, i.e., for $V = L^{\max}(\lambda)$, $i_\lambda(n) : X_n \hookrightarrow \mathbb{P}(V)$ is a closed embedding (cf. 7.1.19), where $i_\lambda(n) = i_{\lambda|X_n}$ (cf. 7.1.13).

Let $S \subset \mathbb{P}(V)$ be the open subset consisting of those lines $[v]$ such that the λ-th weight component of v is nonzero. Then $i_\lambda(n)^{-1}S = \mathcal{U}_n^-$ by Theorem 6.2.8 and (6.2.8.3). In particular,

$$\bar{i}_\lambda(n) : \mathcal{U}_n^- \to V, \quad g \mapsto gv_\lambda,$$

is a closed embedding, where v_λ is a highest weight vector of V. Observe that, by (6.2.8.3), the λ-th weight component of any $v \in \operatorname{Im} \bar{i}_\lambda(n)$ is 1. Choose a large enough $k(n) = k_\lambda(n)$ such that, for all $k \geq k(n)$, $\bar{i}_\lambda(n,k) : \mathcal{U}_n^- \to V^{(k)}$ (obtained by composing $\bar{i}_\lambda(n)$ with the projection) is also a closed embedding. By 7.3.1 the unipotent algebraic group $\hat{\mathcal{U}}^{-(k)}$ acts algebraically on $V^{(k)}$, inducing a closed embedding

$$\hat{i}_\lambda(k) : \hat{\mathcal{U}}^{-(k)}/H \hookrightarrow V^{(k)}, \quad gH \mapsto gv_\lambda,$$

where H is the isotropy subgroup of $\hat{\mathcal{U}}^{-(k)}$ for $v_\lambda \in V^{(k)}$ (cf. proof of Lemma 7.3.2). Consider the commutative diagram (with π the standard projection map):

$$\begin{array}{ccccc} \mathcal{U}_n^- & \xrightarrow{i(n,k)} & \hat{\mathcal{U}}^{-(k)} & \xrightarrow{\pi} & \hat{\mathcal{U}}^{-(k)}/H \\ & \bar{i}_\lambda(n,k) \searrow & & \swarrow \hat{i}_\lambda(k) & \\ & & V^{(k)} & & \end{array}.$$

Since both maps $\bar{i}_\lambda(n,k)$ and $\hat{i}_\lambda(k)$ are closed embeddings, so is the map $\pi \circ i(n,k)$. Note that, so far, we have not proved that $i(n,k)$ is a morphism.

Fix any $k' \geq 1$. We next show that $i(n,k')$ is a morphism: Take any $\mu \in D_\mathbb{Z}$ with $\langle \mu, \alpha_i^\vee \rangle \geq k' - 1$ for all the simple coroots $\alpha_i^\vee$. Then, we have the closed embedding (by Lemma 7.3.2):

$$\theta_{k'}(\mu) : \hat{\mathcal{U}}^{-(k')} \hookrightarrow L^{(k')},$$

making the following diagram commutative (where $L := L^{\max}(\mu)$).

$$\begin{array}{ccc} \mathcal{U}_n^- & \xrightarrow{i(n,k')} & \hat{\mathcal{U}}^{-(k')} \\ & \bar{i}_\mu(n,k') \searrow & \big\downarrow \theta_{k'}(\mu) \\ & & L^{(k')} \end{array}.$$

By Proposition 7.1.15(b), $\bar{i}_\mu(n,k')$ is a morphism and hence so is $i(n,k')$.

Since $\pi \circ i(n,k)$ is a closed embedding (for all $k \geq k(n)$), by [Hartshorne–77, Chap. II, Exercise 3.11(a)], $i(n,k)$ is a closed embedding. This proves the proposition. □

7.3.8 Corollary. *The affine ind-variety $\mathcal{U}^-$ together with its group structure is, in fact, an (affine) ind-group. Further, for any $k \geq 1$, the ind-group morphism $i(k) : \mathcal{U}^- \to \hat{\mathcal{U}}^{-(k)}$ (obtained by the inclusion i followed by the projection) is surjective. In particular, there exists $n(k)$ such that*

$$i(k)\big(\mathcal{U}_{n(k)}^-\big) = \hat{\mathcal{U}}^{-(k)}.$$

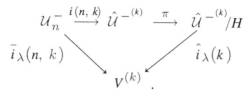

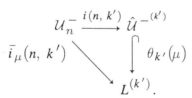

Proof. To prove the first assertion, we need to show that the map $\mu : \mathcal{U}^- \times \mathcal{U}^- \to \mathcal{U}^-$, $(g, h) \mapsto gh^{-1}$, is a morphism. For any $n \in \mathbb{Z}_+$, by Theorem 5.1.3(d), there exists $m = m(n)$ such that $\mu(\mathcal{U}_n^- \times \mathcal{U}_n^-) \subset \mathcal{U}_m^-$. Now, the assertion, that μ is a morphism, follows from Proposition 7.3.7 and Lemma 4.1.2, since $\hat{\mathcal{U}}^{-(k)}$ is an algebraic group for all k.

We next prove that $i(k)$ is surjective: Consider the abstract subgroup H of $\hat{\mathcal{U}}^{-(k)}$ generated by the one-parameter subgroups $\{i(k)\mathcal{U}_{-\alpha_i} = \operatorname{Exp} \mathfrak{g}_{-\alpha_i}; 1 \leq i \leq \ell\}$. Then the Zariski closure $\bar{H} = \hat{\mathcal{U}}^{-(k)}$, since $\hat{\mathcal{U}}^{-(k)}$ is connected and Lie $\bar{H} =$ Lie $(\hat{\mathcal{U}}^{-(k)})$ (as $\{\mathfrak{g}_{-\alpha_i}; 1 \leq i \leq \ell\}$ generates Lie $(\hat{\mathcal{U}}^{-(k)})$ as a Lie algebra). Moreover, by [Borel–91, Chap. I, Proposition 2.2], $H = \bar{H}$. But since $\operatorname{Im} i(k)$ is a subgroup, we get the surjectivity of $i(k)$.

To prove the "In particular" statement, write $\hat{\mathcal{U}}^{-(k)} = \bigcup_{n \geq 0} i(k)\,(\mathcal{U}_n^-)$. By Chevalley's theorem A.10, each $i(k)(\mathcal{U}_n^-)$ is a constructible set in $\hat{\mathcal{U}}^{-(k)}$. So, by Exercise 7.3.E.2, there exists a large enough $n(k)$ such that $\hat{\mathcal{U}}^{-(k)} = i(k)\big(\mathcal{U}_{n(k)}^-\big)$. □

7.3.9 Lemma. *For any integrable highest weight $\mathfrak{g}$-module $V = V(\lambda)$, with $\lambda \in D_{\mathbb{Z}}$, the map $\rho : \mathcal{U}^- \times V(\lambda) \to V(\lambda)$, induced by the restriction of the representation of $\mathcal{G}$ on $V(\lambda)$ (cf. Theorem 6.2.3), is a morphism of ind-varieties, i.e., $V(\lambda)$ is an algebraic representation of the ind-group $\mathcal{U}^-$ (cf. Definition 4.2.3).*

Proof. Define, for $k \geq 1$, $V_k = \bigoplus_{\substack{\beta \in Q^+ \\ |\beta| < k}} V_{\lambda-\beta} \subset V$. It is easy to see that, for any $n \in \mathbb{Z}_+$ and $k \geq 1$, there exists k' such that $\rho(\mathcal{U}_n^- \times V_k) \subset V_{k'}$. Take large enough $N \geq k$, $N \geq k'$ such that $i(n, N) : \mathcal{U}_n^- \to \hat{\mathcal{U}}^{-(N)}$ is an embedding (cf. Proposition 7.3.7). Let $V_k \to V^{(N)}$ be the inclusion of V_k in V followed by projection. Using Lemma 4.1.2, the lemma follows from the following commutative diagram (cf. Lemma 7.3.3):

$$\begin{array}{ccc} \mathcal{U}_n^- \times V_k & \xrightarrow{\ \rho\ } & V_{k'} \\ \cap\big\downarrow & & \cap\big\downarrow \\ \hat{\mathcal{U}}^{-(N)} \times V^{(N)} & \longrightarrow & V^{(N)}, \end{array}$$

since the canonical projection map $V_p \to V^{(N)}$ is a closed embedding for any $p \leq N$ and the bottom horizontal map is a morphism by 7.3.1. □

For any $v \in W$, recall the definition of the unipotent group $\mathcal{U}_v = \mathcal{U}_{\Phi_v}$ from Example 6.1.5(b). As in Definition 7.3.4, both $X_w^Y \cap \mathcal{B}_Y^v$ and $X_w^Y \cap v\mathcal{B}_Y^e$ are locally closed subsets of X_w^Y.

7.3.10 Lemma. *For $v, w \in W'_Y$ with $v \le w$, the map*

$$\theta_{v,w} : \mathcal{U}_v \times \left(X^Y_w \cap \mathcal{B}^v_Y\right) \to X^Y_w \cap v\mathcal{B}^e_Y ,$$

defined by

$$\theta_{v,w}(g, x) = gx, \quad \textit{for } g \in \mathcal{U}_v \textit{ and } x \in X^Y_w \cap \mathcal{B}^v_Y ,$$

is a biregular T-equivariant isomorphism, where T acts by conjugation on $\mathcal{U}_v$ and by left multiplication on the other two factors.

Proof. By Exercise 7.3.E.1, the map

$$\theta_v : \mathcal{U}_v \times \mathcal{B}^v_Y \to v\mathcal{B}^e_Y, \quad (g, x) \mapsto gx,$$

is a bijection (for any $v \in W'_Y$). Restricting the map θ_v to $\mathcal{U}_v \times (X^Y_w \cap \mathcal{B}^v_Y)$,we get the map $\theta_{v,w}$. Observe that Im $\theta_{v,w} = X^Y_w \cap v\,\mathcal{B}^e_Y$, since X^Y_w is $\mathcal{B}$; in particular $\mathcal{U}_v$, stable and θ_v is surjective. Since θ_v is injective, so is $\theta_{v,w}$. Clearly θ_v is T-equivariant.

Now the multiplication map $\mathcal{U}_v \times X^Y_w \to X^Y_w$ being a morphism (cf. 7.1.19), so is the map $\theta_{v,w}$.

Finally, we show that $\theta^{-1}_{v,w}$ is a morphism. For the notational convenience, we assume that $Y = \emptyset$. (The proof for general Y is the same.)

Choose a large enough k (depending upon v and w) such that both the canonical morphisms

$$\pi_1 : v^{-1}X_w \cap \left(\mathcal{U}^-\mathcal{B}/\mathcal{B}\right) \to \hat{\mathcal{U}}^{-^{(k)}} , \text{ and}$$

$$\pi_2 : v^{-1}X_w \cap \left((v^{-1}\mathcal{U}^- v \cap \mathcal{U}^-)\,\mathcal{B}/\mathcal{B}\right) \to \hat{\mathcal{U}}_v^{-^{(k)}}$$

are closed embeddings (cf. Proposition 7.3.7 and its proof), where $\hat{\mathcal{U}}_v^{-^{(k)}}$ is the unipotent group with Lie algebra

$$\Big(\bigoplus_{\alpha\in\Delta^+\cap v^{-1}\Delta^+} \mathfrak{g}_{-\alpha}\Big)\Big/\Big(\bigoplus_{\alpha\in\Delta^+,\,|\alpha|\ge k} \mathfrak{g}_{-\alpha}\Big).$$

(Observe that, for large enough k,

$$\bigoplus_{\alpha\in\Delta^+\cap v^{-1}\Delta^+} \mathfrak{g}_{-\alpha} \supset \bigoplus_{\alpha\in\Delta^+,|\alpha|\ge k} \mathfrak{g}_{-\alpha} ,$$

and, by Exercise 7.3.E.3, $(v^{-1}\mathcal{U}^- v \cap \mathcal{U}^-)\mathcal{B}/\mathcal{B}$ is a closed subset of $\mathcal{U}^-\mathcal{B}/\mathcal{B}$.)

We have the following "negative" analogue of (5.2.3.7) for any $w \in W$:

(1) $$w^{-1}\mathcal{U}^- w = ((w^{-1}\mathcal{U}^- w) \cap \mathcal{U}^-) \odot ((w^{-1}\mathcal{U}^- w) \cap \mathcal{U}).$$

Using the left translation by v^{-1} and (1), the map $\theta_{v,w}$ transforms to the map $\hat{\theta}_{v,w}$ and, moreover, fits in the following commutative diagram:

$$\begin{array}{ccc}
\mathcal{U}^-_{v^{-1}} \times \left(v^{-1}X_w \cap \left((v^{-1}\mathcal{U}^- v \cap \mathcal{U}^-)\mathcal{B}/\mathcal{B}\right)\right) & \xrightarrow{\hat{\theta}_{v,w}} & v^{-1}X_w \cap (\mathcal{U}^-\mathcal{B}/\mathcal{B}) \\
{\scriptstyle I\times\pi_2}\big\downarrow & & \big\downarrow{\scriptstyle \pi_1} \\
\mathcal{U}^-_{v^{-1}} \times \hat{\mathcal{U}}_v^{-(k)} & \xrightarrow{\mu} & \hat{\mathcal{U}}^{-(k)},
\end{array}$$

where $\mathcal{U}^-_{v^{-1}}$ is the unipotent group with Lie algebra $\bigoplus_{\alpha\in\Delta^-\cap v^{-1}\Delta^+}\mathfrak{g}_\alpha$ and μ is induced by the multiplication map. By the analogue of Lemma 6.1.4 for the unipotent group $\hat{\mathcal{U}}^{-(k)}$, μ is a biregular isomorphism, and both the vertical maps are closed embeddings (since so are π_1 and π_2). From this we obtain that $\hat{\theta}^{-1}_{v,w}$ is a regular map and hence so is $\theta^{-1}_{v,w}$. This proves the lemma. □

7.3.E EXERCISES

(1) For any subset $Y \subset \{1, \cdots, \ell\}$ and $v \in W'_Y$, the map

$$\theta_v : \mathcal{U}_v \times \mathcal{B}^v_Y \to v\mathcal{B}^e_Y\,,\ (g, x) \mapsto gx,$$

is a bijection.

(2) Let X be a variety over an uncountable (algebraically closed) field k. Assume that $X = \cup_{n\geq 1} X_n$, where all the $X_n^{'s}$ are constructible subsets. Then prove that X is a finite union $X = \cup_{n=1}^{n_o} X_n$, for some $n_o \geq 1$.
Hint. We can assume that X is irreducible. Write $X = \cup_{n\geq 1}\overline{X_n}$. Then, by [Borho–Jantzen–77], $X = \cup_{n=1}^{m}\overline{X_n}$, for some m. (For $k = \mathbb{C}$, this also follows from the Baire Category Theorem for locally compact spaces.) Further, by Exercise A.E.3, $\cup_{n=1}^{m} X_n$ contains a nonempty open subset U of X. Thus $X\setminus U$ is of dimension $<$ dim X. Now complete the exercise by induction on dim X.

(3) Show that, for any subset $Y \subset \{1, \ldots, \ell\}$ and $v \in W'_Y$, $\mathcal{B}^v_Y$ is a closed subset of the open subset $v\mathcal{B}^e_Y$ of $\mathcal{X}^Y$.
Hint. Use (7.1.21.2) and the injective map $\mathcal{X}^Y \hookrightarrow \mathbb{P}(L^{\max}(\lambda))$, for Y-regular λ.

7.4. Study of the Group $\mathcal{G}^{\min}$ Defined by Kac–Peterson

Let $\mathcal{G}$ be a Kac–Moody group with standard Borel subgroup $\mathcal{B}$ and standard maximal torus $T \subset \mathcal{B}$ (cf. Theorem 6.1.17). For any $Y \subset \{1, \ldots, \ell\}$, let $\mathcal{P}_Y$ be the standard parabolic subgroup of $\mathcal{G}$ (cf. 6.1.18).

7.4.1 Definition. Following Kac–Peterson, let $\mathcal{G}^{\min} \subset \mathcal{G}$ be the subgroup generated by the one-parameter subgroups $\{\mathcal{U}_\alpha\}_{\alpha\in\Delta_{\text{re}}}$ and T (cf. 6.2.7 for the definition of $\mathcal{U}_\alpha$). It should however be mentioned that Kac–Peterson originally defined $\mathcal{G}^{\min}$ intrinsically (without making use of the group $\mathcal{G}$). Define

$$\mathcal{B}^{\min} := \mathcal{B} \cap \mathcal{G}^{\min}.$$

By 6.1.16, $N \subset \mathcal{G}^{\min}$

The following result follows from Theorem 6.1.17, where the notation is as in loc. cit.

7.4.2 Lemma. *($\mathcal{G}^{\min}$, $\mathcal{B}^{\min}$, N, S) is a Tits system. In particular,*

(1)
$$\mathcal{G}^{\min} = \bigsqcup_{w \in W} \mathcal{B}^{\min} w \mathcal{B}^{\min}.$$

More generally, for any $Y, Y' \subset \{1, \dots, \ell\}$,

(2)
$$\mathcal{G}^{\min} = \bigsqcup_{w \in W_Y \backslash W / W_{Y'}} \mathcal{P}_Y^{\min} w \mathcal{P}_{Y'}^{\min}, \text{ where } \mathcal{P}_Y^{\min} := \mathcal{P}_Y \cap \mathcal{G}^{\min}.$$

Proof. By (6.2.7.1), $\mathcal{B}^{\min}$ and N generate $\mathcal{G}^{\min}$ as a group. Hence, to prove the first part of the lemma, it suffices to show that for any $w \in W$,

(3)
$$\mathcal{B} w \mathcal{B} \cap \mathcal{G}^{\min} = \mathcal{B}^{\min} w \mathcal{B}^{\min}.$$

By Exercise 6.2.E.1,

(4)
$$\mathcal{B} w \mathcal{B} = \mathcal{U}_{\Phi_w} w \mathcal{B}.$$

By Lemma 6.1.4,

(5)
$$\mathcal{U}_{\Phi_w} \subset \mathcal{B}^{\min}.$$

By (4) and (5), (3) follows easily. Now (1) and (2) follow from Theorem 5.1.3(c). □

7.4.3 Lemma. *Let H be the subgroup of $\mathcal{G}^{\min}$ generated by T and the one-parameter subgroups $\{\mathcal{U}_\alpha\}_{\alpha \in \Delta^+_{\mathrm{re}}}$. Then*

(1)
$$H = \mathcal{B}^{\min}.$$

Proof. We first show that

(2)
$$\mathcal{G}^{\min} = \bigcup_{w \in W} HwH :$$

For any simple reflection s_i, write $H = \mathcal{U}_{\alpha_i} H^i$, where H^i is a subgroup of H containing T stable under the conjugation by s_i (cf. Lemma 6.1.3). Thus, for any $w \in W$,

$$wHs_i = w\mathcal{U}_{\alpha_i} H^i s_i = w\mathcal{U}_{\alpha_i} s_i H^i.$$

But $w\mathcal{U}_{\alpha_i} s_i H^i = \mathcal{U}_{w\alpha_i} w s_i H^i \subset Hws_iH$, if $w\alpha_i > 0$.

From the SL_2-theory, we have

$$s_i^{-1}\mathcal{U}_{\alpha_i} s_i \subset \big(\mathcal{U}_{\alpha_i} s_i \mathcal{U}_{\alpha_i} \cup \mathcal{U}_{\alpha_i}\big) \cdot T.$$

Hence,

$$\begin{aligned} w\mathcal{U}_{\alpha_i} s_i H^i &\subset \big(ws_i\mathcal{U}_{\alpha_i} s_i \mathcal{U}_{\alpha_i} H^i\big) \cup \big(ws_i\mathcal{U}_{\alpha_i} H^i\big) \\ &\subset \big(\mathcal{U}_{-w\alpha_i} wH\big) \cup \big(\mathcal{U}_{-w\alpha_i} ws_i H\big) \\ &\subset HwH \cup Hws_iH, \qquad \text{if } w\alpha_i < 0. \end{aligned}$$

So, in either case, $wHs_i \subset HwH \cup Hws_iH$. Hence $\bigcup_{w\in W} HwH$ is a subgroup of $\mathcal{G}^{\min}$, which contains $\{\mathcal{U}_\alpha\}_{\alpha\in\Delta_{\mathrm{re}}}$ and T. Thus (2) follows. Now (1) follows from the Bruhat decomposition (7.4.2.1) and (2), since $H \subset \mathcal{B}^{\min}$. □

7.4.4 Definition. Similar to the "maximal" integrable highest weight $\mathfrak{g}$-module $L^{\max}(\lambda)$ (cf. 2.1.5), define the *"maximal" integrable lowest weight $\mathfrak{g}$-module* $L^{\max}(\lambda)^-$, for any $\lambda \in D$, by

$$(1) \qquad L^{\max}(\lambda)^- := \frac{U(\mathfrak{g}) \otimes_{U(\mathfrak{b}^-)} \mathbb{C}_{-\lambda}}{\sum_{i=1}^{\ell} U(\mathfrak{g})\big(e_i^{\lambda(\alpha_i^\vee)+1} \otimes 1\big)}, \qquad \text{where } \mathfrak{b}^- := \mathfrak{n}^- \oplus \mathfrak{h}.$$

It is easy to see that $L^{\max}(\lambda)^-$ is an integrable $\mathfrak{g}$-module generated by the lowest weight vector $1 \otimes 1$ of weight $-\lambda$ (cf. proof of Lemma 2.1.7) and, moreover, it has a unique irreducible $\mathfrak{g}$-module quotient $L(\lambda)^-$ (follow the same proof as that of Lemma 2.1.2). Further, if $\lambda \in D_{\mathbb{Z}}$, the actions of T (via Lemma 6.1.7) and the one-parameter groups $\{\mathcal{U}_\alpha\}_{\alpha\in\Delta_{\mathrm{re}}}$ (obtained by exponentiating the Lie algebra actions of $\mathfrak{g}_\alpha$'s) on $L^{\max}(\lambda)^-$ together give rise to an action of $\mathcal{G}^{\min}$ (on $L^{\max}(\lambda)^-$).

For any finite family $\Lambda := \{\lambda_1, \dots, \lambda_n\} \subset D_{\mathbb{Z}}$, define the map

$$\psi_\Lambda : \mathcal{G}^{\min} \to \bigoplus_{i=1}^{n} \big(L^{\max}(\lambda_i) \oplus L^{\max}(\lambda_i)^-\big)$$

by

$$(2) \qquad \psi_\Lambda(g) = \sum_{i=1}^{n} (gv_{\lambda_i} + gv_{\lambda_i}^*),$$

where v_{λ_i}, resp. $v_{\lambda_i}^*$, is a fixed nonzero highest, resp. lowest, weight vector of $L^{\max}(\lambda_i)$, resp. $L^{\max}(\lambda_i)^-$.

For any $m \in \mathbb{Z}_+$, define $\mathcal{G}_m^{\min} := \tilde{X}_m \cap \tilde{X}_m^-$, where $\tilde{X}_m := \cup_{\ell(w)\le m} \mathcal{B}^{\min} w \mathcal{B}^{\min}$, $\tilde{X}_m^- := \cup_{\ell(w)\le m} \mathcal{B}^- w \mathcal{B}^-$, and $\mathcal{B}^- \subset \mathcal{G}^{\min}$ is the standard opposite Borel subgroup as in 6.2.7. By Lemma 7.4.2 and Exercise 7.4.E.1, $\mathcal{G}^{\min} = \cup_{m\ge 0} \mathcal{G}_m^{\min}$.

7.4.5 Proposition. *Let $\Lambda = \{\lambda_1, \dots, \lambda_n\}$ be as above. Now assume that $(e^{\lambda_1}, \dots, e^{\lambda_n}) : T \to (\mathbb{C}^*)^n$ is surjective, and no $w \neq 1 \in W$ fixes all the λ_i's. Then, for any $m \geq 0$, the image $\psi_\Lambda(\mathcal{G}_m^{\min})$ is a Zariski closed subset of a finite-dimensional vector subspace of $V_\Lambda := \bigoplus_{i=1}^n (L^{\max}(\lambda_i) \oplus L^{\max}(\lambda_i)^-)$. Thus, Im ψ_Λ is Zariski closed in V_Λ.*

Proof. Clearly,
(1)

$$\psi_\Lambda(\mathcal{G}_m^{\min}) \subset V_\Lambda(m) := \sum_{\ell(v),\ell(w)\leq m} \left(\bigoplus_{i=1}^{n} \left(L^{\max}(\lambda_i)_{\geq w\lambda_i} \oplus (L^{\max}(\lambda_i)^-)_{\leq -v\lambda_i} \right) \right),$$

where $L^{\max}(\lambda_i)_{\geq w\lambda_i} := \bigoplus_{\mu \geq w\lambda_i} L^{\max}(\lambda_i)_\mu$, and $(L^{\max}(\lambda_i)^-)_{\leq -v\lambda_i}$ has a similar meaning. In particular, $\psi_\Lambda(\mathcal{G}_m^{\min})$ is contained in a finite-dimensional vector subspace of V_Λ.

By (7.4.2.4) and (7.4.2.5),

$$(2) \qquad \mathcal{G}^{\min}/\mathcal{B}^{\min} \xrightarrow{\sim} \mathcal{G}/\mathcal{B}.$$

For any $\lambda \in D_\mathbb{Z}$, as in Section 7.1, consider the map $i_\lambda : \mathcal{G}^{\min}/\mathcal{B}^{\min} \to \mathbb{P}(L^{\max}(\lambda))$, $g\mathcal{B}^{\min} \mapsto [gv_\lambda]$, and let $\theta_\Lambda : \mathcal{G}^{\min}/\mathcal{B}^{\min} \to \prod_{i=1}^n \mathbb{P}(L^{\max}(\lambda_i))$ be the corresponding product map. Analogously, we have the map $\theta_\Lambda^- : \mathcal{G}^{\min}/\mathcal{B}^- \to \prod_{i=1}^n \mathbb{P}(L^{\max}(\lambda_i)^-)$. Equipping $\mathcal{G}^{\min}/\mathcal{B}^{\min}$ with the stable projective ind-variety structure (cf. 7.1.19), θ_Λ is a morphism of ind-varieties (cf. Proposition 7.1.15(b)). Similarly, θ_Λ^- is a morphism with the analogous ind-variety structure on $\mathcal{G}^{\min}/\mathcal{B}^-$.

For any $\mathfrak{g}$-module $M \in \mathcal{O}$, as in 2.1.1, let $M^\vee \subset M^*$ be the restricted dual, i.e., $M^\vee$ is the direct sum of $\mathfrak{h}$-weight spaces of M^*. Then $M^\vee$ is a $\mathfrak{g}$-submodule of the dual $\mathfrak{g}$-module M^*. It is easy to see that, for $\lambda \in D$, as $\mathfrak{g}$-modules

$$(*) \qquad L(\lambda)^- \simeq L(\lambda)^\vee.$$

Consider the following $\mathfrak{g}$-module maps, for any $\lambda \in D$,

$$L^{\max}(\lambda) \otimes L^{\max}(\lambda)^- \to L^{\max}(\lambda) \otimes L(\lambda)^- \simeq$$
$$L^{\max}(\lambda) \otimes L(\lambda)^\vee \hookrightarrow L^{\max}(\lambda) \otimes L^{\max}(\lambda)^\vee \to \mathbb{C},$$

where the last map is given by $v \otimes f \mapsto f(v)$ and the other maps are the obvious maps. We denote the $\mathfrak{g}$-invariant pairing $L^{\max}(\lambda) \times L^{\max}(\lambda)^- \to \mathbb{C}$, obtained from the composite of the above maps, by $\langle\, , \rangle$. Observe that $\langle\, , \rangle$ depends on the choice of the isomorphism $(*)$ and thus it is determined only up to nonzero scalar multiples.

Define the following Zariski closed subset of $V_\Lambda(m)$:

$$S_m := \Big\{ x = \sum_{i=1}^{n} x_i + x_i^* \in V_\Lambda(m) : x_i \in L^{\max}(\lambda_i), x_i^* \in L^{\max}(\lambda_i)^-, \langle x_i, x_i^* \rangle$$
$$= \langle v_{\lambda_i}, v_{\lambda_i}^* \rangle, ([x_1], \dots, [x_n]) \in \theta_\Lambda(X_m) \text{ and } ([x_1^*], \dots, [x_n^*]) \in \theta_\Lambda^-(X_m^-) \Big\},$$

where $[y]$ denotes the line through y, $X_m := \tilde{X}_m/\mathcal{B}^{\min}$ and $X_m^- := \tilde{X}_m^-/\mathcal{B}^-$. Observe that this X_m is the same as X_m defined in 7.1.13 under the identification (2).

We claim that

(3) $$\psi_\Lambda(\mathcal{G}_m^{\min}) = S_m :$$

Clearly, we have the inclusion

(4) $$\psi_\Lambda(\mathcal{G}_m^{\min}) \subset S_m.$$

Conversely, take $x = \sum_i x_i + x_i^* \in S_m$. Then, by the definition of S_m, there exist $g \in \tilde{X}_m$ and $g' \in \tilde{X}_m^-$ such that, for all $1 \le i \le n$,

$$g v_{\lambda_i} = z_i x_i, \quad g' v_{\lambda_i}^* = z_i' x_i^*,$$

for some $z_i, z_i' \in \mathbb{C}^*$.

Since $\left(e^{\lambda_1}, \dots, e^{\lambda_n}\right) : T \to (\mathbb{C}^*)^n$ is surjective (by assumption), there exist $t, t' \in T$ such that

(5) $$g t v_{\lambda_i} = x_i \text{ and } g' t' v_{\lambda_i}^* = x_i^*.$$

From the Birkhoff decomposition (Theorem 6.2.8), write

(6) $$(g't')^{-1} g t = u' \bar{w} t_o u,$$

for some $u' \in \mathcal{U}^-$, $u \in \mathcal{U}^{\min}$, $t_o \in T$ and $w \in W$, where $\bar{w}$ is a coset representative of w and $\mathcal{U}^{\min} := \mathcal{U} \cap \mathcal{G}^{\min}$ (we choose $\bar{1} = 1$). Then

(7) $$\langle v_{\lambda_i}, v_{\lambda_i}^* \rangle = \langle x_i, x_i^* \rangle = \langle \bar{w} t_o v_{\lambda_i}, v_{\lambda_i}^* \rangle = e^{\lambda_i}(t_o) \langle \bar{w} v_{\lambda_i}, v_{\lambda_i}^* \rangle.$$

For $w \ne 1 \in W$, there exists an $1 \le i_o \le n$ such that $w\lambda_{i_o} \ne \lambda_{i_o}$ (by assumption). Hence, from (7), we get

(8) $$\bar{w} = 1, \text{ and } e^{\lambda_i}(t_o) = 1, \text{ for all } i.$$

So (6) reduces to

(9) $$gtu^{-1} = g't'u't_o.$$

By (5), (8) and (9), for $h = gtu^{-1}$, we get

$$hv_{\lambda_i} = x_i,\ hv^*_{\lambda_i} = x^*_i, \text{ for all } 1 \le i \le n.$$

By (9), $h \in \mathcal{G}^{\min}_m$ and hence $x \in \psi_\Lambda(\mathcal{G}^{\min}_m)$, proving

(10) $$S_m \subset \psi_\Lambda(\mathcal{G}^{\min}_m).$$

Combining (4) and (10), we get (3). This proves the first part of the proposition.

Since no $w \neq 1 \in W$ fixes all the λ_i's, $\sum \lambda_i \in D^o$. From this it is easy to see using Corollary 1.3.22 that, for any finite-dimensional vector subspace $V \subset V_\Lambda$, there exists $m = m_V$ such that

(11) $$\psi_\Lambda^{-1}(V) \subset \mathcal{G}^{\min}_m.$$

In particular,

$$V \cap \psi_\Lambda(\mathcal{G}^{\min}) = V \cap \psi_\Lambda(\mathcal{G}^{\min}_m),$$

which shows that $V \cap \psi_\Lambda(\mathcal{G}^{\min})$ is closed in V. Hence $\psi_\Lambda(\mathcal{G}^{\min})$ is Zariski closed in V_Λ. □

Let us fix a $\mathbb{Z}$-basis $\Lambda_o = \{\omega_1, \dots, \omega_d\} \subset D_\mathbb{Z}$ of $\mathfrak{h}^*_\mathbb{Z}$ such that

(∗) $$\omega_i(\alpha_j^\vee) = \delta_{i,j}, \quad \text{for } 1 \le i \le d, 1 \le j \le \ell,$$

where $\{\alpha_1^\vee, \dots, \alpha_\ell^\vee\}$ is the full set of simple coroots.

7.4.6 Theorem. *For any finite (possibly empty) family $\Lambda \subset D_\mathbb{Z}$, the map $\psi_{\tilde\Lambda} : \mathcal{G}^{\min} \to V_{\tilde\Lambda} := \bigoplus_{\lambda\in\tilde\Lambda}(L^{\max}(\lambda) \oplus L^{\max}(\lambda)^-)$, defined by (7.4.4.2), is injective and, for any $m \ge 0$, $\psi_{\tilde\Lambda}(\mathcal{G}^{\min}_m)$ is a Zariski closed subset of a finite-dimensional vector subspace of $V_{\tilde\Lambda}$, where $\tilde\Lambda := \Lambda_o \sqcup \Lambda$. Thus,* Im $\psi_{\tilde\Lambda}$ *is Zariski closed in $V_{\tilde\Lambda}$. Hence, there is a unique (affine) ind-variety structure on $\mathcal{G}^{\min}$ (denoted by $\mathcal{G}^{\min}(\tilde\Lambda)$) with the filtration $\{\mathcal{G}^{\min}_m\}_{m\ge0}$ making $\psi_{\tilde\Lambda}$ a closed embedding.*

Proof. We first prove the injectivity of $\psi_{\tilde\Lambda}$: From the homogeneity, it suffices to show that $\psi_{\tilde\Lambda}^{-1}(v_o) = \{1\}$, where $v_o := \sum_{\lambda\in\tilde\Lambda}(v_\lambda + v^*_\lambda)$. Using Lemma 7.1.2, since $\tilde\Lambda$ spans $\mathfrak{h}^*_\mathbb{Z}$, we get that

$$\psi_{\tilde\Lambda}^{-1}(v_o) \subset \mathcal{U}^{\min}.$$

Similarly, from the symmetry of $\psi_{\tilde{\Lambda}}$ with respect to $\mathcal{U}^{\min}$ and $\mathcal{U}^-$,

$$\psi_{\tilde{\Lambda}}^{-1}(v_o) \subset \mathcal{U}^-.$$

So $\psi_{\tilde{\Lambda}}^{-1}(v_o) \subset \mathcal{U}^{\min} \cap \mathcal{U}^- = \{1\}$, by (6.2.8.1). This proves the injectivity of $\psi_{\tilde{\Lambda}}$.

We now prove that $\psi_{\tilde{\Lambda}}$ has closed image. Enumerate $\Lambda = \{\lambda_1, \dots, \lambda_q\}$ and write $\lambda_p = \sum_{i=1}^{d} m_i(p)\omega_i$. Since $\lambda_p \in D_{\mathbb{Z}}$, $m_i(p) \in \mathbb{Z}_+$ for all $1 \le i \le \ell$. Further, $L^{\max}(\omega_i)$ is a one-dimensional representation for all $i > \ell$ (cf. Lemma 8.3.2), i.e., a character and hence $L^{\max}(\omega_i)^{\otimes m}$ makes sense for all $m \in \mathbb{Z}$.

Consider the $\mathcal{G}$-module map

$$\theta_{\lambda_p} : L^{\max}(\lambda_p) \to L^{\max}(\omega_1)^{\otimes m_1(p)} \otimes \cdots \otimes L^{\max}(\omega_d)^{\otimes m_d(p)},$$

which takes

$$v_{\lambda_p} \mapsto v_{\omega_1}^{\otimes m_1(p)} \otimes v_{\omega_2}^{\otimes m_2(p)} \otimes \cdots \otimes v_{\omega_d}^{\otimes m_d(p)}$$

(cf. proof of Proposition 7.1.15(b)). Similarly, we have the $\mathcal{G}^{\min}$-module map

$$\theta_{\lambda_p}^- : L^{\max}(\lambda_p)^- \to (L^{\max}(\omega_1)^-)^{\otimes m_1(p)} \otimes \cdots \otimes (L^{\max}(\omega_d)^-)^{\otimes m_d(p)}.$$

Consider the $\mathcal{G}^{\min}$-orbit $\mathcal{G}^{\min} \cdot v_{\lambda_p} \subset L^{\max}(\lambda_p)$ and define, for any $m \ge 0$,

$$\begin{aligned}\tilde{S}_m := \Big\{ x = \sum_{i=1}^{d+q} x_i + x_i^* \in V_{\tilde{\Lambda}}(m) : \sum_{i=1}^{d} x_i + x_i^* \in \psi_{\Lambda_o}(\mathcal{G}_m^{\min}) \text{ and, for} \\ \text{all } 1 \le p \le q,\ x_{d+p} \in \mathcal{G}^{\min} \cdot v_{\lambda_p} \cup \{0\},\ x_{d+p}^* \in \mathcal{G}^{\min} \cdot v_{\lambda_p}^* \cup \{0\}, \\ \theta_{\lambda_p}(x_{d+p}) = x_1^{\otimes m_1(p)} \otimes \cdots \otimes x_d^{\otimes m_d(p)} \text{ and} \\ \theta_{\lambda_p}^-(x_{d+p}^*) = (x_1^*)^{\otimes m_1(p)} \otimes \cdots \otimes (x_d^*)^{\otimes m_d(p)} \Big\},\end{aligned}$$

where $V_{\tilde{\Lambda}}(m)$ is as defined in (7.4.5.1).

By Proposition 7.4.5 and its proof, $\psi_{\Lambda_o}(\mathcal{G}_m^{\min})$ is a closed subset of $V_{\Lambda_o}(m)$. Also the maps θ_{λ_p} and $\theta_{\lambda_p}^-$ (being linear maps) are continuous under the Zariski topologies. Further, $\mathcal{G}^{\min} \cdot v_{\lambda_p} \cup \{0\}$ is a closed subset of $L^{\max}(\lambda_p)$, since $i_\lambda(\mathcal{G}/\mathcal{B}) \subset \mathbb{P}(L^{\max}(\lambda))$ is a closed subset (cf. 7.1.13) and $\mathcal{G}^{\min} \cdot \mathcal{B} = \mathcal{G}$ (cf. Lemma 7.4.2 and its proof). Similarly, $\mathcal{G}^{\min} \cdot v_{\lambda_p}^* \cup \{0\}$ is a closed subset of $L^{\max}(\lambda_p)^-$. From these we obtain that $\tilde{S}_m$ is a closed subset of $V_{\tilde{\Lambda}}(m)$.

We next show that

(1) $$\tilde{S}_m = \psi_{\tilde{\Lambda}}(\mathcal{G}_m^{\min}).$$

The inclusion $\tilde{S}_m \supset \psi_{\tilde{\Lambda}}(\mathcal{G}_m^{\min})$ is clear. Conversely, take $x = \sum_{i=1}^{d+q} x_i + x_i^* \in \tilde{S}_m$. Then, by the definition of $\tilde{S}_m$, there exists $g \in \mathcal{G}_m^{\min}$ such that

(2) $$\psi_{\Lambda_o}(g) = \sum_{i=1}^{d} x_i + x_i^*, \text{ i.e., } x_i = g v_{\omega_i} \text{ and } x_i^* = g v_{\omega_i}^*, \text{ for all } 1 \le i \le d.$$

Thus, for any $1 \le p \le q$,

(3) $$\theta_{\lambda_p}(x_{d+p}) = g \cdot \left(v_{\omega_1}^{\otimes m_1(p)} \otimes \cdots \otimes v_{\omega_d}^{\otimes m_d(p)}\right),$$

and

(4) $$\theta_{\lambda_p}^{-}(x_{d+p}^*) = g \cdot \left((v_{\omega_1}^*)^{\otimes m_1(p)} \otimes \cdots \otimes (v_{\omega_d}^*)^{\otimes m_d(p)}\right).$$

But,

(5) $$\theta_{\lambda_p}(g \cdot v_{\lambda_p}) = g \cdot \left(v_{\omega_1}^{\otimes m_1(p)} \otimes \cdots \otimes v_{\omega_d}^{\otimes m_d(p)}\right).$$

From (3) and (5), we get that $x_{d+p} \neq 0$ and $\theta_{\lambda_p}(x_{d+p}) = \theta_{\lambda_p}(g \cdot v_{\lambda_p})$. Further, $\theta_{\lambda_p | \mathcal{G}^{\min} \cdot v_{\lambda_p}}$ is injective: To see this, define $Q_1 := \left\{g \in \mathcal{G} : \theta_{\lambda_p}(g \cdot v_{\lambda_p}) \in \mathbb{C}^* \theta_{\lambda_p}(v_{\lambda_p})\right\}$ and $Q_2 := \left\{g \in \mathcal{G} : g \cdot v_{\lambda_p} \in \mathbb{C}^* v_{\lambda_p}\right\}$. Then $Q_1 \supset Q_2$ are both standard parabolic subgroups of $\mathcal{G}$. From this it is easy to see that $Q_1 = Q_2$; otherwise we can find $n \in (Q_1 \backslash Q_2) \cap N(T)$ leading to a contradiction, where $N(T)$ is the normalizer of T in $\mathcal{G}$. Now take $g \in \mathcal{G}$ such that $\theta_{\lambda_p}(g \cdot v_{\lambda_p}) = \theta_{\lambda_p}(v_{\lambda_p})$. Then $g \in Q_1 = Q_2$, which immediately gives $g \cdot v_{\lambda_p} = v_{\lambda_p}$, thus proving the assertion that $\theta_{\lambda_p | \mathcal{G}^{\min} \cdot v_{\lambda_p}}$ is injective. Hence

(6) $$x_{d+p} = g \cdot v_{\lambda_p}.$$

Similarly,

(7) $$x_{d+p}^* = g \cdot v_{\lambda_p}^*.$$

Combining (2), (6) and (7), we get that $\psi_{\tilde{\Lambda}}(g) = x$, and thus $\tilde{S}_m \subset \psi_{\tilde{\Lambda}}\left(\mathcal{G}_m^{\min}\right)$ and (1) follows. This proves that $\psi_{\tilde{\Lambda}}\left(\mathcal{G}_m^{\min}\right)$ is a closed subset of $V_{\tilde{\Lambda}}(m)$.

The assertion that Im $\psi_{\tilde{\Lambda}}$ is closed in $V_{\tilde{\Lambda}}$ follows by the same argument as given in the proof of Proposition 7.4.5 for the corresponding fact. □

7.4.7 Definition. Choose any $\mathbb{Z}$-basis Λ_o of $\mathfrak{h}_{\mathbb{Z}}^*$ satisfying $(*)$ just above Theorem 7.4.6. Then for any finite family $\Lambda \subset D_{\mathbb{Z}}$, by Theorem 7.4.6, we have an affine ind-variety structure $\mathcal{G}^{\min}(\tilde{\Lambda})$ on $\mathcal{G}^{\min}$ with the same filtration $\left\{\mathcal{G}_m^{\min}\right\}_{m \ge 0}$, where

$\tilde{\Lambda} := \Lambda_o \sqcup \Lambda$. Moreover, it is easy to see that for any finite family $\Lambda' \subset D_{\mathbb{Z}}$ containing Λ, the identity map $\mathcal{G}^{\min}(\tilde{\Lambda}') \to \mathcal{G}^{\min}(\tilde{\Lambda})$ is a morphism.

By Exercise 7.4.E.3, for any $m \geq 0$, there exists a sufficiently large finite family $\Lambda_m \subset D_{\mathbb{Z}}$ such that the identity map $G_m^{\min}(\tilde{\Lambda}') \to \mathcal{G}_m^{\min}(\tilde{\Lambda}_m)$ is a biregular isomorphism for all finite Λ' with $\Lambda_m \subset \Lambda' \subset D_{\mathbb{Z}}$, where $\mathcal{G}_m^{\min}(\tilde{\Lambda})$ denotes the set $G_m^{\min}$ endowed with the closed (reduced) subvariety structure inherited from $\mathcal{G}^{\min}(\tilde{\Lambda})$. The variety structure $G_m^{\min}(\tilde{\Lambda}_m)$ on $\mathcal{G}_m^{\min}$ is called the *stable* (*affine*) *variety structure* on $\mathcal{G}_m^{\min}$. Clearly, this stable variety structure on $\mathcal{G}_m^{\min}$ does not depend on the choice of Λ_o and Λ_m.

Now we put the *stable* (*affine*) *ind-variety structure* on $\mathcal{G}^{\min}$ by taking the filtration $\{\mathcal{G}_m^{\min}\}_{m \geq 0}$ of $\mathcal{G}^{\min}$ and putting the stable variety structure on each $\mathcal{G}_m^{\min}$. The affine ind-variety $\mathcal{G}^{\min}$, resp. the variety $\mathcal{G}_m^{\min}$, endowed with the stable structure is simply denoted by $\mathcal{G}^{\min}$, resp. $\mathcal{G}_m^{\min}$.

For any finite family $\Lambda \subset D_{\mathbb{Z}}$, the map $\psi_\Lambda : \mathcal{G}^{\min} \to \bigoplus_{\lambda \in \Lambda} L^{\max}(\lambda) \oplus L^{\max}(\lambda)^-$ defined in 7.4.4 is clearly a morphism.

7.4.8 Remark. In the symmetrizable case, the identity map $\mathcal{G}^{\min}(\tilde{\Lambda}') \to \mathcal{G}^{\min}(\tilde{\Lambda})$ is a biregular isomorphism for any finite families $\Lambda \subset \Lambda' \subset D_{\mathbb{Z}}$. This can be seen by observing that, for any $\lambda \in D_{\mathbb{Z}}$, the map

$$\mathcal{G}^{\min}(\Lambda_o) \to L(\lambda) \oplus L(\lambda)^-, \quad g \mapsto g(v_\lambda + v_\lambda^*),$$

is a morphism, which follows since the maps $\theta_\lambda, \theta_\lambda^-$ of the proof of Theorem 7.4.6 are injective and thus are closed embeddings.

7.4.9 Definition. Recall that any complex variety Z is also endowed with the strong (called Hausdorff or analytic as well) topology (cf. [Mumford–88, §10, Chapter I]). Now, given an ind-variety Z over $\mathbb{C}$ with filtration $(Z_n)_{n \geq 0}$, define the *strong topology* on Z by declaring a set $A \subset Z$ open iff $A \cap Z_n$ is open in Z_n under the strong topology on Z_n. A morphism of ind-varieties is continuous in the strong topology (use [loc. cit.]). In particular, two equivalent ind-variety structures on Z give rise to the same strong topology. Endowed with the strong topology, we denote the ind-variety Z by Z_{an} ("an" for analytic).

For a closed embedding of ind-varieties $Z \hookrightarrow Z'$, $Z_{\text{an}} \hookrightarrow Z'_{\text{an}}$ is a homeomorphism onto its image.

For an ind-group H, H_{an} is a topological group. Use [Milnor–Stasheff–74, Lemma 5.5], which asserts that, for two sequences of locally compact Hausdorff spaces $A_1 \subset A_2 \subset \cdots$, and $B_1 \subset B_2 \subset \cdots$, with direct limits A and B respectively, the product topology on $A \times B$ coincides with the direct limit topology induced from the sequence $A_1 \times B_1 \subset A_2 \times B_2 \subset \cdots$.

7.4.10 Lemma. *The canonical map $\pi : \mathcal{G}^{\min} \to \mathcal{X}$ is regular. In particular, the subgroup $\mathcal{B}^{\min} \subset \mathcal{G}^{\min}$ is closed.*

Moreover, the map π admits a regular section on the open subset $\mathcal{U}^- \subset \mathcal{X}$ (cf. 7.3.6). In particular, $\mathcal{X}_{\mathrm{an}}$, resp. $\mathcal{X}$, has the quotient topology induced from π, i.e., a set $V \subset \mathcal{X}_{\mathrm{an}}$, resp. $\mathcal{X}$, is open iff $\pi^{-1}(V)$ is open in $\mathcal{G}^{\min}_{\mathrm{an}}$, resp. $\mathcal{G}^{\min}$.

(This lemma can be generalized to $\mathcal{X}$ replaced by $\mathcal{X}^Y$, cf. Exercise 7.4.E.5.)

Proof. Fix $m \geq 0$. Then $\pi(\mathcal{G}^{\min}_m) \subset X_m$. Pick a large enough $\lambda_m \in D^o$ such that $i_{\lambda_m} : X_m \to \mathbb{P}(L^{\max}(\lambda_m))$ is a closed embedding (cf. 7.1.19). Since $\mathcal{G}^{\min} \to L^{\max}(\lambda_m)$, $g \mapsto g v_{\lambda_m}$, is a morphism by 7.4.7, we get that $\pi_{|\mathcal{G}^{\min}_m}$ is a morphism and hence so is π. In particular, $\mathcal{B}^{\min} = \pi^{-1}(1 \bmod \mathcal{B})$ is Zariski closed in $\mathcal{G}^{\min}$.

Consider the "identity" map $\sigma : \mathcal{U}^- \subset \mathcal{X} \to \mathcal{G}^{\min}$. Then σ is a morphism: First, $\sigma(\mathcal{U}^-_m) \subset \mathcal{G}^{\min}_m$, for any $m \geq 0$. Moreover, for any $\lambda \in D_{\mathbb{Z}}$ and $g \in \mathcal{U}^-$, $g v^*_\lambda = v^*_\lambda$. Thus, it suffices to prove that $\mathcal{U}^- \to L^{\max}(\lambda)$, $g \mapsto g v_\lambda$, is a morphism, where $\mathcal{U}^- \subset \mathcal{X}$ is endowed with the open ind-subvariety structure. But this follows since $i_\lambda : \mathcal{X} \to \mathbb{P}(L^{\max}(\lambda))$ is a morphism (by Proposition 7.1.15(b)), and the component of $g v_\lambda$ in $L^{\max}(\lambda)_\lambda$ is v_λ for any $g \in \mathcal{U}^-$ (use (6.2.8.3)).

The assertion, that the space $\mathcal{X}$ has the quotient topology, follows from the existence of a regular section on $\mathcal{U}^-$ and since the left multiplication by any element of $\mathcal{G}^{\min}$ on $\mathcal{G}^{\min}$ as well as $\mathcal{X}$ is a morphism (in particular, continuous). The same argument shows that $\mathcal{X}_{\mathrm{an}}$ has the quotient topology induced from π.

□

7.4.11 Proposition. *The subgroup $\mathcal{U}^{\min} := \mathcal{U} \cap \mathcal{G}^{\min}$ is closed in $\mathcal{G}^{\min}$. Moreover, the product maps*

$$\mu_1 : \mathcal{U}^{\min} \times T \to \mathcal{B}^{\min}, \quad (u,t) \mapsto u \cdot t, \qquad \textit{and}$$
$$\mu_2 : \mathcal{U}^- \times \mathcal{B}^{\min} \to \pi^{-1}(\mathcal{U}^-), \quad (u,b) \mapsto u \cdot b,$$

are biregular isomorphisms, where $\mathcal{U}^{\min}$ and $\mathcal{B}^{\min}$ have closed ind-subvariety structures coming from $\mathcal{G}^{\min}$, $\mathcal{U}^-$ is open ind-subvariety of $\mathcal{X}$ and $\pi : \mathcal{G}^{\min} \to \mathcal{X}$ is the canonical map. In particular, T is closed in $\mathcal{B}^{\min}$ and hence in $\mathcal{G}^{\min}$.

Proof. To prove that $\mathcal{U}^{\min}$ is closed in $\mathcal{G}^{\min}$, observe that

$$\mathcal{U}^{\min} = \psi^{-1}_{\Lambda_o}\left(\sum_{i=1}^{d}\left(v_{\omega_i} \oplus L^{\max}(\omega_i)^-\right)\right).$$

To prove that μ_1 is a morphism, it suffices to show, with the notation in Theorem 7.4.6, that $\psi_{\bar{\Lambda}} \circ \mu_1 : \mathcal{U}^{\min} \times T \to V_{\bar{\Lambda}}$ is a morphism for any finite family $\Lambda \subset D_{\mathbb{Z}}$. But this follows easily since $\psi_{\bar{\Lambda}}$ is a morphism (cf. 7.4.7).

Clearly, μ_1 is bijective. We next show that $\pi_T \circ \mu_1^{-1}$ is a morphism, where π_T is the projection $\mathcal{U}^{\min} \times T \to T$:

Consider the commutative diagram:

$$\begin{array}{ccc} \mathcal{B}^{\min} & \xrightarrow{\psi_{\Lambda_o}} & \bigoplus_{i=1}^{d} \left(L^{\max}(\omega_i) \oplus L^{\max}(\omega_i)^-\right) \\ \pi_T \circ \mu_1^{-1} \big\downarrow & & \big\downarrow \beta \\ T & \xrightarrow[\theta]{} & \bigoplus_{i=1}^{d} L^{\max}(\omega_i)_{\omega_i}, \end{array}$$

where β is the projection induced from the projections $L^{\max}(\omega_i) \to L^{\max}(\omega_i)_{\omega_i}$ onto the ω_i-weight spaces and $\theta(t) := \sum_{i=1}^{d} e^{\omega_i}(t)\, v_{\omega_i}$. Since θ is an open embedding and $\beta \circ \psi_{\Lambda_o}$ is a morphism, we get that $\pi_T \circ \mu_1^{-1}$ is a morphism.

Now, it is easy to see that the product map $\mu_3 : \mathcal{B}^{\min} \times T \to \mathcal{B}^{\min}$, $(g, t) \mapsto g \cdot t$, is a morphism. (In fact, the product map $\mathcal{G}^{\min} \times T \to \mathcal{G}^{\min}$ is a morphism.) Hence the composite map $\mu_3 \circ \gamma_1$:

$$\mathcal{B}^{\min} \xrightarrow{\gamma_1} \mathcal{B}^{\min} \times T \xrightarrow{\mu_3} \mathcal{B}^{\min}$$

is a morphism, where $\gamma_1(b) := \left(b, \left(\pi_T \circ \mu_1^{-1}(b)\right)^{-1}\right)$ for $b \in \mathcal{B}^{\min}$. But $\mu_3 \circ \gamma_1 = \pi_{\mathcal{U}^{\min}} \circ \mu_1^{-1}$, where $\pi_{\mathcal{U}^{\min}}$ is the projection $\mathcal{U}^{\min} \times T \to \mathcal{U}^{\min}$. Thus μ_1^{-1} is a morphism, proving that μ_1 is a biregular isomorphism.

By Lemma 7.3.9, $\mathcal{U}^-$ acts algebraically on $L^{\max}(\lambda)$ (for any $\lambda \in D_{\mathbb{Z}}$) and a similar proof shows that $\mathcal{U}^-$ acts algebraically on $L^{\max}(\lambda)^-$. From this it is easy to see that the multiplication map $\mu_4 : \mathcal{U}^- \times \mathcal{G}^{\min} \to \mathcal{G}^{\min}$ is a morphism and hence so is the multiplication map μ_2. Moreover, μ_2 is a bijection by (7.4.5.2) and 7.3.6. Observe further that

$$\text{(1)} \qquad \pi_{\mathcal{U}^-} \circ \mu_2^{-1} = \pi_{|\pi^{-1}(\mathcal{U}^-)},$$

and hence $\pi_{\mathcal{U}^-} \circ \mu_2^{-1}$ is a morphism by Lemma 7.4.10. Since $\mathcal{U}^-$ is an ind-group (cf. Corollary 7.3.8), the composite map $\mu_4 \circ \gamma_2$:

$$\pi^{-1}(\mathcal{U}^-) \xrightarrow{\gamma_2} \mathcal{U}^- \times \pi^{-1}(\mathcal{U}^-) \xrightarrow{\mu_4} \mathcal{G}^{\min}$$

is a morphism, where $\gamma_2(g) := (\pi(g)^{-1}, g)$, for $g \in \pi^{-1}(\mathcal{U}^-)$. From (1) we see that $\mu_4 \circ \gamma_2 = \pi_{\mathcal{B}^{\min}} \circ \mu_2^{-1}$, and thus μ_2^{-1} is a morphism. This proves that μ_2 is a biregular isomorphism. □

7.4.12 Remark. Since

$$\mathcal{U}^- = \psi_{\Lambda_o}^{-1}\left(\sum_{i=1}^{d}\left(L^{\max}(\omega_i) \oplus v_{\omega_i}^*\right)\right),$$

$\mathcal{U}^-$ is closed in $\mathcal{G}^{\min}$ In particular, $\mathcal{U}^-$ acquires another ind-variety structure as a closed ind-subvariety of $\mathcal{G}^{\min}$ However, by Lemma 7.4.10, this ind-variety structure on $\mathcal{U}^-$ coincides with the earlier ind-variety structure on $\mathcal{U}^-$, obtained as an open ind-subvariety of $\mathcal{X}$.

As a consequence of Proposition 7.4.11, we obtain the following.

7.4.13 Corollary. *For any $\lambda \in D_{\mathbb{Z}}$, the module maps*

$$m : \mathcal{G}^{\min} \times L^{\max}(\lambda) \to L^{\max}(\lambda), \qquad \textit{and}$$
$$m^- : \mathcal{G}^{\min} \times L^{\max}(\lambda)^- \to L^{\max}(\lambda)^-$$

are morphisms of ind-varieties.

Thus the same result is true for any integrable highest weight $\mathfrak{g}$-module $V(\lambda)$ with highest weight λ and any integrable lowest weight $\mathfrak{g}$-module $V(\lambda)^-$ with lowest weight $-\lambda$.

Proof. We prove that m is a morphism. (The proof for m^- is similar.) By Lemma 7.3.9, $\mathcal{U}^-$ acts algebraically on $L^{\max}(\lambda)$ and a similar proof shows that $\mathcal{U}^{\min}$ acts algebraically on $L^{\max}(\lambda)$ (cf. Exercise 7.4.E.4). It is easy to see that T acts algebraically on $L^{\max}(\lambda)$. Thus, by Proposition 7.4.11, m restricted to the open subset $\pi^{-1}(\mathcal{U}^-) \times L^{\max}(\lambda)$ is a morphism.

Now consider the open cover $\mathcal{G}^{\min} = \bigcup_{g\in\mathcal{G}^{\min}} g\pi^{-1}(\mathcal{U}^-)$. Since the left multiplication by g on $\mathcal{G}^{\min}$ is a morphism and so is the action of g on $L^{\max}(\lambda)$, the restriction of m to $g\pi^{-1}(\mathcal{U}^-) \times L^{\max}(\lambda)$ is again a morphism. This proves that m is a morphism. □

7.4.14 Theorem. *$\mathcal{G}^{\min}$ is an affine ind-group. In particular, $\mathcal{G}^{\min}_{\text{an}}$ is a topological group.*

Proof. The assertion that the multiplication map $\mu : \mathcal{G}^{\min} \times \mathcal{G}^{\min} \to \mathcal{G}^{\min}$ is a morphism follows from Corollary 7.4.13. We now prove that the map $i : \mathcal{G}^{\min} \to \mathcal{G}^{\min}$, $g \mapsto g^{-1}$, is a morphism. Since $\mathcal{U}^-$ and $\mathcal{U}^{\min}$ are ind-groups (cf. Corollary 7.3.8 and Exercise 7.4.E.4) and μ is a morphism, the map $\mathcal{U}^- \times \mathcal{U}^{\min} \times T \to \mathcal{G}^{\min}$, $(g, h, t) \mapsto t^{-1}h^{-1}g^{-1}$, is a morphism with the product ind-variety structure on the domain. Combining this with Proposition 7.4.11, we get that the map i restricted to $\pi^{-1}(\mathcal{U}^-)$ is a morphism. Now, since μ is a morphism, $i_{|g\cdot\pi^{-1}(\mathcal{U}^-)}$ is a morphism as well for any $g \in \mathcal{G}^{\min}$. This proves that i is a morphism on the whole of $\mathcal{G}^{\min}$, proving the theorem. □

$\mathcal{B}^{\min}$, being a closed subgroup of the ind-group $\mathcal{G}^{\min}$, is an ind-group; in particular, $\mathcal{B}^{\min}_{\text{an}}$ is a topological group.

7.4.15 Corollary. *$\pi : \mathcal{G}^{\min}_{\text{an}} \to \mathcal{X}_{\text{an}}$ is a locally trivial principal $\mathcal{B}^{\min}_{\text{an}}$-bundle. In fact, $\pi : \mathcal{G}^{\min} \to \mathcal{X}$ is a Zariski locally trivial principal $\mathcal{B}^{\min}$-bundle.*

Proof. The corollary follows from Proposition 7.4.11. □

Recall from Lemma 7.4.2 that $(\mathcal{G}^{\min}, \mathcal{B}^{\min}, N, S)$ is a Tits system. In fact, one has the following stronger result.

7.4.16 Proposition. *$(\mathcal{G}^{\min}_{\text{an}}, \mathcal{B}^{\min}_{\text{an}}, N, S)$ is a topological Tits system (cf. Definition 5.1.4), with $(\mathcal{P}^{\min}_s)_{\text{an}}/\mathcal{B}^{\min}_{\text{an}}$ homeomorphic to the two-dimensional sphere S^2 for each $s \in S$, where $\mathcal{P}^{\min}_s := \mathcal{P}_s \cap \mathcal{G}^{\min}$.*

In particular, for any subset $Y \subset \{1, \dots, \ell\}$, $\mathcal{X}^Y_{\text{an}} = (\mathcal{G}/\mathcal{P}_Y)_{\text{an}}$ is a CW-complex with cells $\{\mathcal{B}\, w\mathcal{P}_Y/\mathcal{P}_Y\}_{w\in W'_Y}$. Moreover, we have a homeomorphism, in fact, a biregular isomorphism, for any $w \in W'_Y$:

$$(1) \qquad \mathcal{B}w\mathcal{P}_Y/\mathcal{P}_Y \simeq \mathbb{C}^{\ell(w)},$$

where $\mathcal{B}w\mathcal{P}_Y/\mathcal{P}_Y$ is an open subvariety of $\mathcal{X}^Y_w$ under the stable variety structure.

Proof. By Theorem 7.4.14, $\mathcal{G}^{\min}_{\text{an}}$ is a Hausdorff topological group and, by Lemma 7.4.10, $\mathcal{B}^{\min}$ is a closed subgroup. Since T is Zariski closed in $\mathcal{G}^{\min}$ (cf. Proposition 7.4.11), $N_m := \bigcup_{\ell(w)\le m} wT$ is closed in $\mathcal{G}^{\min}$ for any $m \geq 0$ (and so is $N_m \backslash T$). This proves that N is (Zariski) closed in $\mathcal{G}^{\min}$ and T is open in N (under the subspace topology on N induced from $\mathcal{G}^{\min}$). By Remark 7.1.9 and Lemma 7.4.10, the axiom (BN_6) of Definition 5.1.4 is satisfied. This proves that $(\mathcal{G}^{\min}_{\text{an}}, \mathcal{B}^{\min}_{\text{an}}, N, S)$ is a topological Tits system.

The "In particular" part of the proposition follows from Theorem 5.1.5, since $\mathcal{P}^{\min}_Y := \mathcal{P}_Y \cap \mathcal{G}^{\min}$ is closed in $\mathcal{G}^{\min}$ and hence in $\mathcal{G}^{\min}_{\text{an}}$ and the inductive limit topology on $\mathcal{X}^Y$ (cf. Theorem 5.1.5(c)) coincides with the strong topology $\mathcal{X}^Y_{\text{an}}$ (by Exercise 7.4.E.5). Finally, (1) follows from Proposition 7.1.15(a) and the proof of Proposition 7.1.8(b). □

7.4.17 Proposition. *The group $\mathcal{U}^{\min}_{\text{an}}$ is a contractible space.*

Proof. Choose $\rho^\vee \in \mathfrak{h}$ such that $\alpha_i(\rho^\vee) = 1$ for all the simple roots α_i. Define the contraction map $\theta : [0, 1] \times \mathcal{U}^{\min}_{\text{an}} \to \mathcal{U}^{\min}_{\text{an}}$ by

$$\begin{aligned}\theta(s, g) &= \operatorname{Exp}\Big(-\frac{1-s}{s}\rho^\vee\Big) \cdot g \cdot \operatorname{Exp}\Big(\frac{1-s}{s}\rho^\vee\Big), && \text{if } s \neq 0\\ &= 1, && \text{if } s = 0.\end{aligned}$$

Take any finite family $\Lambda \subset D_{\mathbb{Z}}$. Then, with the notation of Theorem 7.4.6, θ is the restriction of the map (under the continuous injective map $\psi_{\bar{\Lambda}} : \mathcal{G}^{\min}_{\text{an}} \to$

$(V_{\tilde{\Lambda}})_{an})$ $\hat{\theta}_{\Lambda} : [0,1] \times (V_{\tilde{\Lambda}})_{an} \to (V_{\tilde{\Lambda}})_{an}$ given by (for $a_\lambda \in L^{\max}(\lambda)$ and $b_\lambda \in L^{\max}(\lambda)^-$)

$$\hat{\theta}_{\Lambda}\Big(s, \sum_{\lambda\in\tilde{\Lambda}} (a_\lambda + b_\lambda)\Big) = \sum_{\lambda\in\tilde{\Lambda}} \Big(v_\lambda \oplus e^{-\frac{1-s}{s}\lambda(\rho^\vee)} \operatorname{Exp}\Big(-\frac{1-s}{s}\rho^\vee\Big)\cdot b_\lambda\Big), \text{ if } s \neq 0$$
$$= \sum_{\lambda}\Big(v_\lambda \oplus \frac{\langle v_\lambda, b_\lambda\rangle}{\langle v_\lambda, v_\lambda^*\rangle} v_\lambda^*\Big), \qquad \text{if } s = 0,$$

where the pairing $\langle\, ,\, \rangle : L^{\max}(\lambda) \times L^{\max}(\lambda)^- \to \mathbb{C}$ is as defined in the proof of Proposition 7.4.5. It is easy to see that $\hat{\theta}_\Lambda$ is continuous under the analytic topology on $V_{\tilde{\Lambda}}$ for any finite family $\Lambda \subset D_{\mathbb{Z}}$ and hence so is θ. This proves the proposition. □

7.4.E EXERCISES

(1) Prove that the commutator subgroup $[\mathcal{G}^{\min}, \mathcal{G}^{\min}]$ of $\mathcal{G}^{\min}$ is isomorphic with the group defined in [Peterson–Kac–83, §2] (in their notation the group G).

Hence, or otherwise, prove that there exists a group involution $\tilde{\omega} : \mathcal{G}^{\min} \to \mathcal{G}^{\min}$ with the property

$$\tilde{\omega}(t) = t^{-1}, \text{ for } t \in T; \quad \tilde{\omega}(\mathcal{U}_\alpha) = \mathcal{U}_{-\alpha}, \text{ for any } \alpha \in \Delta_{\mathrm{re}}.$$

In particular, $\tilde{\omega}(\mathcal{B}^{\min}) = \mathcal{B}^-$. ($\tilde{\omega}$ should be viewed as a group analogue of the Cartan involution ω of $\mathfrak{g}$ as in 1.1.2.)

(2) Show that the Zariski topology on $\mathcal{G}^{\min}(\tilde{\Lambda})$ does not depend on the choices of Λ_o and Λ, i.e., the identity map $\mathcal{G}^{\min}(\Lambda_o \sqcup \Lambda) \to \mathcal{G}^{\min}(\Lambda'_o \sqcup \Lambda')$ is a homeomorphism for any $\mathbb{Z}$-bases Λ_o, Λ'_o of $\mathfrak{h}^*_{\mathbb{Z}}$ satisfying (∗) above Theorem 7.4.6 and any finite families $\Lambda, \Lambda' \subset D_{\mathbb{Z}}$.

Hint. For any $\lambda \in D_{\mathbb{Z}}$, show that the injective map $\theta_{\lambda|\mathcal{G}^{\min}\cdot v_\lambda} : \mathcal{G}^{\min}\cdot v_\lambda \to L^{\max}(\omega_1)^{\otimes m_1} \otimes \cdots \otimes L^{\max}(\omega_d)^{\otimes m_d}$ is a homeomorphism onto its image, where $\lambda = \sum_i m_i\omega_i$, θ_λ is the map defined in the proof of Theorem 7.4.6 and $\mathcal{G}^{\min}\cdot v_\lambda \subset L^{\max}(\lambda)$ is equipped with the subspace topology. To show this, observe first that the associated map $\mathbb{P}(\mathcal{G}^{\min}\cdot v_\lambda) \to \mathbb{P}(L^{\max}(\omega_1)^{\otimes m_1} \otimes \cdots \otimes L^{\max}(\omega_d)^{\otimes m_d})$ is a homeomorphism onto its image, where $\mathbb{P}(\mathcal{G}^{\min}\cdot v_\lambda)$ denotes the quotient space $\mathcal{G}^{\min}\cdot v_\lambda/\mathbb{C}^*$. Now use the homeomorphism $\theta_{\lambda|\mathcal{G}^{\min}\cdot v_\lambda}$ to show that the identity map $\mathcal{G}^{\min}(\tilde{\Lambda}) \to \mathcal{G}^{\min}(\Lambda_o)$ is a homeomorphism.

(3) With the notation as in 7.4.7, for any $m \geq 0$, show that there exists a sufficiently large finite family $\Lambda_m \subset D_{\mathbb{Z}}$ such that the identity map

$$\mathcal{G}^{\min}_m(\tilde{\Lambda}') \to \mathcal{G}^{\min}_m(\tilde{\Lambda}_m)$$

is a biregular isomorphism for any finite family Λ' with $\Lambda_m \subset \Lambda' \subset D_{\mathbb{Z}}$.

Hint. Fix $k, n \geq 0$ and choose $\lambda_k \in D^o$, resp. $\mu_n \in D^o$, such that the stable variety structure on $\mathcal{U}_k^-$, resp. $\mathcal{U}_n^{\min}$, coming from $\mathcal{X}$, resp. $\mathcal{G}^{\min}/\mathcal{B}^-$, is given by λ_k, resp. μ_n. Then show that, for any finite family $\Lambda \subset D_{\mathbb{Z}}$ containing λ_k and μ_n, the product map $\mathcal{U}_k^- \times \mathcal{U}_n^{\min} \times T \to \mathcal{G}^{\min}(\tilde{\Lambda})$ is a biregular isomorphism onto its image (which is the closed subset $\mathcal{U}_k^- \cdot \mathcal{U}_n^{\min} \cdot T$ of the open subset $\mathcal{U}^- \cdot \mathcal{U}^{\min} \cdot T$ of $\mathcal{G}^{\min}$), where the domain is equipped with the product variety structure with respect to the stable variety structures on $\mathcal{U}_k^-$ and $\mathcal{U}_n^{\min}$. Now cover $\mathcal{G}_m^{\min}$ by finitely many open subsets of the form $\left(g_i \cdot \mathcal{U}_k^- \cdot \mathcal{U}_n^{\min} \cdot T\right) \cap \mathcal{G}_m^{\min}$ for some large enough $k, n \geq 0$ and $g_i \in \mathcal{G}_m^{\min}$.

(4) Show that $\mathcal{U}^{\min}$ is an ind-group under the closed ind-subvariety structure coming from $\mathcal{G}^{\min}$ (cf. Proposition 7.4.11). Moreover, for any $\lambda \in D_{\mathbb{Z}}$, the $\mathcal{U}^{\min}$-modules $L^{\max}(\lambda)$ and $L^{\max}(\lambda)^-$ are algebraic. (Do not use Corollary 7.4.13 or Theorem 7.4.14.)

Hence, any integrable highest weight $\mathfrak{g}$-module $V(\lambda)$ with highest weight $\lambda \in D_{\mathbb{Z}}$ is an algebraic $\mathcal{U}^{\min}$-module. Similarly, any integrable lowest weight $\mathfrak{g}$-module $V(\lambda)^-$ is an algebraic $\mathcal{U}^{\min}$-module.
Hint. Prove the analogue of Proposition 7.3.7 for $\mathcal{U}^{\min}$ and then follow the arguments as in the proofs of Corollary 7.3.8 and Lemma 7.3.9.

(5) Prove the analogues of Lemma 7.4.10 and Corollary 7.4.15 for $\mathcal{X}$ replaced by any $\mathcal{X}^Y$, $\mathcal{B}^{\min}$ replaced by $\mathcal{P}_Y^{\min} := \mathcal{P}_Y \cap \mathcal{G}^{\min}$ and $\mathcal{U}^-$ replaced suitably.

(6) Prove that the Lie algebra Lie $(\mathcal{G}^{\min})$ of the ind-group $\mathcal{G}^{\min}$ is canonically isomorphic with $\mathfrak{g}$.
Hint. For any $1 \leq i \leq \ell$, the ind-group morphism $\mathcal{G}_i \to \mathcal{G}^{\min}$ gives rise to the Lie algebra homomorphism $\mathfrak{g}_i \to$ Lie $(\mathcal{G}^{\min})$, where $\mathcal{G}_i = \mathcal{G}_{\{i\}}$ with Lie $\mathcal{G}_i = \mathfrak{g}_i$ is as in 6.1.13. Glue these to construct a Lie algebra homomorphism $\tilde{\mathfrak{g}} \to$ Lie $(\mathcal{G}^{\min})$, where $\tilde{\mathfrak{g}}$ is as in 1.1.2.

(7) For any $\lambda \in D_{\mathbb{Z}}$, prove that the derivative of the algebraic representation of $\mathcal{G}^{\min}$ in any integrable highest weight $\mathfrak{g}$-module $V(\lambda)$ (cf. Corollary 7.4.13) coincides with the representation of the Lie algebra $\mathfrak{g}$ in $V(\lambda)$. Prove the same statement for $V(\lambda)^-$.

(8) We define an alternative ind-variety structure on $\mathcal{G}^{\min}$ as follows:

Let Λ_o be as in Theorem 7.4.6. Then prove that, for any finite family $\Lambda \subset D_{\mathbb{Z}}$, the map

$$\tilde{\psi}_{\tilde{\Lambda}} : \mathcal{G}^{\min} \to \bigoplus_{\lambda \in \tilde{\Lambda}} \left(L(\lambda) \oplus L(\lambda)^\vee\right),$$

defined by $g \mapsto \sum_{\lambda \in \tilde{\Lambda}} (gv_\lambda + gv_\lambda^*)$, is again injective with closed image, where $L(\lambda)^\vee$ is the restricted dual of $L(\lambda)$ (cf. 2.1.1) and v_λ^* is a nonzero vector of $L(\lambda)^\vee$ of weight $-\lambda$.

In particular, we can endow $\mathcal{G}^{\min}$ with a unique (affine) ind-variety structure, denoted $\mathfrak{G}^{\min}(\tilde{\Lambda})$, making $\tilde{\psi}_{\tilde{\Lambda}}$ a closed embedding. Of course, for symmetrizable $\mathfrak{g}$, this coincides with the ind-variety structure $\mathcal{G}^{\min}(\tilde{\Lambda}) = \mathcal{G}^{\min}$ (cf. Remark 7.4.8).

Show further that the ind-variety structures $\mathfrak{G}^{\min}(\Lambda_o)$ and $\mathfrak{G}^{\min}(\tilde{\Lambda})$ are equivalent for any $\Lambda \in D_{\mathbb{Z}}$. Thus, we can drop the qualification $\tilde{\Lambda}$ and denote the corresponding ind-variety $\mathfrak{G}^{\min}(\tilde{\Lambda})$ by $\mathfrak{G}^{\min}$ itself.

Prove that $\mathfrak{G}^{\min}$ is an ind-group with its Lie algebra canonically isomorphic with $\mathfrak{g}/\mathfrak{r}$, where $\mathfrak{r}$ is the radical of $\mathfrak{g}$ (cf. 1.5.6). In particular, if $\mathfrak{r} \neq 0$, the ind-variety structures $\mathcal{G}^{\min}$ and $\mathfrak{G}^{\min}$ on $\mathcal{G}^{\min}$ are *not* equivalent. (Of course, $\mathfrak{r} = 0$ for symmetrizable $\mathfrak{g}$ by Corollary 3.2.10, and is expected to be zero in general.)
Hint. To prove that $\tilde{\psi}_{\tilde{\Lambda}}$ is injective with closed image, follow a similar proof as that of Proposition 7.4.5 and Theorem 7.4.6.

(9) Show that $(\mathcal{G}^{\min}, N, \mathcal{U} \cap \mathcal{G}^{\min}, \mathcal{U}^-, T, S)$ is a refined Tits system.

(10) Recall from 6.1.13 and Theorem 6.1.17 that, for any $1 \leq i \leq \ell$, there exists a connected reductive linear algebraic group $\mathcal{G}_i$ with $\operatorname{Lie}\mathcal{G}_i = \mathfrak{g}_i$ and, moreover, $\mathcal{G}_i$ sits as a subgroup of $\mathcal{G}^{\min}$. Let ω_o be the (unique) conjugate-linear involution of $\mathfrak{g}$ such that $\omega_{o|\mathfrak{g}_{\mathbb{R}}} = \omega_{|\mathfrak{g}_{\mathbb{R}}}$, where ω is the Cartan involution of $\mathfrak{g}$ as in 1.1.2 and $\mathfrak{g}_{\mathbb{R}}$ is a real form of $\mathfrak{g}$ defined in 2.3.9. Then ω_o keeps each $\mathfrak{g}_i$ stable. Let $\mathfrak{k}_i \subset \mathfrak{g}_i$ be its fixed point real Lie subalgebra. Let $K_i \subset \mathcal{G}_i$ be the corresponding connected real Lie subgroup. Define $K \subset \mathcal{G}^{\min}$ as the abstract subgroup generated by the subgroups $\{K_i\}_{1 \leq i \leq \ell}$. Then K is called the *standard unitary form* of $\mathcal{G}$ (or $\mathcal{G}^{\min}$). Prove the following:

(a) Each K_i is compact and the commutator subgroup $[K_i, K_i] \simeq SU(2)$ (as Lie groups). Moreover, $K_i\mathcal{B} = \mathcal{P}_i$ where $\mathcal{P}_i$ is the minimal parabolic subgroup of $\mathcal{G}$.

(b) $K\mathcal{B} = \mathcal{G}$.

(c) K is closed in $\mathcal{G}^{\min}_{\text{an}}$.

Hint. To prove (b), use (a) and Theorem 5.1.3(i_2).

7.C Comments. [Kazhdan–Lusztig–80] and [Lusztig–83] realized the generalized flag varieties, in the affine case, as projective ind-varieties. We give a slight variant of their construction in Section 13.2. The generalized flag varieties associated to the affine SL_n were also studied by [Kac–Peterson–81], where a certain Borel–Weil type result was established in this case. [Tits–82] defined a projective ind-variety structure on the generalized flag variety $\mathcal{G}/\mathcal{B}$ over any field of char. 0. A detailed account of his construction (as well as its extension to an arbitrary $\mathcal{G}/\mathcal{P}_Y$) is given in [Slodowy–84b]. The construction of ind-variety structures on $\mathcal{G}/\mathcal{P}_Y$, as given in Section 7.1 (Subsections 7.1.1–7.1.15), is taken from [Tits–82], [Slodowy–84b]. [Mathieu–88a, §18] realized $\mathcal{G}/\mathcal{P}_Y$ as an ind-scheme over $\mathbb{Z}$.

The Bott–Samelson–Demazure–Hansen varieties were first introduced by [Bott–Samelson–58] in a differential geometric and topological context. [Demazure–74] and [Hansen–73] adapted the construction in algebro-geometric situation and used it to desingularize the Schubert varieties and to determine the Chow group of $\mathcal{G}/\mathcal{B}$ (in the finite case). [Tits–82] extended the definition to the Kac–Moody case. Notion of the stable ind-variety structure on $\mathcal{G}/\mathcal{P}_Y$ (and Corollary 7.1.18) is due to [Kumar–87a]. Proposition 7.1.20 in the finite case was proved by [Bernstein–Gelfand–Gelfand–73]. Exercise 7.1.E.1 generalizes the corresponding result in the finite case due to [Bernstein–Gelfand–Gelfand–73, Theorem 2.9]. Proposition 7.1.21 in the affine case is due to [Kazhdan–Lusztig–80], and in the general case it is proved in [Kac–Peterson–83].

The treatment of homogeneous line bundles as in Section 7.2 is essentially taken from [Slodowy–84b].

Lemma 7.3.10 (in the finite case) is due to [Kazhdan–Lusztig–80]; the general case is similar (cf. [Kumar–90]). Proposition 7.3.7 is taken from [Kumar–96].

As mentioned in (6.C), the "minimum" Kac–Moody group $\mathcal{G}^{\min}$, over a field of char. 0, was introduced in [Kac–Peterson–83] and [Peterson–Kac–83]. Lemma 7.4.2 is due to them. In the symmetrizable case, Theorems 7.4.6, 7.4.14, Corollaries 7.4.13, 7.4.15 and the second part of Proposition 7.4.16 are announced without proofs in [Kac–Peterson–83]. I have not seen any detailed proofs of these in the literature, except that a proof of the result, that $\mathcal{G}/\mathcal{B}$ is a CW-complex, appeared in [Kac–85b]. Exercise 7.4.E.9 is taken from [Kac–Peterson–85a] and the unitary form given in Exercise 7.4.E.10 was studied by [Kac–Peterson–83, 84b, 85a].

VIII

Demazure and Weyl–Kac Character Formulas

We give a proof of the Demazure character formula (in an arbitrary Kac–Moody setting) and, as a consequence, obtain the Weyl–Kac character formula for an arbitrary Kac–Moody Lie algebra. For a dominant integral weight $\lambda \in D_{\mathbb{Z}}$, let $L^{\max}(\lambda)$ be the maximal integrable highest weight $\mathfrak{g}$-module, as defined in 2.1.5, and, for any $w \in W$, let $v_{w\lambda} \in L^{\max}(\lambda)$ be an extremal weight vector of weight $w\lambda$, which is unique up to scalar multiples. Then, the Demazure module $L_w^{\max}(\lambda)$ is, by definition, the $U(\mathfrak{b})$-span of $v_{w\lambda}$. The Demazure character formula determines the character of $L_w^{\max}(\lambda)$ as a T-module. The proof uses algebro-geometric techniques.

For any word $\mathfrak{w} = (s_{i_1}, \dots , s_{i_n})$ of length n, let $Z_{\mathfrak{w}}$ be the Bott–Samelson–Demazure–Hansen variety defined in the last chapter. We begin Section 8.1 by determining the canonical bundle of $Z_{\mathfrak{w}}$. We prove a fundamental, and most crucial, cohomology vanishing theorem for certain line bundles on $Z_{\mathfrak{w}}$. These line bundles arise as the pullback of homogeneous line bundles on $\mathcal{G}/\mathcal{B}$ twisted by the inverse of certain divisors. More specifically, let $1 \leq j \leq k \leq n$ be such that the subword $(s_{i_j}, \dots , s_{i_k})$ is reduced. Then, for any $\lambda \in D_{\mathbb{Z}}$ and $p > 0$, $H^p\Big(Z_{\mathfrak{w}}, \mathcal{L}_{\mathfrak{w}}(\lambda) \otimes \mathcal{O}_{Z_{\mathfrak{w}}}\big(-\sum_{q=j}^{k} Z_{\mathfrak{w}(q)}\big)\Big) = 0$, where $Z_{\mathfrak{w}(q)}$ are the coordinate divisors of $Z_{\mathfrak{w}}$ obtained by taking the q-th coordinate to be 1.

The proof of this vanishing involves the explicit description of the canonical bundle of $Z_{\mathfrak{w}}$ mentioned above, a result of Grauert–Riemenschneider, Serre duality, the Leray spectral sequence, the projection formula and a careful induction. Fix $w \in W$ and take a reduced decomposition $w = s_{i_1} \cdots s_{i_n}$. For the reduced word $\mathfrak{w} = (s_{i_1}, \dots , s_{i_n})$, there is a surjective birational morphism $m_{\mathfrak{w}} : Z_{\mathfrak{w}} \to X_w$, giving rise to the induced morphism $\tilde{m}_{\mathfrak{w}} : Z_{\mathfrak{w}} \to \tilde{X}_w$, where $\tilde{X}_w$ is the normalization of X_w. It is shown that $\tilde{m}_{\mathfrak{w}}$ is a rational resolution, i.e., all the higher direct images of the structure sheaf $\mathcal{O}_{Z_{\mathfrak{w}}}$ of $Z_{\mathfrak{w}}$ via $\tilde{m}_{\mathfrak{w}}$ are zero, and, of course, $\tilde{m}_{\mathfrak{w}*}\mathcal{O}_{Z_{\mathfrak{w}}} = \mathcal{O}_{\tilde{X}_w}$. In fact, a stronger result is proved in Theorem 8.1.13. Given any finite-dimensional pro-representation M of $\mathcal{B}$, we construct, in a functorial manner, an associated vector bundle $\mathcal{L}_{\mathfrak{w}}(M)$ on $Z_{\mathfrak{w}}$. Since $\tilde{m}_{\mathfrak{w}}$ is a rational resolution, we get that $\tilde{m}_{\mathfrak{w}*}\mathcal{L}_{\mathfrak{w}}(M)$ is a locally free sheaf on $\tilde{X}_w$. We next show, using Joseph's functor, that for any $\lambda \in D_{\mathbb{Z}}$, the restricted dual $H^0(Z_\infty, \mathcal{L}_\infty(\lambda))^\vee \simeq L^{\max}(\lambda)$, where $H^0(Z_\infty, \mathcal{L}_\infty(\lambda))$ refers to a certain inverse limit of $H^0(Z_{\mathfrak{w}}, \mathcal{L}_{\mathfrak{w}}(\lambda))$ over the sequences $\mathfrak{w}$. From this we deduce that $H^0(X_w, \mathcal{L}_w(\lambda)) \simeq L_w^{\max}(\lambda)^* \simeq H^0(Z_{\mathfrak{w}}, \mathcal{L}_{\mathfrak{w}}(\lambda))$.

In Section 8.2, we use the above results to prove that any Schubert variety $X_w^Y \subset \mathcal{G}/\mathcal{P}_Y$ is normal, Cohen–Macaulay, has rational singularities, $H^p(X_w^Y, \mathcal{L}_w^Y(\lambda)) = 0$ for all $p > 0$, and the restriction map $H^0(X_w^Y, \mathcal{L}_w^Y(\lambda)) \to H^0(X_v^Y, \mathcal{L}_v^Y(\lambda))$ is surjective whenever $v \leq w$ (for any homogeneous line bundle $\mathcal{L}^Y(\lambda)$ on $\mathcal{G}/\mathcal{P}_Y$ with λ dominant). We also obtain that, in the symmetrizable case, X_w^Y is projectively normal and projectively Cohen–Macaulay in the embedding given by any ample $\mathcal{L}_w^Y(\lambda)$. Moreover, we prove the Demazure character formula, which asserts that, for any $w \in W$, there is an explicit operator D_w on the representation ring $A(T)$ of the torus T such that $\operatorname{ch} L_w^{\max}(\lambda) = D_w(e^\lambda)$.

On taking the appropriate limit, the Demazure character formula leads to a character formula for $L^{\max}(\lambda)$, providing a generalization of the Weyl–Kac character formula to an arbitrary Kac–Moody algebra, proved originally by Kac in the symmetrizable case (cf. the proof given in Chapter 2). There is no purely algebraic proof known so far for this generalization. We extend the Borel–Weil–Bott Theorem to an arbitrary Kac–Moody setup by proving that, for any $\lambda \in D_{\mathbb{Z}}$, $H^p(\mathcal{G}/\mathcal{B}, \mathcal{L}(v * \lambda)) = 0$ unless $p = \ell(v)$ and $H^{\ell(v)}\big(\mathcal{G}/\mathcal{B}, \mathcal{L}(v * \lambda)\big)^\vee \simeq H^0\big(\mathcal{G}/\mathcal{B}, \mathcal{L}(\lambda)\big)^\vee \simeq L^{\max}(\lambda)$, where $*$ is the shifted action of the Weyl group. Moreover, if λ does not belong to $W * D_{\mathbb{Z}}$, then $H^p(\mathcal{G}/\mathcal{B}, \mathcal{L}(\lambda)) = 0$ for all $p \geq 0$. The proof crucially uses a trick by Demazure which relates $H^*(X_w, \mathcal{L}_w(\lambda))$ with $H^*(X_w, \mathcal{L}_w(s_i * \lambda))$ for any simple reflection s_i, as in his simple proof of Bott's theorem [Demazure–76].

In the finite case, a simple proof of the Borel–Weil theorem, using the Peter–Weyl theorem and the Tannaka–Kreĭn duality, is indicated in Exercises 8.3.E. Also, in the exercises, it is indicated how to obtain the Borel–Weil–Bott theorem from Kostant's $\mathfrak{n}$-homology result and vice-versa (in the finite case).

In the non-symmetrizable case it is still not known whether $L^{\max}(\lambda)$ is irreducible. Its irreducibility, for all $\lambda \in D$, will imply the vanishing of the radical of $\mathfrak{g}$ (cf. Remark 8.3.6).

Most of the main results of this chapter are due to Kumar and, independently, to Mathieu.

8.1. Cohomology of Certain Line Bundles on $Z_{\mathfrak{w}}$

In this section $\mathfrak{g}$ is an arbitrary Kac–Moody Lie algebra with the associated group $\mathcal{G}$, as in Theorem 6.1.17. For any $\mathfrak{w} \in \mathfrak{W}$, recall the definition of the Bott–Samelson–Demazure–Hansen variety $Z_{\mathfrak{w}}$ from 7.1.3 and of the line bundle $\mathcal{L}_{\mathfrak{w}}(\lambda) := m_{\mathfrak{w}}^* \mathcal{L}(\lambda)$ on $Z_{\mathfrak{w}}$ for any $\lambda \in \mathfrak{h}_{\mathbb{Z}}^*$ from 7.2.4.

8.1.1 For $\mathfrak{w} = (s_{i_1}, \dots , s_{i_n}) \in \mathfrak{W}$ and $1 \leq q \leq n$, define the subwords of $\mathfrak{w}$:

$$\mathfrak{w}(q) = (s_{i_1}, \dots , \hat{s}_{i_q}, \dots , s_{i_n}), \qquad \text{and}$$
$$\mathfrak{w}[q] = (s_{i_1}, \dots , s_{i_q}).$$

Observe that $\mathfrak{w}(q)$ is nothing but $\hat{\mathfrak{w}}_J$ in the notation of 7.1.3, where $J = 1, 2, \dots , \hat{q}, \dots , n$.

By Corollary 7.1.10(d), $Z_{\mathfrak{w}(q)} \hookrightarrow Z_{\mathfrak{w}}$, via $i_{\mathfrak{w}(q),\mathfrak{w}}$, is a closed irreducible smooth subvariety of codimension one; and the map $\psi_{\mathfrak{w},q} : Z_{\mathfrak{w}} \to Z_{\mathfrak{w}[q]}$, induced from the projection on the first q-factors, is a smooth morphism by Corollary 7.1.10(b).

8.1.2 Proposition. *For any $\mathfrak{w} \in \mathfrak{W}$ of length n, the canonical line bundle $K_{Z_{\mathfrak{w}}}$ of $Z_{\mathfrak{w}}$ is isomorphic with the line bundle*

$$\mathcal{L}_{\mathfrak{w}}(-\rho) \otimes \mathcal{O}_{Z_{\mathfrak{w}}}\Big(-\sum_{q=1}^{n} Z_{\mathfrak{w}(q)}\Big),$$

where $\rho \in \mathfrak{h}^*_{\mathbb{Z}}$ is any element satisfying $\rho(\alpha_i^\vee) = 1$ for all the simple coroots $\alpha_i^\vee$.

Proof. We prove the proposition by induction on $\ell(\mathfrak{w}) = n$: For $n = 1$, $Z_{\mathfrak{w}} = \mathcal{P}_{i_1}/\mathcal{B} \cong \mathbb{P}^1$ and hence $K_{Z_{\mathfrak{w}}} \simeq \mathcal{O}_{\mathbb{P}^1}(-2x_o)$, for any point $x_o \in \mathbb{P}^1$ (cf. [Hartshorne–77, Chap. II, Example 8.20.1]). It can easily be seen that $\mathcal{L}_{\mathfrak{w}}(-\rho) \simeq \mathcal{O}_{\mathbb{P}^1}(-x_o)$. So, the proposition follows in this case.

Recall the $\mathbb{P}^1$-fibration $\psi : Z_{\mathfrak{w}} \to Z_{\mathfrak{w}(n)}$ and the section σ from Corollary 7.1.10(b). By induction, we assume the validity of the proposition for $Z_{\mathfrak{w}(n)}$. Further, it is easy to see that the line bundle $\mathcal{L}_{\mathfrak{w}}(\rho)$ is of degree 1 along the fibers of ψ. Now the proposition follows from Lemmas A.18 and A.16 by observing that

$$(1) \qquad \sigma^*\mathcal{L}_{\mathfrak{w}}(\rho) = \mathcal{L}_{\mathfrak{w}(n)}(\rho). \qquad \square$$

8.1.3 Definition. For any $n \geq 1$, let $\pi_n : \mathcal{B}^n \to \mathcal{B}$, $(b_1, \dots , b_n) \mapsto b_n$, be the projection on the n-th factor. Then π_n is a pro-group morphism. In particular, given a pro-representation M of $\mathcal{B}$, we can view M as a pro-representation of $\mathcal{B}^n$ via π_n. Viewed as a pro-representation of $\mathcal{B}^n$, we denote M by $M(n)$. (If there is no cause for confusion, we simply denote $M(n)$ by M itself.)

For any $\lambda \in \mathfrak{h}^*_{\mathbb{Z}}$, by $\mathbb{C}_\lambda$ we mean the one-dimensional pro-representation of $\mathcal{B}$ such that $\mathcal{U}$ acts trivially and $t \cdot z = e^\lambda(t)\, z$, for $t \in T$ and $z \in \mathbb{C}_\lambda$ (cf. 6.1.6).

Recall the definition of H-equivariant (algebraic) vector bundle for a pro-group H from 7.1.5.

8.1.4 Lemma. *Fix $\mathfrak{w} = (s_{i_1}, \dots, s_{i_n}) \in \mathfrak{W}$ of length n. For any finite-dimensional pro-representation M of the pro-group $\mathcal{B}^n$, there is associated, in a functorial way, a $\mathcal{P}_{i_1}$-equivariant algebraic vector bundle $\mathcal{L}_{\mathfrak{w}}(M)$ on the base $Z_{\mathfrak{w}}$ (which is a $\mathcal{P}_{i_1}$-variety by Proposition 7.1.8(a)) constructed in the proof below.*

In particular, for any $p \geq 0$, $H^p(Z_{\mathfrak{w}}, \mathcal{L}_{\mathfrak{w}}(M))$ is a finite-dimensional pro-representation of $\mathcal{P}_{i_1}$ (and hence; in particular, of $\mathcal{B}$).

Further, for any $\lambda \in \mathfrak{h}^*_{\mathbb{Z}}$, as algebraic line bundles on $Z_{\mathfrak{w}}$,

$$(1) \qquad \mathcal{L}_{\mathfrak{w}}(-\lambda) \simeq \mathcal{L}_{\mathfrak{w}}(\mathbb{C}_\lambda),$$

where $\mathbb{C}_\lambda$ is viewed as a $\mathcal{B}^n$-module as in 8.1.3.

The functor $\mathcal{L}_{\mathfrak{w}}$ is an exact functor on the category of finite-dimensional pro-representations of $\mathcal{B}^n$.

Proof. Since M is a finite-dimensional pro-representation of $\mathcal{B}^n$, there exists a large enough $\mathfrak{k} = (k_1, \dots, k_n)$ (i.e., the k_i's are large enough positive integers) satisfying the condition (7.1.3.7) such that, with the notation as in 7.1.3, the action of $\mathcal{B}^n$ on M factors through the action of $\mathcal{B}^n/\mathcal{U}_{\mathfrak{w}}(\mathfrak{k})$.

Now define

$$\mathcal{L}_{\mathfrak{w}}(M) := \mathcal{P}_{\mathfrak{w}}/\mathcal{U}_{\mathfrak{w}}(\mathfrak{k}) \underset{\mathcal{B}^n/\mathcal{U}_{\mathfrak{w}}(\mathfrak{k})}{\times} M$$

as the vector bundle associated to the locally trivial principal $\mathcal{B}^n/\mathcal{U}_{\mathfrak{w}}(\mathfrak{k})$-bundle $\theta_{\mathfrak{w},\mathfrak{k}} : \mathcal{P}_{\mathfrak{w}}/\mathcal{U}_{\mathfrak{w}}(\mathfrak{k}) \to Z_{\mathfrak{w}}$ (cf. Proposition 7.1.8(b)) via the representation M (cf. A.41). The pro-group $\mathcal{P}_{i_1}$ acts on

$$\mathcal{P}_{\mathfrak{w}}/\mathcal{U}_{\mathfrak{w}}(\mathfrak{k}) := \mathcal{P}_{i_1}/\mathcal{U}_{i_1}(k_1) \times \cdots \times \mathcal{P}_{i_n}/\mathcal{U}_{i_n}(k_n)$$

via the left multiplication on the first factor, and the action is clearly regular (cf. Definition 7.1.5). This induces a regular action of $\mathcal{P}_{i_1}$ on $\mathcal{L}_{\mathfrak{w}}(M)$ by acting only on the $\mathcal{P}_{\mathfrak{w}}/\mathcal{U}_{\mathfrak{w}}(\mathfrak{k})$-factor. This makes $\mathcal{L}_{\mathfrak{w}}(M)$ a $\mathcal{P}_{i_i}$-equivariant algebraic vector bundle.

We now show that the vector bundle $\mathcal{L}_{\mathfrak{w}}(M)$ does not depend on the choice of the sequence $\mathfrak{k}$. So let $\mathfrak{k}'$ be another sequence of positive integers satisfying (7.1.3.7) such that the action of $\mathcal{B}^n$ on M descends to an action of $\mathcal{B}^n/\mathcal{U}_{\mathfrak{w}}(\mathfrak{k}')$, and let $\mathcal{L}'_{\mathfrak{w}}(M)$ be the vector bundle thus obtained. We can assume that $\mathfrak{k} \leq \mathfrak{k}'$, i.e., $k_i \leq k'_i$ for all $1 \leq i \leq n$. Then, we have the canonical $\mathcal{P}_{i_1}$-equivariant quotient map γ making the following diagram commutative:

$$\begin{array}{ccc} \mathcal{P}_{\mathfrak{w}}/\mathcal{U}_{\mathfrak{w}}(\mathfrak{k}') & \xrightarrow{\gamma} & \mathcal{P}_{\mathfrak{w}}/\mathcal{U}_{\mathfrak{w}}(\mathfrak{k}) \\ {\scriptstyle \theta_{\mathfrak{w},\mathfrak{k}'}} \searrow & & \swarrow {\scriptstyle \theta_{\mathfrak{w},\mathfrak{k}}} \\ & Z_{\mathfrak{w}} & \end{array} .$$

From this it is easy to see that there is a $\mathcal{P}_{i_1}$-equivariant bundle morphism α,

$$\begin{array}{ccc} \mathcal{L}'_{\mathfrak{w}}(M) & \xrightarrow{\alpha} & \mathcal{L}_{\mathfrak{w}}(M) \\ & \searrow \quad \swarrow & \\ & Z_{\mathfrak{w}}, & \end{array}$$

which is an isomorphism on the fibers. Thus, α is an isomorphism (cf. Exercise 4.2.E.3). This proves the first part of the lemma.

To prove (1), assume first that $\lambda \in D^\circ$. Fix a nonzero highest weight vector $v_\lambda \in L^{\max}(\lambda)$. Choose a large enough $\mathfrak{k} = (k_1, \dots, k_n)$ as in the proof of Proposition 7.1.12 for $M = \mathbb{C}v_\lambda$ and $V = L^{\max}(\lambda)$, i.e., the map

$$\theta : \mathcal{P}_{i_1} \times \cdots \times \mathcal{P}_{i_n} \times M \to V, (p_1, \dots, p_n; v) \mapsto p_1 \cdots p_n v\,,$$

for $p_j \in \mathcal{P}_{i_j}$ and $v \in M$, has its image in a finite-dimensional subspace $M' \subset V$ and, moreover, θ descends to a morphism

$$\bar{\theta} : \mathcal{P}_{i_1}/\mathcal{U}_{i_1}(k_1) \times \cdots \times \mathcal{P}_{i_n}/\mathcal{U}_{i_n}(k_n) \times M \to M'\,.$$

Now, define the map

$$\beta : \mathcal{L}_{\mathfrak{w}}(\mathbb{C}_\lambda) \to \mathbb{P}(L^{\max}(\lambda)) \times L^{\max}(\lambda),$$

where $\mathcal{L}_{\mathfrak{w}}(\mathbb{C}_\lambda) := \mathcal{P}_{\mathfrak{w}}/\mathcal{U}_{\mathfrak{w}}(\mathfrak{k}) \underset{\mathcal{B}^n/\mathcal{U}_{\mathfrak{w}}(\mathfrak{k})}{\times} \mathbb{C}_\lambda$, by

$$\beta\left((\bar{p}_1, \dots, \bar{p}_n, z) \bmod \mathcal{B}^n/\mathcal{U}_{\mathfrak{w}}(\mathfrak{k})\right) = \left(m^\lambda_{\mathfrak{w}}[p_1, \dots, p_n],\, zp_1 \cdots p_n v_\lambda\right),$$

for $p_j \in \mathcal{P}_{i_j}$ and $z \in \mathbb{C}_\lambda = \mathbb{C}$, where $\bar{p}_j := p_j \bmod \mathcal{U}_{i_j}(k_j)$, and $m^\lambda_{\mathfrak{w}}$ is the map $m_{\mathfrak{w}}(v_\lambda)$ of Proposition 7.1.12. It is easy to see that β induces a morphism $\bar{\beta}$ which is an isomorphism on the fibers:

$$\begin{array}{ccc} \mathcal{L}_{\mathfrak{w}}(\mathbb{C}_\lambda) & \xrightarrow{\bar{\beta}} & \mathcal{L}_{\mathfrak{w}}(-\lambda) \\ & \searrow \quad \swarrow & \\ & Z_{\mathfrak{w}}\,. & \end{array}$$

Thus, $\bar{\beta}$ is an isomorphism of line bundles. Now (1) follows for general $\lambda \in \mathfrak{h}^*_{\mathbb{Z}}$ by writing $\lambda = \lambda_1 - \lambda_2$ with $\lambda_1, \lambda_2 \in D^\circ$, because of the following obvious result:

For any finite-dimensional pro-representations M, N of $\mathcal{B}^n$, there is a canonical isomorphism of vector bundles on $Z_{\mathfrak{w}}$:

$$\mathcal{L}_{\mathfrak{w}}(M) \otimes \mathcal{L}_{\mathfrak{w}}(N) \simeq \mathcal{L}_{\mathfrak{w}}(M \otimes N).$$

The exactness of the functor $\mathcal{L}$ follows trivially from its definition. $\square$

8.1.5 Lemma. *Let M be a finite-dimensional pro-representation of $\mathcal{B}$ and let $\mathfrak{w} = (s_{i_1}, \dots, s_{i_n}) \in \mathfrak{W}$. Fix $1 \le j \le n$. Then, for any $p \ge 0$, the sheaf $R^p\psi_*(\mathcal{L}_{\mathfrak{w}}(M))$ is canonically isomorphic with the vector bundle $\mathcal{L}_{\mathfrak{w}[j]}(N)$, where $\psi = \psi_{\mathfrak{w},j} : Z_{\mathfrak{w}} \to Z_{\mathfrak{w}[j]}$ is as in 8.1.1, N is the pro-representation $H^p(Z_{\mathfrak{v}}, \mathcal{L}_{\mathfrak{v}}(M))$ of $\mathcal{B}$ (cf. Lemma 8.1.4), and $\mathfrak{v} := (s_{i_{j+1}}, \dots, s_{i_n})$.*

Proof. Choose any sequence $\mathfrak{k} = (k_1, \dots, k_n)$ satisfying (7.1.3.7), so that $\theta = \theta_{\mathfrak{w},\mathfrak{k}} : \mathcal{P}_{\mathfrak{w}}/\mathcal{U}_{\mathfrak{w}}(\mathfrak{k}) \to Z_{\mathfrak{w}}$ is a locally trivial principal $\mathcal{B}^n/\mathcal{U}_{\mathfrak{w}}(\mathfrak{k})$-bundle (cf. Proposition 7.1.8(b)). We assume further (choosing k_n large enough) that the action of $\mathcal{B}$ on M descends to an action of $\mathcal{B}/\mathcal{U}_{i_n}(k_n)$.

Consider the locally trivial principal H-bundle $\bar{\theta} : (\mathcal{P}_{\mathfrak{w}}/\mathcal{U}_{\mathfrak{w}}(\mathfrak{k}))/L \to Z_{\mathfrak{w}}$, where $H := \mathcal{B}/\mathcal{U}_{i_j}(k_j)$, $L := L_1 \times 1 \times L_2$, $L_1 := \mathcal{B}/\mathcal{U}_{i_1}(k_1) \times \cdots \times \mathcal{B}/\mathcal{U}_{i_{j-1}}(k_{j-1})$, $L_2 := \mathcal{B}/\mathcal{U}_{i_{j+1}}(k_{j+1}) \times \cdots \times \mathcal{B}/\mathcal{U}_{i_n}(k_n)$, and H is identified canonically with $(\mathcal{B}^n/\mathcal{U}_{\mathfrak{w}}(\mathfrak{k}))/L$. Let

$$Z'_{\mathfrak{w}[j]} := \left(\mathcal{P}_{i_1}/\mathcal{U}_{i_1}(k_1) \times \cdots \times \mathcal{P}_{i_j}/\mathcal{U}_{i_j}(k_j)\right) / L_1 \times 1\,,$$

where the action of $L_1 \times 1$ is as given by (7.1.3.9). Then $(\mathcal{P}_{\mathfrak{w}}/\mathcal{U}_{\mathfrak{w}}(\mathfrak{k}))/L$ can canonically be identified with $Z'_{\mathfrak{w}[j]} \times Z_{\mathfrak{v}}$. We have the following commutative diagram:

$$\begin{array}{ccc}
Z'_{\mathfrak{w}[j]} \times Z_{\mathfrak{v}} & \xrightarrow{\ \bar{\theta}\ } & Z_{\mathfrak{w}} \\
\downarrow{\scriptstyle \pi_1} & & \downarrow{\scriptstyle \psi} \\
Z'_{\mathfrak{w}[j]} & \xrightarrow[\bar{\theta}_j]{} & Z_{\mathfrak{w}[j]}\,,
\end{array}$$

where π_1 is the projection on the first factor and $\bar{\theta}_j$ is nothing but the map $\bar{\theta}$ with $\mathfrak{w}$ replaced by $\mathfrak{w}[j]$. Both of $\bar{\theta}$ and $\bar{\theta}_j$ are locally trivial principal H-bundles. It is easy to see that $\bar{\theta}$ descends to a biregular isomorphism $\hat{\theta}$ making the following diagram commutative:

$$\begin{array}{ccc}
Z'_{\mathfrak{w}[j]} \times_H Z_{\mathfrak{v}} & \xrightarrow[\sim]{\ \hat{\theta}\ } & Z_{\mathfrak{w}} \\
& {\scriptstyle (\bar{\theta}_j)_{Z_{\mathfrak{v}}}} \searrow \quad \swarrow {\scriptstyle \psi} & \\
& Z_{\mathfrak{w}[j]}\,, &
\end{array}$$

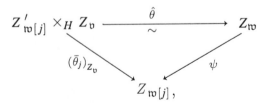

where $(\bar{\theta}_j)_{Z_{\mathfrak{v}}}$ is the fibration with fiber $Z_{\mathfrak{v}}$ associated to the principal H-bundle $\bar{\theta}_j$. It is easy to see that the action of $\mathcal{B}$ on $\mathcal{L}_{\mathfrak{v}}(M)$ (obtained from the restriction of the $\mathcal{P}_{i_{j+1}}$–action) descends to give an H-equivariant vector bundle structure on $\mathcal{L}_{\mathfrak{v}}(M)$ (cf. the proof of Lemma 8.1.4). Moreover, we have

$$(1) \qquad \bar{\theta}^*(\mathcal{L}_{\mathfrak{w}}(M)) \simeq \epsilon_{Z'_{\mathfrak{w}[j]}} \boxtimes \mathcal{L}_{\mathfrak{v}}(M), \text{ as } H\text{-equivariant vector bundles,}$$

where $\epsilon_{Z'_{\mathfrak{w}[j]}}$ is the trivial line bundle on $Z'_{\mathfrak{w}[j]}$ with the obvious H-action, $\boxtimes$ denotes the external tensor product (cf. Example 4.2.7(e)), and $\bar{\theta}$ being an H-equivariant morphism (with the trivial action of H on $Z_{\mathfrak{w}}$), $\bar{\theta}^*(\mathcal{L}_{\mathfrak{w}}(M))$ acquires a canonical H-equivariant vector bundle structure (with the trivial action of H on $\mathcal{L}_{\mathfrak{w}}(M)$). Thus, by Lemma A.42,

$$(2) \qquad \hat{\theta}^*(\mathcal{L}_{\mathfrak{w}}(M)) \simeq \mathcal{L}_{\mathfrak{v}}(M)_{\bar{\theta}_j}, \text{ as vector bundles,}$$

$\mathcal{L}_{\mathfrak{v}}(M)_{\bar{\theta}_j}$ is the unique vector bundle on $Z'_{\mathfrak{w}[j]} \times_H Z_{\mathfrak{v}}$ such that

$$\hat{\theta}_j^*(\mathcal{L}_{\mathfrak{v}}(M)_{\bar{\theta}_j}) \simeq \epsilon_{Z'_{\mathfrak{w}[j]}} \boxtimes \mathcal{L}_{\mathfrak{v}}(M),$$

as H-equivariant vector bundles on $Z'_{\mathfrak{w}[j]} \times Z_{\mathfrak{v}}$ with $\hat{\theta}_j : Z'_{\mathfrak{w}[j]} \times Z_{\mathfrak{v}} \to Z'_{\mathfrak{w}[j]} \times_H Z_{\mathfrak{v}}$ being the orbit map. The lemma follows now by using (2) and Lemma A.42. □

8.1.6 Definition. Let $\mathfrak{w} = (s_{i_1}, \dots, s_{i_n}) \in \mathfrak{W}$ be any word. Recall from 7.1.11 the $\mathcal{B}$-equivariant morphism $m_{\mathfrak{w}} : Z_{\mathfrak{w}} \to X_m$ for some m; in fact, we can take $m = n$ by using Theorem 5.1.3 (d). Since $Z_{\mathfrak{w}}$ is irreducible and projective (cf. Proposition 7.1.8(a) and Corollary 7.1.10(c)), Im $m_{\mathfrak{w}}$ is a closed irreducible $\mathcal{B}$-stable subset of X_m. In particular, Im $m_{\mathfrak{w}} = X_w$ for some $w \in W$.

Let $\tilde{X}_w$ be the normalization of X_w. From the uniqueness of normalization (cf. Proposition A.6), we see that $\tilde{X}_w$ is again a $\mathcal{B}$-variety and, moreover, it is projective (cf. loc. cit.). The morphism $m_{\mathfrak{w}}$ lifts uniquely to a $\mathcal{B}$-equivariant morphism $\tilde{m}_{\mathfrak{w}}$, making the following diagram commutative (cf. Proposition A.7):

$$\begin{array}{ccc} & & \tilde{X}_w \\ & \overset{\tilde{m}_{\mathfrak{w}}}{\nearrow} & \big\downarrow \nu \\ Z_{\mathfrak{w}} & \xrightarrow[m_{\mathfrak{w}}]{} & X_w \,. \end{array}$$

By Proposition 7.1.15, if $\mathfrak{w}$ is reduced, $m_{\mathfrak{w}}$ is a birational map onto X_w, where $w := \pi(\mathfrak{w})$. If $\mathfrak{w}$ is not reduced, w is in general different from $\pi(\mathfrak{w})$.

8.1.7 Proposition. *Let $\mathfrak{w} = (s_{i_1}, \dots, s_{i_n}) \in \mathfrak{W}$ be reduced with $\pi(\mathfrak{w}) = w$. Assume that*

$$(1) \qquad H^i(Z_{\mathfrak{w}}, \mathcal{L}_{\mathfrak{w}}(p\lambda)) = 0, \textit{ for all } p, i > 0 \textit{ and some } \lambda \in D^\circ.$$

Then, for any finite-dimensional pro-representation M of $\mathcal{B}$,

(2) $$R^i \tilde{m}_{\mathfrak{w}^*}(\mathcal{L}_{\mathfrak{w}}(M)) = 0, \quad \text{for all } i > 0, \text{ and}$$

(3) *$\tilde{m}_{\mathfrak{w}^*}(\mathcal{L}_{\mathfrak{w}}(M))$ is a $\mathcal{B}$-equivariant vector bundle on $\tilde{X}_w$ of rank $= \dim M$.*

Moreover, the canonical $\mathcal{O}_{Z_{\mathfrak{w}}}$-module map (cf. Lemma A.8)

(4) $$\tilde{m}^*_{\mathfrak{w}} \tilde{m}_{\mathfrak{w}^*}(\mathcal{L}_{\mathfrak{w}}(M)) \to \mathcal{L}_{\mathfrak{w}}(M) \text{ is an isomorphism.}$$

If $\mathfrak{v} = (s_{j_1}, \dots, s_{j_n}) \in \mathfrak{W}$ is another reduced word with $\pi(\mathfrak{v}) = w$, then there is a canonical isomorphism, as given in the proof below:

(5) $$\tilde{m}_{\mathfrak{v}^*}(\mathcal{L}_{\mathfrak{v}}(M)) \simeq \tilde{m}_{\mathfrak{w}^*}(\mathcal{L}_{\mathfrak{w}}(M))$$

of $\mathcal{B}$-equivariant $\mathcal{O}_{\tilde{X}_w}$-modules on $\tilde{X}_w$. In particular, $\tilde{m}_{\mathfrak{v}^*}(\mathcal{L}_{\mathfrak{v}}(M))$ is a $\mathcal{B}$-equivariant vector bundle. Observe that, a priori, we are not assuming (1) for $Z_{\mathfrak{v}}$.

Further, under the assumption (1) for $\mathfrak{w}$ as well as $\mathfrak{v}$, as $\mathcal{B}$-modules,

(6) $$H^i(Z_{\mathfrak{w}}, \mathcal{L}_{\mathfrak{w}}(M)) \simeq H^i(Z_{\mathfrak{v}}, \mathcal{L}_{\mathfrak{v}}(M)), \text{ for all } i \geq 0.$$

We denote the $\mathcal{B}$-equivariant vector bundle $\tilde{m}_{\mathfrak{w}^*}(\mathcal{L}_{\mathfrak{w}}(M))$ on $\tilde{X}_w$ by $\tilde{\mathcal{L}}_w(M)$.

Proof. We prove (2) and (3) by induction on $\dim M$. If $\dim M = 1$, then $M \simeq \mathbb{C}_\mu$ for some $\mu \in \mathfrak{h}^*_{\mathbb{Z}}$. Further, by Lemma 8.1.4,

$$\mathcal{L}_{\mathfrak{w}}(\mathbb{C}_\mu) \simeq m^*_{\mathfrak{w}}(\mathcal{L}_w(-\mu)) \simeq \tilde{m}^*_{\mathfrak{w}}(\nu^*(\mathcal{L}_w(-\mu))),$$

where $\mathcal{L}_w(-\mu) := \mathcal{L}(-\mu)_{|X_w}$. Then, by the projection formula (cf. Theorem A.22), one has

$$R^i \tilde{m}_{\mathfrak{w}*}(\mathcal{L}_{\mathfrak{w}}(\mathbb{C}_\mu)) \simeq R^i \tilde{m}_{\mathfrak{w}*}(\mathcal{O}_{Z_{\mathfrak{w}}}) \otimes \nu^* \mathcal{L}_w(-\mu).$$

Thus, (2) and (3) will follow for $M = \mathbb{C}_\mu$ if one shows that

$$R^i \tilde{m}_{\mathfrak{w}*}(\mathcal{O}_{Z_{\mathfrak{w}}}) \simeq \begin{cases} 0 & \text{if } i > 0; \\ \mathcal{O}_{\tilde{X}_w} & \text{if } i = 0. \end{cases}$$

First, since $\tilde{X}_w$ is normal, the equality

$$\tilde{m}_{\mathfrak{w}*}(\mathcal{O}_{Z_{\mathfrak{w}}}) = \mathcal{O}_{\tilde{X}_w}$$

follows from Zariski's main theorem (A.9.1). Next, recall that $\tilde{X}_w$ is a projective variety, as noted in 8.1.6. Since the ampleness is preserved by a finite morphism (cf. Proposition A.30), $\nu^*\mathcal{L}_w(-\lambda)$ is an ample line bundle on $\tilde{X}_w$ (cf. Exercise 7.2.E.2). Therefore, the assumption (1) implies, by a lemma due to Kempf (see Lemma A.31), that

$$R^i\tilde{m}_{\mathfrak{w}*}(\mathcal{O}_{Z_{\mathfrak{w}}}) = 0, \text{ for } i > 0.$$

This proves (2) and (3) when $\dim M = 1$.

Now we come to the general M with $\dim M > 1$. Since M is a pro-representation of $\mathcal{B}$, there exists a $\mathcal{U}(q) = \mathcal{U}_\emptyset(q)$ (cf. Lemma 6.1.14) such that M is a module for the solvable group $\mathcal{B}/\mathcal{U}(q)$. In particular, by Lie's theorem, there exists an exact sequence of $\mathcal{B}/\mathcal{U}(q)$-modules (cf. [Humphreys–72, Theorem 4.1]):

$$0 \to N \to M \to Q \to 0,$$

with both N and Q nonzero. Hence we get, abbreviating $\tilde{m}_{\mathfrak{w}}$ by m, the long exact sequence of sheaves:

$$(7) \quad \begin{aligned} &0 \to m_*(\mathcal{L}_{\mathfrak{w}}(N)) \to m_*(\mathcal{L}_{\mathfrak{w}}(M)) \to m_*(\mathcal{L}_{\mathfrak{w}}(Q)) \to \cdots \\ &\to R^i m_*(\mathcal{L}_{\mathfrak{w}}(N)) \to R^i m_*(\mathcal{L}_{\mathfrak{w}}(M)) \to R^i m_*(\mathcal{L}_{\mathfrak{w}}(Q)) \to \cdots. \end{aligned}$$

By induction, we have the validity of (2) and (3) for N and Q. Hence, by (7), we get the validity of (2) and (3) for M. Observe that, since m is a $\mathcal{B}$-equivariant morphism and $\mathcal{L}_{\mathfrak{w}}(M)$ is a $\mathcal{B}$-equivariant vector bundle, $m_*(\mathcal{L}_{\mathfrak{w}}(M))$ is a $\mathcal{B}$-equivariant sheaf.

By (2) and (3) we have the following commutative diagram with exact rows; exactness of the top row follows since m^* is an exact functor on the . of locally free sheaves.

$$\begin{array}{ccccccc} 0 \to & m^*m_*(\mathcal{L}_{\mathfrak{w}}(N)) & \to & m^*m_*(\mathcal{L}_{\mathfrak{w}}(M)) & \to & m^*m_*(\mathcal{L}_{\mathfrak{w}}(Q)) & \to 0 \\ & \downarrow & & \downarrow & & \downarrow & \\ 0 \to & \mathcal{L}_{\mathfrak{w}}(N) & \to & \mathcal{L}_{\mathfrak{w}}(M) & \to & \mathcal{L}_{\mathfrak{w}}(Q) & \to 0. \end{array}$$

Now (4) follows by induction on $\dim M$ and the five lemma [Spanier–66, Lemma 11, Sec. 5, Ch. 4].

We now come to the proof of (5): By making successive use of the Coxeter relations as in Theorem 1.3.11(a), we can assume, by reindexing the s_i's if necessary, that the reduced words $\mathfrak{w}$ and $\mathfrak{v}$ are as follows: $\mathfrak{w} = \mathfrak{w}' * \mathfrak{w}_o * \mathfrak{w}''$ and $\mathfrak{v} = \mathfrak{w}' * \mathfrak{w}'_o * \mathfrak{w}''$, for some (possibly empty) words $\mathfrak{w}' = (s_{i_1}, \ldots, s_{i_q})$, $\mathfrak{w}'' = (s_{i_r}, \ldots, s_{i_n})$, $\mathfrak{w}_o = (s_1, s_2, s_1, s_2, \ldots)$, and $\mathfrak{w}'_o = (s_2, s_1, s_2, s_1, \ldots)$, where the subgroup W_2 of W generated by s_1 and s_2 is finite, $\mathfrak{w}_o$ and $\mathfrak{w}'_o$ both are of length m, and s_1s_2 is of order m. In particular, $\pi(\mathfrak{w}_o) = \pi(\mathfrak{w}'_o)$ is the maximal element of W_2, providing two different decompositions of $\pi(\mathfrak{w}_o)$. Let $\mathcal{P} = \mathcal{P}_Y \subset \mathcal{G}$ be the standard parabolic subgroup (cf. Definition 6.1.18), where $Y := \{1, 2\}$ is of finite type. Choose $\mathcal{U}_{i_1}(k_1) \subset \cdots \subset \mathcal{U}_{i_n}(k_n)$ as in (7.1.3.7) with the following additional conditions: (a) $\mathcal{U}_{i_n}(k_n)$ acts trivially on M, and (b) $\mathcal{U}_{i_{q+1}}(k_{q+1}) = \cdots = \mathcal{U}_{i_{r-1}}(k_{r-1}) = \mathcal{U}_Y(k)$, for some normal pro-subgroup $\mathcal{U}_Y(k)$ of $\mathcal{P}$ (cf. Lemma 6.1.14).

Consider the product algebraic group

$$C := \mathcal{P}_{\mathfrak{w}'}/\mathcal{U}_{\mathfrak{w}'}(\mathfrak{k}') \times \mathcal{P}/\mathcal{U}_Y(k) \times \mathcal{P}_{\mathfrak{w}''}/\mathcal{U}_{\mathfrak{w}''}(\mathfrak{k}''),$$

where $\mathfrak{k}' := (k_1, \ldots, k_q)$, $\mathfrak{k}'' := (k_r, \ldots, k_n)$. The group

$$H := \mathcal{B}^q/\mathcal{U}_{\mathfrak{w}'}(\mathfrak{k}') \times \mathcal{B}/\mathcal{U}_Y(k) \times \mathcal{B}^{n-r+1}/\mathcal{U}_{\mathfrak{w}''}(\mathfrak{k}'')$$

acts on C on the right by the same formula as (7.1.3.9). Moreover, the orbit space $Z := C/H$ acquires a natural smooth variety structure such that $C \to Z$ is a locally trivial principal H-bundle. (The proof is similar to the proof of Proposition 7.1.8 and hence is omitted.) Observe that the $\mathcal{B}$-module M gives rise to a canonical H-module structure on M via Definition 8.1.3 by using the assumption (a) as above. In particular, to M, there is associated a $\mathcal{B}$-equivariant vector bundle $\mathcal{L}_Z(M) := C \underset{H}{\times} M \to Z$ (cf. A.41), where $\mathcal{B}$ acts on $\mathcal{L}_Z(M)$ via the left multiplication on the first factor of C.

Now define the morphism

$$f_1 : \mathcal{P}_{\mathfrak{w}}/\mathcal{U}_{\mathfrak{w}}(\mathfrak{k}) \to C,$$

where $\mathcal{P}_{\mathfrak{w}}/\mathcal{U}_{\mathfrak{w}}(\mathfrak{k}) := \mathcal{P}_{\mathfrak{w}'}/\mathcal{U}_{\mathfrak{w}'}(\mathfrak{k}') \times \mathcal{P}_{\mathfrak{w}_o}/\mathcal{U}_{\mathfrak{w}_o}(\mathfrak{k}_o) \times \mathcal{P}_{\mathfrak{w}''}/\mathcal{U}_{\mathfrak{w}''}(\mathfrak{k}'')$ and $\mathfrak{k}_o := (k_{q+1}, \ldots, k_{r-1})$, by

$$f_1(\bar{p}_1, \ldots, \bar{p}_q, \bar{p}_{q+1}, \ldots, \bar{p}_{r-1}, \bar{p}_r, \ldots, \bar{p}_n) = (\bar{p}_1, \ldots, \bar{p}_q, \overline{p_{q+1} \cdots p_{r-1}}, \bar{p}_r, \ldots, \bar{p}_n),$$

and, similarly, define

$$f_2 : \mathcal{P}_{\mathfrak{v}}/\mathcal{U}_{\mathfrak{v}}(\mathfrak{k}) \to C.$$

These morphisms descend to give $\mathcal{B}$-equivariant morphisms $\hat{f}_1, \hat{f}_2$ and $\bar{f}_1, \bar{f}_2$, making the following diagram commutative:

$$\begin{array}{ccccc} \mathcal{L}_{\mathfrak{w}}(M) & \xrightarrow{\hat{f}_1} & \mathcal{L}_Z(M) & \xleftarrow{\hat{f}_2} & \mathcal{L}_{\mathfrak{v}}(M) \\ \downarrow & & \downarrow & & \downarrow \\ Z_{\mathfrak{w}} & \underset{\bar{f}_1}{\longrightarrow} & Z & \underset{\bar{f}_2}{\longleftarrow} & Z_{\mathfrak{v}}, \end{array}$$

where all the vertical maps are bundle projections. Moreover, $\hat{f}_1$ and $\hat{f}_2$ are linear isomorphisms on the fibers. In particular, as $\mathcal{B}$-equivariant vector bundles,

$$\bar{f}_1^* \mathcal{L}_Z(M) \simeq \mathcal{L}_{\mathfrak{w}}(M), \quad \bar{f}_2^* \mathcal{L}_Z(M) \simeq \mathcal{L}_{\mathfrak{v}}(M). \tag{8}$$

Clearly, $\bar{f}_1$ and $\bar{f}_2$ are both surjective (by Theorem 5.1.3(i_2)).
Define the $\mathcal{B}$-equivariant morphism $\theta : Z \to X_w$ by

$$[p_1, \dots, p_q, p, p_r, \dots, p_n] \mapsto p_1 \cdots p_q p \, p_r \cdots p_n ,$$

for $p_j \in \mathcal{P}_{i_j}$ and $p \in \mathcal{P}$. (By Theorem 5.1.3(i), Im θ indeed lies in X_w.) Then, clearly,

$$\theta \circ \bar{f}_1 = m_{\mathfrak{w}} \text{ and } \theta \circ \bar{f}_2 = m_{\mathfrak{v}}.$$

Since $m_{\mathfrak{w}}$, resp. $m_{\mathfrak{v}}$, is birational, so is $\bar{f}_1$, resp. $\bar{f}_2$. Further, Z being smooth (in particular normal), θ induces a $\mathcal{B}$-equivariant morphism $\tilde{\theta} : Z \to \tilde{X}_w$ and, of course,

$$\tilde{\theta} \, \bar{f}_1 = \tilde{m}_{\mathfrak{w}} \quad \text{and} \quad \tilde{\theta} \, \bar{f}_2 = \tilde{m}_{\mathfrak{v}}. \tag{9}$$

Hence, by (8) and (9), as $\mathcal{B}$-equivariant vector bundles on $\tilde{X}_w$,

$$\begin{aligned} \tilde{m}_{\mathfrak{w}^*}(\mathcal{L}_{\mathfrak{w}}(M)) &\simeq \tilde{\theta}_* \, \bar{f}_{1*} \, \bar{f}_1^*(\mathcal{L}_Z(M)) \\ &\simeq \tilde{\theta}_*(\mathcal{L}_Z(M)), \end{aligned} \tag{10}$$

by the projection formula (A.22.2). Similarly, as $\mathcal{B}$-equivariant sheaves,

$$\tilde{m}_{\mathfrak{v}^*}(\mathcal{L}_{\mathfrak{v}}(M)) \simeq \tilde{\theta}_*(\mathcal{L}_Z(M)). \tag{11}$$

Of course (10) and (11) together prove (5).
By the Leray Spectral Sequence Theorem A.23, (2) and (5), we get

$$\begin{aligned} H^i(Z_{\mathfrak{w}}, \mathcal{L}_{\mathfrak{w}}(M)) &\simeq H^i(\tilde{X}_w, \tilde{m}_{\mathfrak{w}*}\mathcal{L}_{\mathfrak{w}}(M)) \\ &\simeq H^i(\tilde{X}_w, \tilde{m}_{\mathfrak{v}*}\mathcal{L}_{\mathfrak{v}}(M)) \simeq H^i(Z_{\mathfrak{v}}, \mathcal{L}_{\mathfrak{v}}(M)). \end{aligned}$$

This proves (6). □

With these preliminaries, we come to the following most crucial result of this chapter.

8.1.8 Theorem. *Let* $\mathfrak{w} = (s_{i_1}, \dots, s_{i_n}) \in \mathfrak{W}$ *be any word and let* $1 \le j \le k \le n$ *be such that the subword* $\mathfrak{v} = (s_{i_j}, \dots, s_{i_k})$ *is reduced. Then, for any* $\lambda \in D_{\mathbb{Z}}$,

(1) $$H^p\Big(Z_{\mathfrak{w}}, \mathcal{L}_{\mathfrak{w}}(\lambda) \otimes \mathcal{O}_{Z_{\mathfrak{w}}}\big(-\sum_{q=j}^{k} Z_{\mathfrak{w}(q)}\big)\Big) = 0, \textit{ for all } p > 0.$$

We also have

(1′) $$H^p(Z_{\mathfrak{w}}, \mathcal{L}_{\mathfrak{w}}(\lambda)) = 0, \textit{ for all } p > 0.$$

Further, if $k < n$ *and* $(s_{i_j}, \dots, s_{i_k}, s_{i_{k+1}})$ *is* **not** *reduced (and* $\mathfrak{v}$ *is reduced), then, in fact, we have*

(2) $$H^p\Big(Z_{\mathfrak{w}}, \mathcal{L}_{\mathfrak{w}}(\lambda) \otimes \mathcal{O}_{Z_{\mathfrak{w}}}\big(-\sum_{q=j}^{k} Z_{\mathfrak{w}(q)}\big)\Big) = 0 \textit{ for all } p \ge 0.$$

Proof. It is easy to see that the theorem is true for $n = 1$, because then $Z_{\mathfrak{w}} \simeq \mathbb{P}^1$ and $H^1(\mathbb{P}^1, \mathcal{O}(q)) = 0$ for $q \ge -1$. Now we assume, by induction, that (1) is true for any sequence $\mathfrak{u} = (s_{j_1}, \dots, s_{j_{n'}})$ with $n' < n$ and any choice of $1 \le j' \le k' \le n'$ satisfying the assumption of the theorem.

Now we work with the fixed sequence $\mathfrak{w} = (s_{i_1}, \dots, s_{i_n})$. We claim that if we know the validity of (1) for some $1 \le j < k \le n$, then (1) is also valid when we replace k by $k-1$. To prove this, consider the sheaf exact sequence corresponding to the codimension-one subvariety $i : Z_{\mathfrak{w}(k)} \subset Z_{\mathfrak{w}}$ (cf. (A.15.1)):

(3) $$0 \to \mathcal{O}_{Z_{\mathfrak{w}}}(-Z_{\mathfrak{w}(k)}) \to \mathcal{O}_{Z_{\mathfrak{w}}} \to i_*\mathcal{O}_{Z_{\mathfrak{w}(k)}} \to 0.$$

Tensoring this over $\mathcal{O}_{Z_{\mathfrak{w}}}$ with the locally free sheaf

$$\mathcal{L}_{\mathfrak{w}}(\lambda) \otimes \mathcal{O}_{Z_{\mathfrak{w}}}\big(-\sum_{q=j}^{k-1} Z_{\mathfrak{w}(q)}\big),$$

we get the exact sequence

(4) $$0 \to \mathcal{L}_{\mathfrak{w}}(\lambda) \otimes \mathcal{O}_{Z_{\mathfrak{w}}}\big(-\sum_{q=j}^{k} Z_{\mathfrak{w}(q)}\big) \to \mathcal{L}_{\mathfrak{w}}(\lambda) \otimes \mathcal{O}_{Z_{\mathfrak{w}}}\big(-\sum_{q=j}^{k-1} Z_{\mathfrak{w}(q)}\big)$$
$$\to (i_*\mathcal{O}_{Z_{\mathfrak{w}(k)}}) \otimes \Big(\mathcal{L}_{\mathfrak{w}}(\lambda) \otimes \mathcal{O}_{Z_{\mathfrak{w}}}\big(-\sum_{q=j}^{k-1} Z_{\mathfrak{w}(q)}\big)\Big) \to 0.$$

But, as can be easily seen from Proposition 7.1.8, $Z_{\mathfrak{w}(k)}$ intersects transversally (inside $Z_{\mathfrak{w}}$) $Z_{\mathfrak{w}(q)}$ for any $q \neq k$; the intersection being the codimension-one subvariety $Z_{\mathfrak{w}(k)(q)}$ of $Z_{\mathfrak{w}(k)}$. Hence, the last sheaf in the above sequence (4) can be identified with the sheaf

$$i_*\left(\mathcal{L}_{\mathfrak{w}(k)}(\lambda) \otimes \mathcal{O}_{Z_{\mathfrak{w}(k)}}\Big(-\sum_{q=j}^{k-1} Z_{\mathfrak{w}(q)} \cap Z_{\mathfrak{w}(k)}\Big)\right),$$

by Lemma A.17. Now, by the induction hypothesis and the long exact cohomology sequence corresponding to the sheaf sequence (4), the assertion (validity of (1) for $j \leq k-1$) follows since $H^p(Z_{\mathfrak{w}}, i_*\mathcal{F}) \simeq H^p(Z_{\mathfrak{w}(k)}, \mathcal{F})$, for any sheaf $\mathcal{F}$ on $Z_{\mathfrak{w}(k)}$ (cf. Lemma A.21). Also, from the long exact cohomology sequence, corresponding to the sheaf sequence (3) tensored with $\mathcal{L}_{\mathfrak{w}}(\lambda)$, it follows that $H^p(Z_{\mathfrak{w}}, \mathcal{L}_{\mathfrak{w}}(\lambda)) = 0$ for all $p > 0$, provided we know the validity of (1) for the pair $k \leq k$. So, we can assume now that the pair $j \leq k$ is such that the sequence $\mathfrak{v}$ is (as given) reduced, and either (a) $k = n$, or (b) $k < n$ and the sequence $(s_{i_j}, \dots, s_{i_k}, s_{i_{k+1}})$ is not reduced.

We deal with these two cases separately.

Case (a), $k = n$. By an argument exactly similar to the one used above, we can assume that we have two subcases:

(a_1) $j = 1$, or

(a_2) $j > 1$ and $(s_{i_{j-1}}, s_{i_j}, \dots, s_{i_k})$ is not reduced.

In the case (a_1), i.e., $j = 1$, the theorem follows from a result of Grauert–Riemenschneider (see Theorem A.33), since

$$\begin{aligned} H^p\Big(Z_{\mathfrak{w}}, \mathcal{L}_{\mathfrak{w}}(\lambda) \otimes \mathcal{O}_{Z_{\mathfrak{w}}}\Big(-\sum_{q=1}^{n} Z_{\mathfrak{w}(q)}\Big)\Big) &\\ = H^p(Z_{\mathfrak{w}}, \mathcal{L}_{\mathfrak{w}}(\lambda+\rho) \otimes K_{Z_{\mathfrak{w}}}), &\text{ by Proposition 8.1.2} \\ \approx H^{n-p}(Z_{\mathfrak{w}}, \mathcal{L}_{\mathfrak{w}}(-(\lambda+\rho)))^*, &\text{ by Serre duality Theorem A.37} \\ = 0, \qquad &\text{by Theorem A.33,} \end{aligned}$$

since $\mathcal{L}_w(\lambda+\rho)$ is ample on X_w (cf. Exercise 7.2.E.2) and $m_{\mathfrak{w}} : Z_{\mathfrak{w}} \to X_w$ is a birational morphism by Proposition 7.1.15(a).

Case (a_2). Since $v := s_{i_j} \cdots s_{i_n}$ is reduced and $s_{i_{j-1}} s_{i_j} \cdots s_{i_n}$ is not reduced, we can write $v = s_{i_{j-1}} r_{j+1} \cdots r_n$ for some simple reflections $r_{j+1}, \dots, r_n$. Define the following sequences:

$$\mathfrak{u} = (s_{i_1}, \dots, s_{i_{j-1}}),\ \mathfrak{v} = (s_{i_j}, \dots, s_{i_n}),\ \text{and } \mathfrak{v}' = (s_{i_{j-1}}, r_{j+1}, \dots, r_n).$$

We have the canonical projection $\psi = \psi_{\mathfrak{w},j-1} : Z_{\mathfrak{w}} \to Z_{\mathfrak{u}}$. By the induction hypothesis,

$$H^p\left(Z_{\mathfrak{v}}, \mathcal{L}_{\mathfrak{v}}(\lambda) \otimes \mathcal{O}_{Z_{\mathfrak{v}}}\left(-\sum_{q=1}^{n-j+1} Z_{\mathfrak{v}(q)}\right)\right) = 0, \qquad \text{for all } p > 0.$$

Hence, by the Leray spectral sequence Theorem A.23 for the map ψ together with the proof of Lemma 8.1.5 (by observing that, in the notation of loc. cit., $\bar{\theta}^{-1}(Z_{\mathfrak{w}(q)}) = Z'_{\mathfrak{u}} \times Z_{\mathfrak{v}(q-j+1)}$ for any $j \le q \le n$), we get

$$\begin{aligned} H^p\Big(Z_{\mathfrak{w}}, \mathcal{L}_{\mathfrak{w}}(\lambda) \otimes \mathcal{O}_{Z_{\mathfrak{w}}}\big(-\sum_{q=j}^{n} Z_{\mathfrak{w}(q)}\big)\Big) & \\ \approx H^p\Big(Z_{\mathfrak{u}}, \mathcal{L}_{\mathfrak{u}}\big(H^0\big(Z_{\mathfrak{v}}, \mathcal{L}_{\mathfrak{v}}(\lambda) \otimes \mathcal{O}_{Z_{\mathfrak{v}}}\big(-\sum_{q=1}^{n-j+1} Z_{\mathfrak{v}(q)}\big)\big)\big)\Big) & \\ = H^p\big(Z_{\mathfrak{u}}, \mathcal{L}_{\mathfrak{u}}(H^0(Z_{\mathfrak{v}}, \mathcal{L}_{\mathfrak{v}}(\lambda+\rho) \otimes K_{Z_{\mathfrak{v}}}))\big), & \qquad (5) \end{aligned}$$

by Proposition 8.1.2.

Now we claim that the $\mathcal{B}$-module structure on $H^0(Z_{\mathfrak{v}}, \mathcal{L}_{\mathfrak{v}}(\lambda+\rho) \otimes K_{Z_{\mathfrak{v}}})$ admits an extension to the $\mathcal{P}_{i_{j-1}}$-module structure: By the Serre duality Theorem A.37,

$$H^0\big(Z_{\mathfrak{v}}, \mathcal{L}_{\mathfrak{v}}(\lambda+\rho) \otimes K_{Z_{\mathfrak{v}}}\big) \approx H^{n-j+1}\big(Z_{\mathfrak{v}}, \mathcal{L}_{\mathfrak{v}}(-(\lambda+\rho))\big)^*.$$

Observe that the above is a $\mathcal{B}$-equivariant isomorphism (since it is induced by a $\mathcal{B}$-invariant pairing). By the induction hypothesis, (1′) is satisfied for $\mathfrak{v}$ and $\mathfrak{v}'$ and hence, by Proposition 8.1.7,

$$H^{n-j+1}\big(Z_{\mathfrak{v}}, \mathcal{L}_{\mathfrak{v}}(-(\lambda+\rho))\big) \simeq H^{n-j+1}\big(Z_{\mathfrak{v}'}, \mathcal{L}_{\mathfrak{v}'}(-(\lambda+\rho))\big),$$

as $\mathcal{B}$-modules. But $\mathfrak{v}'$ is the sequence $(s_{i_{j-1}}, r_{j+1}, \dots, r_n)$ and hence the claim is established (cf. Lemma 8.1.4).

Now, considering the Leray spectral sequence corresponding to the projection $Z_{\mathfrak{u}} \to Z_{\mathfrak{u}[j-2]}$, we conclude, by Lemma 8.1.5 and Exercise 8.1.E.1, that

$$\begin{aligned} H^p\Big(Z_{\mathfrak{u}}, \mathcal{L}_{\mathfrak{u}}(H^0(Z_{\mathfrak{v}}, \mathcal{L}_{\mathfrak{v}}(\lambda+\rho) \otimes K_{Z_{\mathfrak{v}}}))\Big) & \\ \approx H^p\Big(Z_{\mathfrak{u}[j-2]}, \mathcal{L}_{\mathfrak{u}[j-2]}(H^0(Z_{\mathfrak{v}}, \mathcal{L}_{\mathfrak{v}}(\lambda+\rho) \otimes K_{Z_{\mathfrak{v}}}))\Big) & \\ \approx H^p\Big(Z_{\mathfrak{w}'}, \mathcal{L}_{\mathfrak{w}'}(\lambda) \otimes \mathcal{O}_{Z_{\mathfrak{w}'}}\big(-\sum_{q=j-1}^{n-1} Z_{\mathfrak{w}'(q)}\big)\Big), & \qquad (6) \end{aligned}$$

where $\mathfrak{w}' := \mathfrak{w}(j-1) = (s_{i_1}, \ldots, s_{i_{j-2}}, s_{i_j}, \ldots, s_{i_n})$. (The last isomorphism is obtained by considering the isomorphism (5) for the sequence $\mathfrak{w}$ replaced by $\mathfrak{w}'$.) Hence, by induction and (5), (6), the case (a_2) is taken care of.

Now we come to the case (b), i.e., $k < n$ and the sequence $(s_{i_j}, \ldots, s_{i_k}, s_{i_{k+1}})$ is not reduced. Define the following sequences:

$$\mathfrak{v} = (s_{i_j}, \ldots, s_{i_k}),\ \mathfrak{x} = (s_{i_1}, \ldots, s_{i_k}),\ \mathfrak{y} = (s_{i_{k+1}}, \ldots, s_{i_n}),$$
$$\text{and}\quad \mathfrak{u} = (s_{i_1}, \ldots, s_{i_{j-1}}).$$

Let $\sigma = \psi_{\mathfrak{w},k} : Z_{\mathfrak{w}} \to Z_{\mathfrak{x}}$ be the canonical projection. Then, by Lemma A.16,

$$\mathcal{O}_{Z_{\mathfrak{w}}}\Big(-\sum_{q=j}^{k} Z_{\mathfrak{w}(q)}\Big) \approx \sigma^*\Big(\mathcal{O}_{Z_{\mathfrak{x}}}\Big(-\sum_{q=j}^{k} Z_{\mathfrak{x}(q)}\Big)\Big).$$

Hence, by the projection formula, Lemma 8.1.5 and the Leray spectral sequence corresponding to the map σ (since $H^p(Z_{\mathfrak{y}}, \mathcal{L}_{\mathfrak{y}}(\lambda)) = 0$ for all $p > 0$ by induction), we have

(7)
$$H^p\Big(Z_{\mathfrak{w}}, \mathcal{L}_{\mathfrak{w}}(\lambda)\otimes\mathcal{O}_{Z_{\mathfrak{w}}}\Big(-\sum_{q=j}^{k} Z_{\mathfrak{w}(q)}\Big)\Big) \approx H^p\Big(Z_{\mathfrak{x}}, \mathcal{L}_{\mathfrak{x}}(M)\otimes\mathcal{O}_{Z_{\mathfrak{x}}}\Big(-\sum_{q=j}^{k} Z_{\mathfrak{x}(q)}\Big)\Big),$$

where the $\mathcal{B}$-module $M := H^0(Z_{\mathfrak{y}}, \mathcal{L}_{\mathfrak{y}}(\lambda))$.

By Serre duality,

(8) $$H^p(Z_{\mathfrak{v}}, \mathcal{L}_{\mathfrak{v}}(\rho) \otimes \mathcal{L}_{\mathfrak{v}}(M) \otimes K_{Z_{\mathfrak{v}}}) \approx H^{k-j+1-p}(Z_{\mathfrak{v}}, \mathcal{L}_{\mathfrak{v}}(-\rho) \otimes \mathcal{L}_{\mathfrak{v}}(M^*))^*.$$

Since $(s_{i_j}, \ldots, s_{i_k}, s_{i_{k+1}})$ is not reduced, we can write

$$s_{i_j} \cdots s_{i_k} = r_j \cdots r_{k-1} s_{i_{k+1}},$$

for some simple reflections $r_j, \ldots, r_{k-1}$. But, by (8.1.7.6), for all $p \geq 0$,

(9) $$H^p\big(Z_{\mathfrak{v}}, \mathcal{L}_{\mathfrak{v}}(-\rho) \otimes \mathcal{L}_{\mathfrak{v}}(M^*)\big) \approx H^p\big(Z_{\mathfrak{v}'}, \mathcal{L}_{\mathfrak{v}'}(-\rho) \otimes \mathcal{L}_{\mathfrak{v}'}(M^*)\big),$$

where $\mathfrak{v}'$ is the reduced sequence $(r_j, \ldots, r_{k-1}, s_{i_{k+1}})$.

But M (and hence M^*) is of course a $\mathcal{P}_{i_{k+1}}$-module and hence, by the Leray spectral sequence for the map $Z_{\mathfrak{v}'} \to Z_{\mathfrak{v}'[k-j]}$ and Exercise 8.1.E.1,

$$H^p(Z_{\mathfrak{v}'}, \mathcal{L}_{\mathfrak{v}'}(-\rho) \otimes \mathcal{L}_{\mathfrak{v}'}(M^*)) = 0,$$

for all $p \geq 0$. Hence, by (9),

(10) $$H^p\big(Z_{\mathfrak{v}}, \mathcal{L}_{\mathfrak{v}}(-\rho) \otimes \mathcal{L}_{\mathfrak{v}}(M^*)\big) = 0, \text{ for all } p \geq 0.$$

Combining (8), (10) and Proposition 8.1.2, we get

$$(11)\qquad H^p\Big(Z_{\mathfrak{v}}, \mathcal{L}_{\mathfrak{v}}(M) \otimes \mathcal{O}_{Z_{\mathfrak{v}}}\big(-\sum_{q=1}^{k-j+1} Z_{\mathfrak{v}(q)}\big)\Big) = 0, \text{ for all } p \geq 0.$$

Now, by an argument exactly analogous to the one used to derive (5), replacing the $\mathcal{B}$-module $\mathbb{C}_{-\lambda}$by M, we get

$$H^p\Big(Z_{\mathfrak{x}}, \mathcal{L}_{\mathfrak{x}}(M) \otimes \mathcal{O}_{Z_{\mathfrak{x}}}\big(-\sum_{q=j}^{k} Z_{\mathfrak{x}(q)}\big)\Big)$$

$$\approx H^p\Big(Z_{\mathfrak{u}}, \mathcal{L}_{\mathfrak{u}}\big(H^0\big(Z_{\mathfrak{v}}, \mathcal{L}_{\mathfrak{v}}(M) \otimes \mathcal{O}_{Z_{\mathfrak{v}}}\big(-\sum_{q=1}^{k-j+1} Z_{\mathfrak{v}(q)}\big)\big)\big)\Big)$$

$$(12)\qquad = 0, \text{ for all } p \geq 0, \text{ by (11)}.$$

Combining (7) and (12), we get (2) and hence; in particular, (1). This completes the induction and hence proves the theorem. □

8.1.9 Remark. Taking $\mathfrak{w} = (s_i, s_i)$ and $j = 1, k = 2$, it is easy to see that the assumption in the above theorem that $\mathfrak{v}$ be reduced is essential in general.

Recall the definition of the surjective morphism $m_{\mathfrak{w}} : Z_{\mathfrak{w}} \to X_w$ from 8.1.6.

8.1.10 Corollary. *Let $\mathfrak{w}$ be any word. Then there exists a maximal reduced subword $\mathfrak{v} = \mathfrak{w}_J$ of $\mathfrak{w}$ (i.e., $\mathfrak{v}$ is reduced and if $\mathfrak{u} = \mathfrak{w}_{J'}$ is any reduced subword of $\mathfrak{w}$ such that $J \subseteq J'$, then $\mathfrak{u} = \mathfrak{v}$) such that $m_{\mathfrak{w}}(Z_{\mathfrak{v}}) = X_w$ and the canonical map $H^0(Z_{\mathfrak{w}}, \mathcal{L}_{\mathfrak{w}}(\lambda)) \to H^0(Z_{\mathfrak{v}}, \mathcal{L}_{\mathfrak{v}}(\lambda))$, induced from the inclusion of $Z_{\mathfrak{v}}$ in $Z_{\mathfrak{w}}$ via the map $i_{\mathfrak{v},\mathfrak{w}}$ (cf. 7.1.3), is an isomorphism for any $\lambda \in D_{\mathbb{Z}}$.*

Proof. We first prove that if $\mathfrak{w}_1$ is any reduced sequence and $\mathfrak{w}_2 := (s_i, \mathfrak{w}_1)$ is not a reduced sequence, the canonical map $H^0(Z_{\mathfrak{w}_2}, \mathcal{L}_{\mathfrak{w}_2}(\lambda)) \to H^0(Z_{\mathfrak{w}_1}, \mathcal{L}_{\mathfrak{w}_1}(\lambda))$ is an isomorphism: By the Leray spectral sequence corresponding to $\psi = \psi_{\mathfrak{w}_2,1} : Z_{\mathfrak{w}_2} \to Z_{s_i}$ and Lemma 8.1.5, we get that

$$H^0(Z_{\mathfrak{w}_2}, \mathcal{L}_{\mathfrak{w}_2}(\lambda)) \approx H^0(Z_{s_i}, \mathcal{L}_{s_i}(H^0(Z_{\mathfrak{w}_1}, \mathcal{L}_{\mathfrak{w}_1}(\lambda)))).$$

But $H^0(Z_{\mathfrak{w}_1}, \mathcal{L}_{\mathfrak{w}_1}(\lambda))$ being a $\mathcal{P}_i$-module (by Theorem 8.1.8 and (8.1.7.6)), we get that $H^0(Z_{\mathfrak{w}_2}, \mathcal{L}_{\mathfrak{w}_2}(\lambda)) \xrightarrow{\sim} H^0(Z_{\mathfrak{w}_1}, \mathcal{L}_{\mathfrak{w}_1}(\lambda))$ (by the $p = 0$ case of Exercise 8.1.E.1). Now we prove the corollary by induction on the length n of $\mathfrak{w} = (s_{i_1}, \ldots, s_{i_n})$. Choose (by induction) a maximal reduced subword $\mathfrak{v}'$ of $\mathfrak{w}'$, where $\mathfrak{w}' := (s_{i_2}, \ldots, s_{i_n})$, such that $m_{\mathfrak{w}}(Z_{\mathfrak{w}'}) = m_{\mathfrak{w}}(Z_{\mathfrak{v}'})$ and

$$H^0(Z_{\mathfrak{w}'}, \mathcal{L}_{\mathfrak{w}'}(\lambda)) \to H^0(Z_{\mathfrak{v}'}, \mathcal{L}_{\mathfrak{v}'}(\lambda))$$

is an isomorphism. Now there are two cases to consider:
(a) $(s_{i_1}, \mathfrak{v}')$ is reduced,
(b) $(s_{i_1}, \mathfrak{v}')$ is not reduced.

In the case (a), it is easy to see that $\mathfrak{v} := (s_{i_1}, \mathfrak{v}')$ is a maximal reduced subword of $\mathfrak{w}$. By the Leray spectral sequence, this case is taken care of. In the case (b), it is easy to see that $\mathfrak{v}'$ is a maximal reduced subword of $\mathfrak{w}$. Now $H^0(Z_{\mathfrak{w}'}, \mathcal{L}_{\mathfrak{w}'}(\lambda)) \tilde{\to} H^0(Z_{\mathfrak{v}'}, \mathcal{L}_{\mathfrak{v}'}(\lambda))$ (by the choice of $\mathfrak{v}'$). But, by the argument in the first paragraph of the proof, $H^0(Z_{\mathfrak{w}}, \mathcal{L}_{\mathfrak{w}}(\lambda)) \tilde{\to} H^0(Z_{\mathfrak{w}'}, \mathcal{L}_{\mathfrak{w}'}(\lambda))$. The assertion $m_{\mathfrak{w}}(Z_{\mathfrak{v}}) = X_w$ in both the cases is easy to see from Theorem 5.1.3(d). □

8.1.11 Corollary. *Let $\mathfrak{w}$ be any word of length n. Then, for any $1 \leq j \leq n$ and $\lambda \in D_{\mathbb{Z}}$, the canonical map $H^0(Z_{\mathfrak{w}}, \mathcal{L}_{\mathfrak{w}}(\lambda)) \to H^0(Z_{\mathfrak{w}(j)}, \mathcal{L}_{\mathfrak{w}(j)}(\lambda))$ is surjective.*

Proof. Consider the sheaf exact sequence (corresponding to the codimension-one subvariety $i : Z_{\mathfrak{w}(j)} \subset Z_{\mathfrak{w}}$):

$$0 \to \mathcal{O}_{Z_{\mathfrak{w}}}(-Z_{\mathfrak{w}(j)}) \to \mathcal{O}_{Z_{\mathfrak{w}}} \to i_*\mathcal{O}_{Z_{\mathfrak{w}(j)}} \to 0.$$

On tensoring over $\mathcal{O}_{Z_{\mathfrak{w}}}$ with the (locally free) sheaf $\mathcal{L}_{\mathfrak{w}}(\lambda)$, we get the exact sequence

$$0 \to \mathcal{L}_{\mathfrak{w}}(\lambda) \otimes \mathcal{O}_{Z_{\mathfrak{w}}}(-Z_{\mathfrak{w}(j)}) \to \mathcal{L}_{\mathfrak{w}}(\lambda) \to i_*\mathcal{L}_{\mathfrak{w}(j)}(\lambda) \to 0.$$

Considering the corresponding long exact cohomology sequence, together with Theorem 8.1.8, we get the corollary. □

8.1.12 Remark. Let $\mathfrak{w} \in \mathfrak{W}$ be any word and $\mathfrak{v}$ any maximal reduced subword. I do not know if $m_{\mathfrak{w}}(Z_{\mathfrak{w}}) = m_{\mathfrak{w}}(Z_{\mathfrak{v}})$.

As a consequence of Corollary 8.1.10, we get the following strengthening of both Proposition 8.1.7 and Corollary 8.1.10.

For any (not necessarily reduced) word $\mathfrak{w} = (s_{i_1}, \dots, s_{i_n}) \in \mathfrak{W}$, recall the definition of the surjective morphism $m = \tilde{m}_{\mathfrak{w}} : Z_{\mathfrak{w}} \to \tilde{X}_w$ from 8.1.6.

8.1.13 Theorem. *For any $\mathfrak{w} \in \mathfrak{W}$ as above and any finite-dimensional pro-representation M of $\mathcal{B}$, we have the following:*

(1) *The canonical inclusion $\mathcal{O}_{\tilde{X}_w} \hookrightarrow m_*(\mathcal{O}_{Z_{\mathfrak{w}}})$ is an isomorphism.*

(2) *$R^i m_*(\mathcal{L}_{\mathfrak{w}}(M)) = 0$, for all $i > 0$; and $m_*(\mathcal{L}_{\mathfrak{w}}(M))$ is locally free.*

(3) *$m^*(\tilde{\mathcal{L}}_w(M)) \approx \mathcal{L}_{\mathfrak{w}}(M)$, as $\mathcal{B}$-equivariant vector bundles on $Z_{\mathfrak{w}}$,*

where $\tilde{\mathcal{L}}_w(M)$ is the $\mathcal{B}$-equivariant vector bundle on $\tilde{X}_w$ defined in Proposition 8.1.7. In particular, the canonical map

$$(4)\qquad m^* : H^i(\tilde{X}_w, \tilde{\mathcal{L}}_w(M)) \xrightarrow{\sim} H^i(Z_{\mathfrak{w}}, \mathcal{L}_{\mathfrak{w}}(M))$$

is an isomorphism, for all $i \geq 0$.

Further, if $\mathfrak{u}$ is any subword of $\mathfrak{w}$ such that $m_{|Z_{\mathfrak{u}}}$ is surjective, then the canonical map (induced by $i_{\mathfrak{u},\mathfrak{w}}$)

$$(5)\qquad i^*_{\mathfrak{u},\mathfrak{w}} : H^i(Z_{\mathfrak{w}}, \mathcal{L}_{\mathfrak{w}}(M)) \to H^i(Z_{\mathfrak{u}}, \mathcal{L}_{\mathfrak{u}}(M))$$

is an isomorphism for all $i \geq 0$.

Proof. To prove (1), by Lemma A.32, it suffices to show that the canonical map

$$m^* : H^0(\tilde{X}_w, \nu^*\mathcal{L}_w(\lambda)) \to H^0(Z_{\mathfrak{w}}, \mathcal{L}_{\mathfrak{w}}(\lambda))$$

is an isomorphism for any $\lambda \in D_{\mathbb{Z}}$, where $\nu : \tilde{X}_w \to X_w$ is the normalization. Choose a maximal subword $\mathfrak{v}$ of $\mathfrak{w}$ as in Corollary 8.1.10. Then, we have the commutative diagram

$$\begin{array}{ccc} H^0(\tilde{X}_w, \nu^*\mathcal{L}_w(\lambda)) & \xrightarrow{m^*} & H^0(Z_{\mathfrak{w}}, \mathcal{L}_{\mathfrak{w}}(\lambda)) \\ & {\scriptstyle m'^*}\searrow \quad \swarrow{\scriptstyle i^*_{\mathfrak{v},\mathfrak{w}}} & \\ & H^0(Z_{\mathfrak{v}}, \mathcal{L}_{\mathfrak{v}}(\lambda)), & \end{array}$$

where $m' := m_{|Z_{\mathfrak{v}}}$. By Corollary 8.1.10, $i^*_{\mathfrak{v},\mathfrak{w}}$ is an isomorphism. Further, m' is birational (cf. 8.1.6) and hence m'^* is an isomorphism by the projection formula (A.22.2) coupled with Zariski's main theorem A.9. In particular, from the above diagram, we get that m^* is an isomorphism. This proves (1).

By Lemma A.31, Theorem 8.1.8 and (1), we get that

$$(6)\qquad R^i m_*(\mathcal{O}_{Z_{\mathfrak{w}}}) = 0, \text{ for all } i > 0.$$

Now, by the same argument used in the proof of Proposition 8.1.7, we obtain (2) and also that the following canonical map is an isomorphism:

$$(7)\qquad m^* m_*(\mathcal{L}_{\mathfrak{w}}(M)) \xrightarrow{\sim} \mathcal{L}_{\mathfrak{w}}(M).$$

We next show that the natural $\mathcal{O}_{\tilde{X}_w}$-module map

$$(8)\qquad m_*\mathcal{L}_{\mathfrak{w}}(M) \to m'_*\mathcal{L}_{\mathfrak{v}}(M) \text{ is an isomorphism:}$$

For $M = \mathbb{C}_\lambda$ (with $\lambda \in \mathfrak{h}^*_{\mathbb{Z}}$) this follows from theprojection formula and (1). Now, (8) for general M follows by induction on $\dim M$ and the five lemma, by using (2) and the exact sequence (8.1.7.7). Combining (7) and (8), we get (3).

Combining (1)–(3) with the Leray spectral sequence associated to the morphism m, we get (4). The assertion (5) follows by using (4) for $Z_{\mathfrak{w}}$ and $Z_{\mathfrak{u}}$. This completes the proof of the theorem. □

Recall the definition of a subword $\hat{\mathfrak{w}}_J$ (and the corresponding word $\mathfrak{w}_J$) of a word $\mathfrak{w}$ from 7.1.3.

8.1.14 Lemma. *Let $\mathfrak{w} \in \mathfrak{W}$ be a word with two subwords $\hat{\mathfrak{w}}_J$ and $\hat{\mathfrak{w}}_{J'}$ such that $\mathfrak{w}_J = \mathfrak{w}_{J'}$ (as elements of $\mathfrak{W}$). These give rise to two embeddings*

$$i_{\hat{\mathfrak{w}}_J,\mathfrak{w}},\ i_{\hat{\mathfrak{w}}_{J'},\mathfrak{w}} : Z_{\mathfrak{v}} \to Z_{\mathfrak{w}},$$

where $\mathfrak{v} := \mathfrak{w}_J (= \mathfrak{w}_{J'})$ (cf. Corollary 7.1.10(d)). Let M be a finite-dimensional pro-representation of $\mathcal{B}$. It is easy to see that there is a canonical $\mathcal{B}$-equivariant bundle morphism $\hat{i}_{\hat{\mathfrak{w}}_J,\mathfrak{w}} : \mathcal{L}_{\mathfrak{v}}(M) \to \mathcal{L}_{\mathfrak{w}}(M)$ covering $i_{\hat{\mathfrak{w}}_J,\mathfrak{w}}$. Then, the induced $\mathcal{B}$-module maps

$$\hat{i}^*_{\hat{\mathfrak{w}}_J,\mathfrak{w}},\ \hat{i}^*_{\hat{\mathfrak{w}}_{J'},\mathfrak{w}} : H^*(Z_{\mathfrak{w}}, \mathcal{L}_{\mathfrak{w}}(M)) \to H^*(Z_{\mathfrak{v}}, \mathcal{L}_{\mathfrak{v}}(M))$$

are equal.

For $\mathfrak{v} \leq \mathfrak{w}$, set $\hat{i}^*_{\mathfrak{v},\mathfrak{w}} : H^*(Z_{\mathfrak{w}}, \mathcal{L}_{\mathfrak{w}}(M)) \to H^*(Z_{\mathfrak{v}}, \mathcal{L}_{\mathfrak{v}}(M))$ to be the map $\hat{i}^*_{\hat{\mathfrak{w}}_J,\mathfrak{w}}$ if $\mathfrak{v} = \mathfrak{w}_J$.

Proof. Let $\mathfrak{w} = (s_{i_1}, \dots, s_{i_n})$. We prove the lemma by induction on n. If $n = 1$, there is nothing to prove. Let $J = (j_1, \dots, j_p)$ and $J' = (j'_1, \dots, j'_p)$. Then, by assumption, $s_{i_{j_k}} = s_{i_{j'_k}}$ for any $1 \leq k \leq p$. Assume that $j_1 \leq j'_1$. Define the sequence $J'' := (j_1, j'_2, \dots, j'_p)$. Then $\mathfrak{w}_{J''} = \mathfrak{w}_J$.

Let $\mathfrak{w}'$ be the subword $(s_{i_{j_1+1}}, s_{i_{j_1+2}}, \dots, s_{i_n})$. Then, by induction,

$$(1)\qquad \hat{i}^*_{\hat{\mathfrak{w}}'_K,\mathfrak{w}'} = \hat{i}^*_{\hat{\mathfrak{w}}'_{K'},\mathfrak{w}'},$$

where $K := (j_2, \dots, j_p)$ and $K' := (j'_2, \dots, j'_p)$. By Lemma 8.1.5, the E_2-term of the Leray spectral sequence for the projection $\psi = \psi_{\mathfrak{w},j_1} : Z_{\mathfrak{w}} \to Z_{\mathfrak{w}[j_1]}$ with respect to the bundle $\mathcal{L}_{\mathfrak{w}}(M)$ on $Z_{\mathfrak{w}}$ is given by

$$E_2^{p,q} = H^p(Z_{\mathfrak{w}[j_1]}, \mathcal{L}(H^q(Z_{\mathfrak{w}'}, \mathcal{L}_{\mathfrak{w}'}(M)))).$$

By (1), the bundle morphisms $\hat{i}_{\hat{\mathfrak{w}}_J,\mathfrak{w}}$ and $\hat{i}_{\hat{\mathfrak{w}}_{J''},\mathfrak{w}}$ induce the same map at the E_2-term and hence the same map at the E_∞-term. Thus

$$\hat{i}^*_{\hat{\mathfrak{w}}_{J''},\mathfrak{w}} = \hat{i}^*_{\hat{\mathfrak{w}}_J,\mathfrak{w}}.$$

To prove $\hat{i}^*_{\hat{\mathfrak{w}}_{J''},\mathfrak{w}} = \hat{i}^*_{\hat{\mathfrak{w}}_{J'},\mathfrak{w}}$, again by the Leray spectral sequence and Lemma 8.1.5, we can assume that $p = 1$ since J'' and J' differ only in one place. Thus let $J = (j_1)$, $J' = (j_1')$ with $s_{i_{j_1}} = s_{i_{j_1'}}$. Assume further that $j_1 < j_1'$. By considering the subword $\hat{\mathfrak{w}}_{\tilde{J}}$ corresponding to the sequence $\tilde{J} := (j_1, j_1')$, we are reduced to the case when $\mathfrak{w} = (s_i, s_i)$ and $J = (1)$, $J' = (2)$, for some simple reflection s_i.

Consider the surjective morphism $m_{\mathfrak{w}} : Z_{\mathfrak{w}} \to \mathcal{P}_i/\mathcal{B}$ (cf. 8.1.6). Clearly, $m_{\mathfrak{w}} \circ i_{\hat{\mathfrak{w}}_J,\mathfrak{w}} = m_{\mathfrak{w}} \circ i_{\hat{\mathfrak{w}}_{J'},\mathfrak{w}}$ as maps $Z_{s_i} \to \mathcal{P}_i/\mathcal{B}$ and, moreover,

$$m^*_{\mathfrak{w}} : H^*(\mathcal{P}_i/\mathcal{B}, \mathcal{L}_{s_i}(M)) \to H^*(Z_{\mathfrak{w}}, \mathcal{L}_{\mathfrak{w}}(M))$$

is an isomorphism, which follows from (8.1.13.4). This proves the lemma. □

8.1.15 Definition. For any finite-dimensional pro-representation M of $\mathcal{B}$ and any $p \geq 0$, define the $\mathcal{B}$-module

(1) $$H^p(Z_\infty, \mathcal{L}_\infty(M)) := \underset{\mathfrak{w}\in\mathfrak{W}}{\text{Inv. lt.}}\ H^p(Z_{\mathfrak{w}}, \mathcal{L}_{\mathfrak{w}}(M)),$$

where the inverse limit is taken with respect to the partial order $\leq$ in $\mathfrak{W}$ (cf. 7.1.3) and the homomorphisms $\hat{i}^*_{\mathfrak{v},\mathfrak{w}}$, for $\mathfrak{v} \leq \mathfrak{w}$, defined in the above lemma.

Similarly, define the pro-representation of $\mathcal{B}$:

(2) $$H^p(Z_\infty, \mathcal{L}_\infty(M))^\vee := \underset{\mathfrak{w}\in\mathfrak{W}}{\text{limit}}\ H^p(Z_{\mathfrak{w}}, \mathcal{L}_{\mathfrak{w}}(M))^*.$$

For any $1 \leq i \leq \ell$, let $\mathfrak{W}_i \subset \mathfrak{W}$ be the subset consisting of those words $\mathfrak{w} = (s_{i_1}, \dots, s_{i_n})$ $(n \geq 1)$ such that $s_{i_1} = s_i$. The subset $\mathfrak{W}_i \subset \mathfrak{W}$ is clearly cofinal in the directed set $(\mathfrak{W}, \leq)$. In particular,

(3) $$H^p(Z_\infty, \mathcal{L}_\infty(M)) \cong \underset{\mathfrak{w}\in\mathfrak{W}_i}{\text{Inv. lt.}}\ H^p(Z_{\mathfrak{w}}, \mathcal{L}_{\mathfrak{w}}(M)).$$

By Lemma 8.1.4, for any $\mathfrak{w} \in \mathfrak{W}_i$, $\mathcal{L}_{\mathfrak{w}}(M)$ is a $\mathcal{P}_i$-equivariant algebraic vector bundle on $Z_{\mathfrak{w}}$ and, moreover, for any $\mathfrak{w} \in \mathfrak{W}_i$ and subword $\hat{\mathfrak{w}}_J$ of $\mathfrak{w}$ corresponding to any sequence $J = (1, j_2, \dots, j_p)$, the bundle morphism $\hat{i}_{\hat{\mathfrak{w}}_J,\mathfrak{w}}$ (cf. Lemma 8.1.14) is $\mathcal{P}_i$-equivariant. In particular, the induced map $\hat{i}^*_{\hat{\mathfrak{w}}_J,\mathfrak{w}}$ is $\mathcal{P}_i$-equivariant, and hence the $\mathcal{B}$-module $H^p(Z_\infty, \mathcal{L}_\infty(M))$ has a compatible $\mathcal{P}_i$-module structure for any $1 \leq i \leq \ell$. Similarly, $H^p(Z_\infty, \mathcal{L}_\infty(M))^\vee$ is a pro-representation of $\mathcal{P}_i$ which is compatible with the $\mathcal{B}$-module structure. In fact, later in Section 8.3 we will see that both $H^p(Z_\infty, \mathcal{L}_\infty(M))$ and $H^p(Z_\infty, \mathcal{L}_\infty(M))^\vee$ have $\mathcal{G}$-module structures.

8.1.16 Definition. Let M be a finite-dimensional $\mathfrak{b}$-module which is $\mathfrak{h}$-semisimple. For any simple reflection s_i, $1 \leq i \leq \ell$, define the finite-dimensional $\mathfrak{p}_i$ (in particular a $\mathfrak{b}$)-module $\mathcal{D}_{s_i}(M)$ to be the quotient of the induced module $U(\mathfrak{p}_i) \underset{U(\mathfrak{b})}{\otimes} M$

by the smallest $\mathfrak{p}_i$-submodule K of finite-codimension, where $\mathfrak{p}_i$ is the minimal parabolic subalgebra $\mathfrak{b} \oplus \mathbb{C} f_i$ as in 1.2.2. (As shown below, $U(\mathfrak{p}_i) \underset{U(\mathfrak{b})}{\otimes} M$ indeed has a unique smallest $\mathfrak{p}_i$-submodule of finite-codimension.) As in 1.1.2, let $\mathfrak{g}(i)$ be the Lie subalgebra of $\mathfrak{p}_i$ spanned by $\{e_i, f_i, \alpha_i^\vee\}$ and let $\mathfrak{b}(i)$ be the Lie subalgebra spanned by $\{e_i, \alpha_i^\vee\}$. Then $\mathfrak{g}(i)$ is isomorphic with the Lie algebra sl_2. Define the ideal $\mathfrak{s}(i)$ of $\mathfrak{p}_i$ by $\mathfrak{s}(i) = \mathfrak{u}_i \oplus \{h \in \mathfrak{h} : \alpha_i(h) = 0\}$, where $\mathfrak{u}_i$ is as in 1.2.2. Then

(1) $$\mathfrak{p}_i = \mathfrak{g}(i) \oplus \mathfrak{s}(i)$$

as vector spaces. From the PBW theorem, the canonical map

(2) $$j : U(\mathfrak{g}(i)) \underset{U(\mathfrak{b}(i))}{\otimes} M \xrightarrow{\sim} U(\mathfrak{p}_i) \underset{U(\mathfrak{b})}{\otimes} M$$

is a $\mathfrak{g}(i)$-module isomorphism. In particular, the inverse image of any $\mathfrak{p}_i$-submodule of the right side of (2) is a $\mathfrak{g}(i)$-submodule of the left side. From Exercise 8.1.E.2, $U(\mathfrak{g}(i)) \underset{U(\mathfrak{b}(i))}{\otimes} M$ has only finitely many $\mathfrak{g}(i)$-submodules; in particular, $U(\mathfrak{p}_i) \underset{U(\mathfrak{b})}{\otimes} M$ has only finitely many $\mathfrak{p}_i$-submodules. Hence $U(\mathfrak{g}(i)) \underset{U(\mathfrak{b}(i))}{\otimes} M$, resp. $U(\mathfrak{p}_i) \underset{U(\mathfrak{b})}{\otimes} M$, has a unique smallest $\mathfrak{g}(i)$-submodule $\tilde{K}$, resp. $\mathfrak{p}_i$-submodule K, of finite-codimension. Moreover, $j(\tilde{K}) \subset K$. Thus, there is a surjective $\mathfrak{g}(i)$-module homomorphism

(3) $$\hat{j} : \mathcal{D}^{\mathfrak{g}(i)}(M) \twoheadrightarrow \mathcal{D}_{s_i}(M),$$

induced by j, where $\mathcal{D}^{\mathfrak{g}(i)}(M) := \big(U(\mathfrak{g}(i)) \underset{U(\mathfrak{b}(i))}{\otimes} M\big)/\tilde{K}$ corresponds to $\mathcal{D}_{s_i}(M)$ for $\mathfrak{g} = \mathfrak{g}(i) \simeq sl_2$ and the unique simple reflection s_i of $\mathfrak{g}(i)$. (In the proof of Proposition 8.1.17, we will see that $\hat{j}$ is an isomorphism at least when M is a finite-dimensional pro-representation of $\mathcal{B}$.)

Clearly $\mathcal{D}_{s_i}(M)$ is again $\mathfrak{h}$-semisimple.

Now, for an arbitrary word $\mathfrak{w} = (s_{i_1}, \dots, s_{i_n}) \in \mathfrak{W}$, define the *Joseph functor*

(4) $$\mathcal{D}_{\mathfrak{w}}(M) := \mathcal{D}_{s_{i_1}} \cdots \mathcal{D}_{s_{i_n}}(M).$$

8.1.17 Proposition. *For any finite-dimensional pro-representation M of $\mathcal{B}$ under π and any word $\mathfrak{w} = (s_{i_1}, \dots, s_{i_n}) \in \mathfrak{W}$ (not necessarily reduced), $H^0(Z_{\mathfrak{w}}, \mathcal{L}_{\mathfrak{w}}(M))^*$ is $U(\mathfrak{p}_{i_1})$-isomorphic with $\mathcal{D}_{\mathfrak{w}}(M^*)$, where M is to be regarded as a $\mathfrak{b}$-module via the restriction of $\dot{\pi}$ to $\mathfrak{b}$ (cf. 4.4.23).*

Proof. We first prove the proposition for $\mathfrak{w} = (s_i)$ of length 1: The $\mathcal{B}$-module map obtained by restriction $H^0(Z_{s_i}, \mathcal{L}_{s_i}(M)) \to H^0(Z_\phi, \mathcal{L}_\phi(M)) \simeq M$, on dualizing, gives rise to the $\mathcal{B}$ (and hence $\mathfrak{b}$)-module map $M^* \to H^0(Z_{s_i}, \mathcal{L}_{s_i}(M))^*$. But, since $H^0(Z_{s_i}, \mathcal{L}_{s_i}(M))$ is a $\mathcal{P}_i$-module, we get a $\mathfrak{p}_i$-module map

$$\psi : U(\mathfrak{p}_i) \underset{U(\mathfrak{b})}{\otimes} M^* \to H^0(Z_{s_i}, \mathcal{L}_{s_i}(M))^*,$$

and hence an induced $\mathfrak{p}_i$-module map $\bar{\psi} : \mathcal{D}_{s_i}(M^*) \to H^0(Z_{s_i}, \mathcal{L}_{s_i}(M))^*$. Now we prove that $\bar{\psi}$ is an isomorphism:

We first assume that $\mathfrak{g} = \mathfrak{g}(i)$ and $\mathfrak{b} = \mathfrak{b}(i)$, i.e., we assume that $\mathcal{G} = \mathrm{SL}_2$. In this case, we can assume by Exercise 8.1.E.2(c) that M is a cyclic $\mathfrak{b}$-module $F(p,q)$ (with integers $p \leq q$ such that $q - p$ is even) spanned by $\{v_p, v_{p+2}, v_{p+4}, \dots, v_q\}$ with the action $\alpha_i^\vee v_s := s v_s$ and $e_i v_s := v_{s+2}$ for all $p \leq s \leq q$, where v_{q+2} is interpreted as 0. Thus,

$$F(p,q) \simeq V_{\frac{q-p}{2}} \otimes \mathbb{C}_{\frac{q+p}{2}}, \text{ as } \mathfrak{b}\text{-modules}, \tag{1}$$

where V_r is the irreducible sl_2-module of dimension $r+1$, and $\mathbb{C}_r$ is the one-dimensional $\mathfrak{b}(i)$-module with e_i acting trivially and $\alpha_i^\vee$ acting via the multiplication by r.

Now, by Exercise 8.1.E.2(b), for $M = F(p,q)$,

$$\mathcal{D}^{\mathfrak{g}(i)}(M^*) \simeq V^*_{\frac{q-p}{2}} \otimes \mathcal{D}^{\mathfrak{g}(i)}\big(\mathbb{C}^*_{\frac{q+p}{2}}\big), \tag{2}$$

where $\mathcal{D}^{\mathfrak{g}(i)}$ is defined in 8.1.16. Further, by Exercise 8.1.E.1, for $M = F(p,q)$,

$$H^0\big(Z_{s_i}, \mathcal{L}_{s_i}(M)\big) \simeq V_{\frac{q-p}{2}} \otimes H^0\big(Z_{s_i}, \mathcal{L}_{s_i}\big(\mathbb{C}_{\frac{q+p}{2}}\big)\big), \tag{3}$$

and hence, to prove that $\bar{\psi}$ is an isomorphism for the case $\mathcal{G} = \mathrm{SL}_2$, we can assume that $M = \mathbb{C}_r$ for some $r \in \mathbb{Z}$. It is not difficult to see that the isomorphisms (2) and (3) are compatible. If $r > 0$,

$$H^0(Z_{s_i}, \mathcal{L}_{s_i}(\mathbb{C}_r)) = 0 = \mathcal{D}^{\mathfrak{g}(i)}(\mathbb{C}_r^*);$$

and if $r \leq 0$, $H^0(Z_{s_i}, \mathcal{L}_{s_i}(\mathbb{C}_r))$ is the irreducible sl_2-module of dimension $1 - r$ and so is $\mathcal{D}^{\mathfrak{g}(i)}(\mathbb{C}_r^*)$ (cf. Exercise 8.1.E.2(d)). But, by Corollary 8.1.11, the sl_2-module map $\bar{\psi}$ is nonzero and hence an isomorphism. So, this proves the proposition in the case of $\mathcal{G} = \mathrm{SL}_2$ and $\ell(\mathfrak{w}) = 1$.

Now we come to the case of general $\mathfrak{g}$ (but still $\ell(\mathfrak{w}) = 1$). We have the following commutative diagram:

$$\begin{array}{ccc}
\mathcal{D}^{\mathfrak{g}(i)}(M^*) & \xrightarrow[\sim]{\bar{\psi}^{\mathfrak{g}(i)}} & H^0(Z_{s_i}^{\mathfrak{g}(i)}, \mathcal{L}_{s_i}^{\mathfrak{g}(i)}(M))^* \\
\Big\downarrow \hat{j} & & \Big\downarrow \wr\, \theta^* \\
\mathcal{D}_{s_i}(M^*) & \xrightarrow{\bar{\psi}} & H^0(Z_{s_i}, \mathcal{L}_{s_i}(M))^*,
\end{array}$$

where $\hat{j}$ is the surjective map as in (8.1.16.3), $\bar{\psi}^{\mathfrak{g}(i)}$ is the map $\bar{\psi}$ for $\mathfrak{g} = \mathfrak{g}(i)$, $Z_{s_i}^{\mathfrak{g}(i)}$ is the Bott–Samelson–Demazure–Hansen variety Z_{s_i} for the case of $\mathfrak{g} = \mathfrak{g}(i)$ and $\mathcal{L}_{s_i}^{\mathfrak{g}(i)}(M)$ is the homogenous bundle on $Z_{s_i}^{\mathfrak{g}(i)}$ associated to the $\mathfrak{b}(i)$-module M. The right vertical isomorphism θ^* is induced by the canonical isomorphism $\theta : Z_{s_i}^{\mathfrak{g}(i)} \simeq Z_{s_i}$. (Observe that the bundle $\mathcal{L}_{s_i}^{\mathfrak{g}(i)}(M)$ is the pullback of the bundle $\mathcal{L}_{s_i}(M)$ via θ.) Moreover, $\bar{\psi}^{\mathfrak{g}(i)}$ is an isomorphism as proved above. Thus, from the commutativity of the above diagram, $\hat{j}$ is injective and hence an isomorphism. This gives that $\bar{\psi}$ is an isomorphism as well.

Now the general case, i.e., $\mathfrak{w} = (s_{i_1}, \dots, s_{i_n})$ follows easily by induction on the length n of $\mathfrak{w}$, by using the morphism $\psi_{\mathfrak{w},1} : Z_{\mathfrak{w}} \to Z_{s_{i_1}}$ and Lemma 8.1.5. □

8.1.18 Remark. As a consequence of (8.1.7.6), Proposition 8.1.17 and Theorem 8.1.8, for any finite-dimensional pro-representation M of $\mathcal{B}$, $\mathcal{D}_{\mathfrak{w}}(M)$ is $\mathfrak{b}$-isomorphic with $\mathcal{D}_{\mathfrak{v}}(M)$ for any reduced words $\mathfrak{v}$ and $\mathfrak{w}$ with $\pi(\mathfrak{v}) = \pi(\mathfrak{w})$. Joseph gave an algebraic proof of this in the finite case [Joseph–85, Proposition 2.15].

8.1.19 Lemma. *Let $\mathfrak{w}$ be a reduced word. Then, $H^0(Z_{\mathfrak{w}}, \mathcal{L}_{\mathfrak{w}}(\lambda))^*$ is a cyclic $U(\mathfrak{b})$-module generated by an element of weight $w\lambda$, for any $\lambda \in D_{\mathbb{Z}}$, where $w := \pi(\mathfrak{w})$.*

Proof. Let $X_w^o := \mathcal{B}w\mathcal{B}/\mathcal{B}$ and let $Z_{\mathfrak{w}}^o$ be the $\mathcal{B}$-stable (dense) open subset of $Z_{\mathfrak{w}}$ defined in (7.1.3.3). Then, by Proposition 7.1.15(a), $m_{\mathfrak{w}|Z_{\mathfrak{w}}^o} : Z_{\mathfrak{w}}^o \to X_w^o$ is a biregular isomorphism. This lifts to a bundle isomorphism $\mathcal{L}_{\mathfrak{w}}(\lambda)_{|Z_{\mathfrak{w}}^o} \to \mathcal{L}(\lambda)_{|X_w^o}$ (cf. Definition 7.2.4). In particular, we get an injective $\mathcal{B}$-module map of pro-representations of $\mathcal{B}$:

$$i : H^0(Z_{\mathfrak{w}}, \mathcal{L}_{\mathfrak{w}}(\lambda)) \hookrightarrow H^0(X_w^o, \mathcal{L}(\lambda)_{|X_w^o}),$$

and hence, on dualizing, we get the surjective $\mathcal{B}$-module map

$$\delta : M \twoheadrightarrow H^0(Z_{\mathfrak{w}}, \mathcal{L}_{\mathfrak{w}}(\lambda))^*,$$

where $M := H^0\big(X_w^o, \mathcal{L}(\lambda)_{|X_w^o}\big)^*$. So, to prove the lemma, it suffices to show that M is a cyclic $U(\mathfrak{b})$-module generated by an element of weight $w\lambda$.

Recall the definition of the unipotent algebraic group $\mathcal{U}_w = \mathcal{U}_{\Phi_w}$ from Example 6.1.5(b), and define the map $\theta : \mathcal{U}_w \to X_w^o$ by $\theta(g) = g\bar{w}\mathcal{B}$, where $\bar{w}$ is a representative of w in N. By Exercise 6.2.E.1, θ is a bijection. Further, it is easy to see that θ is a morphism and hence it is a biregular isomorphism by Theorem A.11. Observe that X_w^o is smooth, and hence normal, by Proposition 7.1.15(a). Moreover, θ is clearly $\mathcal{U}_w$-equivariant under the action of $\mathcal{U}_w$ on itself via the left multiplication, and its action on X_w^o via the standard action of $\mathcal{B}$. In particular, as $\mathcal{U}_w$-modules,

$$H^0\big(X_w^o, \mathcal{L}(\lambda)_{|X_w^o}\big) \approx H^0\big(\mathcal{U}_w, \theta^*(\mathcal{L}(\lambda)_{|X_w^o})\big). \tag{1}$$

The bundle $\theta^*\big(\mathcal{L}(\lambda)_{|X_w^o}\big)$ admits a nowhere vanishing section σ given by $\sigma(g)(g\bar{w}v_\lambda) = 1$, for $g \in \mathcal{U}_w$, where v_λ is a fixed nonzero highest weight vector of $L^{\max}(\lambda)$. Hence, $\theta^*(\mathcal{L}(\lambda)_{|X_w^o})$ is a trivial line bundle giving rise to an isomorphism of $\mathcal{U}_w$-modules:

$$H^0\big(\mathcal{U}_w, \theta^*(\mathcal{L}(\lambda)_{|X_w^o})\big) \approx \mathbb{C}[\mathcal{U}_w] \otimes \sigma, \tag{2}$$

where $\mathcal{U}_w$ acts on $\mathbb{C}[\mathcal{U}_w]$ via the left regular representation and acts trivially on σ. Under the identifications (1)–(2), the map δ can be rewritten as the map

$$\bar{\delta} : \mathbb{C}[\mathcal{U}_w]^* \otimes \sigma^* \twoheadrightarrow H^0(Z_{\mathfrak{w}}, \mathcal{L}_{\mathfrak{w}}(\lambda))^*.$$

Let $\mathrm{Dist}(\mathcal{U}_w) \subset \mathbb{C}[\mathcal{U}_w]^*$ denote the space of distributions on $\mathcal{U}_w$ with support in 1 (cf. [Jantzen–87, Part I, §7.1]). Then, we prove that $\bar{\delta}_{|(\mathrm{Dist}(\mathcal{U}_w)\otimes\sigma^*)}$ is still surjective:

Let $\bar{i} : H^0(Z_{\mathfrak{w}}, \mathcal{L}_{\mathfrak{w}}(\lambda)) \hookrightarrow \mathbb{C}[\mathcal{U}_w] \otimes \sigma$ be the injective map i under the identifications (1)–(2). Since $H^0(Z_{\mathfrak{w}}, \mathcal{L}_{\mathfrak{w}}(\lambda))$ is finite-dimensional, there exists a large enough n such that the composite map $\pi_n \circ \bar{i} : H^0(Z_{\mathfrak{w}}, \mathcal{L}_{\mathfrak{w}}(\lambda)) \to (\mathbb{C}[\mathcal{U}_w]/\mathfrak{m}^n) \otimes \sigma$ still remains injective, where $\mathfrak{m} \subset \mathbb{C}[\mathcal{U}_w]$ is the maximal ideal consisting of functions vanishing at 1 and $\pi_n : \mathbb{C}[\mathcal{U}_w] \otimes \sigma \to (\mathbb{C}[\mathcal{U}_w]/\mathfrak{m}^n) \otimes \sigma$ is the canonical projection map. Dualizing the injective map $\pi_n \circ \bar{i}$, we get the surjectivity of $\bar{\delta}_{|(\mathrm{Dist}(\mathcal{U}_w)\otimes\sigma^*)}$. Now the lemma follows from the following result of Cartier (cf. [Demazure–Gabriel–70, II, §6, n°1]). Observe that σ^* is of weight $w\lambda$. □

8.1.20 Theorem. *For any algebraic group G over an algebraically closed field k of char. 0, there is a natural isomorphism*

$$\psi : U(\mathrm{Lie}\, G) \xrightarrow{\sim} \mathrm{Dist}\, G$$

commuting with the $U(\mathrm{Lie}\, G)$*-module structures, where* $U(\mathrm{Lie}\, G)$ *acts on itself by the left multiplication and its action on Dist G is induced from the dual of the left regular representation of G on* $k[G]$. □

8.1.21 The Borel–Weil homomorphism. For any countable-dimensional vector space V, define the linear map

$$\beta_V : V^* \to H^0(\mathbb{P}(V), \mathcal{L}_V^*)$$

by

(1) $$\beta_V(f)(\delta) = (\delta, f_{|\delta}), \text{ for } f \in V^* \text{ and } \delta \in \mathbb{P}(V),$$

where $\mathcal{L}_V$ is the tautological line bundle on $\mathbb{P}(V)$ as in Example 4.2.7(c). Recall that the fiber of $\mathcal{L}_V^*$ over a line $\delta \in \mathbb{P}(V)$ is the dual vector space $\delta^* \simeq V^*/\delta^\perp$, and the notation H^0 is as in (4.2.6.1).

Now take $\lambda \in D_{\mathbb{Z}}$ and define a linear map

$$\beta = \beta(\lambda) : L^{\max}(\lambda)^* \to H^0(\mathcal{X}, \mathcal{L}(\lambda))$$

as the composite $i_\lambda^* \circ \beta_V$, where $i_\lambda^* : H^0(\mathbb{P}(V), \mathcal{L}_V^*) \to H^0(\mathcal{X}, \mathcal{L}(\lambda))$ is induced from the morphism $i_\lambda : \mathcal{X} \to \mathbb{P}(V)$ (cf. 7.2.1 and Exercise 7.2.E.1), and $V := L^{\max}(\lambda)$.

Let $Y \subset \{1, \dots, \ell\}$ be any subset. Recall the definition of $\mathfrak{h}^*_{\mathbb{Z},Y}$ from 7.2.1. For any $\lambda \in \mathfrak{h}^*_{\mathbb{Z},Y}$, by 7.2.3, there is a natural action of $\mathcal{G}$ on the line bundle $\mathcal{L}^Y(\lambda)$ and hence, by 4.2.6, $\mathcal{G}$ acts naturally on $H^p(\mathcal{X}^Y, \mathcal{L}^Y(\lambda))$. Under this action, the map $\beta(\lambda)$ is a $\mathcal{G}$-module map, since $\mathcal{G}$ acts linearly on V (and hence acts on $\mathbb{P}(V)$) and i_λ is $\mathcal{G}$-equivariant.

For any $w \in W$ and simple reflection s_i such that $s_i w < w$, the left multiplication of the minimal parabolic subgroup $\mathcal{P}_i$ on $\mathcal{X}^Y$ keeps X_w^Y stable. Thus, we get a natural action of $\mathcal{P}_i$ on $H^p(X_w^Y, \mathcal{L}_w^Y(\lambda))$ making it into a pro-representation of $\mathcal{P}_i$. For $\lambda \in D_{\mathbb{Z}}$, let

$$\beta_w(\lambda) : L^{\max}(\lambda)^* \to H^0(X_w, \mathcal{L}_w(\lambda))$$

be the map $\beta(\lambda)$ followed by the canonical restriction map

$$H^0(\mathcal{X}, \mathcal{L}(\lambda)) \to H^0(X_w, \mathcal{L}_w(\lambda)).$$

Then $\beta_w(\lambda)$ is a $\mathcal{P}_i$-module map (under the assumption $s_i w < w$).

8.1.22 Definition. Let V be an integrable highest weight $\mathfrak{g}$-module with highest weight λ. For $w \in W$, let V_w be the $U(\mathfrak{b})$-submodule of V generated by any nonzero vector $v_{w\lambda}$ of V of weight $w\lambda$. Observe that $v_{w\lambda}$ is unique up to scalar multiples by Lemma 1.3.5(a). The $\mathfrak{b}$-submodule $V_w \subset V$ is called the *Demazure submodule* of V corresponding to the Weyl group element w.

Assume now that $\lambda \in D_{\mathbb{Z}}$. By Proposition 4.4.26, it is easy to see that V_w is $\mathcal{U}$-stable. Moreover, V_w is T-stable since it is a weight module (with weights $\in D_{\mathbb{Z}}$). Thus V_w is $\mathcal{B}$-stable; in fact, V_w is the $\mathcal{B}$-submodule of V generated by $v_{w\lambda}$ (use 4.4.23).

8.1.23 Lemma. *For any $\lambda \in D_{\mathbb{Z}}$, $w \in W$, and simple reflection s_i such that $s_i w < w$, the $\mathcal{P}_i$-module map $\beta_w(\lambda) : L^{\max}(\lambda)^* \to H^0(X_w, \mathcal{L}_w(\lambda))$, defined in 8.1.21, has kernel precisely equal to the subspace $[L^{\max}(\lambda)/ L_w^{\max}(\lambda)]^*$, where $L_w^{\max}(\lambda) = L^{\max}(\lambda)_w$ is the Demazure submodule of $L^{\max}(\lambda)$ corresponding to w.*

In particular, $L_w^{\max}(\lambda)$ is $\mathcal{P}_i$-stable and we get an injective $\mathcal{P}_i$-module map $\bar{\beta}_w(\lambda) : L_w^{\max}(\lambda)^* \hookrightarrow H^0(X_w, \mathcal{L}_w(\lambda))$.

Proof. Since $\mathcal{B}w\mathcal{B}/\mathcal{B} \subset X_w$ is dense open (cf. Proposition 7.1.15), the restriction map $H^0(X_w, \mathcal{L}_w(\lambda)) \to H^0\big(\mathcal{B}w\mathcal{B}/\mathcal{B}, \mathcal{L}_w(\lambda)_{|(\mathcal{B}w\mathcal{B}/\mathcal{B})}\big)$ is injective. In particular,

$$\begin{aligned}\operatorname{Ker} \beta_w(\lambda) &= \big\{f \in L^{\max}(\lambda)^* : \beta(\lambda)(f)_{|(\mathcal{B}w\mathcal{B}/\mathcal{B})} \equiv 0\big\}\\ &= \big\{f \in L^{\max}(\lambda)^* : f_{|(\mathcal{B}w\mathcal{B}\, v_\lambda)} \equiv 0\big\}\\ &= \big\{f \in L^{\max}(\lambda)^* : f_{|L_w^{\max}(\lambda)} \equiv 0\big\}.\end{aligned}$$

This proves the lemma. □

8.1.24 Definition. For any word $\mathfrak{w} = (s_{i_1}, \dots, s_{i_n})$ and $\lambda \in D_{\mathbb{Z}}$, the surjective morphism $m_{\mathfrak{w}} : Z_{\mathfrak{w}} \to X_w$ (cf. Definition 8.1.6) induces the injective map

$$(1) \qquad \bar{m}_{\mathfrak{w}}(\lambda) : H^0(X_w, \mathcal{L}_w(\lambda)) \hookrightarrow H^0(Z_{\mathfrak{w}}, \mathcal{L}_{\mathfrak{w}}(\lambda)),$$

and hence, taking duals, we get the surjective map

$$m_{\mathfrak{w}}(\lambda) : H^0(Z_{\mathfrak{w}}, \mathcal{L}_{\mathfrak{w}}(\lambda))^* \twoheadrightarrow H^0(X_w, \mathcal{L}_w(\lambda))^*.$$

Composing this with the dual of the injective map $\bar{\beta}_w(\lambda)$ (cf. Lemma 8.1.23), we get the surjective map

$$(2) \qquad \phi_{\mathfrak{w}}(\lambda) : H^0(Z_{\mathfrak{w}}, \mathcal{L}_{\mathfrak{w}}(\lambda))^* \twoheadrightarrow L_w^{\max}(\lambda) \subset L^{\max}(\lambda).$$

Moreover, $\phi_{\mathfrak{w}}(\lambda)$ is a $\mathcal{P}_{i_1}$-module map. Since $m_{\mathfrak{w}}$ is surjective and $Z_{\mathfrak{w}}$ is $\mathcal{P}_{i_1}$-stable, so are X_w and $L_w^{\max}(\lambda)$.

Taking the direct limit over $\mathfrak{w}$, we get the map

$$(3) \qquad \phi(\lambda) : H^0(Z_\infty, \mathcal{L}_\infty(\lambda))^\vee \to L^{\max}(\lambda).$$

Moreover, $\phi(\lambda)$ is a $\mathcal{P}_i$-module map for every $\mathcal{P}_i$, $1 \le i \le \ell$ (cf. 8.1.15).

8.1.25 Theorem. *For any $\lambda \in D_{\mathbb{Z}}$, the map*

$$\phi(\lambda) : H^0(Z_\infty, \mathcal{L}_\infty(\lambda))^\vee \to L^{\max}(\lambda),$$

defined above, is an isomorphism. In particular, the weight spaces of $H^0(Z_\infty, \mathcal{L}_\infty(\lambda))^\vee$ are finite dimensional.

Proof. Since $N := H^0(Z_\infty, \mathcal{L}_\infty(\lambda))^\vee$ is a pro-representation of $\mathcal{P}_i$ (cf. 8.1.15), on differentiation, we get a pro-representation π_i of $\hat{\mathfrak{p}}_i$ on N (cf. 4.4.23 and Lemma 6.1.14). Hence, by Corollary 1.3.10, we obtain a $\mathfrak{g}$-module structure π on the same space N such that

(1) $$\pi_{|\mathfrak{p}_i} = \pi_{i|\mathfrak{p}_i}.$$

Since N is a locally finite T-module, it is a weight module. In particular, (N, π) is an integrable $\mathfrak{g}$-module by (1). Moreover, $\phi(\lambda)$ being a $\mathcal{P}_i$-module map; in particular, it is a $\mathfrak{p}_i$-module map for all $1 \leq i \leq \ell$ (cf. Proof of Proposition 4.4.26). Hence, $\phi(\lambda)$ is a $\mathfrak{g}$-module map.

We next show that (N, π) is a highest weight module with highest weight λ. From Proposition 8.1.17, it is easy to see that N has at most one vector $v_\lambda \neq 0$ of weight λ (up to a nonzero scalar multiple) and, moreover, the weights of N are contained in the cone $\lambda - \sum_{i=1}^{\ell} \mathbb{Z}_+\alpha_i$. Further, by Corollary 8.1.11, N has at least one vector $v_\lambda \neq 0$ of weight λ. Let $M := U(\mathfrak{g}) \cdot v_\lambda \subset N$. Fix $w \in W$ and a reduced decomposition $w = s_{i_1} \cdots s_{i_n}$. Denote by $\mathfrak{w}$ the word $(s_{i_1}, \dots, s_{i_n})$. Since N, resp. M, is integrable, the weight space $N_{w\lambda}$, resp. $M_{w\lambda}$, corresponding to the weight $w\lambda$ is one-dimensional (by Lemma 1.3.5(a)) and hence $N_{w\lambda} \subset M$. But, by Lemma 8.1.19, $A_{\mathfrak{w}} := H^0(Z_{\mathfrak{w}}, \mathcal{L}_{\mathfrak{w}}(\lambda))^*$ is a cyclic $U(\mathfrak{b})$-module generated by an element of weight $w\lambda$. Further, by Corollary 8.1.11, $A_{\mathfrak{w}}$ injects inside N, and hence we get $A_{\mathfrak{w}} \subset M$.

But now let $\mathfrak{w} = (s_{i_1}, \dots, s_{i_n})$ be any (not necessarily reduced) word. Then there exists a (maximal) reduced subword $\mathfrak{v}$ of $\mathfrak{w}$ such that the canonical map $H^0(Z_{\mathfrak{w}}, \mathcal{L}_{\mathfrak{w}}(\lambda)) \to H^0(Z_{\mathfrak{v}}, \mathcal{L}_{\mathfrak{v}}(\lambda))$ is an isomorphism (cf. Corollary 8.1.10). Now, since N is the direct limit of $H^0(Z_{\mathfrak{w}}, \mathcal{L}_{\mathfrak{w}}(\lambda))^*$, the assertion that (N, π) is a highest weight module with highest weight λ, follows.

By (8.1.24.2), Im $\phi(\lambda)$ contains the highest weight vector of $L^{\max}(\lambda)$ and hence $\phi(\lambda)$ is surjective (since it is a $\mathfrak{g}$-module map). Finally, since N is an integrable highest weight $\mathfrak{g}$-module with highest weight λ, there is a surjective $\mathfrak{g}$-module map $\theta : L^{\max}(\lambda) \to N$ (by Lemma 2.1.7). But since $\operatorname{Hom}_{\mathfrak{g}}(L^{\max}(\lambda), L^{\max}(\lambda))$ is one-dimensional, we obtain that $\phi(\lambda)$ is an isomorphism. □

8.1.26 Corollary. *For any $\lambda \in D_{\mathbb{Z}}$ and any word $\mathfrak{w} = (s_{i_1}, \dots, s_{i_n})$, the $\mathcal{P}_{i_1}$-module maps (cf. (8.1.24.1) and Lemma 8.1.23)*

$$\bar{m}_{\mathfrak{w}}(\lambda) : H^0(X_w, \mathcal{L}_w(\lambda)) \to H^0(Z_{\mathfrak{w}}, \mathcal{L}_{\mathfrak{w}}(\lambda))$$

$$\bar{\beta}_w(\lambda) : L_w^{\max}(\lambda)^* \to H^0(X_w, \mathcal{L}_w(\lambda))$$

are both isomorphisms, where $w \in W$ is such that $m_{\mathfrak{w}} : Z_{\mathfrak{w}} \to X_w$ is surjective (cf. Definition 8.1.6).

Proof. It clearly suffices to show that the surjective map

$$\phi_{\mathfrak{w}}(\lambda) : H^0(Z_{\mathfrak{w}}, \mathcal{L}_{\mathfrak{w}}(\lambda))^* \to L_w^{\max}(\lambda)$$

(cf. (8.1.24.2)) is injective. Since the subset $\mathfrak{W}_{\geq \mathfrak{w}} := \{\mathfrak{v} \in \mathfrak{W} : \mathfrak{v} \geq \mathfrak{w}\}$ is cofinal in $\mathfrak{W}$, we obtain that there is a canonical map

$$\delta_{\mathfrak{w}} : H^0(Z_{\mathfrak{w}}, \mathcal{L}_{\mathfrak{w}}(\lambda))^* \to H^0(Z_\infty, \mathcal{L}_\infty(\lambda))^\vee.$$

Moreover, by Corollary 8.1.11, $\delta_{\mathfrak{w}}$ is injective.

We have the following commutative diagram:

$$\begin{array}{ccc} H^0(Z_{\mathfrak{w}}, \mathcal{L}_{\mathfrak{w}}(\lambda))^* & \xrightarrow{\phi_{\mathfrak{w}}(\lambda)} & L_w^{\max}(\lambda) \\ \downarrow \delta_{\mathfrak{w}} & & \downarrow \\ H^0(Z_\infty, \mathcal{L}_\infty(\lambda))^\vee & \xrightarrow{\phi(\lambda)} & L^{\max}(\lambda). \end{array}$$

The injectivity of $\phi_{\mathfrak{w}}(\lambda)$ follows from the injectivity of $\phi(\lambda)$ (cf. Theorem 8.1.25). This proves the corollary. □

8.1.E EXERCISES

(1) For a finite-dimensional pro-representation M of $\mathcal{P}_i$ (for a minimal parabolic subgroup $\mathcal{P}_i$ of $\mathcal{G}$) and a finite-dimensional pro-representation N of $\mathcal{B}$, show that

$$H^p(Z_i, \mathcal{L}_{Z_i}(M \otimes N)) \simeq M \otimes H^p(Z_i, \mathcal{L}_{Z_i}(N)), \text{ as } \mathcal{P}_i\text{-modules},$$

where $Z_i := \mathcal{P}_i/\mathcal{B}$. Further, show that

$$H^p(Z_i, \mathcal{L}_{Z_i}(-\rho)) = 0, \text{ for all } p \geq 0.$$

(This exercise is essentially a particular case of [Jantzen–87, Part I, Propositions 4.8 and 5.12(a)].)

(2) In the following, we take $\mathfrak{g} = sl_2$ with the standard Borel subalgebra $\mathfrak{b}$ consisting of the upper triangular (trace 0) matrices. Show the following:

(a) For any finite-dimensional $\mathfrak{b}$-module M which is $\mathfrak{h}$-semisimple, $U(\mathfrak{g}) \underset{U(\mathfrak{b})}{\otimes} M$ has only finitely many $\mathfrak{g}$-submodules.

(b) For any finite-dimensional $\mathfrak{g}$-module M, resp. $\mathfrak{b}$-module N,

$$\mathcal{D}_s(M \otimes N) \simeq M \otimes \mathcal{D}_s(N), \text{ as } \mathfrak{g}\text{-modules},$$

where s is the (simple) reflection in the Weyl group of sl_2.

(c) Any finite-dimensional $\mathfrak{b}$-module M which is $\mathfrak{h}$-semisimple, is a direct sum of cyclic modules $F(p, q)$, for integers $p \leq q$ such that $q - p$ is even, where $F(p, q)$ is as defined in the proof of Proposition 8.1.17.

(d) For $r \in \mathbb{Z}$, let $\mathbb{C}_r$ be the one-dimensional $\mathfrak{b}$-module as in the proof of Proposition 8.1.17. Then $\mathcal{D}_s(\mathbb{C}_r) = 0$, if $r < 0$, and $\mathcal{D}_s(\mathbb{C}_r)$ is the irreducible sl_2-module of dimension $r + 1$, for $r \geq 0$.

Prove the same result for $\mathcal{D}_s(\mathbb{C}_r)$ replaced by $H^0(SL(2)/B,\ \mathcal{L}(\mathbb{C}_r^*))$, where B is the Borel subgroup of SL_2 with Lie algebra $\mathfrak{b}$.

8.2. Normality of Schubert Varieties and the Demazure Character Formula

In this section $\mathfrak{g}$ is an arbitrary Kac–Moody Lie algebra (associated to a $\ell \times \ell$ GCM), and let $\mathcal{G}$ be the associated group. For any subset $Y \subset \{1, \dots, \ell\}$ (Y not necessarily of finite type), let $\mathcal{P} = \mathcal{P}_Y$ be the associated standard parabolic subgroup of $\mathcal{G}$ (cf. Definition 6.1.18). For any $w \in W$, recall the definition of the Schubert variety $X_w^Y \subset \mathcal{G}/\mathcal{P}_Y$ from 7.1.13, which is equipped with the stable variety structure to make it an irreducible projective variety (cf. 7.1.19).

8.2.1 Definition. Recall the definition of the $\mathcal{P}_i$-module map (for any $w \in W$, simple reflection s_i such that $s_i w < w$, and $\lambda \in D_{\mathbb{Z}}$) $\bar{\beta}_w(\lambda) : L_w^{\max}(\lambda)^* \hookrightarrow H^0(X_w, \mathcal{L}_w(\lambda))$ from Lemma 8.1.23. If $\lambda \in D_{\mathbb{Z}} \cap \mathfrak{h}_{\mathbb{Z},Y}^*$ (cf. 7.2.1), in exactly the same way, we can define the map

$$\bar{\beta}_w^Y(\lambda) : L_w^{\max}(\lambda)^* \hookrightarrow H^0\big(X_w^Y, \mathcal{L}_w^Y(\lambda)\big),$$

making the following diagram commutative:

$$\text{(D)} \qquad \begin{array}{ccc} L_w^{\max}(\lambda)^* & \xrightarrow{\bar{\beta}_w^Y(\lambda)} & H^0(X_w^Y, \mathcal{L}_w^Y(\lambda)) \\ & \searrow^{\bar{\beta}_w(\lambda)} & \downarrow \alpha^* \\ & & H^0(X_w, \mathcal{L}_w(\lambda)), \end{array}$$

where α^* is induced from the projection $\alpha : X_w \to X_w^Y$ and $\mathcal{L}_w^Y(\lambda)$ is the line bundle on X_w^Y defined in 7.2.1. Since α is surjective, α^* is injective.

With the above notation, we have the following:

8.2.2 Theorem. *For any $v \leq w \in W$ and any $\lambda \in D_{\mathbb{Z}} \cap \mathfrak{h}^*_{\mathbb{Z},Y}$, we have the following:*

(a) *The maps*

$$\bar{\beta}^Y_w(\lambda) : L^{\max}_w(\lambda)^* \to H^0\big(X^Y_w, \mathcal{L}^Y_w(\lambda)\big)$$

and

$$\alpha^* : H^0\big(X^Y_w, \mathcal{L}^Y_w(\lambda)\big) \to H^0\big(X_w, \mathcal{L}_w(\lambda)\big)$$

are both isomorphisms.

(b) *X^Y_w is a normal variety.*

(c) *For any (not necessarily reduced) word $\mathfrak{w} \in \mathfrak{W}$ such that $m(Z_{\mathfrak{w}}) = X^Y_w$, the morphism $m : Z_{\mathfrak{w}} \to X^Y_w$ is a trivial morphism (cf. A.24), where $m := m^Y_{\mathfrak{w}}$ (cf. 7.1.11).*

In particular, for any algebraic vector bundle $\mathcal{S}$ on X^Y_w, we have the following isomorphisms (for any $p \geq 0$):

(1) $$m^* : H^p\big(X^Y_w, \mathcal{S}\big) \tilde{\to} H^p\big(Z_{\mathfrak{w}}, m^*\mathcal{S}\big),$$

(2) $$\alpha^* : H^p\big(X^Y_w, \mathcal{S}\big) \tilde{\to} H^p\big(X_w, \alpha^*\mathcal{S}\big).$$

Taking $\mathcal{S} = \mathcal{L}^Y_w(\lambda)$ for λ as above, we get

(3) $$H^p\big(X^Y_w, \mathcal{L}^Y_w(\lambda)\big) = 0, \text{ for all } p > 0.$$

Taking $w' \in wW_Y \cap W'_Y$ and reduced $\mathfrak{w}'$ with $\pi(\mathfrak{w}') = w'$, we get that $m^Y_{\mathfrak{w}'}$ is a rational resolution and hence X^Y_w is Cohen–Macaulay.

(d) *The canonical restriction map $H^0\big(X^Y_w, \mathcal{L}^Y_w(\lambda)\big) \to H^0\big(X^Y_v, \mathcal{L}^Y_v(\lambda)\big)$ is surjective.*

(e) *Assume that $\mathfrak{g}$ is symmetrizable. Then, for $\lambda \in D^o_Y$, the linear system on X^Y_w, given by $\mathcal{L}^Y_w(\lambda)$, embeds X^Y_w as a projectively normal and projectively Cohen–Macaulay variety (for an explanation see the proof of this part).*

Proof. (a): By Corollary 8.1.26 the map $\bar{\beta}_w(\lambda)$ is an isomorphism and, hence, from the diagram (D) of 8.2.1, α^* is surjective and thus is an isomorphism. Hence $\bar{\beta}^Y_w(\lambda)$ also is an isomorphism. This proves (a).

(b): Let $w' \in W'_Y$ be the unique (minimal) element in the coset wW_Y and choose any reduced word $\mathfrak{w}'$ with $\pi(\mathfrak{w}') = w'$. Then the morphism $m^Y_{\mathfrak{w}'} : Z_{\mathfrak{w}'} \to X^Y_w$ is surjective and birational by Proposition 7.1.15(a). By Corollary 8.1.26 and the (a) part applied to w', we get that the induced map $(m^Y_{\mathfrak{w}'})^*$:

$H^0(X_w^Y, \mathcal{L}_w^Y(\lambda)) \to H^0(Z_{\mathfrak{w}'}, \mathcal{L}_{\mathfrak{w}'}(\lambda))$ is an isomorphism. Taking any $\lambda^o \in D_Y^o$ and applying Lemma A.32 to the ample line bundle $\mathcal{L} = \mathcal{L}_w^Y(\lambda^o)$, we get that $m_{\mathfrak{w}'*}^Y \mathcal{O}_{Z_{\mathfrak{w}'}} = \mathcal{O}_{X_w^Y}$. (By Exercise 7.2.E.2, $\mathcal{L}_w^Y(\lambda^o)$ is indeed ample.) This gives that X_w^Y is normal, proving (b).

(c): The assertion that $m_{\mathfrak{w}}^Y : Z_{\mathfrak{w}} \to X_w^Y$ is a trivial morphism follows from the (a) part, Theorem 8.1.8 (specifically (1′)), Corollary 8.1.26, and Lemmas A.31–A.32. (1) follows from Lemma A.25. (2) follows from (1) by applying it to X_w^Y and X_w. (3) follows from (1) and Theorem 8.1.8. Since $m_{\mathfrak{w}'}^Y$ is a birational map (cf. the proof of the (b) part), and is a trivial morphism, from the definition (cf. A.24), it is a rational resolution. In particular, from Lemma A.38, X_w^Y is Cohen–Macaulay.

(d): Since $L_v^{\max}(\lambda) \subseteq L_w^{\max}(\lambda)$ (as can be easily seen, e.g., by using Lemma 8.1.23), we obtain the (d) part by using the isomorphisms $\bar{\beta}_w^Y(\lambda)$ and $\bar{\beta}_v^Y(\lambda)$ from the (a) part.

(e): Since $\lambda \in D_Y^o$, we have the closed embedding $i_\lambda : X_w^Y \hookrightarrow \mathbb{P}(L^{\max}(\lambda))$ (cf. 7.1.13 and Proposition 7.1.15(c); observe that we have used the symmetrizability of $\mathfrak{g}$ here) such that $i_\lambda^*(\mathcal{L}_V^*) \simeq \mathcal{L}_w^Y(\lambda)$, where $V := L^{\max}(\lambda)$. By Exercise 8.2.E.1, there is a closed embedding $\hat{i}_\lambda : X_w^Y \hookrightarrow \mathbb{P}(L^*)$, where $L := H^0(X_w^Y, \mathcal{L}_w^Y(\lambda))$ and, moreover, $\hat{i}_\lambda^*(\mathcal{O}_{L^*}(1)) \cong \mathcal{L}_w^Y(\lambda)$. Now the assertion of the (e) part means that the embedding $\hat{i}_\lambda$ is projectively normal and projectively Cohen–Macaulay. We first prove the projective normality of $\hat{i}_\lambda$:

First of all, by [Hartshorne–77, Chap. III, Theorem 5.1], there is an isomorphism

$$S^n\big(H^0(X_w^Y, \mathcal{L}_w^Y(\lambda))\big) \cong H^0(\mathbb{P}(L^*), \mathcal{O}(n)),$$

where S^n denotes the n-th symmetric power and $\mathcal{O}(n) := \mathcal{O}_{L^*}(1)^{\otimes n}$. So, by Proposition A.39 and the (b) part, it suffices to show that for $\lambda, \lambda' \in D_Y^o$ the canonical map

$$\gamma^Y : H^0\big(X_w^Y, \mathcal{L}_w^Y(\lambda)\big) \otimes H^0\big(X_w^Y, \mathcal{L}_w^Y(\lambda')\big) \to H^0\big(X_w^Y, \mathcal{L}_w^Y(\lambda + \lambda')\big)$$

is surjective. In view of the (a) part, to prove the surjectivity of γ^Y, we need to show that, for $\lambda, \lambda' \in D_{\mathbb{Z}}$, the canonical map

$$\gamma : H^0(X_w, \mathcal{L}_w(\lambda)) \otimes H^0(X_w, \mathcal{L}_w(\lambda')) \to H^0(X_w, \mathcal{L}_w(\lambda + \lambda'))$$

is surjective.

We have the following commutative diagram:

$$\text{(F)}\qquad \begin{array}{ccc} H^0(X_w, \mathcal{L}_w(\lambda)) \otimes H^0(X_w, \mathcal{L}_w(\lambda')) & \longrightarrow & H^0(X_w, \mathcal{L}_w(\lambda+\lambda')) \\ \uparrow \wr & & \uparrow \wr \\ [L_w^{\max}(\lambda)]^* \otimes [L_w^{\max}(\lambda')]^* & & [L_w^{\max}(\lambda+\lambda')]^* \\ \uparrow\!\!\!\!\!\uparrow & & \uparrow\!\!\!\!\!\uparrow \\ [L^{\max}(\lambda) \otimes L^{\max}(\lambda')]^* & \longrightarrow & [L^{\max}(\lambda+\lambda')]^* , \end{array}$$

where the two top vertical maps are induced by $\bar{\beta}_w$ and areisomorphisms by Corollary 8.1.26, the two bottom vertical maps are the canonical restriction maps and hence are surjective, and the bottom horizontal map is obtained by dualizing the unique $\mathfrak{g}$-module map (cf. proof of Proposition 7.1.15(b))

$$\theta = \theta_{\lambda,\lambda'} : L^{\max}(\lambda+\lambda') \to L^{\max}(\lambda) \otimes L^{\max}(\lambda'),$$

which takes $v_{\lambda+\lambda'} \mapsto v_\lambda \otimes v_{\lambda'}$, where v_λ is some fixed nonzero highest weight vector of $L^{\max}(\lambda)$.

But θ is injective for symmetrizable $\mathfrak{g}$ since, in this case, $L^{\max}(\lambda+\lambda')$ is an irreducible $\mathfrak{g}$-module by Corollaries 2.2.6 and 3.2.10. In fact, θ is always injective (cf. Remark 7.1.16). So the bottom horizontal map in the above diagram (F) is surjective and hence so is the top horizontal map. This proves the projective normality of X_w^Y.

To prove that X_w^Y is projectively Cohen–Macaulay under the embedding $\hat{i}_\lambda$, since it is projectively normal, it suffices to show (cf. [Eisenbud–95, Exercise 18.16]) that

$$\text{(4)}\qquad H^p\big(X_w^Y, \mathcal{L}_w^Y(n\lambda)\big) = 0, \text{ for all } n \in \mathbb{Z} \text{ and } 0 < p < \dim X_w^Y.$$

For any $p > 0$ and $n \geq 0$, (4) follows from (3). For any $p < \dim X_w^Y$ and $n < 0$, (4) follows from the (c) part of the theorem and Lemma A.38.

This completes the proof of the theorem. □

By the (b) part of the above theorem, we get the following corollary.

8.2.3 Corollary. *Theorem 8.1.13 and Proposition 8.1.7 are true with $\tilde{X}_w$ replaced by X_w. In particular, for any finite-dimensional pro-representation M of $\mathcal{B}$, we have the $\mathcal{B}$-equivariant vector bundle $\mathcal{L}_w(M)$ on X_w defined as $\tilde{\mathcal{L}}_w(M)$ in Proposition 8.1.7. Moreover, by (8.1.7.5), $\mathcal{L}_w(M)$ is in fact a $\mathcal{P}_j$-equivariant vector bundle (extending the $\mathcal{B}$-structure) for any minimal parabolic $\mathcal{P}_j$ such that $\mathcal{P}_j$ keeps X_w stable under the left multiplication, i.e., if $s_j w < w$.* □

8.2.4 Lemma. *For any $v \leq w \in W$ and any finite-dimensional pro-representation M of $\mathcal{B}$, there is a canonical isomorphism*

$$i^* \mathcal{L}_w(M) \simeq \mathcal{L}_v(M)$$

of $\mathcal{B}$-equivariant vector bundles, where $i : X_v \hookrightarrow X_w$ is the canonical inclusion.

Moreover, for a simple reflection s_j such that both $s_j v < v$ and $s_j w < w$, the above isomorphism (described in the proof) is $\mathcal{P}_j$-equivariant.

Proof. Take a reduced word $\mathfrak{w}$ with $\pi(\mathfrak{w}) = w$, and a reduced subword $\mathfrak{v} = \mathfrak{w}_J$ of $\mathfrak{w}$ such that $\pi(\mathfrak{v}) = v$. (This is guaranteed by Lemma 1.3.16.) Consider the commutative diagram:

$$\begin{array}{ccc} Z_{\mathfrak{v}} & \xrightarrow{i_{\mathfrak{v},\mathfrak{w}}} & Z_{\mathfrak{w}} \\ {\scriptstyle m_{\mathfrak{v}}}\downarrow & & \downarrow{\scriptstyle m_{\mathfrak{w}}} \\ X_v & \xrightarrow{i} & X_w , \end{array}$$

where $i_{\mathfrak{v},\mathfrak{w}}$ is the inclusion given in (7.1.3.2) and $m_{\mathfrak{v}}$ is the morphism as in Proposition 7.1.15. By the definition,

(1) $$m_{\mathfrak{w}_*}(\mathcal{L}_{\mathfrak{w}}(M)) = \mathcal{L}_w(M), \qquad \text{and}$$

(2) $$m_{\mathfrak{v}_*}(\mathcal{L}_{\mathfrak{v}}(M)) = \mathcal{L}_v(M).$$

Further,

(3) $$i^*_{\mathfrak{v},\mathfrak{w}}(\mathcal{L}_{\mathfrak{w}}(M)) \simeq \mathcal{L}_{\mathfrak{v}}(M), \qquad \text{cf. Lemma 8.1.14, and}$$

(4) $$m^*_{\mathfrak{w}}(\mathcal{L}_w(M)) \simeq \mathcal{L}_{\mathfrak{w}}(M), \qquad \text{by (8.1.7.4).}$$

So,

$$m^*_{\mathfrak{v}}\, i^*(\mathcal{L}_w(M)) \simeq i^*_{\mathfrak{v},\mathfrak{w}}\, m^*_{\mathfrak{w}}(\mathcal{L}_w(M))$$

(5) $$\simeq \mathcal{L}_{\mathfrak{v}}(M), \qquad \text{by (3)–(4).}$$

Taking $m_{\mathfrak{v}*}$ of (5), and using the projection formula and the normality of X_v, we get

(6) $$i^*(\mathcal{L}_w(M)) \simeq \mathcal{L}_v(M),$$

proving the first part of the lemma.

To prove the second part, observe that we can take a reduced word $\mathfrak{w}$ with $\pi(\mathfrak{w}) = w$ such that $\mathfrak{w} = (s_j, s_{i_1}, \dots, s_{i_n})$, and we can choose a reduced subword $\mathfrak{v} = (s_j, s_{i_1}, \dots, \hat{s}_{i_{j_1}}, \dots, \hat{s}_{i_{j_p}}, \dots, s_{i_n})$ such that $\pi(\mathfrak{v}) = v$. (Use the fact that $s_j v \leq s_j w$, which follows from Corollary 1.3.19.) With these choices, all the bundles involved in the above proof are $\mathcal{P}_j$-equivariant and, moreover, all the maps are $\mathcal{P}_j$-equivariant, proving the second part of the lemma. □

8.2.5 Corollary. *For any finite-dimensional pro-representation M of $\mathcal{B}$, there exists a $\mathcal{B}$-equivariant algebraic vector bundle $\mathcal{L}(M)$ on the ind-variety $\mathcal{X} = \mathcal{G}/\mathcal{B}$ (cf. Definition 7.1.5) such that, for any $w \in W$, there is a canonical isomorphism of $\mathcal{B}$-equivariant vector bundles:*

$$\mathcal{L}(M)_{|X_w} \simeq \mathcal{L}_w(M).$$

In fact, $\mathcal{L}(M)$ is a $\mathcal{P}_j$-equivariant algebraic vector bundle for any minimal parabolic subgroup $\mathcal{P}_j$, extending the $\mathcal{B}$-structure.

Proof. Let $X_0 \subset X_1 \subset \cdots \subset X_n \subset \cdots$ be the filtration of the ind-variety $\mathcal{X}$ (cf. Definition 7.1.19). For any n, there exists a large enough $w_n \in W$ such that $X_n \subset X_{w_n}$. Let $\mathcal{L}_n(M) \to X_n$ be the restriction of the vector bundle $\mathcal{L}_{w_n}(M)$ to X_n. By Lemma 8.2.4, there exists a canonical isomorphism of $\mathcal{B}$-equivariant vector bundles $i_n : \mathcal{L}_n(M) \simeq \mathcal{L}_{n+1}(M)_{|X_n}$, for any $n \geq 0$. Thus, from Example 4.2.7(f), we get an algebraic vector bundle $\mathcal{L}(M) \to \mathcal{X}$ together with bundle isomorphisms $\mathcal{L}(M)_{|X_n} \simeq \mathcal{L}_n(M)$ for all n. Moreover, since each $\mathcal{L}_n(M)$ is a $\mathcal{B}$-equivariant vector bundle and i_n is a $\mathcal{B}$-equivariant isomorphism, we get that $\mathcal{L}(M)$ is a $\mathcal{B}$-equivariant vector bundle.

Now, for any minimal parabolic subgroup $\mathcal{P}_j$ of $\mathcal{G}$, we put a compatible $\mathcal{P}_j$-equivariant algebraic vector bundle structure on $\mathcal{L}(M)$ as follows: Define the subset $W^j := \{w \in W : s_j w < w\}$ of W. Then, clearly, W^j is cofinal in W under the partial order $\leq$. For any $v, w \in W^j$ such that $v \leq w$, by Lemma 8.2.4, the canonical isomorphism $\mathcal{L}_v(M) \simeq \mathcal{L}_w(M)_{|X_v}$ is $\mathcal{P}_j$-equivariant. This turns $\mathcal{L}(M)$ into a $\mathcal{P}_j$-equivariant vector bundle. □

8.2.6 Remark. Observe that by (8.1.4.1) and the normality of X_w's, for any $\lambda \in \mathfrak{h}^*_{\mathbb{Z}}$, the algebraic line bundle $\mathcal{L}(-\lambda)$ on $\mathcal{X}$ (cf. 7.2.1) is canonically isomorphic with the line bundle $\mathcal{L}(\mathbb{C}_\lambda)$, where $\mathbb{C}_\lambda$ is the one-dimensional pro-representation of $\mathcal{B}$ as in 8.1.3.

8.2.7 Demazure operators. For any simple reflection s_i, $1 \leq i \leq \ell$, following Demazure, define the $\mathbb{Z}$-linear operator $D_{s_i} : A(T) \to A(T)$ by

$$D_{s_i}(e^\lambda) = \frac{e^\lambda - e^{s_i\lambda - \alpha_i}}{1 - e^{-\alpha_i}}, \quad \text{for } e^\lambda \in X(T),$$

where $A(T) := \mathbb{Z}[X(T)]$ is the group algebra of the character group $X(T)$ (cf. 6.1.6) and α_i is the i-th simple (positive) root. It is easy to see that $D_{s_i}(e^\lambda) \in A(T)$. In fact, one has the following simple lemma.

Now for any word $\mathfrak{w} = (s_{i_1}, \ldots, s_{i_n}) \in \mathfrak{W}$, define $D_{\mathfrak{w}} = D_{s_{i_1}} \circ \cdots \circ D_{s_{i_n}} : A(T) \to A(T)$.

8.2.8 Lemma.

$$\begin{aligned} D_{s_i}(e^\lambda) &= e^\lambda + e^{\lambda-\alpha_i} + \cdots + e^{s_i\lambda}, && \text{if } \lambda(\alpha_i^\vee) \geq 0 \\ &= 0, && \text{if } \lambda(\alpha_i^\vee) = -1 \\ &= -(e^{\lambda+\alpha_i} + \cdots + e^{s_i\lambda-\alpha_i}), && \text{if } \lambda(\alpha_i^\vee) < -1. \quad \square \end{aligned}$$

The ring $A(T)$ admits an involution defined by $\overline{e^\lambda} = e^{-\lambda}$. We denote $\overline{D_{\mathfrak{w}}(e^\lambda)}$ by $\bar{D}_{\mathfrak{w}}(e^\lambda)$.

Now we are ready to prove the following Demazure character formula in an arbitrary Kac–Moody setting.

8.2.9 Theorem. *For any (not necessarily reduced) word $\mathfrak{w}$ and any finite-dimensional pro-representation M of $\mathcal{B}$, we have*

(1) $$\chi(Z_{\mathfrak{w}}, \mathcal{L}_{\mathfrak{w}}(M)) = \bar{D}_{\mathfrak{w}}(\overline{\operatorname{ch} M}), \qquad \textit{as elements of } A(T),$$

where $\chi(Z_{\mathfrak{w}}, \mathcal{L}_{\mathfrak{w}}(M)) := \sum_p (-1)^p \operatorname{ch} H^p(Z_{\mathfrak{w}}, \mathcal{L}_{\mathfrak{w}}(M)) \in A(T)$ and, for any finite-dimensional T-module N, $\operatorname{ch} N$ denotes its formal T-character (cf. 2.1.1).

In particular, for $\lambda \in \mathfrak{h}_{\mathbb{Z}}^*$ and any reduced word $\mathfrak{w}$ with $\pi(\mathfrak{w}) = w$,

(2) $$\chi(X_w, \mathcal{L}_w(\lambda)) = \bar{D}_{\mathfrak{w}}(e^\lambda).$$

Hence, if $\lambda \in D_{\mathbb{Z}}$,

(3) $$\operatorname{ch} H^0(X_w, \mathcal{L}_w(\lambda)) = \bar{D}_{\mathfrak{w}}(e^\lambda), \qquad \textit{and}$$

(4) $$\operatorname{ch} L_w^{\max}(\lambda) = D_{\mathfrak{w}}(e^\lambda).$$

Proof. For any exact sequence

$$0 \to M_1 \to M \to M_2 \to 0$$

of finite-dimensional pro-representations of $\mathcal{B}$, we have from the corresponding long exact cohomology sequence ($\mathcal{L}_{\mathfrak{w}}$ being an exact functor)

(5) $$\chi(Z_{\mathfrak{w}}, \mathcal{L}_{\mathfrak{w}}(M)) = \chi(Z_{\mathfrak{w}}, \mathcal{L}_{\mathfrak{w}}(M_1)) + \chi(Z_{\mathfrak{w}}, \mathcal{L}_{\mathfrak{w}}(M_2)).$$

We prove (1) by induction on the length n of $\mathfrak{w} = (s_{i_1}, \ldots, s_{i_n})$. If $n = 1$, (1) follows for any one-dimensional pro-representation M of $\mathcal{B}$ from Exercise 8.2.E.2. (Observe that any one-dimensional pro-representation M of $\mathcal{B}$ is of the form $M = \mathbb{C}_\lambda$, for some $\lambda \in \mathfrak{h}_{\mathbb{Z}}^*$.) Now, by (5) and Lie's theorem (cf.

Proof of Proposition 8.1.7), we get the validity of (1) for general M (in the case $n = 1$). Assume the validity of (1) for $\mathfrak{w}[n-1]$ by induction (and any M). In view of Lemma 8.1.5, the Leray spectral sequence for the fibration $\psi = \psi_{\mathfrak{w},n-1} : Z_{\mathfrak{w}} \to Z_{\mathfrak{w}[n-1]}$ takes the form

$$E_2^{p,q} = H^p\Big(Z_{\mathfrak{w}[n-1]}, \mathcal{L}_{\mathfrak{w}[n-1]}(H^q(\mathcal{P}_{i_n}/\mathcal{B}, \mathcal{L}_{s_{i_n}}(M)))\Big)$$

and it converges to $H^{p+q}(Z_{\mathfrak{w}}, \mathcal{L}_{\mathfrak{w}}(M))$.

From this it is easy to see that

$$\sum_{p,q}(-1)^{p+q} \operatorname{ch} H^p\Big(Z_{\mathfrak{w}[n-1]}, \mathcal{L}_{\mathfrak{w}[n-1]}(H^q(\mathcal{P}_{i_n}/\mathcal{B}, \mathcal{L}_{s_{i_n}}(M)))\Big)$$
$$= \chi(Z_{\mathfrak{w}}, \mathcal{L}_{\mathfrak{w}}(M)). \tag{6}$$

But, by the induction hypothesis and the case $n = 1$, the left side of (6) is given by

$$\begin{aligned}\sum_q(-1)^q \chi\Big(Z_{\mathfrak{w}[n-1]}, \mathcal{L}_{\mathfrak{w}[n-1]}(H^q(\mathcal{P}_{i_n}/\mathcal{B}, \mathcal{L}_{s_{i_n}}(M)))\Big)& \\ = \bar{D}_{\mathfrak{w}[n-1]}(\bar{D}_{s_{i_n}}(\overline{\operatorname{ch} M}))& \\ = \bar{D}_{\mathfrak{w}}(\overline{\operatorname{ch} M}).&\end{aligned}$$

This, together with (6), proves (1) for $\mathfrak{w}$ and thereby completes the induction.

(2) and (3) follow from (1) by Theorem 8.2.2(c). (4) follows from (3) and Theorem 8.2.2(a). □

The following corollary follows immediately from (8.2.9.2).

8.2.10 Corollary. *For any reduced word $\mathfrak{w}$, the operator $D_{\mathfrak{w}} : A(T) \to A(T)$ depends only on the Weyl group element $\pi(\mathfrak{w})$.*

For $w \in W$, we set $D_w = D_{\mathfrak{w}}$ for any reduced word $\mathfrak{w}$ with $\pi(\mathfrak{w}) = w$. □

(This corollary also admits a purely algebraic proof similar to the proof of Theorem 11.1.2(b).)

8.2.E EXERCISES

(1) Let V be a finite-dimensional vector space and let X be a projective variety with a closed embedding $i : X \hookrightarrow \mathbb{P}(V)$. Let $\mathcal{O}(1) = \mathcal{O}_V(1)$ be the dual of the tautological line bundle $\mathcal{L}_V$ on $\mathbb{P}(V)$ (cf. 4.2.7(c)) and let $\mathcal{O}_X(1)$ be the pullback $i^*\mathcal{O}(1)$. (Then $\mathcal{O}_X(1)$ are precisely the very ample line bundles on X as we vary i and V, cf. [Hartshorne–77, Chap. II, §5].)

Prove the following:

(a) The restriction map $H^0(X, \mathcal{O}_X(1)) \to H^0(x, \mathcal{O}_X(1)_{|x})$ (for any $x \in X$) is surjective. Hence, we get a line $l_x := H^0(x, \mathcal{O}_X(1)_{|x})^* \hookrightarrow W^*$, where $W := H^0(X, \mathcal{O}_X(1))$.

(b) The map $x \mapsto l_x$ provides a (closed) embedding $\hat{i} : X \hookrightarrow \mathbb{P}(W^*)$ such that $\hat{i}^*(\mathcal{O}_{W^*}(1)) \cong \mathcal{O}_X(1)$.

(2) Prove (8.2.9.1) for $M = \mathbb{C}_\lambda$ ($\lambda \in \mathfrak{h}^*_{\mathbb{Z}}$) and $\mathfrak{w} = (s_i)$, for any simple reflection s_i.*Hint*. Use Lemma 8.2.8 and Exercise 8.1.E.2.

(3) Extend the Demazure character formula (8.2.9.4) for any $\lambda \in D$, i.e., prove that for any $\lambda \in D$ and $w \in W$,

$$\operatorname{ch} L_w^{\max}(\lambda) = D_{\mathfrak{w}}(e^\lambda),$$

where $\mathfrak{w}$ is any reduced word with $\pi(\mathfrak{w}) = w$.

8.3. Extension of the Weyl–Kac Character Formula and the Borel–Weil–Bott Theorem

In this section $\mathfrak{g}$ is an arbitrary Kac–Moody algebra.

Recall that a purely algebraic proof of the Weyl–Kac character formula for symmetrizable Kac–Moody algebras was given in Section 2.2. We now give a geometric proof of this theorem. An advantage of this geometric proof is that it works for an arbitrary (not necessarily symmetrizable) Kac–Moody algebra.

Fix any $\rho \in \mathfrak{h}^*_{\mathbb{Z}}$ satisfying $\langle \rho, \alpha_i^\vee \rangle = 1$ for all the simple coroots $\alpha_i^\vee$. Recall the definition of the algebra of formal characters $\mathcal{A}$ from 2.1.1, the sign representation ε of W (where W is the Weyl group associated to $\mathfrak{g}$) from Definition 1.3.12, and the integrable highest weight $\mathfrak{g}$-module $L^{\max}(\lambda)$ for any $\lambda \in D$ from 2.1.5. Then, the formal character $\operatorname{ch}(L^{\max}(\lambda)) \in \mathcal{A}$ (cf. (2.1.1.3)). Also, recall the notation $D_{\mathbb{Z}} := D \cap \mathfrak{h}^*_{\mathbb{Z}}$ from 6.1.6.

8.3.1 Theorem. *For an arbitrary Kac–Moody algebra* $\mathfrak{g}$ *and any* $\lambda \in D$, *we have the character formula:*

(1)
$$\Big(\sum_{w\in W} \varepsilon(w)\, e^{w\rho}\Big) \cdot \operatorname{ch}(L^{\max}(\lambda)) = \sum_{w\in W} \varepsilon(w)\, e^{w(\lambda+\rho)}$$

in the ring $\mathcal{A}$. *It is easy to see that, for any* $\lambda \in D$, $\sum_{w\in W} \varepsilon(w)\, e^{w(\lambda+\rho)} \in \mathcal{A}$.

Moreover, we have the "denominator" formula:

(2)
$$\sum_{w\in W} \varepsilon(w)\, e^{w\rho-\rho} = \prod_{\beta\in\Delta^+} (1 - e^{-\beta})^{\operatorname{mult}\beta},$$

where, as earlier, $\operatorname{mult}\beta$ *denotes the dimension of the* β*-th root space.*

We first begin with the following preparatory lemmas.

8.3.2 Lemma. *Let $\lambda, \lambda' \in D$ be such that $\lambda(\alpha_i^\vee) = \lambda'(\alpha_i^\vee)$, for all the simple coroots $\alpha_i^\vee$. Then $L^{\max}(\lambda)$ and $L^{\max}(\lambda')$ are isomorphic as $\mathfrak{g}'$-modules (cf. Corollary 1.2.3 for the notation $\mathfrak{g}'$).*

Moreover,

$$\text{(1)} \qquad \operatorname{ch} L^{\max}(\lambda) = e^{\lambda-\lambda'} \operatorname{ch} L^{\max}(\lambda').$$

Proof. The inclusion $U(\mathfrak{g}') \hookrightarrow U(\mathfrak{g})$ induces an isomorphism of $\mathfrak{g}'$-modules (by Corollary 1.2.3)

$$M(\lambda) \xrightarrow{\sim} M(\lambda').$$

This induces an isomorphism of $\mathfrak{g}'$-modules $L^{\max}(\lambda) \xrightarrow{\sim} L^{\max}(\lambda')$ (using Lemma 2.1.6). Assertion (1) is easy to verify from the above isomorphism. □

Recall the definition of the Demazure module $L_w^{\max}(\lambda)$ from 8.1.22. Express, for any $w \in W$,

$$\operatorname{ch} L_w^{\max}(\lambda) = \sum_\mu m_\mu^\lambda(w)\, e^\mu \text{ , and}$$

$$\operatorname{ch} L^{\max}(\lambda) = \sum_\mu m_\mu^\lambda\, e^\mu.$$

8.3.3 Lemma. *For any integrable highest weight $\mathfrak{g}$-module $V(\lambda)$ with highest weight $\lambda \in D_{\mathbb{Z}}$, we have*

$$\text{(1)} \qquad V_v(\lambda) \subset V_w(\lambda), \qquad \textit{for } v \leq w \in W.$$

In particular,

$$\text{(2)} \qquad v\lambda \geq w\lambda \qquad \textit{(cf. 2.1.1(a))}.$$

Further,

$$\text{(3)} \qquad \cup_{w\in W} V_w(\lambda) = V(\lambda).$$

Hence, for any μ , there exists $w = w_\mu \in W$ such that $m_\mu^\lambda = m_\mu^\lambda(w)$ (since $m_\mu^\lambda < \infty$).

Proof. For any subset $S \subset V(\lambda)$, let $\langle S\rangle \subset V(\lambda)$ denote the vector subspace spanned by S. By definition, the $\mathfrak{b}$-submodule of $V(\lambda)$ generated by $v_{w\lambda}$ is $V_w(\lambda)$ and hence, by 8.1.22,

$$\text{(4)} \qquad \langle \mathcal{B}w\mathcal{B}v_\lambda\rangle = V_w(\lambda).$$

We first take $V(\lambda) = L^{\max}(\lambda)$. Looking at the morphism $i_\lambda : \mathcal{G}/\mathcal{B} \to \mathbb{P}(L^{\max}(\lambda))$, we see that $i_\lambda^{-1}(\mathbb{P}(L_w^{\max}(\lambda)))$ is closed in $\mathcal{G}/\mathcal{B}$ and hence, by Proposition 7.1.15(a),

(5) $$\left\langle \cup_{\theta \le w} \mathcal{B}\theta\mathcal{B}v_\lambda \right\rangle = L_w^{\max}(\lambda).$$

Since there exists a surjective $\mathcal{G}$-module map $L^{\max}(\lambda) \twoheadrightarrow V(\lambda)$ (cf. Lemma 2.1.7), we get (5) for an arbitrary $V(\lambda)$, i.e.,

(6) $$\left\langle \cup_{\theta \le w} \mathcal{B}\theta\mathcal{B}v_\lambda \right\rangle = V_w(\lambda).$$

In particular, for $v \le w$,

$$V_v(\lambda) \subseteq V_w(\lambda), \quad \text{proving (1).}$$

Since the Bruhat–Chevalley partial order on W is directed (cf. Lemma 1.3.20), from (1), we obtain that $M := \bigcup_{w \in W} V_w(\lambda)$ is indeed a vector subspace of $V(\lambda)$. We next show that M is a $\mathcal{P}_i$-submodule for any minimal parabolic subgroup $\mathcal{P}_i$:

By Theorem 5.1.3(d), and (4), (6),

(7) $$\begin{aligned} \langle \mathcal{P}_i \cdot V_w(\lambda) \rangle &= V_w(\lambda), && \text{if } s_i w < w \\ &= V_{s_i w}(\lambda), && \text{if } s_i w > w. \end{aligned}$$

In particular, M is stable under $\mathcal{P}_i$ and hence M is a $\mathcal{G}$-submodule of $V(\lambda)$. But M contains the highest weight space $V(\lambda)_\lambda$ and hence $M = V(\lambda)$, proving the lemma. □

Recall that, as in 1.3.1, the "shifted" action of W on $\mathfrak{h}^*$ is defined by $w * \mu := w(\mu + \rho) - \rho$, for $w \in W$ and $\mu \in \mathfrak{h}^*$.

Fix any $v \in W$, $\mu \in \mathfrak{h}^*$, $\lambda \in D_{\mathbb{Z}}$. Then we have the following:

8.3.4 Lemma.

(a) $$\sum_{w \in W} \varepsilon(w)\, m^\lambda_{w*\mu}(v) = \begin{cases} \varepsilon(w_o), \textit{ if } \mu = w_o * \lambda, & \textit{for some (and hence, by Lemma 3.2.5, unique) } w_o \in W \\ 0, & \textit{if } \mu \notin W * \lambda. \end{cases}$$

Hence,

(b) $$\sum_{w \in W} \varepsilon(w)\, m^\lambda_{w*\mu} = \begin{cases} \varepsilon(w_o), & \textit{if } \mu = w_o * \lambda \\ 0, & \textit{if } \mu \notin W * \lambda. \end{cases}$$

(The left side in (b) *and hence in* (a) *is a finite sum, as proved below.)*

Proof. We abbreviate m^λ_ν by m_ν and $m^\lambda_\nu(v)$ by $m_\nu(v)$. For any $\mu \in \mathfrak{h}^*$, there exist only finitely many $w_1, \dots, w_n \in W$ (depending on μ) such that $m_{w_i * \mu} \neq 0$, for $1 \leq i \leq n$. (In particular, $m_{w*\mu}(v) = 0$ for any $v \in W$ and any $w \notin \{w_1, \dots, w_n\}$.) To prove this observe first that, $L^{\max}(\lambda)$ being integrable, $m_{w*\mu} = m_{\mu+\rho-w^{-1}\rho}$. Next, the weights of $L^{\max}(\lambda)$ are contained in the "cone" $\lambda - \sum_i \mathbb{Z}_+\alpha_i$ and, for any fixed μ, $\mu+\rho-w^{-1}\rho$ does not belong to this cone for all sufficiently large $\ell(w)$ as $\rho - w^{-1}\rho$ is a sum of $\ell(w)$ positive roots by (1.3.22.3). Thus, if $\ell(w)$ is sufficiently large, $\mu+\rho-w^{-1}\rho$ is not a weight of $L^{\max}(\lambda)$. Consequently the sums in the lemma make sense.

We prove (a) by induction on $\ell(v)$: It is obvious for $\ell(v) = 0$. So, take $\ell(v) > 0$ and write $v = s_i v'$ with $\ell(v') < \ell(v)$. By (8.2.9.4), we get the following:

$$\begin{aligned}\sum_\nu m_\nu(v)e^\nu &= D_{s_i}\Big(\sum_\nu m_\nu(v')e^\nu\Big)\\ &= \sum_\nu m_\nu(v')\left(\frac{e^\nu - e^{-\alpha_i}e^{s_i\nu}}{1-e^{-\alpha_i}}\right)\\ &= \sum_\nu m_\nu(v')\left(e^\nu\sum_{k=0}^{\infty}e^{-k\alpha_i} - e^{s_i\nu}\sum_{k=1}^{\infty}e^{-k\alpha_i}\right).\end{aligned}$$

Hence, for any $\nu \in \mathfrak{h}^*$,

$$m_\nu(v) = \sum_{k=0}^{\infty} m_{\nu+k\alpha_i}(v') - \sum_{k=1}^{\infty} m_{s_i\nu-k\alpha_i}(v').$$

We can (and will) assume that the set $\{w_1, \dots, w_n\}$, if necessary by enlarging, is stable under the left multiplication by s_i. Thus

$$\begin{aligned}\sum_{w\in W}\varepsilon(w)m_{w*\mu}(v) &= \sum_{j=1}^{n}\varepsilon(w_j)\,m_{w_j*\mu}(v)\\ &= \sum_j \varepsilon(w_j)\sum_{k=0}^{\infty} m_{w_j*\mu+k\alpha_i}(v')\\ &\quad -\sum_j \varepsilon(w_j)\sum_{k=0}^{\infty} m_{(s_iw_j)*\mu-k\alpha_i}(v')\\ &\quad \text{(since } s_i\nu = s_i*\nu+\alpha_i)\\ &= \sum_j \varepsilon(w_j)\sum_{k=0}^{\infty} m_{w_j*\mu+k\alpha_i}(v')\\ &\quad +\sum_j \varepsilon(w_j)\sum_{k=0}^{\infty} m_{w_j*\mu-k\alpha_i}(v')\\ &\quad \text{(replacing } w \text{ by } s_iw \text{ in the second sum)}\end{aligned}$$

(1) $$= \sum_j \varepsilon(w_j) m_{w_j * \mu}(v') + \sum_j \varepsilon(w_j) \sum_{k \in \mathbb{Z}} m_{w_j * \mu + k\alpha_i}(v').$$

But

(2) $$\sum_j \varepsilon(w_j) \sum_{k \in \mathbb{Z}} m_{w_j * \mu + k\alpha_i}(v') = 0.$$

The validity of (2) is clear if $\mu \notin \mathfrak{h}^*_{\mathbb{Z}}$ (since $\lambda \in \mathfrak{h}^*_{\mathbb{Z}}$ by assumption). In fact, in this case, each $m_{w_j * \mu + k\alpha_i}(v') = 0$. So, assume that $\mu \in \mathfrak{h}^*_{\mathbb{Z}}$. In this case, the left side of (2) is equal to

$$-\sum_j \varepsilon(w_j) \sum_{k \in \mathbb{Z}} m_{(s_i w_j) * \mu + k\alpha_i}(v') = -\sum_j \varepsilon(w_j) \sum_{k \in \mathbb{Z}} m_{w_j * \mu + k\alpha_i}(v').$$

This proves (2) and, hence, by induction (1) gives (a).

To prove (b), for any w_j $(1 \leq j \leq n)$ choose v_j such that $m_{w_j * \mu} = m_{w_j * \mu}(v'_j)$ for all $v'_j \geq v_j$. If we now choose any $v \in W$ such that $v \geq v_j$ for all $1 \leq j \leq n$, then $m_{w_j * \mu} = m_{w_j * \mu}(v)$ for all $1 \leq j \leq n$. So (b) follows from (a). □

8.3.5 *Proof of Theorem 8.3.1.* We first assume that $\lambda \in D_{\mathbb{Z}}$. Then, writing $\operatorname{ch} L^{\max}(\lambda) = \sum_\mu m_\mu e^\mu$,

$$\begin{aligned}
\Big(\sum_{w \in W} \varepsilon(w)\, e^{w\rho - \rho}\Big) \cdot \operatorname{ch} L^{\max}(\lambda) &= \sum_w \varepsilon(w)\, e^{w\rho-\rho} \cdot \sum_\mu m_\mu\, e^\mu \\
&= \sum_{w,\mu} \varepsilon(w)\, m_\mu e^{\mu + w\rho - \rho} \\
&= \sum_{w,\mu} \varepsilon(w)\, m_{\mu - w\rho + \rho}\, e^\mu \\
&= \sum_{w,\mu} \varepsilon(w)\, m_{w^{-1}\mu - \rho + w^{-1}\rho}\, e^\mu \\
&\quad (\text{since } m_\nu = m_{w\nu}) \\
&= \sum_{w,\mu} \varepsilon(w^{-1})\, m_{w * \mu}\, e^\mu \\
&= \sum_\mu \Big(\sum_w \varepsilon(w)\, m_{w*\mu}\Big) e^\mu \\
&= \sum_{w_o \in W} \varepsilon(w_o)\, e^{w_o * \lambda}, \qquad \text{by Lemma 8.3.4(b).}
\end{aligned}$$

This proves the first part of the theorem in the case when $\lambda \in D_{\mathbb{Z}}$. The case of an arbitrary $\lambda \in D$ follows from this case and Lemma 8.3.2 by taking $\lambda' \in D_{\mathbb{Z}}$ such that $\lambda(\alpha_i^\vee) = \lambda'(\alpha_i^\vee)$ for all $1 \leq i \leq \ell$ (which is possible by 6.1.6(c_4)).

We come to the proof of the denominator formula (8.3.1.2). Take any $\lambda \in D$ and $\mu \in \mathfrak{h}^*$ satisfying

$$(*) \qquad \mu = \lambda - \sum_i n_i\alpha_i, \qquad \text{where } 0 \le n_i \le \lambda(\alpha_i^\vee), \text{ for all } 1 \le i \le \ell.$$

Then the μ-weight spaces in the Verma module $M(\lambda)$ and $L^{\max}(\lambda)$ are the same (by the definition of $L^{\max}(\lambda)$). Hence the coefficients of e^μ in

$$(1) \qquad e^\lambda\Big(\prod_{\beta\in\Delta^+}(1-e^{-\beta})^{-\operatorname{mult}\beta}\Big)\Big(\sum_{w\in W}\varepsilon(w)\,e^{w\rho-\rho}\Big) \text{ and } \sum_w \varepsilon(w)\,e^{w*\lambda}$$

are the same by Lemma 2.1.13 and the first part of the theorem, since $\sum_w \varepsilon(w)\,e^{w\rho-\rho} = \sum n_\beta e^\beta$, for some $n_\beta \in \mathbb{Z}$, where the summation runs over $\beta \in -\sum_i \mathbb{Z}_+\alpha_i$.

For any $y \in W$ and simple reflection s such that $sy > y$, by Lemma 1.3.13,

$$(2) \qquad (sy) * \lambda \le y * \lambda.$$

For any $w \neq 1 \in W$, take a simple reflection s_i such that $ws_i < w$. Then, by making repeated use of (2), we get

$$w * \lambda \le s_i * \lambda = \lambda - (\lambda(\alpha_i^\vee) + 1)\alpha_i.$$

Thus $\mu = w*\lambda$, for $w \neq 1$, does not satisfy $(*)$. This shows that any μ satisfying $(*)$ can occur in the second term of (1) if and only if $\mu = \lambda$.

Now fix any $0 \neq \gamma \in \sum_{i=1}^{\ell} \mathbb{Z}_+\alpha_i$. To prove the denominator formula, it suffices to show that $e^{\lambda-\gamma}$ does not occur in the first term of (1) for some choice of $\lambda \in D$ (the choice of λ is allowed to depend on the choice of γ). Now choose a sufficiently large λ (i.e., $\lambda(\alpha_i^\vee)$ is sufficiently large for each simple coroot $\alpha_i^\vee$) such that $\mu := \lambda - \gamma$ satisfies $(*)$ with respect to λ. With this choice of λ, the coefficients of e^μ in the two terms of (1) are the same. But, by the above paragraph, the coefficient of e^μ in the second term of (1) is zero. This completes the proof of the denominator formula as well. □

8.3.6 Remark. In the symmetrizable case, for any $\lambda \in D$, $L^{\max}(\lambda)$ is irreducible (cf. Corollary 2.2.6). But, in the non-symmetrizable case, the irreducibility of $L^{\max}(\lambda)$ is not known. The irreducibility of $L^{\max}(\lambda)$, for all $\lambda \in D$, for any fixed Kac–Moody algebra $\mathfrak{g}$ would imply the vanishing of the radical of $\mathfrak{g}$ (cf. Corollary 3.2.10 in the symmetrizable case).

8.3.7 Proposition. *Take a simple reflection s_i and $\lambda \in \mathfrak{h}^*_{\mathbb{Z}}$ such that $\lambda(\alpha_i^\vee) \geq -1$. Then, for any $p \in \mathbb{Z}$ and any $w \in W$ such that $ws_i < w$, there is an isomorphism of $\mathcal{B}$-modules*

(1) $$\beta_w : H^p(X_w, \mathcal{L}_w(\lambda)) \xrightarrow{\sim} H^{p+1}(X_w, \mathcal{L}_w(s_i * \lambda)).$$

In fact, this isomorphism is a $\mathcal{P}_j$-module isomorphism for all those $1 \leq j \leq \ell$ such that X_w is $\mathcal{P}_j$-stable under left multiplication, i.e., those j satisfying $s_j w < w$.

Moreover, for any $v \geq w$ such that $vs_i < v$, we have the following commutative diagram:

(∗) $$\begin{array}{ccc} H^p(X_v, \mathcal{L}_v(\lambda)) & \xrightarrow[\sim]{\beta_v} & H^{p+1}(X_v, \mathcal{L}_v(s_i * \lambda)) \\ \downarrow & & \downarrow \\ H^p(s_w, \mathcal{L}_w(\lambda)) & \xrightarrow[\sim]{\beta_w} & H^{p+1}(X_w, \mathcal{L}_w(s_i * \lambda)), \end{array}$$

where the vertical maps are the canonical restriction maps.

Proof. Let V be the pro-representation of $\mathcal{P}_i$ defined by

$$V := H^0(Z_{s_i}, \mathcal{L}_{s_i}(\lambda + \rho)).$$

The inclusion $Z_\emptyset \hookrightarrow Z_{s_i}$ induces a surjective $\mathcal{B}$-module map (cf. the proof of Proposition 8.1.17)

$$p : V \twoheadrightarrow \mathbb{C}_{-(\lambda+\rho)}.$$

Thus, we get the exact sequence of $\mathcal{B}$-modules:

$$0 \to K \to V \xrightarrow{p} \mathbb{C}_{-(\lambda+\rho)} \to 0,$$

where $K := \operatorname{Ker} p$. Tensoring this with the $\mathcal{B}$-module $\mathbb{C}_\rho$, we get the exact sequence of $\mathcal{B}$-modules:

(2) $$0 \to K \otimes \mathbb{C}_\rho \to V \otimes \mathbb{C}_\rho \to \mathbb{C}_{-\lambda} \to 0.$$

If $(\lambda + \rho)(\alpha_i^\vee) > 0$, by the proof of Proposition 8.1.17 again, we have an injective $\mathcal{B}$-module map $\mathbb{C}_{-s_i(\lambda+\rho)} \hookrightarrow K$. Let Q be the quotient. Then Q is, in fact, a $\mathcal{P}_i$-module (since Q is obtained as a subquotient from the irreducible pro-representation V of $\mathcal{P}_i$ by stripping off its highest and lowest weight spaces) and we get the exact sequence of $\mathcal{B}$-modules:

(3) $$0 \to \mathbb{C}_{-s_i(\lambda+\rho)+\rho} \to K \otimes \mathbb{C}_\rho \to Q \otimes \mathbb{C}_\rho \to 0\cdot$$

By Exercise 8.1.E.1, abbreviating $\mathcal{L}_{\mathcal{P}_i/\mathcal{B}}$ by $\mathcal{L}$,

$$(4) \quad H^p(\mathcal{P}_i/\mathcal{B}, \mathcal{L}(V \otimes \mathbb{C}_\rho)) = 0 = H^p(\mathcal{P}_i/\mathcal{B}, \mathcal{L}(Q \otimes \mathbb{C}_\rho)), \text{ for all } p \geq 0\,.$$

We next show that, for all $p \geq 0$,

$$(5) \qquad H^p(X_w, \mathcal{L}_w(V \otimes \mathbb{C}_\rho)) = H^p(X_w, \mathcal{L}_w(Q \otimes \mathbb{C}_\rho)) = 0\,.$$

Take a reduced word $\mathfrak{w}$ with $\pi(\mathfrak{w}) = w$ such that $\mathfrak{w}$ ends in s_i. By Theorem 8.2.2(b)–(c) and (8.1.13.3),

$$H^p(X_w, \mathcal{L}_w(V \otimes \mathbb{C}_\rho)) \tilde{\to} H^p(Z_{\mathfrak{w}}, \mathcal{L}_{\mathfrak{w}}(V \otimes \mathbb{C}_\rho)), \text{ and}$$

$$H^p(X_w, \mathcal{L}_w(Q \otimes \mathbb{C}_\rho)) \tilde{\to} H^p(Z_{\mathfrak{w}}, \mathcal{L}_{\mathfrak{w}}(Q \otimes \mathbb{C}_\rho)).$$

Using the Leray spectral sequence for the projection $Z_{\mathfrak{w}} \to Z_{\mathfrak{w}[n-1]}$ (where $n = \ell(w)$), Lemma 8.1.5 and (4), together with the above isomorphisms, we get (5).

Using (5) in the long exact cohomology sequences associated to (2)–(3), we obtain:

$$(6) \qquad H^p(X_w, \mathcal{L}_w(\lambda)) \simeq H^{p+1}(X_w, \mathcal{L}_w(K \otimes \mathbb{C}_\rho)), \text{ and}$$

$$(7) \qquad H^{p+1}(X_w, \mathcal{L}_w(s_i * \lambda)) \simeq H^{p+1}(X_w, \mathcal{L}_w(K \otimes \mathbb{C}_\rho)).$$

Further, both of these isomorphisms are $\mathcal{P}_j$-module isomorphisms for any j such that $s_j w < w$ (by Corollary 8.2.3). Combining (6)–(7), we obtain (1) in the case $(\lambda + \rho)(\alpha_i^\vee) > 0$. If $(\lambda + \rho)(\alpha_i^\vee) = 0$, then $\lambda = s_i * \lambda$ and, in this case, $H^p(X_w, \mathcal{L}_w(\lambda)) = 0$ for all $p \geq 0$ (as can be seen by the same argument which is used to prove (5)). So (1) follows in this case as well.

The commutativity of $(*)$ is easy to see from the functoriality of the long exact cohomology sequences (since, for any finite-dimensional pro-representation M of $\mathcal{B}$, $\mathcal{L}_v(M)_{|X_w} \cong \mathcal{L}_w(M)$ as $\mathcal{B}$-equivariant vector bundles on X_w, cf. Lemma 8.2.4). □

8.3.8 Definition. For any finite-dimensional pro-representation M of $\mathcal{B}$, recall the definition of the $\mathcal{B}$-equivariant (in fact $\mathcal{P}_j$-equivariant, for any minimal parabolic subgroup $\mathcal{P}_j$) algebraic vector bundle $\mathcal{L}(M)$ on $\mathcal{X} = \mathcal{G}/\mathcal{B}$ from Corollary 8.2.5. For any $n \geq 0$, there exists $w_n \in W$ such that $X_{w_n} \supset X_n$ (by Lemma 1.3.20). From this it is easy to see that (for any $p \geq 0$)

$$(1) \qquad H^p(\mathcal{X}, \mathcal{L}(M)) = \underset{w \in W}{\text{Inv. lt.}}\ H^p(X_w, \mathcal{L}_w(M)),$$

where the cohomology $H^*(\mathcal{X}, \mathcal{L}(M))$ is as defined in 4.2.6 and W is the directed set under the Bruhat–Chevalley partial order $\leq$. By 4.2.6, $H^p(\mathcal{X}, \mathcal{L}(M))$ is canonically a $\mathcal{P}_j$-module for each $\mathcal{P}_j$.

We also define

(2) $$H^p(\mathcal{X}, \mathcal{L}(M))^\vee := \operatorname*{limit}_{w\in W} H^p(X_w, \mathcal{L}_w(M))^*.$$

Then, similarly, $H^p(\mathcal{X}, \mathcal{L}(M))^\vee$ is canonically a $\mathcal{P}_j$-module (for each $\mathcal{P}_j$). In fact, it is a pro-representation of $\mathcal{P}_j$ since, for any $w \in W$, $H^p(X_w, \mathcal{L}_w(M))^*$ is a pro-representation of those $\mathcal{P}_j$ such that $s_j w < w$ (by Corollary 8.2.3).

8.3.9 Lemma. *For any finite-dimensional pro-representation M of $\mathcal{B}$, there exists a (unique) $\mathcal{G}$-module structure on $H^p(\mathcal{X}, \mathcal{L}(M))^\vee$ (as well as on $H^p(\mathcal{X}, \mathcal{L}(M))$) extending the $\mathcal{P}_j$-module structures (for every minimal parabolic subgroup $\mathcal{P}_j$) given in 8.3.8. Hence, $H^p(\mathcal{X}, \mathcal{L}(M))^\vee \in \mathfrak{M}(\mathcal{G})$ (cf. 6.2.1 for the notation $\mathfrak{M}(\mathcal{G})$). Further, there is a canonical $\mathcal{G}$-module isomorphism*

(1) $$\left(H^p(\mathcal{X}, \mathcal{L}(M))^\vee\right)^* \simeq H^p(\mathcal{X}, \mathcal{L}(M)).$$

Proof. We first put a $\mathcal{G}$-module structure on $N := H^p(\mathcal{X}, \mathcal{L}(M))^\vee$. Since N is a pro-representation of $\mathcal{P}_j$, on differentiation, we get a pro-representation of $\hat{\mathfrak{p}}_j \simeq \operatorname{Lie} \mathcal{P}_j$ on N (cf. 4.4.23 and Lemma 6.1.14). On restriction, we get a (locally finite) representation of each $\mathfrak{p}_j$ on N and hence, by Corollary 1.3.10, there is a compatible $\mathfrak{g}$-module structure on N. By Exercise 6.1.E.1, we get a $\hat{\mathfrak{g}}$-module structure on N extending the $\mathfrak{g}$-module and $\hat{\mathfrak{p}}_j$-module structures on it. Since N is a pro-representation of $\mathcal{P}_j$; in particular, it is a locally finite algebraic T-module and hence $N \in \mathfrak{M}_T(\hat{\mathfrak{g}})$ (cf. 6.2.1). So, by Theorem 6.2.3 and Lemma 4.4.24, we get a $\mathcal{G}$-module structure on N extending the $\mathcal{P}_j$-module structures for every $\mathcal{P}_j$.

Consider the dual θ_w^* of the natural map $\theta_w : H^p(X_w, \mathcal{L}_w(M))^* \to N$ for any $w \in W$. Taking the inverse limit of the maps $\{\theta_w^*\}_{w\in W}$, we get the map

$$\gamma : N^* \to H^p(\mathcal{X}, \mathcal{L}(M)),$$

by (8.3.8.1); identifying $(H^p(X_w, \mathcal{L}_w(M))^*)^* \simeq H^p(X_w, \mathcal{L}_w(M))$. Since the dual of a direct limit of vector spaces is isomorphic with the inverse limit of the dual vector spaces, it is easy to see that γ is an isomorphism and, moreover, it is a $\mathcal{P}_j$-module morphism for all the minimal parabolics $\mathcal{P}_j$. This proves (1). Since N is a $\mathcal{G}$-module, so is N^* and, hence, transporting the $\mathcal{G}$-module structure via γ, we get that $H^p(\mathcal{X}, \mathcal{L}(M))$ is a $\mathcal{G}$-module extending the $\mathcal{P}_j$-module structures. This proves the lemma. □

8.3.10 Remark. By the same proof as above, we obtain that, for any finite-dimensional pro-representation M of $\mathcal{B}$, $H^p(Z_\infty, \mathcal{L}_\infty(M))^\vee$ and $H^p(Z_\infty, \mathcal{L}_\infty(M))$ are both $\mathcal{G}$-modules extending the $\mathcal{P}_j$-module structures on these spaces (cf. Definition 8.1.15). Moreover, from the general property of direct and inverse limits mentioned in the proof of Lemma 8.3.9, together with the finite dimensionality of $H^p(Z_{\mathfrak{w}}, \mathcal{L}_{\mathfrak{w}}(M))$, we get a canonical isomorphism

$$H^p(Z_\infty, \mathcal{L}_\infty(M))^{\vee *} \simeq H^p(Z_\infty, \mathcal{L}_\infty(M)).$$

In fact, by (8.2.2.1), there are canonical isomorphisms

$$H^p(Z_\infty, \mathcal{L}_\infty(M))^\vee \stackrel{\sim}{\to} H^p(\mathcal{X}, \mathcal{L}(M))^\vee, \text{ and}$$

$$H^p(\mathcal{X}, \mathcal{L}(M)) \stackrel{\sim}{\to} H^p(Z_\infty, \mathcal{L}_\infty(M)).$$

Now we are ready to prove the following theorem, which is a generalization of the Borel–Weil–Bott theorem to an arbitrary Kac–Moody situation.

8.3.11 Theorem. *Let $\mathcal{G}$ be an arbitrary Kac–Moody group with Borel subgroup $\mathcal{B}$ and Weyl group W. Then, for any $\lambda \in \mathfrak{h}^*_{\mathbb{Z}}$ such that $\lambda + \rho \in D$, $v \in W$, and $p \in \mathbb{Z}$, we have*

$$H^p(\mathcal{G}/\mathcal{B}, \mathcal{L}(\lambda))^\vee \approx H^{p+\ell(v)}(\mathcal{G}/\mathcal{B}, \mathcal{L}(v * \lambda))^\vee, \quad \textit{as } \mathcal{G}\textit{-modules}.$$

Proof. We prove the theorem by induction on $\ell(v)$. For $v = 1$ there is nothing to prove; so take $v = s_i v'$ with $\ell(v') < \ell(v)$. We first observe that

$$(v' * \lambda)(\alpha_i^\vee) = v'(\lambda + \rho)(\alpha_i^\vee) - 1 = (\lambda + \rho)(v'^{-1}\alpha_i^\vee) - 1 \geq -1,$$

since $\lambda + \rho \in D$ (by assumption) and $v'^{-1}\alpha_i^\vee = \sum n_j \alpha_j^\vee$ with $n_j \geq 0$ (by Lemma 1.3.13). Hence, by Proposition 8.3.7, $H^p(X_w, \mathcal{L}_w(v' * \lambda)) \approx H^{p+1}(X_w, \mathcal{L}_w(v * \lambda))$ as $\mathcal{B}$-modules, provided $ws_i < w$. But since the set W_i, consisting of all those Weyl group elements w such that $ws_i < w$, is cofinal in W, we get that $H^p(\mathcal{G}/\mathcal{B}, \mathcal{L}(v' * \lambda))^\vee \approx H^{p+1}(\mathcal{G}/\mathcal{B}, \mathcal{L}(v * \lambda))^\vee$, as $\mathcal{B}$-modules (using the commutativity of $(*)$ in 8.3.7). It remains to show that the isomorphism is indeed a $\mathcal{G}$-module isomorphism.

For any simple reflection s_j, define $W_{j,i} = \{w \in W_i : s_j w < w\}$. In view of Proposition 8.3.7, it suffices to show that $W_{j,i}$ also is cofinal in W. Pick any $w \in W_i$. If $s_j w > w$, we claim that $s_j w \in W_{j,i}$, i.e., $s_j w s_i < s_j w$: If not, we have $s_j w s_i > s_j w > w$. But $\ell(s_j w s_i) \leq \ell(w)$, which is a contradiction. This proves $s_j w s_i < s_j w$. Hence $W_{j,i}$ is cofinal in W_i and hence in W. This proves the theorem. □

8.3.12 Corollary. *With the notation as in the above Theorem 8.3.11, we have the following:*

(a) *For* $\lambda \in D_{\mathbb{Z}}$ *and any* $v \in W$,

$$H^p(\mathcal{G}/\mathcal{B}, \mathcal{L}(v * \lambda))^\vee = 0, \text{ unless } p = \ell(v), \text{ and}$$

$$H^{\ell(v)}(\mathcal{G}/\mathcal{B}, \mathcal{L}(v * \lambda))^\vee \approx H^0(\mathcal{G}/\mathcal{B}, \mathcal{L}(\lambda))^\vee \approx L^{\max}(\lambda), \quad \text{as } \mathcal{G}\text{-modules.}$$

(b) *For any* $\lambda \in \mathfrak{h}^*_{\mathbb{Z}}$ *but* $\lambda \notin W * D$, $H^p(\mathcal{G}/\mathcal{B}, \mathcal{L}(\lambda))^\vee = 0$, *for all* $p \geq 0$.

Proof. (a): By Theorem 8.3.11 and the definition of $H^i(\mathcal{G}/\mathcal{B}, \mathcal{L}(M))^\vee$ (cf. (8.3.8.2)), one has

$$\begin{aligned} H^p(\mathcal{G}/\mathcal{B}, \mathcal{L}(v * \lambda))^\vee &\simeq H^{p-\ell(v)}(\mathcal{G}/\mathcal{B}, \mathcal{L}(\lambda))^\vee \\ &\simeq \operatorname*{limit}_{w \in W} \; H^{p-\ell(v)}(X_w, \mathcal{L}_w(\lambda))^*. \end{aligned}$$

Further, by Theorem 8.2.2(a) and (8.2.2.3),

$$H^{p-\ell(v)}(X_w, \mathcal{L}_w(\lambda))^* \simeq \begin{cases} L_w^{\max}(\lambda), & \text{if } p = \ell(v); \\ 0, & \text{otherwise.} \end{cases}$$

Since $L^{\max}(\lambda)$ is the direct limit of $\{L_w^{\max}(\lambda)\}_w$ (by Lemma 8.3.3), this proves (a).

(b): We prove by induction on $p \geq -1$ that for all $q \leq p$,

(1) $$H^q(\mathcal{G}/\mathcal{B}, \mathcal{L}(\lambda))^\vee = 0, \quad \text{for all } \lambda \in \mathfrak{h}^*_{\mathbb{Z}} \backslash W * D.$$

Of course (1) is true for $p = -1$, so the induction starts. Since $\lambda \notin D$, there exists a simple coroot $\alpha_i^\vee$ such that $\lambda(\alpha_i^\vee) \leq -1$. Then $\mu := s_i * \lambda$ satisfies $\mu(\alpha_i^\vee) \geq -1$. Hence, by Proposition 8.3.7, for any $w \in W_i$,

$$H^{p-1}(X_w, \mathcal{L}_w(s_i * \lambda))^* \approx H^p(X_w, \mathcal{L}_w(\lambda))^*.$$

Thus, since W_i is cofinal in W, we get

$$H^{p-1}(\mathcal{G}/\mathcal{B}, \mathcal{L}(s_i * \lambda))^\vee \approx H^p(\mathcal{G}/\mathcal{B}, \mathcal{L}(\lambda))^\vee.$$

Since $s_i * \lambda \in \mathfrak{h}^*_{\mathbb{Z}} \backslash W * D$, we get by induction that $H^p(\mathcal{G}/\mathcal{B}, \mathcal{L}(\lambda))^\vee = 0$, completing the induction. This proves (b). □

8.3.E EXERCISES

(1) Generalize Lemma 8.3.3 to an arbitrary $\lambda \in D$.
Hint. Use Lemma 8.3.2.

(2) Let $Y \subset \{1, \dots, \ell\}$ be a subset of finite type and let $\mathcal{P}_Y$ be the associated standard parabolic subgroup of $\mathcal{G}$. Let M be a finite-dimensional pro-representation of $\mathcal{P}_Y$. Then construct a $\mathcal{B}$-equivariant algebraic vector bundle $\mathcal{L}^Y(M)$ on $\mathcal{X}^Y := \mathcal{G}/\mathcal{P}_Y$ of rank $= \dim M$ such that, under the canonical morphism $\pi : \mathcal{X} \to \mathcal{X}^Y$,

$$\pi^* \mathcal{L}^Y(M) \simeq \mathcal{L}(M)$$

as $\mathcal{B}$-equivariant algebraic vector bundles (cf. Corollary 8.2.5).

Show further that, in fact, $\mathcal{L}^Y(M)$ is a $\mathcal{P}_j$-equivariant algebraic vector bundle (for each minimal parabolic subgroup $\mathcal{P}_j$) such that the above isomorphism is $\mathcal{P}_j$-equivariant. Moreover, show that (under π^*)

$$H^p(\mathcal{X}^Y, \mathcal{L}^Y(M)) \simeq H^p(\mathcal{X}, \mathcal{L}(M))$$

as representations of each $\mathcal{P}_j$. In particular, $H^p(\mathcal{X}^Y, \mathcal{L}^Y(M))$ is a $\mathcal{G}$-module, making the above isomorphism a $\mathcal{G}$-module isomorphism.

In the following exercises (3)–(6), assume that $\mathcal{G} = G$ is of finite type, i.e., G is a (finite-dimensional) semisimple simply-connected complex algebraic group. By the Peter–Weyl theorem and the Tannaka–Kreĭn duality [Brocker-Dieck–85, Chap. III], the affine coordinate ring $\mathbb{C}[G]$, as a $G \times G$-module, is given by

$$(*) \qquad \mathbb{C}[G] \simeq \bigoplus_{\lambda \in D} L(\lambda)^* \otimes L(\lambda),$$

where $G \times G$ acts on $\mathbb{C}[G]$ via $((g, h)f)x = f(g^{-1}xh)$ and $G \times G$ acts on $L(\lambda)^* \otimes L(\lambda)$ factorwise.

(3) Use the above decomposition $(*)$ to obtain a simple proof of the Borel–Weil theorem: For $\lambda \in D$, $H^0(G/B, \mathcal{L}(\lambda)) \simeq L(\lambda)^*$, as G-modules.

(4) Prove Bott's result that for any parabolic subgroup P and any finite-dimensional algebraic P-module M,

$$H^p(G/P, \mathcal{L}(M)) \simeq \sum_{\lambda \in D} L(\lambda)^* \otimes H^p(\mathfrak{u}, L(\lambda) \otimes M)^{\mathfrak{r}}$$

as G-modules, where $\mathfrak{u}$ is the nil-radical of the parabolic subalgebra $\mathfrak{p} := \operatorname{Lie} P$ and $\mathfrak{r}$ is a Levi component of $\mathfrak{p}$. (The superscript $\mathfrak{r}$ denotes the space of $\mathfrak{r}$-invariants.)

Hint. Realize both $H^*(G/P, -)$ and $H^*(\mathfrak{u}, -)$ as right derived functors of appropriate functors on certain abelian categories and then prove the above isomorphism for H^0.

(5) Use the above exercise (4) and Kostant's result on thecohomology of $\mathfrak{n}$ (cf. Exercise 3.2.E.1(d)) to obtain the Borel–Weil–Bott theorem; which asserts that for any dominant integral weight λ,

$$H^p(G/B, \mathcal{L}(w * \lambda)) = 0, \quad \text{unless } p = \ell(w), \text{ and}$$
$$H^{\ell(w)}(G/B, \mathcal{L}(w * \lambda)) \simeq H^0(G/B, \mathcal{L}(\lambda)) \simeq L(\lambda)^*, \quad \text{as } G\text{-modules.}$$

(6) Conversely, use Exercise (4) and the Borel–Weil–Bott theorem (given above) to obtain Kostant's result on $H^*(\mathfrak{n}, L(\lambda))$.

8.C Comments. Most of the main results in this chapter for the Kac–Moody case are proved by [Kumar–87a] and also, independently, by [Mathieu–88a].

More specifically, Proposition 8.1.2 (in the finite case) is due to [Ramanathan–85] and the functor $D_{\mathfrak{w}}(M)$ defined in 8.1.16 was introduced by [Joseph–85]. Results of Exercise 8.1.E.2 are proved in [Joseph–85]. Theorem 8.1.8 is proved by [Kumar–87a]. (Theorem 8.1.8(1′) is also proved by [Mathieu–88a].) Among the other main results, Proposition 8.1.7, Theorem 8.1.13, Proposition 8.1.17, Theorems 8.1.25, 8.2.2, 8.2.9, 8.3.1, 8.3.11 and Corollary 8.3.12 are all proved in [Kumar–87a] and [Mathieu–88a] in the Kac–Moody case. Kumar's work is over char. 0 (without using any char. p methods at all). On the other hand, Mathieu uses char. p methods in his work, more specifically, the Frobenius splitting methods introduced by [Mehta–Ramanathan–85] and [Ramanan–Ramanathan–85]. In particular, some of Mathieu's results are available over any field (even over $\mathbb{Z}$). Even though we have the symmetrizability assumption in Theorem 8.2.2(e) on the projectively normal and projectively Cohen–Macaulay property of the Schubert Varieties X_w^Y, it remains true in general, i.e., the symmetrizability assumption can be removed. The projective normality in general is proved in [Mathieu–89, Corollary 1] and can also be easily deduced from [Kumar–89a, Theorem 2.7]. Once we have the projective normality of X_w^Y, its projective Cohen–Macaulay property follows by the same proof as that of Theorem 8.2.2(e) given in the book.

[Foda–Misra–Okado–98], [Kuniba–Misra–Okado–Takagi–Uchiyama–98], [Sanderson–00] have given some explicit expressions for the characters of certain Demazure modules (in some affine cases).

In the following paragraphs we assume that we are in the finite case, i.e., $\mathcal{G} = G$ is a finite-dimensional semisimple simply-connected algebraic group.

As is well known, Demazure's original proof of his character formula [Demazure–74] has a serious gap (as pointed out by V. Kac). [Joseph–85] gave a proof of the Demazure character formula for any sufficiently large dominant λ, thus obtaining the normality of the Schubert varieties $X_w \subset G/B$ in char. 0 as a corollary; since the normality of X_w is equivalent to the validity of the Demazure character formula for all large multiples of a dominant regular λ. [Seshadri–87] proved the normality of X_w over an arbitrary field. Thus, the work [Mehta–Ramanathan–85] introducing the remarkable "Frobenius splitting" method, together with the normality of X_w, gives the Demazure character formula for any dominant λ (over an arbitrary field). [Andersen–85] and [Ramanan–Ramanathan–85] proved the surjectivity of $H^0(G/B, \mathcal{L}(\lambda)) \to H^0(X_w, \mathcal{L}_w(\lambda))$ for any dominant λ, and thus obtained the normality of X_w and the Demazure character formula over an arbitrary field.

The projective normality of X_w is proved in [Ramanan–Ramanathan–85]; the property that X_w has rational singularities (thus Cohen–Macaulay) and is projectively Cohen–Macaulay is proved in [Ramanathan–85]. Some of these results in special cases have been obtained by Seshadri and his school via Standard Monomial Theory (cf. the survey article [Lakshmibai–Seshadri–91]).

[Kashiwara–93] has given a proof of the Demazure character formula (valid in the symmetrizable Kac–Moody case) by using his crystal base, and [Littelmann–98] has given a proof of the Demazure character formula using his Lakshmibai–Seshadri path model.

Recently, [Kumar–Littelmann–02] have obtained several of these results algebraically via the quantum groups at roots of unity.

The famous Borel–Weil theorem (cf. [Borel–54]), that $H^0(G/B, \mathcal{L}(\lambda)) \simeq L(\lambda)^*$ for dominant λ (in char. 0), was extended by [Bott–57] who proved that $H^p(G/B, \mathcal{L}(w * \lambda)) \simeq L(\lambda)^*$, for $p = \ell(w)$ and 0 otherwise. (Refer to the papers [Demazure–68, 76] for a simple proof of Bott's theorem.) Theorem 8.3.11 is proved in [Demazure–76] (in the finite case).

There is a recent generalization of the Borel–Weil–Bott theorem by [Kostant–99] applicable to any compact simply-connected homogeneous space of positive Euler characteristic, which is further extended to the affine case by [Landweber–2001]. In another direction, [Teleman–98] has obtained a generalization of the Borel–Weil–Bott theorem for the moduli stack of G-bundles over a smooth projective curve (for a semisimple group G).

IX

BGG and Kempf Resolutions

The aim of this chapter is to obtain the BGG resolution and the dual Kempf resolution in an arbitrary Kac–Moody situation.

In Section 9.1, we assume that $\mathfrak{g}$ is symmetrizable and show that there exists a resolution of the integrable module $L(\lambda)$ for any $\lambda \in D$:

$$(*) \qquad \cdots \longrightarrow F_p \xrightarrow{\delta^p} \cdots \to F_1 \xrightarrow{\delta^1} F_0 \xrightarrow{\delta^0} L(\lambda) \longrightarrow 0,$$

where $F_p := \bigoplus_{\substack{w \in W \\ \ell(w)=p}} M(w * \lambda)$ and δ^p are $\mathfrak{g}$-module maps. The proof proceeds by taking the standard resolution $U(\mathfrak{g}) \otimes_{U(\mathfrak{b})} \Lambda^*(\mathfrak{g}/\mathfrak{b})$ of the trivial $\mathfrak{g}$-module for the pair $(\mathfrak{g}, \mathfrak{b})$, as in the proof of the $\mathfrak{n}^-$-homology result of Garland–Lepowsky in Section 3.2, and tensoring it with $L(\lambda)$ to get a resolution of $L(\lambda)$. Next, it is shown that all the terms $U(\mathfrak{g}) \otimes_{U(\mathfrak{b})} (\Lambda^p(\mathfrak{g}/\mathfrak{b}) \otimes L(\lambda))$ admit a filtration by $\mathfrak{g}$-submodules such that the successive quotients are Verma modules. Now, we decompose the resolution as a direct sum of generalized eigenspaces of the Casimir–Kac operator Ω and consider the direct summand $\mathcal{F}_\bullet$ corresponding to the eigenvalue $d_\lambda := \langle \lambda+\rho, \lambda+\rho\rangle - \langle \rho, \rho\rangle$, which is the constant by which Ω acts on $L(\lambda)$. It is further shown that each $\mathcal{F}_p$ has a filtration by $\mathfrak{g}$-submodules such that the associated gr is isomorphic with F_p. To prove that $\mathcal{F}_p$ itself is isomorphic with F_p, we need a certain Ext vanishing between two Verma modules as in Proposition 9.1.7. In fact, we prove a more general resolution similar to $(*)$ that is valid for the parabolic case.

In Section 9.2, which no longer requires the symmetrizability assumption on $\mathfrak{g}$, we define a chain complex $\mathcal{S}$ (which we call the BGG complex) with the same F_p's (as above), but now the chain maps s^p are explicitly given in terms of the combinatorics of the Weyl group. However, as explained below, the exactness of this new complex is obtained by using the exactness of $(*)$ in the symmetrizable case and, in general, by using the exactness of the Kempf complex obtained in Section 9.3.

The combinatorial description of the maps s^p crucially relies on the following basic result on $\mathfrak{g}$-module maps between Verma modules: For $\lambda \in D$ and $v, w \in$

W, $\operatorname{Hom}_{\mathfrak{g}}(M(v * \lambda), M(w * \lambda)) \neq 0$ iff $v \geq w$, and in this case it is one-dimensional. This result is obtained by using Enright's completion functor. We give a self-contained construction of this functor. We next show that given any chain complex $\mathcal{A}$ with the same components F_p and $\mathfrak{g}$-module chain maps b^p, under some very mild restrictions on b^p, the chain complex $\mathcal{A}$ is isomorphic with the chain complex $\mathcal{S}$ in the category of chain complexes of $\mathfrak{g}$-modules. In particular, the chain complex $\mathcal{S}$ is exact iff $\mathcal{A}$ is exact. Thus, the exactness of $\mathcal{S}$ follows, in the symmetrizable case, from the exactness of $(*)$.

In Section 9.3, we define the Kempf resolution for an arbitrary $\mathfrak{g}$ due to Kumar. To define the Kempf resolution, we first recall from B.7 that, given a topological space Z with an abelian sheaf $\mathcal{L}$ and a filtration of Z by closed subspaces: $Z = Z_0 \supset Z_1 \supset \cdots$, there is associated the following cochain complex consisting of local cohomology groups (called the Grothendieck–Cousin complex):

$$(\mathcal{G}) \qquad 0 \to H^0(Z, \mathcal{L}) \to H^0_{Z_0/Z_1}(Z, \mathcal{L}) \to H^1_{Z_1/Z_2}(Z, \mathcal{L}) \to \cdots .$$

Taking Z to be the Schubert variety X_w with the line bundle $\mathcal{L}_w(\lambda)$, for any $w \in W$ and $\lambda \in D_{\mathbb{Z}}$, and considering the filtration of X_w by $\mathcal{F}_p(w) := \cup_{\ell(v) \geq p} X_w \cap X^v$, we get the Grothendieck–Cousin complex $\mathcal{C}(w)$ in this case, where X^v is the closure of the $\mathcal{U}^-$-orbit $\mathcal{U}^- v\mathcal{B}/\mathcal{B}$ in $\mathcal{G}/\mathcal{B}$. Further, it is shown that $\mathcal{C}(w)$ is exact, by using the Cohen–Macaulay property of X_w. Dualizing $\mathcal{C}(w)$ and then taking the limit over $w \in W$, we get the following exact complex (which we refer to as the Kempf complex):

$$(\mathcal{K}) \quad 0 \leftarrow H^0(\mathcal{X}, \mathcal{L}(\lambda))^\vee \leftarrow H^0_{\mathcal{F}_0/\mathcal{F}_1}(\mathcal{X}, \mathcal{L}(\lambda))^\vee \leftarrow H^1_{\mathcal{F}_1/\mathcal{F}_2}(\mathcal{X}, \mathcal{L}(\lambda))^\vee \leftarrow \cdots ,$$

where $H^i_{\mathcal{F}_p/\mathcal{F}_{p+1}}(\mathcal{X}, \mathcal{L}(\lambda))^\vee := \operatorname{limit}_{w \in W} \left(H^i_{\mathcal{F}_p(w)/\mathcal{F}_{p+1}(w)}(X_w, \mathcal{L}_w(\lambda))^\vee\right)$. Moreover, the terms $H^p_{\mathcal{F}_p/\mathcal{F}_{p+1}}(\mathcal{X}, \mathcal{L}(\lambda))^\vee$ are identified with the direct sum

$$\oplus_{\ell(v)=p} H^p_{X^v/X^v \setminus (\mathcal{U}^- v\mathcal{B}/\mathcal{B})}(\mathcal{X}, \mathcal{L}(\lambda))^\vee .$$

Finally, by a somewhat long and involved argument, we show that

$$H^{\ell(v)}_{X^v/X^v \setminus (\mathcal{U}^- v\mathcal{B}/\mathcal{B})}(\mathcal{X}, \mathcal{L}(\lambda))^\vee$$

is $\mathfrak{g}$-module isomorphic with the Verma module $M(v * \lambda)$. Thus, for any $p \geq 0$, the p-th term of the above complex $\mathcal{K}$ is isomorphic with F_p as a $\mathfrak{g}$-module. In particular, taking the exact chain complex $\mathcal{K}$ for $\mathcal{A}$, from the uniqueness result described above, we get that the BGG complex $\mathcal{S}$ itself is exact for an arbitrary Kac–Moody algebra and, in fact, the geometrically defined chain complex $\mathcal{K}$ is isomorphic with the combinatorially defined chain complex $\mathcal{S}$. The exactness of the Kempf complex $\mathcal{K}$ provides, as a corollary, an alternative proof of the generalization of the Weyl–Kac character formula for an arbitrary Kac–Moody algebra obtained in Chapter 8. As another important corollary, we extend the $\mathfrak{n}^-$-homology result of Garland–Lepowsky (proved in Chapter 3 in the symmetrizable case) to an arbitrary Kac–Moody case. Yet another consequence is the vanishing of $\mathrm{Ext}^*_{(\mathfrak{g},\mathfrak{h})}(L^{\max}(\lambda), L^{\max}(\mu)^\sigma)$ for $\lambda \neq \mu \in D$.

It will be interesting to generalize the Kempf complex to cover the case of any $Y \subset \{1, \dots, \ell\}$.

9.1. BGG Resolution: An Algebraic Proof in the Symmetrizable Case

In this section $\mathfrak{g} = \mathfrak{g}(A)$ is the (symmetrizable) Kac–Moody Lie algebra, associated to any symmetrizable $\ell \times \ell$ GCM A.

For any subset $Y \subset \{1, \dots, \ell\}$ (not necessarily of finite type) and $\lambda \in D_Y$, let $L_Y(\lambda)$ be the irreducible, highest weight, integrable $\mathfrak{g}_Y$-module with highest weight λ (cf. Lemma 3.2.1).

9.1.1 Definition. Let $\mathfrak{p}_Y \subset \mathfrak{g}$ be the standard parabolic subalgebra (cf. 1.2.2). For $\lambda \in D_Y$, we define the *Y-generalized Verma module*

$$M_Y(\lambda) := U(\mathfrak{g}) \otimes_{U(\mathfrak{p}_Y)} L_Y(\lambda), \tag{1}$$

where $U(\mathfrak{p}_Y)$ acts on $U(\mathfrak{g})$ by right multiplication and the $\mathfrak{g}_Y$-module $L_Y(\lambda)$ is considered as a $\mathfrak{p}_Y := \mathfrak{g}_Y \oplus \mathfrak{u}_Y$ module by letting $\mathfrak{u}_Y$ act trivially on $L_Y(\lambda)$. Then $M_Y(\lambda)$ is a $U(\mathfrak{g})$-module under the left multiplication on the $U(\mathfrak{g})$-factor. By the PBW theorem, $M_Y(\lambda)$ is a free $U(\mathfrak{u}_Y^-)$-module, where $\mathfrak{u}_Y^-$ is as in 1.2.2. Further, by Lemma 1.3.3(b), $M_Y(\lambda)$ is an integrable $\mathfrak{g}_Y$-module. Moreover, $M_Y(\lambda)$ is canonically a $\mathfrak{g}$-module quotient of $M(\lambda)$. In particular, $M_Y(\lambda) \in \mathcal{O}$ (cf. 2.1.1). Observe that, for $Y = \emptyset$, $M_Y(\lambda)$ is nothing but the Verma module $M(\lambda)$ defined in 2.1.1.

A (possibly finite) filtration of a $\mathfrak{g}$-module $M \in \mathcal{O}$ by $\mathfrak{g}$-submodules: $M_{-1} = 0 \subset M_0 \subset M_1 \subset \cdots$ is called a *Y-generalized Verma filtration* if it satisfies the following two conditions:

(a) $\bigcup_{i \geq 0} M_i = M$, and
(b) for any $i \geq 0$, as $\mathfrak{g}$-modules,

$$M_i/M_{i-1} \simeq \bigoplus_{\lambda \in D_Y} n_i(\lambda)\, M_Y(\lambda), \quad \text{for some } n_i(\lambda) \geq 0. \tag{2}$$

(Since $M \in \mathcal{O}$, $n_i(\lambda) < \infty$.)

Since each $M_Y(\lambda)$ is a free $U(\mathfrak{u}_Y^-)$-module, by (2), so is M if M admits a Y-generalized Verma filtration. Since $\mathfrak{g}_Y$ normalizes $\mathfrak{u}_Y^-$ (cf. 1.2.2) and M is a

$\mathfrak{g}$-module, $\mathbb{C} \otimes_{U(\mathfrak{u}_Y^-)} M$ is a $\mathfrak{g}_Y$-module under its action on the M-factor. Further, from (2), it is easy to see that

$$(3) \qquad \mathbb{C} \otimes_{U(\mathfrak{u}_Y^-)} M \simeq \bigoplus_{\lambda \in D_Y} \Big(\sum_{i \geq 0} n_i(\lambda)\Big) L_Y(\lambda), \qquad \text{as } \mathfrak{g}_Y\text{-modules.}$$

Since all the weight spaces of M are finite dimensional (as $M \in \mathcal{O}$), we have that $\sum_{i \geq 0} n_i(\lambda) < \infty$ for any $\lambda \in D_Y$.

For any M as above and $\lambda \in D_Y$, define the *multiplicity* $(M : M_Y(\lambda))$ of $M_Y(\lambda)$ in M by

$$(4) \qquad (M : M_Y(\lambda)) := \sum_{i \geq 0} n_i(\lambda).$$

By (3), the multiplicity $(M : M_Y(\lambda))$ is well defined, i.e., it does not depend on the choice of the Y-generalized Verma filtration of M. For any $M \in \mathcal{O}$ and $z \in \mathbb{C}$, recall the definition of the generalized z-eigenspace M^z for the Casimir operator Ω_M from 2.1.17.

9.1.2 Lemma. *Let M be a module in the category $\mathcal{O}$ which admits a Y-generalized Verma filtration. Then, for any $z \in \mathbb{C}$, M^z again admits a Y-generalized Verma filtration.*

(In fact, a more general result is true, cf. Exercise 9.1.E.)

Proof. Let $M_{-1} = 0 \subset M_0 \subset M_1 \subset \cdots$ be a Y-generalized Verma filtration of M. Then, clearly,

$$(1) \qquad M_i^z = M^z \cap M_i.$$

We claim that $M_{-1}^z = 0 \subset M_0^z \subset M_1^z \subset \cdots$ is a Y-generalized Verma filtration of M^z:

By (1), we have $\cup_{i \geq 0} M_i^z = M^z$. Further, $M \rightsquigarrow M^z$ begin an exact functor from the category $\mathcal{O}$ to itself (cf. Lemma 2.1.18), we obtain that for any $i \geq 0$,

$$(2) \qquad M_i^z / M_{i-1}^z \simeq (M_i / M_{i-1})^z.$$

But, by assumption,

$$(3) \qquad M_i / M_{i-1} \simeq \bigoplus_{\lambda \in D_Y} n_i(\lambda)\, M_Y(\lambda),$$

and, by Lemma 2.1.16, $\Omega_{M_Y(\lambda)} = \langle \lambda, 2\rho + \lambda \rangle I$ (since $M_Y(\lambda)$ is a quotient of $M(\lambda)$). Thus, by (2)–(3), we get

$$(4) \qquad M_i^z / M_{i-1}^z \simeq \bigoplus_{\lambda} n_i(\lambda)\, M_Y(\lambda),$$

where the summation runs over those $\lambda \in D_Y$ such that $\langle \lambda, 2\rho + \lambda \rangle = z$. This proves the lemma. $\square$

The following theorem is the main result of this section, known as the *generalized Bernstein–Gelfand–Gelfand* (for short BGG) *resolution*.

9.1.3 Theorem. *Let $\mathfrak{g}$ be symmetrizable. For any subset $Y \subset \{1, \dots, \ell\}$ of finite type and any $\lambda \in D$, there exists an exact sequence*

$$(1) \qquad \cdots \to F_p \xrightarrow{\delta^p} \cdots \to F_1 \xrightarrow{\delta^1} F_0 \xrightarrow{\delta^0} L(\lambda) \to 0,$$

where $F_p = F_p^Y := \bigoplus_{\substack{w \in W'_Y \text{ and} \\ \ell(w)=p}} M_Y(w^{-1} * \lambda)$ and each δ^p $(p \geq 0)$ is a $\mathfrak{g}$-module map.

Recall that W'_Y is defined by (1.3.17.1) and the shifted action $*$ is defined in 1.3.1. Observe that for $w \in W'_Y$ and $\lambda \in D$, $w^{-1} * \lambda \in D_Y$.

Proof. Recall the resolution (3.2.7.2) (abbreviating $L(\lambda)$ by L):

$$(2) \qquad \cdots \to X_p \to \cdots \to X_0 \to L \to 0,$$

consisting of $\mathfrak{g}$-modules $X_p \in \mathcal{O}$ and $\mathfrak{g}$-module maps between them, where $X_p := U(\mathfrak{g}) \otimes_{U(\mathfrak{p}_Y)} (\Lambda^p(\mathfrak{g}/\mathfrak{p}_Y) \otimes_{\mathbb{C}} L)$.

For any $z \in \mathbb{C}$, from the exact sequence (2), we get the exact sequence

$$(3) \qquad \cdots \to X_p^z \to \cdots \to X_0^z \to L^z \to 0.$$

But, by Lemma 2.1.16, $L^z = 0$ unless $z = \langle \lambda, 2\rho + \lambda \rangle$ and, in this case, $L^z = L$. Hence, setting $z_o = \langle \lambda, 2\rho + \lambda \rangle$, we get the exact sequence of $\mathfrak{g}$-modules:

$$(4) \qquad \cdots \to X_p^{z_o} \to \cdots \to X_0^{z_o} \to L \to 0.$$

We next determine $X_p^{z_o}$. To this end, we first prove the following:

9.1.4 Lemma. *For any $p \geq 0$, the $\mathfrak{p}_Y$-module $M = \Lambda^p(\mathfrak{g}/\mathfrak{p}_Y) \otimes_{\mathbb{C}} L$ (where $L = L(\lambda)$ is as above) has a (possibly finite) filtration by $\mathfrak{p}_Y$-submodules*

$$(\mathcal{M}) \qquad M^{-1} = 0 \subset M^0 \subset M^1 \subset \cdots,$$

with the following properties:

(a) $\cup_i M^i = M$.

(b) *For all $i \geq 0$, the action of $\mathfrak{u}_Y$ on (the $\mathfrak{p}_Y$-module) M^i/M^{i-1} is trivial.*

(c) *As a $\mathfrak{g}_Y$-module,*

$$M \simeq \oplus_{i \geq 0} M^i/M^{i-1}.$$

(Of course, $M \simeq \Lambda^p(\mathfrak{u}_Y^-) \otimes_{\mathbb{C}} L$ as a $\mathfrak{g}_Y$-module.)

(d) *As a $\mathfrak{g}_Y$-module,*

$$(1) \qquad M^i/M^{i-1} \simeq \oplus_{\mu \in D_Y} n_\mu^i L_Y(\mu), \quad \text{for some } n_\mu^i \in \mathbb{Z}_+.$$

In particular, as a $\mathfrak{g}_Y$-module,

(2) $$M \simeq \oplus_{\mu \in D_Y} n_\mu \, L_Y(\mu),$$

where $n_\mu := \sum_{i \geq 0} n^i_\mu \in \mathbb{Z}_+$.

Proof. For any $\beta \in \mathfrak{h}^*$, let M_β be the β-th weight space of M. Now, for any $i \geq 0$, set
$$M^i = \oplus M_{\lambda-\alpha} \, ,$$
where the summation is taken over all those $\alpha \in \sum_j \mathbb{Z}_+ \alpha_j$ such that $\mathrm{ht}_Y(\alpha) \leq i$ (ht_Y is as in (6.1.10.1)). Then, clearly, (a) and (b) are satisfied.

Observe that $\Lambda^p(\mathfrak{u}_Y^-) \otimes_{\mathbb{C}} L$ is an integrable $\mathfrak{g}_Y$-module by Corollary 1.3.4, and, moreover, it belongs to the category $\mathcal{O}_Y$ (cf. 2.1.1). Hence, by Remark 3.2.2, (c) and (d) follow. The assertion that $n_\mu \in \mathbb{Z}_+$ follows since the weight spaces of M are finite dimensional. This proves the lemma. □

9.1.5 Corollary. *For any* $p \geq 0$, $U(\mathfrak{g}) \otimes_{U(\mathfrak{p}_Y)} (\Lambda^p(\mathfrak{g}/\mathfrak{p}_Y) \otimes_{\mathbb{C}} L)$ *admits a* Y-*generalized Verma filtration.*

Proof. Take any filtration $\{M^i\}_{i \geq -1}$ of $M := \Lambda^p(\mathfrak{g}/\mathfrak{p}_Y) \otimes_{\mathbb{C}} L$ by $\mathfrak{p}_Y$-submodules, given in Lemma 9.1.4. This gives rise to the filtration $\tilde{M}^i$ of $X_p := U(\mathfrak{g}) \otimes_{U(\mathfrak{p}_Y)} M$ by $\mathfrak{g}$-submodules defined by $\tilde{M}^i := U(\mathfrak{g}) \otimes_{U(\mathfrak{p}_Y)} M^i \subset X_p$. (We are using the fact that $M \rightsquigarrow U(\mathfrak{g}) \otimes_{U(\mathfrak{p}_Y)} M$ is an exact functor from the category of $\mathfrak{p}_Y$-modules to the category of $\mathfrak{g}$-modules, since $U(\mathfrak{g})$ is $U(\mathfrak{p}_Y)$-free by the PBW theorem.)

By (9.1.4.1) and 9.1.4(b), as $\mathfrak{g}$-modules,
$$\tilde{M}^i / \tilde{M}^{i-1} \simeq \oplus_{\mu \in D_Y} \, n^i_\mu \, M_Y(\mu).$$

This proves the corollary. □

9.1.6 *Proof of Theorem 9.1.3* (continued). By Corollary 9.1.5 and Lemma 9.1.2, $X^{z_o}_p$ admits a Y-generalized Verma filtration. We next claim that, for $\mu \in D_Y$,

(1) $$\begin{aligned} \left(X^{z_o}_p : M_Y(\mu)\right) &= 1, \text{ if } \mu = w^{-1} * \lambda, \text{ for some } w \in W'_Y \text{ with } \ell(w) = p \\ &= 0, \quad \text{otherwise.} \end{aligned}$$

To prove this, by (9.1.1.3)–(9.1.1.4), it suffices to show that (as $\mathfrak{g}_Y$-modules)

(2) $$\mathbb{C} \otimes_{U(\mathfrak{u}_Y^-)} X^{z_o}_p \simeq \bigoplus_{\substack{w \in W'_Y, \text{ and} \\ \ell(w) = p}} L_Y(w^{-1} * \lambda).$$

By (3.2.7.9), following the same notation as in the proof of Theorem 3.2.7, we get

(3) $$\mathbb{C} \otimes_{U(\mathfrak{u}_Y^-)} X_p^{z_o} = Z_p^{z_o},$$

where $Z_p := \Lambda^p(\mathfrak{u}_Y^-) \otimes L \simeq \mathbb{C} \otimes_{U(\mathfrak{u}_Y^-)} X_p$ as $\mathfrak{g}_Y$-modules and $Z_p^{z_o}$ is the generalized z_o-eigenspace of Z_p for the operator Ω^Y. Since $Z_p \in \mathcal{O}_Y$ and it is an integrable $\mathfrak{g}_Y$-module, by Remark 3.2.2,

(4) $$Z_p^{z_o} = \oplus_{\mu \in D_Y} n_\mu L_Y(\mu), \qquad \text{for some } n_\mu \in \mathbb{Z}_+,$$

and, moreover,

(5) $$\Omega^Y_{L_Y(\mu)} = \langle \mu, \mu + 2\rho \rangle I, \quad \text{by (3.2.7.7)}.$$

From (4)–(5) we get that

(6) $$n_\mu = 0, \qquad \text{for } \mu \in D_Y \text{ with } \langle \mu, \mu + 2\rho \rangle \neq z_o.$$

So take $\mu \in D_Y$ with $\langle \mu, \mu + 2\rho \rangle = z_o$. Assume further that μ is a weight of Z_p. Then, by the proof of Theorem 3.2.7, we obtain that such a μ is necessarily of the form $\mu = w^{-1} * \lambda$, for $w \in W_Y'$ with $\ell(w) = p$, and, Moreover,

(7) $$n_{w^{-1} * \lambda} = 1\,, \qquad \text{for } w \in W_Y' \text{ with } \ell(w) = p.$$

Combining (3), (4), (6) and (7), we obtain (2), and thus assertion (1) is established. Thus, for any $p \geq 0$, there exists a filtration of $X_p^{z_o}$ by $\mathfrak{g}$-submodules:

$$A^{-1} = 0 \subset A^0 \subset A^1 \subset \cdots \subset A^N = X_p^{z_o},$$

such that

(8) $$A^i / A^{i-1} \simeq M_Y(w_i^{-1} * \lambda), \qquad \text{for } 0 \leq i \leq N,$$

where $\{w_0, \ldots, w_N\}$ is an enumeration (without repetition) of the elements of W_Y' of length p. Now the theorem follows from (9.1.3.4) and (8) by using the following proposition and the interpretation of $\mathrm{Ext}^1_{(\mathfrak{g}, \mathfrak{g}_Y)}$ as the group of equivalence classes of $\mathfrak{g}_Y$-split extensions (cf. D.7). Observe that since each $M_Y(\mu)$, $\mu \in D_Y$, is $\mathfrak{g}_Y$-integrable by 9.1.1, so is each A^i, and hence we can use the complete reducibility (Remark 3.2.2) to obtain the $\mathfrak{g}_Y$-splitting. □

9.1.7 Proposition. *Let Y be of finite type. For any $\mathfrak{g}$-module $A \in \mathcal{O}$ which is $\mathfrak{g}_Y$-integrable, $v \in W'_Y$, $\lambda \in D$ and $p \geq 0$, if $Ext^p_{(\mathfrak{g},\mathfrak{g}_Y)}(M_Y(v^{-1} * \lambda), A) \neq 0$, there exist elements $\mu_1 > \cdots > \mu_p > \mu_{p+1} = v^{-1} * \lambda$ in D_Y such that*

$$(1) \qquad [A : L(\mu_1)][M_Y(\mu_1) : L(\mu_2)] \cdots [M_Y(\mu_p) : L(\mu_{p+1})] \neq 0\,,$$

where the notation $[\,.\,:\,.\,]$ is as in 2.1.11.

In particular, for $v \neq w \in W'_Y$ of the same length and $\lambda \in D$,

$$(2) \qquad \mathrm{Ext}^p_{(\mathfrak{g},\mathfrak{g}_Y)}\big(M_Y(v^{-1} * \lambda),\ M_Y(w^{-1} * \lambda)\big) = 0, \quad \text{for all } p \in \mathbb{Z}_+.$$

We first prove the following:

9.1.8 Lemma. *Let Y be of finite type. For any $\mu \in D_Y$ and any $\mathfrak{g}$-module $M \in \mathcal{O}$ which is $\mathfrak{g}_Y$-integrable,*

$$\mathrm{Ext}^p_{(\mathfrak{g},\mathfrak{g}_Y)}(M_Y(\mu), M) \simeq \mathrm{Hom}_{\mathfrak{g}_Y}\big(H_p(\mathfrak{u}_Y^-, M^\sigma),\ L_Y(\mu)\big)\,.$$

(Since $\mathfrak{g}_Y$ normalizes $\mathfrak{u}_Y^-$, there is a natural $\mathfrak{g}_Y$-module structure on $H_p(\mathfrak{u}_Y^-, M^\sigma)$ by 3.1.1.)

Proof. Since Y is of finite type, $\mathfrak{g}$ is finitely semisimple as a $\mathfrak{g}_Y$-module. Thus, by Remark 3.3.6,

$$(1) \qquad \mathrm{Ext}^p_{(\mathfrak{g},\mathfrak{g}_Y)}(M_Y(\mu), M) \simeq \mathrm{Tor}_p^{(\mathfrak{g},\mathfrak{g}_Y)}((M_Y(\mu)^\omega)^t, M^\sigma)^*.$$

Next observe that, as a $\mathfrak{g}$-module,

$$M_Y(\mu)^\omega \simeq U(\mathfrak{g}) \otimes_{U(\mathfrak{p}_Y^-)} (L_Y(\mu)^\omega).$$

Hence, by (3.1.14.1), (3.1.9.1) and (3.1.13.1),

$$\begin{aligned} \mathrm{Tor}_p^{(\mathfrak{g},\mathfrak{g}_Y)}\big((M_Y(\mu)^\omega)^t, M^\sigma\big) &\simeq \mathrm{Tor}_p^{(\mathfrak{p}_Y^-,\mathfrak{g}_Y)}\big((L_Y(\mu)^\omega)^t, M^\sigma\big) \\ (2) \qquad &\simeq H_p(\mathfrak{p}_Y^-, \mathfrak{g}_Y, L_Y(\mu)^\omega \otimes M^\sigma). \end{aligned}$$

By 3.1.3, $H_*(\mathfrak{p}_Y^-, \mathfrak{g}_Y, L_Y(\mu)^\omega \otimes M^\sigma)$ is the homology of the complex

$$0 \leftarrow C_0 \leftarrow C_1 \leftarrow \cdots \xleftarrow{\partial'_p} C_p \leftarrow \cdots,$$

where $C_p := [L_Y(\mu)^\omega \otimes \Lambda^p(\mathfrak{u}_Y^-) \otimes M^\sigma]_{\mathfrak{g}_Y}$, $[N]_{\mathfrak{g}_Y}$ denotes the space of coinvariants $N/\mathfrak{g}_Y \cdot N$ for a $\mathfrak{g}_Y$-module N, the differential ∂'_* is induced from the differential $I \otimes \partial_*$ (since $\mathfrak{u}_Y^-$ acts trivially on $L_Y(\mu)^\omega$), the $\mathfrak{g}_Y$-module map

$\partial_* : \Lambda^*(\mathfrak{u}_Y^-) \otimes M^\sigma \to \Lambda^{*-1}(\mathfrak{u}_Y^-) \otimes M^\sigma$ is the differential asin 3.1.1 for the Lie algebra $\mathfrak{u}_Y^-$ with coefficients in the $\mathfrak{u}_Y^-$-module M^σ and $I := I_{L_Y(\mu)^\omega}$.

Since $\Lambda^p(\mathfrak{u}_Y^-) \in \mathcal{O}_Y$ is an integrable $\mathfrak{g}_Y$-module by Corollary 1.3.4 and $M^\sigma \in \mathcal{O}_Y$ is $\mathfrak{g}_Y$-integrable by Lemma 2.1.14, by Remark 3.2.2 we can decompose (as a direct sum of $\mathfrak{g}_Y$-modules)

$$\Lambda^p(\mathfrak{u}_Y^-) \otimes M^\sigma = B_p \oplus H_p \oplus F_p,$$

where $B_p := \operatorname{Im} \partial_{p+1}$, H_p is, by definition, complementary to B_p in $\operatorname{Ker} \partial_p$ and F_p is complementary to $\operatorname{Ker} \partial_p$. In particular,

$$\text{(3)} \qquad H_p \simeq H_p(\mathfrak{u}_Y^-, M^\sigma), \qquad \text{as } \mathfrak{g}_Y\text{-modules.}$$

This gives rise to the decomposition (again as $\mathfrak{g}_Y$-modules):

$$L_Y(\mu)^\omega \otimes \Lambda^p(\mathfrak{u}_Y^-) \otimes M^\sigma = (L_Y(\mu)^\omega \otimes B_p) \oplus (L_Y(\mu)^\omega \otimes H_p) \oplus (L_Y(\mu)^\omega \otimes F_p).$$

Since $\partial_{p|F_p} : F_p \to B_{p-1}$ is a $\mathfrak{g}_Y$-module isomorphism, so is

$$I \otimes \partial_{p|F_p} : L_Y(\mu)^\omega \otimes F_p \to L_Y(\mu)^\omega \otimes B_{p-1}\,.$$

From this it is easy to see that

$$\text{(4)} \qquad H_p(\mathfrak{p}_Y^-, \mathfrak{g}_Y, L_Y(\mu)^\omega \otimes M^\sigma) \simeq [L_Y(\mu)^\omega \otimes H_p]_{\mathfrak{g}_Y}.$$

Now,

$$\begin{aligned}
\big([L_Y(\mu)^\omega \otimes H_p]_{\mathfrak{g}_Y}\big)^* &\simeq \operatorname{Hom}_{\mathfrak{g}_Y}(L_Y(\mu)^\omega \otimes H_p, \mathbb{C}) \\
&\simeq \operatorname{Hom}_{\mathfrak{g}_Y}(H_p, (L_Y(\mu)^\omega)^*), \quad \text{by (3.1.13.9)} \\
&\simeq \operatorname{Hom}_{\mathfrak{g}_Y}(H_p, (L_Y(\mu)^\omega)^\vee), \text{ since } H_p \text{ is } \mathfrak{h}\text{-semisimple} \\
\text{(5)} \qquad &\simeq \operatorname{Hom}_{\mathfrak{g}_Y}(H_p, L_Y(\mu)), \qquad \text{by (3.3.8.1).}
\end{aligned}$$

Combining (1)–(5), we get the lemma. □

9.1.9 *Proof of Proposition 9.1.7.* We first prove (1): Using Lemma 9.1.8, we obtain that

$$\operatorname{Hom}_{\mathfrak{g}_Y}(H_p(\mathfrak{u}_Y^-, M), L_Y(v^{-1} * \lambda)) \neq 0,$$

where $M := A^\sigma$. Observe that M is $\mathfrak{g}_Y$-integrable by Lemma 2.1.14. Let $(M_i)_{i \geq 0}$ be an increasing filtration of M as given by Lemma 2.1.10. Since homology commutes with direct limits, there exists i_o such that

$$\text{(1)} \qquad \operatorname{Hom}_{\mathfrak{g}_Y}(H_p(\mathfrak{u}_Y^-, M_{i_o}), L_Y(v^{-1} * \lambda)) \neq 0.$$

Let $i_o \geq 1$ be the smallest integer satisfying (1). By condition (ii) of Lemma 2.1.10, $N := M_{i_o}/M_{i_o-1}$ is a highest weight $\mathfrak{g}$-module with highest weight λ_{i_o} (which is $\mathfrak{g}_Y$-integrable) and hence $\lambda_{i_o} \in D_Y$ (cf. Lemma 3.2.1). Let L be the (highest weight) $\mathfrak{g}_Y$-submodule of N generated by the highest weight vector of N (of weight λ_{i_o}). Then, by Corollary 2.2.6 (for $\mathfrak{g}_Y$) and Lemma 3.2.1, $L \simeq L_Y(\lambda_{i_o})$. Moreover, L is, in fact, $\mathfrak{p}_Y$-stable, such that $\mathfrak{u}_Y$ acts trivially in L (as can be easily seen since $\mathfrak{u}_Y \subset \mathfrak{p}_Y$ is an ideal). In particular, we get a surjective $\mathfrak{g}$-module map $q : M_Y(\lambda_{i_o}) \twoheadrightarrow N$. Since i_o is (by choice) the smallest integer satisfying (1), from the long exact homology sequence for $H_*(\mathfrak{u}_Y^-, -)$, corresponding to the coefficient sequence $0 \to M_{i_o-1} \to M_{i_o} \to N \to 0$, we get that

(2) $$\operatorname{Hom}_{\mathfrak{g}_Y}(H_p(\mathfrak{u}_Y^-, N), L_Y(v^{-1} * \lambda)) \neq 0.$$

Now, consider the long exact homology sequence:

(3) $$\cdots \to H_p(\mathfrak{u}_Y^-, M_Y(\lambda_{i_o})) \to H_p(\mathfrak{u}_Y^-, N) \to H_{p-1}(\mathfrak{u}_Y^-, K) \\ \to H_{p-1}(\mathfrak{u}_Y^-, M_Y(\lambda_{i_o})) \to \cdots,$$

corresponding to the sequence

$$0 \to K \to M_Y(\lambda_{i_o}) \xrightarrow{q} N \to 0, \qquad \text{where } K := \operatorname{Ker} q.$$

Since $M_Y(\lambda_{i_o})$ is $U(\mathfrak{u}_Y^-)$-free, we get from (3) that

(4) $$H_p(\mathfrak{u}_Y^-, N) \xrightarrow{\sim} H_{p-1}(\mathfrak{u}_Y^-, K), \qquad \text{for } p \geq 1:$$

Clearly (4) is true if $p \geq 2$. From (3) above, $L_Y(\lambda_{i_o}) \simeq H_0(\mathfrak{u}_Y^-, M_Y(\lambda_{i_o})) \to H_0(\mathfrak{u}_Y^-, N)$ is surjective; in particular, nonzero and hence injective, being a $\mathfrak{g}_Y$-module map. Hence (4) follows for $p = 1$ as well.

We prove (9.1.7.1) by induction on p. If $p = 0$, it is clear since, by Exercise D.E,

$$\operatorname{Ext}^0_{(\mathfrak{g},\mathfrak{g}_Y)}(M_Y(v^{-1} * \lambda), A) \simeq \operatorname{Hom}_{\mathfrak{g}}(M_Y(v^{-1} * \lambda), A).$$

So take $p \geq 1$ and assume the validity of (9.1.7.1) for $p - 1$ and $A = K^\sigma$. Observe that by 9.1.1, K and hence K^σ indeed is $\mathfrak{g}_Y$-integrable. By (2) and (4) we get that $\operatorname{Hom}_{\mathfrak{g}_Y}(H_{p-1}(\mathfrak{u}_Y^-, K), L_Y(v^{-1} * \lambda)) \neq 0$ and hence, by Lemma 9.1.8, $\operatorname{Ext}^{p-1}_{(\mathfrak{g},\mathfrak{g}_Y)}(M_Y(v^{-1} * \lambda), K^\sigma) \neq 0$. Thus, by induction, there exist $\mu_2 > \cdots > \mu_{p+1} = v^{-1} * \lambda$ in D_Y such that

$$[K : L(\mu_2)][M_Y(\mu_2) : L(\mu_3)] \cdots [M_Y(\mu_p) : L(\mu_{p+1})] \neq 0$$

(since $[K : L(\mu_2)] = [K^\sigma : L(\mu_2)]$ by (2.1.1.6) and Lemma 2.1.12). But, K being a proper submodule of $M_Y(\lambda_{i_o})$, we have that $[M_Y(\lambda_{i_o}) : L(\mu_2)] \neq 0$ and $\lambda_{i_o} > \mu_2$.

Now take $\mu_1 = \lambda_{i_o}$. Then $\mu_1 > \mu_2 > \cdots > \mu_p > \mu_{p+1} = v^{-1} * \lambda$ satisfies (9.1.7.1), since $[A : L(\lambda_{i_o}] = [M : L(\lambda_{i_o}] \neq 0$ (as N is a quotient of $M_Y(\lambda_{i_o})$ and hence of $M(\lambda_{i_o})$). This completes the induction and hence proves (9.1.7.1).

To prove (9.1.7.2), assume that (if possible)

$$\mathrm{Ext}^p_{(\mathfrak{g}, \mathfrak{g}_Y)}(M_Y(v^{-1} * \lambda), M_Y(w^{-1} * \lambda)) \neq 0.$$

Then, by (9.1.7.1) for $A = M_Y(w^{-1} * \lambda)$ and Exercise 2.3.E.1, we get that there exist Weyl group elements $w^{-1} \leq w_1 < \cdots < w_p < w_{p+1} = v^{-1}$. (Observe that $M_Y(\mu)$ is a quotient of $M(\mu)$ for any $\mu \in D_Y$ by 9.1.1.) But this is a contradiction, since $\ell(v) = \ell(w)$ (by assumption). Hence (9.1.7.2) is established. □

9.1.10 Remark. Even though we prove Lemma 9.1.8 under the assumption that Y is of finite type, it can be proved (by the same proof) for an arbitrary Y using Remark 3.1.15 and Exercise 3.1.E.5. Its validity for any subset Y implies the validity of Proposition 9.1.7 (and thus Theorem 9.1.3) for any Y, since in this section 9.1, the assumption on Y to be of finite type is not used anywhere else.

9.1.E EXERCISE. For $M, N \in \mathcal{O}$, $M \oplus N$ admits a Y-generalized Verma filtration iff both M and N admit Y-generalized Verma filtrations.

9.2. A Combinatorial Description of the BGG Resolution

We give a purely combinatorial description of differentials in the BGG resolution (cf. Theorem 9.1.3), and show their "uniqueness."

In this section $\mathfrak{g} = \mathfrak{g}(A)$ is any (not necessarily symmetrizable) Kac–Moody Lie algebra with the associated Weyl group W, and $Y \subset \{1, \ldots, \ell\}$ is any subset.

9.2.1 Definition. Let $\mathcal{W}$ be any Coxeter group. For $w_1, w_2 \in \mathcal{W}$ denote $w_1 \to w_2$ to mean that $w_1 \leq w_2$ and $\ell(w_2) = \ell(w_1) + 1$.

A quadruple (w_1, w_2, w_3, w_4) of elements of $\mathcal{W}$ is called a *square* if $w_1 \to w_2 \to w_4$, $w_1 \to w_3 \to w_4$ and $w_2 \neq w_3$. A square (w_1, w_2, w_3, w_4) is pictorially denoted as

$$\begin{array}{ccccc} & \nearrow & w_2 & \searrow & \\ w_1 & & & & w_4. \\ & \searrow & w_3 & \nearrow & \end{array}$$

The following lemma is due to [Bernstein–Gelfand–Gelfand–75, Lemmas 10.3–10.4]. (Even though they prove the result in the case where $\mathcal{W}$ is finite, the same proof with minor modifications works for general $\mathcal{W}$.) A more precise result than the (a) part of the following lemma can be found in [Björner–Wachs–82, Lemma 4.3].

9.2.2 Lemma. *(a) Let $w_1, w_2 \in \mathcal{W}$ be such that $\ell(w_2) = \ell(w_1) + 2$. Then the number of $w \in \mathcal{W}$ such that $w_1 \to w \to w_2$ is equal to either 0 or 2. Moreover, this number is 0 iff $w_1 \not\leq w_2$.*

(b) To each arrow $w_1 \to w_2$ in $\mathcal{W}$, one can assign a number $\epsilon(w_1, w_2) \in \{\pm 1\}$ such that, for every square (w_1, w_2, w_3, w_4) as above, the product of the numbers assigned to the four arrows occurring in it is equal to -1. □

9.2.3 Theorem. *Let $\mathfrak{g}$ be an arbitrary Kac–Moody Lie algebra. For $\lambda \in D$, $v, w \in W$,*

(a) $$\mathrm{Hom}_{\mathfrak{g}}(M(v * \lambda), M(w * \lambda)) \neq 0 \textit{ iff } v \geq w.$$

Moreover, in this case, i.e., $v \geq w$,

(b) $$\dim \mathrm{Hom}_{\mathfrak{g}}(M(v * \lambda), M(w * \lambda)) = 1.$$

Before we can prove the above theorem, we need some preparatory work.

9.2.4 Definition. For $\lambda \in \mathfrak{h}^*$ and any simplereflection s_i, set

$$\underline{s}_i * \lambda = \begin{cases} s_i * \lambda, & \text{if } s_i * \lambda \leq \lambda \\ \lambda, & \text{otherwise,} \end{cases}$$

and

$$\bar{s}_i * \lambda = \begin{cases} s_i * \lambda, & \text{if } s_i * \lambda \geq \lambda \\ \lambda, & \text{otherwise.} \end{cases}$$

9.2.5 Lemma. *For $\lambda \in \mathfrak{h}^*$ and any simple reflection s_i,*

$$\dim \mathrm{Hom}_{\mathfrak{g}}(M(\underline{s}_i * \lambda), M(\lambda)) = 1.$$

Proof. We can of course assume that $s_i * \lambda < \lambda$. Then $\lambda - s_i * \lambda = n\alpha_i$, for $n := \langle \lambda, \alpha_i^\vee \rangle + 1$ a positive integer. Let $v_\lambda \in M(\lambda)$ be a nonzero highest weight vector. Then we claim that

(1) $$e_j f_i^n v_\lambda = 0, \qquad \text{for all } 1 \leq j \leq \ell :$$

This is clearly true for $j \neq i$. Further, by Exercise 1.2.E.3,

(2) $$e_i f_i^n = f_i^n e_i + n f_i^{n-1} \big(\alpha_i^\vee - n + 1\big).$$

From (2), (1) follows for $j = i$ as well. This proves that $\mathrm{Hom}_{\mathfrak{g}}(M(s_i * \lambda), M(\lambda)) \neq 0$. Since the weight space $M(\lambda)_{s_i * \lambda}$ is one-dimensional, the lemma follows. □

9.2.6 Lemma. *For* $\lambda, \mu \in \mathfrak{h}^*$ *and any simple reflection* s_i,

$$\mathrm{Hom}_{\mathfrak{g}}\big(M(\underline{s}_i * \lambda), M(\underline{s}_i * \mu)\big) \geq \mathrm{Hom}_{\mathfrak{g}}(M(\lambda), M(\mu)).$$

Proof. Since $U(\mathfrak{n}^-)$ has no zero divisors, any nonzero $\mathfrak{g}$-module map $M(\lambda) \to M(\mu)$ is injective. Fix $\mathfrak{g}$-module embeddings (cf. Lemma 9.2.5) $j_\lambda : M(\underline{s}_i * \lambda) \hookrightarrow M(\lambda)$ and $j_\mu : M(\underline{s}_i * \mu) \hookrightarrow M(\mu)$. We identify $M(\underline{s}_i * \lambda)$ as a submodule of $M(\lambda)$ via j_λ and, similarly, $M(\underline{s}_i * \mu)$ as a submodule of $M(\mu)$. To prove the lemma, it suffices to show that, for any $\mathfrak{g}$-module map $\theta : M(\lambda) \to M(\mu)$,

$$\theta(M(\underline{s}_i * \lambda)) \subset M(\underline{s}_i * \mu). \tag{1}$$

Let $\bar{\theta} : M(\lambda) \to M(\mu)/M(\underline{s}_i * \mu)$ be the composition of θ followed by the projection. To prove (1), we just need to show that $\bar{\theta}(v_o) = 0$, for v_o a nonzero highest weight vector of $M(\underline{s}_i * \lambda)$.

By Lemma 1.3.3(b), f_i acts locally nilpotently on $Q := M(\mu)/M(\underline{s}_i * \mu)$ and, of course (from weight considerations), e_i acts locally nilpotently on Q. In particular, Q is an integrable $\mathfrak{g}(i)$-module, where $\mathfrak{g}(i)$ is as in 1.1.2. If $\bar{\theta}(v_o) \neq 0$, from Exercise 1.2.E.4 (since $e_i \cdot v_o = 0$), we get that $(\underline{s}_i * \lambda)(\alpha_i^\vee) \in \mathbb{Z}_+$, which is a contradiction! Hence $\bar{\theta}(v_o) = 0$. This proves the lemma. □

We next recall Enright's completion functor, but we need some definitions beforehand.

9.2.7 Definition. Let $\mathfrak{s}$ be any Lie algebra (not necessarily finite dimensional) together with an embedding $i : s\ell_2 \hookrightarrow \mathfrak{s}$. We identify $s\ell_2$ as a subalgebra of $\mathfrak{s}$ via i. Let $\mathcal{I}_{\mathfrak{s}}(i)$ or just $\mathcal{I}_{\mathfrak{s}}$ (when the reference to i is clear) be the full subcategory of the category of $\mathfrak{s}$-modules consisting of all the $\mathfrak{s}$-modules M satisfying:

(a) M is a weight module under the action of H,
(b) Y acts injectively on M, and
(c) X acts locally nilpotently on M,

where $\{X, Y, H\}$ is the standard basis of $s\ell_2$.

An $\mathfrak{s}$-module M is said to be *complete* (with respect to i) if M is a weight module under the action of H and, moreover, for each $n \in \mathbb{Z}_+$,

$$Y^{n+1} : M_n^X \xrightarrow{\sim} M_{-n-2}^X \quad \text{is an isomorphism,} \tag{1}$$

where M_n denotes the eigenspace of M corresponding to the eigenvalue n for the H-action and M_n^X denotes $\{v \in M_n : Xv = 0\}$. (Observe that by Exercise 1.2.E.3, $Y^{n+1}(M_n^X) \subset M_{-n-2}^X$.)

For $\mathfrak{s}$-modules M, N, it is clear that $M \oplus N$ is complete if and only if both M and N are complete.

Let M be any $\mathfrak{s}$-module. By a *completion* $\bar{M}$ of M (with respect to i), we mean a $\mathfrak{s}$-module $\bar{M}$ together with an $\mathfrak{s}$-module embedding $j : M \hookrightarrow \bar{M}$ such that

(c_1) $\bar{M}/j(M)$ is locally finite under $s\ell_2$, and
(c_2) $\bar{M}$ is complete.

9.2.8 Lemma. *Let M be an $\mathfrak{s}$-module which is X locally nilpotent and is complete. Then $M \in \mathcal{I}_{\mathfrak{s}}$. In particular, for $M \in \mathcal{I}_{\mathfrak{s}}$, any completion $\bar{M} \in \mathcal{I}_{\mathfrak{s}}$.*

Proof. Define $N = \{m \in M : Y^p m = 0$, for some $p = p(m) \in \mathbb{Z}_+\}$. Then, by Lemma 1.3.3(b), N is a $s\ell_2$-submodule of M. Moreover, since X acts locally nilpotently on M and M is a weight module for H, N is a locally finite $s\ell_2$-module by Exercise 9.2.E.4. In particular, if $N \neq 0$, there exists a nonzero $v_o \in N$ such that $v_o \in N_n^X$ for some $n \in \mathbb{Z}_+$. Further, $Y^{n+1} v_o \in N_{-n-2}^X$. But, N being locally $s\ell_2$-finite, $N_{-n-2}^X = 0$ and hence $Y^{n+1} v_o = 0$. Since M is complete (by assumption), the map $Y^{n+1} : M_n^X \xrightarrow{\sim} M_{-n-2}^X$ is an isomorphism, which is a contradiction. Hence $N = 0$, i.e., Y acts injectively on M, which proves that $M \in \mathcal{I}_{\mathfrak{s}}$. $\square$

9.2.9 Theorem. *Let $\mathfrak{s}$ be a Lie algebra together with an embedding $i : s\ell_2 \hookrightarrow \mathfrak{s}$. Assume further that $\mathfrak{s}$ is an integrable $s\ell_2$-module under the adjoint action. Then we have the following:*

(a) *For any $M \in \mathcal{I}_{\mathfrak{s}}$, there exists a completion $\bar{M}$. Further, $\bar{M}$ is unique in the sense that if $j' : M \hookrightarrow \bar{M}'$ is another completion, then there exists a unique $\mathfrak{s}$-module isomorphism $\theta : \bar{M} \xrightarrow{\sim} \bar{M}'$ making the following diagram commutative:*

$$\begin{array}{ccc} & M & \\ {}^{j}\swarrow & & \searrow^{j'} \\ \bar{M} & \xrightarrow[\theta]{\sim} & \bar{M}'. \end{array}$$

(b) *For $M, N \in \mathcal{I}_{\mathfrak{s}}$ and an $\mathfrak{s}$-module morphism $f : M \to N$, there exists a unique $\mathfrak{s}$-module map $\bar{f} : \bar{M} \to \bar{N}$ making the following diagram commutative:*

$$(*) \qquad \begin{array}{ccc} M & \xrightarrow{f} & N \\ \downarrow & & \downarrow \\ \bar{M} & \xrightarrow[\bar{f}]{} & \bar{N} \end{array}$$

(c) *Let $M \in \mathcal{I}_{\mathfrak{s}}$ and F a finite-dimensional $\mathfrak{s}$-module. Then $M \otimes F \in \mathcal{I}_{\mathfrak{s}}$ and, moreover, $\overline{(M \otimes F)} = \bar{M} \otimes F$.*

(d) *Let* $\mathfrak{s} = \mathfrak{g}$ *be any Kac–Moody Lie algebra and let* $\mathfrak{g}(i) \hookrightarrow \mathfrak{g}$ *be the subalgebra defined in 1.1.2, for any* $1 \leq i \leq \ell$. *Then, for any* $\lambda \in \mathfrak{h}^*$, $M(\lambda) \in \mathcal{I}_{\mathfrak{g}}(\mathfrak{g}(i))$ *and, moreover,* $\overline{M(\lambda)} \simeq M(\bar{s}_i * \lambda)$ *as* $\mathfrak{g}$*-modules.*

Proof. (a): Fix a basis $\{v_i\}_{i\in\Lambda}$ of $\bigoplus_{n\in\mathbb{Z}_+} M^X_{-n-2}$ such that $v_i \in M^X_{-n_i-2}$ for some $n_i \in \mathbb{Z}_+$. For any $z \in \mathbb{C}$, let $V(z)$ be the Verma module for $s\ell_2$ with highest weight z (i.e., H acts by z). Any v_i determines a unique $\mathfrak{s}$-module homomorphism

$$\theta_i : \tilde{V}(-n_i - 2) \simeq U(\mathfrak{s}) \otimes_{U(\mathfrak{b}_o)} \mathbb{C}_{-n_i-2} \to M,$$

taking $1 \otimes 1_{-n_i-2} \mapsto v_i$, where $\tilde{V}(z) := U(\mathfrak{s}) \otimes_{U(s\ell_2)} V(z)$, and $\mathfrak{b}_o$ is the Borel subalgebra of $s\ell_2$ spanned by $\{X, H\}$. Consider the $s\ell_2$-module embedding $\gamma_i : V(-n_i - 2) \hookrightarrow V(n_i)$, taking $1 \otimes 1_{-n_i-2} \mapsto Y^{n_i+1} \otimes 1_{n_i}$ (cf. proof of Lemma 9.2.5). (Then γ_i has finite-dimensional cokernel.) This induces a $\mathfrak{s}$-module embedding $\tilde{\gamma}_i : \tilde{V}(-n_i - 2) \hookrightarrow \tilde{V}(n_i)$. Set $L' := M \oplus \big(\oplus_i \tilde{V}(-n_i - 2)\big)$ and $L := M \oplus \big(\oplus_i \tilde{V}(n_i)\big)$. Then, there is a canonical $\mathfrak{s}$-module embedding $\tilde{\gamma} : L' \to L$ induced by the embeddings $\tilde{\gamma}_i$. Further, there is a (surjective) $\mathfrak{s}$-module map $\theta : L' \to M$, given by $\theta_{|M} = I_M$ and $\theta_{|\tilde{V}(-n_i-2)} = \theta_i$. Let $K' := \operatorname{Ker}\theta$ and let $K = \tilde{\gamma}(K')$. Finally, let $\tilde{K} := \{v \in L : Y^p v \in K$, for some $p = p(v) \in \mathbb{Z}_+\}$. By Lemma 1.3.3(b), $\tilde{K}$ is a $\mathfrak{s}$-submodule of L. Set

$$\bar{M} := L/\tilde{K},$$

and define the map $j : M \to \bar{M}$ by taking M to the M-factor via the identity map and via the zero map to all the other factors $\tilde{V}(n_i)$. We claim that j is a completion of M:

Clearly $\bar{M}$ is a weight module under H and j is injective (since Y acts injectively on M by assumption). Further, $\bar{M}/j(M)$ is locally finite under the $s\ell_2$-action. To prove this, observe that

$$\bigoplus_i \big(\tilde{V}(n_i)/\tilde{\gamma}_i \tilde{V}(-n_i - 2)\big) \twoheadrightarrow L/(K + M),$$

as $\mathfrak{s}$-modules and, moreover, by Lemma 1.3.3(b), $\tilde{V}(n_i)/\tilde{\gamma}_i \tilde{V}(-n_i - 2)$ is locally nilpotent under the actions of X and Y and hence it is a locally finite $s\ell_2$-module. But, since $\bar{M}/j(M)$ is a quotient of $L/(K + M)$, $\bar{M}/j(M)$ is a locally finite $s\ell_2$-module as well. In particular, j induces an isomorphism $M^X_{-n} \simeq \bar{M}^X_{-n}$, for all the integers $n > 0$.

We next show that, for all $n \in \mathbb{Z}_+$, $Y^{n+1} : \bar{M}^X_n \to \bar{M}^X_{-n-2} \approx M^X_{-n-2}$ is an isomorphism. By the definition of $\tilde{K}$, Y acts injectively on $\bar{M}$ and, by the construction of L, $Y^{n+1}(\bar{M}_n^X) = M_{-n-2}^X$. This completes the proof that $j : M \to \bar{M}$ is a completion.

We now prove (b) and revert to the uniqueness of $\bar{M}$ after proving (b).

(b): Fix any completions $j_M : M \hookrightarrow \bar{M}$ and $j_N : N \hookrightarrow \bar{N}$. Then we construct an $\mathfrak{s}$-module map $\bar{f} : \bar{M} \to \bar{N}$, making the diagram (∗) of (b) commutative:

Let $\Gamma \subset M \oplus N \subset \bar{M} \oplus \bar{N}$ be the graph of f. Define

$$\tilde{\Gamma} = \{v \in \bar{M} \oplus \bar{N} : Y^p v \in \Gamma, \text{ for some } p = p(v) \in \mathbb{Z}_+\}.$$

Then, by Lemma 1.3.3(b), $\tilde{\Gamma}$ is a $\mathfrak{s}$-submodule of $\bar{M} \oplus \bar{N}$. We next prove that the projection on the first factor induces an isomorphism $\pi_1 : \tilde{\Gamma} \xrightarrow{\sim} \bar{M}$:

First of all, the map π_1 is injective. Let $(0, w) \in \tilde{\Gamma}$. Then, for some $p \in \mathbb{Z}_+$, $(0, Y^p w) \in \Gamma$, and hence $Y^p w = 0$. But since $\bar{N} \in \mathcal{I}_{\mathfrak{s}}$ (by Lemma 9.2.8), $Y^p w = 0 \Rightarrow w = 0$, proving the injectivity of π_1.

We next prove the surjectivity of π_1. By Lemma 9.2.10, $\bar{M}$ is generated (as a $s\ell_2$-module) by M and $\bar{M}_n^X$ ($n \in \mathbb{Z}_+$). Since $\pi_1(\tilde{\Gamma}) \supset M$, it suffices to show that, for any $n \in \mathbb{Z}_+$, $\bar{M}_n^X \subset \pi_1(\tilde{\Gamma})$. So take $v \in \bar{M}_n^X$. Then $Y^{n+1} v \in \bar{M}_{-n-2}^X = M_{-n-2}^X$. In particular, for $w := f(Y^{n+1}v) \in N_{-n-2}^X = \bar{N}_{-n-2}^X$, we have $(Y^{n+1}v, w) \in \Gamma$. But, $\bar{N}$ being complete, there exists $w' \in \bar{N}_n^X$ such that $w = Y^{n+1} w'$, and hence $(v, w') \in \tilde{\Gamma}$. This proves that π_1 is surjective as well, and hence gives rise to the $\mathfrak{s}$-module map $\bar{f}$, as asserted.

We next prove that $\bar{f}$ is unique making (∗) commutative. It suffices to show that if $\bar{f}_{|M} \equiv 0$, then $\bar{f}$ itself is 0. Such an $\bar{f}$ gives rise to an $\mathfrak{s}$-module map $\bar{f}' : \bar{M}/M \to \bar{N}$. But, since Y acts locally nilpotently on $\bar{M}/M$ and acts injectively on $\bar{N}$ (by Lemma 9.2.8), $\bar{f}' \equiv 0$. This proves the uniqueness and hence completes the proof of (b).

Now, we come to the uniqueness of the completion in the (a) part: Let $j' : M \to \bar{M}'$ be another completion. From the functorial property proved above, we have $\mathfrak{s}$-module morphisms $\bar{M} \to \bar{M}'$, $\bar{M}' \to \bar{M}$ and, from the uniqueness, they are inverses of each other. This completes the uniqueness part of (a).

(c): By the complete reducibility theorem [Humphreys–72, Theorem 6.3] (cf. Exercise 3.2.E.2), F is completely reducible as an $s\ell_2$-module and, moreover, any finite-dimensional irreducible module of $s\ell_2$ is a direct summand of $\otimes^n V_2$ for some $n \in \mathbb{Z}_+$, where V_2 is the standard two-dimensional representation of $s\ell_2$. So, to prove the (c) part, it suffices to show that, for $M \in \mathcal{I}_{s\ell_2}$, Y acts injectively on $M \otimes V_2$ and, moreover, if $M \in \mathcal{I}_{s\ell_2}$ is complete, then so is $M \otimes V_2$.

Let e_1, e_2 be the standard basis of V_2 so that $Xe_1 = 0$ and $Ye_1 = e_2$. Write any element $v \in M \otimes V_2$ as $v = w_1 \otimes e_1 + w_2 \otimes e_2$. Then $Yv = 0$ implies that $Yw_1 = 0$ and $(Yw_2) + w_1 = 0$. In particular, if $M \in \mathcal{I}_{s\ell_2}$, we get $w_1 = 0 = w_2$. This proves that Y acts injectively on $M \otimes V_2$.

Assume now that $M \in \mathcal{I}_{s\ell_2}$ is complete. To prove that $\tilde{M} := M \otimes V_2$ is complete, it suffices to show that, for all $n \in \mathbb{Z}_+$,

$$(1) \qquad Y^{n+1} : \tilde{M}^X_n \to \tilde{M}^X_{-n-2} \text{ is surjective.}$$

It is easy to see that

$$(2) \qquad \tilde{M}^X = \{w_1 \otimes e_1 + w_2 \otimes e_2 \in \tilde{M} : w_2 = -Xw_1 \text{ and } X^2 w_1 = 0\}.$$

Moreover, for $w_1 \otimes e_1 + w_2 \otimes e_2 \in \tilde{M}$,

$$(3) \qquad \begin{aligned} &Y^{n+1}(w_1 \otimes e_1 + w_2 \otimes e_2) \\ &\quad = \left(Y^{n+1} w_1\right) \otimes e_1 + \left((n+1)Y^n w_1 + Y^{n+1} w_2\right) \otimes e_2. \end{aligned}$$

In view of (2)–(3), to prove the surjectivity of Y^{n+1} in (1), we need to show the following: For $w_1' \in M_{-n-3}$ such that $X^2 w_1' = 0$, there exists $w_1 \in M_{n-1}$ satisfying

$$(4) \qquad X^2 w_1 = 0, \quad \text{and } Y^{n+1} w_1 = w_1'.$$

(Since $Y^{n+1}(\tilde{M}^X_n) \subset \tilde{M}^X_{-n-2}$, $Y^{n+1}(w_1 \otimes e_1 - Xw_1 \otimes e_2)$ will automatically be equal to $w_1' \otimes e_1 - Xw_1' \otimes e_2$ by (2)–(4).)

Now $XYXw_1' + (n+1)Xw_1' = HXw_1' + (n+1)Xw_1' = 0$, since $X^2 w_1' = 0$. Hence, M being complete, there exist $\theta_1 \in M^X_{n+1}$ and $\theta_2 \in M^X_{n-1}$ such that

$$(5) \qquad Y^{n+2}\theta_1 = YXw_1' + (n+1)w_1', \quad \text{and}$$

$$(6) \qquad Y^n \theta_2 = Xw_1'.$$

Setting $w_1 = \frac{1}{n+1}(Y\theta_1 - \theta_2)$, from (5)–(6), we see that (4) is satisfied. This proves (1) and hence the (c) part is established.

(d): Since $\underline{s}_i * (\bar{s}_i * \lambda) = \lambda$, by Lemma 9.2.5 there is an embedding $M(\lambda) \hookrightarrow M(\bar{s}_i * \lambda)$. Clearly, $M(\lambda)$ belongs to $\mathcal{I}_{\mathfrak{g}}(\mathfrak{g}(i))$. Further, by Lemma 1.3.3(b) and the proof of Lemma 9.2.5, $Y = f_i$ acts locally nilpotently on $M(\bar{s}_i * \lambda)/M(\lambda)$. So, to prove (d), by Exercise 9.2.E.4 it suffices to show that $M(\bar{s}_i * \lambda)$ is complete.As a $\mathfrak{g}(i)$-module we can write

$$(7) \qquad M(\bar{s}_i * \lambda) \cong U(\mathfrak{u}_i^-) \otimes V\left(\bar{s}_i * \lambda(\alpha_i^\vee)\right),$$

where $\mathfrak{u}_i^-$ is as in 1.2.2, $U(\mathfrak{u}_i^-)$ is a $\mathfrak{g}(i)$-module induced by the adjoint action of $\mathfrak{g}(i)$ on $\mathfrak{u}_i^-$, and we put the tensor product $\mathfrak{g}(i)$-module structure on the right side. By Corollary 1.3.4 and Exercise 9.2.E.4, $\mathfrak{u}_i^-$ (and hence $U(\mathfrak{u}_i^-)$) is a locally finite $\mathfrak{g}(i)$-module. Further, by Exercise 9.2.E.2, $V(\bar{s}_i * \lambda(\alpha_i^\vee))$ is a complete $\mathfrak{g}(i)$-module since $\bar{s}_i * \lambda(\alpha_i^\vee) \notin \{-2,-3,-4,\cdots\}$. Hence, by the (c) part and (7), $M(\bar{s}_i * \lambda)$ is complete. This proves (d), and hence completes the proof of the theorem, modulo the following Lemma 9.2.10. □

Even though, for the proof of Theorem 9.2.9, we need the following lemma only for $\mathfrak{g} = s\ell_2$, we prove it for any semisimple Lie algebra.

Recall that any (finite-dimensional) semisimple Lie algebra $\mathfrak{g}$ is an example of a symmetrizable Kac–Moody Lie algebra (cf. [Serre–87, Chapter 6, Appendix]). In particular, various notions such as integrable $\mathfrak{g}$-modules, category $\mathcal{O}$, Casimir operator etc. have the same meanings as given in Chapters 1–2 in the Kac–Moody case.

9.2.10 Lemma. *Let $\mathfrak{g}$ be a semisimple Lie algebra and $\mathfrak{b} \subset \mathfrak{g}$ a Borel subalgebra with nilradical $\mathfrak{n}$, and Cartan subalgebra $\mathfrak{h} \subset \mathfrak{b}$. Let $M \subset N$ be $\mathfrak{g}$-modules such that N is a weight module under the $\mathfrak{h}$-action and is locally $U(\mathfrak{n})$-finite and N/M is an integrable $\mathfrak{g}$-module. Then, N is generated (as a $\mathfrak{g}$-module) by M and $N_\lambda^{\mathfrak{n}}$ ($\lambda \in D$), where $N_\lambda^{\mathfrak{n}} := \{v \in N : \mathfrak{n} \cdot v = 0$ and $h \cdot v = \lambda(h)v$, for $h \in \mathfrak{h}\}$.*

Proof. From the complete reducibility theorem and Exercise 9.2.E.4, it suffices to show that for any $v \in (N/M)_\lambda^{\mathfrak{n}}$ (for $\lambda \in D$), there exists a vector $v_o \in N_\lambda^{\mathfrak{n}}$ such that $\pi(v_o) = v$, where $\pi : N \to N/M$ is the projection. Choose any vector $w \in N_\lambda$ with $\pi(w) = v$, and let $L \subset N$, resp. $V \subset N/M$, be the $\mathfrak{g}$-submodule generated by w, resp. v. Since N is locally $U(\mathfrak{n})$-finite, $L \in \mathcal{O}$.

As in 2.1.17, let L^z (for $z \in \mathbb{C}$) be the generalized z-eigenspace for the Casimir operator Ω acting on L. Taking $z_o = \langle \lambda, 2\rho + \lambda \rangle$, by Lemmas 2.1.16 and 2.1.18, π restricts to a surjective map $L^{z_o} \twoheadrightarrow V$. Choose a vector $v_o' \in (L^{z_o})_\lambda$ with $\pi(v_o') = v$, and also choose a nonzero vector $v_o \in (L^{z_o})_\mu^{\mathfrak{n}}$ such that $v_o \in U(\mathfrak{n}) \cdot v_o'$ (for some $\mu \in \mathfrak{h}^*$). (This is possible since N is locally $U(\mathfrak{n})$-finite.) Then

(1) $$\mu = \lambda + \sum_i n_i \alpha_i, \quad \text{for some } n_i \in \mathbb{Z}_+.$$

By Lemma 2.1.16,

$$\Omega\, v_o = \langle \mu, 2\rho + \mu \rangle v_o.$$

But, since $v_o \in L^{z_o}$, we get

(2) $$\langle \mu, 2\rho + \mu \rangle = \langle \lambda, 2\rho + \lambda \rangle.$$

Combining (1) and (2), since $\lambda \in D$, we get $\mu = \lambda$ and hence $v_o = v_o'$ (up to a nonzero scalar multiple). This proves the lemma. □

9.2.11 Corollary. *Let $\mathfrak{g}$ be any Kac–Moody algebra. Then, for any $\lambda, \mu \in \mathfrak{h}^*$ and $1 \leq i \leq \ell$,*

$$\dim \operatorname{Hom}_{\mathfrak{g}}(M(\lambda), M(\mu)) \leq \dim \operatorname{Hom}_{\mathfrak{g}}(M(\bar{s}_i * \lambda), M(\bar{s}_i * \mu)).$$

Proof. Apply Theorem 9.2.9 for the Lie algebra $\mathfrak{s} = \mathfrak{g}$ and the subalgebra $\mathfrak{g}(i)$. By the (b) part of the theorem, for any $\mathfrak{g}$-module map $f : M(\lambda) \to M(\mu)$, we get the $\mathfrak{g}$-module map $\bar{f}$ making the following diagram commutative:

$$\begin{array}{ccc} M(\lambda) & \xrightarrow{\ f\ } & M(\mu) \\ \downarrow & & \downarrow \\ \overline{M(\lambda)} & \xrightarrow[\bar{f}]{} & \overline{M(\mu)} \end{array}$$

Now the corollary follows from the (d) part of the theorem. □

With these preparations, we can now prove Theorem 9.2.3.

9.2.12 *Proof of Theorem 9.2.3.* We first consider the case $v \geq w$ and prove by induction on $\ell(v)$ that

$$\text{(1)} \qquad \operatorname{Hom}_{\mathfrak{g}}(M(v * \lambda), M(w * \lambda)) \neq 0\,.$$

For $\ell(v) = 0$, (1) is obvious. So take $\ell(v) > 0$ and choose a simple reflection s_i such that $v' := s_i v < v$. Set

$$\text{(2)} \qquad w' = \begin{cases} w\,, & \text{if } s_i w > w \\ s_i w\,, & \text{otherwise.} \end{cases}$$

Then, by Lemma 9.2.5 in the first case (i.e., $s_i w > w$) and Lemma 9.2.6 in the second case,

$$\text{(3)} \quad \dim \operatorname{Hom}_{\mathfrak{g}}(M(v * \lambda), M(w * \lambda)) \geq \dim \operatorname{Hom}_{\mathfrak{g}}(M(v' * \lambda), M(w' * \lambda)).$$

But, by induction, the right side of the above inequality is nonzero and hence the induction is complete, proving (1). Observe that $v' \geq w'$ by Corollary 1.3.19 and, moreover, $s_i * (v' * \lambda) \leq v' * \lambda$ by Lemma 1.3.13 and, similarly, in the second case $s_i * (w' * \lambda) \leq w' * \lambda$.

We next show that, for any $v, w \in W$ and $\lambda \in D$,

$$\text{(4)} \qquad \dim \operatorname{Hom}_{\mathfrak{g}}(M(v * \lambda), M(w * \lambda)) \leq 1 :$$

Fix an embedding $M(w * \lambda) \hookrightarrow M(\lambda)$ (guarnteed by (1)). Then, for a simple reflection s_i such that $v' := s_i v < v$,

$$\text{(5)} \qquad \begin{aligned} \dim \operatorname{Hom}_{\mathfrak{g}}(M(v * \lambda), M(w * \lambda)) &\le \dim \operatorname{Hom}_{\mathfrak{g}}(M(v * \lambda), M(\lambda)) \\ &\le \dim \operatorname{Hom}_{\mathfrak{g}}(M(v' * \lambda), M(\lambda)), \end{aligned}$$

where the last inequality follows from Corollary 9.2.11. By (5) and induction on $\ell(v)$, we get

$$\dim \operatorname{Hom}_{\mathfrak{g}}(M(v * \lambda), M(w * \lambda)) \le \dim \operatorname{Hom}_{\mathfrak{g}}(M(\lambda), M(\lambda)) = 1.$$

This proves (4).

Combining (1) and (4), we get the (b) part of Theorem 9.2.3. To complete the proof of the theorem, we need to show that

$$\text{(6)} \qquad \operatorname{Hom}_{\mathfrak{g}}\big(M(v * \lambda), M(w * \lambda)\big) = 0, \text{ if } v \not\geq w.$$

Again we prove (6) by induction on $\ell(v)$. For $v = e$, (6) is clear since λ is not a weight of $M(w * \lambda)$ for $w \neq e$. So, assume that $\ell(v) > 0$ and take a simple reflection s_i such that $v' := s_i v < v$. Then, by Corollary 9.2.11 and Lemma 1.3.13,

$$\text{(7)} \quad \dim \operatorname{Hom}_{\mathfrak{g}}\big(M(v * \lambda), M(w * \lambda)\big) \le \dim \operatorname{Hom}_{\mathfrak{g}}\big(M(v' * \lambda), M(w' * \lambda)\big),$$

where w' is given by (2).

It is easy to see using Corollary 1.3.19 that $w' \not\leq v'$. By induction, the right side of (7) is 0 and hence so is the left side. This completes the induction and hence finishes the proof of the theorem. □

9.2.13 Definition. Recall that, for a symmetrizable Kac–Moody algebra $\mathfrak{g}$, any subset $Y \subset \{1, \ldots, \ell\}$ and any $\lambda \in D_Y$, the Y-generalized Verma module $M_Y(\lambda)$ was defined in 9.1.1. For an arbitrary (not necessarily symmetrizable) $\mathfrak{g}$, we extend the definition of $M_Y(\lambda)$ (again for $\lambda \in D_Y$) by

$$M_Y(\lambda) := U(\mathfrak{g}) \otimes_{U(\mathfrak{p}_Y)} L_Y^{\max}(\lambda),$$

where (cf. Definition 2.1.5)

$$L_Y^{\max}(\lambda) := U(\mathfrak{g}_Y)/\mathcal{I}\langle e_i, f_i^{\langle \lambda, \alpha_i^\vee \rangle + 1}, h - \lambda(h); h \in \mathfrak{h}, i \in Y \rangle,$$

and $\mathfrak{u}_Y$ acts trivially on $L_Y^{\max}(\lambda)$, where $\mathcal{I}$ denotes the left ideal generated by the elements inside the brackets.

As in 9.1.1, $M_Y(\lambda)$ is canonically a $\mathfrak{g}$-module quotient of $M(\lambda)$; in particular, $M_Y(\lambda) \in \mathcal{O}$. Further, $M_Y(\lambda)$ is a free $U(\mathfrak{u}_Y^-)$-module and, moreover, it is an integrable $\mathfrak{g}_Y$-module by Lemma 1.3.3(b).

9.2.14 Lemma. *For $\lambda \in \mathfrak{h}^*$ and $\mu \in D_Y$, let $f : M(\lambda) \to M(\mu)$ be a $\mathfrak{g}$-module homomorphism such that $\pi_\mu \circ f$ is nonzero, where $\pi_\mu : M(\mu) \to M_Y(\mu)$ is the canonical (surjective) $\mathfrak{g}$-module homomorphism. Then $\lambda \in D_Y$ and, moreover, $\pi \circ f$ descends to give a (nonzero) $\mathfrak{g}$-module homomorphism $f_Y : M_Y(\lambda) \to M_Y(\mu)$.*

In particular, for $v, w \in W'_Y$ with $w \to v$ and $\lambda \in D$, any $\mathfrak{g}$-module homomorphism $\theta : M(v^{-1} * \lambda) \hookrightarrow M(w^{-1} * \lambda)$ descends to a nonzero $\mathfrak{g}$-module homomorphism

$$M_Y(v^{-1} * \lambda) \to M_Y(w^{-1} * \lambda).$$

Recall that, by Theorem 9.2.3, such a θ exists and is unique up to a nonzero scalar multiple.

Proof. By 9.2.13, $M_Y(\mu)$ is an integrable $\mathfrak{g}_Y$-module. Since, for the highest weight vector $e_\lambda \in M(\lambda)$, $\pi_\mu \circ f(e_\lambda) \neq 0$, we get by Lemma 2.1.7 that $\lambda \in D_Y$.

There exists a $\mathfrak{g}_Y$-module homomorphism $\tilde{f}_Y : L_Y^{\max}(\lambda) \to M_Y(\mu)$ (by Lemma 2.1.7), such that the highest weight vector e_λ goes to $\pi_\mu \circ f(e_\lambda)$. Further, since $\mathfrak{g}_Y$ normalizes $\mathfrak{u}_Y$, it is easy to see that $\mathfrak{u}_Y$ acts trivially on $\operatorname{Im} \tilde{f}_Y$. Thus, $\tilde{f}_Y$ extends uniquely to a $\mathfrak{g}$-module homomorphism $f_Y : M_Y(\lambda) \to M_Y(\mu)$. Clearly $f_Y \circ \pi_\lambda = \pi_\mu \circ f$, since they coincide on e_λ.

As observed in Theorem 9.1.3, for $\lambda \in D$ and $w \in W'_Y$, $w^{-1} * \lambda \in D_Y$. So, to prove the "In particular" statement, it suffices to show that

(1) $$\pi_{w^{-1}*\lambda} \circ \theta \neq 0.$$

From the definition of $M_Y(\lambda)$, it is clear that

(2) $$\begin{aligned}\operatorname{Ker} \pi_{w^{-1}*\lambda} &= \sum_{i\in Y} U(\mathfrak{g})\, f_i^{w^{-1}*\lambda(\alpha_i^\vee)+1} e_o \\ &= \sum_{i\in Y} M\big((s_i w^{-1}) * \lambda\big) \subset M(w^{-1} * \lambda),\end{aligned}$$

by Lemma 9.2.5 and its proof, where e_o is the highest weight vector of $M(w^{-1} * \lambda)$. If (1) were false, by (2) we would get

$$\operatorname{Im} \theta \subset \sum_{i\in Y} M\big((s_i w^{-1}) * \lambda\big) \subset M(w^{-1} * \lambda).$$

In particular, this would give (cf. 2.1.11)

(3) $$\Big[M\big((s_{i_o} w^{-1}) * \lambda\big) : L(v^{-1} * \lambda)\Big] \neq 0, \text{ for some } i_o \in Y.$$

For symmetrizable $\mathfrak{g}$ by Exercise 2.3.E.1, and for general $\mathfrak{g}$ by [Neidhardt–84, Theorem 4.1] along with Theorem 9.2.3(a), (3) would give

$$v \geq ws_{i_o}. \tag{4}$$

Since $w \to v$ and $w \in W'_Y$ (by assumption), (4) forces $v = ws_{i_o}$, which is a contradiction since $v \in W'_Y$ by assumption. This proves (1) and hence completes the proof of the lemma. □

9.2.15 Definition. For $\mu, \mu' \in D_Y$, a $\mathfrak{g}$-module homomorphism $f : M_Y(\mu) \to M_Y(\mu')$ is called *standard* if f lifts to a $\mathfrak{g}$-module homomorphism $\hat{f} : M(\mu) \to M(\mu')$, making the following diagram commutative:

$$\begin{array}{ccc} M(\mu) & \xrightarrow{\hat{f}} & M(\mu') \\ {\scriptstyle \pi_\mu}\big\downarrow & & \big\downarrow{\scriptstyle \pi_{\mu'}} \\ M_Y(\mu) & \xrightarrow{f} & M_Y(\mu'). \end{array}$$

By Lemma 9.2.14, for $\lambda \in D$ and $v, w \in W'_Y$ with $w \to v$, the space of standard $\mathfrak{g}$-module maps $M_Y(v^{-1} * \lambda) \to M_Y(w^{-1} * \lambda)$ is one-dimensional. For $Y = \emptyset$, any $\mathfrak{g}$-module map $M(\mu) \to M(\mu')$ is of course standard.

Let $\lambda \in D$. For any $w \in W$, fix a $\mathfrak{g}$-module embedding (cf. Theorem 9.2.3)

$$i_w : M(w^{-1} * \lambda) \hookrightarrow M(\lambda). \tag{1}$$

These embeddings give rise to a unique $\mathfrak{g}$-module embedding (again by Theorem 9.2.3), for any $w \leq v$,

$$i_{w,v} : M(v^{-1} * \lambda) \hookrightarrow M(w^{-1} * \lambda),$$

satisfying

$$i_w \circ i_{w,v} = i_v. \tag{2}$$

For $p \in \mathbb{Z}_+$ set $W'_Y(p) = \{v \in W'_Y : \ell(v) = p\}$. As in Theorem 9.1.3, set (for $\lambda \in D$)

$$F_p^Y := \bigoplus_{v \in W'_Y(p)} M_Y(v^{-1} * \lambda).$$

For any $p \geq 1$ and $\lambda \in D$, any $\mathfrak{g}$-module homomorphism $b = b^p : F_p^Y \to F_{p-1}^Y$ is given by

$$b_{|M_Y(v^{-1} * \lambda)} = \oplus_{w \in W'_Y(p-1)} b_{w,v}, \tag{3}$$

for some $\mathfrak{g}$-module homomorphisms

$$b_{w,v} : M_Y(v^{-1} * \lambda) \to M_Y(w^{-1} * \lambda).$$

Conversely, any collection of $\mathfrak{g}$-module homomorphisms

$$(b_{w,v})_{w \in W'_Y(p-1), v \in W'_Y(p)}$$

gives rise to a $\mathfrak{g}$-module homomorphism $b : F_p^Y \to F_{p-1}^Y$ defined by (3). Such a b is called *standard* if each $b_{w,v}$ is standard. By Theorem 9.2.3, any standard $b_{w,v} = 0$ if $w \nrightarrow v$.

For $w \in W'_Y(p-1)$ and $v \in W'_Y(p)$ with $w \to v$, let $\bar{i}_{w,v} : M_Y(v^{-1} * \lambda) \to M_Y(w^{-1} * \lambda)$ be the standard homomorphism lifting to $i_{w,v}$. Then the standard $\mathfrak{g}$-module homomorphisms $b : F_p^Y \to F_{p-1}^Y$ are given precisely by $\left(b_{w,v} = z_{w,v}\bar{i}_{w,v}\right)_{w \to v}$ with $w \in W'_Y(p-1)$ and $v \in W'_Y(p)$, where $z_{w,v}$ are any complex numbers.

9.2.16 Lemma. *For $p \geq 1$, let $b^{p+1} : F_{p+1}^Y \to F_p^Y$ and $b^p : F_p^Y \to F_{p-1}^Y$ be (any) standard $\mathfrak{g}$-module homomorphisms given respectively by $\left(b_{v,u}^{p+1} = z_{v,u}\bar{i}_{v,u}\right)_{v \to u}$ and $\left(b_{w,v}^p = z_{w,v}\bar{i}_{w,v}\right)_{w \to v}$. Then $b^p \circ b^{p+1} = 0$ iff for all the squares (cf. Definition 9.2.1) (w, v, v', u) of elements of W with each $w, v, v', u \in W'_Y$ and $\ell(w) = p - 1$,*

(1) $$z_{w,v}\, z_{v,u} + z_{w,v'}\, z_{v',u} = 0.$$

Proof. The lemma follows easily from Lemma 9.2.2(a) and the identity

(2) $$\bar{i}_{w,v}\, \bar{i}_{v,u} = \bar{i}_{w,v'}\, \bar{i}_{v',u}. \qquad \square$$

9.2.17 Definition (BGG complex). For $\lambda \in D$, and any subset $Y \subset \{1, \dots, \ell\}$, define the chain complex

$(\mathcal{S})$ $$\cdots \longrightarrow F_p^Y \xrightarrow{s^p} \cdots \longrightarrow F_1^Y \xrightarrow{s^1} F_0^Y \xrightarrow{\hat{\epsilon}} L^{\max}(\lambda) \longrightarrow 0,$$

where F_p^Y is as in 9.2.15 and $s^p : F_Y^p \to F_Y^{p-1}$, for $p \geq 1$, are the standard $\mathfrak{g}$-module homomorphisms given by

$$s_{w,v}^p := \epsilon(w, v)\, \bar{i}_{w,v},$$

for $w \to v$, $w \in W'_Y(p-1)$, $v \in W'_Y(p)$, and $\epsilon(w, v) \in \{\pm 1\}$ is as in Lemma 9.2.2(b). The $\mathfrak{g}$-module map $\hat{\epsilon} : F_0^Y = M_Y(\lambda) \to L^{\max}(\lambda)$ is the standard quotient map.

By Lemmas 9.2.16 and 9.2.2(b), $\mathcal{S}$ is indeed a chain complex. (Observe that $\hat{\epsilon} \circ s^1 = 0$ by (9.2.14.2).) This complex will be referred to as the *BGG complex*. Clearly,

(1) $$s^p_{|M_Y(v^{-1} * \lambda)} \neq 0,$$

for any $v \in W'_Y(p)$, $p \geq 1$ (since there exists $w \in W'_Y(p-1)$ such that $w \to v$).

9.2.18 Theorem. *Let λ, Y, $\hat{\epsilon}$ and F_p^Y be as in 9.2.17. Consider any chain complex*

$$(\mathcal{A}) \qquad \cdots \longrightarrow F_p^Y \xrightarrow{b^p} \cdots \longrightarrow F_1^Y \xrightarrow{b^1} F_0^Y \xrightarrow{\hat{\epsilon}} L^{\max}(\lambda) \longrightarrow 0,$$

consisting of standard $\mathfrak{g}$-module maps b^p such that $b^p_{|M_Y(v^{-1}*\lambda)} \neq 0$ for any $v \in W'_Y(p)$, $p \geq 1$. Then, there exists an isomorphism of chain complexes $\varphi : \mathcal{A} \to \mathcal{S}$ such that $\varphi_{|L^{\max}(\lambda)}$ is the identity map and each $\varphi^p : F_p^Y \to F_p^Y$ is a standard $\mathfrak{g}$-module isomorphism, where $\mathcal{S}$ is the BGG complex defined in 9.2.17.

Proof. Assume, by induction on $p \geq 0$, that we have constructed standard $\mathfrak{g}$-module isomorphisms $\varphi^i : F_i^Y \to F_i^Y$ for all $0 \leq i \leq p$ such that

$$(1) \qquad s^i\varphi^i = \varphi^{i-1}b^i, \quad \text{for all } 0 < i \leq p, \text{ and } \hat{\epsilon} \circ \varphi^0 = \hat{\epsilon}.$$

Then, by Theorem 9.2.3(a), φ^i takes $M_Y(v^{-1} * \lambda)$ to itself for any $v \in W'_Y(i)$. Let φ^i_v be the restriction of φ^i to $M_Y(v^{-1} * \lambda)$, so that

$$(2) \qquad \varphi^i_v = c_v I, \quad \text{for some } c_v \in \mathbb{C}\backslash\{0\}.$$

The relations (1) are equivalent to the following relations:

(3)

$$\varphi^0 = I \text{ and } c_v = \frac{c_w z_{w,v}}{\epsilon(w,v)}, \quad \text{for all } w \to v \text{ with } w, v \in W'_Y \text{ and } 1 \leq \ell(v) \leq p,$$

where $b_{w,v} = z_{w,v}\bar{i}_{w,v}$ for $w \to v$ as in 9.2.15.

We start the induction by taking $\varphi^0 = I$. Now, we need to define a standard $\mathfrak{g}$-module isomorphism $\varphi^{p+1} : F_{p+1}^Y \to F_{p+1}^Y$ satisfying

$$(4) \qquad s^{p+1}\varphi^{p+1} = \varphi^p b^{p+1}.$$

Take $v \in W'_Y(p+1)$ and choose a simple reflection s_i such that $s_i v < v$; in particular, $s_i v \in W'_Y$. Then, for any $w \in W'_Y(p)$ with $w \to v$ and $w \neq s_i v$, by Corollary 1.3.19, we have $s_i w < w$; in particular, $s_i w \in W'_Y$ and $s_i w \to s_i v$. Applying (9.2.16.1) to the square $(s_i w, w, s_i v, v)$, we get

$$(5) \qquad z_{s_i w,w}\, z_{w,v} = -z_{s_i w, s_i v}\, z_{s_i v, v}.$$

We next prove that, for any $v \in W'_Y$ and simple reflection s_i such that $s_i v < v$,

$$(6) \qquad z_{s_i v,v} \neq 0.$$

Since, if $z_{s_i v,v} = 0$, by (5) we get $z_{w,v} = 0$ for all $w \to v$ and $w \in W'_Y$ (since, by induction, we can assume that $z_{s_i w,w} \neq 0$). But this is a contradiction since, by assumption, $b_{|M_Y(v^{-1}*\lambda)} \neq 0$.

Now, for $v \in W'_Y(p+1)$, consider the map $c_v I : M_Y(v^{-1} * \lambda) \to M_Y(v^{-1} * \lambda)$, where

(7) $$c_v := \frac{c_{s_i v}\, z_{s_i v, v}}{\epsilon(s_i v, v)},$$

and s_i is a simple reflection such that $s_i v < v$. Since $c_{s_i v} \neq 0$, so is c_v by (6).

We next assert that, for any $w \to v$, $w \in W'_Y$, $v \in W'_Y(p+1)$,

(8) $$c_v = \frac{c_w\, z_{w,v}}{\epsilon(w, v)}.$$

To prove (8), in view of (7), we can assume that $w \neq s_i v$. Now use (3) (applied to $s_i w \to s_i v$ and $s_i w \to w$) together with (5), and Lemma 9.2.2(b). Observe that by (8), c_v defined by (7) does not depend upon the choice of s_i such that $s_i v < v$.

Define the standard $\mathfrak{g}$-module isomorphism $\varphi^{p+1} : F^Y_{p+1} \to F^Y_{p+1}$ by

$$\varphi^{p+1} = \bigoplus_{v \in W'_Y(p+1)} c_v I_{M_Y(v^{-1} * \lambda)},$$

where c_v is defined by (7). Then, by (8), we get (4).

This completes the induction and hence proves the theorem. □

9.2.19 Lemma. *Let λ, Y and F^Y_p be as in 9.2.17, and let $\mathcal{A}$ be an exact complex*

$$\cdots \longrightarrow F^Y_p \xrightarrow{b^p} \cdots \longrightarrow F^Y_1 \xrightarrow{b^1} F^Y_0 \xrightarrow{b^0} L^{\max}(\lambda) \longrightarrow 0$$

consisting of any $\mathfrak{g}$-module maps b^p, $p \geq 0$. Then

$$b^p_{|M_Y(w^{-1} * \lambda)} \neq 0, \qquad \textit{for any } w \in W'_Y(p).$$

Proof. Assume, if possible, that there exists a $w \in W'_Y(p)$ such that $b^p_{|M_Y(w^{-1} * \lambda)} = 0$. Then, by the exactness of $\mathcal{A}$, there exists a weight vector $e' \in F^Y_{p+1}$ such that $b^{p+1}(e') = e_o$, where e_o is a nonzero highest weight vector of $M_Y(w^{-1} * \lambda)$. In particular, there exists a weight vector $e'' \in M_Y(v^{-1} * \lambda)$, for some $v \in W'_Y(p+1)$, such that the component c_o of $b^{p+1}(e'')$ in $M_Y(w^{-1} * \lambda)$ equals e_o. But since $v^{-1} * \lambda \neq w^{-1} * \lambda$ (by Lemma 3.2.5), weight $e'' = w^{-1} * \lambda < v^{-1} * \lambda$. Now, for any $w_1, v_1 \in W'_Y$, any $\mathfrak{g}$-homomorphism $M_Y(v_1^{-1} * \lambda) \to M_Y(w_1^{-1} * \lambda)$ is 0 unless $v_1 \geq w_1$ by [Neidhardt–84, Theorem 4.1] together with Theorem 9.2.3(a). Thus $w \to v$ (and $w^{-1} * \lambda < v^{-1} * \lambda$), which contradicts Lemma 1.3.13. This forces e'' and hence $c_o = 0$, a contradiction, proving the lemma. □

9.2.20 Corollary. *Let the notation and assumptions be as in Theorem 9.2.18. Assume further that $\mathcal{A}$ is exact. Then so is $\mathcal{S}$.*

In particular, if $\mathfrak{g}$ is symmetrizable and Y is a subset of finite type, $\mathcal{S}$ is exact and, moreover, any $\mathcal{A}$ satisfying the assumptions of Theorem 9.2.18 is exact.

(In the next section, we will see that, for $Y = \emptyset$, $\mathcal{S}$ is exact for an arbitrary $\mathfrak{g}$.)

Proof. The first part follows immediately from Theorem 9.2.18.

To prove that $\mathcal{S}$ is exact, in view of Lemma 9.2.19 and Theorem 9.2.18, it suffices to show that (when $\mathfrak{g}$ is symmetrizable and Y is of finite type) there exists an exact complex $\mathcal{A}$ consisting of standard $\mathfrak{g}$-module maps b^p, $p \geq 1$. For $\mathcal{A}$ take the exact chain complex as given in the proof of Theorem 9.1.3. By Exercise 9.2.E.3, all the differentials δ^p $(p \geq 1)$ are standard. This proves the corollary. □

9.2.E EXERCISES

(1) Extend Theorem 9.2.3 to cover those $\lambda \in \mathfrak{h}^*$ such that $\lambda + \rho \in D$; more specifically, prove the following: For $v \geq w$,

$$\mathrm{Hom}_{\mathfrak{g}}\big(M(v * \lambda), M(w * \lambda)\big) \text{ is one-dimensional}$$

and, for any $v, w \in W$,

$$\mathrm{Hom}_{\mathfrak{g}}\big(M(v * \lambda), M(w * \lambda)\big) = 0 \text{ if } v_o \not\geq w_o,$$

where v_o, resp. w_o, is the smallest coset representative of the coset $vW_{\lambda+\rho}$, resp. $wW_{\lambda+\rho}$, and $W_{\lambda+\rho} \subset W$ is the isotropy of $\lambda + \rho$.

Hint. Follow the same proof as in 9.2.12 and use the fact that, for $v \in W$ which is the smallest coset representative of $vW_{\lambda+\rho}$ and any simple reflection s_i such that $s_i v < v$, $s_i v$ is the smallest coset representative of $s_i v W_{\lambda+\rho}$.

(2) Show that for any $z \in \mathbb{C}\backslash\{-2, -3, -4, \cdots\}$, the $s\ell_2$-module $V(z)$ is complete, where $V(z)$ is as in the proof of Theorem 9.2.9.

(3) Let $\mathfrak{g}$ be symmetrizable. Then show that, for any $\lambda \in D$ and subset $Y \subset \{1, \ldots, \ell\}$ of finite type, the differentials δ^p $(p \geq 1)$, given in the proof of Theorem 9.1.3, are all standard $\mathfrak{g}$-module maps.

(4) Let $\mathfrak{g}$ be a (finite-dimensional) semisimple Lie algebra. Then prove that a representation V of $\mathfrak{g}$ is integrable iff it is locally finite.

Hint. Use [Humphreys–72, §6.4].)

9.3. Kempf Resolution

In this section $\mathcal{G}$ is an arbitrary Kac–Moody group with the standard Borel subgroup $\mathcal{B}$ and Weyl group W (cf. Section 6.1). Also, for any $w \in W$, recall the definitions of the subsets $\mathcal{B}^w$ and X^w of $\mathcal{G}/\mathcal{B}$ from 7.1.21 and the Schubert variety X_w from 7.1.13.

9.3.1 Definition. For any $w \in W$, define a decreasing filtration $\{\mathcal{F}_p(w)\}_{p\geq 0}$ of X_w consisting of closed subsets by

$$\mathcal{F}_p(w) := \bigcup_{\substack{v\in W \\ \ell(v)\geq p}} (X_w \cap X^v).$$

Clearly $\mathcal{F}_0(w) = X_w$. By Lemma 7.1.22(b), we can take $v \leq w$ in the above union. In particular, $\mathcal{F}_p(w)$ is indeed closed in X_w and $\mathcal{F}_p(w) = \emptyset$ if $p > \ell(w)$.

Fix $\lambda \in D_{\mathbb{Z}}$ and consider the line bundle $\mathcal{L}_w(\lambda)$ on X_w (cf. 7.2.1). By B.7, the filtration $\{\mathcal{F}_p(w)\}_{p\geq 0}$ of X_w and the line bundle $\mathcal{L}_w(\lambda)$ together give rise to the Grothendieck–Cousin complex

$$\begin{aligned} \mathcal{C}(w): \qquad & 0 \to H^0\big(X_w, \mathcal{L}_w(\lambda)\big) \xrightarrow{\epsilon(w)} H^0_{\mathcal{F}_0(w)/\mathcal{F}_1(w)}\big(X_w, \mathcal{L}_w(\lambda)\big) \\ & \xrightarrow{d_1(w)} H^1_{\mathcal{F}_1(w)/\mathcal{F}_2(w)}\big(X_w, \mathcal{L}_w(\lambda)\big) \longrightarrow \cdots . \end{aligned}$$

Let $\mathcal{P}_j$ be a minimal parabolic subgroup such that $\mathcal{P}_j$ keeps X_w stable, i.e., $s_j w < w$. By Definitions 7.1.19 and 7.2.3, there exists a large enough k such that the action of $\mathcal{P}_j$ on X_w descends to an algebraic action of the (finite-dimensional) affine algebraic group $\mathcal{P}_j/\mathcal{U}_j(k)$ (cf. Lemma 6.1.14 for the notation $\mathcal{U}_j(k)$) and, moreover, $\mathcal{L}_w(\lambda)$ is a $\mathcal{P}_j/\mathcal{U}_j(k)$-equivariant line bundle. Further, since $\mathcal{F}_p(w)$ is T-stable, by Lemma B.10, each $H^i_{\mathcal{F}_p(w)/\mathcal{F}_{p+1}(w)}(X_w, \mathcal{L}_w(\lambda))$ is canonically a $(\mathfrak{p}_j, T)$-module and all the differentials $\epsilon(w)$ and $d_i(w)$ in the complex $\mathcal{C}(w)$ are $\mathfrak{p}_j$-module maps. (Recall that a $\mathfrak{p}_j$-module M is called a $(\mathfrak{p}_j, T)$*-module* if the $\mathfrak{h}$-module structure on M, got by restricting the $\mathfrak{p}_j$-module structure, integrates to give a locally finite algebraic T-module, cf. 6.2.1.)

9.3.2 Theorem. *For any $w \in W$ and $\lambda \in D_{\mathbb{Z}}$, the above complex $\mathcal{C}(w)$ is exact.*

Proof. We apply Theorem B.9. By Theorem 8.2.2(c), X_w is Cohen–Macaulay. Further, $\mathcal{F}_p(w)\backslash\mathcal{F}_{p+1}(w) = \bigsqcup_{v\in W(p)}(X_w \cap \mathcal{B}^v)$, and each $X_w \cap \mathcal{B}^v$ is open in $\mathcal{F}_p(w)\backslash\mathcal{F}_{p+1}(w)$ by Proposition 7.1.21, where $W(p)$ is the set of length p elements in W. In particular, by Lemma 7.3.5, as a locally closed subvariety of X_w, $\mathcal{F}_p(w)\backslash\mathcal{F}_{p+1}(w)$ is an affine variety. To prove that, for any $v \in W$, the inclusion $i : X_w \cap \mathcal{B}^v \hookrightarrow X_w$ is an affine morphism, by virtue of Lemmas 7.3.5 and 7.3.10, it suffices to observe the following:

(a) For any two affine open subsets U and V in a variety Z, $U \cap V$ is again affine (cf. [Hartshorne–77, Chap. II, Exercise 4.3]), and

(b) for any two affine varieties Z_1, Z_2, the inclusion $Z_2 \hookrightarrow Z_1 \times Z_2$, $z \mapsto (z_1, z)$, for any fixed $z_1 \in Z_1$, is an affine morphism by definition (cf. Definition A.13).

By Lemma 7.3.10, $\text{codim}_{X_w}(X_w \cap \mathcal{B}^v) = \dim U_v = \ell(v)$. Finally, by (8.2.2.3), $H^n\big(X_w, \mathcal{L}_w(\lambda)\big) = 0$, for all $n > 0$. So, all the hypotheses of Theorem B.9 are satisfied, proving the theorem. $\square$

9.3.3 Definition. Recall that a locally finite algebraic T-module V is a weight module, i.e., $V = \bigoplus_{\lambda \in \mathfrak{h}^*_{\mathbb{Z}}} V_\lambda$, where V_λ is the weight space corresponding to the weight λ. As in 2.1.1, consider the restricted dual $V^\vee := \bigoplus_{\lambda \in \mathfrak{h}^*_{\mathbb{Z}}} V^*_\lambda$, where V^*_λ is the full vector space dual of V_λ. Then $V^\vee$ is canonically a (locally finite algebraic) T-module.

A weight module V is called *admissible* if all the weight spaces V_λ are finite dimensional. For an admissible module V, we define its *formal character* (cf. (2.1.1.3))

$$\text{ch}\, V = \sum_\lambda \dim V_\lambda \, e^\lambda.$$

For any $v, w \in W$, $\lambda \in D_{\mathbb{Z}}$ and $i \geq 0$, set

$$(1) \qquad H^i_v(X_w, \mathcal{L}_w(\lambda)) := H^i_{X_w \cap X^v / X_w \cap \partial X^v}(X_w, \mathcal{L}_w(\lambda)),$$

where ∂X^v is as defined in 7.3.4. Then $H^i_v(X_w, \mathcal{L}_w(\lambda))$ is canonically a $(\mathfrak{p}_j, T)$-module for any minimal parabolic $\mathcal{P}_j$ such that $\mathcal{P}_j$, under the left multiplication, keeps X_w stable (cf. 9.3.1).

For any $w' \in W$ with $w \leq w'$, we have the inclusion $X_w \subset X_{w'}$ (cf. (7.1.13.1)) and hence (for any $p \geq 0$) $\mathcal{F}_p(w') \cap X_w \subset \mathcal{F}_p(w)$; in fact, the equality holds. This gives rise to the canonical $\mathfrak{b}$-module maps (via Lemmas B.2 and B.10)

$$\phi^p_{w,w'} : H^i_{\mathcal{F}_p(w')/\mathcal{F}_{p+1}(w')}(X_{w'}, \mathcal{L}_{w'}(\lambda)) \to H^i_{\mathcal{F}_p(w)/\mathcal{F}_{p+1}(w)}(X_w, \mathcal{L}_w(\lambda)),$$

and also

$$\phi^v_{w,w'} : H^i_v(X_{w'}, \mathcal{L}_{w'}(\lambda)) \to H^i_v(X_w, \mathcal{L}_w(\lambda)).$$

Further, if $\mathcal{P}_j$ is a minimal parabolic which keeps both X_w and $X_{w'}$ stable, then $\phi^p_{w,w'}$ and $\phi^v_{w,w'}$ are $\mathfrak{p}_j$-module maps.

Taking the restricted duals, we get the $\mathfrak{b}$-module maps

$$\psi^p_{w,w'} : H^i_{\mathcal{F}_p(w)/\mathcal{F}_{p+1}(w)}(X_w, \mathcal{L}_w(\lambda))^\vee \to H^i_{\mathcal{F}_p(w')/\mathcal{F}_{p+1}(w')}(X_{w'}, \mathcal{L}_{w'}(\lambda))^\vee,$$

and

$$\psi^v_{w,w'} : H^i_v(X_w, \mathcal{L}_w(\lambda))^\vee \to H^i_v(X_{w'}, \mathcal{L}_{w'}(\lambda))^\vee.$$

Define the following $(\mathfrak{b}, T)$-modules, for any $p \geq 0$ and $v \in W$:

$$H^i_{\mathcal{F}_p/\mathcal{F}_{p+1}}(\mathcal{X}, \mathcal{L}(\lambda))^\vee := \operatorname*{limit}_{w \in W} \big(H^i_{\mathcal{F}_p(w)/\mathcal{F}_{p+1}(w)}(X_w, \mathcal{L}_w(\lambda))^\vee\big),$$

and

$$H^i_v(\mathcal{X}, \mathcal{L}(\lambda))^\vee := \operatorname*{limit}_{w \in W} \big(H^i_v(X_w, \mathcal{L}_w(\lambda))^\vee\big),$$

where the limits are taken via the maps $\psi^p_{w,w'}$ and $\psi^v_{w,w'}$ respectively.

The subset $W_j := \{w \in W : s_j w < w\}$ is clearly cofinal in W. In particular, $H^i_{\mathcal{F}_p/\mathcal{F}_{p+1}}(\mathcal{X}, \mathcal{L}(\lambda))^\vee$ and $H^i_v(\mathcal{X}, \mathcal{L}(\lambda))^\vee$ are both $\mathfrak{p}_j$-modules for any minimal parabolic $\mathfrak{p}_j$, and hence are $\tilde{\mathfrak{g}}$-modules satisfying, in addition, the relation (R_4) of 1.1.2, where $\tilde{\mathfrak{g}}$ is as defined in 1.1.2.

The maps $\psi^p_{w,w'}$ give rise to the chain map $\psi_{w,w'}$ from the restricted dual $\mathcal{K}(w)$ of the cochain complex $\mathcal{C}(w)$ to $\mathcal{K}(w')$. Taking the direct limit of the chain complexes $\mathcal{K}(w)$ (over $w \in W$) via the chain maps $\psi_{w,w'}$, we get the fundamental chain complex (consisting of $\tilde{\mathfrak{g}}$-modules and $\tilde{\mathfrak{g}}$-module differentials $\epsilon^\vee$ and δ^i)

$$\begin{aligned}\mathcal{K}: \quad 0 \leftarrow H^0(\mathcal{X}, \mathcal{L}(\lambda))^\vee &\xleftarrow{\epsilon^\vee} H^0_{\mathcal{F}_0/\mathcal{F}_1}(\mathcal{X}, \mathcal{L}(\lambda))^\vee \\ &\xleftarrow{\delta^1} \cdots \xleftarrow{\delta^p} H^p_{\mathcal{F}_p/\mathcal{F}_{p+1}}(\mathcal{X}, \mathcal{L}(\lambda))^\vee \leftarrow \cdots,\end{aligned}$$

where, as defined in (8.3.8.2), $H^i(\mathcal{X}, \mathcal{L}(\lambda))^\vee := \operatorname*{limit}_{w \in W} \big(H^i(X_w, \mathcal{L}_w(\lambda))^*\big)$.

We shall refer to the complex $\mathcal{K}$ as the *Kempf complex* (who first studied this special example of the dual of the Grothendieck–Cousin complex in the finite case in [Kempf–78]).

As an immediate consequence of Theorem 9.3.2, we get the following:

9.3.4 Corollary. *The Kempf complex $\mathcal{K}$ defined above is exact for any $\lambda \in D_{\mathbb{Z}}$.* □

We now determine $H^p_{\mathcal{F}_p/\mathcal{F}_{p+1}}(\mathcal{X}, \mathcal{L}(\lambda))^\vee$.

9.3.5 Lemma. *For any $w \in W$, $p \geq 0$ and minimal parabolic $\mathcal{P}_j$ such that $\mathcal{P}_j$ keeps X_w stable, there is a canonical $\mathfrak{p}_j$-module isomorphism*

$$H^i_{\mathcal{F}_p(w)/\mathcal{F}_{p+1}(w)}(X_w, \mathcal{L}_w(\lambda)) \simeq \bigoplus_{v \in W(p)} H^i_v(X_w, \mathcal{L}_w(\lambda)),$$

where $W(p)$ is as in the proof of Theorem 9.3.2.

In particular, taking limits of the restricted duals, we get a canonical $\tilde{\mathfrak{g}}$-module isomorphism:

$$(1) \qquad H^i_{\mathcal{F}_p/\mathcal{F}_{p+1}}(\mathcal{X}, \mathcal{L}(\lambda))^\vee \simeq \bigoplus_{v \in W(p)} H^i_v(\mathcal{X}, \mathcal{L}(\lambda))^\vee.$$

In the sequel, we will make this identification.

Proof. For any $v \in W(p)$, there is a canonical map given by Lemma B.2:

$$H^i_v(X_w, \mathcal{L}_w(\lambda)) \to H^i_{\mathcal{F}_p(w)/\mathcal{F}_{p+1}(w)}(X_w, \mathcal{L}_w(\lambda))$$

(induced from the inclusions $X_w \cap X^v \subset \mathcal{F}_p(w)$ and $X_w \cap \partial X^v \subset \mathcal{F}_{p+1}(w)$). Moreover, by Lemma B.10, this is a $\mathfrak{p}_j$-module map. These maps give rise to the $\mathfrak{p}_j$-module map

$$\theta : \bigoplus_{v \in W(p)} H^i_v(X_w, \mathcal{L}_w(\lambda)) \to H^i_{\mathcal{F}_p(w)/\mathcal{F}_{p+1}(w)}(X_w, \mathcal{L}_w(\lambda)).$$

From the following commutative diagram, θ is an isomorphism.

$$\begin{array}{ccc}
\bigoplus_{v \in W(p)} H^i_v(X_w, \mathcal{L}_w(\lambda)) & \xrightarrow{\ \theta\ } & H^i_{\mathcal{F}_p(w)/\mathcal{F}_{p+1}(w)}(X_w, \mathcal{L}_w(\lambda)) \\
{\scriptstyle \theta_1} \searrow & & \nearrow {\scriptstyle \theta_2} \\
& \bigoplus_{v \in W(p)} H^i_{X_w \cap \mathcal{B}^v}(X_w \backslash \mathcal{F}_{p+1}(w), \mathcal{L}_w(\lambda)), &
\end{array}$$

where θ_1 is the isomorphism induced by Lemmas B.4 and B.3, and θ_2 is the isomorphism induced by Lemmas B.4 and B.6 together with (B.1.13), since $\mathcal{F}_p(w)\backslash\mathcal{F}_{p+1}(w)$ is the disjoint union $\bigsqcup_{v \in W(p)}(X_w \cap \mathcal{B}^v)$ of open subsets of $\mathcal{F}_p(w)\backslash\mathcal{F}_{p+1}(w)$ (cf. proof of Theorem 9.3.2). This proves the lemma. □

9.3.6 Definition. For any $v \in W(p+1)$ and $v' \in W(p)$, define the $\tilde{\mathfrak{g}}$-module map $\delta^{v,v'} : H^{p+1}_v(\mathcal{X}, \mathcal{L}(\lambda))^\vee \to H^p_{v'}(\mathcal{X}, \mathcal{L}(\lambda))^\vee$ as the restriction of the map δ^{p+1} of the Kempf complex $\mathcal{K}$ to $H^{p+1}_v(\mathcal{X}, \mathcal{L}(\lambda))^\vee$ followed by the projection onto the $H^p_{v'}(\mathcal{X}, \mathcal{L}(\lambda))^\vee$-factor under the decomposition (9.3.5.1).

For any $w \in W$, one can similarly define the map

$$\delta^{v,v'}(w) : H^{p+1}_v(X_w, \mathcal{L}_w(\lambda))^\vee \to H^p_{v'}(X_w, \mathcal{L}_w(\lambda))^\vee$$

by using Lemma 9.3.5 and the Kempf complex $\mathcal{K}(w)$, where recall that $\mathcal{K}(w)$ is the restricted dual of the cochain complex $\mathcal{C}(w)$ (cf. 9.3.3).

The following proposition determines the structure of $H^{\ell(v)}_v(\mathcal{X}, \mathcal{L}(\lambda))^\vee$.

9.3.7 Proposition. *The $\tilde{\mathfrak{g}}$-module structure on $H_v^{\ell(v)}(\mathcal{X}, \mathcal{L}(\lambda))^\vee$ (cf. Definition 9.3.3) descends to give a $\mathfrak{g}$-module structure and, moreover, it is isomorphic with the Verma module $M(v * \lambda)$ as $\mathfrak{g}$-modules.*

As a preparation for the proof of the above proposition, we prove the following five Lemmas 9.3.8–9.3.12.

9.3.8 Lemma. *Let V be a $\mathfrak{g}$-submodule of the Verma module $M(\mu)$, for any $\mu \in \mathfrak{h}^*$. Then V itselfis isomorphic to the Verma module $M(\mu')$ if and only if* $\operatorname{ch} V = \operatorname{ch} M(\mu')$ *(cf. (2.1.1.3)).*

Proof. It suffices to prove the implication '$\Leftarrow$'. Choose a nonzero vector $v_o \in V$ of weight μ', which exists and is unique up to a nonzero scalar multiple by the assumption on $\operatorname{ch} V$. Clearly, $U(\mathfrak{b})v_o = \mathbb{C}v_o$. Let V' be the $\mathfrak{g}$-submodule of V generated by v_o. Since any nonzero homomorphism of one Verma module into another is injective (cf. Proof of Lemma 9.2.6), V' is isomorphic with the Verma module $M(\mu')$. Hence

$$\begin{aligned}\operatorname{ch} V' &= \operatorname{ch} M(\mu')\\ &= \operatorname{ch} V \text{ (by assumption).}\end{aligned}$$

But since V' is a submodule of V, this is possible only if $V' = V$, proving the lemma. □

9.3.9 Lemma. *For any $v, w \in W$ we have:*

(a) $H_v^p(X_w, \mathcal{L}_w(\lambda)) = 0$, *unless $v \le w$ and $p = \ell(v)$, and*
(b) *if $v \le w$,*

$$H_v^{\ell(v)}(X_w, \mathcal{L}_w(\lambda)) \approx H_{\{e\}}^{\ell(v)}(\mathcal{U}_v, \mathcal{O}_{\mathcal{U}_v}) \otimes H^0\big(X_w \cap \mathcal{B}^v, \mathcal{L}_w(\lambda)|_{X_w \cap \mathcal{B}^v}\big),$$

as T-modules, where the unipotent group $\mathcal{U}_v = \mathcal{U}_{\Phi_v}$ is as in Example 6.1.5 (b), $H_{\{e\}}^*(\cdot, \cdot)$ denotes the local cohomology with support in the singleton $\{e\}$, and T acts diagonally on the right side of (b).

Proof. Since $X_w \cap \mathcal{B}^v$ is closed in the open subset $X_w \cap (v\mathcal{U}^- \mathcal{B}/\mathcal{B})$ of X_w by Lemma 7.3.10,

$$\begin{aligned}H_v^p(X_w, \mathcal{L}_w(\lambda)) &\approx H^p_{X_w \cap \mathcal{B}^v}\big(X_w \cap (v\mathcal{U}^- \mathcal{B}/\mathcal{B}), \mathcal{L}_w(\lambda)\big),\\ &\qquad \text{by Lemmas B.4 and B.3}\\ &\approx H^p_{\{e\}\times X_w \cap \mathcal{B}^v}\big(\mathcal{U}_v \times (X_w \cap \mathcal{B}^v), \theta^*_{v,w}\, \mathcal{L}_w(\lambda)\big),\\ &\qquad \text{by Lemma 7.3.10}\\ &\approx H^p_{\{e\}}(\mathcal{U}_v, \mathcal{O}_{\mathcal{U}_v}) \otimes H^0\big(X_w \cap \mathcal{B}^v, \mathcal{L}_w(\lambda)|_{X_w \cap \mathcal{B}^v}\big),\\ &\qquad \text{as } T\text{-modules,}\end{aligned}$$

where the last isomorphism is obtained from Lemma B.12 together with (A.23.2) by considering the projection $\mathcal{U}_v \times (X_w \cap \mathcal{B}^v) \to \mathcal{U}_v$ and by observing that $\theta^*_{v,w}\mathcal{L}_w(\lambda) \simeq \mathcal{O}_{\mathcal{U}_v} \boxtimes (\mathcal{L}_w(\lambda)_{|X_w \cap \mathcal{B}^v})$ as T-equivariant line bundles (cf. 7.2.3).

By Lemma B.11, $H^p_{\{e\}}(\mathcal{U}_v, \mathcal{O}_{\mathcal{U}_v}) = 0$, unless $p = \dim \mathcal{U}_v = \ell(v)$, since $\mathcal{U}_v$ is biregular isomorphic with its Lie algebra. Now the lemma follows by observing that $X_w \cap \mathcal{B}^v = \emptyset$ unless $v \leq w$ (cf. Lemma 7.1.22(b)). □

9.3.10 Lemma. *For any $v, w \in W$ and any simple reflection s_i such that $vs_i > v$ and $ws_i < w$, we have*

$$H^{\ell(v)+1}_{X_w \cap X^v / X_w \cap ((\partial X^v) \setminus \mathcal{B}^{vs_i})}(X_w, \mathcal{L}_w(\lambda)) = 0.$$

(Observe that by Proposition 7.1.21, $\mathcal{B}^{vs_i}$ is an open subset of ∂X^v.)

Proof. Let $\pi_i : \mathcal{G}/\mathcal{B} \to \mathcal{G}/\mathcal{P}_i$ be the projection. Set $X^i_w := \pi_i(X_w)$. Then the restriction of π_i to X_w is a smooth morphism onto X^i_w (cf. Exercise 9.3.E.1). The map π_i on restriction gives rise to the surjective morphism $\tilde{\pi}_i : X_w \cap (v\mathcal{U}^-\mathcal{P}_i/\mathcal{B}) \to X^i_w \cap (v\mathcal{U}^-\mathcal{P}_i/\mathcal{P}_i)$ which is smooth and projective by the base change property (cf. A.14 and [Hartshorne–77, Chap. II, Exercise 4.9]). By Lemma 7.3.10, $X_w \cap (\mathcal{U}^- v\mathcal{P}_i/\mathcal{B})$ is closed in the open subspace $X_w \cap (v\mathcal{U}^-\mathcal{P}_i/\mathcal{B})$ of X_w. Hence, by Lemmas B.3 and B.4,

$$(1) \qquad H^*_{X_w \cap X^v / X_w \cap ((\partial X^v) \setminus \mathcal{B}^{vs_i})}(X_w, \mathcal{L}_w(\lambda)) \approx H^*_{X_w \cap (\mathcal{U}^- v\mathcal{P}_i/\mathcal{B})}\big(X_w \cap (v\mathcal{U}^-\mathcal{P}_i/\mathcal{B}), \mathcal{L}_w(\lambda)\big).$$

(Observe that $\mathcal{B}^v \cup \mathcal{B}^{vs_i} = \mathcal{U}^- v\mathcal{P}_i/\mathcal{B}$ by Lemma 5.2.2.)

By the semicontinuity theorem A.29, the direct images $R^q\tilde{\pi}_{i*}(\mathcal{L}_w(\lambda)) = 0$, for $q > 0$, since λ is (by assumption) dominant and the fibers of $\tilde{\pi}_i$ are left translates of $\mathcal{P}_i/\mathcal{B} \approx \mathbb{P}^1$. Hence, by Lemma B.12,

$$(2) \qquad H^*_{X_w \cap (\mathcal{U}^- v\mathcal{P}_i/\mathcal{B})}\big(X_w \cap (v\mathcal{U}^-\mathcal{P}_i/\mathcal{B}), \mathcal{L}_w(\lambda)\big) \approx H^*_{X^i_w \cap (\mathcal{U}^- v\mathcal{P}_i/\mathcal{P}_i)}\big(X^i_w \cap (v\mathcal{U}^-\mathcal{P}_i/\mathcal{P}_i), \tilde{\pi}_{i*}\mathcal{L}_w(\lambda)\big).$$

Finally, by Lemmas B.12 and 7.3.10 (cf. Proof of Lemma 9.3.9),

$$(3) \qquad H^*_{X^i_w \cap (\mathcal{U}^- v\mathcal{P}_i/\mathcal{P}_i)}\big(X^i_w \cap (v\mathcal{U}^-\mathcal{P}_i/\mathcal{P}_i), \tilde{\pi}_{i*}\mathcal{L}_w(\lambda)\big) \approx H^*_{\{e\}}(\mathcal{U}_v, \mathcal{S}),$$

for some quasi-coherent sheaf $\mathcal{S}$ on $\mathcal{U}_v$. But, by [Grothendieck–67, Proposition 1.12], $H^p_{\{e\}}(\mathcal{U}_v, \mathcal{S}) = 0$ for all $p > \dim \mathcal{U}_v = \ell(v)$. Combining (1)–(3), the lemma follows. □

9.3.11 Lemma. *The $\tilde{\mathfrak{g}}$-module structure on $H^0_e(\mathcal{X}, \mathcal{L}(\lambda))^\vee$, defined in 9.3.3, descends to give a $\mathfrak{g}$-module structure and, moreover, as a $\mathfrak{g}$-module it is isomorphic with the Verma module $M(\lambda)$.*

Proof. As in 9.3.3, by the definition, writing $H^0_e(\mathcal{X}, \mathcal{L}(\lambda))^\vee = N$,

$$\begin{aligned} N &= \operatorname*{limit}_{w\in W} \left(H^0_e(X_w, \mathcal{L}_w(\lambda))^\vee\right) \\ &= \operatorname*{limit}_{w\in W} \left(H^0(X_w \cap \mathcal{B}^e, \mathcal{L}_w(\lambda))^\vee\right). \end{aligned} \tag{1}$$

Define a regular section σ of the line bundle $\mathcal{L}(\lambda)_{|\mathcal{B}^e}$ by

$$\sigma(g\mathcal{B})x = v^*_\lambda(g^{-1}x), \tag{2}$$

for $g \in \mathcal{U}^-$ and x in the line $i_\lambda(g\mathcal{B})$, where $i_\lambda : \mathcal{X} \to \mathbb{P}(L^{\max}(\lambda))$ is the morphism $g\mathcal{B} \mapsto [gv_\lambda]$, v_λ is a fixed highest weight vector in $L^{\max}(\lambda)$ and $v^*_\lambda : L^{\max}(\lambda) \to \mathbb{C}$ is the linear form which takes $v_\lambda \mapsto 1$ and takes any other weight vector to 0. (By Lemma 7.3.9, σ is indeed regular.) Clearly, σ never vanishes on $\mathcal{B}^e$. Hence

$$H^0(X_w \cap \mathcal{B}^e, \mathcal{L}_w(\lambda)) \simeq \mathbb{C}[X_w \cap \mathcal{B}^e] \cdot \sigma_{|X_w \cap \mathcal{B}^e}. \tag{3}$$

Further, it is easy to see that

$$\operatorname*{limit}_{w\in W} \left(\mathbb{C}[X_w \cap \mathcal{B}^e]^\vee\right) \simeq \operatorname*{limit}_{n\in\mathbb{N}} \left(\mathbb{C}[\mathcal{U}^-_n]^\vee\right), \tag{4}$$

where $\mathcal{U}^-_n$ is defined in (7.3.6.1).

It is easy to see that the T-character of σ under the canonical action of T on $H^0(\mathcal{B}^e, \mathcal{L}(\lambda))$ is given by

$$\operatorname{ch}\sigma = e^{-\lambda}. \tag{5}$$

In particular, by (1), (3)–(4), as T-modules,

$$N \simeq \mathbb{C}[\mathcal{U}^-]^\vee \otimes \sigma^*, \tag{6}$$

where $\mathbb{C}[\mathcal{U}^-]^\vee := \operatorname*{limit}_{n\in\mathbb{N}} \left(\mathbb{C}[\mathcal{U}^-_n]^\vee\right)$.

By Proposition 7.3.7, for any n there exists $k(n)$ such that for $k \geq k(n)$,

$$\mathbb{C}[\mathcal{U}^-_n]^\vee \hookrightarrow \mathbb{C}[\hat{\mathcal{U}}^{-(k)}]^\vee. \tag{7}$$

(Observe that T acts on $\hat{\mathcal{U}}^{-(k)}$ via the adjoint action and $\mathbb{C}[\hat{\mathcal{U}}^{-(k)}]$ is a weight module under the induced T-action and hence $\mathbb{C}[\hat{\mathcal{U}}^{-(k)}]^\vee$ makes sense.) Further, for any k, by Corollary 7.3.8, there exists $n(k)$ such that for all $n \geq n(k)$,

$$\mathbb{C}[\mathcal{U}^-_n]^\vee \twoheadrightarrow \mathbb{C}[\hat{\mathcal{U}}^{-(k)}]^\vee. \tag{8}$$

From (7)–(8), $\mathbb{C}[\mathcal{U}^-]^\vee$ is an admissible T-module and, moreover,

$$(9) \qquad \operatorname{ch} \mathbb{C}[\mathcal{U}^-]^\vee = \prod_{\alpha \in \Delta^+} (1 - e^{-\alpha})^{-\operatorname{mult} \alpha} \in \mathcal{A},$$

where $\mathcal{A}$ is as in 2.1.1. Hence, by (5)–(6) and (9), N is an admissible T-module with

$$(10) \qquad \operatorname{ch} N = e^\lambda \prod_{\alpha \in \Delta^+} (1 - e^{-\alpha})^{-\operatorname{mult} \alpha} = \operatorname{ch} M(\lambda), \text{ by Lemma 2.1.13.}$$

For any $k \geq 1$, the left regular representation ρ_ℓ of $\hat{\mathcal{U}}^{-(k)}$ on $\mathbb{C}[\hat{\mathcal{U}}^{-(k)}]$ induces a representation of its Lie algebra $\hat{\mathfrak{n}}^{-(k)} := \hat{\mathfrak{n}}^- / \hat{\mathfrak{n}}^-(k)$ (cf. 7.3.1) on the restricted dual $M^k := \mathbb{C}[\hat{\mathcal{U}}^{-(k)}]^\vee$. By [Jantzen–87, Part I, 7.10], there is a $U(\hat{\mathfrak{n}}^{-(k)})$-module isomorphism

$$(11) \qquad \theta : U\big(\hat{\mathfrak{n}}^{-(k)}\big) \underset{\sim}{\longrightarrow} M^k,$$

given by $X \mapsto X \cdot \delta_1$, where δ_1 is the delta function at 1 ($\delta_1 f := f(1)$). Moreover, θ commutes with the T-actions (since T acts trivially on δ_1).

The $\hat{\mathfrak{n}}^{-(k)}$-module structure on M^k of course gives rise to a $\mathfrak{n}^-$ (and hence $\tilde{\mathfrak{n}}^-$)-module structure (where $\tilde{\mathfrak{n}}^-$ is as in 1.1.2).

For any fixed $k \geq 1$, the morphisms $i(n,k) : \mathcal{U}_n^- \to \hat{\mathcal{U}}^{-(k)}$ (cf. Proposition 7.3.7) induce the T-module map

$$\psi^k : \mathbb{C}[\mathcal{U}^-]^\vee \to M^k.$$

It is easy to see that ψ^k is a $\tilde{\mathfrak{n}}^-$-module map, where the action of $\tilde{\mathfrak{n}}^-$ on the left side comes from its action on N via its identification with $\mathbb{C}[\mathcal{U}^-]^\vee$ as in (6). Since the $\tilde{\mathfrak{n}}^-$-action on M^k descends to an action of $\mathfrak{n}^-$, by (7), we get that the $\tilde{\mathfrak{n}}^-$-action on N descends to an action of $\mathfrak{n}^-$, i.e., the $\tilde{\mathfrak{g}}$-action on N descends to an action of $\mathfrak{g}$.

Finally, by (10), all the weights of N are $\leq \lambda$ and hence we have a nonzero $\mathfrak{g}$-module map $\beta : M(\lambda) \to N$. Composing β with $\theta^{-1}\psi^k$, we see that $\theta^{-1}\psi^k\beta(v_\lambda) = 1$ (from the weight considerations) for some nonzero highest weight vector $v_\lambda \in M(\lambda)$. Hence, for any $k \geq 1$, $\theta^{-1}\psi^k\beta$ is nothing but the unique $\mathfrak{n}^-$-module map

$$U(\mathfrak{n}^-) \to U\big(\hat{\mathfrak{n}}^{-(k)}\big), \quad \text{taking } 1 \mapsto 1.$$

This forces β to be injective. Since $\operatorname{ch} M(\lambda) = \operatorname{ch} N$ by (10), β is surjective as well. This completes the proof of the lemma. □

9.3.12 Lemma. *For any $v \in W$, $H_v^{\ell(v)}(\mathcal{X}, \mathcal{L}(\lambda))^\vee$ is an admissible T-module. Further,*

$$\mathrm{ch}\Big(H_v^{\ell(v)}(\mathcal{X}, \mathcal{L}(\lambda))^\vee\Big) = e^{v*\lambda} \prod_{\alpha\in\Delta^+} \big(1 - e^{-\alpha}\big)^{-\mathrm{mult}\,\alpha} = \mathrm{ch}\, M(v * \lambda).$$

Proof. By Lemma 9.3.9(b), for any $v \leq w$, as T-modules

(1) $$H_v^{\ell(v)}(X_w, \mathcal{L}_w(\lambda)) \simeq H_{\{e\}}^{\ell(v)}\big(\mathcal{U}_v, \mathcal{O}_{\mathcal{U}_v}\big) \otimes H^0\big(X_w \cap \mathcal{B}^v, \mathcal{L}_w(\lambda)|_{X_w\cap\mathcal{B}^v}\big).$$

Similar to the section σ (cf. (9.3.11.2)), define the regular section σ_v of the line bundle $\mathcal{L}(\lambda)_{|v\mathcal{B}^e}$ by

(2) $$\sigma_v(\bar{v}g\mathcal{B})\,x = v_\lambda^*(g^{-1}\bar{v}^{-1}x),$$

for $g \in \mathcal{U}^-$ and x in the line $i_\lambda(\bar{v}g\mathcal{B})$ (where $\bar{v} \in N$ is a fixed representative of v). Then

(3) $$\mathrm{ch}\,\sigma_v = e^{-v\lambda},$$

and σ_v never vanishes on $v\mathcal{B}^e$. Hence

(4) $$H^0\Big(X_w \cap \mathcal{B}^v, \mathcal{L}_w(\lambda)_{|X_w\cap\mathcal{B}^v}\Big) \simeq \mathbb{C}[X_w \cap \mathcal{B}^v] \cdot \sigma_{v|X_w\cap\mathcal{B}^v}.$$

In particular, from (1), (3) and (4) (since tensor product commutes with direct limits, cf. [Godement–58, 1.6.]), as T-modules

(5) $$H_v^{\ell(v)}(\mathcal{X}, \mathcal{L}(\lambda))^\vee \simeq H_{\{e\}}^{\ell(v)}\big(\mathcal{U}_v, \mathcal{O}_{\mathcal{U}_v}\big)^\vee \otimes \mathbb{C}[\mathcal{B}^v]^\vee \otimes \sigma_v^*,$$

where

$$\mathbb{C}[\mathcal{B}^v]^\vee := \underset{w\in W}{\mathrm{limit}}\ \big(\mathbb{C}[X_w \cap \mathcal{B}^v]^\vee\big).$$

(Observe that to deduce (5), we have used the fact that $H_{\{e\}}^{\ell(v)}\big(\mathcal{U}_v, \mathcal{O}_{\mathcal{U}_v}\big)^\vee$ and $\mathbb{C}[\mathcal{B}^v]^\vee$ both are admissible T-modules with characters $\in \mathcal{A}$ to be established in (8) and (12).)

On the other hand, by Lemma 7.3.10, there is a T-equivariant isomorphism

$$(v^{-1}\mathcal{U}_v v) \times v^{-1}(X_w \cap \mathcal{B}^v) \approx (v^{-1}X_w) \cap \mathcal{B}^e,$$

and hence, as T-modules,

(6) $$\mathbb{C}[v^{-1}\mathcal{U}_v v] \otimes \mathbb{C}[v^{-1}(X_w \cap \mathcal{B}^v)] \simeq \mathbb{C}[(v^{-1}X_w) \cap \mathcal{B}^e].$$

Taking limits, we get (as T-modules)

$$\mathbb{C}[v^{-1}\mathcal{U}_v v]^\vee \otimes \mathbb{C}[v^{-1}\mathcal{B}^v]^\vee \simeq \mathbb{C}[\mathcal{U}^-]^\vee, \tag{7}$$

where

$$\mathbb{C}[v^{-1}\mathcal{B}^v]^\vee := \underset{w\in W}{\text{limit}} \left(\mathbb{C}[v^{-1}(X_w \cap \mathcal{B}^v)]^\vee\right),$$

and (as in the proof of Lemma 9.3.11) $\mathbb{C}[\mathcal{U}^-]^\vee := \underset{n\in\mathbb{N}}{\text{limit}} \left(\mathbb{C}[\mathcal{U}_n^-]^\vee\right)$. Again, to deduce (7), we have used that $\mathbb{C}[v^{-1}\mathcal{U}_v v]^\vee$ and $\mathbb{C}[v^{-1}\mathcal{B}^v]^\vee$ are admissible T-modules with characters $\in \mathcal{A}$ (by (10)–(11)). (It is easy to see that $\underset{w\in W}{\text{limit}}\ \mathbb{C}[(v^{-1}X_w) \cap \mathcal{B}^e]^\vee \simeq \mathbb{C}[\mathcal{U}^-]^\vee$, as T-modules.)

By Lemma B.11, $\operatorname{ch} H_e^{\ell(v)}(\mathcal{U}_v, \mathcal{O}_{\mathcal{U}_v})^\vee \in \mathcal{A}$ and, moreover, as elements of $\mathcal{A}$,

$$\operatorname{ch} H_e^{\ell(v)}(\mathcal{U}_v, \mathcal{O}_{\mathcal{U}_v})^\vee = \prod_{\alpha\in\Delta^+\cap v\Delta^-} e^{-\alpha}(1-e^{-\alpha})^{-1}. \tag{8}$$

Also, by (9.3.11.9),

$$\operatorname{ch} \mathbb{C}[\mathcal{U}^-]^\vee = \prod_{\alpha\in\Delta^+} (1-e^{-\alpha})^{-\operatorname{mult}\alpha} \in \mathcal{A}. \tag{9}$$

Further,

$$\operatorname{ch} \mathbb{C}[v^{-1}\mathcal{U}_v v]^\vee = \prod_{\alpha\in\Delta^+\cap v^{-1}\Delta^-} (1-e^{-\alpha})^{-1}. \tag{10}$$

In particular, from (7), (9)–(10), and Exercise 9.3.E.4,

$$\operatorname{ch} \mathbb{C}[v^{-1}\mathcal{B}^v]^\vee = \prod_{\alpha\in\Delta^+\cap v^{-1}\Delta^+} (1-e^{-\alpha})^{-\operatorname{mult}\alpha}. \tag{11}$$

This gives

$$\operatorname{ch} \mathbb{C}[\mathcal{B}^v]^\vee = \prod_{\alpha\in\Delta^+\cap v\Delta^+} (1-e^{-\alpha})^{-\operatorname{mult}\alpha} \in \mathcal{A}. \tag{12}$$

Combining (3), (5), (8) and (12), we get that $H_v^{\ell(v)}(\mathcal{X}, \mathcal{L}(\lambda))^\vee$ is an admissible T-module and

$$\begin{aligned}
\operatorname{ch} H_v^{\ell(v)}(\mathcal{X}, \mathcal{L}(\lambda))^\vee &= e^{v\lambda} \cdot e^{-\sum_{\alpha\in\Delta^+\cap v\Delta^-}\alpha} \cdot \prod_{\alpha\in\Delta^+} (1-e^{-\alpha})^{-\operatorname{mult}\alpha} \in \mathcal{A} \\
&= e^{v*\lambda} \prod_{\alpha\in\Delta^+} (1-e^{-\alpha})^{-\operatorname{mult}\alpha}, \quad \text{by (1.3.22.3)} \\
&= \operatorname{ch} M(v*\lambda), \qquad \text{by Lemma 2.1.13.}
\end{aligned}$$

This proves the lemma. $\square$

9.3.13 *Proof of Proposition 9.3.7.* We prove, by induction on $\ell(v)$, that the $\tilde{\mathfrak{g}}$-module structure on $H_v^{\ell(v)}(\mathcal{X}, \mathcal{L}(\lambda))^\vee$ descends to a $\mathfrak{g}$-module structure and, moreover, as a $\mathfrak{g}$-module it is isomorphic with the Verma module $M(v * \lambda)$.

The case $\ell(v) = 0$ (i.e., $v = e$) is the content of Lemma 9.3.11. So, take $\ell(v) > 0$ and write $v = v's_i$, for some simple reflection s_i such that $v > v'$. We first assert that the $\tilde{\mathfrak{g}}$-module map

$$\delta^{v,v'} : H_v^{p+1}(\mathcal{X}, \mathcal{L}(\lambda))^\vee \to H_{v'}^p(\mathcal{X}, \mathcal{L}(\lambda))^\vee \qquad \text{(defined in 9.3.6)}$$

is injective, where $p := \ell(v')$. In fact, we prove that the map $\delta^{v,v'}(w) : H_v^{p+1}(X_w, \mathcal{L}_w(\lambda))^\vee \to H_{v'}^p(X_w, \mathcal{L}_w(\lambda))^\vee$ (cf. 9.3.6) is injective, for any $w \in W$ such that $ws_i < w$. Consider the triple $X_w \cap X^{v'} \supset X_w \cap (\partial X^{v'}) \supset X_w \cap \big((\partial X^{v'})\backslash \mathcal{B}^v\big)$ of closed subspaces of X_w (call these subspaces $Y_1 \supset Y_2 \supset Y_3$, respectively). This gives rise to the long exact sequence as in (B.1.15):

$$\cdots \longrightarrow H_{v'}^p(X_w, \mathcal{L}_w(\lambda)) \xrightarrow{\delta} H_{Y_2/Y_3}^{p+1}(X_w, \mathcal{L}_w(\lambda))$$
$$\longrightarrow H_{Y_1/Y_3}^{p+1}(X_w, \mathcal{L}_w(\lambda)) \longrightarrow \cdots.$$

But, by Lemma 9.3.10, $H_{Y_1/Y_3}^{p+1}(X_w, \mathcal{L}_w(\lambda)) = 0$. Further, by Lemmas B.3 and B.4, $H_{Y_2/Y_3}^{p+1}(X_w, \mathcal{L}_w(\lambda))$ can canonically be identified with $H_v^{p+1}(X_w, \mathcal{L}_w(\lambda))$ and, further, the dual map $\delta^\vee : H_{Y_2/Y_3}^{p+1}(X_w, \mathcal{L}_w(\lambda))^\vee \to H_{v'}^p(X_w, \mathcal{L}_w(\lambda))^\vee$ can easily be seen to be the same map as $\delta^{v,v'}(w)$. This establishes the assertion that $\delta^{v,v'}(w)$ is injective.

This; in particular, implies (by the induction hypothesis) that the $\tilde{\mathfrak{g}}$-module structure on $H_v^{\ell(v)}(\mathcal{X}, \mathcal{L}(\lambda))^\vee$ descends to a $\mathfrak{g}$-module structure and, moreover, by Lemmas 9.3.8 and 9.3.12, it is isomorphic (as a $\mathfrak{g}$-module) with the Verma module $M(v * \lambda)$. $\square$

Finally, we come to the main result of this section. This generalizes Theorem 9.1.3 for $Y = \emptyset$ to an arbitrary (not necessarily symmetrizable) Kac–Moody algebra.

9.3.14 Theorem. *Let $\mathfrak{g}$ be an arbitrary Kac–Moody Lie algebra and let $\lambda \in D_{\mathbb{Z}}$. Then there is an exact sequence of $\mathfrak{g}$-module maps:*

$$(\mathcal{C}) \qquad \cdots \to F_p \xrightarrow{\delta^p} \cdots \to F_1 \xrightarrow{\delta^1} F_0 \xrightarrow{\delta^0} L^{\max}(\lambda) \to 0,$$

obtained from the Kempf complex $\mathcal{K}$ (cf. 9.3.3) via the identification (9.3.5.1) together with Corollary 8.3.12(a) and Proposition 9.3.7, where $F_p := \bigoplus_{v \in W(p)} M(v * \lambda)$.

Moreover, let $\mathcal{A}$ be any chain complex of the following form consisting of $\mathfrak{g}$-module maps:

$$(\mathcal{A}) \qquad \cdots \longrightarrow F_p \xrightarrow{b^p} \cdots \to F_1 \xrightarrow{b^1} F_0 \xrightarrow{b^0} L^{\max}(\lambda) \longrightarrow 0,$$

such that $b^p_{|M(v*\lambda)} \neq 0$ (for any $v \in W(p)$, $p \geq 0$). Then, there exists a $\mathfrak{g}$-module isomorphism $\varphi : \mathcal{C} \to \mathcal{A}$ of the chain complexes. In particular, any such $\mathcal{A}$ is exact.

Taking $\mathcal{A}$ to be the (combinatorially defined) BGG complex $\mathcal{S}$ for $Y = \emptyset$ (cf. Definition 9.2.17), we get that $\mathcal{S}$ is exact and is $\mathfrak{g}$-isomorphic with the (geometrically defined) complex $\mathcal{C}$.

Proof. The first part follows from Corollaries 9.3.4 and 8.3.12(a) together with (9.3.5.1) and Proposition 9.3.7.

The second part follows from Theorem 9.2.18 and Lemma 9.2.19. (Observe that the assertion $\delta^p_{|M(v*\lambda)} \neq 0$, for any $v \in W(p)$, $p \geq 1$, has also been geometrically established in 9.3.13.) □

9.3.15 Remark. The above theorem can easily be extended to any $\lambda \in D$ (from $\lambda \in D_{\mathbb{Z}}$) by taking $\lambda' \in D_{\mathbb{Z}}$ such that $\lambda(\alpha_i^\vee) = \lambda'(\alpha_i^\vee)$ for all the simple coroots $\alpha_i^\vee$ and using Lemma 8.3.2.

As an immediate consequence of Theorem 9.3.14 (and Remark 9.3.15), we obtain the following generalization of the Weyl–Kac character formula for an arbitrary Kac–Moody algebra. Recall that this generalization was proved by a different method in Section 8.3.

9.3.16 Corollary. *For an arbitrary $\mathfrak{g}$ and $\lambda \in D$,*

$$\operatorname{ch} L^{\max}(\lambda) = \Big(\prod_{\alpha\in\Delta^+} (1 - e^{-\alpha})^{-\operatorname{mult}\alpha}\Big)\Big(\sum_{w\in W} \epsilon(w)\, e^{w*\lambda}\Big)$$

in the ring $\mathcal{A}$. In particular, taking $\lambda = 0$,

$$\prod_{\alpha\in\Delta^+} (1 - e^{-\alpha})^{\operatorname{mult}\alpha} = \sum_{w\in W} \epsilon(w)\, e^{w\rho-\rho}. \qquad \square$$

As an easy consequence of Theorem 9.3.14, we obtain the following generalization of Theorem 3.2.7 for an arbitrary Kac–Moody algebra $\mathfrak{g}$ (but $Y = \emptyset$). Since $\mathfrak{n}^-$ is an ideal in $\mathfrak{b}^-$, for any $\mathfrak{b}^-$-module (in particular, for a $\mathfrak{g}$-module) V, $H_*(\mathfrak{n}^-, V)$ has a natural $\mathfrak{h}$-module structure (cf. 3.1.1).

9.3.17 Theorem. *Let $\mathfrak{g}$ be an arbitrary Kac–Moody algebra and let $\lambda \in D$. Then, for any $p \geq 0$, as $\mathfrak{h}$-modules,*

$$H_p(\mathfrak{n}^-, L^{\max}(\lambda)) \simeq \bigoplus_{v \in W(p)} \mathbb{C}_{v*\lambda},$$

where $\mathbb{C}_\mu$ is the one-dimensional $\mathfrak{h}$-module with weight μ.

Proof. From the definition of the complex (3.1.1.1), which computes the Lie algebra homology, it is easy to see that $H_p(\mathfrak{n}^-, L^{\max}(\lambda))$ is a weight module. So, it suffices to show that, for any $\mu \in \mathfrak{h}^*$, the μ-weight space $H_p(\mathfrak{n}^-, L^{\max}(\lambda))_\mu = 0$ unless $\mu = v * \lambda$ for some $v \in W(p)$ and in this case it is one-dimensional.

From the definition of the Lie algebra homology of a pair (cf. 3.1.3), it is easy to see that, for any $\mathfrak{b}^-$-module V which is a weight module under the $\mathfrak{h}$-action, the Lie algebra homology

$$H_*(\mathfrak{b}^-, \mathfrak{h}, V) \simeq H_*(\mathfrak{n}^-, V)_0. \tag{1}$$

For any $\mu \in \mathfrak{h}^*$, consider $\mathbb{C}_\mu$ as a $\mathfrak{b}^-$-module where $\mathfrak{n}^-$ acts trivially (and $\mathfrak{h}$ acts by weight μ). Then, by (1),

$$H_*(\mathfrak{b}^-, \mathfrak{h}, L^{\max}(\lambda) \otimes \mathbb{C}_\mu) \simeq H_*(\mathfrak{n}^-, L^{\max}(\lambda))_{-\mu}. \tag{2}$$

The resolution $\mathcal{C}$ of Theorem 9.3.14 gives rise to the resolution (for any $\mu \in \mathfrak{h}^*$):

$$\cdots \to F_p \otimes \mathbb{C}_\mu \to \cdots \to F_0 \otimes \mathbb{C}_\mu \to L^{\max}(\lambda) \otimes \mathbb{C}_\mu \to 0.$$

By Proposition 3.1.10, as $\mathfrak{b}^-$-modules,

$$\begin{aligned} F_p \otimes \mathbb{C}_\mu &= \bigoplus_{v \in W(p)} (M(v * \lambda) \otimes \mathbb{C}_\mu) \\ &\simeq \bigoplus_{v \in W(p)} U(\mathfrak{b}^-) \otimes_{U(\mathfrak{h})} \mathbb{C}_{v*\lambda+\mu}. \end{aligned} \tag{3}$$

In particular, $F_p \otimes \mathbb{C}_\mu$ is $(U(\mathfrak{b}^-), U(\mathfrak{h}))$-projective by Lemma 3.1.7, and $\operatorname{Tor}_*^{(U(\mathfrak{b}^-),U(\mathfrak{h}))}\left(\mathbb{C}_0, L^{\max}(\lambda) \otimes \mathbb{C}_\mu\right)$ is the homology of the complex (since each $F_p \otimes \mathbb{C}_\mu$ is a finitely semisimple $\mathfrak{h}$-module):

$$\cdots \to \mathbb{C}_0 \otimes_{U(\mathfrak{b}^-)} (F_p \otimes \mathbb{C}_\mu) \to \cdots \to \mathbb{C}_0 \otimes_{U(\mathfrak{b}^-)} (F_0 \otimes \mathbb{C}_\mu) \to 0. \tag{$*$}$$

But, by (3) and (3.1.7.2),

$$\mathbb{C}_0 \otimes_{U(\mathfrak{b}^-)} (F_p \otimes \mathbb{C}_\mu) \simeq \bigoplus_{v \in W(p)} \mathbb{C}_0 \otimes_{U(\mathfrak{h})} \mathbb{C}_{v*\lambda+\mu}\,. \tag{4}$$

Now, for any $\theta \in \mathfrak{h}^*$,

(5) $$\mathbb{C}_0 \otimes_{U(\mathfrak{h})} \mathbb{C}_\theta \simeq \mathbb{C}, \quad \text{if } \theta = 0$$
$$= 0, \quad \text{otherwise.}$$

Hence, if $\mu = -w * \lambda$ for some $w \in W(p)$, then the complex $(*)$ is one-dimensional in the p-th component and 0 elsewhere. Moreover, if $-\mu \notin W * \lambda$, the complex $(*)$ is identically 0. So the same is true of its homology. But, by (3.1.9.1),

$$\mathrm{Tor}_*^{(U(\mathfrak{b}^-),U(\mathfrak{h}))}\big(\mathbb{C}_0, L^{\max}(\lambda) \otimes \mathbb{C}_\mu\big) \simeq H_*\big(\mathfrak{b}^-, \mathfrak{h}, L^{\max}(\lambda) \otimes \mathbb{C}_\mu\big).$$

So the theorem follows from (2). □

We derive the following corollary of the above theorem. This generalizes Corollary 3.3.10(b_3) from symmetrizable to an arbitrary $\mathfrak{g}$.

9.3.18 Corollary. *Let $\mathfrak{g}$ be an arbitrary Kac–Moody algebra. Then, for $\lambda \neq \mu \in D$,*

$$\mathrm{Ext}^*_{(\mathfrak{g},\mathfrak{h})}\big(L^{\max}(\lambda), L^{\max}(\mu)^\sigma\big) = 0,$$

where M^σ is as in 2.1.1.

(By Exercise 9.3.E.2, $\mathrm{Ext}^*_{\mathfrak{g}}\big(L^{\max}(\lambda), L^{\max}(\mu)^\sigma\big) = 0$ as well.)

Proof. For any $\theta \in \mathfrak{h}^*$,

$$\mathrm{Ext}^p_{(\mathfrak{g},\mathfrak{h})}\big(M(\theta), L^{\max}(\mu)^\sigma\big) \simeq \big[H_p\big(\mathfrak{n}^-, L^{\max}(\mu)\big)_\theta\big]^*, \quad \text{by Lemma 9.1.8}$$

(1) $$\simeq \begin{cases} 0, & \text{if } \theta \notin W(p) * \mu \\ \mathbb{C}, & \text{if } \theta = w * \mu \text{ for some } w \in W(p), \end{cases}$$

where the last isomorphism of course follows from Theorem 9.3.17. (Although Lemma 9.1.8 was proved under the assumption that $\mathfrak{g}$ is symmetrizable, the same proof works to give the first isomorphism.)

In particular, for any $w \in W$,

$$\mathrm{Ext}^*_{(\mathfrak{g},\mathfrak{h})}\big(M(w * \lambda), L^{\max}(\mu)^\sigma\big) = 0,$$

since $w * \lambda \notin W * \mu$ by Exercise 1.4.E.2. This gives the vanishing

(2) $$\mathrm{Ext}^*_{(\mathfrak{g},\mathfrak{h})}\big(F_p, L^{\max}(\mu)^\sigma\big) = 0,$$

for all the terms F_p $(p \geq 0)$ of the resolution $\mathcal{C}$ of Theorem 9.3.14:

$$\cdots \to F_p \xrightarrow{\delta^p} \cdots \xrightarrow{\delta^1} F_0 \xrightarrow{\delta^0} L^{\max}(\lambda) \to 0.$$

Let $I_p := \operatorname{Im} \delta^{p+1} \subset F_p$ for $p \geq 0$ and set $I_{-1} = L^{\max}(\lambda)$. Then, for any $p \geq 0$, we get the short exact sequence:

$$0 \to I_p \longrightarrow F_p \xrightarrow{\delta^p} I_{p-1} \to 0.$$

Considering the corresponding long exact sequence for the functor $\operatorname{Ext}_{(\mathfrak{g},\mathfrak{h})}(-, L^{\max}(\mu)^\sigma)$ (cf. (D.12.4)) and using (2), we get, for all $p \geq 0$ and $i \in \mathbb{Z}$,

$$\text{(3)} \qquad \operatorname{Ext}^{i-1}_{(\mathfrak{g},\mathfrak{h})}\big(I_p, L^{\max}(\mu)^\sigma\big) \simeq \operatorname{Ext}^{i}_{(\mathfrak{g},\mathfrak{h})}\big(I_{p-1}, L^{\max}(\mu)^\sigma\big).$$

Hence, for all $p \geq 0$,

$$\operatorname{Ext}^{i}_{(\mathfrak{g},\mathfrak{h})}\big(L^{\max}(\lambda), L^{\max}(\mu)^\sigma\big) \simeq \operatorname{Ext}^{i-p}_{(\mathfrak{g},\mathfrak{h})}\big(I_{p-1}, L^{\max}(\mu)^\sigma\big).$$

Taking $p > i$ in the above, we get the corollary. □

9.3.19 Open Problem. Generalize the complex $\mathcal{C}$ obtained from the Kempf complex $\mathcal{K}$, as in Theorem 9.3.14, to cover the case of any subset $Y \subset \{1, \dots, \ell\}$ (and any Kac–Moody Lie algebra $\mathfrak{g}$). Recall that such a complex is obtained in the symmetrizable case by algebraic methods (cf. Theorem 9.1.3 and Remark 9.1.10).

9.3.E EXERCISES

(1) For any $w \in W$ and $1 \leq i \leq \ell$, show that the set theoretic fibers of the morphism $\pi_i^w : X_w \to X_w^i$ are $\mathbb{P}^1$ or single points (where π_i^w is the restriction of the morphism $\pi_i : \mathcal{G}/\mathcal{B} \to \mathcal{G}/\mathcal{P}_i$, cf. 7.1.19, to X_w and $X_w^i := \pi_i(X_w)$).

If $ws_i < w$, then prove that π_i^w is a locally trivial $\mathbb{P}^1$-fibration (in the Zariski topology). In particular, it is a smooth morphism.

(2) With the notation and assumptions as in Corollary 9.3.18, prove that $\operatorname{Ext}^*_{\mathfrak{g}}\big(L^{\max}(\lambda), L^{\max}(\mu)^\sigma\big) = 0$.
Hint. Use Lemma 3.3.5, Remark 3.3.6 and Theorem E.13.

(3) For any Kac–Moody algebra $\mathfrak{g}$ and $\lambda \in \mathfrak{h}^*$, $H_p(\mathfrak{g}, \mathfrak{h}, M(\lambda)) = 0$ unless $\lambda = w * 0$ for some $w \in W(p)$ and in this case it is one-dimensional.
Hint. Use (3.1.14.1) and (9.3.17.1) together with Theorem 9.3.17.

Use this to calculate $H_*(\mathfrak{g}, M(\lambda))$.

(4) Prove that the ring $\mathcal{A}$ defined in 2.1.1 is an integral domain.

(5) Use Theorem 9.3.17 to generalize it for an arbitrary subset $Y \subset \{1, \dots, \ell\}$.

9.C Comments. As is well known, a resolution of $L(\lambda)$ as in Theorem 9.1.3 for $Y = \emptyset$ was first obtained by [Bernstein–Gelfand–Gelfand–75] in the finite case, by making crucial use of the center of $U(\mathfrak{g})$. This was generalized by [Lepowsky–77] to cover all Y (in the finite case). Again, in the finite case, [Kempf–78] realized such a resolution (for $Y = \emptyset$) as the dual of a particular Grothendieck–Cousin complex, which we refer to as the Kempf complex. (Actually, he did not explicitly identify the terms F_p of his complex as the sum of Verma modules; this was done by [Brylinski–81].)

[Garland–Lepowsky–76] proved the existence of the BGG resolution for symmetrizable Kac–Moody algebras (for Y of finite type). But they proved only a weaker version, in which the modules F_p (a priori) only admitted filtrations such that the set of successive quotients coincides with the Y-generalized Verma modules $\{M_Y(v^{-1} * \lambda)\}_{v \in W'_Y, \ell(w)=p}$. It was later shown by [Rocha–Caridi–Wallach–82], by proving an appropriate Ext vanishing result (a certain variant of Proposition 9.1.7, cf. [loc. cit.] and [Rocha–Caridi–80]), to be actually the direct sum of these generalized Verma modules. The proof of Garland–Lepowsky followed in spirit the work of Bernstein et. al. cited above. They (Garland–Lepowsky) also crucially used the center of $U(\mathfrak{g})$ (rather its one "special" element — the Casimir operator, which of course exists only in the symmetrizable case).

The BGG complex (cf. 9.2.17) for $Y = \emptyset$ was defined by [Bernstein–Gelfand–Gelfand–75] in the finite case; its extension to the general case is straightforward once we have the basic Theorem 9.2.3. Theorem 9.2.3 in the finite case is proved in [Verma–68] and [Bernstein–Gelfand–Gelfand–71]; in the symmetrizable case in [Rocha–Caridi–Wallach–82]; and for the general Kac–Moody case in [Neidhardt–87].

The treatment of Enright's completion functor (cf. Definition 9.2.7, Lemma 9.2.8 and Theorem 9.2.9) is taken from [Enright–79, §3] (though the proof of Theorem 9.2.9(c) is somewhat different). As in [Neidhardt–87] this functor is crucially used in the proof of Theorem 9.2.3. Lemma 9.2.14 and Definition 9.2.15 are due to [Lepowsky–77]. Theorem 9.2.18 and Lemma 9.2.19 are due to [Rocha–Caridi–80] and [Rocha–Caridi–Wallach–82].

The results in Section 9.3 are due to [Kumar–90], including the extension of the Kempf resolution to the general Kac–Moody case Theorem 9.3.14, and its consequences: Theorem 9.3.17 and Corollary 9.3.18.

X

Defining Equations of $\mathcal{G}/\mathcal{P}$ and Conjugacy Theorems

Fix a subset $Y \subset \{1, \dots, \ell\}$ and a Y-regular weight $\Lambda \in D_{\mathbb{Z}}$, i.e., Λ is dominant totally integral and $\Lambda(\alpha_i^\vee) = 0$ iff $i \in Y$. The orbit through the highest weight vector gives rise to an embedding $i_\Lambda : \mathcal{G}/\mathcal{P}_Y \hookrightarrow \mathbb{P}(L(\Lambda))$. In Section 10.1, we are concerned with two problems: the first, to determine the precise image of i_Λ, and the second, to find defining equations of the image. Both of these questions were answered by Kostant in the finite case by beautifully using "convex geometry." His results remain valid in the symmetrizable Kac–Moody case (proved by Kac–Peterson), which we now describe. For any $0 \neq v \in L(\Lambda)$, the line $[v]$ through v lies in the image of i_Λ iff v belongs to the Kostant cone $K(\Lambda) := \{v \in L(\Lambda) : v \otimes v \in L(2\Lambda)\}$. Moreover, taking a basis $\{x_i\}$ of $\mathfrak{g}$ (consisting of root vectors, apart from the elements of $\mathfrak{h}$), $v \in K(\Lambda)$ iff $\langle \Lambda, \Lambda \rangle v \otimes v = \sum_i x_i v \otimes y_i v$, where $\{y_i\}$ is the dual basis of $\mathfrak{g}$ with respect to a nondegenerate invariant form $\langle \cdot, \cdot \rangle$ on $\mathfrak{g}$. This follows by considering the action of the Casimir–Kac operator on $L(\Lambda) \otimes L(\Lambda)$. From this, a set of defining equations for the image of i_Λ can be read off. Specifically, the image of i_Λ is defined by the collection of the quadratic equations $\{P_{\phi,\phi'}\}_{\phi,\phi' \in \Gamma}$, where

$$P_{\phi,\phi'} := \langle \Lambda, \Lambda \rangle v_\phi^* v_{\phi'}^* - \sum_i (x_i(v_\phi^*))(y_i(v_{\phi'}^*)) \in S^2(L(\Lambda)^*),$$

$\{v_\phi\}_{\phi \in \Gamma}$ is a basis of $L(\Lambda)$ consisting of weight vectors and $\{v_\phi^*\}$ is the dual basis of the restricted dual $L(\Lambda)^\vee$. These equations generalize the classical Plücker relations for the embedding of the Grassmannian of k-planes in $\mathbb{C}^n$ inside $\mathbb{P}(\Lambda^k \mathbb{C}^n)$. In Section 10.1 we give a complete proof of these results.

In Section 10.2, we use the above results to characterize certain special subalgebras of a symmetrizable Kac–Moody algebra $\mathfrak{g}$ and also prove some conjugacy theorems. Let (V, π) be a representation of $\mathfrak{g}$. A Lie subalgebra $\mathfrak{s} \subset \mathfrak{g}$ is called π-triangular if there exists a $\mathfrak{s}$-stable flag $V_0 = 0 \subset V_1 \subset V_2 \subset \cdots$ such that $\cup V_i = V$ and V_{i+1}/V_i is one-dimensional for each $i \geq 0$. Similarly, $\mathfrak{s}$ is called

π-diagonalizable, resp. π-semisimple, if V is a direct sum of one-dimensional, resp. finite-dimensional, irreducible $\mathfrak{s}$-submodules. For symmetrizable $\mathfrak{g}$, we characterize subalgebras $\mathfrak{s}$ which are $\mathrm{ad}_{\mathfrak{g}}$-triangular (cf. Theorem 10.2.5). In particular, we show that such subalgebras are characterized by the property that there exist $g \in \mathcal{G}^{\min}$ and $w \in W$ such that $(\mathrm{Ad}\, g)\mathfrak{s} \subset \mathfrak{b} \cap w\mathfrak{b}^-$. Moreover, such an $\mathfrak{s}$ (more generally, any $\mathrm{ad}_{\mathfrak{g}}$-locally finite subalgebra) lies in $\mathfrak{g}_{\mathrm{fin}}$ and $\mathrm{Exp}\, \mathfrak{s} \subset \mathcal{G}^{\min}$. As a corollary of the above characterization, we show that $\mathrm{ad}_{\mathfrak{g}}$-diagonalizable subalgebras $\mathfrak{s}$ of $\mathfrak{g}$ are precisely those satisfying $(\mathrm{Ad}\, g)\mathfrak{s} \subset \mathfrak{h}$, for some $g \in \mathcal{G}^{\min}$.

As another corollary, we show the existence of the Jordan–Chevalley decomposition of any $\mathrm{ad}_{\mathfrak{g}}$-locally finite element x of $\mathfrak{g}$ as $x = x_s + x_n$, with $[x_s, x_n] = 0$ and x_s, resp.x_n, is $\mathrm{ad}_{\mathfrak{g}}$-diagonalizable, resp. $\mathrm{ad}_{\mathfrak{g}}$-locally nilpotent (cf. Theorem 10.2.12).

These results culminate in the result that the GCM A itself is an invariant of the underlying Kac–Moody Lie algebra $\mathfrak{g}(A)$. More specifically, for two symmetrizable GCM's A and B with $\mathfrak{g}(A) \simeq \mathfrak{g}(B)$ (as Lie algebras), we have $A = B$ (up to a permutation of the indexing set $\{1, \dots, \ell\}$). In addition, we determine the group $\mathrm{Aut}(\mathfrak{g}(A))$ of the Lie algebra automorphisms of $\mathfrak{g}(A)$ and show that, for A indecomposable, any $\theta \in \mathrm{Aut}(\mathfrak{g}(A))$ can be written as $\theta = \gamma\omega^{\epsilon}\, \mathrm{Ad}\, g$, for some $\epsilon \in \{0, 1\}$, $g \in \mathcal{G}^{\min}$ and "diagram automorphism" γ, where ω is the Cartan involution of $\mathfrak{g}$.

In Section 10.3, we prove the group analogues of some of the results of Section 10.2 by using the Hilbert–Mumford theory of unstable points (although indirectly). We do not require the symmetrizability assumption on $\mathfrak{g}$ anymore. We first construct a strictly convex W-invariant real analytic function on the interior C^o of the Tits cone. This function is used to construct a function $\Phi_v : \mathcal{G} \to C^o$ for any nonzero $v \in V$, where V is a highest weight $\mathfrak{g}$-module $V(\lambda)$ with $\lambda \in C^o \cap D_{\mathbb{Z}}$.

The main properties of the function Φ_v are summarized in Proposition 10.3.8. In particular, we prove that Φ_v is equivariant with respect to the left multiplication of N on $\mathcal{G}$ and the Weyl group action on C^o and, moreover, there exists $g_o \in \mathcal{G}$ such that $F(\Phi_v(g_o)) \geq F(\Phi_v(g))$ for all $g \in \mathcal{G}$. This proposition is used to characterize subgroups S of a parabolic subgroup of finite type. It is shown that such subgroups S are precisely those which are contained in a finite union of double cosets $\mathcal{B}w\mathcal{B}$ (cf. Theorem 10.3.9). We also characterize those subgroups $S \subset \mathcal{G}$ such that the adjoint action of S on $\mathfrak{g}$ is locally finite. In particular, we show that such subgroups S are precisely those of the form $gSg^{-1} \subset \mathcal{P}_Y \cap w\mathcal{P}^-_{Y'}w^{-1}$, for some $g \in \mathcal{G}^{\min}$, $w \in W$ and finite type subsets $Y, Y' \subset \{1, \dots, \ell\}$, where $\mathcal{P}^-_{Y'}$ is the standard negative parabolic subgroup (cf. Theorem 10.3.12). As a corollary, we get a characterization of $\mathrm{ad}_{\mathfrak{g}}$-locally finite subalgebras of symmetrizable $\mathfrak{g}$ as subalgebras $\mathfrak{s}$ of the form $(\mathrm{Ad}\, g)\, \mathfrak{s} \subset \mathfrak{p}_Y \cap (\mathrm{Ad}\, n)\, \mathfrak{p}^-_{Y'}$, for some $g \in \mathcal{G}^{\min}$, $n \in N$ and finite type subsets $Y, Y' \subset \{1, \dots, \ell\}$.

We also study subgroups $S \subset \mathcal{G}^{\min}$ which are reductive in $\mathcal{G}^{\min}$ (and similarly subalgebras $\mathfrak{s} \subset \mathfrak{g}$ reductive in $\mathfrak{g}$) in one of the exercises.

The results of Sections 10.2 and 10.3 are due to Kac–Peterson.

10.1. Quadratic Generation of Defining Ideals of $\mathcal{G}/\mathcal{P}$ in Projective Embeddings

In this section $\mathfrak{g}$ is a symmetrizable Kac–Moody Lie algebra.

We recall some of the basic definitions and elementary facts from the geometry of polyhedra.

10.1.1 Definition. Let V be a finite-dimensional real vector space and let $S = \{v_1, \dots, v_n\} \subset V$ be a nonempty finite subset. Then the convex hull $[S]$ of S, i.e., the smallest convex subset of V containing S, is the (compact) polyhedron given by

$$[S] = \Big\{\sum_i t_i v_i : \text{each } t_i \geq 0 \text{ and } \sum t_i = 1\Big\}.$$

For an affine hyperplane $H \subset V$, let H^+ and H^- be the two open half spaces separated by H. A point $v \in [S]$ is called a *vertex* of $[S]$ if there exists an affine hyperplane $H \subset V$ through v such that $[S]\backslash\{v\}$ lies entirely in one of $H^{\pm}$.

A pair of distinct vertices $\{v, v'\}$ of $[S]$ (or the closed line segment $[v, v']$ joining v, v') is called an *edge* of $[S]$ if there exists an affine hyperplane H containing $[v, v']$ such that $[S]\backslash[v, v']$ lies entirely in one of $H^{\pm}$.

The following lemma is easy to establish and is left as an exercise. A proof is given, e.g., in [Moody–Pianzola–95, §7.3, Lemma 4].

10.1.2 Lemma. *Let the notation be as above. In addition, let $X \subset V$ be a countable subset. Then we have the following:*

(a) Any vertex $v \in [S]$ actually lies in S itself.

(b) For any vertex $v \in [S]$ such that $v \in X$, there exists an affine hyperplane H containing v such that $[S]\backslash\{v\}$ lies entirely in one of $H^{\pm}$ and $H \cap X = \{v\}$.

10.1.3 Definition. For $\Lambda \in D$, let $L(\Lambda)$ be the integrable highest weight (irreducible) $\mathfrak{g}$-module. Then the tensor product $L(\Lambda) \otimes L(\Lambda)$, of course, contains a unique copy of $L(2\Lambda)$. Define the *Kostant cone* $K(\Lambda) \subset L(\Lambda)$ by

$$K(\Lambda) := \{v \in L(\Lambda) : v \otimes v \in L(2\Lambda)\}.$$

If $\Lambda \in D_{\mathbb{Z}}$, the associated Kac–Moody group $\mathcal{G}$ acts on $L(\Lambda)$ and $L(2\Lambda)$ (cf. Theorem 6.2.3) and $K(\Lambda)$ is clearly $\mathcal{G}$-stable.

Fix a $\mathfrak{g}$-invariant bilinear form $\langle\, ,\, \rangle$ on $\mathfrak{g}$ as in Theorem 1.5.4. For any $\alpha \in \Delta \cup \{0\}$, choose dual bases $\{e_\alpha^1, \cdots, e_\alpha^{n_\alpha}\}$ and $\{f_\alpha^1, \ldots, f_\alpha^{n_\alpha}\}$, with respect to $\langle\, ,\, \rangle$, of $\mathfrak{g}_\alpha$ and $\mathfrak{g}_{-\alpha}$ respectively, where $n_\alpha := \dim \mathfrak{g}_\alpha$. We also denote by $\langle\, ,\, \rangle$ the induced W-invariant nondegenerate symmetric bilinear form on $\mathfrak{h}^*$ (cf. 1.5.3).

10.1.4 Proposition. *For $\Lambda \in D$, $K(\Lambda) \subset L(\Lambda)$ is precisely the set of vectors $v \in L(\Lambda)$ which satisfy the following equation in $L(\Lambda) \otimes L(\Lambda)$:*

(1) $$\langle \Lambda, \Lambda \rangle\, v \otimes v = \sum_{\alpha \in \Delta \cup \{0\}} \sum_{j=1}^{n_\alpha} e_\alpha^j v \otimes f_\alpha^j v.$$

Observe that, for any $v \in L(\Lambda)$, all but finitely many terms in the above sum are zero. Further, by the proof of Lemma 1.5.9, the sum on the right side, for any $v \in L(\Lambda)$, is independent of the choice of the dual bases e_α^j and f_α^j.

Proof. Recall the definition of the Casimir–Kac element in $\hat{U}(\mathfrak{g})$ from 1.5.8:

$$\Omega = 2\nu^{-1}(\rho) + \Omega_0 + 2 \sum_{\alpha \in \Delta^+} \Omega_\alpha\,, \quad \text{where}$$

$$\Omega_\alpha := \sum_{j=1}^{n_\alpha} f_\alpha^j e_\alpha^j\,, \quad \text{for } \alpha \in \Delta^+ \cup \{0\}.$$

For $M \in \mathcal{O}$, let Ω_M be the Casimir–Kac operator acting on M (cf. 2.1.15). We omit the subscript M from Ω_M when no confusion is likely. We make a choice of the dual bases $\{e_\alpha^j\}$, $\{f_\alpha^j\}$ so that $e_{-\alpha}^j = f_\alpha^j$ for all $\alpha \in \Delta$. For any $v \in L(\Lambda)$, it is easy to see that

(2) $$\Omega(v \otimes v) = (\Omega v) \otimes v + v \otimes \Omega v + 2 \sum_{\alpha \in \Delta \cup \{0\}} \sum_{j=1}^{n_\alpha} e_\alpha^j v \otimes f_\alpha^j v,$$

since $\sum_j e_0^j v \otimes f_0^j v = \sum_j f_0^j v \otimes e_0^j v$.

By Lemma 2.1.16,

(3) $$\Omega_{L(2\Lambda)} = 4\langle \Lambda, \Lambda + \rho \rangle I, \text{ and } \Omega_{L(\Lambda)} = \langle \Lambda, \Lambda + 2\rho \rangle I\,,$$

and, moreover, for any irreducible component $L(\lambda)$ of $L(\Lambda) \otimes L(\Lambda)$ such that $\lambda \neq 2\Lambda$, $\Omega_{L(\lambda)}$ acts by a scalar different from $4\langle \Lambda, \Lambda + \rho \rangle$ (use Lemma 2.2.4).

In particular, for $v \in L(\Lambda)$,

(4) $$v \otimes v \in L(2\Lambda) \text{ iff } \Omega(v \otimes v) = 4\langle \Lambda, \Lambda + \rho \rangle (v \otimes v).$$

Combining (2)–(4), the proposition follows. □

We now come to the main result of this section.

10.1.5 Theorem. *For* $0 \neq \Lambda \in D_{\mathbb{Z}}$,

$$K(\Lambda)\backslash\{0\} = \mathcal{G} \cdot v_\Lambda,$$

where v_Λ *is a nonzero highest weight vector of* $L(\Lambda)$.

Proof. To prove the theorem, we need the following lemmas.

10.1.6 Lemma. *Let* $\mathfrak{g} = \bigoplus_{r\in\mathbb{R}} \mathfrak{g}_r$ *be a Lie algebra gradation of* $\mathfrak{g}$, *i.e.,* $[\mathfrak{g}_r, \mathfrak{g}_s] \subset \mathfrak{g}_{r+s}$, *which satisfies* $\mathfrak{g}_0 = \mathfrak{h}$. *In addition, for a fixed* $\Lambda \in D$, *let* $L(\Lambda) = \bigoplus_{r\in\mathbb{R}} L(\Lambda)_r$ *be a gradation compatible with the gradation of* $\mathfrak{g}$, *i.e.,* $\mathfrak{g}_r \cdot L(\Lambda)_s \subset L(\Lambda)_{r+s}$. *Then*

(a) *any* $v \in K(\Lambda) \cap L(\Lambda)_r$ *(for any* $r \in \mathbb{R}$*) belongs to* $L(\Lambda)_{w\Lambda}$ *for some* $w \in W$, *where* $L(\Lambda)_{w\Lambda}$ *denotes the weight space of* $L(\Lambda)$ *corresponding to the weight* $w\Lambda$, *and*

(b) *for* $v \in K(\Lambda)$, *write* $v = \sum_{r\leq s} v_r$ *with* $v_r \in L(\Lambda)_r$, *then* $v_s \in K(\Lambda)$.

Proof. Since $\mathfrak{h} = \mathfrak{g}_0$, we have $[\mathfrak{h}, \mathfrak{g}_r] \subset \mathfrak{g}_r$. In particular, for any simple root α_i, $\mathfrak{g}_{\pm\alpha_i}$ being one-dimensional, $\mathfrak{g}_{\alpha_i} \subset \mathfrak{g}_{r_i}$ and $\mathfrak{g}_{-\alpha_i} \subset \mathfrak{g}_{-r_i}$, for some $r_i \in \mathbb{R}\backslash\{0\}$. From this we see that, for any $\alpha = \sum_i n_i\alpha_i \in \Delta$,

$$\mathfrak{g}_\alpha \subset \mathfrak{g}_{r(\alpha)}, \tag{1}$$

where $r(\alpha) := \sum_i n_i r_i$. In particular, $r(\alpha) \neq 0$.

We now prove (a): Since $v \in K(\Lambda)$, by Proposition 10.1.4,

$$\langle\Lambda, \Lambda\rangle v \otimes v = \sum_{\alpha\in\Delta\cup\{0\}} \sum_{j=1}^{n_\alpha} \big(e_\alpha^j v \otimes f_\alpha^j v\big). \tag{2}$$

Since $v \in L(\Lambda)_r$, equating the components of the two sides of (2) in $L(\Lambda)_r \otimes L(\Lambda)_r$, we get (using (1))

$$\langle\Lambda, \Lambda\rangle v \otimes v = \sum_{j=1}^{n_0} e_0^j v \otimes f_0^j v. \tag{3}$$

Now consider the weight decomposition:

$$v = \sum_\lambda v_\lambda, \quad 0 \neq v_\lambda \in L(\Lambda)_\lambda,$$

where λ runs over a finite subset $S \subset \mathfrak{h}^*_{\leq\Lambda}$ (cf. 2.1.1(a) for the notation $\mathfrak{h}^*_{\leq\Lambda}$). Then (3) takes the form

$$\begin{aligned}\sum_{\lambda,\mu\in S} \langle\Lambda, \Lambda\rangle\, v_\lambda \otimes v_\mu &= \sum_{\lambda,\mu\in S} \sum_{j=1}^{n_0} \lambda(e_0^j)\, \mu(f_0^j)\, v_\lambda \otimes v_\mu \\ &= \sum_{\lambda,\mu\in S} \langle\lambda, \mu\rangle\, v_\lambda \otimes v_\mu.\end{aligned}$$

Since $\{v_\lambda \otimes v_\mu\}_{\lambda,\mu\in S}$ are linearly independent, this gives that

$$(4) \qquad \langle \Lambda, \Lambda\rangle = \langle \lambda, \mu\rangle, \qquad \text{for all } \lambda, \mu \in S.$$

By [Kac–90, Proposition 11.4(a)], (4) implies that $\lambda = \mu \in W \cdot \Lambda$. In particular, S is the singleton $\{w\Lambda\}$ for some $w \in W$. This proves (a).

Considering the component of total degree $2s$ in (2) and using Proposition 10.1.4, we immediately obtain (b). (An element of $L(\Lambda)_r \otimes L(\Lambda)_s$ is said to be of total degree $r + s$.) □

10.1.7 Definition. For any nonzero v in a weight $\mathfrak{h}$-module V, define the *support of* v to be

$$\operatorname{supp} v := \{\lambda \in \mathfrak{h}^* : v_\lambda \neq 0\},$$

where $v = \sum v_\lambda$ is the weight space decomposition.

Let $S(v)$ be the convex hull $[\operatorname{supp} v] \subset \mathfrak{h}^*$.

10.1.8 Lemma. *For any $\Lambda \in D \cap \mathfrak{h}^*_{\mathbb{R}}$ (cf. 1.4.1) and $v \in K(\Lambda)\backslash\{0\}$,*

(a) $\mathcal{V}(S(v)) = (W \cdot \Lambda) \cap S(v)$, where $\mathcal{V}(S(v))$ is the set of vertices of the polyhedron $S(v)$,

(b) if $\{\lambda, \mu\}$ is an edge of $S(v)$, then $\mu = s_\alpha\lambda$ for some real root α, where s_α is the reflection defined in 2.3.7.

Proof. (a): Fix $\lambda \in \mathcal{V}(S(v))$. By Lemma 10.1.2, $\lambda \in \operatorname{supp} v$ and, moreover, there exists an affine hyperplane $H_\lambda \subset \mathfrak{h}^*_{\mathbb{R}}$ through λ such that $S(v)\backslash\{\lambda\}$ lies entirely in one of the half spaces, say H^+_λ, separated by H_λ and $H_\lambda \cap (\lambda + Q) = \{\lambda\}$, where $Q \subset \mathfrak{h}^*_{\mathbb{R}}$ is the root lattice. Let H^0_λ be the hyperplane $H_\lambda - \lambda$ passing through 0. Choose a (nonzero) vector $\mu \in \mathfrak{h}^*_{\mathbb{R}}$ such that $\lambda + \mu \notin H_\lambda \cup H^+_\lambda$. Then; in particular,

$$(1) \qquad \mathfrak{h}^*_{\mathbb{R}} = H^0_\lambda \oplus \mathbb{R}\mu.$$

Let $\pi : \mathfrak{h}^*_{\mathbb{R}} \to \mathbb{R} \simeq \mathbb{R}\mu$ be the projection on the second factor under the decomposition (1). This gives rise to a Lie algebra gradation of $\mathfrak{g}$ by

$$(2) \qquad \mathfrak{g}_r := \bigoplus_{\alpha\in\Delta\cup\{0\} \text{ with } \pi(\alpha)=r} \mathfrak{g}_\alpha, \quad \text{for } r \in \mathbb{R}.$$

From the choice of H_λ, it is easy to see that $\mathfrak{g}_0 = \mathfrak{h}$. Similarly, define a gradation of $L(\Lambda)$, compatible with the above gradation of $\mathfrak{g}$, by

$$(3) \qquad L(\Lambda)_r = \bigoplus_{\alpha\in Q, \pi(\alpha)=r} L(\Lambda)_{\lambda+\alpha}.$$

Then, by the choices of H_λ and μ,

$$v \in \bigoplus_{r \le 0} L(\Lambda)_r \quad \text{and} \quad L(\Lambda)_0 = L(\Lambda)_\lambda.$$

In particular, by Lemma 10.1.6(b), $0 \neq v_0 \in K(\Lambda) \cap L(\Lambda)_0$ (where v_0 is the component of v in $L(\Lambda)_0$) and hence, by Lemma 10.1.6(a), $\lambda \in W \cdot \Lambda$. This proves the inclusion

$$\mathcal{V}(S(v)) \subset (W \cdot \Lambda) \cap S(v).$$

To prove the reverse inclusion, let $w\Lambda \in S(v)$ for some $w \in W$. Then $\Lambda \in w^{-1}S(v) = S(n^{-1}v)$, where $n \in N$ is such that $n \mod T = w$. Clearly, $\Lambda \in \mathcal{V}(S(n^{-1}v))$ so that $w\Lambda \in \mathcal{V}(Sv)$. This proves the reverse inclusion and hence completes the proof of (a).

(b): Since $v \in K(\Lambda)$, by Proposition 10.1.4,

$$\langle \Lambda, \Lambda \rangle\, v \otimes v = \sum_{\alpha \in \Delta \cup \{0\}} \sum_{j=1}^{n_\alpha} e_\alpha^j v \otimes f_\alpha^j v. \tag{4}$$

Collecting the components in (4) corresponding to $L(\Lambda)_\lambda \otimes L(\Lambda)_\mu$, we get

$$\langle \Lambda, \Lambda \rangle\, v_\lambda \otimes v_\mu = \langle \lambda, \mu \rangle\, v_\lambda \otimes v_\mu + \sum_{\alpha \in \Delta} \sum_{j=1}^{n_\alpha} (e_\alpha^j v_{\lambda-\alpha}) \otimes f_\alpha^j v_{\mu+\alpha}, \tag{5}$$

where $v = \sum_{\theta \in \operatorname{supp} v} v_\theta$ is the weight space decomposition of v. By Lemma 10.1.2(a), $\lambda, \mu \in \operatorname{supp} v$; in particular, they are weights of $L(\Lambda)$ and hence, by [Kac–90, Proposition 11.4],

$$\langle \Lambda, \Lambda \rangle \neq \langle \lambda, \mu \rangle, \quad \text{since } \lambda \neq \mu. \tag{6}$$

By (5) and (6) we obtain that there exists an $\alpha \in \Delta$ such that

$$\{\lambda - \alpha, \mu + \alpha\} \subset \operatorname{supp} v.$$

Since $\{\lambda, \mu\}$ is an edge of $S(v)$, there exists an affine hyperplane $H \subset \mathfrak{h}_{\mathbb{R}}^*$ such that the closed line segment $[\lambda, \mu]$ is contained in H and $S(v) \backslash [\lambda, \mu] \subset H^+$, for an open half space $H^+ \subset \mathfrak{h}_{\mathbb{R}}^*$ determined by H. If at least one of $\lambda - \alpha$, $\mu + \alpha$ does not belong to $[\lambda, \mu]$, then the midpoint $\frac{1}{2}(\lambda - \alpha + \mu + \alpha) = \frac{1}{2}(\lambda + \mu) \in H^+$, a contradiction! Hence $\{\lambda - \alpha, \mu + \alpha\} \subset [\lambda, \mu]$. Thus $\lambda - \mu = t\alpha$, for some $0 \neq t \in \mathbb{R}$. Write, by (a), $\lambda = w\Lambda$ and $\mu = w'\Lambda$, for some $w, w' \in W$. Then $\Lambda - w^{-1}w'\Lambda = t(w^{-1}\alpha)$ and $(w'^{-1}w\Lambda - \Lambda) = tw'^{-1}\alpha$. In particular, $w^{-1}\alpha$ and $w'^{-1}\alpha$ are of opposite signs, so $\alpha \in \Delta_{\text{re}}$ (by Lemma 1.3.14). Finally, since $\lambda = \mu + t\alpha$ and $\lambda, \mu \in W \cdot \Lambda$, we get $\langle \lambda, \lambda \rangle = \langle \lambda, \lambda \rangle + t^2 \langle \alpha, \alpha \rangle + 2t \langle \mu, \alpha \rangle$. Hence $t = -\langle \mu, \alpha^\vee \rangle$, since $t \neq 0$. Thus $\lambda = s_\alpha \mu$, proving (b). ☐

10.1.9 Lemma. *For $\Lambda \in D$ and $\lambda \in W \cdot \Lambda$, the set $\Delta^+(\lambda) := \{\alpha \in \Delta^+_{\mathrm{re}} : \langle \lambda, \alpha^\vee \rangle < 0\}$ is finite.*

Proof. Write $\lambda = w\Lambda$, for some $w \in W$. Then

$$\begin{aligned}\Delta^+(\lambda) &= \left\{\alpha \in \Delta^+_{\mathrm{re}} : \langle \Lambda, w^{-1}\alpha^\vee \rangle < 0\right\} \\ &\subseteq \left\{\alpha \in \Delta^+ : w^{-1}\alpha \in \Delta^-\right\} = \Phi_w.\end{aligned}$$

But Φ_w is finite by Lemma 1.3.14. $\square$

With these preparations, we are ready now to prove Theorem 10.1.5.

10.1.10 *Proof of Theorem 10.1.5.* For any $g \in \mathcal{G}$, clearly $gv_\Lambda \in K(\Lambda)\backslash\{0\}$ from the definition of $K(\Lambda)$. Conversely, take $v \in K(\Lambda)\backslash\{0\}$. For $\mu \in \mathcal{V}(S(v))$, let $\Phi_\mu(v) := \{\beta \in \Delta^+_{\mathrm{re}} : \{\mu, s_\beta\mu\}$ is an edge of $S(v)\}$ be the set of 'edges through μ' (cf. Lemma 10.1.8(b)). Recall the definition of $\Delta^+(\mu)$ from Lemma 10.1.9. Define $\Phi'_\mu(v) = \{\alpha \in \Delta^+(\mu) : \mu + t\alpha \in S(v)$ for some $0 \neq t \in \mathbb{R}\}$. Then

$$\Phi_\lambda(v) \subset \Phi'_\lambda(v)\,, \tag{1}$$

for any $\lambda \in \mathcal{V}(S(v))$ such that $|\lambda - \Lambda| \leq |\mu - \Lambda|$, for all $\mu \in \mathcal{V}(S(v))$, where $|\cdot|$ is the principal gradation defined by (1.2.2.3). Note that for such a λ, one has $|\mu - \Lambda| \geq |\lambda - \Lambda|$, for all $\mu \in S(v)$. We choose such a λ.

If $\Phi_\lambda(v)$ is empty, then $S(v) = \{\lambda\}$ and hence $v \in L(\Lambda)_\lambda$ and, of course, $\lambda = w\Lambda$ for some $w \in W$ by Lemma 10.1.8(a). In particular, we can find $g \in N$ such that $g^{-1}v = v_\Lambda$, i.e., $v \in \mathcal{G} \cdot v_\Lambda$ in this case. (We have used here the hypothesis that $\Lambda \neq 0$.)

So assume that $\Phi_\lambda(v) \neq \emptyset$. Take $\alpha \in \Phi_\lambda(v)$ and nonzero $e_\alpha \in \mathfrak{g}_\alpha$. Since $|\lambda - \Lambda|$ is minimal among $\mu \in \mathcal{V}(S(v))$, $s_\alpha\lambda = \lambda + k\alpha$ for some positive integer k (and $\alpha \in \Delta^+(\lambda)$). We prove that there exists $z_0 \in \mathbb{C}$ such that $\lambda \in \mathcal{V}(S(\mathrm{Exp}(z_0e_\alpha) \cdot v))$, $|\lambda - \Lambda|$ is minimal among all $\nu - \Lambda$, where ν runs through the vertices of $S(\mathrm{Exp}(z_0e_\alpha) \cdot v)$ and, moreover,

$$\Phi'_\lambda(\mathrm{Exp}(z_0e_\alpha) \cdot v) \subset \Phi'_\lambda(v)\backslash\{\alpha\}\,. \tag{2}$$

Since $|\lambda - \Lambda|$ is minimal among $\mu \in \mathcal{V}(S(v))$, it is easy to see that, for any $z \in \mathbb{C}$, $\lambda \in \mathrm{supp}\,(\mathrm{Exp}(ze_\alpha) \cdot v)$ and hence, by Lemma 10.1.8(a), $\lambda \in \mathcal{V}(S(\mathrm{Exp}(ze_\alpha) \cdot v))$. Further,

$$S(\mathrm{Exp}(ze_\alpha) \cdot v) \subset S(v) + \mathbb{R}_+\alpha, \tag{3}$$

and hence $|\lambda - \Lambda|$ is minimal among all $\mu \in \mathcal{V}((\mathrm{Exp}(ze_\alpha) \cdot v))$.

We next prove that, for any $z \in \mathbb{C}$,

(4) $$\Phi'_\lambda(\operatorname{Exp}(ze_\alpha) \cdot v) \subset \Phi'_\lambda(v) :$$

Take $\beta \in \Phi'_\lambda(\operatorname{Exp}(ze_\alpha) \cdot v)$. Then there exists $0 \neq t_o \in \mathbb{R}$ such that

$$\lambda + t_o\beta \in S(\operatorname{Exp}(ze_\alpha) \cdot v).$$

By (3) write

$$\lambda + t_o\beta = \gamma + s\alpha, \quad \text{for } \gamma \in S(v) \text{ and } s \in \mathbb{R}_+.$$

Then, for any $s' \in [0, 1]$,

(5) $$\lambda + s't_o\beta - s's\alpha = (1 - s')\lambda + s'\gamma\,,$$

and hence, since $S(v)$ is convex, $\lambda + s't_o\beta - s's\alpha \in S(v)$. Further, since $\lambda + k\alpha \in S(v)$ for some $k > 0$,

(6) $$\lambda + s's\alpha \in S(v), \quad \text{for all small enough } s' > 0.$$

Hence, from (5) and (6),

$$\lambda + \frac{1}{2}s't_o\beta \in S(v), \text{ thus } \beta \in \Phi'_\lambda(v), \text{ proving (4).}$$

Finally, we show that for an appropriate choice of z_0, $\alpha \notin \Phi'_\lambda(\operatorname{Exp}(z_0e_\alpha) \cdot v)$, which will complete the proof of (2).

Since $\{\lambda, s_\alpha(\lambda) = \lambda + k\alpha\}$ is an edge of $S(v)$, $\lambda + \ell\alpha \notin \operatorname{supp} v$ for any $\ell < 0$; thus the component of $\operatorname{Exp}(ze_\alpha) \cdot v$ in $L(\Lambda)_{\lambda+k\alpha}$ is given by

(7) $$z^k\frac{e_\alpha^k}{k!}v_\lambda + z^{k-1}\frac{e_\alpha^{k-1}}{(k-1)!}v_{\lambda+\alpha} + \cdots + v_{\lambda+k\alpha},$$

where $v = \sum_\mu v_\mu$ is the weight decomposition. Since λ and $\lambda + k\alpha$ are both vertices of $S(v)$, neither v_λ nor $v_{\lambda+k\alpha}$ is zero. Since $L(\Lambda)_{\lambda+k\alpha}$ is one-dimensional by Lemma 1.3.5(a), write

$$\frac{e_\alpha^{k-i}}{(k-i)!}v_{\lambda+i\alpha} = c_i v_{\lambda+k\alpha}, \text{ for some } c_i \in \mathbb{C}.$$

So (7) becomes

(8) $$(z^kc_0 + z^{k-1}c_1 + \cdots + zc_{k-1} + 1)v_{\lambda+k\alpha},$$

and $c_0 \neq 0$ by Exercise 10.1.E. Now choose $z = z_0$ so that (8) is 0. With this choice of z_0,

$$\lambda + k\alpha \notin \operatorname{supp}(\operatorname{Exp}(z_0 e_\alpha) \cdot v). \tag{9}$$

We now show that $\alpha \notin \Phi'_\lambda(\operatorname{Exp}(z_0 e_\alpha) \cdot v)$: Choose a hyperplane $H \subset \mathfrak{h}^*_{\mathbb{R}}$ such that $[\lambda, \lambda + k\alpha] \subset H$ and $S(v)\backslash[\lambda, \lambda + k\alpha] \subset H^+$, where H^+ is an open half space separated by H. Then for any $z \in \mathbb{C}$, using (3), $S(\operatorname{Exp}(ze_\alpha) \cdot v) \subset H \cup H^+$ and, moreover, $S(\operatorname{Exp}(ze_\alpha) \cdot v) \cap H = [\lambda, \lambda + s_o\alpha]$, for some $s_o = s_o(z) \in \mathbb{R}_+$. In particular, if $s_o > 0$, $\{\lambda, \lambda + s_o\alpha\}$ would be an edge of $S(\operatorname{Exp}(ze_\alpha) \cdot v)$, forcing $s_o = k$ (by Lemma 10.1.8(b)). So, if $z = z_0$, this gives $s_o = 0$ (by (9) and Lemma 10.1.8(a)) and hence $\alpha \notin \Phi'_\lambda(\operatorname{Exp}(z_0 e_\alpha) \cdot v)$. This proves (2).

Since $\Phi'_\lambda(v)$ is finite (by Lemma 10.1.9), by successively using (2) we find $z_0, \dots, z_p \in \mathbb{C}$ and $\alpha_0 = \alpha, \alpha_1, \cdots, \alpha_p \in \Delta^+(\lambda)$ such that, for all $0 \le i \le p$,

$$\alpha_i \in \Phi_\lambda\big(\operatorname{Exp}(z_{i-1} e_{\alpha_{i-1}}) \cdots \operatorname{Exp}(z_0 e_{\alpha_0}) v\big)$$

and $\Phi_\lambda\big(\operatorname{Exp}(z_p e_{\alpha_p}) \cdots \operatorname{Exp}(z_0 e_{\alpha_0}) v\big) = \emptyset$. Hence, by the first part of the proof, $\operatorname{Exp}(z_p e_{\alpha_p}) \cdots \operatorname{Exp}(z_0 e_{\alpha_0}) v \in \mathcal{G} \cdot v_\Lambda$. This completes the proof of the theorem. □

Combining Proposition 10.1.4 and Theorem 10.1.5, we get the following:

10.1.11 Corollary. *For any $\Lambda \in D_{\mathbb{Z}}$, let $Y \subset \{1, \dots, \ell\}$ be the subset consisting of those i such that $\langle \Lambda, \alpha_i^\vee \rangle = 0$. Then the ind-subvariety*

$$i_\Lambda : \mathcal{G}/\mathcal{P}_Y \hookrightarrow \mathbb{P}(L(\Lambda)),$$

obtained as the orbit through the highest weight vector, is given by the vanishing of a certain collection of equations $P_{\phi,\phi'}$ in $S^2(L(\Lambda)^*)$ given below in the proof. Observe that, since $\mathfrak{g}$ is symmetrizable, i_Λ is a closed embedding by Proposition 7.1.15(c).

Proof. Let $\{v_\phi\}_{\phi \in \Gamma}$ be a basis of $L(\Lambda)$ consisting of weight vectors and let $\{v_\phi^*\}$ be the dual basis of $L(\Lambda)^\vee := \bigoplus_\lambda (L(\Lambda)_\lambda)^*$. For any $\phi, \phi' \in \Gamma$, consider the element

$$P_{\phi,\phi'} := \langle \Lambda, \Lambda \rangle v_\phi^* v_{\phi'}^* - \sum_{\alpha \in \Delta \cup \{0\}} \sum_{j=1}^{n_\alpha} \big(e_\alpha^j(v_\phi^*)\big)\big(f_\alpha^j(v_{\phi'}^*)\big) \in S^2(L(\Lambda)^*). \tag{1}$$

(It is easy to see that, for any fixed ϕ, ϕ', the sum in the right side is finite.)

By Proposition 10.1.4 and Theorem 10.1.5, it follows immediately that $i_\Lambda(\mathcal{G}/\mathcal{P}_Y)$ is given by the vanishing of the equations $\{P_{\phi,\phi'}\}_{\phi,\phi' \in \Gamma}$. □

10.1.12 Remark. Kostant, in fact, proved that (in the finite case) the full ideal of $i_\Lambda(\mathcal{G}/\mathcal{P}_Y)$ in $\mathbb{P}(L(\Lambda))$ is generated by the collection of the above equations $P_{\phi,\phi'}$. (The proof appeared in [Garfinkle–82].)

10.1.E EXERCISE. Take $\Lambda \in D$, $\lambda \in W \cdot \Lambda$ and $\alpha \in \Delta^+_{\mathrm{re}}$ such that $\langle \lambda, \alpha^\vee \rangle = -k$, where k is a positive integer. Then, show that $e^k_\alpha v_\lambda \neq 0$, where $0 \neq v_\lambda \in L(\Lambda)_\lambda$ and $0 \neq e_\alpha \in \mathfrak{g}_\alpha$.
Hint. Show that $\lambda - \alpha$ is not a weight of $L(\Lambda)$ by proving $\langle \lambda - \alpha, \lambda - \alpha \rangle > \langle \Lambda, \Lambda \rangle$ and then consider the $s\ell_2(\alpha)$-submodule generated by v_λ, where $s\ell_2(\alpha) = \mathfrak{g}_\alpha \oplus \mathbb{C}\alpha^\vee \oplus \mathfrak{g}_{-\alpha}$.)

10.2. Conjugacy Theorems for Lie Algebras

In this section $\mathfrak{g}$ is any symmetrizable Kac–Moody Lie algebra (unless specified otherwise).

10.2.1 Definition. Let $\mathfrak{s}$ be any Lie algebra and let (V, π) be an $\mathfrak{s}$-module. Then $\mathfrak{s}$ is called

(a) *π-triangular* if there exists a flag of $\mathfrak{s}$-submodules: $V_0 = (0) \subset V_1 \subset V_2 \subset \cdots$, such that $\cup V_i = V$ and, for each $i \geq 0$, V_{i+1}/V_i is one-dimensional.

(b) *π-diagonalizable*, resp. *π-finitely semisimple*, if V is a direct sum of one-dimensional, resp. finite-dimensional, irreducible $\mathfrak{s}$-submodules (cf. Definition 3.1.6).

(c) *π-locally finite* if any $v \in V$ is contained in a finite-dimensional $\mathfrak{s}$-submodule of V.

Clearly, for a countable-dimensional $\mathfrak{s}$-module V: $\mathfrak{s}$ is π-diagonalizable $\Rightarrow$ $\mathfrak{s}$ is π-triangular $\Rightarrow$ $\mathfrak{s}$ is π-locally finite. It is easy to see that if $\mathfrak{s}$ is π-diagonalizable and $W \subset V$ is $\mathfrak{s}$-stable, then $\mathfrak{s}$ is $\pi_{|W}$-diagonalizable (use Exercise 1.2.E.2). The same property is, of course, true for π-triangular and π-locally finite $\mathfrak{s}$.

An element $x \in \mathfrak{s}$ is called *π-triangular*, *π-diagonalizable*, *π-finitely semisimple* if the Lie subalgebra $\mathbb{C}x \subset \mathfrak{s}$ satisfies the same for the representation π restricted to $\mathbb{C}x$, respectively.

The adjoint representation $\mathrm{ad}_\mathfrak{g}$ of $\mathfrak{g}$ gives rise to the dual representation $\mathrm{ad}^*_\mathfrak{g}$ on the restricted dual space $\mathfrak{g}^\vee := \mathfrak{h}^* \bigoplus (\oplus_{\alpha\in\Delta}\, \mathfrak{g}^*_\alpha)$. Observe that, for symmetrizable $\mathfrak{g}$, from the nondegenerate invariant pairing $\mathfrak{g} \otimes \mathfrak{g} \to \mathbb{C}$ (cf. Corollaries 1.5.7 and 3.2.10), $\mathfrak{g} \simeq \mathfrak{g}^\vee$ (as $\mathfrak{g}$-modules).

10.2.2 Lemma. *Let $\mathfrak{s}$ be a subalgebra of $\mathfrak{g}$ such that $\mathfrak{s} + \mathfrak{n}$, resp. $\mathfrak{s} + \mathfrak{n}^-$, has finite-codimension in $\mathfrak{g}$. Then, for $\lambda \in D$, $L(\lambda)/\mathfrak{s} \cdot L(\lambda)$, resp. $L(\lambda)^{\mathfrak{s}}$, is finite dimensional.*

Proof. First assume that $\mathfrak{s} + \mathfrak{n}$ has finite-codimension in $\mathfrak{g}$. Choose finitely many $x_i \in \mathfrak{g}$ such that x_i's act locally finitely on $L(\lambda)$ and $\mathfrak{s} + \sum_i \mathbb{C}x_i + \mathfrak{n} = \mathfrak{g}$. (This is possible by Lemma 1.3.3(c).) In particular, by the PBW Theorem,

$$L(\lambda) = U(\mathfrak{g}) \cdot v_\lambda = U(\mathfrak{s})F,$$

for some finite-dimensional subspace $F \subset L(\lambda)$. Hence $L(\lambda)/\mathfrak{s} \cdot L(\lambda)$ is finite dimensional.

Now we come to the case when $\mathfrak{s} + \mathfrak{n}^-$ has finite codimension in $\mathfrak{g}$. Then, it is easy to see that $L(\lambda)^{\mathfrak{s}} \hookrightarrow \big(L(\lambda)^\vee/\mathfrak{s} \cdot (L(\lambda)^\vee)\big)^*$, where $L(\lambda)^\vee$ is the restricted dual as in 2.1.1(c). Since $L(\lambda)^\vee$ is a lowest weight irreducible integrable $\mathfrak{g}$-module (by Lemma 2.1.14), the lemma in this case follows from the previous case by replacing $\mathfrak{n}$ with $\mathfrak{n}^-$. □

10.2.3 Remark. The above lemma is true (by the same proof) for an arbitrary Kac–Moody algebra $\mathfrak{g}$. In fact, the same proof shows that for any integrable highest weight $\mathfrak{g}$-module $V(\lambda)$, $V(\lambda)/\mathfrak{s} \cdot V(\lambda)$ is finite dimensional for any subalgebra $\mathfrak{s}$ such that $\mathfrak{s} + \mathfrak{n}$, resp. $\mathfrak{s} + \hat{\mathfrak{n}}$, has finite codimension in $\mathfrak{g}$, resp. $\hat{\mathfrak{g}}$, where $\hat{\mathfrak{n}}$, $\hat{\mathfrak{g}}$ are as in 6.1.1.

10.2.4 Lemma. *For any $\lambda \in \mathfrak{h}^*$ such that $\lambda(\alpha_i^\vee) \neq 0$ for all the simple coroots $\alpha_i^\vee$, and any highest weight $\mathfrak{g}$-module $V(\lambda)$ with highest weight λ, the stabilizer $\mathfrak{s}$ of the highest weight line $V(\lambda)_\lambda$ in $\mathfrak{g}$ is precisely $\mathfrak{b}$.*

Proof. Clearly $\mathfrak{b} \subset \mathfrak{s}$. If possible, let $\mathfrak{s} \cap \mathfrak{g}_{-\alpha} \neq 0$ for some $\alpha \in \Delta^+$, and choose $\alpha \in \Delta^+$ such that $|\alpha|$ is minimum with this property. Then, since $\lambda(\alpha_i^\vee) \neq 0$, α can not be a simple root. From the minimality of $|\alpha|$, for any $1 \leq i \leq \ell$, $[e_i, \mathfrak{s} \cap \mathfrak{g}_{-\alpha}] = 0$. Hence, by Theorem 1.2.1(d), $[\mathfrak{n}, \mathfrak{s} \cap \mathfrak{g}_{-\alpha}] = 0$. In particular, the ideal I of $\mathfrak{g}$ generated by $\mathfrak{s} \cap \mathfrak{g}_{-\alpha}$ satisfies $I \cap \mathfrak{h} = (0)$. But then, by Corollary 3.2.10, $I = (0)$, a contradiction! This proves the lemma. □

Recall that a weight $\lambda \in D$ is called *regular* if $\lambda(\alpha_i^\vee) \neq 0$ for all the simple coroots $\alpha_i^\vee$, where D is as in (2.1.5.1).

10.2.5 Theorem. *The following conditions are equivalent for a subalgebra $\mathfrak{s} \subset \mathfrak{g}$:*

(a) $\mathfrak{s}$ is $\mathrm{ad}_{\mathfrak{g}}$-triangular.

(b) $\mathfrak{s}$ is a finite-dimensional solvable Lie algebra which is $\mathrm{ad}_{\mathfrak{g}}$-locally finite.

(c) $\mathfrak{s}$ is triangular for every integrable highest weight (irreducible) $\mathfrak{g}$-module $L(\lambda)$ and its restricted dual $L(\lambda)^\vee$.

(d) $\mathfrak{s}$ is triangular for the integrable highest weight (irreducible) $\mathfrak{g}$-module $L(\lambda)$ and its restricted dual $L(\lambda)^\vee$, for some regular λ.

(e) There exist $g \in \mathcal{G}^{\min}$ and $w \in W$ such that $(\mathrm{Ad}\ g)\mathfrak{s} \subset \mathfrak{b} \cap w\mathfrak{b}^-$, where $\mathcal{G}^{\min} \subset \mathcal{G}$ is the subgroup as in 7.4.1.

(By 6.2.10 and Exercise 6.2.E.2, for any $g \in \mathcal{G}^{\min}$, $\mathrm{Ad}\, g$ is a Lie algebra automorphism of $\mathfrak{g}$.)

Proof. (a) ⇒ (b): Since $\mathfrak{g}$ is a finitely generated Lie algebra, there exists a large enough i such that the subspace V_i of $\mathfrak{g}$ (cf. Definition 10.2.1(a)) contains a full set of Lie algebra generators of $\mathfrak{g}$. Hence, the Lie algebra homomorphism (induced by the adjoint action) $\mathfrak{s}/\mathfrak{s} \cap \mathfrak{c} \to \mathrm{End}_{\mathbb{C}}\, V_i$ is injective, where $\mathfrak{c}$ is the center of $\mathfrak{g}$. In particular, by Corollary 1.2.3(b), $\mathfrak{s}$ is finite dimensional. Moreover, since there is a complete flag in V_i stable under $\mathfrak{s}$ (by assumption), $\mathfrak{s}/\mathfrak{s} \cap \mathfrak{c}$ (and hence $\mathfrak{s}$) is a solvable Lie algebra. This proves (b).

(b) ⇒ (c): We first show that $\mathfrak{s}$ is $L(\lambda)$-triangular: Let $\mathfrak{b}^{-\vee} := \{f \in \mathfrak{g}^\vee : f_{|\mathfrak{n}} \equiv 0\}$. Then, clearly, the orthogonal complement of $\mathfrak{b}^{-\vee}$ in $\mathfrak{g}$, under the standard pairing, equals $\mathfrak{n}$. Since $\mathfrak{s}$ is finite dimensional, from the weight consideration, $(\mathrm{ad}^*_{\mathfrak{g}}\ \mathfrak{s})(\mathfrak{b}^{-\vee}) \subset \mathfrak{b}^{-\vee} + V$, for some finite-dimensional subspace V of $\mathfrak{g}^\vee$. Now let $\tilde{V} := \mathrm{ad}^*_{\mathfrak{g}}(U(\mathfrak{s}))\, V$. Then, by (b) (since $\mathfrak{g} \simeq \mathfrak{g}^\vee$ as $\mathfrak{g}$-modules), $\tilde{V}$ is finite dimensional and, moreover, $\tilde{V} + \mathfrak{b}^{-\vee}$ is $\mathrm{ad}^*_{\mathfrak{g}}(\mathfrak{s})$-stable. Define the subspace Z of $\mathfrak{g}$ by

$$Z = \{x \in \mathfrak{g} : (\tilde{V} + \mathfrak{b}^{-\vee})(x) = 0\}.$$

Then Z is a $\mathrm{ad}_{\mathfrak{g}}(\mathfrak{s})$-stable subspace of $\mathfrak{n}$ of finite codimension. Let $\tilde{Z} \subset \mathfrak{n}$ be the Lie subalgebra generated by Z. By Lemma 10.2.2, the subspace $L(\lambda)^{\tilde{Z}}$ is finite dimensional and, of course, contains the highest weight vector v_λ. Moreover, $\tilde{Z}$ being $\mathrm{ad}_{\mathfrak{g}}(\mathfrak{s})$-stable, $L(\lambda)^{\tilde{Z}}$ is stable under the action of $\mathfrak{s}$. Now, since $\mathfrak{s}$ is $\mathrm{ad}_{\mathfrak{g}}$-locally finite (by (b)), it is easy to see that $\{v \in L(\lambda) : U(\mathfrak{s})v$ is finite dimensional$\}$ is a $\mathfrak{g}$-submodule of $L(\lambda)$. Hence $\mathfrak{s}$ is locally finite for $L(\lambda)$. Since $\mathfrak{s}$ is solvable, by Lie's theorem (cf. [Serre–87, Chap. I, Theorem 3]), we get that $\mathfrak{s}$ is $L(\lambda)$-triangular.

Since $L(\lambda)^\vee$ is a lowest weight irreducible integrable $\mathfrak{g}$-module (by Lemma 2.1.14), and the assumptions of (b) are symmetric with respect to $\mathfrak{n}$ and $\mathfrak{n}^-$, we get that$\mathfrak{s}$ is $L(\lambda)^\vee$-triangular as well. This proves the (c) part.

(c) ⇒ (d): In fact, (d) is a special case of (c).

(d) ⇒ (e): By Lemma 8.3.2, we can assume that $\lambda \in D^o$, where D^o is as defined in 7.1.1. Recall the inclusion $i_L : \mathcal{G}/\mathcal{B} \hookrightarrow \mathbb{P}(L)$ from Lemma 7.1.2, where $L = L(\lambda)$. Let $V := U(\mathfrak{s}) \cdot v_\lambda$. Then, by (d), V is a finite-dimensional $\mathfrak{s}$-submodule of L. For any $p \geq 1$, let $I^p := \{f \in (S^p(L))^* : f\big((gv_\lambda)^p\big) = 0$, for all $g \in \mathcal{G}\}$, where $S^p(L)$ is the p-th symmetric power of L and $v_\lambda \in L$ is a highest weight vector. Then, clearly, I^p is a $\mathcal{G}$-submodule, resp. $\mathfrak{g}$-submodule, of $(S^p(L))^*$ under the canonical action of $\mathcal{G}$, resp. $\mathfrak{g}$, on $(S^p(L))^*$ (use Proposition 6.2.11 for the case of $\mathfrak{g}$). Consider the canonical $\mathfrak{s}$-module map $\theta : (S^p(L))^* \to (S^p(V))^*$, obtained by restriction, where the $\mathfrak{s}$-module structure on L is obtained via restricting its $\mathfrak{g}$-module structure. Define $I^p_V := \theta(I^p)$. Then I^p_V is a $\mathfrak{s}$-submodule of $(S^p(V))^*$. Since $\mathfrak{s}$ is triangular for its action on V (by assumption), $\mathfrak{s}$ is triangular on $(S^p(V))^*$ as well. Choose a $\mathfrak{s}$-stable complete flag:

$$(0) \subset V_0 \subset V_1 \subset \cdots \subset V_{N_p} = (S^p(V))^*, \text{ for } N_p := \dim S^p(V).$$

Consider the (solvable) algebraic subgroup H_p of Aut V defined by

$$H_p = \{g \in \operatorname{Aut} V : \hat{g}(I_V^p) = I_V^p \text{ and } \hat{g} \text{ keeps each } V_i \text{ stable}\},$$

where $\hat{g}$ is the induced automorphism of $(S^p(V))^*$. Then the Lie algebra homomorphism $\pi : \mathfrak{s} \to \operatorname{End} V$ (induced by the $\mathfrak{s}$-module V) satisfies

(1) $$\pi(\mathfrak{s}) \subset \operatorname{Lie} H_p, \text{ for any } p \geq 1.$$

Let $Z_p(V) \subset \mathbb{P}(V)$ be the projective subvariety obtained as the zero set of I_V^p, i.e.,

$$Z_p(V) := \{[v] \in \mathbb{P}(V) : f(v^p) = 0 \text{ for all } f \in I_V^p\},$$

where $[v]$ is the line through $v \in V$. Then $Z_p(V)$ is stable under the standard action of H_p on $\mathbb{P}(V)$. Similarly, let $Z_p \subset \mathbb{P}(L)$ be the closed ind-subvariety obtained as the zero set of I^p (cf. Example 4.1.3(6)). Then

(2) $$Z_p(V) = Z_p \cap \mathbb{P}(V).$$

By the Borel's fixed point theorem [Borel–91, Chap. III, Theorem 10.4], the identity component H_p^o has a common fixed point $x_p \in Z_p(V)$ if $Z_p(V)$ is nonempty. By Corollary 10.1.11,

(3) $$Z_2 = i_L(\mathcal{G}/\mathcal{B}).$$

In particular, by (2) and (3),

(4) $$Z_2(V) = i_L(\mathcal{G}/\mathcal{B}) \cap \mathbb{P}(V).$$

So we get $x_2 = [gv_\lambda] \in \mathbb{P}(V)$, for some $g \in \mathcal{G}^{\min}$ by (7.4.5.2), and (by choice) H_2^o fixes the line $x_2 \subset V$. In particular, by (1), $\pi(\mathfrak{s})x_2 \subset x_2$, i.e., $\big((\operatorname{Ad} g^{-1})\,\mathfrak{s}\big) \cdot v_\lambda \subset \mathbb{C}v_\lambda$ (cf. (6.2.10.3)). By Lemma 10.2.4 and Exercise 6.2.E.2, this gives

$$(\operatorname{Ad} g^{-1})\,\mathfrak{s} \subset \mathfrak{b}.$$

By a similar argument, using $L(\lambda)^\vee$ instead of $L(\lambda)$, we get $(\operatorname{Ad} g_1^{-1})\mathfrak{s} \subset \mathfrak{b}^-$, for some $g_1 \in \mathcal{G}^{\min}$. Write $g_1^{-1}g = u_- n u_+$ for some $u_- \in \mathcal{U}^-$, $n \in N$ and $u_+ \in \mathcal{U} \cap \mathcal{G}^{\min}$ (cf. the Birkhoff decomposition Theorem 6.2.8). Then

$$(\operatorname{Ad}(u_+ g^{-1}))\,\mathfrak{s} \subset \mathfrak{b} \cap w\mathfrak{b}^-, \text{ for } w := n^{-1} \mod T.$$

This proves (e).

(e) $\Rightarrow$ (a): By the analogue of (6.2.10.3) for π replaced by the adjoint representation Ad of $\mathcal{G}^{\min}$ in $\mathfrak{g}$, it suffices to show that, for any $w \in W$, $\mathfrak{b} \cap w\mathfrak{b}^-$ is $\operatorname{ad}_\mathfrak{g}$-triangular. It is easy to see that $\mathfrak{b} \cap w\mathfrak{b}^-$ is a finite-dimensional solvable Lie algebra. Thus, by Lie's theorem, we need to show that $\mathfrak{b} \cap w\mathfrak{b}^-$ is $\operatorname{ad}_\mathfrak{g}$-locally finite. Since all the roots in $\Delta^+ \cap w\Delta^-$ are real (cf. Lemma 1.3.14), by Lemma 1.3.3(c), we get that $\mathfrak{b} \cap w\mathfrak{b}^-$ is $\operatorname{ad}_\mathfrak{g}$-locally finite. This proves (a) and hence finishes the proof of the theorem. $\square$

10.2.6 Corollary. *Let $\mathfrak{s}$ be an $\mathrm{ad}_{\mathfrak{g}}$-locally finite subalgebra of $\mathfrak{g}$. Then $\mathfrak{s} \subset \mathfrak{g}_{\mathrm{fin}}$ (cf. Proposition 6.2.11). Moreover,* $\operatorname{Exp} \mathfrak{s} \subset \mathcal{G}^{\min}$.

Proof. For any nonzero $x \in \mathfrak{s}$, apply Theorem 10.2.5 to the one-dimensional subalgebra $\mathbb{C}x$. Then

(1) $\quad (\mathrm{Ad}\ g)x \in \mathfrak{b} \cap w\mathfrak{b}^-$, for some $g = g(x) \in \mathcal{G}^{\min}$ and $w = w(x) \in W$.

In particular, $x \in (\mathrm{Ad}\ g^{-1})\,\mathfrak{b}$ and hence $x \in \mathfrak{g}_{\mathrm{fin}}$ (from the definition of $\mathfrak{g}_{\mathrm{fin}}$ as in 6.2.10).

The second part of the corollary follows from (1), by using (6.2.11.1) and observing that

(2) $$\operatorname{Exp}(\mathfrak{b} \cap w\mathfrak{b}^-) \subset \mathcal{G}^{\min}. \qquad \square$$

In fact, it is equal to the subgroup of $\mathcal{G}$ generated by T and $\{U_\beta\}_{\beta\in\Phi_w}$ (cf. Lemma 6.1.4 and Example 6.1.5(b)).

10.2.7 Remark. The implication '(d) $\Rightarrow$ (e)' in Theorem 10.2.5 holds good (by the same proof) for any Kac–Moody group $\mathcal{G}$ such that $i_L(\mathcal{G}/\mathcal{B}) = Z_p$ for some p, provided we replace $\mathfrak{g}$, $\mathfrak{b}$, $\mathfrak{b}^-$ respectively by $\mathfrak{g}/\mathfrak{r}$, $\mathfrak{b}/\mathfrak{r}^+$, $\mathfrak{b}^-/\mathfrak{r}^-$ and take $\mathfrak{s} \subset \mathfrak{g}/\mathfrak{r}$ (satisfying (d)), where $\mathfrak{r}$ is the radical defined in 1.5.6 and $\mathfrak{r}^\pm := \mathfrak{r} \cap \mathfrak{n}^\pm$.

Recall [Humphreys–72, §15.3] that a nilpotent subalgebra $\mathfrak{a}$ of a Lie algebra $\mathfrak{s}$ is called a *Cartan subalgebra* if $\mathfrak{a}$ equals its own normalizer $N(\mathfrak{a})$. Then, by [loc. cit., Theorem 15.3], Cartan subalgebras exist for any finite-dimensional $\mathfrak{s}$.

10.2.8 Lemma. *For $w \in W$, consider the Lie subalgebra $\mathfrak{b}_w := \mathfrak{b} \cap w\mathfrak{b}^- \subset \mathfrak{g}$. Then any $\mathrm{ad}_{\mathfrak{b}_w}$-diagonalizable subalgebra $\mathfrak{a}$ of $\mathfrak{b}_w$ satisfies:*

$$(\mathrm{Ad}\, g)\, \mathfrak{a} \subset \mathfrak{h}\,,$$

for some g in the subgroup $\mathcal{U}_{\Phi_w}$ of $\mathcal{G}^{\min}$ generated by $\{\mathcal{U}_\alpha;\ \alpha \in \Delta^+ \cap w\Delta^-\}$ (cf. Example 6.1.5(b)).

Proof. Clearly, the center $\mathfrak{z}$ of $\mathfrak{b}_w$ is contained in $\mathfrak{h}$. Moreover, since $\mathfrak{a}$ is $\mathrm{ad}_{\mathfrak{b}_w}$-diagonalizable, $[\mathfrak{a}, \mathfrak{a}] \subset \mathfrak{z} \subset \mathfrak{h}$. Further, it is easy to see that $[\mathfrak{b}_w, \mathfrak{b}_w] \cap \mathfrak{h} = 0$, and hence $\mathfrak{a}$ is abelian.

Let $\mathfrak{s}$ be the centralizer of $\mathfrak{a}$ in $\mathfrak{b}_w$ and let $\mathfrak{c}$ be a Cartan subalgebra of $\mathfrak{s}$. Then, $\mathfrak{a}$ being central in $\mathfrak{s}$, $\mathfrak{a} \subset \mathfrak{c} \subset \mathfrak{s}$. We now show that $\mathfrak{c}$ is also a Cartan subalgebra of $\mathfrak{b}_w$. Let $\mathfrak{d}$ be the normalizer of $\mathfrak{c}$ in $\mathfrak{b}_w$ and write $\mathfrak{d} = \mathfrak{c} \oplus \mathfrak{c}^\perp$, where $\mathfrak{c}^\perp$ is an $\mathfrak{a}$-stable subspace of $\mathfrak{d}$ under the adjoint action. (This is possible since, by assumption, $\mathfrak{a}$ is $\mathrm{ad}_{\mathfrak{b}_w}$-diagonalizable; in particular, semisimple.) Then

$$[\mathfrak{a}, \mathfrak{c}^\perp] \subset [\mathfrak{a}, \mathfrak{d}] \cap \mathfrak{c}^\perp \subset \mathfrak{c} \cap \mathfrak{c}^\perp = (0),$$

i.e., $\mathfrak{c}^\perp \subset \mathfrak{s}$. But since $\mathfrak{d} \cap \mathfrak{s} = \mathfrak{c}$, we get $\mathfrak{d} = \mathfrak{c}$, thus $\mathfrak{c}$ is a Cartan subalgebra of $\mathfrak{b}_w$. It is easy to see that $\mathfrak{h}$ is a Cartan subalgebra of $\mathfrak{b}_w$. But then, by [Humphreys–72, Theorem 16.2], $(\mathrm{Ad}\ g)\ \mathfrak{c} = \mathfrak{h}$ for some $g \in \mathcal{U}_{\Phi_w}$, proving the lemma. □

As a consequence of Theorem 10.2.5 and Lemma 10.2.8, we obtain the following.

10.2.9 Corollary. *Let $\mathfrak{s}$ be an $\mathrm{ad}_\mathfrak{g}$-diagonalizable subalgebra of $\mathfrak{g}$. Then there exists $g \in \mathcal{G}^{\min}$ such that*

$$(1) \qquad (\mathrm{Ad}\, g)\, \mathfrak{s} \subset \mathfrak{h}.$$

Thus, such an $\mathfrak{s}$ is diagonalizable for any integrable highest weight $\mathfrak{g}$-module $L(\lambda)$ and its restricted dual $L(\lambda)^\vee$.

Proof. Since $\mathfrak{s}$ is $\mathrm{ad}_\mathfrak{g}$-diagonalizable; in particular, it is $\mathrm{ad}_\mathfrak{g}$-triangular. Hence, by Theorem 10.2.5, there exist $g' \in \mathcal{G}^{\min}$ and $w \in W$ such that $(\mathrm{Ad}\, g')\mathfrak{s} \subset \mathfrak{b} \cap w\mathfrak{b}^-$. Since $\mathrm{Ad}\, g'$ is a Lie algebra automorphism of $\mathfrak{g}$ (cf. 6.2.10), $(\mathrm{Ad}\, g')\mathfrak{s}$ is $\mathrm{ad}_\mathfrak{g}$-diagonalizable as well. Now (1) follows from Lemma 10.2.8. □

We get the following from Theorem 10.2.5 and the above Corollary 10.2.9.

10.2.10 Corollary. *Let $\mathfrak{s}$ be a $\mathrm{ad}_\mathfrak{g}$-locally finite, resp. $\mathrm{ad}_\mathfrak{g}$-finitely semisimple, subalgebra of $\mathfrak{g}$. Then $\mathfrak{s}$ is locally finite, resp. finitely semisimple, for any integrable highest weight $\mathfrak{g}$-module $L(\lambda)$ and its restricted dual $L(\lambda)^\vee$.*

Proof. By the same argument as used in the proof of Theorem 10.2.5 for the implication '(a) ⇒ (b)', we get that $\mathfrak{s}$ is finite dimensional. Let $\{s_1, \dots, s_N\}$ be a basis of $\mathfrak{s}$. Then, by Theorem 10.2.5, s_i is $L(\lambda)$ and $L(\lambda)^\vee$ locally finite for each $1 \le i \le N$. So, by the PBW Theorem, $\mathfrak{s}$ is $L(\lambda)$ and $L(\lambda)^\vee$ locally finite.

Now assume that $\mathfrak{s}$ is $\mathrm{ad}_\mathfrak{g}$-finitely semisimple. Then, by Exercise 10.2.E.1, $\mathfrak{s}$ is $\mathrm{ad}_\mathfrak{s}$-semisimple. Further, by the Schur's lemma [Humphreys–72, §6.1], the center $Z(\mathfrak{s})$ of $\mathfrak{s}$ is $\mathrm{ad}_\mathfrak{g}$-diagonalizable. Hence, by Corollary 10.2.9, $Z(\mathfrak{s})$ is diagonalizable for $L(\lambda)$ and $L(\lambda)^\vee$. Thus, by the previous part of the corollary and [Humphreys–72, §6.4, Exercise 5(c)–(d)], $\mathfrak{s}$ is finitely semisimple for $L(\lambda)$ and $L(\lambda)^\vee$. This proves the corollary. □

For a finite-dimensional vector space V and $x \in \mathrm{End}\, V$, recall the Jordan–Chevalley (for short JC) decomposition $x = x_s + x_n$ from [Borel–91, Chap. I, §4.2]. Its extension to an arbitrary V and $x \in \mathrm{End}_{\mathrm{fin}}(V)$, where (as in 6.2.10) $\mathrm{End}_{\mathrm{fin}}(V)$ denotes the set of all the locally finite endomorphisms of V, is straightforward and well known (cf. loc. cit. or [Springer–98, §2.4.7]). However, we include its proof for convenience of the reader.

10.2.11 Proposition. *Let V be any vector space over an algebraically closed field and let $x \in \mathrm{End}_{\mathrm{fin}}(V)$. Then there exists a decomposition:*

$$x = x_s + x_n\,, \tag{1}$$

satisfying the following:

(a) *$x_s, x_n \in \mathrm{End}_{\mathrm{fin}}(V)$ and they commute.*
(b) *x_s, resp. x_n, is diagonalizable, resp. locally nilpotent, for V.*
(c) *For any x-stable subspace $M \subset V$, M is x_s and x_n-stable. Moreover, if M is finite dimensional and x-stable, $x_{|M} = x_{s_{|M}} + x_{n_{|M}}$ is the JC decomposition.*
(d) *Any $y \in \mathrm{End}(V)$ (not necessarily locally finite) which commutes with x also commutes with x_s and x_n.*

Further, any decomposition (1) satisfying (a) and (b) is unique.

This decomposition is again called the *JC decomposition* of x.

Proof. For any finite-dimensional x-stable subspace $M \subset V$, define $x_s^M, x_n^M \in \mathrm{End}\, M$ via the JC decomposition $x_{|M} = x_s^M + x_n^M$. By the properties of the JC decomposition (cf. [Borel–91, Chap. I, Proposition 4.2]), for any other x-stable finite-dimensional subspace $M' \subset V$, x_s^M and x_n^M both keep $M \cap M'$ stable and, moreover, $(x_s^M)_{|M\cap M'} = (x_s^{M'})_{|M\cap M'}$, $(x_n^M)_{|M\cap M'} = (x_n^{M'})_{|M\cap M'}$. Since $x \in \mathrm{End}_{\mathrm{fin}}(V)$ (by assumption), we get well defined $x_s, x_n \in \mathrm{End}_{\mathrm{fin}}(V)$ such that, for any x-stable finite-dimensional subspace M, x_s and x_n keep M stable and, moreover,

$$x_{s_{|M}} = x_s^M, \quad x_{n_{|M}} = x_n^M. \tag{2}$$

From the above construction and the corresponding properties of the JC decomposition, we obtain (1) and the properties (a), (b) and (c). So we come to the property (d). Take $y \in \mathrm{End}(V)$ which commutes with x. We prove that y commutes with x_s (and hence commutes with x_n by (1)): By one of the properties of the JC decomposition (cf. loc. cit.), for any x-stable finite-dimensional subspace $M \subset V$, there exists a polynomial $P_M(t) \in \mathbb{C}[t]$ such that

$$x_{s_{|M}} = P_M(x)_{|M}. \tag{3}$$

Since $yP_M(x) = P_M(x)y$ on the whole of V; in particular, we get by (3)

$$y x_s = x_s y \text{ as maps from } M \cap y^{-1}M \to V. \tag{4}$$

For any $v \in V$, choose a finite-dimensional x-stable subspace M containing $\{v, yv\}$. Then (4) gives $yx_s(v) = x_s\, y(v)$. This proves that x_s commutes with y, and hence the property (d) is established.

For the uniqueness, let $x = x'_s + x'_n$ be another decomposition satisfying (a) and (b). By (a), x'_s commutes with x and hence, by the property (d) for the original decomposition (1), x'_s commutes with x_s and x_n. Similarly, x'_n commutes with x_s and x_n. Hence, for any $v \in V$, we can find a finite-dimensional subspace $M \subset V$ containing v such that M is stable under all of x_s, x_n, x'_s, x'_n. But then $x_{|M} = x_{s_{|M}} + x_{n_{|M}} = x'_{s_{|M}} + x'_{n_{|M}}$. From the uniqueness of theJC decomposition, we get $x_{s_{|M}} = x'_{s_{|M}}, x_{n_{|M}} = x'_{n_{|M}}$. This proves the uniqueness part. □

10.2.12 Theorem. *Let $x \in \mathfrak{g}$ be $\mathrm{ad}_{\mathfrak{g}}$-locally finite. Then x can be decomposed as*

(1) $$x = x_s + x_n, \qquad x_s, x_n \in \mathfrak{g},$$

satisfying the following:

(a) $[x_s, x_n] = 0$.

(b) *x_s, resp. x_n, is π-diagonalizable, resp. π-locally nilpotent, for each of the $\mathfrak{g}$-modules $\pi = \mathrm{ad}_{\mathfrak{g}}$, $L(\lambda)$, $L(\lambda)^\vee$, where $L(\lambda)$ is any integrable highest weight $\mathfrak{g}$-module.*

(c) *For any π as in (b) and any $\pi(x)$-stable subspace M, $\pi(x_s)$ and $\pi(x_n)$ both keep M stable.*

(d) *$\mathfrak{g}^x = \mathfrak{g}^{x_s} \cap \mathfrak{g}^{x_n}$, where $\mathfrak{g}^x$ denotes the centralizer of x in $\mathfrak{g}$.*

Further, the decomposition (1) satisfying (a) and (b) is unique.

The decomposition of any $\mathrm{ad}_{\mathfrak{g}}$-locally finite element x as above is called the *Jordan–Chevalley decomposition.*

Proof. We first prove the existence. By Theorem 10.2.5, there exist $g \in \mathcal{G}^{\min}$ and $w \in W$ such that $(\mathrm{Ad}\, g)\, x \in \mathfrak{b}_w$. Hence, using (6.2.10.3), we can assume that $x \in \mathfrak{b}_w$. Take the pro-subgroup $\mathcal{B}_w := \mathcal{B} \cap w\mathcal{B}^- w^{-1}$ of $\mathcal{B}$ (cf. Lemma 6.1.14), which is, in fact, a (finite-dimensional) affine algebraic group, where $\mathcal{B}^-$ is as defined in 6.2.7. Then $\mathrm{Lie}\, \mathcal{B}_w = \mathfrak{b}_w$. Consider the Jordan–Chevalley decomposition $x = x_s + x_n$ in the Lie algebra $\mathfrak{b}_w$ (cf. [Borel–91, Chap. I, Theorem 4.4]). In particular, it satisfies (a). We now prove that this decomposition satisfies all the properties (b)–(d) by using the properties in loc. cit. For any π as in (b), $\mathfrak{b}_w$ is π-locally finite (cf. Theorem 10.2.5). Take any $\mathfrak{b}_w$-stable finite-dimensional subspace M. Then the Lie algebra action of $\mathfrak{b}_w$ on M integrates to give an algebraic representation of $\mathcal{B}_w$ (cf. Theorem 6.2.3 for $\pi = L(\lambda)$; the case of $L(\lambda)^\vee$ is similar; and Subsection 6.2.10 for $\pi = \mathrm{ad}_{\mathfrak{g}}$). Hence the properties (b) and (c) follow from [Borel–91, Chap. I, Theorem 4.4]. To prove (d), take $y \in \mathfrak{g}^x$. Then $\pi(y)$ commutes with $\pi(x)$ and hence, by Proposition 10.2.11(d), $\pi(y)$ commutes with $\pi(x_s)$ and $\pi(x_n)$, i.e., $\pi([y, x_s]) = \pi([y, x_n]) = 0$. Since this is true for all π as in (b), we get $[y, x_s] = [y, x_n] = 0$ (cf. Lemma 6.2.5).

This gives the inclusion $\mathfrak{g}^x \subset \mathfrak{g}^{x_s} \cap \mathfrak{g}^{x_n}$. The reverse inclusion follows from the decomposition (1). This proves (d).

To prove the uniqueness of the decomposition (1), let $x = x_s' + x_n'$ be another decomposition satisfying (a) and (b). Then, by Proposition 10.2.11, $\pi(x_s) = \pi(x_s')$ and $\pi(x_n) = \pi(x_n')$ for all π as in (b). Hence, $x_s = x_s'$ and $x_n = x_n'$. This completes the uniqueness part and hence the theorem is proved. □

10.2.13 Proposition. *Let $\mathfrak{g} = \mathfrak{g}(A)$ be any Kac–Moody Lie algebra such thatthe GCM A is indecomposable (but not necessarily symmetrizable), and let Δ be the set of roots. Let $\beta = \{\beta_1, \dots, \beta_\ell\}$ be a $\mathbb{R}$-linearly independent set of real roots in Δ such that $\Delta = \Delta_\beta^+ \cup \Delta_\beta^-$, where $\Delta_\beta^+ := \left(\sum_{i=1}^{\ell} \mathbb{Z}_+\beta_i\right) \cap \Delta$ and $\Delta_\beta^- := -\Delta_\beta^+$. Then there exists $w \in W$ (the Weyl group associated to $\mathfrak{g}$) such that*

$$w\Delta_\beta^+ = \Delta^+ \text{ or } w\Delta_\beta^+ = \Delta^-.$$

Proof. By the classification theorem of indecomposable GCM's [Kac–90, Theorem 4.3], A is of one of the following three types: finite, affine or indefinite.

In the first case, the proposition is well known (cf. [Humphreys–72, Theorem 10.3]).

Next, we consider the affine case. Choose $\lambda \in \mathfrak{h}_{\mathbb{R}}^*$ such that

$$(1) \qquad \langle \lambda, \beta_i^\vee \rangle > 0 \quad \text{for all } 1 \le i \le \ell.$$

(This is possible since β_i, and hence $\beta_i^\vee$, are linearly independent.) Let C be the Tits cone for $\mathfrak{g}(A)$ (cf. 1.4.1) and $\bar{C}$ its closure in the Hausdorff topology on $\mathfrak{h}_{\mathbb{R}}^*$. Then $\mathfrak{h}_{\mathbb{R}}^* = \bar{C} \cup (-\bar{C})$ (cf. [Kac–90, Proposition 5.8(b)]). Hence we can choose $\lambda \in C \cup (-C)$ satisfying (1), since $C \cup (-C)$ is dense in $\mathfrak{h}_{\mathbb{R}}^*$ and the set S of λ's satisfying (1) is open and nonempty. Assume that $\lambda \in C$. Then, by the definition of C, there exists $w \in W$ such that $\langle w\lambda, \alpha_i^\vee \rangle \ge 0$, for all the simple coroots $\alpha_i^\vee$. By (1), we get that $\langle w\lambda, \alpha_i^\vee \rangle \ne 0$ and hence

$$(2) \qquad \langle w\lambda, \alpha_i^\vee \rangle > 0, \quad \text{for all } \alpha_i^\vee.$$

By (1) and (2), we get that $w\Delta_\beta^+ \subset \Delta^+$ and hence $w\Delta_\beta^- \subset \Delta^-$. Thus, we have the equality in the above inclusions. Similarly, if $\lambda \in -C$, we get $w\Delta_\beta^+ = \Delta^-$. This proves the proposition in the affine case.

Finally, we come to the case when A is indefinite. In this case, by [Kac–90, Theorem 5.6(c)], there exists an imaginary root $\delta \in \Delta^+$ such that

$$(3) \qquad \langle \delta, \alpha_i^\vee \rangle < 0, \quad \text{for all the simple coroots } \alpha_i^\vee.$$

Assume (if necessary replacing Δ_β^+ by Δ_β^-) that $\delta \in \Delta_\beta^+$. Let W_β be the subgroup of W generated by $\{s_{\beta_i}\}_{1 \le i \le \ell}$, where $s_{\beta_i}\chi := \chi - \langle \chi, \beta_i^\vee \rangle \beta_i$. Since δ is imaginary, it is easy to see that $W_\beta \delta \subset \Delta_\beta^+$. For any $\gamma = \sum_{i=1}^{\ell} n_i \beta_i \in \Delta_\beta^+$, define $|\gamma|_\beta = \sum n_i$. Take $w \in W_\beta$ such that $|w\delta|_\beta$ is minimum. Then $\langle w\delta, \beta_i^\vee \rangle \leq 0$, for all $1 \leq i \leq \ell$. But, by (3), $\langle w\delta, \beta_i^\vee \rangle \neq 0$ and hence

$$\langle w\delta, \beta_i^\vee \rangle < 0, \quad \text{for all } 1 \leq i \leq \ell. \tag{4}$$

From (3) and (4), we get

$$w^{-1}\Delta_\beta^+ \subset \Delta^+ \quad \text{and hence } w^{-1}\Delta_\beta^- \subset \Delta^-.$$

This proves $w^{-1}\Delta_\beta^+ = \Delta^+$, proving the proposition. □

10.2.14 Definition. Any maximal $\mathrm{ad}_{\mathfrak{g}}$-diagonalizable subalgebra $\mathfrak{s} \subset \mathfrak{g}$ is called a *maximal toral subalgebra* of $\mathfrak{g}$.

Let $\mathrm{Aut}_{\mathrm{Lie}}\,\mathfrak{g}$ be the group of all the Lie algebra automorphisms of $\mathfrak{g}$ and let $\Gamma \subset \mathrm{Aut}_{\mathrm{Lie}}\,\mathfrak{g}$ be the subgroup consisting of all those $\theta \in \mathrm{Aut}_{\mathrm{Lie}}\,\mathfrak{g}$ such that θ keeps each of the subsets $\mathfrak{h}$, $\{e_i; 1 \leq i \leq \ell\}$ and $\{f_i; 1 \leq i \leq \ell\}$ separately stable.

10.2.15 Theorem. *Let $\mathfrak{g} = \mathfrak{g}(A)$ be a symmetrizable Kac–Moody Lie algebra. Then we have the following:*

(a) A subalgebra $\mathfrak{s} \subset \mathfrak{g}$ is a maximal toralsubalgebra iff there exists $g \in \mathcal{G}^{\min}$ such that

$$(\mathrm{Ad}\, g)\, \mathfrak{s} = \mathfrak{h}.$$

(b) For another symmetrizable Kac–Moody algebra $\mathfrak{g}(B)$, there exists a Lie algebra isomorphism $\varphi : \mathfrak{g}(A) \xrightarrow{\sim} \mathfrak{g}(B)$ iff A and B are both of the same size, say $\ell \times \ell$, and there exists a permutation σ of $\{1, \ldots, \ell\}$ such that $b_{i,j} = a_{\sigma(i),\sigma(j)}$, for all $1 \leq i, j \leq \ell$.

In the latter case, we say that B is obtained from A by *permuting its indices.*

(c) Assume that A is indecomposable. Then, any $\theta \in \mathrm{Aut}_{\mathrm{Lie}}\,\mathfrak{g}$ can be written as $\theta = \gamma \omega^\varepsilon\, \mathrm{Ad}\, g$, for some $g \in \mathcal{G}^{\min}$, $\gamma \in \Gamma$ and $\varepsilon \in \{0, 1\}$, where ω is the Cartan involution of $\mathfrak{g}$ defined in 1.1.2.

(See Exercise 10.2.E.3 for an extension of this result to decomposable A.)

Proof. (a) follows immediately from Corollary 10.2.9 since $\mathfrak{h}$ is an $\mathrm{ad}_{\mathfrak{g}}$-diagonalizable subalgebra.

(b): If B is obtained from A by permuting its indices, then $\mathfrak{g}(B) \simeq \mathfrak{g}(A)$ (cf. Exercise 10.2.E.2). So, we prove the reverse implication, i.e., if $\varphi : \mathfrak{g}(A) \simeq \mathfrak{g}(B)$ is a Lie algebra isomorphism, we prove that B is obtained from A by permuting its indices. Let $\mathfrak{h}(A) \subset \mathfrak{g}(A)$ be the standard Cartan subalgebra. Then $\varphi^{-1}(\mathfrak{h}(B))$ is a maximal toral subalgebra and hence, by (a), there exists $g \in \mathcal{G}^{\min}(A)$ such that $\varphi_1 := \varphi \circ \mathrm{Ad}\, g$ satisfies $\varphi_1(\mathfrak{h}(A)) = \mathfrak{h}(B)$. Let φ_1^* be the dual map $\mathfrak{h}(B)^* \to \mathfrak{h}(A)^*$. It is easy to see that $\varphi_1^*\Delta(B) = \Delta(A)$. By Corollary 1.3.6(a) and (1.3.5.4),

$\varphi_1^* \Delta(B)_{\mathrm{re}} = \Delta(A)_{\mathrm{re}}$. Hence, by Proposition 10.2.13, there exists $n \in N(A)$ such that $\varphi_2 := \varphi_1 \circ (\mathrm{Ad}\, n) \circ \Omega$ satisfies $\varphi_2^*(\Delta(B)^+) = \Delta(A)^+$, where $N(A)$ is the normalizer of $T(A)$ in $\mathcal{G}^{\min}(A)$, $\Delta(A)^+$ is the standard set of positive roots and $\Omega : \mathfrak{g}(A) \to \mathfrak{g}(A)$ is a certain Lie algebra automorphism such that, for any indecomposable component A_i of A, $\Omega_{|\mathfrak{g}(A_i)}$ is either the identity map or the Cartan involution $\omega(A_i)$ (cf. 1.1.2). In particular, φ_2^* takes the standard simple roots in $\Delta(B)^+$ bijectively to the standard simple roots in $\Delta(A)^+$. Hence, there exist $\lambda_i, \mu_i \in \mathbb{C}^*$ and a permutation σ of $\{1, \dots, \ell\}$ such that, for $e_i := e_i(A)$ and $\bar{e}_i := e_i(B)$,

$$\text{(1)} \qquad \varphi_2(e_i) = \lambda_i \bar{e}_{\sigma(i)} \text{ and } \varphi_2(f_i) = \mu_i \bar{f}_{\sigma(i)}, \quad \text{for all } 1 \leq i \leq \ell.$$

Hence, for any $h \in \mathfrak{h}(A)$,

$$\lambda_i \alpha_i(h)\, \bar{e}_{\sigma(i)} = \varphi_2[h, e_i] = [\varphi_2 h, \varphi_2 e_i] = \lambda_i \bar{\alpha}_{\sigma(i)}(\varphi_2(h)) \bar{e}_{\sigma(i)},$$

giving

$$\text{(2)} \qquad \alpha_i(h) = \bar{\alpha}_{\sigma(i)}(\varphi_2(h)), \quad \text{for all } h \in \mathfrak{h},$$

where α_i, resp. $\bar{\alpha}_i$, are the simple roots of $\mathfrak{g}(A)$, resp. $\mathfrak{g}(B)$. Now

$$\text{(3)} \qquad \varphi_2(\alpha_i^\vee) = \varphi_2[e_i, f_i] = [\varphi_2 e_i, \varphi_2 f_i] = \lambda_i \mu_i \bar{\alpha}^\vee_{\sigma(i)}.$$

Combining (2) for $h = \alpha_i^\vee$, and (3), we get

$$\text{(4)} \qquad 1 = \lambda_i \mu_i.$$

By (2) (for $h = \alpha_j^\vee$), (3) and (4), we get $\alpha_i(\alpha_j^\vee) = \bar{\alpha}_{\sigma(i)}(\bar{\alpha}^\vee_{\sigma(j)})$. From this (b) follows.

(c): To prove (c), we follow the above argument for the proof of the (b) part (taking $A = B$). Thus, there exist $\varepsilon \in \{0, 1\}$ and $g \in \mathcal{G}^{\min}$ such that $\theta_2 := \theta \circ \mathrm{Ad}\, g \circ \omega^\varepsilon$ satisfies $\theta_2(\mathfrak{h}) = \mathfrak{h}$, $\theta_2(e_i) = \lambda_i e_{\sigma(i)}$ and $\theta_2(f_i) = \lambda_i^{-1} f_{\sigma(i)}$, for a permutation σ of $\{1, \dots, \ell\}$ and some $\lambda_i \in \mathbb{C}^*$.

Now take $t \in T := \mathrm{Hom}_{\mathbb{Z}}(\mathfrak{h}^*_{\mathbb{Z}}, \mathbb{C}^*)$ satisfying $t(\alpha_i) = \lambda_i^{-1}$. (This is possible since $\mathbb{C}^*$ is an injective $\mathbb{Z}$-module, cf. [Matsumura–89, Theorem B3].) Then $\gamma := \theta_2 \circ \mathrm{Ad}(t) \in \Gamma$. But since, for any $t \in T$,

$$\text{(5)} \qquad \mathrm{Ad}\, t \circ \omega = \omega \circ \mathrm{Ad}(t^{-1}),$$

the (c) part follows. This proves the theorem. □

10.2.E EXERCISES

(1) Let A be an associative algebra. Then an A-module V is called *semisimple* if $V = \oplus_i V_i$, where V_i's are irreducible submodules of V.

Prove that an A-module V is semisimple iff V is the sum (not necessarily direct sum) of its irreducible A-submodules.

Prove further that any A-submodule W of a semisimple A-module V is semisimple.

(Compare this exercise with Exercise 3.1.E.3.)

(2) For any two GCM's A and B such that B is obtained from A by permuting its indices, prove that $\mathfrak{g}(A) \simeq \mathfrak{g}(B)$ as Lie algebras.

(3) Extend Theorem 10.2.15(c) for any (possibly decomposable) symmetrizable GCM A by replacing ω^{ε} with an automorphism Ω of $\mathfrak{g}(A)$ such that $\Omega_{|\mathfrak{g}(A_i)} = \omega(A_i)^{\varepsilon_i}$ for any indecomposable component A_i of A (for $\varepsilon_i \in \{0, 1\}$).

(4) For any GCM $A = (a_{i,j})_{1 \le i,j \le \ell}$ let $\mathrm{Aut}(A)$ be the group of permutations σ of $\{1, \dots, \ell\}$ such that $a_{i,j} = a_{\sigma(i),\sigma(j)}$, for all $1 \le i, j \le \ell$. Show that the subgroup $\Gamma \subset \mathrm{Aut}_{\mathrm{Lie}}\, \mathfrak{g}$, defined in Definition 10.2.14, fits into an exact sequence of groups:

$$0 \to \mathrm{Hom}_{\mathbb{C}}(\mathfrak{g}/[\mathfrak{g}, \mathfrak{g}], Z(\mathfrak{g})) \to \Gamma \to \mathrm{Aut}(A) \to 1,$$

where $Z(\mathfrak{g})$ is the center of $\mathfrak{g}$.

(5) For any $\mathrm{ad}_{\mathfrak{g}}$-locally finite element $x \in \mathfrak{g}$, prove that $\mathrm{Ad}\,(\mathrm{Exp}\, x) = \exp\,(\mathrm{ad}\, x)$ as automorphisms of $\mathfrak{g}$.

10.3. Conjugacy Theorems for Groups

In this section $\mathfrak{g}$ is any Kac–Moody Lie algebra.

We recall some of the basic definitions and elementary properties of convex functions (cf. [Roberts–Varberg-73] for a detailed theory).

10.3.1 Definition. Let $U \subset \mathbb{R}^n$ be a nonempty convex subset. A function $f : U \to \mathbb{R}$ is called *convex* if, for all $x, y \in U$ and $t \in [0, 1]$,

$$f(tx + (1-t)y) \le tf(x) + (1-t)\, f(y).$$

If the above inequality is strict whenever $x \ne y$ and $t \in (0, 1)$, the function f is called *strictly convex*.

We will use the following properties of convex functions which are easy to establish.

(a) The function $x \mapsto e^x$ is strictly convex on $\mathbb{R}$. Thus, the function $x \mapsto e^{Tx}$ is convex on $\mathbb{R}^n$, where $T : \mathbb{R}^n \to \mathbb{R}$ is any linear function.

(b) Let $U \subset \mathbb{R}^n$ be a nonempty convex subset and let $\{f_n\}_{n\geq 1}$ be a sequence of convex functions on U such that $f = \sum f_n$ converges pointwise on U. Then f is convex on U and, moreover, f is strictly convex if at least one of f_n is strictly convex.

(c) Let f be a convex function on a convex open subset $U \subset \mathbb{R}^n$ and suppose that f is differentiable at $x_o \in U$. Then, for $\alpha \in \mathbb{R}^n \backslash \{0\}$ and $t > 0$ such that $x_o + t\alpha \in U$,

$$f(x_o + t\alpha) - f(x_o) \geq t(\partial_\alpha f)(x_o),$$

where ∂_α denotes the directional derivative of f in the direction of α. Furthermore, the inequality is strict if f is strictly convex.

To prove this, observe that, for any $s \in [0, 1]$, $f(x_o + ts\alpha) = f(s(x_o + t\alpha) + (1 - s)x_o) \leq sf(x_o + t\alpha) + (1 - s) f(x_o)$.

(d) Let V be a finite-dimensional real vector space and let $S = \{v_1, \dots , v_d\} \subset V$ a finite nonempty subset. As in 10.1.1, let $[S]$ be the convex hull of S. Let $f : [S] \to \mathbb{R}$ be a convex function. Then, for any $v \in [S]$,

$$f(v) \leq \max_i \{f(v_i)\}.$$

This is easy to establish by induction on d.

(e) Let $f : U \to \mathbb{R}$ be a strictly convex function on a convex subset $U \subset \mathbb{R}^n$. If f attains an absolute minimum at some point $x \in U$, then x is unique.

For, if x and y are two such distinct points, then, for any $t \in (0, 1)$,

$$f(tx + (1 - t)y) < tf(x) + (1 - t) f(y) = f(x).$$

A contradiction, since f attains a minimum at x.

Recall the definition of the Tits cone $C \subset \mathfrak{h}^*_{\mathbb{R}}$ from 1.4.1. By Proposition 1.4.2(c), C is a (W-invariant) convex cone. Let C^o denote the interior of C. By Exercise 1.4.E.1, C^o is a (W-invariant) convex open subset of $\mathfrak{h}^*_{\mathbb{R}}$. A subset $S \subset C^o$ is called *admissible* if it is W-invariant and there exists a finite subset $A = A_S \subset \mathfrak{h}^*_{\mathbb{R}}$ such that $S \subseteq C^o \cap (A - Q^+)$, where $Q^+ := \oplus_{i=1}^{\ell} \mathbb{Z}_+\alpha_i$.

10.3.2 Theorem. *There exists a strictly convex W-invariant real analytic function $F : C^o \to (0, \infty)$ such that, for every admissible subset $S \subset C^o$, the series $\sum_{\lambda\in S,\, \mu\in S\cap D_{\mathbb{R}}} F(\frac{1}{2}(\lambda + \mu))$ converges, where the dominant chamber $D_{\mathbb{R}}$ is as in (1.4.1.1).*

Proof. Pick a basis $\{h_i\}$ of $\mathfrak{h}_{\mathbb{R}}$ such that $\langle \alpha_j, h_i \rangle$ is a positive integer for all the simple roots α_j. Define $F = F_{\{h_i\}} : C^o \to (0, \infty)$ by

$$F(\lambda) = \sum_i \sum_{w \in W} e^{\langle \lambda, wh_i \rangle}. \tag{1}$$

To prove that the series is convergent, we prove the following: For any $h \in \mathfrak{h}_{\mathbb{R}}$ such that $\langle \alpha_j, h \rangle$ is a positive integer for all α_j, and $\lambda \in D^o_{\mathbb{R}}$ (the interior of $D_{\mathbb{R}}$),

$$\sum_{w \in W} e^{\langle \lambda, wh \rangle} \leq e^{\langle \lambda, h \rangle} \sum_{(m_1, \dots, m_\ell) \in \mathbb{Z}^\ell_+} e^{-\delta(\lambda)\,(m_1 + \cdots + m_\ell)}, \tag{2}$$

where $\delta(\lambda) := \min\{\langle \lambda, \alpha_j^\vee \rangle;\ 1 \leq j \leq \ell\} > 0$.

By (1.3.22.2),

$$h - wh \in Q^\vee_+, \text{ where } Q^\vee_+ := \oplus_j \mathbb{Z}_+ \alpha_j^\vee. \tag{3}$$

Hence,

$$\begin{aligned} \sum_{w \in W} e^{\langle \lambda, wh \rangle} &= e^{\langle \lambda, h \rangle} \sum_{w \in W} e^{\langle \lambda, wh - h \rangle} \\ &\leq e^{\langle \lambda, h \rangle} \sum_{(m_1, \dots, m_\ell) \in \mathbb{Z}^\ell_+} e^{-\langle \lambda, m_1 \alpha_1^\vee + \cdots + m_\ell \alpha_\ell^\vee \rangle}, \end{aligned} \tag{4}$$

since the W-isotropy of h is trivial (by Proposition 1.4.2(a)). From (4), we get (2).

From (2), we see that the series in (1) is convergent for all $\lambda \in D^o_{\mathbb{R}}$ and, since F is (clearly) W-invariant, the series in (1) is convergent for all $\lambda \in W \cdot D^o_{\mathbb{R}}$. From the last part of Exercise 1.4.E.1 and the convexity of the functions $\lambda \mapsto e^{\langle \lambda, wh_i \rangle}$ (cf. 10.3.1(a)), we get that the series in (1) is convergent for all $\lambda \in C^o$. By 10.3.1(a)–(b), F is a convex function.

Since $\{h_i\}$ is a basis of $\mathfrak{h}_{\mathbb{R}}$ and $x \mapsto e^x$ is strictly convex on $\mathbb{R}$, it is easy to see that $\lambda \mapsto \sum_i e^{\langle \lambda, h_i \rangle}$ is strictly convex on $\mathfrak{h}^*_{\mathbb{R}}$ (and hence on C^o). In particular, by 10.3.1(b), F is strictly convex.

To prove that F is real analytic on C^o, we extend F to $\hat{C}^o := C^o + \sqrt{-1}\,\mathfrak{h}^*_{\mathbb{R}}$ by the same definition as (1). Then the series defining F is absolutely convergent on $\hat{C}^o$. Further, since each term in the series of F is convex on C^o, for any $\Gamma = \{\lambda_1, \dots, \lambda_d\} \subset C^o$ the absolute convergence of F on $\hat{C}^o$ is uniform on $[\Gamma] + \sqrt{-1}\,\mathfrak{h}^*_{\mathbb{R}}$; use 10.3.1(d). In particular, F is holomorphic on $\hat{C}^o$ and hence real analytic on C^o.

Finally, we prove the convergence of $\sum_{\lambda\in S,\,\mu\in S\cap D_{\mathbb{R}}} F(\frac{1}{2}(\lambda+\mu))$. Let $N := \max\{\#W_Y : Y \subset \{1,\dots,\ell\}$ is of finite type$\}$. For any h_i as above, using Proposition 1.4.2, we get:

$$\begin{aligned}
\sum_{\lambda\in S}\sum_{\mu\in S\cap D_{\mathbb{R}}}\sum_{w\in W} e^{\frac{1}{2}\langle\lambda+\mu, wh_i\rangle} &= \sum_{\lambda\in S}\sum_{\mu\in S\cap D_{\mathbb{R}},\, w\in W} e^{\frac{1}{2}\langle\lambda+w\mu, h_i\rangle}\\
&\le N\sum_{\lambda\in S}\sum_{\mu\in S} e^{\frac{1}{2}\langle\lambda+\mu, h_i\rangle}\\
&= N\Big(\sum_{\lambda\in S} e^{\frac{1}{2}\langle\lambda, h_i\rangle}\Big)^2\\
&\le N\Big(\sum_{\theta\in A} e^{\frac{1}{2}\langle\theta, h_i\rangle}\Big(\sum_{(n_1,\dots,n_\ell)\in\mathbb{Z}_+^\ell} e^{-\frac{1}{2}\sum_{j=1}^{\ell} n_j\langle\alpha_j, h_i\rangle}\Big)\Big)^2\\
&< \infty,
\end{aligned}$$

where $A \subset \mathfrak{h}_{\mathbb{R}}^*$ is a finite subset such that $S \subseteq C^o \cap (A - Q^+)$. This proves the theorem. □

10.3.3 Remark. In the above proof, to define $F = F_{\{h_i\}}$, we chose a basis $\{h_i\}$ of $\mathfrak{h}_{\mathbb{R}}$ such that $\langle\alpha_j, h_i\rangle$ is a positive integer for all the simple roots α_j. In fact, the theorem holds for the function $F = F_{\{h_i\}}$ defined by (1) for any basis $\{h_i\}$ of $\mathfrak{h}_{\mathbb{R}}$ such that $h_i \in (C^\vee)^o$ (with a slightly modified proof), where $(C^\vee)^o$ is the interior of the Tits cone $C^\vee \subset \mathfrak{h}_{\mathbb{R}}$.

10.3.4 Definition. Fix any $F : C^o \to (0,\infty)$ as in Theorem 10.3.2. For any compact convex subset $K \subset C^o$, F being strictly convex, $F_{|K}$ attains its absolute minimum on a unique point denoted μ_K of K (cf. 10.3.1(e)).

From now on we shall fix one function $F : C^o \to (0,\infty)$ satisfying the properties of Theorem 10.3.2.

As a consequence of Theorem 10.3.2, we obtain the following.

10.3.5 Corollary. *Let $S \subset C^o$ be an admissible subset. Fix $\varepsilon > 0$. Then the set of all the finite subsets $Z \subset S$ such that $F(\mu_{[Z]}) \ge \varepsilon$ and $Z \cap D_{\mathbb{R}} \neq \emptyset$ is finite.*

Proof. By Theorem 10.3.2, the set

$$E_\varepsilon = \{(\lambda,\mu) \in S \times (S \cap D_{\mathbb{R}}) : F(\frac{1}{2}(\lambda+\mu)) \ge \varepsilon\}$$

is finite. Further, for any $\lambda, \mu \in Z$,

$$F(\frac{1}{2}(\lambda+\mu)) \ge F(\mu_{[Z]}) \ge \varepsilon.$$

In particular, since $Z \cap D_{\mathbb{R}} \neq \emptyset$, $Z \subset \pi_1(E_\varepsilon)$, where π_1 is the projection to the first factor. This proves the corollary. □

10.3.6 Lemma. *Let $K \subset C^o$ be a nonempty compact convex subset such that $\mu_K \in D_{\mathbb{R}}$. Then, for any $x \in (K + \sum_{i=1}^{\ell} \mathbb{R}_+\alpha_i) \cap C^o$,*

(1) $$F(x) \geq F(\mu_K).$$

Moreover, for $x \neq \mu_K$, $F(x) > F(\mu_K)$.

Proof. Write $x = \mu_K + \lambda + \beta$, for $\mu_K + \lambda \in K$ and $\beta \in \sum \mathbb{R}_+\alpha_i$. Since, for $0 \leq t \leq 1$,

$$F(\mu_K + t\lambda) = F(t(\mu_K + \lambda) + (1-t)\,\mu_K) \geq F(\mu_K),$$

we get

(2) $$(\partial_\lambda F)(\mu_K) \geq 0.$$

For any simple coroot $\alpha_i^\vee$, set $t_i := \langle \mu_K, \alpha_i^\vee \rangle$. Then $F(\mu_K) = F(s_i\mu_K) = F(\mu_K - t_i\alpha_i)$. In particular, for any $0 \leq t \leq 1$,

$$F(\mu_K - tt_i\alpha_i) = F(t(\mu_K - t_i\alpha_i) + (1-t)\,\mu_K) \leq F(\mu_K).$$

From this inequality, we get

(3) $$(\partial_{\alpha_i} F)(\mu_K) \geq 0, \quad \text{if } t_i > 0.$$

If $t_i = 0$, since F is W-invariant,

$$F(\mu_K - t\alpha_i) = F(\mu_K + t\alpha_i),$$

for any $t \in \mathbb{R}$ such that $\mu_K + t\alpha_i \in C^o$. In particular,

$$F(\mu_K) = F\Big(\frac{1}{2}(\mu_K - t\alpha_i + \mu_K + t\alpha_i)\Big) \leq F(\mu_K + t\alpha_i).$$

So, again,

(4) $$(\partial_{\alpha_i} F)(\mu_K) \geq 0, \quad \text{for } t_i = 0.$$

(In fact, though we do not need it, in this case $(\partial_{\alpha_i} F)(\mu_K) = 0$.) Combining (2)–(4), we get (since $\mu_K \in D_{\mathbb{R}}$ by assumption):

$$(\partial_{\lambda+\beta} F)(\mu_K) \geq 0.$$

Now (1) follows from 10.3.1(c). Since $\tilde{K} := (K + \sum_{i=1}^{\ell} \mathbb{R}_+\alpha_i) \cap C^o$ is a convex subset of C^o containing K, by (1) and 10.3.1(e), $F(x) > F(\mu_K)$ if $x \neq \mu_K \in \tilde{K}$. □

10.3.7 Definition. Recall the definition of the category $\mathfrak{M}(\mathcal{G})$ from 6.2.1. For any V in $\mathfrak{M}(\mathcal{G})$, let $P(V)$ be the set of T-weights of V. Define the full subcategory $\mathfrak{M}^o(\mathcal{G})$ to consist of those $V \in \mathfrak{M}(\mathcal{G})$ such that $P(V) \subset C^o$. Then, just as $\mathfrak{M}(\mathcal{G})$, $\mathfrak{M}^o(\mathcal{G})$ also is closed under taking direct sums, tensor products, submodules and quotient modules.

For any $\lambda \in C^o \cap D_{\mathbb{Z}}$, the module $V = L^{\max}(\lambda)$ belongs to $\mathfrak{M}^o(\mathcal{G})$, and hence so does its any quotient. This follows from Exercise 2.3.E.3, since $P(V) \subset [W \cdot \lambda] \subset C^o$. For any $V \in \mathfrak{M}^o(\mathcal{G})$ such that V is a finitely generated $\mathcal{G}$-module, it is easy to see that $P(V)$ is an admissible set.

Fix any function $F : C^o \to (0, \infty)$ which satisfies the properties of Theorem 10.3.2. For any $V \in \mathfrak{M}^o(\mathcal{G})$ and nonzero $v \in V$, define the function $\Phi_v : \mathcal{G} \to C^o$ by

$$\Phi_v(g) = \mu_{[\text{supp}\, gv]},$$

where supp gv is as defined in 10.1.7. Observe that Φ_v depends on the choice of F.

The following proposition summarizes some of the basic properties of the function Φ_v.

10.3.8 Proposition. *(a) For any $n \in N$,*

$$\Phi_v(ng) = w \cdot \Phi_v(g), \tag{1}$$

where $w := n \mod T$ (cf. Corollary 6.1.8).

(b) For any $g \in \mathcal{G}$ such that $\Phi_v(g) \in D_{\mathbb{R}}$,

$$F(\Phi_v(g)) \leq F(\Phi_v(bg)), \text{ for all } b \in \mathcal{B}. \tag{2}$$

Moreover, the equality occurs in (2) for some $b \in \mathcal{B}$ iff

$$\Phi_v(g) = \Phi_v(bg).$$

(c) There exists $g_o \in \mathcal{G}$ such that $\Phi_v(g_o) \in D_{\mathbb{R}}$ and

$$F(\Phi_v(g_o)) \geq F(\Phi_v(g)), \text{ for all } g \in \mathcal{G}. \tag{3}$$

Further, let $Z := \{g_o' \in \mathcal{G} : \Phi_v(g_o') \in D_{\mathbb{R}} \text{ and } F(\Phi_v(g_o')) = F(\Phi_v(g_o))\}$. Then $Z = \mathcal{P}_{g_o} g_o$, where $\mathcal{P}_{g_o}$ is the standard parabolic subgroup $\mathcal{B}W_{\Phi_v(g_o)}\mathcal{B}$ and

$W_{\Phi_v(g_o)} \subset W$ is the isotropy subgroup of $\Phi_v(g_o)$. (Since $\Phi_v(g_o) \in C^o$, by Proposition 1.4.2(f), $\mathcal{P}_{g_o}$ is of finite type.)

Moreover, for $g'_o \in Z$, $\Phi_v(g_o) = \Phi_v(g'_o)$.

Proof. Since F is W-invariant, (1) follows from Lemma 1.3.5(a). Further, (2) follows readily from Lemma 10.3.6. We now prove (c): Let $V' \subset V$ be the $\mathcal{G}$-submodule generated by v. Then $P(V')$ is an admissible set (cf. 10.3.7). Now, the existence of g_o satisfying (3) is guaranteed by Corollary 10.3.5 and (1). We next take $g'_o \in Z$. By the Bruhat decomposition, write $g'_o g_o^{-1} = bnb'$, for some $b, b' \in \mathcal{B}$ and $n \in N$. Let $\bar{g}_o = b' g_o$. Then, by the (b) part, $\Phi_v(\bar{g}_o) = \Phi_v(g_o)$. Similarly, $\Phi_v(b^{-1}g'_o) = \Phi_v(g'_o)$. Hence, $\Phi_v(n\bar{g}_o) = \Phi_v(b^{-1}g'_o) = \Phi_v(g'_o)$. But, by (1), $\Phi_v(n\bar{g}_o) = w \cdot \Phi_v(\bar{g}_o) = w \cdot \Phi_v(g_o)$ (where $w := n \mod T \in W$). Thus, we get $w \cdot \Phi_v(g_o) = \Phi_v(g'_o)$. But since, by assumption, $\Phi_v(g_o), \Phi_v(g'_o)$ both belong to $D_{\mathbb{R}}$, by Proposition 1.4.2(b), $\Phi_v(g_o) = \Phi_v(g'_o)$ and $w \in W_{\Phi_v(g_o)}$. This proves that $Z \subset \mathcal{P}_{g_o} g_o$. Conversely, for $g'_o \in \mathcal{P}_{g_o} g_o$, by (a)–(b) it is easy to see that $g'_o \in Z$. This proves the (c) part. □

10.3.9 Theorem. *Let $\mathcal{G}$ be any Kac–Moody group. The following conditions on a subgroup S of $\mathcal{G}$ are equivalent:*

(a) $S \subset g^{-1}\mathcal{P}_Y g$, for some $g \in \mathcal{G}$ and some standard parabolic subgroup $\mathcal{P}_Y$ of finite type.

(b) $S \subset \cup_w \mathcal{B}w\mathcal{B}$, where the union runs over a finite subset of the Weyl group W.

(c) Any $\mathcal{G}$-module $V \in \mathfrak{M}^o(\mathcal{G})$ is locally finite under the action of S.

(d) There exists a $\mathcal{G}$-module $V \in \mathfrak{M}^o(\mathcal{G})$ and a nonzero finite-dimensional S-stable subspace $V_1 \subset V$.

A subgroup S satisfying any of the above four equivalent properties is called a *bounded* subgroup of $\mathcal{G}$.

Proof. (a) $\Rightarrow$ (b): By the Bruhat decomposition Corollary 6.1.20, $g \in \mathcal{B}z\mathcal{B}$ (for some $z \in W$) and hence, Y being of finite type, $g^{-1}\mathcal{P}_Y g$ is contained in a finite union of the cosets $\mathcal{B}w\mathcal{B}$. (Use the axiom (BN_4) of Tits system as in 5.1.1.) This proves (b).

The implication (b) $\Rightarrow$ (c) is clear from the definition of the category $\mathfrak{M}^o(\mathcal{G})$; in fact (b) $\Rightarrow$ (c) for any $V \in \mathfrak{M}(\mathcal{G})$. Assertion (d) is a special case of (c). So, finally, we prove the implication (d) $\Rightarrow$(a). Replacing V by $\wedge^{\dim V_1} V$, we can assume that V_1 is one-dimensional, i.e., S fixes the line $\mathbb{C}v$, for some nonzero $v \in V$. As in Proposition 10.3.8(c), choose $g_o \in \mathcal{G}$ such that $\Phi_v(g_o) \in D_{\mathbb{R}}$ and $F(\Phi_v(g_o)) \geq F(\Phi_v(g))$ for all $g \in \mathcal{G}$. Then, by Proposition 10.3.8(c), $S \subset g_o^{-1}\mathcal{P}_{g_o} g_o$, where $\mathcal{P}_{g_o}$ is the standard parabolic subgroup $\mathcal{B}W_{\Phi_v(g_o)}\mathcal{B}$ which is of finite type. This proves (a). □

10.3.10 Remarks. (a) The above theorem is true (by the same proof) for any subsemigroup S of $\mathcal{G}$.

(b) By (7.4.5.2), for any $g \in \mathcal{G}$, there exists $g_1 \in \mathcal{G}^{\min}$ such that $g\mathcal{P}_Y g^{-1} = g_1\mathcal{P}_Y g_1^{-1}$.

10.3.11 Definition. For any subset $Y \subset \{1, \dots, \ell\}$, similarly to the definition of $\mathcal{P}_Y$ as in 6.1.18, define the *standard negative parabolic subgroup* $\mathcal{P}_Y^-$ of $\mathcal{G}$ by

$$\mathcal{P}_Y^- := \mathcal{B}^- W_Y \mathcal{B}^-,$$

where $\mathcal{B}^-$ is as defined in 6.2.7. (It is easy to see that $\mathcal{P}_Y^-$ is indeed a group.) Clearly $\mathcal{P}_Y^- \subset \mathcal{G}^{\min}$.

A subgroup $S \subset \mathcal{G}^{\min}$ is called *antibounded* if there exist $g \in \mathcal{G}^{\min}$ and a standard negative parabolic subgroup $\mathcal{P}_Y^-$ (where Y is of finite type) such that $S \subset g\mathcal{P}_Y^- g^{-1}$.

Antibounded subgroups are characterized by the four equivalent conditions, similar to the ones in Theorem 10.3.9.

10.3.12 Theorem. *Let $S \subset \mathcal{G}^{\min}$ be a subgroup. Then the following conditions are equivalent:*

(a) S is bounded and antibounded.

(b) There exist finite type subsets Y and Y' of $\{1, \dots, \ell\}$, $g \in \mathcal{G}^{\min}$, and $w \in W$, such that

$$gSg^{-1} \subset \mathcal{P}_Y \cap w\mathcal{P}_{Y'}^- w^{-1}.$$

(c) The adjoint action of S on $\mathfrak{g}$ is locally finite (cf. 6.2.10 and Exercise 6.2.E.2).

Proof. (a)⇒(b): By the definitions and Remark 10.3.10(b), there exist $g_1, g_2 \in \mathcal{G}^{\min}$ and finite type subsets Y, Y' such that

$$S \subset g_1^{-1}\mathcal{P}_Y g_1 \cap g_2^{-1}\mathcal{P}_{Y'}^- g_2.$$

Write, by the Birkhoff decomposition, $g_2 g_1^{-1} = u_- n^{-1} u$, for $u_- \in \mathcal{U}^-$, $u \in \mathcal{U}$ and $n \in N$. Putting $g = ug_1$, we get $gSg^{-1} \subset \mathcal{P}_Y \cap w\mathcal{P}_{Y'}^- w^{-1}$, where $w = n$ mod T. (Since $g_1, g_2 \in \mathcal{G}^{\min}$, so is u and hence $g \in \mathcal{G}^{\min}$.)

(b)⇒(c): We can, of course, assume that $S \subset \mathcal{P}_Y \cap w\mathcal{P}_{Y'}^- w^{-1}$. Using the identity $\mathcal{B}z\mathcal{B} = \mathcal{B}z\mathcal{U}_{\Phi_{z^{-1}}}$ for any $z \in W$ (cf. Exercise 6.2.E.1), we get that, for any $x \in \mathfrak{g}$,

$$(1) \qquad \operatorname{Ad}(\mathcal{P}_Y \cap \mathcal{G}^{\min}) \cdot x \subset \operatorname{Ad}(\mathcal{U} \cap \mathcal{G}^{\min}) \cdot M,$$

for some finite-dimensional subspace $M \subset \mathfrak{g}$. But, by Lemma 7.4.3 and Exercise 10.2.E.5,

$$\text{(2)} \qquad \operatorname{Ad}(\mathcal{U} \cap \mathcal{G}^{\min}) \cdot M \subset (\operatorname{ad} U(\mathfrak{n})) \cdot M.$$

(In fact, even though we do not need it, the equality occurs in (2).)

Combining (1) and (2), we get

$$\text{(3)} \qquad \operatorname{Ad}(\mathcal{P}_Y \cap \mathcal{G}^{\min}) \cdot x \subset (\operatorname{ad} U(\mathfrak{n})) \cdot M.$$

Similarly, using Lemma 1.3.5,

$$\text{(4)} \qquad \operatorname{Ad}(w\mathcal{P}_{Y'}^{-}w^{-1}) \cdot x \subset (\operatorname{ad} U(\mathfrak{n}^{-})) \cdot M',$$

for some finite-dimensional subspace $M' \subset \mathfrak{g}$.

Combining (3) and (4), we obtain that the span of $\operatorname{Ad}(\mathcal{P}_Y \cap w\mathcal{P}_{Y'}^{-}w^{-1}) \cdot x$ is finite dimensional. This proves (c).

(c) $\Rightarrow$(a): We prove that S is bounded. (The proof that S is antibounded is identical.) Assuming that S is not bounded, by Theorem 10.3.9(b), we can choose an infinite sequence of elements in S:

$$a_k = u_k n_k u'_k, \text{ with } u_k,\ u'_k \in \mathcal{G}^{\min} \cap \mathcal{U},\ n_k \in N$$

such that $w_k := n_k \mod T \in W$ are all distinct.

For any $\alpha \in \Delta_{\text{re}}$, choose $\beta_k(\alpha)$ with minimal $|\beta_k(\alpha)|$ in $\operatorname{supp}((\operatorname{Ad} a_k)\,\mathfrak{g}_\alpha)$. Then, by Exercise 10.2.E.5,

$$\text{(5)} \qquad |w_k \cdot \alpha| \geq |\beta_k(\alpha)|.$$

Replacing α by $-\alpha$ in (5), we get

$$\text{(6)} \qquad |w_k \cdot \alpha| \leq -|\beta_k(-\alpha)|.$$

But, since S acts locally finitely on $\mathfrak{g}$ (by assumption), $-c(\alpha) \leq |\beta_k(\alpha)| \leq c(\alpha)$, for some constant $c(\alpha)$ depending only upon α (and not on k). In particular, by (5)–(6),

$$\text{(7)} \qquad -c'(\alpha) \leq |w_k \cdot \alpha| \leq c'(\alpha),$$

for $c'(\alpha) := \max\{c(\alpha), c(-\alpha)\}$. Taking $\alpha = \alpha_1, \cdots, \alpha_\ell$ in (7), we arrive at a contradition, since the map $W \to Q^{\oplus \ell}$, $w \mapsto (w\alpha_1, \cdots, w\alpha_\ell)$, is injective (by Lemma 1.3.13), where Q is the root lattice.

This proves (a), proving the theorem. □

As a corollary of the above theorem we prove its infinitesimal version.

10.3.13 Corollary. *Let $\mathfrak{g}$ be a symmetrizable Kac–Moody algebra and $\mathfrak{s} \subset \mathfrak{g}$ a subalgebra. Then $\mathfrak{s}$ is $\mathrm{ad}_{\mathfrak{g}}$-locally finite iff there exist $g \in \mathcal{G}^{\min}$, $n \in N$ and finite type subsets $Y, Y' \subset \{1, \ldots, \ell\}$ such that $(\mathrm{Ad}\, g)\, \mathfrak{s} \subset \mathfrak{p}_Y \cap (\mathrm{Ad}\, n)\mathfrak{p}^-_{Y'}$ (cf. 1.2.2 for the definition of $\mathfrak{p}_Y$, $\mathfrak{p}^-_Y$).*

Proof. We first assume that $\mathfrak{s}$ is $\mathrm{ad}_{\mathfrak{g}}$-locally finite. By Corollary 10.2.6, $\mathfrak{s} \subset \mathfrak{g}_{\mathrm{fin}}$ and $\mathrm{Exp}\, \mathfrak{s} \subset \mathcal{G}^{\min}$. (This is where we use the symmetrizability assumption on $\mathfrak{g}$.) Let S be the subgroup of $\mathcal{G}^{\min}$ generated by $\mathrm{Exp}\, \mathfrak{s}$. By Exercise 10.2.E.5, $\mathfrak{g}$ is locally finite under the adjoint action of S since $\mathfrak{s}$ is $\mathrm{ad}_{\mathfrak{g}}$-locally finite. Hence, by Theorem 10.3.12, there exist $g \in \mathcal{G}^{\min}$, $n \in N$ and finite type subsets $Y, Y' \subset \{1, \ldots, \ell\}$ such that $gSg^{-1} \subset \mathcal{P}_Y \cap n\mathcal{P}^-_{Y'}n^{-1}$. From this it is easy to see that $(\mathrm{Ad}\, g)\, \mathfrak{s}$ normalizes $\mathfrak{p} := \mathfrak{p}_Y \cap ((\mathrm{Ad}\, n)\, \mathfrak{p}^-_{Y'})$. But the normalizer of $\mathfrak{p}$ in $\mathfrak{g}$ is $\mathfrak{p}$ itself (as can be seen by using the action of $\mathfrak{h}$). This proves that $(\mathrm{Ad}\, g)\, \mathfrak{s} \subset \mathfrak{p}$, thereby establishing the "only if" part.

We now come to the "if" part: Let $\mathfrak{p}$ be as above. Since $\mathfrak{g}$ is an integrable $\mathfrak{g}$-module and $\mathfrak{p}$ is a finite-dimensional Lie algebra spanned by $\mathfrak{h}$ and certain real root vectors, by the PBW theorem, $\mathfrak{p}$ is $\mathrm{ad}_{\mathfrak{g}}$-locally finite. In particular, $\mathfrak{s}$ is $\mathrm{ad}_{\mathfrak{g}}$-locally finite, proving the "if" part. ☐

10.3.E EXERCISES

(1) In the finite case, i.e., when $\mathfrak{g}(A)$ is finite dimensional, prove that the function $F : \mathfrak{h}^*_{\mathbb{R}} \to (0, \infty)$ defined by $F(\lambda) = \langle \lambda, \lambda \rangle + 1$ satisfies the properties of Theorem 10.3.2, where $\langle\ ,\ \rangle$ is a W-invariant nondegenerate symmetric bilinear form. (Observe that, in the finite case, $C^o = \mathfrak{h}^*_{\mathbb{R}}$ by Proposition 1.4.2(e).)

In this case prove that, for a finite-dimensional $\mathfrak{g}(A)$-module V and nonzero $v \in V$ such that $\mathrm{supp}\, v$ is contained in an open half space, the isotropy subgroup $\mathcal{G}(A)_{\mathbb{C}v}$ of the line $\mathbb{C}v$ is contained in a proper parabolic subgroup.

(2) A subgroup $S \subset \mathcal{G}^{\min}$, resp. subalgebra $\mathfrak{s} \subset \mathfrak{g}$, is called *reductive* in $\mathcal{G}^{\min}$, resp. in $\mathfrak{g}$, if, under the adjoint action of S, resp. $\mathfrak{s}$, $\mathfrak{g}$ decomposes as a direct sum of finite-dimensional irreducible S-modules, resp. $\mathfrak{s}$-modules.

(a) Prove that for any subgroup $S \subset \mathcal{G}^{\min}$ reductive in $\mathcal{G}^{\min}$, there exists a subset $Y \subset \{1, \ldots, \ell\}$ of finite type and $g \in \mathcal{G}^{\min}$ such that $gSg^{-1} \subset \mathcal{G}_Y$, where $\mathcal{G}_Y$ is the (finite-dimensional) reductive group defined in 6.1.13.

Hint. Show that the group $\mathcal{P}_Y \cap w\mathcal{P}^-_{Y'}w^{-1}$, for $w \in W$ and any finite type subsets Y, Y', is generated by T and $\{\mathcal{U}_\beta\}_{\beta \in \Delta_Y \cap (-w\Delta_{Y'})}$, where Δ_Y is as in 1.2.2. Now use Theorem 10.3.12.

(b) From (a) deduce that any $\mathrm{Ad}_{\mathfrak{g}}$-diagonalizable connected subgroup H of $\mathcal{G}^{\min}$ can be conjugated by $\mathcal{G}^{\min}$ into T.

(An infinitesimal analogue of this result for symmetrizable $\mathfrak{g}$ is proved in Theorem 10.2.15(a).)

(c) Use (a) to show that, for symmetrizable connected $\mathfrak{g}$, a subalgebra $\mathfrak{s} \subset \mathfrak{g}$ reductive in $\mathfrak{g}$ can be conjugated by $\mathcal{G}^{\min}$ into $\mathfrak{g}_Y$, for some finite type subset Y.

(d) Use Theorem 10.3.12 to show that an $\mathrm{Ad}_{\mathfrak{g}}$-triangular subgroup S of $\mathcal{G}^{\min}$ can be conjugated by $\mathcal{G}^{\min}$ into $\mathcal{B} \cap w\mathcal{B}^- w^{-1}$ for some $w \in W$. (See Theorem 10.2.5 for an infinitesimal analogue of this result for symmetrizable $\mathfrak{g}$.)

(e) Prove that any finite-dimensional semisimple subalgebra of an affine Kac–Moody algebra $\mathfrak{g}$ is reductive in $\mathfrak{g}$. Construct a counter example (in the nonaffine case).

10.C Comments. As mentioned in the beginning of this chapter, Proposition 10.1.4, Theorem 10.1.5, and Corollary 10.1.11 in the finite case are due to Kostant. His proof appeared in [Garfinkle–82]. In fact, Kostant proved in the finite case that the full ideal of $\mathcal{G}/\mathcal{P}_Y$ inside $\mathbb{P}(L(\Lambda))$ (with the notation as in Corollary 10.1.11) is generated by the collection of equations given in the proof of Corollary 10.1.11. (There is another char. p proof of Corollary 10.1.11 in the finite case due to [Ramanathan–87] using the Frobenius splitting methods, though he does not give the precise equations.) The generalization of these results to symmetrizable Kac–Moody algebras is due to [Peterson–Kac–83] following a similar argument as that of Kostant. As given in [Kac–90, ch. 14], for $\mathfrak{g}$ of affine type and $\Lambda = \omega_0$ (the zeroth fundamental weight), the quadratic equations defining $\mathcal{G}/\mathcal{P}_{\{1,\dots,\ell\}}$ in $\mathbb{P}(L(\omega_0))$ are equivalent to the hierarchy of Hirota bilinear forms studied in [Date–Jimbo–Kashiwara–Miwa–82b, 83]. (The case $\mathfrak{g} = \widehat{s\ell_2}$ corresponds to the KdV hierarchy.) Further connections of Hirota bilinear forms with infinite flag varieties were established in [Kac–Peterson–86] (cf. also [Kac–Raina–87]).

Most of the results in Subchapter 10.2 are announced in [Peterson–Kac–83] with a broad outline of their proofs. (A more detailed account of the proofs is given in [Moody–Pianzola–95, §7.4], e.g., we have taken the proof of Lemma 10.2.8 from loc. cit.) Lemma 10.2.2 is taken from [Kac–90, Chapter 11, Exercise 11.10].

The results in Subchapter 10.3 (including Exercises 10.3.E) are taken from [Kac–Peterson–87]. (Also see [Slodowy–84a, 85b].)

XI

Topology of Kac–Moody Groups and Their Flag Varieties

The aim of this chapter is to study the T-equivariant cohomology $H_T^*(\mathcal{X}^Y)$ of the flag varieties $\mathcal{X}^Y = \mathcal{G}/\mathcal{P}_Y$ in terms of the nil-Hecke ring and also prove a certain positivity result for the cup product in $H_T^*(\mathcal{X}^Y)$. In addition, we study the degeneracy of the Leray–Serre spectral sequence for the fibration $\mathcal{G}^{\min} \to \mathcal{G}^{\min}/T$.

Let $\mathfrak{g}$ be any Kac–Moody Lie algebra with Cartan subalgebra $\mathfrak{h}$ and the associated Weyl group W, and let $Q = Q(\mathfrak{h}^*)$ be the quotient field of the symmetric algebra $S = S(\mathfrak{h}^*)$. Then the action of W on $\mathfrak{h}^*$ extends to an action on Q via the field automorphisms. We take the Q-vector space Q_W with basis $\{\delta_w\}_{w\in W}$ and define a multiplication in Q_W by $(q\delta_v)(q'\delta_w) = q(vq')\delta_{vw}$. Thus, we get an associative ring, which is a certain twisted analogue of the group algebra $Q[W]$. In addition, Q_W has a commutative coproduct $\Delta(\delta_v) = \delta_v \otimes \delta_v$, although it is not a Hopf algebra. The ring Q_W contains, for each simple reflection s_i, a remarkable element $x_{s_i} = \frac{1}{\alpha_i}(\delta_{s_i} - \delta_e)$. These elements satisfy the braid property, and thus we can define the elements $x_w \in Q_W$ for any $w \in W$. Define the fundamental subring, called the nil-Hecke ring, $R = R_W$ of Q_W as $R = \bigoplus_{w\in W} Sx_w$.

Now, the dual vector space $\Omega_W := \operatorname{Hom}_Q(Q_W, Q)$ acquires a commutative product (coming from the coproduct in Q_W). Let $\{\xi^w\}_{w\in W}$ be the "basis" of $\Omega = \Omega_W$ dual to the Q-basis $\{x_w\}$ of Q_W, and let $\Lambda \subset \Omega$ be the S-subalgebra which is the "restricted" S-dual of the nil-Hecke ring R. Alternatively, Λ can be described as the S-span of the elements $\{\xi^w\}_{w\in W}$. The elements ξ^w give rise to the basic "upper triangular" matrix $D = (d_{v,w})_{v,w\in W}$, where $d_{v,w} := \xi^v(\delta_w)$. Various properties of the rings R and Λ and the elements x_w, ξ^w are discussed in Section 11.1. We particularly mention Theorems 11.1.2, 11.1.7, 11.1.13 and Propositions 11.1.11, 11.1.15 and 11.1.19. As an immediate corollary of Proposition 11.1.19, we obtain the following result, conjectured by Deodhar for the case $u = v$. For any $u \le v \le w \in W$, $\#\{\beta \in \Delta_{\mathrm{re}}^+ : u \le s_\beta v \le w\} \ge \ell(w) - \ell(u)$.

The ring Q_W has a natural representation in Q defined by $(q\delta_w)\bullet q' = q(wq')$. We prove in Section 11.2 that the nil-Hecke ring R is precisely the stabilizer of $S \subset Q$, although this result is not used in the subsequent study of the T-equivariant cohomology of $\mathcal{G}/\mathcal{P}_Y$.

There is also a left Q_W-module structure on Ω (such that R keeps Λ stable) defined by $(x \bullet f)a = f(ax)$, for $a, x \in Q_W$ and $f \in \Omega$.

In Section 11.3, we study the T-equivariant and the singular cohomologies of $\mathcal{X}^Y$ for any subset $Y \subset \{1, \dots, \ell\}$. In fact, more generally, we study the T-equivariant and the singular cohomologies of any closed $\mathcal{B}$-stable subset $\mathcal{X}^Y_\Theta$ of $\mathcal{X}^Y$. The Bruhat cells provide a CW-complex structure on $\mathcal{X}^Y$, with only even-dimensional cells. Thus the singular cohomology with integral coefficients $H^*(\mathcal{X}^Y)$ in odd degrees is 0 and $H^{2i}(\mathcal{X}^Y) \simeq \bigoplus_{\substack{w\in W'_Y \text{with} \\ \ell(w)=i}} \mathbb{Z}\, \varepsilon^Y_w$, where $\{\varepsilon^Y_w\}_{w\in W'_Y}$ is the Schubert basis. (For $Y = \emptyset$, we denote ε^Y_w by ε_w itself.)

The cohomologies $H^*(\mathcal{X})$ and $H^*_T(\mathcal{X})$ both admit canonical W-module structures obtained from the T-equivariant homotopy equivalence $\mathcal{G}^{\min}/T \to \mathcal{X}$ (cf. Definition 11.3.4). Moreover, for any simple reflection s_i, there is an analogue of the BGG–Demazure operator $D_{s_i} : H^*(\mathcal{X}) \to H^{*-2}(\mathcal{X})$, obtained by "integration" along the fibers of the $\mathbb{P}^1$-fibration $\mathcal{G}/\mathcal{B} \to \mathcal{G}/\mathcal{P}_i$, where $\mathcal{P}_i$ is the minimal parabolic subgroup corresponding to $Y = \{i\}$. We similarly obtain the operators $\hat{D}_{s_i} : H^*_T(\mathcal{X}) \to H^{*-2}_T(\mathcal{X})$. The main result of Section 11.3 asserts that there is a canonical graded $S_{\mathbb{Z}}$-algebra isomorphism $\nu : H^*_T(\mathcal{X}) \xrightarrow{\sim} \Lambda_{\mathbb{Z}}$, such that the (Weyl group) action of s_i, resp. $\hat{D}_{s_i}$, corresponds under ν to the $\bullet$ action of δ_{s_i}, resp. x_{s_i}, where $S_{\mathbb{Z}} := S(\mathfrak{h}^*_{\mathbb{Z}})$ and $\Lambda_{\mathbb{Z}} := \bigoplus_{w\in W} S_{\mathbb{Z}}\xi^w$.

Moreover, the $S_{\mathbb{Z}}$-basis $\{\xi^w\}_{w\in W}$ of $\Lambda_{\mathbb{Z}}$ corresponds under ν to an equivariant analogue $\{\hat{\varepsilon}_w\}_w$ of the Schubert basis. A characterization of the basis $\{\hat{\varepsilon}_w\}_w$ is given in Proposition 11.3.10. The map ν is obtained from the localization map in equivariant cohomology induced from the inclusion $W \simeq (\mathcal{G}/\mathcal{B})^T \hookrightarrow \mathcal{G}/\mathcal{B}$. The localization theorem in equivariant cohomology, equivariant Borel homomorphism and the operators $\hat{D}_{s_i}$ play a central role in proving that ν is an isomorphism. The isomorphism ν easily generalizes to the parabolic case, more generally, to any $\mathcal{B}$-stable closed subset $\mathcal{X}^Y_\Theta \subset \mathcal{X}^Y$ (cf. Corollaries 11.3.14 and 11.3.16). In particular, we get an isomorphism of $\mathbb{Z}$-algebras ν_0: $H^*(\mathcal{X}) \simeq \mathbb{Z} \otimes_{S_{\mathbb{Z}}} \Lambda_{\mathbb{Z}}$ by "evaluating" ν at 0.

These identifications allow us to translate various results proved in Section 11.1 about the S-algebra Λ into results on $H^*_T(\mathcal{X})$ and $H^*(\mathcal{X})$. In particular, the action of $\hat{D}_{s_i}$ on $H^*_T(\mathcal{X})$ takes a particularly simple form: $\hat{D}_{s_i}(\hat{\varepsilon}_w) = \hat{\varepsilon}_{ws_i}$ if $ws_i < w$, and 0 otherwise. Moreover, the product formula in Λ (cf. Theorem 11.1.13) gives rise to the cup product formula in $H^*_T(\mathcal{X})$ multiplying any two Schubert classes $\hat{\varepsilon}_u$ and $\hat{\varepsilon}_v$ purely in terms of the matrix D defined above (cf. Corollary 11.3.17). "Evaluating" this cup product formula at 0, we obtain a general expression for the product of any two Schubert classes $\varepsilon_u, \varepsilon_v \in H^*(\mathcal{X})$, thus generalizing the classical Chevalley formula.

In Section 11.4, we prove a "positivity" result for the cup product in $H_T^*(\mathcal{X})$ in the Schubert basis. More specifically, write (for any $u, v \in W$)

$$\hat{\varepsilon}_u \, \hat{\varepsilon}_v = \sum_w p_{u,v}^w \, \hat{\varepsilon}_w.$$

Then we prove that each $p_{u,v}^w$ is a homogenous polynomial in the simple roots $\{\alpha_1, \dots, \alpha_\ell\}$ with nonnegative integral coefficients. In particular, the cup product $\varepsilon_u \varepsilon_v$ written in terms of the Schubert basis $\{\varepsilon_w\}_w$ of $H^*(\mathcal{X})$ has nonnegative coefficients. (In fact, we prove a more general result in Theorem 11.4.11.) The main technical result needed to prove the above positivity is Theorem 11.4.2, which roughly asserts that for a complete B-variety X and a T-stable closed subvariety $Z \subset X$, there exists a B-stable closed (not necessarily reduced) subscheme $Z' \subset X$ which is rationally equivalent to Z, where B is a connected solvable algebraic group and $T \subset B$ is a maximal torus. (Actually Theorem 11.4.2 is an equivariant version of it.)

Finally, in Section 11.5, we study the Leray–Serre cohomology spectral sequence for the fibration $\mathcal{G}^{\min} \to \mathcal{G}^{\min}/T$ over any field coefficient k. It is shown that this spectral sequence collapses at E_3, i.e., $E_3 = E_\infty$, by making crucial use of the BGG–Demazure operators D_{s_i} and the ideal of generalized invariants (cf. 11.5.2 for its definition). This important result, due to Kac–Peterson, is used to determine $H^*(\mathcal{G}^{\min}, k)$ (cf. Theorem 11.5.12 and Corollary 11.5.13). Their basic idea is that the operators D_{s_i}, the description of the differential d_2 in the above spectral sequence, and collapsing of the spectral sequence at E_3 determine the product in $H^*(\mathcal{G}^{\min}, k)$. It is conjectured that the above spectral sequence degenerates at E_3 even over $\mathbb{Z}$.

As an easy consequence of the surjectivity of the Borel homomorphism $\beta : S \to H^*(\mathcal{X}, \mathbb{C})$ in the finite case, it is shown that there is a graded S-algebra isomorphism (in the finite case) $S \otimes_{S^W} S \to H_T(\mathcal{X}, \mathbb{C})$, where S^W is the subring of W-invariants. Some results on the cohomology of $\mathcal{G}^{\min}$ and $\mathcal{X}$ in the finite and affine cases are described in Exercises 11.5.E, including a result of Kostant in the finite case identifying $H^*(\mathcal{X}, \mathbb{C})$ with the affine coordinate ring $\mathbb{C}[\mathcal{N} \cap \mathfrak{h}]$ (where $\mathcal{N} \subset \mathfrak{g}$ is the nilpotent cone and $\mathcal{N} \cap \mathfrak{h}$ is the scheme theoretic intersection), and also an explicit realization of the double Schubert polynomials for sl_n and its generalization to an arbitrary finite type $\mathfrak{g}$.

11.1. The Nil-Hecke Ring

In this section $\mathfrak{g}$ is any Kac–Moody Lie algebra with the standard Cartan subalgebra $\mathfrak{h}$. Then $\mathfrak{h}^*$ (and hence $\mathfrak{h}$) is a W-module for the associated Weyl group W (cf. 1.3.1).

11.1.1 Definition. Let $Q = Q(\mathfrak{h}^*)$ be the quotient field of the symmetric algebra $S = S(\mathfrak{h}^*)$. Then the action of W on $\mathfrak{h}^*$ extends to an action on Q via the field automorphisms. Let Q_W be the Q-vector space with basis $\{\delta_w\}_{w\in W}$, i.e., Q_W as a Q-vector space is nothing but the vector space underlying the group algebra of W over the field Q. Define a "twisted" product in Q_W by

$$(1)\qquad \Big(\sum_{v\in W} q_v\delta_v\Big)\cdot\sum_{w} q'_w\delta_w = \sum_{v,w} q_v(vq'_w)\,\delta_{vw},\quad \text{for } q_v, q'_w \in Q\,.$$

Of course, all but finitely many q_v's are assumed to be 0 and similarly for q'_w.

This product makes Q_W into an associative (noncommutative) ring with multiplicative identity $1 = \delta_e$. Even though the ring Q_W is also a vector space over Q, it is *not* an algebra over Q since $Q\delta_e$ is not central in Q_W. But clearly Q_W is an algebra over the subfield $Q^W \subset Q$ of W-invariants. We will always refer to Q_W as the ring under (1) (and not the group algebra).

The ring Q_W admits an involutary antiautomorphism t defined by

$$(2)\qquad (q\delta_w)^t = (w^{-1}q)\delta_{w^{-1}},\quad \text{for } w \in W \text{ and } q \in Q.$$

By the definition, Q_W has a left Q-vector space structure. In addition, the ring structure equips Q_W with a right Q-vector space structure obtained by the right multiplication:

$$(3)\qquad x\cdot q = x(q\delta_e),\quad \text{for } x \in Q_W \text{ and } q \in Q.$$

(We often drop the "dot" in the above.)Define the Q-linear coproduct $\Delta : Q_W \to Q_W \otimes_Q Q_W$ by

$$(4)\qquad \Delta(q\delta_w) = (q\delta_w)\otimes\delta_w = \delta_w\otimes q\delta_w,\quad \text{for } w \in W \text{ and } q \in Q,$$

where both the copies of Q_W in the tensor product are equipped with the left Q-vector space structure.

It is easy to see that the coproduct Δ is associative and commutative with a counit $\varepsilon : Q_W \to Q$ defined by

$$(5)\qquad \varepsilon(q\delta_w) = q,\quad \text{for } q \in Q,\, w \in W.$$

For the definition of associativity, commutativity and counit for a coproduct, see, e.g., [Spanier–66, Chap. 5, §8]. As observed earlier, Q_W is not an algebra over Q and hence, even though it is equipped with a coalgebra structure, it is *not* a Hopf algebra. However, $Q_W \otimes_Q Q_W$ admits an associative product $\odot$ (defined below) making Δ a ring homomorphism.

$$(6)\quad (q_v\delta_v\otimes q_w\delta_w)\odot(q_{v'}\delta_{v'}\otimes q_{w'}\delta_{w'}) := (q_vq_w)\big(v(q_{v'}q_{w'})\big)\,\delta_{vv'}\otimes\delta_{vw'v^{-1}w}\,.$$

As in 1.1.2, let $\{\alpha_1, \dots, \alpha_\ell\}$ be the simple roots of $\mathfrak{g}$ and let $\{s_1, \dots, s_\ell\}$ be the corresponding simple reflections in W (cf. 1.3.1). For any $1 \le i \le \ell$, define the element $x_i = x_{s_i} \in Q_W$ by

(7) $$x_i = \frac{1}{\alpha_i}\big(\delta_{s_i} - \delta_e\big).$$

Denote the polynomial ring over $\mathbb{Q}$ in the variables $\{\alpha_1, \dots, \alpha_\ell\}$ by $\mathbb{Q}[\underline{\alpha}] \subset S(\mathfrak{h}^*)$, i.e., $\mathbb{Q}[\underline{\alpha}] = \sum_{I \in \mathbb{Z}_+^\ell} \mathbb{Q}\alpha^I$, where $\alpha^I := \alpha_1^{i_1} \cdots \alpha_\ell^{i_\ell}$ for $I = (i_1, \dots, i_\ell)$. Similarly, define $\mathbb{Z}[\underline{\alpha}]$ and $\mathbb{Z}_+[\underline{\alpha}]$.

The following result summarizes some of the basic important properties of the elements x_i.

11.1.2 Theorem. *(a) For any* $1 \le i \le \ell$, $x_i^2 = 0$.

(b) For any $w \in W$*, pick a reduced decomposition* $w = s_{i_1} \cdots s_{i_n}$*. Then the element*

(1) $$x_w := x_{i_1} \cdots x_{i_n} \in Q_W$$

does not depend on the choice of the reduced decomposition of w*. (Hence the notation* x_w *is justified.) We set* $x_e = \delta_e$.

(c) For any $\chi \in \mathfrak{h}^*$ *and* $w \in W$,

$$\chi\, x_w = x_w(w^{-1}\chi) - \sum_{v \xrightarrow{\beta} w} \langle \chi, \beta^\vee \rangle\, x_v,$$

where the notation $v \xrightarrow{\beta} w$ *means* $v \le w$, $\ell(w) = \ell(v) + 1$, $\beta \in \Delta^+_{\mathrm{re}}$ *and* $w = s_\beta v$*;* $s_\beta \in W$ *being defined by* $s_\beta \chi = \chi - \langle \chi, \beta^\vee \rangle \beta$ *for* $\chi \in \mathfrak{h}^*$.

(d) For any $v, w \in W$,

$$\begin{aligned} x_v x_w &= x_{vw}, && \text{if } \ell(vw) = \ell(v) + \ell(w) \\ &= 0, && \text{otherwise.} \end{aligned}$$

(e) Write

(2) $$x_w = \sum_{v \in W} c_{w,v}\, \delta_v, \text{ for } c_{w,v} \in Q.$$

Then

(e₁) $$c_{w,v} = 0\,, \text{ unless } v \le w, \quad \text{and}$$

(e₂) $$c_{w,w} = \prod_{\beta \in \Delta^+ \cap w\Delta^-} \beta^{-1}.$$

(f) Fix a reduced decomposition $w = s_{i_1} \cdots s_{i_n}$. Then, for any $v \leq w$,

$$c_{w,v} = (-1)^n \sum \Big(\big(s_{i_1}^{\varepsilon_1} \alpha_{i_1}\big)\big(s_{i_1}^{\varepsilon_1} s_{i_2}^{\varepsilon_2} \alpha_{i_2}\big) \cdots \big(s_{i_1}^{\varepsilon_1} \cdots s_{i_n}^{\varepsilon_n} \alpha_{i_n}\big)\Big)^{-1},$$

where the summation runs over all those $(\varepsilon_1, \dots, \varepsilon_n) \in \{0, 1\}^n$ satisfying $s_{i_1}^{\varepsilon_1} \cdots s_{i_n}^{\varepsilon_n} = v$. (The notation s_i^0 means the identity element.)

(g) For $w \in W$ and any simple reflection s_i,

$$\begin{aligned} x_w \delta_{s_i} &= -x_w, && \text{if } ws_i < w, \text{ and} \\ &= (w\alpha_i) x_{ws_i} - x_w + \sum_{v \xrightarrow{\beta} ws_i} \langle w\alpha_i, \beta^\vee \rangle x_v, && \text{otherwise.} \end{aligned}$$

(h) For any $w \in W$, we have

$$\Delta(x_w) = \sum_{u,v \leq w} p_{u,v}^w \, x_u \otimes x_v, \tag{3}$$

for some homogeneous polynomials $p_{u,v}^w \in S$ of degree $\ell(u) + \ell(v) - \ell(w)$. In particular, $p_{u,v}^w = 0$ unless $\ell(u) + \ell(v) \geq \ell(w)$.

In fact, $p_{u,v}^w \in \mathbb{Z}[\underline{\alpha}]$.

(By 11.1.3, $\{x_w\}_w$ is a Q-basis of Q_W and hence $p_{u,v}^w$ are uniquely determined by (3).)

Proof. (a) follows from a simple calculation.

We next prove the validity of (b), (c) and (e) simultaneously by induction on $\ell(w)$. If $\ell(w) = 1$, (b) and (e) are clear. To prove (c) in this case, take $1 \leq i \leq \ell$. Then

$$\begin{aligned} \chi x_i &= \frac{\chi}{\alpha_i}(\delta_{s_i} - \delta_e) \\ &= \frac{1}{\alpha_i}(\delta_{s_i} - \delta_e)\, s_i\chi + \frac{s_i\chi - \chi}{\alpha_i}\delta_e \\ &= x_i(s_i\chi) - \langle \chi, \alpha_i^\vee \rangle \delta_e, \end{aligned}$$

which proves (c) in this case.

We now take $w = us_i \in W$ with $\ell(w) \geq 2$ and $u < w$. We assume the validity of (b), (c) and (e) for u by induction. Then

$$\begin{aligned} \chi x_u x_i &= \Big(x_u(u^{-1}\chi) - \sum_{u_o \xrightarrow{\beta_o} u} \langle \chi, \beta_o^\vee \rangle x_{u_o} \Big) x_i \\ &= x_u x_i (w^{-1}\chi) - \langle u^{-1}\chi, \alpha_i^\vee \rangle x_u - \sum_{u_o \xrightarrow{\beta_o} u} \langle \chi, \beta_o^\vee \rangle x_{u_o} x_i, \end{aligned} \tag{4}$$

where the second equality follows from (c) for $\ell(w) = 1$. Further, by Corollary 1.3.19,

$$\{u_o : u_o \xrightarrow{\beta_o} u \text{ and } u_o s_i > u_o\} \xrightarrow{\sim} \{v : v \xrightarrow{\beta} w \text{ and } v \neq u\},$$

under the map $u_o \mapsto u_o s_i$. Moreover, under this correspondence, $\beta_o \mapsto \beta_o$. Further, $x_{u_o} x_i = x_{u_o s_i}$ if $u_o s_i > u_o$ and zero otherwise, which follows from (b) (by the induction hypothesis) and (a), since both of u_o and $u_o s_i$ have lengths $\leq \ell(u)$.

Using the above, (4) reduces to

(5) $$\chi x_u x_i = x_u x_i \big(w^{-1}\chi\big) - \sum_{v \xrightarrow{\beta} w} \langle \chi, \beta^\vee \rangle \, x_v.$$

Now let $w = u' s_j$ be another decomposition such that $u' < w$. Then by (5)

(6) $$\chi x_u x_i - x_u x_i \big(w^{-1}\chi\big) = \chi x_{u'} x_j - x_{u'} x_j \big(w^{-1}\chi\big).$$

Write

(7) $$x_u x_i = \sum_{\theta \in W} q_\theta \delta_\theta \text{ and } x_{u'} x_j = \sum q'_\theta \delta_\theta, \text{ for } q_\theta, q'_\theta \in Q.$$

Then, using (e_1) for u and u' and Corollary 1.3.19, we get

(8) $$q_\theta = q'_\theta = 0, \text{ unless } \theta \leq w;$$

and, by (e_2) for $c_{u,u}$ and Lemma 1.3.14,

(9) $$q_w = c_{u,u}(u\alpha_i)^{-1} = \prod_{\beta \in \Delta^+ \cap w\Delta^-} \beta^{-1}.$$

Similarly,

(10) $$q'_w = \prod_{\beta \in \Delta^+ \cap w\Delta^-} \beta^{-1} = q_w.$$

From (6) and (7) we get, for all $\theta \in W$,

(11) $$(\chi - \theta w^{-1}\chi) q_\theta = (\chi - \theta w^{-1}\chi) q'_\theta.$$

Since (11) is true for all $\chi \in \mathfrak{h}^*$ and $\mathfrak{h}^*$ is a faithful representation of W, we get

(12) $$q_\theta = q'_\theta \quad \text{for all } \theta \neq w.$$

Combining (10) and (12), we get

$$x_u x_i = x_{u'} x_j.$$

This proves (b) for w and the relation (5), resp. (8), (9), is nothing but (c), resp. (e_1), (e_2). The first part of (d) follows from (b). For the second part of (d), take a reduced decomposition $w = s_{i_1} \cdots s_{i_n}$ and let $w_k := s_{i_1} \cdots s_{i_k}$ (for $0 \le k \le n$). Choose k such that $\ell(vw_k) = \ell(v) + \ell(w_k)$ and $vw_k s_{i_{k+1}} < vw_k$. Then, by (b),

$$\begin{aligned} x_v x_w &= x_v x_{w_k} x_{i_{k+1}} \cdots x_{i_n} \\ &= x_{vw_k s_{i_{k+1}}} x_{i_{k+1}} x_{i_{k+1}} \cdots x_{i_n} \\ &= 0, \text{ by (a)}. \end{aligned}$$

This proves the second part of (d). The proof of (f) is easy from the definition of x_w.

We now prove (g). By (d),

$$x_w \delta_{s_i} = -x_w(x_i \alpha_i + \delta_e) = \begin{cases} -x_w, & \text{if } ws_i < w \\ -x_{ws_i}\alpha_i - x_w, & \text{otherwise,} \end{cases}$$

so (g) follows from (c).

(h): We prove this by induction on $\ell(w)$. For $w = e$, (h) is clearly true. So take $w = s_i$. Then

$$\begin{aligned} \Delta(x_i) &= \frac{1}{\alpha_i}\big(\delta_{s_i} \otimes \delta_{s_i} - \delta_e \otimes \delta_e\big) \\ &= \big(\delta_{s_i} - \delta_e\big) \otimes \frac{1}{\alpha_i}\big(\delta_{s_i} - \delta_e\big) + \delta_e \otimes \frac{1}{\alpha_i}\big(\delta_{s_i} - \delta_e\big) + \frac{1}{\alpha_i}\big(\delta_{s_i} - \delta_e\big) \otimes \delta_e \\ &= \alpha_i x_i \otimes x_i + x_e \otimes x_i + x_i \otimes x_e. \end{aligned} \tag{13}$$

This proves (h) for $\ell(w) = 1$. Now take an arbitrary $w = s_i w'$ with $\ell(w) > \ell(w') \ge 1$. Then, Δ being a ring homomorphism, by (d) and the induction hypothesis, there are homogeneous polynomials $p^{w'}_{u',v'}$ of degree $\ell(u') + \ell(v') - \ell(w')$ such that

$$\begin{aligned} \Delta(x_w) &= \Delta(x_i) \odot \Delta(x_{w'}) \\ &= \left(\frac{1}{\alpha_i}\delta_{s_i} \otimes \delta_{s_i} - \frac{1}{\alpha_i}\delta_e \otimes \delta_e\right) \odot \left(\sum_{u',v' \le w'} p^{w'}_{u',v'}\, x_{u'} \otimes x_{v'}\right). \end{aligned} \tag{14}$$

From the definition of $\odot$ it is easy to see that, for any $x, y \in Q_W$, $w_o \in W$ and $q \in Q$,

$$(q\delta_{w_o} \otimes \delta_{w_o}) \odot (x \otimes y) = q\delta_{w_o}x \otimes \delta_{w_o}y. \tag{15}$$

Hence, (14) can be rewritten as

$$\begin{aligned}
\Delta(x_w) &= \sum_{u',v' \leq w'} \left[\frac{1}{\alpha_i}(s_i p^{w'}_{u',v'})\delta_{s_i}x_{u'} \otimes \delta_{s_i}x_{v'} - \frac{1}{\alpha_i}p^{w'}_{u',v'}\, x_{u'} \otimes x_{v'} \right] \\
&= \sum_{u',v' \leq w'} \Big[\alpha_i (s_i p^{w'}_{u',v'}) \frac{1}{\alpha_i}(\delta_{s_i} - \delta_e)x_{u'} \otimes \frac{1}{\alpha_i}(\delta_{s_i} - \delta_e)x_{v'} \\
&\quad + (s_i\, p^{w'}_{u',v'})x_{u'} \otimes \frac{1}{\alpha_i}(\delta_{s_i} - \delta_e)x_{v'} \\
&\quad + (s_i p^{w'}_{u',v'})\frac{1}{\alpha_i}(\delta_{s_i} - \delta_e)x_{u'} \otimes x_{v'} + \frac{s_i p^{w'}_{u',v'} - p^{w'}_{u',v'}}{\alpha_i} x_{u'} \otimes x_{v'} \Big] \\
&= \sum_{u',v' \leq w'} \Big[\alpha_i (s_i p^{w'}_{u',v'}) x_i x_{u'} \otimes x_i x_{v'} + (s_i p^{w'}_{u',v'}) x_{u'} \otimes x_i x_{v'} \\
&\quad + (s_i p^{w'}_{u',v'}) x_i x_{u'} \otimes x_{v'} + (x_i \bullet p^{w'}_{u',v'}) x_{u'} \otimes x_{v'} \Big],
\end{aligned}$$

where $x_i \bullet p := \frac{s_i p - p}{\alpha_i}$ (cf. (11.1.3.1)).

By (d), the above expression is of the same form as asserted by (3), and hence the induction is complete, proving (h). (The assertion that $p^w_{u,v} \in \mathbb{Z}[\underline{\alpha}]$ follows easily by induction and the above expression for $\Delta(x_w)$, since $x_i \bullet \mathbb{Z}[\underline{\alpha}] \subset \mathbb{Z}[\underline{\alpha}]$.) This finishes the proof of the theorem. $\square$

11.1.3 Definition. Define the "change of basis" $W \times W$-matrix $C = (c_{v,w})_{v,w \in W}$, where $c_{v,w} \in Q$ is defined by (11.1.2.2).

By Theorem 11.1.2(e), C is a lower triangular matrix with nonzero diagonal entries (with respect to the Bruhat partial order on W). In particular, $\{x_w\}_{w \in W}$ is a Q-basis of Q_W under the left Q-vector space structure on Q_W. Similarly, it is a Q-basis under the right Q-vector space structure on Q_W.

Now Q has the structure of a left Q_W-module under

$$(q\delta_w) \bullet q' = q(wq'), \text{ for } q, q' \in Q \text{ and } w \in W. \tag{1}$$

Define the subring $\bar{R} = \bar{R}_W \subset Q_W$ by

$$\bar{R} = \{a \in Q_W : a \bullet S \subset S\}, \tag{2}$$

where $S := S(\mathfrak{h}^*) \subset Q$.

The ring $\bar{R}$ is clearly left and right S-stable. Moreover, $x_i \in \bar{R}$ for any $1 \leq i \leq \ell$, and hence any $x_w \in \bar{R}$.

Let $R = R_W \subset \bar{R}$ be the subring defined by

$$(3) \qquad R := \bigoplus_{w \in W} x_w S = \bigoplus_{w \in W} S x_w.$$

By Theorem 11.1.2(c) and (d), R is indeed a ring and, moreover, the second equality follows. Further, $\delta_w \in R$ for any $w \in W$, since each $\delta_{s_i} \in R$.

As we will see later in Section 11.3, R plays a basic role in determining the T-equivariant cohomology of the generalized flag varieties. The ring R is referred to as the *nil-Hecke ring* by Kostant–Kumar in view of the properties (a) and (d) of Theorem 11.1.2.

In Section 11.2 we will prove that, in fact, $R = \bar{R}$.

11.1.4 The dual of R. Let

$$(1) \qquad \Omega_W := \mathrm{Hom}_Q(Q_W, Q)$$

be the full vector space dual of the Q-vector space Q_W (under the left Q-vector structure). Since $\{\delta_w\}_{w \in W}$ is a Q-basis of Q_W, we will often identify

$$(2) \qquad \Omega_W \cong \{f : W \to Q, \, f \text{ any map}\}$$

under $f \mapsto \theta_f$, where $\theta_f : Q_W \to Q$ is the Q-linear map defined by $\theta_f(\delta_w) := f(w)$. Under the above identification, the Q-vector space structure on Ω_W corresponds to the pointwise addition and scalar multiplication of functions, i.e.,

$$(3) \qquad (f + g)(w) = f(w) + g(w), \ (qf)w = q\, f(w),$$

for $f, g : W \to Q$, $w \in W$ and $q \in Q$.

The commutative coproduct Δ defined by (11.1.1.4) induces a commutative product in Ω_W. Explicitly, it is given by

$$(4) \qquad fg(w) := f(w)\, g(w).$$

The function 1, taking value 1 on all the $w \in W$, serves as the identity of the commutative Q-algebra Ω_W.

The ring Q_W acts on Ω_W from the left via Q-linear maps under

$$(5) \qquad (x \bullet f)a = f(ax), \ \text{ for } a, x \in Q_W, \, f \in \Omega_W.$$

The actions of the elements x_i $(1 \le i \le \ell)$, $\alpha\delta_e$ $(\alpha \in Q)$, δ_w $(w \in W)$ on Ω_W will play an important role later in the chapter. So we write them explicitly. For $f \in \Omega_W$ and $v \in W$,

(6) $$\big((\alpha\delta_e) \bullet f\big)(v) = (v\alpha) f(v),$$

(7) $$(\delta_w \bullet f)(v) = f(vw), \qquad \text{and}$$

(8) $$(x_i \bullet f)v = \frac{f(vs_i) - f(v)}{v\,\alpha_i}.$$

We will refer to the action of δ_w's as the *Weyl group action* and the action of x_i's as the *BGG–Demazure operators*. [Bernstein–Gelfand–Gelfand–73] and [Demazure–74] independently introduced analogous operators A_{s_i} on $S(\mathfrak{h}^*)$ (cf. Exercise 11.3.E.2).

From (6), for $\alpha \in Q^W$ and $f \in \Omega_W$,

(9) $$\alpha f = (\alpha\delta_e) \bullet f,$$

where Q^W is the subfield of W-invariants. However, for general $\alpha \in Q$,

$$\alpha f \neq (\alpha\delta_e) \bullet f.$$

Introduce the "restricted" $S(\mathfrak{h}^*)$-dual $\Lambda = \Lambda_W$ of the ring R (cf. (11.1.3.3)) as follows:

(10) $$\Lambda := \{f \in \Omega_W : f(R) \subset S \text{ and } f(x_w) = 0 \text{ for all but finitely many } w \in W\}.$$

We have the following simple

11.1.5 Lemma. *With the notation as above, Λ is a S-subalgebra of the Q-algebra Ω_W. Moreover, Λ is free as an S-module with basis $\{\xi^w\}_{w\in W}$ defined (uniquely) by*

(1) $$\xi^w(x_v) := \delta_{w,v}, \quad v \in W.$$

Further, Λ is stable under the left action $\bullet$ of $R \subset Q_W$ (cf. (11.1.4.5)).

Proof. Since $\{x_v\}_{v\in W}$ is a left Q-basis of Q_W (cf. 11.1.3), there are unique elements $\xi^w \in \Omega_W$ defined by (1). Moreover, by (11.1.3.3), $\xi^w \in \Lambda$ and $\{\xi^w\}_{w\in W}$ is a S-basis of Λ. By Theorem 11.1.2(c)–(d), Λ is stable under R (with respect to the $\bullet$ action) and, by Theorem 11.1.2(h), Λ is stable under multiplication. This proves the lemma. ☐

11.1.6 Definition. Let $\mathcal{D} = \mathcal{D}_W$ be the space of all the functions $\varphi : W \times W \to Q$ with the property that there exists some $n_\varphi \geq 0$ such that

$$\text{(1)} \qquad \varphi(v, w) = 0 \ \text{ whenever } \ell(v) - \ell(w) > n_\varphi.$$

Then $\mathcal{D}$ is an associative Q-algebra under pointwise addition, scalar multiplication and convolution as multiplication, i.e.,

$$\text{(2)} \quad (\varphi_1\varphi_2)(v, w) = \sum_{u \in W} \varphi_1(v, u)\,\varphi_2(u, w), \ \text{ for } v, w \in W \text{ and } \varphi_1, \varphi_2 \in \mathcal{D}.$$

The above sum is finite because of the condition (1) on φ_2.

We can of course think of $\mathcal{D}$ as a certain subspace of the space of all matrices over Q parametrized by $W \times W$. Under this identification, the convolution in $\mathcal{D}$ is nothing but the matrix multiplication.

Now define the matrix $D = (d_{v,w})_{v,w \in W} \in \mathcal{D}$ by

$$\text{(3)} \qquad d_{v,w} := \xi^v(w).$$

This matrix will play a fundamental role later in the chapter.

We collect various properties of the basis $\{\xi^w\}$ in the following theorem. These are obtained essentially by dualizing the results in Theorem 11.1.2.

11.1.7 Theorem. *Take any $v, w \in W$ and $1 \leq i \leq \ell$. Then we have the following:*

(a) $d_{w,v} = 0$, unless $w \leq v$, and $d_{w,w} = \prod_{\beta \in \Delta^+ \cap w\Delta^-} \beta$.

In particular, $d_{w,w} \neq 0$ and D is an invertible element of the ring $\mathcal{D}$. In fact, $D^{-1} = C^t$, where C is the matrix defined in 11.1.3 and C^t denotes its transpose.

(b)

$$\begin{aligned} x_i \bullet \xi^w &= \xi^{ws_i}, && \text{if } ws_i < w \\ &= 0, && \text{otherwise.} \end{aligned}$$

In particular,

$$\text{(1)} \qquad \begin{aligned} d_{w,vs_i} - d_{w,v} &= (v\alpha_i)\, d_{ws_i,v}\,, && \text{if } ws_i < w \\ &= 0, && \text{otherwise.} \end{aligned}$$

(c) ξ^e is the multiplicative identity of Λ, i.e.,

$$\text{(2)} \qquad d_{e,v} = 1, \ \text{for all } v \in W.$$

Further,

$$d_{s_i,v} = \omega_i - v\omega_i, \tag{3}$$

where $\omega_i \in \mathfrak{h}^*$ is any element satisfying $\omega_i(\alpha_j^\vee) = \delta_{i,j}$, for all the simple coroots $\alpha_j^\vee$.

(d) $\xi^u \xi^v = \sum_{u,v\le w} p_{u,v}^w \xi^w$,

where $p_{u,v}^w \in \mathbb{Z}[\underline{\alpha}]$ is as in (11.1.2.3). Recall that it is a homogeneous polynomial of degree $\ell(u)+\ell(v)-\ell(w)$.

(e) For any $\chi \in \mathfrak{h}^*$,

$$(\chi\delta_e) \bullet \xi^w = (w\chi)\xi^w - \sum_{w \xrightarrow{\beta} v} \langle w\chi, \beta^\vee\rangle\, \xi^v.$$

(f)

$$\begin{aligned} \delta_{s_i} \bullet \xi^w &= \xi^w, && \text{if } ws_i > w \\ &= -(w\alpha_i)\xi^{ws_i} + \xi^w + \sum_{ws_i \xrightarrow{\beta} v} \langle w\alpha_i, \beta^\vee\rangle\, \xi^v, && \text{otherwise.} \end{aligned}$$

(g) $d_{w,v} \in S^{\ell(w)}(\underline{\alpha})$, where $S^{\ell(w)}(\underline{\alpha})$ denotes the homogeneous component of $\mathbb{Z}[\underline{\alpha}]$ of degree $\ell(w)$ (in the variables $\alpha_1, \dots, \alpha_\ell$).

(h) x_i acts on Ω_W as a twisted derivation, i.e.,

$$x_i \bullet (\psi_1\psi_2) = (x_i \bullet \psi_1)\psi_2 + (\delta_{s_i} \bullet \psi_1)(x_i \bullet \psi_2), \text{ for any } \psi_1, \psi_2 \in \Omega_W. \tag{4}$$

Let $w = s_{i_1}\cdots s_{i_n}$ be a reduced decomposition. Then, more generally,

$$x_w \bullet (\psi_1\psi_2) = \sum_{\substack{0\le p\le n\\ 1\le j_1<\cdots<j_p\le n}} \Big(\big(x_{i_1}\cdots \hat{\hat{x}}_{i_{j_1}} \cdots \hat{\hat{x}}_{i_{j_p}} \cdots x_{i_n}\big) \bullet \psi_1\Big) \Big(\big(x_{i_{j_1}} \cdots x_{i_{j_p}}\big) \bullet \psi_2\Big), \tag{5}$$

where the notation $\hat{\hat{x}}_i$ means that the element x_i is to be replaced by the element δ_{s_i}.

In particular, for any $u, v, w \in W$, taking a reduced decomposition $w = s_{i_1}\cdots s_{i_n}$, we get

$$p_{u,v}^w = \sum \Big(\big(x_{i_1}\cdots \hat{\hat{x}}_{i_{k_1}} \cdots \hat{\hat{x}}_{i_{k_m}} \cdots x_{i_n}\big) \bullet \xi^u\Big)(e), \tag{6}$$

where the summation runs over those $1 \le k_1 < \cdots < k_m \le n$ such that $s_{i_{k_1}}\cdots s_{i_{k_m}}$ is a reduced decomposition of v. (The notation $p_{u,v}^w$ is as in (d), and we take $p_{u,v}^w = 0$ unless $u, v \le w$.)

(i) Analogue of the Chevalley formula:

$$\xi^{s_i}\xi^w = d_{s_i,w}\xi^w + \sum_{w \xrightarrow{\beta} v} \langle w\omega_i, \beta^\vee\rangle \xi^v.$$

Proof. (a): By Theorem 11.1.2(e), write

(7) $$x_v = \Big(\prod_{\beta\in\Delta^+\cap v\Delta^-} \beta^{-1}\Big)\delta_v + \sum_{u<v} c_{v,u}\delta_u.$$

Since $x_e := \delta_e$, we have $\xi^w(\delta_e) = \delta_{w,e}$ (by the definition). Thus, (a) is true for any w and $v = e$. By (7), $\delta_{w,v} = \xi^w(x_v) = \Big(\prod_{\beta\in\Delta^+\cap v\Delta^-} \beta^{-1}\Big)d_{w,v} + \sum_{u<v} c_{v,u}d_{w,u}$. Fixing w and using the induction on $\ell(v)$, we obtain (a).

(b) follows from Theorem 11.1.2(d).

(2) follows from (1) since $d_{e,e} = 1$ by (a).

We prove (3) by induction on $\ell(v)$. If $v = e$, it follows from (a). So, take $v \neq e$ and write $v = v's_j$ such that $v' < v$. Assuming the validity of (3) for v' (by induction) and applying (1) and (2), (3) follows for v.

(d) follows from Theorem 11.1.2(h).

(e) $$\begin{aligned}(\chi\delta_e)\bullet\xi^w &= \sum_v \big((\chi\delta_e)\bullet\xi^w\big)(x_v)\xi^v\\ &= \sum_v \xi^w(x_v\chi)\xi^v\\ &= \sum_v \xi^w\Big((v\chi)x_v + \sum_{u\xrightarrow{\beta}v}\langle v\chi, \beta^\vee\rangle x_u\Big)\xi^v,\\ &\qquad \text{by Theorem 11.1.2(c)}\\ &= (w\chi)\xi^w - \sum_{w\xrightarrow{\beta}v}\langle w\chi, \beta^\vee\rangle\xi^v.\end{aligned}$$

This proves (e).

(f): $\delta_{s_i}\bullet\xi^w = (\alpha_i\delta_e)\bullet(x_i\bullet\xi^w) + \xi^w.$

Using (b) and (e), (f) follows from the above identity.

(g): We prove this by induction on $\ell(w)$. If $w = e$, we get (g) from (2). So, assume $w \neq e$ and write $w = w's_j$ with $w' < w$. Applying (1) we get, for any $v \in W$,

(8) $$d_{w,vs_j} - d_{w,v} = (v\alpha_j)d_{w',v}.$$

By induction, $d_{w',v} \in S^{\ell(w')}(\underline{\alpha})$ and hence, by (8),

$$(9) \qquad d_{w,vs_j} - d_{w,v} \in S^{\ell(w)}(\underline{\alpha}) \qquad \text{(for any } s_j \text{ such that } ws_j < w).$$

Similarly, again applying (1), we get

$$(10) \qquad d_{w,vs_k} = d_{w,v}, \qquad \text{for any } s_k \text{ such that } ws_k > w.$$

Combining (9) and (10) and inducting on $\ell(v)$, we get $d_{w,v} - d_{w,e} \in S^{\ell(w)}(\underline{\alpha})$. But then $d_{w,v} \in S^{\ell(w)}(\underline{\alpha})$ from (a). This completes the induction and proves (g).

(h): To prove (4), it suffices of course to show that, for any $v \in W$,

$$(\psi_1\psi_2)(\delta_v x_i) = \psi_1(\delta_v x_i)\, \psi_2(\delta_v) + \psi_1(\delta_{vs_i})\, \psi_2(\delta_v x_i).$$

But this is an easy calculation. The identity (5) follows easily from (4) and the induction on $\ell(w)$.

To prove the "In particular" statement, apply $x_w \bullet$ to the identity (d) to get

$$x_w \bullet (\xi^u \xi^v) = \sum_{u,v \le w'} p^{w'}_{u,v} (x_w \bullet \xi^{w'}).$$

Evaluating the above at e, we get

$$\big(x_w \bullet (\xi^u \xi^v)\big)(e) = p^w_{u,v}.$$

Applying (5) to the above and using Theorem 11.1.2(d), we get (6).

(i): It is easy to see from (3) that

$$(11) \qquad \xi^{s_i} \xi^w = \omega_i \xi^w - (\omega_i \delta_e) \bullet \xi^w.$$

The right side of the above identity is equal to $(\omega_i - w\omega_i)\xi^w + \sum_{w \xrightarrow{\beta} v} \langle w\omega_i, \beta^\vee \rangle \xi^v$ by (e). This and (3) together prove (i) and thus the proof of the theorem is completed. □

The following lemma characterizes the matrix D.

11.1.8 Lemma. *Any matrix $\hat{D} = (\hat{d}_{v,w})_{v,w \in W}$ with entries in Q, which satisfies (11.1.7.1) and such that $\hat{d}_{v,e} = \delta_{v,e}$, is unique, i.e., $\hat{D} = D$.*

Proof. We prove, by induction on $\ell(w)$, that

$$(1) \qquad \hat{d}_{v,w} = d_{v,w}\,.$$

For $w = e$, its validity is part of the assumption and Theorem 11.1.7(a). So take $w > e$ and choose a simple reflection s_i such that $ws_i < w$. By (11.1.7.1), we get

$$\begin{aligned} d_{v,w} &= d_{v,ws_i} - (w\alpha_i)\, d_{vs_i,ws_i}, && \text{if } vs_i < v \\ &= d_{v,ws_i}, && \text{otherwise.} \end{aligned}$$

The same identity is true for d replaced by $\hat{d}$. So, by induction on $\ell(w)$, we get

$$\hat{d}_{v,w} = d_{v,w}. \qquad \square$$

11.1.9 Definition. Thinking of Q_W as an algebra over $\mathbb{C} = \mathbb{C}\delta_e \subset Q_W$ (where $\mathbb{C}$ of course corresponds to the constant functions on $\mathfrak{h}$), extend the scalars from $\mathbb{C}$ to $S(\mathfrak{h}^*)$ to get the algebra $\hat{Q}_W$ over $S(\mathfrak{h}^*)$, i.e.,

(1) $$\hat{Q}_W := S(\mathfrak{h}^*) \otimes_{\mathbb{C}} Q_W.$$

(The $S(\mathfrak{h}^*)$-algebra structure of $\hat{Q}_W$ should be distinguished from the noncentral copy of $S(\mathfrak{h}^*) = S(\mathfrak{h}^*)\delta_e$ sitting inside Q_W.)

For any $w \in W$, take a reduced decomposition $w = s_{i_1} \cdots s_{i_p}$, and define the element

(2) $$\hat{x}_w = (1 + \beta_1 \otimes x_{i_1})(1 + \beta_2 \otimes x_{i_2}) \cdots (1 + \beta_p \otimes x_{i_p}) \in \hat{Q}_W,$$

where, for any $1 \le j \le p$, $\beta_j := s_{i_1} \cdots s_{i_{j-1}}\alpha_{i_j}$ and 1 is the multiplicative identity of $\hat{Q}_W$. Observe that, by Lemma 1.3.14, $\Delta^+ \cap w\Delta^- = \{\beta_1, \dots, \beta_p\}$. Then, by Exercise 11.1.E.1, $\hat{x}_w$ does not depend on the choice of the reduced decomposition of w, thereby justifying the notation $\hat{x}_w$.

By Theorem 11.1.2(d), we can write

(3) $$\hat{x}_w = \sum_{v \in W} R(v, w) \otimes x_v,$$

for some (unique) $R(v, w) \in S(\mathfrak{h}^*)$. Moreover, $R(v, w) = 0$ unless $v \le w$. (The uniqueness of $R(v, w)$ follows since $\{x_v\}_{v \in W}$ are $\mathbb{C}$-linearly independent elements of Q_W.)

The following lemma follows easily from Theorem 11.1.2(d).

11.1.10 Lemma. *With the notation as above, for any $v, w \in W$,*

$$R(v, w) = \sum \beta_{j_1} \cdots \beta_{j_m},$$

where the summation runs over all those $1 \leq j_1 < \cdots < j_m \leq p$ such that $s_{i_{j_1}} \cdots s_{i_{j_m}}$ is a reduced decomposition of v. In particular, the right side is independent of the choice of the reduced decomposition of w. □

Now we are ready to give a close expression for the entries $d_{v,w}$ of the fundamental matrix D (introduced in (11.1.6.3)). Recall that an inductive formula to calculate $d_{v,w}$ is given by (11.1.7.1).

11.1.11 Proposition. *For any $v, w \in W$,*

$$d_{v,w} = R(v, w). \tag{1}$$

Proof. We first prove an inductive formula for $R(v, w)$ the same as (11.1.7.1): For any simple reflection s_i such that $ws_i < w$, taking a reduced expression of w ending in s_i, we get

$$\begin{aligned} \hat{x}_w &= \hat{x}_{ws_i}(1 - w\alpha_i \otimes x_i) \\ &= \sum_v R(v, ws_i) \otimes x_v - \sum_v R(v, ws_i) w\alpha_i \otimes x_v x_i \\ &= \sum_v R(v, ws_i) \otimes x_v - \sum_{v > vs_i} R(vs_i, ws_i) w\alpha_i \otimes x_v\,. \end{aligned} \tag{2}$$

From (2), we get (under the assumption $ws_i < w$),

$$\begin{aligned} R(v, w) &= R(v, ws_i) - R(vs_i, ws_i) w\alpha_i, && \text{if } vs_i < v \\ &= R(v, ws_i), && \text{otherwise}\,. \end{aligned} \tag{3}$$

For $ws_i > w$, replacing w by ws_i in (3) and using the second equality of (3) for v replaced by vs_i, we get the same relation as (3). Hence (3) is true for any $v, w \in W$. Again replacing w by ws_i in (3), we can rewrite (3) (valid for any $w \in W$) as

$$\begin{aligned} R(v, ws_i) - R(v, w) &= (w\alpha_i)\, R(vs_i, w), && \text{if } vs_i < v \\ &= 0, && \text{otherwise.} \end{aligned} \tag{4}$$

Now the proposition follows from Lemma 11.1.8. □

As an immediate consequence of the above proposition and Lemmas 11.1.10 and 1.3.16, we get the following:

11.1.12 Corollary. *For any $v, w \in W$,*

$$d_{v,w} = 0 \quad \textit{iff}\ v \nleq w.$$

Moreover,

$$d_{v,w} \in \mathbb{Z}_+[\underline{\alpha}],$$

where $\mathbb{Z}_+[\underline{\alpha}]$ is defined above Theorem 11.1.2. □

The (a) part of the following theorem is a vast generalization of the Chevalley formula Theorem 11.1.7(i). As we will see in Section 11.3, this part gives rise to an expression for the cup product of any two cohomology classes of $\mathcal{G}/\mathcal{B}$ in the Schubert basis, and the (b) part gives an expression for the Weyl group action on $H^*(\mathcal{G}/\mathcal{B})$ of an arbitrary $w \in W$, both in terms of the matrix D, for any Kac–Moody group $\mathcal{G}$. In fact, we obtain expressions in the T-equivariant cohomology $H_T^*(\mathcal{G}/\mathcal{B})$. Recall that the Chevalley formula gives an expression for the product of an arbitrary element of $H^*(\mathcal{G}/\mathcal{B})$ with an element in $H^2(\mathcal{G}/\mathcal{B})$ for the finite case.

Recall the definition of the space $\mathcal{D}$ from 11.1.6.

11.1.13 Theorem. *Fix $w \in W$.*

(a) Define the matrix $P^w \in \mathcal{D}$ by $P^w(u, v) = p^v_{w,u}$ (for $u, v \in W$), where $p^v_{w,u}$ is as in Theorem 11.1.7(d).

Also define the diagonal matrix $D^w \in \mathcal{D}$ by

$$D^w(u, v) = \delta_{u,v}\, d_{w,v}.$$

Then

(1) $$P^w = D \cdot D^w \cdot D^{-1}.$$

(b) For $u \in W$, write

(2) $$\delta_w \bullet \xi^u = \sum_v g^w_{u,v}\, \xi^v, \qquad \textit{for some (unique)}\ g^w_{u,v} \in S(\mathfrak{h}^*).$$

By Theorem 11.1.7(f),

(3) $$g^w_{u,v} = 0, \qquad \textit{unless}\ \ell(u) \geq \ell(v) \geq \ell(u) - \ell(w),$$

and, moreover, $g^w_{u,v} \in S^{\ell(u)-\ell(v)}(\underline{\alpha})$.

Now define two matrices $G^w, S^w \in \mathcal{D}$ by $G^w(u, v) = g^w_{u,v}$ and $S^w(u, v) = \delta_{uw^{-1},v}$. Then

(4) $$G^w = D \cdot S^w \cdot D^{-1}.$$

(Recall that $D^{-1} = C^t$ by Theorem 11.1.7(a).)

Proof. Since $p^v_{w,u} = 0$ unless $u \leq v$, we get that $P^w \in \mathcal{D}$. Now, for any $u, v \in W$,

$$\begin{aligned}(P^w \cdot D)(u, v) &= \sum_{w' \in W} p^{w'}_{w,u} d_{w',v} \\ &= d_{w,v} d_{u,v}, \text{ by Theorem 11.1.7(d)} \\ &= (D \cdot D^w)(u, v).\end{aligned}$$

This proves (1).

Clearly $S^w \in \mathcal{D}$ and, by (3), $G^w \in \mathcal{D}$. By (2) (for any $u, v \in W$),

$$\sum_{w'} G^w(u, w')\, d_{w',v} = d_{u,vw}.$$

This proves (4). □

We give another characterization of the S-algebra $\Lambda = \Lambda_W$ defined by (11.1.4.10), where $S = S(\mathfrak{h}^*)$. Recall from 11.1.4–11.1.5 that Λ is a S-subalgebra of the Q-algebra Ω_W, where Ω_W can be identified with the space of all the maps $f : W \to Q$.

11.1.14 Definition. Define the S-subalgebra $\tilde{\Lambda} \subset \Omega_W$ by

$$\tilde{\Lambda} = \Big\{ W \xrightarrow{f} S : f(s_\beta v) - f(v) \text{ is divisible by } \beta \text{ for all } \beta \in \Delta^+_{\text{re}} \text{ and } v \in W \text{ and, moreover, } f(x_w) = 0 \text{ for all but finitely many } w \in W \Big\}.$$

11.1.15 Proposition. $\Lambda = \tilde{\Lambda}$.

Proof. For any $\beta \in \Delta^+_{\text{re}}$, define the Q-linear operator $A_\beta : \Omega_W \to \Omega_W$ by

(1) $$(A_\beta f)\, v = \frac{f(vs_\beta) - f(v)}{v\beta}, \text{ for } f \in \Omega_W \text{ and } v \in W.$$

For any simple root α_i, A_{α_i} is nothing but the operator $x_i \bullet$ (cf. (11.1.4.8)).

Write $\beta = w\alpha_i$ for some $w \in W$ and simple root α_i. Then, with the notation as in 11.1.4, we have

(2) $$A_\beta = (\delta_w x_i \delta_{w^{-1}}).$$

In particular, by Lemma 11.1.5, A_β keeps Λ stable.

We first show that $\Lambda \subset \tilde{\Lambda}$: Take $f \in \Lambda$. Then, as shown above, $A_\beta f \in \Lambda$ (for all $\beta \in \Delta^+_{\mathrm{re}}$). In particular, $f(vs_\beta) - f(v)$ is divisible by $v\beta$, for all $v \in W$. But since $vs_\beta = vs_\beta v^{-1}v = s_{v\beta}v$, $f \in \tilde{\Lambda}$.

Now we prove the reverse inclusion. Take $f \in \tilde{\Lambda}$ and write $f = \sum_{w\in W} q_w \xi^w$ with $q_w \in Q$, where all but finitely many q_w's are 0 by assumption. This is possible since $\{x_w\}_{w\in W}$ is a Q-basis of Q_W under the left Q-vector space structure (cf. 11.1.3). To show that $f \in \Lambda$, it suffices to show, by Lemma 11.1.5, that each $q_w \in S$. Let $V := \{w \in W : q_w \notin S\}$. Assume, if possible, that $V \neq \emptyset$. Take $w_o \in V$ of minimal length. Replacing f by $g := f - \sum_{w\in W\setminus V} q_w \xi^w$, we can assume that, for $w \notin V$, $q_w = 0$. (Observe that $g \in \tilde{\Lambda}$ since $\Lambda \subset \tilde{\Lambda}$.) By Theorem 11.1.7(a),

$$f(w_o) = q_{w_o} \prod_{\beta\in\Delta^+\cap w_o\Delta^-} \beta. \tag{3}$$

Now we can easily deduce from Lemma 1.3.13 that

$$\Delta^+ \cap w_o\Delta^- = \{\beta \in \Delta^+_{\mathrm{re}} : s_\beta w_o < w_o\}. \tag{4}$$

Hence, for any $\beta \in \Delta^+ \cap w_o\Delta^-$,

$$f(s_\beta w_o) = 0.$$

In particular, by the definition of $\tilde{\Lambda}$, $f(w_o)$ is divisible by any $\beta \in \Delta^+ \cap w_o\Delta^-$. Thus $\prod_{\beta\in\Delta^+\cap w_o\Delta^-} \beta$ divides $f(w_o)$, since any two distinct elements of Δ^+_{re} are not multiples of each other, i.e., $q_{w_o} \in S$. A contradiction to the assumption that $V \neq \emptyset$. This proves that $V = \emptyset$ and hence $f \in \Lambda$. □

Now, we define a "parabolic" analogue of the algebra Λ:

11.1.16 Definition. Let $Y \subset \{1, \dots, \ell\}$ be any subset and, as in 1.3.17, let $W_Y \subset W$ be the subgroup generated by $\{s_i\}_{i\in Y}$. Define $\Lambda^Y = \Lambda^{W_Y}$ as the S-subalgebra of Λ consisting of W_Y-invariant elements,where W (in particular, W_Y) acts on Λ via the action $\bullet$ given by (11.1.4.7).

The following lemma determines Λ^Y.

11.1.17 Lemma. *$\Lambda^Y = \sum_{w\in W'_Y} S\xi^w$, where W'_Y is defined in 1.3.17. In particular, Λ^Y is a free S-submodule of Λ.*

Proof. By Theorem 11.1.7(f), for any $w \in W'_Y$, $\xi^w \in \Lambda^Y$. Conversely, take $\xi = \sum_{w\in W} p^w\xi^w \in \Lambda^Y$, where all but finitely many p^w's are 0 (cf. Lemma 11.1.5). Take $i \in Y$. Since $\delta_{s_i} \bullet \xi = \xi$, we get $x_i \bullet \xi = 0$ (by (11.1.4.7-8)).

In particular, by Theorem 11.1.7(b), $p^w = 0$ for $ws_i < w$. This proves that $\Lambda^Y \subset \sum_{w\in W'_Y} S\xi^w$, completing the proof of the lemma. □

11.1.18 Definition. For $v \le w$ in W, define

$$S(w, v) = \{\beta \in \Delta^+_{\rm re} : s_\beta v \le w\}.$$

11.1.19 Proposition. *For any $v \le w$, there exists a nonzero polynomial $c'_{w,v} \in \mathbb{Q}[\underline{\alpha}]$ such that*

$$(1) \qquad c_{w,v} = (-1)^{\ell(w)-\ell(v)} c'_{w,v} \prod_{\beta\in S(w,v)} \beta^{-1}.$$

In particular, $c_{w,v} \neq 0$ and $c'_{w,v}$ is a homogeneous polynomial with

$$(2) \qquad \deg c'_{w,v} = \#S(w, v) - \ell(w).$$

Similarly, for any $v \le w$, there exists a nonzero polynomial $d'_{v,w} \in \mathbb{Q}[\underline{\alpha}]$ such that

$$(3) \qquad d_{v,w} = d'_{v,w} \prod_{\beta\in \hat{S}(w,v)} \beta,$$

where $\hat{S}(w, v) := \{\beta \in \Delta^+_{\rm re} : v \not\le s_\beta w\}$. In particular, $d'_{v,w}$ is homogeneous with

$$(4) \qquad \deg d'_{v,w} = \ell(v) - \#\hat{S}(w, v).$$

(By Exercise 11.1.E.4, $c'_{w,v}, d'_{v,w} \in \mathbb{Z}[\underline{\alpha}]$. In fact, I conjecture that $c'_{w,v}, d'_{v,w} \in \mathbb{Z}_+[\underline{\alpha}]$, for any $v \le w$.)

Proof. We first show that $c_{w,v} \neq 0$: Equating the coefficients of δ_v in Theorem 11.1.2(c) we obtain, for any $\chi \in \mathfrak{h}^*$,

$$(5) \qquad \chi c_{w,v} = (vw^{-1}\chi)c_{w,v} - \sum_{u\xrightarrow{\beta} w} \chi(\beta^\vee)c_{u,v}.$$

Now take $\chi \in D^o := \{\lambda \in \mathfrak{h}^*_{\mathbb{Z}} : \lambda(\alpha_i^\vee)$ is a strictly positive integer for all $i\}$. By Proposition 1.4.2(a), χ has trivial W-isotropy. If $v \neq w$, rewrite (5) as

$$(6) \qquad c_{w,v} = -\sum_{u\xrightarrow{\beta} w} \chi(\beta^\vee)\, \chi_1^{-1} c_{u,v},$$

where $\chi_1 = \chi_1(v, w) := \chi - vw^{-1}\chi$.

By (6) and induction on $\ell(w) - \ell(v)$, for any $v \leq w$, we can write (using (1.3.22.2) and the fact from 1.3.15 that, for $v < w$, there exists $u \to w$ such that $v \leq u$)

$$c_{w,v} = (-1)^{\ell(w)-\ell(v)} c_{v,v} h_{w,v}/g_{w,v},$$

where $h_{w,v}, g_{w,v}$ are both nonzero homogeneous polynomials $\in \mathbb{Z}_+[\underline{\alpha}]$ such that $\deg g_{w,v} - \deg h_{w,v} = \ell(w) - \ell(v)$. In particular, $c_{w,v} \neq 0$ if $v \leq w$ (by Theorem 11.1.2(e$_2$)).

Now we prove (1) for $v \leq w$ by induction on $\ell(w) - \ell(v)$: For $w = v$, $c_{v,v} = \prod_{\beta \in S(v,v)} \beta^{-1}$ by (11.1.15.4) and Theorem 11.1.2(e$_2$), and hence (1) is satisfied with $c'_{v,v} = 1$.

By (6) and the induction hypothesis, for any $\chi \in D^o$ and $w > v$,

$$(7) \qquad \begin{aligned} c_{w,v} &= (-1)^{\ell(w)-\ell(v)} \chi_1^{-1} \sum_{v \leq u \xrightarrow{\beta} w} \chi(\beta^\vee)\Big(c'_{u,v} \prod_{\gamma \in S(u,v)} \gamma^{-1}\Big) \\ &= (-1)^{\ell(w)-\ell(v)} \chi_1^{-1} f_\chi \prod_{\substack{\gamma \in S(w,v) \text{ and} \\ s_\gamma \neq wv^{-1}}} \gamma^{-1}, \end{aligned}$$

for some $f_\chi \in \mathbb{Q}[\underline{\alpha}]$ which is nonzero since $c_{w,v} \neq 0$.

Let r be the rank of $1 - vw^{-1}$ acting on $\mathfrak{h}^*$. Since $v \neq w$, we get $r \geq 1$. We need to consider the following two cases separately.

Case I: $r = 1$. In this case, by (a subsequent) Lemma 11.2.2, $vw^{-1} = s_\nu$ for some $\nu \in \Delta^+_{\text{re}}$. Hence $\chi_1 := \chi - vw^{-1}\chi = \langle \chi, \nu^\vee \rangle \nu$, so that, by (7),

$$c_{w,v} = (-1)^{\ell(w)-\ell(v)} \frac{f_\chi}{\langle \chi, \nu^\vee \rangle} \prod_{\gamma \in S(w,v)} \gamma^{-1}.$$

Thus, (1) is satisfied by taking $c'_{w,v} = \frac{f_\chi}{\langle \chi, \nu^\vee \rangle}$ for any $\chi \in D^o$.

Case II: $r > 1$. Choose $\chi', \chi'' \in D^o$ such that χ'_1 and χ''_1 are $\mathbb{C}$-linearly independent. Then, by (7), $\chi''_1 f_{\chi'} = \chi'_1 f_{\chi''}$ and hence χ'_1 divides $f_{\chi'}$. Taking $c'_{w,v} = f_{\chi'}/\chi'_1$, we see that (1) is satisfied.

This completes the proof of (1). By Theorem 11.1.2(f), we obtain that $c'_{w,v}$ is homogeneous of degree $\#S(w, v) - \ell(w)$. This proves (2).

The proof of (3) is exactly similar to that of the proof of (1) provided we use Theorem 11.1.7(e), resp. 11.1.7(a), instead of 11.1.2(c), resp. 11.1.2(e). Finally (4) follows from (3) and Theorem 11.1.7(g). □

The following result in the case when $u = v$ was conjectured by Deodhar. For finite W, this more general result is, in fact, equivalent to his conjecture.

11.1.20 Corollary. *For any $u \leq v \leq w \in W$,*

$$\#\{\beta \in \Delta_{\text{re}}^+ : u \leq s_\beta v \leq w\} \geq \ell(w) - \ell(u).$$

Proof. We have, by (11.1.19.2) and (11.1.19.4),

$$\begin{aligned} 0 \leq \deg c'_{w,v} + \deg d'_{u,v} &= \# S(w,v) - \ell(w) + \ell(u) - \# \hat{S}(v,u) \\ &= \#\left\{\beta \in \Delta_{\text{re}}^+ : u \leq s_\beta v \leq w\right\} + \ell(u) - \ell(w). \quad \square \end{aligned}$$

11.1.21 Remark. One can easily extend the above corollary to an arbitrary Coxeter group by considering the standard reflection representation of W (cf. Theorem 1.3.11(b)).

In the case when W is finite, we relate $c_{w,v}$'s and $d_{v,w}$'s directly.

11.1.22 Lemma. *Let W be finite, i.e., $\mathfrak{g}$ is of finite type. Then, for any $v, w \in W$,*

$$d_{v,w} = \left(\prod_{\beta \in \Delta^+} \beta\right) w\left(c_{v^{-1}w_o, w^{-1}w_o}\right),$$

where w_o is the longest element of W.

Proof. Define the matrix $\bar{C} = (\bar{c}_{v,w})_{v,w \in W}$ by

$$\bar{c}_{v,w} := \Big(\prod_{\beta \in \Delta^+} \beta\Big) w(c_{v^{-1}w_o, w^{-1}w_o}).$$

By Theorem 11.1.7(a), it suffices to show that

$$\bar{C} \cdot C^t = I, \text{ i.e.,} \tag{1}$$

for any $v, w \in W$,

$$\sum_u \bar{c}_{v,u} c_{w,u} = \delta_{v,w}. \tag{2}$$

Consider

$$x_w x_{v^{-1}w_o} = \sum_u c_{w,u} u(c_{v^{-1}w_o, u^{-1}w_o})\, \delta_{w_o} + \sum_{\theta < w_o} q_\theta \delta_\theta, \tag{3}$$

for some $q_\theta \in Q$. By Theorem 11.1.2(d)–(e), $x_w x_{v^{-1}w_o}$ has a nonzero component in δ_{w_o} iff $w = v$ (since $\ell(v^{-1}w_o) = \ell(w_o) - \ell(v)$). So, by (3) and Theorem 11.1.2(e_2),

$$\sum_u u(c_{v^{-1}w_o, u^{-1}w_o})\, c_{w,u} = \delta_{v,w} \left(\prod_{\beta \in \Delta^+} \beta\right)^{-1}.$$

This proves (2) and hence the lemma is proved. □

11.1.E EXERCISES

(1) Prove that the element $\hat{x}_w \in \hat{Q}_W$ defined by (11.1.9.2) does not depend on the choice of the reduced decomposition of w, i.e., the elements $\{\hat{x}_{s_i}\}_{1 \leq i \leq \ell}$ satisfy the braid property.
Hint. One can assume that W is a finite Weyl group generated by $\{s_1, s_2\}$. In particular, by Proposition 1.3.21, $s_1 s_2$ is of order $m \in \{2, 3, 4, 6\}$. It suffices to show that $\hat{x}_{s_1}\hat{x}_{s_2}\cdots = \hat{x}_{s_2}\hat{x}_{s_1}\cdots$, where both the sides have m factors.

(2) Determine $c_{w,v}$ and $d_{v,w}$ for $v \leq w$ with $\ell(w) = \ell(v) + 1$.

(3) Deduce the analogue of the Chevalley's formula Theorem 11.1.7(i) from Theorem 11.1.13(a).

(4) In the notation of Proposition 11.1.19, prove that, for any $v \leq w, c'_{w,v}, d'_{v,w} \in \mathbb{Z}[\underline{\alpha}]$.
Hint. Use Proposition 11.1.19 and Theorems 11.1.2(f) and 11.1.7(g).

11.2. Determination of $\bar{R}$

Recall the definition of the rings $R \subset \bar{R}$ from 11.1.3, and the involution t from (11.1.1.2).

11.2.1 Theorem. *(a)* $\bar{R} = R$.

(b) $\bar{R} \cap \bar{R}^t = \bigoplus_{w \in W} S\delta_w$.

In particular, $\bar{R}$ is a finitely generated algebra over $\mathbb{C}$.

Proof. We need the following preparatory lemmas.

11.2.2 Lemma. *Let $e \neq w \in W$ be such that w fixes pointwise a hyperplane $H \subset \mathfrak{h}$. Then $w = v s_i v^{-1}$ for some $v \in W$ and simple reflection s_i. In particular, H is the real-root hyperplane* $\operatorname{Ker}(v\alpha_i)$.

Proof. We first show that w is semisimple as an element of $\operatorname{Aut}\mathfrak{h}$. Since w fixes a hyperplane, the only other possibility for $w \in \operatorname{Aut}\mathfrak{h}$ is that it is unipotent and, in fact, $(w-1)^2 = 0$, i.e., $w^2 + 1 = 2w$. Multiplying by w^{-1}, we get

$$w - 1 = 1 - w^{-1}. \tag{1}$$

Taking $h \in \mathfrak{h}_{\mathbb{R}}$ (cf. 1.4.1) such that $\alpha_j(h) > 0$ for all the simple roots α_j and applying (1) to h, we get $wh - h = h - w^{-1}h$, which contradicts Corollary 1.3.22. Hence w is semisimple. Moreover, since w keeps the integral lattice $\mathfrak{h}_{\mathbb{Z}} \subset \mathfrak{h}_{\mathbb{R}}$ stable (cf. 6.1.6), the remaining eigenvalue of w is -1.

We next show that H is a real-root hyperplane. Recall the definition of the Tits cone $C^\vee \subset \mathfrak{h}_\mathbb{R}$ from 1.4.1. Also define

$$D_\mathbb{R}^{o\vee} = \{h \in \mathfrak{h}_\mathbb{R} : \alpha_i(h) > 0 \text{ for all the simple roots } \alpha_i\}$$

and set

$$X^\vee = \bigcup_{w\in W} w D_\mathbb{R}^{o\vee} \subset C^\vee.$$

Let $H^\pm \subset \mathfrak{h}_\mathbb{R}$ be two open half spaces separated by $H_\mathbb{R} := H \cap \mathfrak{h}_\mathbb{R}$. (Since w keeps $\mathfrak{h}_\mathbb{R}$ stable, $H_\mathbb{R}$ is indeed a hyperplane in $\mathfrak{h}_\mathbb{R}$.) Since $X^\vee$ is nonempty and open in $\mathfrak{h}_\mathbb{R}$ (under the Hausdorff topology), $X^\vee$ intersects at least one of $H^\pm$, say $X^\vee \cap H^+ \neq \emptyset$. Since $X^\vee$ is W-stable; in particular, w-stable, and w has one eigenvalue -1, $X^\vee \cap H^- \neq \emptyset$. Since both of the $X^\vee \cap H^\pm$ are open in $\mathfrak{h}_\mathbb{R}$ and nonempty and $C^\vee$ is convex (cf. Proposition 1.4.2(c)), there exists a nonempty Hausdorff open subset N of $H_\mathbb{R}$ such that $N \subset C^\vee$. Since any point in H has w in its isotropy and any point of $X^\vee$ has trivial isotropy (cf. Proposition 1.4.2(a)), we get

$$N \subset C^\vee \backslash X^\vee \subset \bigcup_{\substack{v\in W,\\ 1\leq j\leq \ell}} v(\operatorname{Ker}\alpha_j).$$

But then, from Baire's category theorem ([Rudin–73, §2.2]), $N \subset v(\operatorname{Ker}\alpha_i)$ for some $v \in W$ and some simple root α_i; which in turn, from the Zariski density of N in H, proves that $H \subset \operatorname{Ker}(v\alpha_i)$. But, H being a hyperplane (by assumption), we get $H = \operatorname{Ker}(v\alpha_i)$.

Finally, we show that $w = vs_iv^{-1}$. The element $v^{-1}wv$ fixes the hyperplane $\operatorname{Ker}\alpha_i$ pointwise. Taking $h_o \in \mathfrak{h}_\mathbb{R}$ such that $\alpha_i(h_o) = 0$ and $\alpha_j(h_o) > 0$ for all $j \neq i$, and applying Proposition 1.4.2(a), we get $v^{-1}wv = s_i$. This proves the lemma. □

11.2.3 Lemma. *Let $p \in S$ be an irreducible polynomial and let $x = \sum_w p_w x_w \in R$ be such that at least one of p_w is coprime to p. Assume further that*

(1) $$x \bullet S \subset pS.$$

Then p is a scalar multiple of a real root.

Proof. We can assume, without loss of generality, that each nonzero p_w is coprime to p. Let k be the maximum integer such that there exists $w \in W$ with $\ell(w) = k$ and $p_w \neq 0$. Rewrite

(2) $$x = \sum_{\ell(w)\leq k} p_w x_w = \sum_{\ell(w)\leq k} q_w \delta_w,$$

where $q_w \in Q$. In fact, each q_w has in its denominator only a product of real roots (possibly with repetitions). By Theorem 11.1.2(e), we have

(3) $$q_w = p_w c_{w,w}, \qquad \text{for } \ell(w) = k,$$

where $c_{w,w}$ is as in Theorem 11.1.2(e$_2$).

Define $V := \bigcup_{v \in W \setminus \{e\}} \operatorname{Ker}(v-1)$, where $v-1$ acts on $\mathfrak{h}$. We first prove that $Z(p) \subset V$, where $Z(p)$ denotes the zero set of p. If not, pick any $h_o \in Z(p)\setminus V$. In particular, h_o has no nontrivial W-isotropy. Pick w_o of length k such that $p_{w_o} \neq 0$. There exists a polynomial $p_o \in S$ such that $p_o(w_o^{-1}h_o) = 1$ and $p_o(w^{-1}h_o) = 0$ for all those (finitely many) $w \neq w_o$ satisfying $q_w \neq 0$. Evaluating $x \bullet p_o$ at h_o we get, from (1)–(3), $p_{w_o}(h_o)c_{w_o,w_o}(h_o) = 0$. (Since h_o has no nontrivial W-isotropy, $q_w(h_o)$ makes sense and, moreover, $c_{w_o,w_o}(h_o) \neq 0$.) Hence $p_{w_o}(h_o) = 0$. So, by the Baire category theorem, p divides p_{w_o}, which is a contradiction to the assumption. So $Z(p) \subset V$.

Since p is irreducible, again by Baire's category theorem, $Z(p) \subset \operatorname{Ker}(v_o - 1)$ for some $v_o \in W\setminus\{e\}$. Since $Z(p)$ is a hypersurface in $\mathfrak{h}$, $\operatorname{Ker}(v_o - 1)$ is a hyperplane. The lemma follows now by Lemma 11.2.2. □

11.2.4 Lemma. *Let $\{p_w\}_{\ell(w)\leq k}$ be polynomials in S such that $\left(\sum_{\ell(w)\leq k} p_w x_w\right) \bullet S \subset \alpha_i S$, for some simple root α_i. Then α_i divides all the p_w's.*

Proof. Consider the element $x := \frac{1}{\alpha_i}\sum_{\ell(w)\leq k} p_w x_w \in Q_W$ and write

(1) $$\alpha_i x = \sum_{\ell(w)\leq k} p_w x_w = \sum_{\ell(w)\leq k} q_w \delta_w \quad (\text{for some } q_w \in Q),$$

where, as in (11.2.3.3),

(2) $$q_w = p_w c_{w,w}, \qquad \text{if } \ell(w) = k.$$

Fix w_o of length k such that $p_{w_o} \neq 0$ and rewrite (1) as

(3) $$\alpha_i x = x_o + q_{w_o}\delta_{w_o} + q_{s_i w_o}\delta_{s_i w_o},$$

where $q_{s_i w_o}$ is interpreted as 0 if $s_i w_o > w_o$ and

$$x_o := \sum_{w \notin \{s_i w_o, w_o\}} q_w \delta_w.$$

Applying (3) to any $p \in S$ and $(w_o^{-1}\alpha_i)p$, we get the following two identities in Q:

(4) $$(\alpha_i x) \bullet p = \left(\sum_{w \notin \{s_i w_o, w_o\}} q_w(wp)\right) + p_{w_o}c_{w_o,w_o}(w_o p) + q_{s_i w_o}(s_i w_o p),$$

and

(5)
$$(\alpha_i x) \bullet \big((w_o^{-1}\alpha_i)p\big) = \Bigg(\sum_{w\notin\{s_i w_o, w_o\}} q_w(wp)(ww_o^{-1}\alpha_i)\Bigg) + p_{w_o} c_{w_o,w_o}(w_o p)\alpha_i - q_{s_i w_o}(s_i w_o p)\alpha_i .$$

Multiplying (4) by α_i and adding with (5), we get (in Q)

(6)
$$(\alpha_i^2 x) \bullet p + (\alpha_i x) \bullet \big((w_o^{-1}\alpha_i)p\big) = 2p_{w_o} c_{w_o,w_o}(w_o p)\alpha_i + \sum_{w\notin\{s_i w_o, w_o\}} q_w(wp)(\alpha_i + ww_o^{-1}\alpha_i).$$

Fix any point $h_o \in \mathfrak{h}_{\mathbb{R}}$ such that $\alpha_i(h_o) = 0$ and $\alpha_j(h_o) > 0$ for all $j \neq i$. By Proposition 1.4.2(a), the W-isotropy of h_o is precisely equal to $\{1, s_i\}$. Choose a polynomial $p_o \in S$ such that $p_o(w_o^{-1}h_o) = 1$ and p_o has a zero of sufficiently high multiplicity at $w^{-1}h_o$ for any (finitely many) $w \notin \{s_i w_o, w_o\}$ such that $q_w \neq 0$. (Since W-isotropy of h_o is $\{1, s_i\}$, this is possible.)

If $s_i w_o > w_o$, we have $\alpha_i \notin \Delta^+ \cap w_o\Delta^-$ and hence $c_{w_o,w_o}(h_o)$ is well defined (and is nonzero). In this case, evaluating (4) for $p = p_o$ at h_o, we get $0 = p_{w_o}(h_o)$. Observe that, by assumption, $(\alpha_i x) \bullet p \in \alpha_i S$ and hence, for any $p \in S$, $((\alpha_i x) \bullet p)\, h_o = 0$.

If $s_i w_o < w_o$, by (1.3.14.1) and Theorem 11.1.2(e_2),

(7)
$$c_{w_o,w_o} = \frac{1}{\alpha_i}\big(s_i c_{s_i w_o, s_i w_o}\big),$$

and, from the previous case, $c_{s_i w_o, s_i w_o}(h_o)$ is well defined (and is nonzero). Thus, evaluating (6) for $p = p_o$ at h_o and using (7), we get $0 = p_{w_o}(h_o)$.

So, in either case, $p_{w_o}(h_o) = 0$. Hence α_i divides p_{w_o}. So, by induction on $\#\{w : p_w \neq 0\}$, the lemma follows. □

11.2.5 *Proof of Theorem 11.2.1.* (a): Let $x \in \bar{R}$. By 11.1.3, we can write $x = \frac{1}{p}\sum_{\ell(w)\le k} p_w x_w$, for some $p, p_w \in S$. We prove that p divides p_w for every w. We can assume, of course, that p is irreducible. By Lemma 11.2.3, if p does not divide some p_w, then p has to be a scalar multiple of a real root (say) $v\alpha_i$, for some $v \in W$ and simple root α_i. Since $\delta_{v^{-1}}\bar{R} = \bar{R}$ and $\delta_v(R) = R$, we can assume that $p = \alpha_i$. But then Lemma 11.2.4 proves that $x \in R$.

(b): Fix $x \in \bar{R} \cap \bar{R}^t$. Since $x \in \bar{R}$, we can write by (a), $x = \sum p_w x_w$, for some $p_w \in S$. Express $x = \sum_w q_w \delta_w$, where $q_w \in Q$. Upon multiplying by a suitable polynomial we can assume, without loss of generality, that all the q_w's have only one fixed real root, say $v\alpha_i$, in their denominators. Since $\delta_{v^{-1}}\bar{R} = \bar{R}$,

$\delta_{v^{-1}} \bar{R}^t = \bar{R}^t$, and $\delta_{v^{-1}}(\oplus_w S\delta_w) = \oplus_w S\delta_w$, we can further assume that all the q_w's have only one fixed α_i in their denominators, i.e.,

$$(1) \qquad x = \frac{1}{\alpha_i} \sum_w \bar{p}_w \delta_w, \text{ for some } \bar{p}_w \in S.$$

We prove that all the $\bar{p}_w$'s are divisible by α_i. By (1), taking t, we get

$$(2) \qquad x^t = \frac{\bar{p}_e}{\alpha_i}\delta_e - \frac{(s_i \bar{p}_{s_i})}{\alpha_i}\delta_{s_i} + \sum_{w \neq e, s_i} w^{-1}\left(\frac{\bar{p}_w}{\alpha_i}\right)\delta_{w^{-1}}.$$

For any $p \in S$, from (1)–(2), we get the following identities in Q:

$$(3) \qquad (\alpha_i x) \bullet p = \bar{p}_e p + \bar{p}_{s_i}(s_i p) + \sum_{w \neq e, s_i} \bar{p}_w (wp), \text{ and}$$

$$(4) \qquad (\alpha_i x^t) \bullet p = \bar{p}_e p - (s_i \bar{p}_{s_i})(s_i p) + \sum_{w \neq e, s_i} \alpha_i w\left(\frac{\bar{p}_{w^{-1}}}{\alpha_i}\right)(wp).$$

As in the proof of Lemma 11.2.4, fix any point $h_o \in \mathfrak{h}_{\mathbb{R}}$ such that $\alpha_i(h_o) = 0$ and $\alpha_j(h_o) > 0$ for all $j \neq i$. Further, choose $p_o \in S$ such that $p_o(h_o) = 1$ and $p_o(w^{-1}h_o) = 0$ for all $w \notin \{e, s_i\}$ such that $\bar{p}_w$ or $\bar{p}_{w^{-1}}$ is nonzero. For such an h_o and any $w \in W$, the rational function $\alpha_i / w\alpha_i$ is well defined at h_o.

Evaluating (3)–(4) for $p = p_o$ at h_o, we get

$$(5) \qquad 0 = \bar{p}_e(h_o) + \bar{p}_{s_i}(h_o), \qquad \text{since } x \bullet S \subset S, \text{ and}$$

$$(6) \qquad 0 = \bar{p}_e(h_o) - \bar{p}_{s_i}(h_o), \qquad \text{since } x^t \bullet S \subset S.$$

Adding (5)–(6), we get

$$\bar{p}_e(h_o) = 0, \text{ i.e., } \alpha_i \text{ divides } \bar{p}_e.$$

To prove that α_i divides $\bar{p}_w$ for general w, consider $x\delta_{w^{-1}} \in \bar{R} \cap \bar{R}^t$. This proves (b) and hence completes the proof of the theorem. □

11.2.E EXERCISE Using W-harmonic functions (or otherwise) give a simpler proof of Theorem 11.2.1(a) in the finite case.

11.3. T-equivariant Cohomology of $\mathcal{G}/\mathcal{B}$

In this section $\mathcal{G}$ is any Kac–Moody group and $\mathcal{B}$, resp. T, is its standard Borel subgroup, resp. standard maximal torus. For any subset $Y \subset \{1, \ldots, \ell\}$, let $\mathcal{P}_Y$

be the associated standard parabolic subgroup (cf. 6.1.18). As in 7.4.9, endow the ind-variety $\mathcal{X}^Y := \mathcal{G}/\mathcal{P}_Y$ (under the stable ind-variety structure) with the strong topology and denote $\mathcal{X}^Y$ endowed with this topology by $\mathcal{X}^Y_{\text{an}}$. (Alhough we will suppress "an" when it is clear from the context which topology we are considering, strong or Zariski.) By Theorem 7.4.14 and Lemma 7.4.10, $\mathcal{X}^Y$ is a $\mathcal{G}^{\min}$-ind-variety (cf. 4.2.1) under the left multiplication.

For a topological space Z, we denote by $H^*(Z)$, resp. $H_*(Z)$, the singular cohomology, resp. singular homology, of Z with integral coefficients. Recall that, for an irreducible complete variety Z of dimension d, $H_{2d}(Z_{\text{an}})$ is a free $\mathbb{Z}$-module of rank 1 generated by the fundamental class $[Z]$ (see, e.g., [Fulton–98, §19.1]).

11.3.1 Definition. Let $\Theta \subset W'_Y$ be any subset satisfying the following condition:

(1) For any $w \in \Theta$ and $v \in W'_Y$ such that $v \leq w$, $v \in \Theta$.

To any such Θ, associate the subset $\mathcal{X}^Y_\Theta \subset \mathcal{X}^Y$ defined by

$$\mathcal{X}^Y_\Theta := \bigsqcup_{w\in\Theta} \mathcal{B}w\mathcal{P}_Y/\mathcal{P}_Y.$$

By Proposition 7.1.15, $\mathcal{X}^Y_\Theta$ is a Zariski closed subset of $\mathcal{X}^Y$; in particular, $\mathcal{X}^Y_\Theta$ is closed in $\mathcal{X}^Y$ under the strong topology. Clearly, $\mathcal{X}^Y_\Theta$ is $\mathcal{B}$-stable under the left multiplication. Conversely, for any $\mathcal{B}$-stable Zariski closed subset $Z \subset \mathcal{X}^Y$, there exists a $\Theta = \Theta(Z) \subset W'_Y$ satisfying (1) such that $Z = \mathcal{X}^Y_\Theta$ (use Proposition 7.1.15).

11.3.2 Proposition. *For any $\Theta \subset W'_Y$ as above (i.e., satisfying (11.3.1.1)) and any $i \in \mathbb{Z}_+$,*

(a) $H_{2i+1}(\mathcal{X}^Y_\Theta) = 0$, and

(b) $H_{2i}(\mathcal{X}^Y_\Theta) = \bigoplus_{w\in\Theta(i)} \mathbb{Z}[X^Y_w]$, where $\Theta(i) := \{w \in \Theta : \ell(w) = i\}$, and (abusing notation) $[X^Y_w]$ denotes the image of the fundamental class $[X^Y_w] \in H_{2\ell(w)}(X^Y_w)$ under the canonical map $H_{2\ell(w)}(X^Y_w) \to H_{2\ell(w)}(\mathcal{X}^Y_\Theta)$.

Thus, $H^{2i+1}(\mathcal{X}^Y_\Theta) = 0$ and $H^{2i}(\mathcal{X}^Y_\Theta) = \bigoplus_{w\in\Theta(i)} \mathbb{Z}\,\varepsilon^Y_w$, where $\varepsilon^Y_w \in H^{2\ell(w)}(\mathcal{X}^Y_\Theta)$ is defined by

(1) $$\varepsilon^Y_w[X^Y_v] = \delta_{v,w}, \text{ for } v, w \in \Theta.$$

Further, the natural restriction map

$$\gamma : H^*(\mathcal{X}^Y) \to H^*(\mathcal{X}^Y_\Theta)$$

is surjective.

For $Y = \emptyset$ and $\Theta = W$, so that $\mathcal{X}^Y_\Theta = \mathcal{X}$, the basis $\{\varepsilon_w := \varepsilon^{\emptyset}_w\}_{w\in W}$ is called the *Schubert basis* of $H^*(\mathcal{X})$.

Proof. Assume, by induction on $n \geq 0$, that

$$\begin{aligned} H_{2i+1}(Z_n) &= 0, \quad \text{for any } i \in \mathbb{Z}_+, \text{ and} \\ H_{2i}(Z_n) &= \bigoplus_{w\in\Theta(i)} \mathbb{Z}[X^Y_w], \text{ for } i \leq n, \text{ and} \\ &= 0, \quad \text{for } i > n, \end{aligned} \tag{2}$$

where $Z_n := \mathcal{X}^Y_\Theta \cap X^Y_n$ and $\{X^Y_n\}_{n\geq 0}$ is the filtration of $\mathcal{X}^Y$ as in 7.1.13. Of course, (2) is trivially true for $n = 0$.

For any $n \geq 0$, we assert the following:

$$\begin{aligned} H_i(Z_{n+1}, Z_n) &= 0, \qquad\qquad \text{for } i \neq 2(n+1), \text{ and} \\ &= \bigoplus_{w\in\Theta(n+1)} \mathbb{Z}\,\theta_w[X^Y_w], \text{ if } i = 2(n+1), \end{aligned} \tag{3}$$

where $\theta_w : H_*(X^Y_w) \to H_*(Z_{n+1}, Z_n)$ is the natural map. To prove (3), consider the map φ (induced from the inclusions) from the disjoint union

$$\varphi : \bigsqcup_{w\in\Theta(n+1)} \left(X^Y_w, X^Y_w \cap X^Y_n\right) \to (Z_{n+1}, Z_n).$$

Since φ induces a homeomorphism of the difference spaces

$$\bigsqcup_{w\in\Theta(n+1)} (X^Y_w \backslash X^Y_n) \xrightarrow{\sim} Z_{n+1}\backslash Z_n,$$

we get that

$$\varphi_* : \bigoplus_{w\in\Theta(n+1)} H_*\left(X^Y_w, X^Y_w \cap X^Y_n\right) \xrightarrow{\sim} H_*(Z_{n+1}, Z_n)$$

is an isomorphism. (Use [Spanier–66, Chap. 6, §6, Theorem 5] to conclude that the induced map φ^* in cohomology is an isomorphism, and then use the Universal Coefficient Theorem C.1.) (Observe that, with the notation in [Spanier–66], $\bar{H}^*$ can be identified with H^* by using [Spanier–66, Chap. 6, §9, Corollary 6 and Example 2].)

Now, for any $w \in W'_Y$ of length $n+1$, using [Spanier–66, Chap. 6, §6, Lemma 11 and Example 13] and the Universal Coefficient Theorem, $H_i(X^Y_w, X^Y_w \cap X^Y_n) = 0$ for $i \neq 2(n+1)$ since $X^Y_w \backslash X^Y_n \simeq \mathbb{C}^{\ell(w)}$, by (7.4.16.1). Moreover,

$H_{2(n+1)}(X_w^Y) \to H_{2(n+1)}(X_w^Y, X_w^Y \cap X_n^Y)$ is an isomorphism. This follows from the long exact homology sequence for the pair $(X_w^Y, X_w^Y \cap X_n^Y)$, since $H_i(X_w^Y \cap X_n^Y) = 0$ for $i > 2n$ by the induction hypothesis applied to $\Theta' := \{v \in W'_Y : v \leq w\}$. Combining the above, together with the fact that, for any $w \in W'_Y$, X_w^Y is an irreducible variety of dimension $\ell(w)$, we get (3).

Now (2) for Z_{n+1} follows from the induction hypothesis (i.e., the validity of (2) for Z_n) and (3). Since the homology commutes with direct limits (cf. [Spanier–66, Chap. 4, §4, Theorem 6]), we get the (a) and (b) parts of the proposition from (2).

The assertion for $H^*(\mathcal{X}_\Theta^Y)$ follows from (a) and (b) and the Universal Coefficient Theorem C.1.

The surjectivity of γ follows, since the canonical map $H_*(\mathcal{X}_\Theta^Y) \to H_*(\mathcal{X}^Y)$ is injective and the image is a direct summand (by the (a) and (b) parts). □

Observe that the above proof can be simplified if we use Proposition 7.4.16.

11.3.3 Lemma. *For any subset $Y \subset \{1, \ldots, \ell\}$, let $\pi = \pi^Y : \mathcal{X} \to \mathcal{X}^Y$ be the canonical morphism. Then the induced map π_* in homology is given by*

(1) $$\pi_*[X_w] = [X_w^Y], \qquad \text{if } w \in W'_Y$$

(2) $$= 0, \qquad \text{otherwise,}$$

where $[X_w]$, resp. $[X_w^Y]$, is the fundamental homology class of X_w, resp. X_w^Y, considered as an element of $H_*(\mathcal{X})$, resp. $H_*(\mathcal{X}^Y)$.

Thus, the induced map π^* in cohomology is given by

(3) $$\pi^*(\varepsilon_w^Y) = \varepsilon_w, \quad \text{for } w \in W'_Y,$$

where ε_w^Y and ε_w are as in Proposition 11.3.2. In particular, π^* is injective.

Proof. By 7.1.13, for any $w \in W$,

$$\pi(X_w) = X_{w'}^Y,$$

where $w' \in W'_Y$ is the unique element such that $w = w'\,w''$ for some $w'' \in W_Y$. By (1.3.17.2) and Proposition 7.1.15(a),

(4) $\pi_{|X_w} : X_w \to X_{w'}^Y$ is surjective and birational if $w \in W'_Y$, and

(5) $\dim \pi(X_w) < \dim X_w$, otherwise.

Now applying [Fulton–98, Lemma 19.1.2 and 1.4], (1) and (2) follow. □

For any subset $Y \subset \{1, \dots, \ell\}$, consider the algebraic action of T on $\mathcal{X}^Y = \mathcal{G}/\mathcal{P}_Y$ via the left multiplication. In particular, under the strong topology on T and $\mathcal{X}^Y$, T acts continuously on $\mathcal{X}^Y$. Let $H_T^*(\mathcal{X}^Y)$ be the T-equivariant cohomology of $\mathcal{X}^Y$ with the integral coefficients (cf. Appendix C).

Now we define the Weyl group actions on $H^*(\mathcal{G}/\mathcal{B})$ and $H_T^*(\mathcal{G}/\mathcal{B})$.

11.3.4 Definition. Since $\mathcal{G}^{\min}$ is an ind-group (cf. Theorem 7.4.14); in particular, the action of N on $\mathcal{G}^{\min}$ via the right multiplication is continuous under the strong topology on $\mathcal{G}^{\min}$ and the subspace topology on N. This action descends to give a (right) action of the Weyl group $W \simeq N/T$ (cf. Corollary 6.1.8) on the coset space $\mathcal{G}^{\min}/T$. Moreover, this action is continuous with respect to the discrete topology on W and the quotient topology on $\mathcal{G}^{\min}/T$. (By the proof of Proposition 7.4.16, T and N are Zariski closed in $\mathcal{G}^{\min}$ and T is Zariski open in N, hence the quotient topology on N/T is discrete.) In particular, this gives rise to a left action of W on the cohomology $H^*(\mathcal{G}^{\min}/T)$, where $w \in W$ acts on $H^*(\mathcal{G}^{\min}/T)$ as R_w^* (R_w being the right action of w on $\mathcal{G}^{\min}/T$).

Consider the continuous (left) action of T on $\mathcal{G}^{\min}/T$ via the left multiplication. Since the right action of W on $\mathcal{G}^{\min}/T$ commutes with the left T-action, we get a left action of W on $H_T^*(\mathcal{G}^{\min}/T)$ as well.

By Corollary 7.4.15 and [Steenrod–51, Theorem 7.4], the canonical map $\pi : \mathcal{G}^{\min}/T \to \mathcal{X}$ is a fiber bundle with fiber $\mathcal{B}_{\text{an}}^{\min}/T_{\text{an}}$, where $\mathcal{X} = \mathcal{G}/\mathcal{B}$ is equipped with the strong topology. It is easy to see that, under the quotient topology, $\mathcal{B}_{\text{an}}^{\min}/T_{\text{an}}$ is homeomorphic with $\mathcal{U}_{\text{an}}^{\min}$. But since $\mathcal{U}_{\text{an}}^{\min}$ is contractible (cf. Proposition 7.4.17), by the Leray–Serre spectral sequence Theorem E.11, $\pi^* : H^*(\mathcal{X}) \to H^*(\mathcal{G}^{\min}/T)$ is an isomorphism. So the action of W on $H^*(\mathcal{G}^{\min}/T)$ gives rise to an action of W on $H^*(\mathcal{X})$ (under this identification). By a similar argument, we also get a $H^*(B(T))$-linear action of W on $H_T^*(\mathcal{X})$.

The canonical map $\eta : H_T^*(\mathcal{X}) \to H^*(\mathcal{X})$ is clearly W-equivariant (cf. (C.6.4)).

11.3.5 Definition (Equivariant Borel homomorphism). We freely follow the notation from Appendix C. For $\lambda \in \mathfrak{h}_{\mathbb{Z}}^*$, let $\mathcal{L}(\lambda)$ be the (T-equivariant) line bundle on $\mathcal{X}$, defined in Section 7.2, and consider the T-equivariant topological line bundle $\pi_2^*(\mathcal{L}(\lambda))$ on $E(T) \times \mathcal{X}$, where $\pi_2 : E(T) \times \mathcal{X} \to \mathcal{X}$ is the (T-equivariant) projection on the second factor. Define the line bundle on $\mathcal{X}_T$:

$$\mathcal{L}(\lambda)_T := T \backslash \pi_2^*(\mathcal{L}(\lambda)) \to \mathcal{X}_T.$$

By the construction, for any base point $e \in E(T)$, the pullback line bundle $i_e^*(\mathcal{L}(\lambda)_T)$ satisfies the following:

(1) $$i_e^*(\mathcal{L}(\lambda)_T) \simeq \mathcal{L}(\lambda), \text{ as line bundles,}$$

where $i_e : \mathcal{X} \hookrightarrow \mathcal{X}_T$ is the inclusion defined in C.6. Define the *equivariant Borel homomorphism*

(2) $$\beta_T : S(\mathfrak{h}^*_{\mathbb{Z}}) \to H_T(\mathcal{X}),$$

as the unique $\mathbb{Z}$-algebra homomorphism such that

(3) $$\beta_T(\lambda) = c_1(\mathcal{L}(\lambda)_T), \text{ for } \lambda \in \mathfrak{h}^*_{\mathbb{Z}},$$

where c_1 is the first Chern class and $S(\mathfrak{h}^*_{\mathbb{Z}})$ is the symmetric algebra over $\mathbb{Z}$ of the weight lattice $\mathfrak{h}^*_{\mathbb{Z}}$. Moreover, β_T is a graded homomorphism provided we assign degree 2 to the elements of $\mathfrak{h}^*_{\mathbb{Z}}$.

We define the *Borel homomorphism*

$$\beta : S(\mathfrak{h}^*_{\mathbb{Z}}) \to H(\mathcal{X})$$

as the composite $\eta \circ \beta_T$, where the evaluation $\eta : H^*_T(\mathcal{X}) \to H^*(\mathcal{X})$ is the standard $\mathbb{Z}$-algebra homomorphism.

11.3.6 Definition. For any simple reflection s_i $(1 \leq i \leq \ell)$, we define a certain analogue of the BGG–Demazure operator

$$D_{s_i} : H^*(\mathcal{X}) \to H^{*-2}(\mathcal{X})$$

as follows:

By Exercise 7.4.E.5, $\pi_i : \mathcal{X} \to \mathcal{X}^i$ is a locally trivial fibration with fiber $\mathcal{P}_i/\mathcal{B} \simeq \mathbb{P}^1$. By Proposition 11.3.2, the restriction map $\gamma : H^*(\mathcal{X}) \to H^*(\mathcal{P}_i/\mathcal{B})$ is surjective (with ε_{s_i} mapping to the generator of $H^2(\mathcal{P}_i/B)$). Choose a graded $\mathbb{Z}$-module splitting $\sigma : H^*(\mathcal{P}_i/\mathcal{B}) \to H^*(\mathcal{X})$ of γ. Then by the Leray–Hirsch Theorem C.5, the map

$$\Phi' : H^*(\mathcal{X}^i) \otimes_{\mathbb{Z}} H^*(\mathcal{P}_i/\mathcal{B}) \to H^*(\mathcal{X}), \; u \otimes v \mapsto (\pi_i^* u) \cup \sigma(v),$$

is an isomorphism. Hence, π_i^* is injective and $H^*(\mathcal{X})$ is a free module over $H^*(\mathcal{X}^i)$ (under π_i^*) with basis 1 and $\sigma(\varepsilon)$, where $\varepsilon := \varepsilon_{s_i|(\mathcal{P}_i/\mathcal{B})} \in H^2(\mathcal{P}_i/\mathcal{B})$ is the generator, i.e.,

(1) $$H^n(\mathcal{X}) \cong H^n(\mathcal{X}^i) \oplus \sigma(\varepsilon)\, H^{n-2}(\mathcal{X}^i), \quad \text{for any } n \geq 0.$$

Write, for any $\alpha \in H^*(\mathcal{X})$,

(2) $$\alpha = \pi_i^* \alpha_1 + \sigma(\varepsilon)\, \pi_i^* \alpha_2,$$

where $\alpha_1 \in H^*(\mathcal{X}^i)$ and $\alpha_2 \in H^{*-2}(\mathcal{X}^i)$ are uniquely determined by (2).

Now define

$$D_{s_i}\alpha := \pi_i^*\alpha_2 \in H^{*-2}(\mathcal{X}). \tag{3}$$

It is easy to see that D_{s_i} does not depend upon the choice of the splitting σ. Clearly,

$$D_{s_i}^2 = 0. \tag{4}$$

Let $\pi : E(T) \to B(T)$ be the classifying bundle for T (under the strong topology on T). Then the T-equivariant $\mathbb{P}^1$-fibration $\pi_i : \mathcal{X} \to \mathcal{X}^i$ gives rise to the $\mathbb{P}^1$-fibration $\hat{\pi}_i : E(T) \times_T \mathcal{X} \to E(T) \times_T \mathcal{X}^i$. We claim that the restriction $\gamma_T : H^*(\mathcal{X}_T) \to H^*(\mathcal{P}_i/\mathcal{B})$ is surjective, where $\mathcal{P}_i/\mathcal{B}$ is the fiber of $\hat{\pi}_i$ over $[x, e\mathcal{P}_i]$ for any $x \in E(T)$. To prove this, consider $\beta_T(\omega_i) \in H^2(\mathcal{X}_T)$, where $\omega_i \in \mathfrak{h}_{\mathbb{Z}}^*$ is defined by $(*)$ just above Theorem 7.4.6. By Exercise 11.3.E.1, $\gamma_T(\beta_T(\omega_i)) = \beta(\omega_i)_{|(\mathcal{P}_i/\mathcal{B})}$ is the standard positive generator of $H^2(\mathcal{P}_i/\mathcal{B})$, proving that γ_T is surjective. (The surjectivity of γ_T also follows from the Leray–Serre spectral sequence corresponding to the $\mathbb{P}^1$-fibration $\hat{\pi}_i$, which degenerates at the E_2-term since the base and the fiber both have cohomologies concentrated in even degrees.) Take a graded $\mathbb{Z}$-module splitting σ_T of γ_T. Then, by the Leray–Hirsch Theorem, similar to (1), we get that $H_T^*(\mathcal{X})$ is a free module over $H_T^*(\mathcal{X}_i)$ with basis 1 and $\sigma_T(\varepsilon)$, i.e.,

$$H_T^n(\mathcal{X}) \simeq H_T^n(\mathcal{X}_i) \oplus \sigma_T(\varepsilon)\, H_T^{n-2}(\mathcal{X}_i), \qquad \text{for any } n \geq 0. \tag{5}$$

This gives rise to the $H^*(B(T))$-linear (in fact, $H_T^*(\mathcal{X}_i)$-linear) operator

$$\hat{D}_{s_i} : H_T^n(\mathcal{X}) \to H_T^{n-2}(\mathcal{X}). \tag{6}$$

A different choice of σ_T gives the same operator $\hat{D}_{s_i}$.

Similar to (4), we have

$$\hat{D}_{s_i}^2 = 0. \tag{7}$$

It is easy to see that the following diagram is commutative:

$$\begin{array}{ccc} H_T^*(\mathcal{X}) & \xrightarrow{\hat{D}_{s_i}} & H_T^{*-2}(\mathcal{X}) \\ \downarrow{\scriptstyle \eta} & & \downarrow{\scriptstyle \eta} \\ H^*(\mathcal{X}) & \xrightarrow[D_{s_i}]{} & H^{*-2}(\mathcal{X}). \end{array}$$

We will see later in Corollary 11.3.12 that the $\hat{D}_{s_i}$'s satisfy the braid property (and hence so do the D_{s_i}'s).

We have the following T-equivariant analogue of Proposition 11.3.2, which follows from the Leray–Hirsch Theorem (using Proposition 11.3.2).

11.3.7 Proposition. *Let $\Theta \subset W'_Y$ be as in Proposition 11.3.2. Then the evaluation map $\eta : H_T^*(\mathcal{X}_\Theta^Y) \to H^*(\mathcal{X}_\Theta^Y)$ is surjective.*

Take any graded $\mathbb{Z}$-module splitting

$$\sigma : H^*(\mathcal{X}_\Theta^Y) \to H_T^*(\mathcal{X}_\Theta^Y).$$

Then the map

$$\Phi' : H^*(B(T)) \otimes_{\mathbb{Z}} H^*(\mathcal{X}_\Theta^Y) \to H_T^*(\mathcal{X}_\Theta^Y), \ a \otimes b \mapsto (\pi^* a) \cup \sigma(b),$$

for $a \in H^*(B(T))$ and $b \in H^*(\mathcal{X}_\Theta^Y)$, is an isomorphism, where $\pi = \pi_{\mathcal{X}_\Theta^Y} : E(T) \times_T \mathcal{X}_\Theta^Y \to B(T)$ is the canonical fibration as in Subsection C.6. In particular, $H_T^*(\mathcal{X}_\Theta^Y)$ is a free graded $H^*(B(T))$-module and any graded $\mathbb{Z}$-basis of $H^*(\mathcal{X}_\Theta^Y)$ upon taking σ provides a (graded) $H^*(B(T))$-module basis of $H_T^*(\mathcal{X}^Y)$.

Hence, η induces a graded algebra isomorphism $\bar{\eta} : \mathbb{Z} \otimes_{H^*(B(T))} H_T^*(\mathcal{X}_\Theta^Y) \xrightarrow{\sim} H^*(\mathcal{X}_\Theta^Y)$, where $H^*(B(T)) \to H^0(B(T)) = \mathbb{Z}$ is the augmentation map. Further, the restriction map $\gamma_T : H_T^*(\mathcal{X}^Y) \to H_T^*(\mathcal{X}_\Theta^Y)$ is surjective.

Proof. The Leray–Serre cohomology spectral sequence corresponding to the fibration $\pi : E(T) \times_T \mathcal{X}_\Theta^Y \to B(T)$ with fiber $\mathcal{X}_\Theta^Y$ degenerates at the E_2-term, since $\mathcal{X}_\Theta^Y$ has cohomology only in even degrees (cf. Proposition 11.3.2) and so is $B(T)$ (as is well known). This gives the surjectivity of η.

Now the Leray–Hirsch Theorem C.5 proves that Φ' is an isomorphism. (Even though the hypothesis in the Leray–Hirsch Theorem, that the total homology $H_*(\mathcal{X}_\Theta^Y)$ be finitely generated over $\mathbb{Z}$, is not satisfied in general, we can still apply the theorem by considering the T-stable filtration Z_n of $\mathcal{X}_\Theta^Y$ defined in the proof of Proposition 11.3.2.)

Since Φ' is an isomorphism, we get that $\bar{\eta}$ is an isomorphism (of graded algebras). The surjectivity of $\gamma_T : H_T^*(\mathcal{X}^Y) \to H_T^*(\mathcal{X}_\Theta^Y)$ follows from the corresponding surjectivity $\gamma : H^*(\mathcal{X}^Y) \to H^*(\mathcal{X}_\Theta^Y)$ (cf. Proposition 11.3.2) and the surjectivity of Φ'. □

11.3.8 Localization map. By Lemma 7.1.22(a), the set of T-fixed points $\mathcal{X}^T = \{\dot{w}\}_{w \in W}$, where $\dot{w} := w\mathcal{B} \in \mathcal{X}$. Moreover, the map $i : W \xrightarrow{\sim} \mathcal{X}^T$, $w \mapsto \dot{w}$, is a homeomorphism with the discrete topology on W. Often we will identify $\mathcal{X}^T$ with W (under i).

As in 11.3.5, let $S_{\mathbb{Z}} = S(\mathfrak{h}_{\mathbb{Z}}^*)$ be the symmetric algebra over $\mathbb{Z}$ of the weight lattice $\mathfrak{h}_{\mathbb{Z}}^*$ graded such that $\mathfrak{h}_{\mathbb{Z}}^*$ has degree 2. Define the map

$$c : \mathfrak{h}_{\mathbb{Z}}^* \to H^2(B(T)) \text{ by } \chi \mapsto c_1(\hat{\chi}),$$

where $\hat{\chi}$ denotes the line bundle on $B(T)$ associated to the universal principal T-bundle $E(T) \to B(T)$ (cf. C.6) via the character e^χ of T (cf. 6.1.6) and

c_1 is the first Chern class. Extend c as a graded algebra homomorphism (again denoted by)

$$c : S_{\mathbb{Z}} \to H^*(B(T)). \tag{1}$$

Then, as is well known, c is an isomorphism. (In particular, $H^{\mathrm{odd}}(B(T)) = 0$.) From now on, we will identify $H^*(B(T))$ with $S_{\mathbb{Z}}$ via c.

The inclusion $W \overset{i}{\simeq} \mathcal{X}^T \hookrightarrow \mathcal{X}$ gives rise to the graded $S_{\mathbb{Z}}$-algebra homomorphism (cf. C.6)

$$i^* : H_T^*(\mathcal{X}) \to H_T^*(W) \simeq S_{\mathbb{Z}} \otimes_{\mathbb{Z}} H^0(W), \tag{2}$$

where the last isomorphism results from Lemma C.7. But, for any $j \geq 0$, $S_{\mathbb{Z}}^j \otimes_{\mathbb{Z}} H^0(W)$ can canonically be identified with $\{f : W \to S_{\mathbb{Z}}^j,\ f \text{ any map}\}$ (since $S_{\mathbb{Z}}^j$ is finitely generated $\mathbb{Z}$-module), where $S_{\mathbb{Z}}^j$ is the j-th symmetric power of $\mathfrak{h}_{\mathbb{Z}}^*$. In particular, i^* gives rise to the $S_{\mathbb{Z}}$-algebra homomorphism

$$\nu : H_T^*(\mathcal{X}) \to \Omega_W, \tag{3}$$

where $\Omega_W := \{f : W \to Q,\ f \text{ any map}\}$ (cf. (11.1.4.2)).

Recall the definition of the $S(\mathfrak{h}^*)$-subalgebra $\Lambda \subset \Omega_W$ from (11.1.4.10). Define the $S_{\mathbb{Z}}$-subalgebra $\Lambda_{\mathbb{Z}} \subset \Lambda$ by (cf. Lemma 11.1.5)

$$\Lambda_{\mathbb{Z}} := \bigoplus_{w \in W} S_{\mathbb{Z}}\, \xi^w.$$

(By Theorem 11.1.7(d), $\Lambda_{\mathbb{Z}}$ is indeed an $S_{\mathbb{Z}}$-algebra.) Moreover, by Theorem 11.1.7 (b) and (f), $\Lambda_{\mathbb{Z}}$ is stable under the left action $\bullet$ of the elements $\delta_w, x_w \in R$, for any $w \in W$, and the action is $S_{\mathbb{Z}}$-linear (cf. 11.1.4). In particular, they induce the actions $\bar{\delta}_w := 1 \otimes \delta_w$ and $\bar{x}_w := 1 \otimes x_w$ on $\mathbb{Z} \otimes_{S_{\mathbb{Z}}} \Lambda_{\mathbb{Z}}$.

We put an even grading on $\Lambda_{\mathbb{Z}}$ defined by

$$\Lambda_{\mathbb{Z}}^{2d} := \bigoplus_{w \in W} S_{\mathbb{Z}}^{d-\ell(w)}\, \xi^w, \quad \text{for any } d \geq 0. \tag{4}$$

By Theorem 11.1.7(d), this grading turns $\Lambda_{\mathbb{Z}}$ into a graded $S_{\mathbb{Z}}$-algebra, where we put the grading on $S_{\mathbb{Z}}$ so that $S_{\mathbb{Z}}^d$ acquires degree $2d$.

Define an analogue of the equivariant Borel homomorphism

$$\bar{\beta}_T : S_{\mathbb{Z}} \to \Omega_W$$

by $(\bar{\beta}_T(p))w = (-1)^d wp$ for $w \in W$ and $p \in S_{\mathbb{Z}}^d$. Then, $\bar{\beta}_T$ is clearly a $\mathbb{Z}$-algebra homomorphism.

By Theorem 11.1.7(c), for $1 \leq i \leq \ell$,

$$\bar{\beta}_T(\omega_i) = \xi^{s_i} - \omega_i\, \xi^e, \tag{5}$$

where $\{\omega_1, \dots, \omega_d\}$ is a basis of $\mathfrak{h}^*_{\mathbb{Z}}$ as in $(*)$ just above Theorem 7.4.6. Moreover,

$$\bar{\beta}_T(\omega_i) = -\omega_i \xi^e, \quad \text{for } \ell+1 \leq i \leq d. \tag{6}$$

By (5) and (6),

$$\operatorname{Im} \bar{\beta}_T \subset \Lambda_{\mathbb{Z}},$$

and, moreover, $\bar{\beta}_T$ is a graded $\mathbb{Z}$-algebra homomorphism.

Now we come to the following basic theorem which determines the T-equivariant cohomology $H_T(\mathcal{X})$.

Recall that the operators $\hat{D}_{s_i}$, resp. D_{s_i}, acting on $H_T^*(\mathcal{X})$, resp. $H^*(\mathcal{X})$, are defined in 11.3.6 and the $\bullet$ action of Q_W on Ω_W is defined by (11.1.4.5). For $w \in W$ and $x \in H_T^*(\mathcal{X})$ or $H^*(\mathcal{X})$, let wx denote R_w^* (cf. 11.3.4).

11.3.9 Theorem. *The $S_{\mathbb{Z}}$-algebra homomorphism $\nu : H_T^*(\mathcal{X}) \to \Omega_W$ (defined above) is injective. Moreover, for any $x \in H_T^*(\mathcal{X})$, simple reflection s_i and $w \in W$,*

$$\nu(wx) = \delta_w \bullet (\nu x), \tag{1}$$

$$\nu \circ \beta_T = \bar{\beta}_T, \text{ where } \beta_T \text{ is as in (11.3.5.2)}, \tag{2}$$

$$\nu(\hat{D}_{s_i} x) = x_i \bullet (\nu x). \tag{3}$$

Further,

$$\operatorname{Im} \nu = \Lambda_{\mathbb{Z}}, \tag{4}$$

and $\nu : H_T^*(\mathcal{X}) \xrightarrow{\sim} \Lambda_{\mathbb{Z}}$ is a graded $S_{\mathbb{Z}}$-algebra isomorphism.

In particular, $\bar{\nu} : H^*(\mathcal{X}) \xrightarrow{\sim} \mathbb{Z} \otimes_{S_{\mathbb{Z}}} \Lambda_{\mathbb{Z}}$ is a graded $\mathbb{Z}$-algebra isomorphism, where $\bar{\nu} := 1 \otimes \nu$ under the identification $\bar{\eta} : \mathbb{Z} \otimes_{S_{\mathbb{Z}}} H_T^*(\mathcal{X}) \xrightarrow{\sim} H^*(\mathcal{X})$ (cf. Proposition 11.3.7). Further,

$$\bar{\nu}(wx) = \bar{\delta}_w \bullet (\bar{\nu} x) \text{ and } \bar{\nu}(D_{s_i} x) = \bar{x}_i \bullet (\bar{\nu} x), \quad \textit{for any } x \in H^*(\mathcal{X}).$$

In view of (11.3.10.2), $\{\hat{\varepsilon}_w := \nu^{-1}(\xi^w)\}_{w \in W}$ is called the *Schubert basis* of $H_T^*(\mathcal{X})$. Thus

$$H_T^*(\mathcal{X}) = \bigoplus_{w \in W} S_{\mathbb{Z}}\, \hat{\varepsilon}_w.$$

Proof. Let $\{X_n\}_{n\geq 0}$ be the filtration of $\mathcal{X}$ as in 7.1.13 and let $i_n : X_n^T \hookrightarrow X_n$ be the inclusion. For any $n \geq 0$, by the Localization Theorem (cf. Theorem C.8), the map $i_n^* : H_T^*(X_n) \to S_{\mathbb{Z}} \otimes_{\mathbb{Z}} H^0(X_n^T)$ induces an isomorphism

$$\hat{i}_n^* : Q \otimes_{S_{\mathbb{Z}}} H_T^*(X_n) \xrightarrow{\sim} Q \otimes_{\mathbb{Z}} H^0(X_n^T),$$

where Q is the quotient field of $S(\mathfrak{h}^*)$ as in 11.1.1. (Observe that $X_n^T = X_n^{T_o}$, where $T_o \subset T$ is a compact form of T, since the action of T on X_n is algebraic and $T_o \subset T$ is Zariski dense.)

By Proposition 11.3.7, $H_T^*(X_n)$ is $S_{\mathbb{Z}}$-free; in particular, the natural map $H_T^*(X_n) = S_{\mathbb{Z}} \otimes_{S_{\mathbb{Z}}} H_T^*(X_n) \to Q \otimes_{S_{\mathbb{Z}}} H_T^*(X_n)$ is injective. Hence, we get that $i_n^* : H_T^*(X_n) \to S_{\mathbb{Z}} \otimes_{\mathbb{Z}} H^0(X_n^T)$ is injective. Now use the fact that, for any fixed p, there exists a large enough $n = n(p)$ such that the restriction map $H_T^p(\mathcal{X}) \to H_T^p(X_n)$ is an isomorphism (cf. Propositions 11.3.7 and 11.3.2), to conclude that ν is injective.

To prove (1), consider the commutative diagram:

$$\begin{array}{ccc} H_T^*(\mathcal{G}^{\min}/T) & \xrightarrow{R_w^*} & H_T^*(\mathcal{G}^{\min}/T) \\ \downarrow & & \downarrow \\ H_T^*(W) & \xrightarrow{r_w^*} & H_T^*(W), \end{array}$$

where the vertical maps are induced by the inclusion $W \xrightarrow{\sim} N/T \hookrightarrow \mathcal{G}^{\min}/T$, and r_w^* is induced from the right multiplication $r_w : W \to W$ by w. From the above diagram (1) follows easily.

We now prove (2). Since ν, β_T and $\bar{\beta}_T$ are all $\mathbb{Z}$-algebra homomorphisms, it suffices to show that, for $\lambda \in \mathfrak{h}_{\mathbb{Z}}^*$,

$$\nu(\beta_T\lambda) = \bar{\beta}_T(\lambda). \tag{5}$$

For any $w \in W$, the T-equivariant line bundle $\mathcal{L}(\lambda)$ on $\mathcal{X}$ (cf. 7.2.1) restricted to the T-fixed point $\dot{w}$ corresponds to the character $e^{-w\lambda}$. Hence the line bundle $\mathcal{L}(\lambda)_T$ on $\mathcal{X}_T$ restricts to the line bundle $\widehat{(-w\lambda)}$ on $E(T) \times_T \{\dot{w}\} \simeq B(T)$ (cf. 11.3.8). But since $c(-w\lambda) = c_1(\widehat{(-w\lambda)})$ by the definition of c (cf. 11.3.8), we get (5) and hence (2).

To prove (3), for $x \in H_T(\mathcal{X})$ write (by virtue of (11.3.6.5))

$$x = \hat{\pi}_i^* x_1 + \beta_T(\omega_i)\, \hat{\pi}_i^* x_2, \qquad \text{for } x_1, x_2 \in H_T^*(\mathcal{X}^i),$$

where $\hat{\pi}_i$ and $\beta_T(\omega_i)$ are as in 11.3.6. Then

$$\nu(\hat{D}_{s_i} x)(w) = \nu(\hat{\pi}_i^* x_2)(w), \quad \text{by the definition of } \hat{D}_{s_i}. \tag{6}$$

On the other hand,

$$\nu(\hat{\pi}_i^* y)(ws_i) = \nu(\hat{\pi}_i^* y)(w), \quad \text{for } y \in H_T^*(\mathcal{X}^i). \tag{7}$$

Hence, by (11.1.4.8),

$$\begin{aligned}
(x_i \bullet (\nu x))w &= \frac{(\nu x)(ws_i) - (\nu x)(w)}{w\,\alpha_i} \\
&= \frac{\nu(\hat{\pi}_i^* x_2)(w)\big[\nu(\beta_T(\omega_i))(ws_i) - \nu(\beta_T(\omega_i))(w)\big]}{w\alpha_i}, \quad \text{by (7)} \\
&= \frac{\nu(\hat{\pi}_i^* x_2)(w)\big[\bar{\beta}_T(\omega_i)(ws_i) - \bar{\beta}_T(\omega_i)(w)\big]}{w\alpha_i}, \quad \text{by (2)} \\
&= \nu(\hat{\pi}_i^* x_2)(w), \qquad \text{by (11.3.8.5) and (11.1.7.2-3).}
\end{aligned} \tag{8}$$

Combining (6) and (8), we get (3).

We next show that

$$\operatorname{Im} \nu \subset \Lambda_{\mathbb{Z}}. \tag{9}$$

As a first step we show that

$$\operatorname{Im} \nu \subset \Lambda, \quad \text{where } \Lambda \text{ is defined by (11.1.4.10).} \tag{10}$$

Take $x \in H_T^{2n}(\mathcal{X})$ (any $n \geq 0$) and $w \in W$. Fix a reduced decomposition $w = s_{i_1} \cdots s_{i_p}$. Then, by (3),

$$\nu\big(\hat{D}s_{i_1} \circ \cdots \circ \hat{D}s_{i_p}(x)\big) = x_w \bullet (\nu x). \tag{11}$$

Evaluating the above at e, we get (from the definition (11.1.4.5))

$$\nu\big(\hat{D}s_{i_1} \circ \cdots \circ \hat{D}s_{i_p}(x)\big)(e) = (\nu x)(x_w). \tag{12}$$

Since $\hat{D}s_i$ decreases the degree by 2,

$$\hat{D}s_{i_1} \circ \cdots \circ \hat{D}s_{i_p}(x) = 0, \text{ if } p = \ell(w) > n.$$

In particular, by (12),

$$(\nu x)(x_w) = 0, \quad \text{if } \ell(w) > n, \text{ and} \tag{13}$$

$$(\nu x)(x_w) \in S_{\mathbb{Z}} \subset S(\mathfrak{h}^*), \quad \text{for any } w \in W. \tag{14}$$

This proves (10).

We now show (9). By (10) and Lemma 11.1.5, write

$$\nu(x) = \sum_{v \in W} p^v \xi^v, \quad \text{for some } p^v \in S(\mathfrak{h}^*),$$

where all but finitely many p^v's are 0. Then, for any $w \in W$,

$$\nu(x)(x_w) = p^w,$$

by the definition of ξ^v (cf. (11.1.5.1)). Hence, by (14), $p^w \in S_{\mathbb{Z}}$. This proves (9).

Now, the assertion that $\nu : H_T^*(\mathcal{X}) \to \Lambda_{\mathbb{Z}}$ is a graded homomorphism follows easily from the definition of the grading on $\Lambda_{\mathbb{Z}}$ as in (11.3.8.4) and Theorem 11.1.7(g). Finally, by the following proposition, $\xi^w \in \operatorname{Im} \nu$. Hence, ν being $S_{\mathbb{Z}}$-linear, $\operatorname{Im} \nu = \Lambda_{\mathbb{Z}}$. This completes the proof of the theorem (modulo the following proposition). □

11.3.10 Proposition. *For any $w \in W$ of length n, there exists a unique class $\hat{\varepsilon}_w \in H_T^{2n}(\mathcal{X})$ such that*

(1) $$\nu(\hat{\varepsilon}_w) = \xi^w.$$

Moreover, $\hat{\varepsilon}_w$ satisfies

(2) $$\eta(\hat{\varepsilon}_w) = \varepsilon_w \in H^{2n}(\mathcal{X}),$$

where ε_w is defined by (11.3.2.1) (taking $Y = \emptyset$ and $\mathcal{X}_\Theta^Y = \mathcal{X}$). In fact, $\hat{\varepsilon}_w \in H_T^{2n}(\mathcal{X})$ is characterized by the properties

(3) $$\eta(\hat{\varepsilon}_w) = \varepsilon_w \quad \text{and} \quad \hat{\varepsilon}_{w|Z^w} = 0,$$

where Z^w is defined as $\bigcup_{\substack{\ell(v) \leq n \\ v \neq w}} \mathcal{B}v\mathcal{B}/\mathcal{B}$.

Proof. By Proposition 7.1.15, Z^w is Zariski closed in $\mathcal{X}$ (in particular, closed under the strong topology). By Proposition 11.3.2, ε_w can be lifted to a class $\varepsilon'_w \in H^{2n}(\mathcal{X}, Z^w)$. Consider the commutative diagram:

(D) $$\begin{array}{ccccccccc} 0 \to & \mathbb{Z} \otimes_{S_{\mathbb{Z}}} H_T^*(\mathcal{X}, Z^w) & \to & \mathbb{Z} \otimes_{S_{\mathbb{Z}}} H_T^*(\mathcal{X}) & \to & \mathbb{Z} \otimes_{S_{\mathbb{Z}}} H_T^*(Z^w) & \to 0 \\ & \downarrow \bar{\eta} & & \wr \downarrow \bar{\eta} & & \wr \downarrow \bar{\eta} & \\ 0 \to & H^*(\mathcal{X}, Z^w) & \to & H^*(\mathcal{X}) & \to & H^*(Z^w) & \to 0, \end{array}$$

in which both the horizontal sequences are exact by Propositions 11.3.7 and 11.3.2, respectively. The two right extreme vertical maps are isomorphisms by

Proposition 11.3.7. Hence, by the Five Lemma (cf. [Spanier–66, Chap. 4, §5, Lemma 11]), the left vertical map again is an isomorphism. In particular, there exists a class $c_w \in H_T^{2n}(\mathcal{X}, Z^w)$ such that

(4) $$\eta(c_w) = \varepsilon'_w,$$

and hence

(5) $$\eta(\gamma(c_w)) = \varepsilon_w,$$

where $\gamma : H_T(\mathcal{X}, Z^w) \to H_T(\mathcal{X})$ is the restriction map.Set $\hat{\varepsilon}_w := \gamma(c_w)$. We now prove that $\nu(\hat{\varepsilon}_w) = \xi^w$: Since $\hat{\varepsilon}_{w|Z^w} = 0$,

(6) $$\nu(\hat{\varepsilon}_w)(v) = 0, \quad \text{for } \ell(v) \leq n \text{ and } v \neq w.$$

We next calculate $\nu(\hat{\varepsilon}_w)(w)$:

Let c_w^n be the restriction of c_w to the pair (X_n, Z^w). By the definition, $X_n \backslash Z^w = \mathcal{B}w\mathcal{B}/\mathcal{B}$ and, moreover, there is a T- equivariant homeomorphism $\mathcal{B}w\mathcal{B}/\mathcal{B} \simeq \mathfrak{n}_w$ (cf. Exercise 6.2.E.1 and 6.1.5(b)), where $\mathfrak{n}_w = \hat{\mathfrak{n}}_{\Phi_w} := \bigoplus_{\beta\in\Delta^+\cap w\Delta^-} \mathfrak{g}_\beta$ is a T-module under the adjoint action. It is easy to see that $H^*(X_n, Z^w) \simeq H^*(\mathfrak{n}_w, \mathfrak{n}_w\backslash\{0\})$ and $H_T^*(X_n, Z^w) \simeq H_T^*(\mathfrak{n}_w, \mathfrak{n}_w\backslash\{0\})$. By (4), under the above identification, c_w^n is nothing but the Thom class of the complex vector bundle $\pi : E(T) \times_T \mathfrak{n}_w \to B(T)$ (cf. [Milnor–Stasheff–74, Theorem 9.1]). Consider $c^n_{w_{|\mathfrak{n}_w}} \in H^{2n}(E(T) \times_T \mathfrak{n}_w)$ and let $d_w \in H^{2n}(B(T))$ be defined by $d_w = \sigma^*\left(c^n_{w_{|\mathfrak{n}_w}}\right)$, where $\sigma : B(T) \to E(T) \times_T \mathfrak{n}_w$ is the zero section. Then, by [Milnor–Stasheff–74, §§9, 14], d_w is the n-th Chern class of the rank n vector bundle π. In particular, by the Whitney product formula for Chern classes (cf. [loc. cit., 14.7]), $d_w = \prod_{\beta\in\Delta^+\cap w\Delta^-} c(\beta)$, where c is as in (11.3.8.1). From this it is easy to see that

(7) $$\nu(\hat{\varepsilon}_w)(w) = \prod_{\beta\in\Delta^+\cap w\Delta^-} \beta.$$

Since ν is graded, we can write

(8) $$\nu(\hat{\varepsilon}_w) = \sum_{\ell(v)\leq n} p^v \xi^v, \quad \text{for } p^v \in S_{\mathbb{Z}}^{n-\ell(v)}.$$

Take $v = v_o$ of minimal length such that $p^{v_o} \neq 0$. Assume $v_o \neq w$. Then, by Theorem 11.1.7(a),

$$\nu(\hat{\varepsilon}_w)(v_o) = p^{v_o}\xi^{v_o}(v_o) \neq 0,$$

which is a contradiction in view of (6). So, the above sum (8) reduces to $\nu(\hat{\varepsilon}_w) = p^w \xi^w$. Evaluating this at w and using (7) and Theorem 11.1.7(a), we get $p^w = 1$, i.e., $\nu(\hat{\varepsilon}_w) = \xi^w$. This proves (1), and (2) follows from (5). From the above proof, (3) follows easily. This completes the proof of the proposition and hence the proof of Theorem 11.3.9 is completed. □

11.3.11 Remark. By the same proof (as in the beginning of the proof of Proposition 11.3.10), we see that, for any two subsets $\Theta \subset \Theta' \subset W'_Y$ satisfying the condition (11.3.1.1),

$$\bar{\eta} : \mathbb{Z} \otimes_{S_\mathbb{Z}} H^*_T(\mathcal{X}^Y_{\Theta'}, \mathcal{X}^Y_{\Theta}) \underset{\sim}{\to} H^*(\mathcal{X}^Y_{\Theta'}, \mathcal{X}^Y_{\Theta})$$

is a graded algebra isomorphism.

Theorem 11.3.9 has several corollaries.

11.3.12 Corollary. *The operators $\hat{D}_{s_i} : H_T(\mathcal{X}) \to H_T(\mathcal{X})$, defined in 11.3.6, satisfy the braid property, i.e., for any $w \in W$, take a reduced decomposition $w = s_{i_1} \dots s_{i_n}$. Then $\hat{D}_w := \hat{D}_{s_{i_1}} \circ \dots \circ \hat{D}_{s_{i_n}} : H_T(\mathcal{X}) \to H_T(\mathcal{X})$ does not depend upon the choice of the reduced decomposition. In particular, the operators $D_{s_i} : H(\mathcal{X}) \to H(\mathcal{X})$ satisfy the braid property.*

Further, for any $1 \le i \le \ell$, $\hat{D}_{s_i}$ satisfies the twisted derivation property, i.e.,

$$\hat{D}_{s_i}(xy) = (\hat{D}_{s_i}x)y + (s_i x)(\hat{D}_{s_i} y),$$

for $x, y \in H_T(\mathcal{X})$. In particular, D_{s_i} satisfies the twisted derivation property.

Proof. Since $\nu : H_T(\mathcal{X}) \to \Omega_W$ is injective (cf. Theorem 11.3.9) and the x_i's satisfy the braid property (cf. Theorem 11.1.2(b)), the first part of the Corollary follows from (11.3.9.3). To prove the twisted derivation property, use Theorem 11.1.7(h). □

Recall the notation $\Lambda^Y = \Lambda^{W_Y}$ from 11.1.16. This is determined in Lemma 11.1.17.

11.3.13 Corollary. *The Weyl group invariants are given by*

$$H^*_T(\mathcal{X})^W = S_\mathbb{Z} \cdot H^0_T(\mathcal{X}), \quad \text{and} \tag{1}$$

$$H^*(\mathcal{X})^W = H^0(\mathcal{X}). \tag{2}$$

Proof. (1) follows immediately from Theorem 11.3.9 and Lemma 11.1.17. To prove (2), using Theorem 11.3.9, we get

$$H^*(\mathcal{X})^W \simeq (\mathbb{Z} \otimes_{S_\mathbb{Z}} \Lambda)^W.$$

Now the same proof as that of Lemma 11.1.17 for $Y = \{1, \dots, \ell\}$ gives (2). □

11.3.14 Corollary. *Let $Y \subset \{1, \ldots, \ell\}$ be any subset. Then the canonical T-equivariant morphism $\pi = \pi^Y : \mathcal{X} \to \mathcal{X}^Y$ induces an injective map $\pi_T^* : H_T^*(\mathcal{X}^Y) \hookrightarrow H_T^*(\mathcal{X})$ with image precisely equal to the W_Y-invariants $H_T^*(\mathcal{X})^{W_Y}$. In particular, as graded $S_{\mathbb{Z}}$-algebras,*

(1)
$$H_T^*(\mathcal{X}^Y) \simeq \Lambda_{\mathbb{Z}}^{W_Y}.$$

Further,

(2)
$$\Lambda_{\mathbb{Z}}^{W_Y} = \bigoplus_{w \in W'_Y} S_{\mathbb{Z}}\, \xi^w.$$

Similarly, the injective map (cf. Lemma 11.3.3) $\pi^* : H^*(\mathcal{X}^Y) \hookrightarrow H^*(\mathcal{X})$ has

(3)
$$\operatorname{Im}(\pi^*) = H^*(\mathcal{X})^{W_Y}.$$

Hence,

(4)
$$H^*(\mathcal{X}^Y) \simeq \mathbb{Z} \otimes_{S_{\mathbb{Z}}} \left(\Lambda_{\mathbb{Z}}^{W_Y}\right).$$

Proof. By Proposition 11.3.7, the evaluation maps

$$\eta_{\mathcal{X}} : H_T^*(\mathcal{X}) \twoheadrightarrow H^*(\mathcal{X}) \quad \text{and} \quad \eta_{\mathcal{X}^Y} : H_T^*(\mathcal{X}^Y) \twoheadrightarrow H^*(\mathcal{X}^Y)$$

are surjective. Moreover, by Lemma 11.3.3, $\pi^* : H^*(\mathcal{X}^Y) \to H^*(\mathcal{X})$ is injective and $\mathbb{Z}$-split, i.e., $\operatorname{Im} \pi^*$ is a graded $\mathbb{Z}$-module direct summand. So, for any graded $\mathbb{Z}$-module splitting $\sigma_{\mathcal{X}^Y}$ of $\eta_{\mathcal{X}^Y}$, we can extend the splitting $\pi_T^* \circ \sigma_{\mathcal{X}^Y}$ of $\eta_{\mathcal{X}}$ over $H^*(\mathcal{X}^Y)$ to a graded $\mathbb{Z}$-module splitting $\sigma_{\mathcal{X}}$ on the whole of $H^*(\mathcal{X})$ (under the identification π^*).

Now, consider the commutative diagram:

$$\begin{array}{ccc}
S_{\mathbb{Z}} \otimes_{\mathbb{Z}} H^*(\mathcal{X}^Y) & \xrightarrow{1 \otimes \pi^*} & S_{\mathbb{Z}} \otimes_{\mathbb{Z}} H^*(\mathcal{X}) \\
\wr \big\downarrow \Phi'_{\mathcal{X}^Y} & & \wr \big\downarrow \Phi'_{\mathcal{X}} \\
H_T^*(\mathcal{X}^Y) & \xrightarrow{\pi_T^*} & H_T^*(\mathcal{X}),
\end{array}$$

where the vertical $S_{\mathbb{Z}}$-module isomorphisms are as defined in Proposition 11.3.7, using the splittings $\sigma_{\mathcal{X}^Y}$ and $\sigma_{\mathcal{X}}$ respectively. Since π^* is injective and $\mathbb{Z}$-split, the top horizontal map is injective and hence so is π_T^*. (The injectivity of π_T^* can also be proved by the Leray–Serre spectral sequence corresponding to the fibration $\mathcal{X}_T \to \mathcal{X}_T^Y$, since it degenerates at E_2.)

For any $w \in W_Y$, from the commutative diagram

$$\begin{array}{ccc} \mathcal{G}^{\min}/T & \xrightarrow{R_w} & \mathcal{G}^{\min}/T \\ & \searrow \quad \swarrow & \\ & \mathcal{X}^Y, & \end{array}$$

we readily see that

(5) $$\operatorname{Im} \pi_T^* \subset H_T^*(\mathcal{X})^{W_Y},$$

where $\mathcal{G}^{\min}/T \to \mathcal{X}^Y$ is the standard projection map. Conversely, for any $w \in W'_Y$, we show that

(6) $$\hat{\varepsilon}_w = \nu^{-1}(\xi^w) \in \operatorname{Im} \pi_T^* :$$

Define the Zariski closed subset

$$Z_Y^w := \bigcup_{\substack{v \in W'_Y, \\ \ell(v) \le n, \\ v \ne w}} Bv\mathcal{P}_Y/\mathcal{P}_Y \subset \mathcal{X}^Y,$$

where $n = \ell(w)$.

Analogous to the diagram (D) as in the proof of Proposition 11.3.10, we have the commutative diagram with exact rows:

$$\text{(D}'\text{)} \quad \begin{array}{ccccccccc} 0 \to & \mathbb{Z} \otimes_{S_\mathbb{Z}} H_T^*(\mathcal{X}^Y, Z_Y^w) & \to & \mathbb{Z} \otimes_{S_\mathbb{Z}} H_T^*(\mathcal{X}^Y) & \to & \mathbb{Z} \otimes_{S_\mathbb{Z}} H_T^*(Z_Y^w) & \to 0 \\ & \wr\downarrow \bar{\eta} & & \wr\downarrow \bar{\eta} & & \wr\downarrow \bar{\eta} & \\ 0 \to & H^*(\mathcal{X}^Y, Z_Y^w) & \to & H^*(\mathcal{X}^Y) & \to & H^*(Z_Y^w) & \to 0. \end{array}$$

From the above, as well as Proposition 11.3.2 and Lemma 11.3.3, we get a class $c_w^Y \in H_T^{2n}(\mathcal{X}^Y, Z_Y^w)$ such that $\eta(\varepsilon'_w) = \varepsilon_w$ and $\varepsilon'_{w|Z^w} = 0$, where $\varepsilon'_w := \pi_T^*\big(c^Y_{w|\mathcal{X}^Y}\big) \in H_T^{2n}(\mathcal{X})$ and Z^w is as defined in Proposition 11.3.10. Hence, by the characterization (11.3.10.3), $\varepsilon'_w = \hat{\varepsilon}_w$, proving (6). This gives

(7) $$\operatorname{Im} \pi_T^* \supset \bigoplus_{w \in W'_Y} S_\mathbb{Z} \hat{\varepsilon}_w.$$

Lemma 11.1.17 easily gives (2). Now (1) follows from (2), (5) and (7) together with Theorem 11.3.9. By the same proof as that of Lemma 11.1.17, we obtain that $\big(\mathbb{Z} \otimes_{S_\mathbb{Z}} \Lambda_\mathbb{Z}\big)^{W_Y}$ is spanned by $\{1 \otimes \xi^w\}_{w \in W'_Y}$. In particular,

(8) $$\big(\mathbb{Z} \otimes_{S_\mathbb{Z}} \Lambda_\mathbb{Z}\big)^{W_Y} \simeq \mathbb{Z} \otimes_{S_\mathbb{Z}} \big(\Lambda_\mathbb{Z}^{W_Y}\big).$$

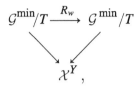

Finally (3) and (4) follow from (1), (8), Proposition 11.3.7 and Theorem 11.3.9. □

11.3.15 Definition. For any $\Theta \subset W'_Y$ satisfying (11.3.1.1), let $\Theta' \subset W$ be defined as

(1) $$\Theta' = \Theta \cdot W_Y.$$

Now, define

$$\Omega_{\Theta'} = \{f : \Theta' \to Q,\ f \text{ any map}\}.$$

Then there is the canonical restriction map

$$\theta_{\Theta'} : \Omega_W \to \Omega_{\Theta'}, \quad \text{taking } f \mapsto f_{|\Theta'}.$$

Define $\Lambda_{\mathbb{Z}}(\Theta) := \theta_{\Theta'}\big(\Lambda_{\mathbb{Z}}^{W_Y}\big)$.

Using Theorem 11.1.7(a), (11.3.14.2) and Lemma 1.3.18, it is easy to see that (since $\xi^w_{|\Theta'} = 0$ if $w \in W'_Y \backslash \Theta$) $\Lambda_{\mathbb{Z}}(\Theta)$ is a free $S_{\mathbb{Z}}$-module with basis $\{\xi^w_{|\Theta'}\}_{w\in\Theta}$, i.e.,

(2) $$\Lambda_{\mathbb{Z}}(\Theta) = \bigoplus_{w\in\Theta} S_{\mathbb{Z}}\big(\xi^w_{|\Theta'}\big).$$

Similar to the grading on $\Lambda_{\mathbb{Z}}$ (cf. (11.3.8.4)), we define the grading on $\Lambda_{\mathbb{Z}}(\Theta)$ by

(3) $$\big(\Lambda_{\mathbb{Z}}(\Theta)\big)^{2d} = \bigoplus_{w\in\Theta} S^{d-\ell(w)}(\mathfrak{h}^*_{\mathbb{Z}})\big(\xi^w_{|\Theta'}\big).$$

11.3.16 Corollary. *For any $\Theta \subset W'_Y$ satisfying (11.3.1.1), there is a unique graded $S_{\mathbb{Z}}$-algebra isomorphism $\nu_\Theta : H^*_T(\mathcal{X}^Y_\Theta) \xrightarrow{\sim} \Lambda_{\mathbb{Z}}(\Theta)$ making the following diagram commutative:*

(A) $$\begin{array}{ccc} H^*_T(\mathcal{X}^Y) & \xrightarrow[\sim]{\nu^Y} & \Lambda_{\mathbb{Z}}^{W_Y} \\ \gamma_T \downarrow & & \downarrow \theta_{\Theta'} \\ H^*_T(\mathcal{X}^Y_\Theta) & \xrightarrow[\nu_\Theta]{\sim} & \Lambda_{\mathbb{Z}}(\Theta), \end{array}$$

where γ_T is induced from the inclusion $\mathcal{X}^Y_\Theta \hookrightarrow \mathcal{X}^Y$ and ν^Y is the restriction of ν. Observe that, by Corollary 11.3.14, ν^Y is an isomorphism.

In particular, there is a unique graded $\mathbb{Z}$-algebra isomorphism

$$H^*(\mathcal{X}^Y_\Theta) \xrightarrow{\sim} \mathbb{Z} \otimes_{S_\mathbb{Z}} \Lambda_\mathbb{Z}(\Theta)$$

making the following diagram commutative:

(A′)
$$\begin{array}{ccc} H^*(\mathcal{X}^Y) & \xrightarrow{\sim} & \mathbb{Z} \otimes_{S_\mathbb{Z}} (\Lambda_\mathbb{Z}^{W_Y}) \\ \gamma \downarrow & & \downarrow 1 \otimes \theta_{\Theta'} \\ H^*(\mathcal{X}^Y_\Theta) & \xrightarrow[\sim]{} & \mathbb{Z} \otimes_{S_\mathbb{Z}} \Lambda_\mathbb{Z}(\Theta). \end{array}$$

Proof. By the definition of the map $\nu : H_T(\mathcal{X}) \to \Omega_W$, it is easy to see that for $x \in \operatorname{Ker} \gamma_T$, $\theta_{\Theta'}\, \nu^Y(x) = 0$. Further, by Proposition 11.3.7, γ_T is surjective. Hence, there exists a unique map ν_Θ making the diagram (A) commutative. Moreover, ν_Θ is a graded $S_\mathbb{Z}$-algebra homomorphism from the corresponding property of all the other maps in the diagram. Since $H_T^*(\mathcal{X}^Y_\Theta)$ is $S_\mathbb{Z}$-free (by Proposition 11.3.7), from the Localization Theorem C.8 and Lemma 7.1.22(a) (cf. proof of Theorem 11.3.9), we get the injectivity of ν_Θ. The surjectivity of ν_Θ follows from the surjectivity of $\theta_{\Theta'}$. This proves the first part of the corollary.

The "In particular" statement follows from the above and Proposition 11.3.7. □

As a vast generalization of the classical Chevalley cup product formula, we obtain, as a corollary of Theorem 11.3.9, the following general cup product formula in the cohomology rings $H_T^*(\mathcal{X})$ and $H^*(\mathcal{X})$. Recall the definition of the Schubert basis $\{\hat{\varepsilon}_w\}_{w\in W}$ of $H_T^*(\mathcal{X})$ from Theorem 11.3.9. Similarly, there is the Schubert basis $\{\varepsilon_w\}_{w\in W}$ of $H^*(\mathcal{X})$ (cf. Proposition 11.3.2).

11.3.17 Corollary. *For any $u, v \in W$, the cup product in $H_T^*(\mathcal{X})$ is given by*

(1)
$$\hat{\varepsilon}_u\, \hat{\varepsilon}_v = \sum_{u,v \leq w} p^w_{u,v}\, \hat{\varepsilon}_w,$$

where $p^w_{u,v} \in S_\mathbb{Z}^{\ell(u)+\ell(v)-\ell(w)}$ is as in (11.1.2.3) (cf. also (11.1.7.6)). In fact, $p^w_{u,v} \in \mathbb{Z}[\underline{\alpha}]$.

An "explicit" expression for $p^w_{u,v}$ is given in Theorem 11.1.13(a) purely in terms of the matrix D.

For any simple reflection s_i, the operator $\hat{D}_{s_i}$ on $H_T^*(\mathcal{X})$ defined in 11.3.6 is given by:

(2)
$$\begin{aligned} \hat{D}_{s_i}\, \hat{\varepsilon}_w &= \hat{\varepsilon}_{ws_i}, \quad \textit{if } ws_i < w \\ &= 0, \qquad \textit{otherwise}. \end{aligned}$$

In particular, the cup product in $H^*(\mathcal{X})$ is given by

(3)
$$\varepsilon_u\, \varepsilon_v = \sum_{\substack{u,v\leq w\\ \ell(w)=\ell(u)+\ell(v)}} p^w_{u,v}\, \varepsilon_w,$$

and the operators D_{s_i} on $H^*(\mathcal{X})$ are given by

(4)
$$\begin{aligned} D_{s_i}\, \varepsilon_w &= \varepsilon_{ws_i}, && \textit{if } ws_i < w \\ &= 0, && \textit{otherwise.} \end{aligned}$$

Proof. (1), resp. (2), follows from Theorem 11.3.9 together with Theorem 11.1.7(d), resp. Theorem 11.1.7(b). Now (3) and (4) follow from (1) and (2). □

11.3.18 Remarks. (a) For a simple reflection u, the product in (11.3.17.1) takes the form given by Theorem 11.1.7(i) (with ξ^w replaced by $\hat{\varepsilon}_w$). Thus, in this case, the product formula (11.3.17.3) in the singular cohomology of $\mathcal{X}$ is nothing but the classical Chevalley cup product formula (cf. [Bernstein-Gelfand-Gelfand–73, Theorem 3.17]).

(b) The Weyl group action of the simple reflections s_i on $H^*_T(\mathcal{X})$ (and also on $H^*(\mathcal{X})$) can explicitly be given by using Theorems 11.1.7(f) and 11.3.9. For general $w \in W$, use Theorem 11.1.13(b).

11.3.E EXERCISES

(1) Recall the definition of the Borel homomorphism $\beta : S(\mathfrak{h}^*_{\mathbb{Z}}) \to H^*(\mathcal{X})$ from11.3.5. Then, for any fundamental weight ω_i, $1 \leq i \leq \ell$ (cf. $(*)$ above Theorem 7.4.6), $\beta(\omega_i)_{|(\mathcal{P}_i/\mathcal{B})}$ is the standard positive generator of $H^2(\mathcal{P}_i/\mathcal{B})$.

(2) For any simple reflection s_i, define the standard BGG–Demazure operator $A_{s_i} : S(\mathfrak{h}^*_{\mathbb{Z}}) \to S(\mathfrak{h}^*_{\mathbb{Z}})$ by $A_{s_i} P = \frac{P - s_i P}{\alpha_i}$, for any $P \in S(\mathfrak{h}^*_{\mathbb{Z}})$. Show that indeed $A_{s_i}(S(\mathfrak{h}^*_{\mathbb{Z}})) \subset S(\mathfrak{h}^*_{\mathbb{Z}})$. Clearly A_{s_i} decreases the degree by 1.

Prove that A_{s_i} satisfies the twisted derivation property:

$$A_{s_i}(PQ) = (A_{s_i} P)Q + (s_i P)(A_{s_i} Q),$$

for $P, Q \in S(\mathfrak{h}^*_{\mathbb{Z}})$.

Recall the definition of the Borel homomorphisms $\beta_T : S(\mathfrak{h}^*_{\mathbb{Z}}) \to H_T(\mathcal{X})$ and $\bar{\beta}_T : S(\mathfrak{h}^*_{\mathbb{Z}}) \to \Lambda_{\mathbb{Z}}$ from 11.3.5 and 11.3.8 respectively. Show that, for any $P \in S(\mathfrak{h}^*_{\mathbb{Z}})$, $\beta_T(s_i P) = s_i(\beta_T(P))$ and $\beta_T(A_{s_i} P) = \hat{D}_{s_i}(\beta_T(P))$. Prove a similar result for $\bar{\beta}_T$.

Hence or otherwise, prove that the A_{s_i}'s satisfy the braid property.

(3) Prove that $\mathcal{G}^{\min} \to \mathcal{G}^{\min}/T$ is a locally trivial principal T-bundle, where $\mathcal{G}^{\min}/T$ is equipped with the quotient topology, and $\mathcal{G}^{\min}$ is equipped with the strong topology $\mathcal{G}^{\min}_{\text{an}}$ (cf. 7.4.9).

Prove further that $\mathcal{G}^{\min}/T$ is connected and simply-connected.

(4) Let A be the GCM $\left(\begin{smallmatrix} 2 & -2 \\ -2 & 2 \end{smallmatrix}\right)$ and let $\mathfrak{g}(A)$ be the associated Kac–Moody algebra. Then $\mathfrak{g}(A)$ is the affine Kac–Moody algebra $\hat{\mathcal{L}}(sl_2)$ corresponding to sl_2 (cf. Section 13.1). Let $\mathcal{P}_Y$ be the standard maximal parabolic subgroup of the associated group $\mathcal{G}$ (corresponding to $Y = \{1\}$). Then show that, for any $i \geq 0$, $H^{2i}(\mathcal{G}/\mathcal{P}_Y, \mathbb{Z})$ is a free $\mathbb{Z}$-module of rank 1 generated by the Schubert class ε_i. Moreover,

$$\varepsilon_i\, \varepsilon_j = \binom{i+j}{i} \varepsilon_{i+j}.$$

Hint. Use (11.3.17.3) for u the simple reflection s_0.

11.4. Positivity of the Cup Product in the Cohomology of Flag Varieties

11.4.1 Definition. Let T be a complex (connected) torus with the character group $X(T)$ (cf. 6.1.6) and let $\pi : E \to A$ be an algebraic principal T-bundle. For $\lambda \in X(T)$, let $\pi_\lambda : E_\lambda \to A$ be the line bundle associated to the principal T-bundle π via the character λ of T. We denote the first Chern class of E_λ by $\lambda_E \in H^2(A, \mathbb{Z})$.

For a T-variety Y, let $\pi_Y : E_Y \to A$ be the associated fiber bundle with fiber Y. If A and Y are complete (e.g., projective), A irreducible, and $Z \subset Y$ is a T-stable closed irreducible subvariety, then the image of the fundamental class $[E_Z]$ in $H_{2(\dim A + \dim Z)}(E_Y)$ is denoted by $[Z]_E$.

If A and Y are not necessarily complete, we can still define the fundamental class $[Z]_E$ provided we replace the singular homology by the Borel–Moore homology (cf. [Fulton–97, Appendix B.2]).

Let B be a connected solvable affine algebraic group over $\mathbb{C}$ with unipotent radical U, and let $T \subset B$ be a maximal torus so that B is the semidirect product UT (cf. [Borel–91, Theorem 10.6(4) and 11.21]). Let $\{\alpha_1, \dots, \alpha_p\}$ be the set of weights of T acting on $\mathfrak{u} := \operatorname{Lie} U$.

11.4.2 Theorem. *Let B and T be as above, and let $\pi : E \to A$ be a principal algebraic T-bundle with irreducible and complete variety A as base. Let Y be a complete B-variety. Then, for any T-stable closed irreducible subvariety $Z \subset Y$, there exist B-stable closed irreducible subvarieties $Z_k \subset Y$ such that*

$$(1) \qquad [Z]_E = \sum_k \pi_Y^*(P_k) \cap [Z_k]_E,$$

as elements of $H_*(E_Y)$, *where* P_k*'s are homogeneous polynomials in* $\{(\alpha_i)_E\}_{1\le i\le p}$ *with nonnegative integral coefficients. Moreover,*

(2) $$\text{degree } P_k \le \dim U.$$

Proof. We prove the theorem by induction on $\dim U$. Assume first that U is one-dimensional and set $\alpha_1 = \alpha$. Choose any biregular isomorphism $\varphi : G_a \xrightarrow{\sim} B/T$, taking 0 to the base point eT, where G_a is the additive group $\mathbb{A}^1$. The left multiplication of T on B/T under φ corresponds to the linear action of T on $\mathbb{A}^1$ with weight α. Embed $G_a \hookrightarrow \mathbb{P}^1 = G_a \cup \{\infty\}$. Then the action of B on B/T (via the left multiplication) extends (under φ) to an algebraic action on $\mathbb{P}^1$, where ∞ is fixed by B. This follows easily since the automorphism group Aut $\mathbb{A}^1$ is a subgroup of Aut $\mathbb{P}^1$, as can be seen by noting that any automorphism f of $\mathbb{A}^1$ is of the form $f(z) = az + b$ for some $a, b \in \mathbb{C}$. Now, putting $f(\infty) = \infty$, f extends to an automorphism of $\mathbb{P}^1$.

Let B act on $B \times_T Y$ via the left multiplication on the first factor and on $B/T \times Y$ via the diagonal action. Then

$$\theta : B \times_T Y \to B/T \times Y, \ \ (b, y) \mapsto (bT, by)\,,$$

is a B-equivariant isomorphism. Let D be the Zariski closure of the image $\theta(B \times_T Z)$ in $\mathbb{P}^1 \times Y$. Then D is B-stable and irreducible, since Z is irreducible by assumption. Let $p_D : D \to \mathbb{P}^1$ be the projection on the first factor. Then p_D is B-equivariant and

(3) $$p_D^{-1}(0) = 0 \times Z \simeq Z.$$

Since $\infty \in \mathbb{P}^1$ is B-fixed (and p_D is B-equivariant), we get that the fiber $D_\infty := p_D^{-1}(\infty)$ is B-stable. Moreover, since $\dim D = n + 1$ (where $n := \dim Z$) and each irreducible component Z_i of D_∞ is of dimension at least n (cf. [Šafarevič–94, Chapter I, §6.3, Theorem 7]), we have $\dim Z_i = n$. Further, D_∞ being B-stable, each Z_i is B-stable (since B is connected). We next prove that, as elements of $H_*(E_Y)$,

(4) $$[Z]_E - \sum_i n_i\,[Z_i]_E = \pi_Y^*(\alpha_E) \cap (\tilde{p}'_{D*}([E_D])),$$

where $p'_D : D \to Y$ is the B-equivariant projection on the Y-factor which induces the (proper) map $\tilde{p}'_D : E_D \to E_Y$, and $n_i \in \mathbb{Z}_+$ is the geometric multiplicity of the irreducible component Z_i in the scheme theoretic fiber $p_D^{-1}(\infty)$.

Let T act on $E \times D$ by $t \cdot (e, d) = (et^{-1}, td)$. The composition f of the T-equivariant projection $E \times D \to D$ with the B (in particular, T)-equivariant map $p_D : D \to \mathbb{P}^1$ is a meromorphic function of the weight α, i.e., $f(t \cdot x) = e^\alpha(t) f(x)$, for all $t \in T$ and $x \in E \times D$ (since the T-action on $\mathbb{A}^1$ is linear of weight α). In particular, f gives rise to a meromorphic section of the pullback line bundle $\pi_D^*(E_\alpha)$, where $\pi_D : E_D \to A$ is the projection. By [Fulton–98, Theorem 3.2(f) and Proposition 19.1.2],

$$\pi_D^*(\alpha_E) \cap [E_D] = [E_{0\times Z}] - \sum_i n_i [E_{\infty \times Z_i}], \tag{5}$$

as elements of $H_*(E_D)$.

Applying the induced map $\tilde{p}'_{D*}$ in homology to (5) and using (C.3.1), we get (4). But, by [Fulton–98, Lemma 19.1.2],

$$\tilde{p}'_{D*}[E_D] = \deg(E_D/E_{\operatorname{Im} p'_D})\, [\operatorname{Im} p'_D]_E. \tag{6}$$

Substituting (7) in (6), we get

$$[Z]_E = \sum_i n_i [Z_i]_E + \deg(D/\operatorname{Im} p'_D)\, \pi_Y^*(\alpha_E) \cap [\operatorname{Im} p'_D]_E, \tag{7}$$

since $\deg(E_D/E_{\operatorname{Im} p'_D}) = \deg(D/\operatorname{Im} p'_D)$. Now (1) and (2) follow from (8) in this case (i.e., when U is one-dimensional).

Now we come to the case of general B. There exists a connected closed subgroup $U' \subset U$ of codimension-1 such that U' is normal in B. Set $B' := U' \cdot T \subset B$. Then B' is a connected (solvable) closed subgroup of B (of codimension-1). By induction, we assume the validity of the theorem for B replaced by B', i.e., we can find B'-stable irreducible Z_k's satisfying

$$[Z]_E = \sum_k \pi_Y^*(P_k) \cap [Z_k]_E, \tag{8}$$

where P_k are homogeneous with $\deg P_k \leq \dim U'$. Fix any B'-stable closed irreducible $Z_k \subset Y$. Replace θ by $\theta' : B \times_{B'} Y \xrightarrow{\sim} B/B' \times Y, (b, y) \mapsto (bB', by)$, and Z by Z_k and choose a biregular isomorphism $G_a \xrightarrow{\sim} B/B'$, taking 0 to the base point eB'. This allows us to B-equivariantly embed $B/B' \hookrightarrow \mathbb{P}^1$ with ∞ fixed by B. By the same argument as above, for any k, we get irreducible B-stable closed subvarieties $Z_{k,k'} \subset Y$ such that

$$[Z_k]_E = \sum_{k'} \pi_Y^*(P_{k,k'}) \cap [Z_{k,k'}]_E, \tag{9}$$

where $\deg P_{k,k'} \leq 1$.

Recall from (C.3.2) that, for any topological space X,

(10) $$a \cap (b \cap c) = (a \cup b) \cap c, \quad \text{for } a, b \in H^*(X), \text{ and } c \in H_*(X).$$

Hence we get (1) and (2) from combining (8)–(10). This completes the induction and hence the theorem is proved. □

11.4.3 Remarks. (1) An analogue of the above theorem for the Chow group (replacing $H_*(E_Y)$) can be proved similarly (without even the completeness assumption on A and Y).

(2) The assumption on A and Y to be complete (in the above theorem) can be removed by replacing the singular homology $H_*(E_Y)$ of E_Y by its Borel–Moore homology (cf. Exercise 11.4.E.1).

Recall that, for a topological space X, $H_*(X)$ is a graded $H^*(X)$-module under $a \cdot x = a \cap x \in H_{n-p}(X)$, for $a \in H^p(X)$, $x \in H_n(X)$. More generally, for a subspace $X' \subset X$, the cap product defines maps (cf. Appendix C):

$$H^p(X, X') \otimes H_n(X, X') \to H_{n-p}(X),$$

and

$$H^p(X) \otimes H_n(X, X') \to H_{n-p}(X, X').$$

11.4.4 Definition. Let $\pi : X \to A$ be a continuous map between topological spaces and $X' \subset X$ any subspace. Define the pairing

$$\langle\, ,\, \rangle'_\pi : H^*(X, X') \times H_*(X, X') \to H_*(A)$$

by $\langle a, x\rangle'_\pi = \pi_*(a \cap x)$, for $a \in H^*(X, X')$ and $x \in H_*(X, X')$. Then, by (C.3.1) and (C.3.2), $\langle\, ,\, \rangle'_\pi$ satisfies

(1) $$\langle b \cdot a, x\rangle'_\pi = (-1)^{\deg a \deg b}\, \langle a, b \cdot x\rangle'_\pi = b \cdot \langle a, x\rangle'_\pi$$

for $b \in H^*(A)$, $a \in H^*(X, X')$ and $x \in H_*(X, X')$. (Here $H^*(X, X')$ and $H_*(X, X')$ are viewed as $H^*(A)$-modules via the homomorphism $\pi^* : H^*(A) \to H^*(X)$.)

Assume further that A is a compact connected oriented topological manifold of dimension d and let $\mu : H^*(A) \xrightarrow{\sim} H_{d-*}(A)$, $a \mapsto a \cap [A]$, be the Poincaré duality isomorphism, where $[A]$ is the fundamental class of A. Then the pairing

$$\langle\, ,\rangle_\pi := \mu^{-1}\langle\, ,\, \rangle'_\pi : H^*(X, X') \times H_*(X, X') \to H^*(A)$$

again satisfies (1), since μ is a $H^*(A)$-linear isomorphism. Moreover,

$$\langle H^p(X, X'), H_q(X, X')\rangle_\pi \subset H^{d+p-q}(A)\,.$$

The following proposition is an easy consequence of the Leray–Hirsch theorem (cf. Theorem C.5).

11.4.5 Proposition. *Let $(E, \dot{E})$ be the total pair of a fiber bundle pair with fiber pair $(F, \dot{F})$ and projection $\pi : E \to A$ onto a base which is a compact connected oriented topological manifold A of dimension d. Assume that $H_*(F, \dot{F})$ is a free and finitely generated $\mathbb{Z}$-module and the restriction map $H^*(E, \dot{E}) \to H^*(F, \dot{F})$ is surjective. Then $H^*(E, \dot{E})$ and $H_*(E, \dot{E})$ are both free and finitely generated as $H^*(A)$-modules.*

Moreover, $\langle\ ,\ \rangle_\pi : H^*(E, \dot{E}) \times H_*(E, \dot{E}) \to H^*(A)$ is nonsingular. In fact, there exists a graded $H^*(A)$-basis $\{a_i\}_{i\in I}$ of $H^*(E, \dot{E})$ and a graded $H^*(A)$-basis $\{x_i\}_{i\in I}$ of $H_*(E, \dot{E})$ such that $\deg x_i = \deg a_i + d$ *and*

$$\text{(1)} \qquad \langle a_i, x_j\rangle_\pi = \delta_{i,j}, \text{ for } i, j \in I.$$

Proof. Choose a graded $\mathbb{Z}$-module splitting $\theta : H^*(F, \dot{F}) \to H^*(E, \dot{E})$. Also choose a graded $\mathbb{Z}$-basis $\{e_i\}_{i\in I}$ of $H_*(F, \dot{F})$ and let $\{e_i^*\}$ be the dual basis of $H^*(F, \dot{F})$, i.e.,

$$e_i^*(e_j) = \delta_{i,j}.$$

Then, by the Leray–Hirsch theorem (C.5) (coupled with C.4), we have the isomorphisms:

$$\text{(2)} \qquad \Phi : H_*(E, \dot{E}) \xrightarrow{\sim} H^*(A) \otimes H_*(F, \dot{F}),$$

defined by $\Phi(x) = \sum_{i\in I} (-1)^{\deg x \deg e_i} \langle \theta(e_i^*), x\rangle_\pi \otimes e_i$, for $x \in H_*(E, \dot{E})$, and

$$\text{(3)} \qquad \Phi' : H^*(A) \otimes H^*(F, \dot{F}) \xrightarrow{\sim} H^*(E, \dot{E}),$$

defined by $\Phi'(b \otimes c) = \pi^*b \cup \theta(c)$, for $b \in H^*(A)$ and $c \in H^*(F, \dot{F})$.

It is easy to see using (11.4.4.1) that, for $b \in H^*(A)$ and $x \in H_*(E, \dot{E})$,

$$\text{(4)} \qquad \Phi(b \cdot x) = b \cdot \Phi(x),$$

where b acts on $H^*(A) \otimes H_*(F, \dot{F})$ via the left multiplication on the $H^*(A)$-factor. From (2) and (4), we obtain that $H_*(E, \dot{E})$ is $H^*(A)$-free. Similarly, from (3), we get the $H^*(A)$-freeness of $H^*(E, \dot{E})$.

Define $a_i := \theta(e_i^*) \in H^{\deg e_i}(E, \dot{E})$, and

$$x_i := (-1)^{(\deg e_i + d)\deg e_i}\, \Phi^{-1}(1 \otimes e_i) \in H_{\deg e_i + d}(E, \dot{E}).$$

Then, by (2)–(4), $\{a_i\}$, resp. $\{x_i\}$, is a $H^*(A)$-basis of $H^*(E, \dot{E})$, resp. $H_*(E, \dot{E})$. Moreover, by the definition of the map Φ, $\langle a_i, x_j\rangle_\pi = \delta_{i,j}$, proving (1). This proves the proposition. □

11.4.6 Definition. A variety X is said to be *affinely paved* if X has a filtration by closed subvarieties:

(1) $$X = X_n \supset X_{n-1} \supset \cdots \supset X_0 \supset X_{-1} = \emptyset,$$

such that each locally closed subvariety $X_p\backslash X_{p-1}$ (of X) is a (finite) disjoint union of varieties $C_{p,q}$ isomorphic to affine spaces $\mathbb{A}^{n_{p,q}}$. The locally closed subvarieties $C_{p,q}$ of X will be referred to as the *cells* of X. These depend of course on the choice of the filtration (1). Let us reenumerate them as $\{C_i\}_{i\in I}$. Also let $d_i := \dim_{\mathbb{C}} C_i$.

We call a closed subvariety $Y \subset X$ *compatibly paved* if Y is a union of certain C_i's. Let $I^Y := \{i \in I : C_i \subset X\backslash Y\}$.

If we assume that X is complete, then the fundamental classes $[\bar{C}_i] \in H_{2d_i}(X)$ form a $\mathbb{Z}$-basis of $H_*(X)$. More generally, for $Y \subset X$ compatibly paved, $\{\alpha_*[\bar{C}_i]\}_{i\in I^Y}$ form a $\mathbb{Z}$-basis of $H_*(X, Y)$, where $\alpha_* : H_*(X) \to H_*(X, Y)$ is the canonical map. (This can be proved for the pair $(X_k \cup Y, Y)$ by induction on $-1 \leq k \leq n$, using the long exact homology sequence for the triple $(X_{k+1} \cup Y, X_k \cup Y, Y)$ and using the arguments as in the proof of Proposition 11.3.2.) In particular, $H_*(X, Y)$ is a free $\mathbb{Z}$-module concentrated only in even degrees, and hence so is $H^*(X, Y)$.

Let $\{\varepsilon_i = \varepsilon_i(Y)\}_{i\in I^Y}$ be the dual basis of $H^*(X, Y)$, i.e.,

$$\varepsilon_i(\alpha_*[\bar{C}_j]) = \delta_{i,j}\ , \quad \text{for } i, j \in I^Y .$$

Recall the notation from 11.4.1.

11.4.7 Proposition. *Let T be a complex (connected) torus and let $\pi : E \to A$ be an algebraic principal T-bundle with a smooth irreducible complete variety A as base. Let X be a complete affinely paved T-variety such that each term X_k as in (11.4.6.1) is T-stable. Assume further that, for any cell C_j in X,*

(∗) $$\bar{C}_j\backslash C_j \subset \bigcup_{d_i<d_j} C_i .$$

Then, for any compatibly paved (in particular, T-stable) subvariety Y, $H_*(E_X, E_Y)$ and $H^*(E_X, E_Y)$ are $H^*(A)$-free with basis $\{\bar{\alpha}_*([\bar{C}_i]_E)\}_{i\in I^Y}$ and $\{\varepsilon_i^E(Y)\}_{i\in I^Y}$ respectively, where $\bar{\alpha}_* : H_*(E_X) \to H_*(E_X, E_Y)$ is the canonical map and $\varepsilon_i^E(Y) \in H^{2d_i}(E_X, E_Y)$ is the unique class satisfying

(1) $$\varepsilon_i^E(Y)_{|(X,Y)} = \varepsilon_i(Y)\ , \text{ and}$$

(2) $$\varepsilon_i^E(Y)_{|(E_{Z^i\cup Y}, E_Y)} = 0\,, \quad \text{where } Z^i := \bigcup_{\substack{d_j\leq d_i\\ j\neq i}} C_j .$$

(By the condition (*) on X, Z^i is closed.) Moreover,

(3) $$\langle \varepsilon_i^E(Y), \bar{\alpha}_*([\bar{C}_j]_E)\rangle_{\pi_X} = \delta_{i,j}, \quad \text{for } i, j \in I^Y.$$

Proof. Let $p : E(T) \to B(T)$ be the universal principal T-bundle (cf. C.6). Then, as is well known, $H^*(B(T))$ is a free $\mathbb{Z}$-module concentrated only in the even degrees (and $B(T)$ is simply-connected). Hence, by the degenerate Leray–Serre spectral sequence for the fiber bundle pair $(E(T)_X, E(T)_Y)$ over the base space $B(T)$ and fiber pair (X, Y), we get that the restriction map $\bar{\gamma}_Y : H^*(E(T)_X, E(T)_Y) \to H^*(X, Y)$ is surjective. (The degeneracy of the spectral sequence at E_2 follows since $H^*(X, Y)$ is $\mathbb{Z}$-free and concentrated in even degrees and so is $H^*(B(T))$.)

Now choose a T-bundle map

$$\begin{array}{ccc} E & \xrightarrow{\varphi} & E(T) \\ \downarrow{\scriptstyle \pi} & & \downarrow{\scriptstyle p} \\ A & \longrightarrow & B(T). \end{array}$$

From this, and the surjectivity of $\bar{\gamma}_Y$, we immediately obtain that the restriction map

$$\gamma_Y : H^*(E_X, E_Y) \twoheadrightarrow H^*(X, Y) \text{ is surjective.}$$

Hence, by Proposition 11.4.5, we get that $H_*(E_X, E_Y)$ and $H^*(E_X, E_Y)$ are both free $H^*(A)$-modules.

We now prove the existence and uniqueness of $\varepsilon_i^E(Y)$ satisfying (1) and (2).

From 11.4.6, it is easy to see that the class $\varepsilon_i(Y) \in H^{2d_i}(X, Y)$ is the restriction of the class $\varepsilon_i(Z^i \cup Y) \in H^{2d_i}(X, Z^i \cup Y)$. In particular, from the surjectivity of $\gamma_{Z^i \cup Y}$, the existence of $\varepsilon_i^E(Y)$ satisfying (1) and (2) follows.

Choose the graded $\mathbb{Z}$-module splitting $\theta : H^*(X, Y) \to H^*(E_X, E_Y)$ defined by $\theta(\varepsilon_i(Y)) = \varepsilon_i^E(Y)$, and consider the isomorphism

$$\Phi' : H^*(A) \otimes H^*\big(X, Z^i \cup Y\big) \xrightarrow{\sim} H^*\big(E_X, E_{Z^i \cup Y}\big),$$

(cf. (11.4.5.3)). From the dimensional consideration, we see that

$$\gamma_{Z^i \cup Y} : H^{2d_i}\big(E_X, E_{Z^i \cup Y}\big) \xrightarrow{\sim} H^{2d_i}\big(X, Z^i \cup Y\big)$$

is an isomorphism. From this the uniqueness of $\varepsilon_i^E(Y)$ (satisfying (1) and (2)) follows. The assertion that $\{\varepsilon_i^E(Y)\}_{i \in I^Y}$ is a $H^*(A)$-basis of $H^*(E_X, E_Y)$ follows from (11.4.5.3).

We next prove (3): First take $i, j \in I^Y$ such that $d_i < d_j$. By 11.4.4,

$$\langle \varepsilon_i^E(Y), \bar{\alpha}_*([\bar{C}_j]_E)\rangle_{\pi_X} \in H^{2d_i - 2d_j}(A) = 0.$$

Next, assume that $i \neq j$ and $d_i \geq d_j$. Since $\varepsilon_i^E(Y) = \varepsilon_i^E(Z^i \cup Y)_{|(E_X, E_Y)}$ and $\bar{\beta}_*\bar{\alpha}_*([\bar{C}_j]_E) = 0$, where $\bar{\beta} : (E_X, E_Y) \to (E_X, E_{Z^i \cup Y})$, we obtain again, using (C.3.1), $\langle \varepsilon_i^E(Y), \bar{\alpha}_*([\bar{C}_j]_E)\rangle_{\pi_X} = 0$ (in fact, $\varepsilon_i^E(Y) \cap \bar{\alpha}_*([\bar{C}_j]_E) = 0$). So (3) follows from Exercise 11.4.E.2. From (3), we see that $\Phi\big(\bar{\alpha}_*([\bar{C}_i]_E)\big) = 1 \otimes \alpha_*[\bar{C}_i]$ (if we choose $\theta(\varepsilon_i(Y)) = \varepsilon_i^E(Y)$), where Φ is the isomorphism as in (11.4.5.2). In particular, by the proof of Proposition 11.4.5, $\big\{\bar{\alpha}_*([\bar{C}_i]_E)\big\}_{i \in I^Y}$ is a $H^*(A)$-basis of $H_*(E_X, E_Y)$. This completes the proof of the proposition. □

If the varieties X and Y are affinely paved with cells $\{C_i\}$and $\{D_j\}$ respectively, then the product variety $Z := X \times Y$ is affinely paved under the filtration

$$Z_k := \bigcup_{k'+k'' \leq k} X_{k'} \times Y_{k''}.$$

The cells of Z are precisely $\{C_i \times D_j\}$.

Let $\delta : X \to X \times X$ be the diagonal map which is, of course, T-equivariant (under the diagonal action of T on $X \times X$), and let $\bar{\delta} : E_X \to E_{X \times X}$ be the induced map. Observe that if X satisfies $(*)$ then so does $X \times X$.

11.4.8 Corollary. *Let the notation and assumptions be as in Proposition 11.4.7 and take $Y = \emptyset$. Write, for $k \in I$,*

(1) $$\bar{\delta}_*([\bar{C}_k]_E) = \sum_{i,j \in I} p_{i,j}^k \cdot \big[\,\overline{C_i \times C_j}\,\big]_E,$$

for some unique $p_{i,j}^k \in H^*(A)$. Then, for any $i, j \in I$,

(2) $$\varepsilon_i^E \, \varepsilon_j^E = \sum_{k \in I} p_{i,j}^k \cdot \varepsilon_k^E.$$

(For $Y = \emptyset$, we abbreviate $\varepsilon_i^E = \varepsilon_i^E(\emptyset)$.)

Proof. Consider the commutative diagram:

$$\begin{array}{ccc} E_X & \xrightarrow{\ \Delta\ } & E_X \times E_X \\ & {\scriptstyle\bar{\delta}}\searrow \quad \nearrow{\scriptstyle\psi} & \\ & E_{X \times X}, & \end{array}$$

where Δ is the diagonal map and $\psi\,[e,(x,y)] := ([e,x],[e,y])$, for $e \in E$ and $x, y \in X$. Now, by the characterizing properties (1) and (2) of Proposition 11.4.7, we get that

(3) $$\psi^*\big(\varepsilon_i^E \boxtimes \varepsilon_j^E\big) = (\varepsilon_i \boxtimes \varepsilon_j)^E.$$

Hence,

(4) $$\begin{aligned} \varepsilon_i^E\,\varepsilon_j^E &= \Delta^*\big(\varepsilon_i^E \boxtimes \varepsilon_j^E\big) \\ &= \bar{\delta}^*\big((\varepsilon_i \boxtimes \varepsilon_j)^E\big). \end{aligned}$$

Now, from the functoriality of the cap product,

$$\begin{aligned} \langle \bar{\delta}^*\big((\varepsilon_i \boxtimes \varepsilon_j)^E\big), [\bar{C}_k]_E\rangle_{\pi_X} &= \langle (\varepsilon_i \boxtimes \varepsilon_j)^E, \bar{\delta}_*[\bar{C}_k]_E\rangle_{\pi_{X\times X}} \\ &= \langle (\varepsilon_i \boxtimes \varepsilon_j)^E, \sum_{i',j'\in I} p_{i',j'}^k \cdot \big[\overline{C_{i'} \times C_{j'}}\,\big]_E\rangle_{\pi_{X\times X}}, \\ &\qquad \text{by (1)} \\ &= p_{i,j}^k, \qquad \text{by (11.4.4.1) and (11.4.7.3).} \end{aligned}$$

Writing $\varepsilon_i^E\varepsilon_j^E = \sum_{k'} q_{i,j}^{k'} \cdot \varepsilon_{k'}^E$ and applying (4), (11.4.4.1) and (11.4.7.3) again, we get $q_{i,j}^k = p_{i,j}^k$. This proves (2). □

11.4.9 Definition. Let T be a complex torus and let X be a complete affinely paved T-variety satisfying the condition $(*)$ of Proposition 11.4.7 and such that each X_k (cf. (11.4.6.1)) is T-stable. Let $\{C_1, \dots, C_N\}$ be all the cells of X and let $\dim C_i = d_i$. Then $[\bar{C}_1], \dots, [\bar{C}_N]$ is a $\mathbb{Z}$-basis of $H_*(X)$. Let $\{\varepsilon_1, \dots, \varepsilon_N\}$ be the dual basis of $H^*(X)$ (cf. 11.4.6).

Then, following the same proof given in the proof of Proposition 11.4.7, there exist unique elements $\hat{\varepsilon}_i \in H_T^{2d_i}(X)$ satisfying the following two properties:

(P_1) $\eta(\hat{\varepsilon}_i) = \varepsilon_i$, where $\eta : H_T^*(X) \to H^*(X)$ is the canonical map (cf. C.6.4), and

(P_2) $\gamma_i(\hat{\varepsilon}_i) = 0$, where $\gamma_i : H_T^*(X) \to H_T^*(Z^i)$ is the restriction map, where (as in (11.4.7.2)) $Z^i := \bigcup_{\substack{j\neq i \\ d_j \le d_i}} C_j$.

Moreover, $\{\hat{\varepsilon}_i\}_{1\le i\le N}$ forms a $H^*(B(T))$-basis of $H_T^*(X)$. In particular, we can write

(1) $$\hat{\varepsilon}_i\,\hat{\varepsilon}_j = \sum p_{i,j}^k\,\hat{\varepsilon}_k$$

for some (unique) $p_{i,j}^k \in H^{2(d_i+d_j-d_k)}(B(T))$.

11.4.10 Lemma. *Let $B = UT$ be as in 11.4.1. Then, for any B-variety X such that it has only finitely many U-orbits $\{C_1, \dots, C_N\}$, any U-orbit in X is B-stable. Further, any such X is affinely paved with B (in particular, T)-stable*

$$X_k := \bigcup_{\dim C_i \leq k} C_i ,$$

and $\{C_1, \dots, C_N\}$ are precisely the cells. Moreover, X satisfies the condition $(*)$ of Proposition 11.4.7.

Proof. Take a B-orbit $V \subset X$. Since X has only finitely many U-orbits, so does V. In particular, V being irreducible (since B is connected), there exists a U-orbit V' dense in V. But V is affine (cf. [Hochschild–81, Chap. 12, Theorem 4.3]), and hence V' is closed in V [loc. cit., Chap. 12, Lemma 4.1]. This forces $V' = V$. This proves the first part.

Since the U-orbits are affine spaces, the second part follows from [Humphreys–95, Proposition 8.3]. □

Let $\{\alpha_1, \dots, \alpha_p\}$ be the set of weights of T acting on $\mathfrak{u} := \operatorname{Lie} U$.

11.4.11 Theorem. *Let X and $B = UT$ be as in the above Lemma 11.4.10. Assume further that X is complete and let $\{\hat{\varepsilon}_i\}_{1 \leq i \leq N}$ be the $S_{\mathbb{Z}}$-basis of $H_T^*(X)$ defined in 11.4.9, where we identify the symmetric algebra $S_{\mathbb{Z}}$ over $\mathbb{Z}$ of the character group $X(T)$ with $H^*(B(T))$ under c as in 11.3.8. As in (11.4.9.1), write*

$$\hat{\varepsilon}_i \, \hat{\varepsilon}_j = \sum_k p_{i,j}^k \, \hat{\varepsilon}_k ,$$

for some (unique) $p_{i,j}^k \in S_{\mathbb{Z}}$. Then each $p_{i,j}^k$ is a homogeneous polynomial of degree $d_i + d_j - d_k$ in $\{\alpha_1, \dots, \alpha_p\}$ with nonnegative integral coefficients.

In particular, writing $\varepsilon_i \varepsilon_j = \sum_k c_{i,j}^k \, \varepsilon_k$ in $H^*(X)$, we have $c_{i,j}^k \in \mathbb{Z}_+$.

Proof. Decomposing $T = \mathbb{C}^* \times \cdots \times \mathbb{C}^*$, we can take $B(T) = \mathbb{CP}^\infty \times \cdots \times \mathbb{CP}^\infty$, where $\mathbb{CP}^\infty := \mathbb{P}(V)$ for a countable-dimensional complex vector space V (cf. 4.1.3(4)). For any positive m, define $B(T)_m := \mathbb{CP}^m \times \cdots \times \mathbb{CP}^m$. Then the classifying bundle $p : E(T) \to B(T)$ restricted to $B(T)_m$ is an algebraic T-bundle, with smooth irreducible projective variety $B(T)_m$ as its base, denoted $p_m : E(T)_m \to B(T)_m$.

Having fixed i, j, we prove that $p_{i,j}^k$ are polynomials in α_n's with nonnegative integral coefficients: Choose any $m \geq d_i + d_j$. Now apply Theorem 11.4.2 by taking $B' := (U \times U) \propto T \subset B \times B$ (T acting diagonally on $U \times U$), $B' \subset B \times B$ acting on $Y := X \times X$ componentwise, and taking the T-bundle $\pi = p_m$. In conjunction with Corollary 11.4.8, for $\pi = p_m$, we get that $p_{i,j}^k$

are homogeneous polynomials in the α_n's with nonnegative integral coefficients. This proves the theorem. $\square$

Let $\mathcal{G}$ be any Kac–Moody group, $\mathcal{B} \subset \mathcal{G}$ the standard Borel subgroup, and $T \subset \mathcal{B}$ the standard maximal torus. Let W be the associated Weyl group. Recall the definition of the $S_{\mathbb{Z}} := S(\mathfrak{h}_{\mathbb{Z}}^*)$-basis (called the Schubert basis) $\{\hat{\varepsilon}^w\}_{w\in W}$ of $H_T^*(\mathcal{G}/\mathcal{B})$ from Theorem 11.3.9 and the Schubert basis $\{\varepsilon^w\}_{w\in W}$ of $H^*(\mathcal{G}/\mathcal{B})$ (cf. Proposition 11.3.2). Let $\{\alpha_1, \dots, \alpha_\ell\}$ be the simple roots of the Kac–Moody Lie algebra $\mathfrak{g}$.

11.4.12 Corollary. *With the notation and assumptions as above, write, for $u, v \in W$,*

$$\hat{\varepsilon}^u \cdot \hat{\varepsilon}^v = \sum_{w\in W} p_{u,v}^w \, \hat{\varepsilon}^w,$$

for some (unique) $p_{u,v}^w \in S_{\mathbb{Z}}$. Then each $p_{u,v}^w$ is a homogeneous polynomial in $\{\alpha_1, \dots, \alpha_\ell\}$ of degree $\ell(u)+\ell(v)-\ell(w)$ with nonnegative integral coefficients. In particular, writing

$$\varepsilon^u \, \varepsilon^v = \sum_{w\in W} c_{u,v}^w \, \varepsilon^w,$$

we have $c_{u,v}^w \in \mathbb{Z}_+$.

Proof. Fix u, v and take any $m \geq \ell(u) + \ell(v)$. Consider $X_m \subset \mathcal{G}/\mathcal{B}$ as defined in 7.1.13. Then $\mathcal{B}$ acts on X_m with only finitely many orbits under $\mathcal{U}$, and X_m is a projective variety. By 7.1.19, the action of $\mathcal{B}$ on $\mathcal{X}$ is regular; in particular, there exists a connected solvable affine algebraic group B_m, which is a quotient group of $\mathcal{B}$, such that the action of $\mathcal{B}$ on X_m factors through an algebraic action of B_m onX_m.Let $\hat{\varepsilon}_m^w := \gamma_m(\hat{\varepsilon}^w)$, where $\gamma_m : H_T^*(\mathcal{X}) \to H_T^*(X_m)$ is the restriction map. Then the B_m-variety X_m satisfies the assumptions of Theorem 11.4.11. Moreover, from the characterizing properties of $\hat{\varepsilon}^w$ (cf. Proposition 11.3.10 and 11.4.9), we see that $\{\hat{\varepsilon}_m^w\}_{\ell(w)\leq m}$ is the $S_{\mathbb{Z}}$-basis of $H_T^*(X_m)$ as defined in 11.4.9. Hence the Corollary follows from Theorem 11.4.11. $\square$

11.4.E EXERCISES

(1) Remove the restriction in Theorem 11.4.2 that A and Y are complete by replacing the singular homology $H_*(E_Y)$ by the Borel–Moore homology.

(2) With the notation as in the proof of Proposition 11.4.7, prove that, for all $i \in I^Y$,

$$\langle \varepsilon_i^E(Y), \bar{\alpha}_*([\bar{C}_i]_E)\rangle_{\pi_X} = 1 \in H^0(A).$$

Hint. There exists a Zariski open subset V of A such that the bundle $\pi : E \to A$ restricted to V is trivial. Now use the Borel–Moore homology.

(3) Remove the restriction in Theorem 11.4.11 that X is complete by replacing $H_T^*(X)$ with the compactly supported T-equivariant cohomology.

(4) Consider the blow-up X of $\mathbb{P}^2$ at a point. Then PGL_3 acts on X with finitely many orbits and the orbit closures give rise to a homology basis of X. But the cup products in the dual cohomology basis *do* have negative coefficients. (Contrast this with Theorem 11.4.11, where we assumed that B is solvable and X has finitely many B-orbits.)

(5) Give an example of an affinely paved projective variety X such that X does **not** satisfy the condition $(*)$ of Proposition 11.4.7.

11.5 Degeneracy of the Leray–Serre Spectral Sequence for the Fibration $\mathcal{G}^{\min} \to \mathcal{G}^{\min}/T$

In this section $\mathcal{G}$ is any Kac–Moody group with the standard maximal torus T and $\mathcal{G}^{\min} \subset \mathcal{G}$ is the subgroup introduced by Kac–Peterson (cf. 7.4.1). We endow $\mathcal{G}^{\min}$ with the strong topology $\mathcal{G}^{\min}_{\text{an}}$ and suppress the subscript "an". Then $\mathcal{G}^{\min}$ is a topological group and $T \subset \mathcal{G}^{\min}$ is a closed subgroup (cf. Theorem 7.4.14 and Proposition 7.4.11). Further, $\mathcal{G}^{\min} \to \mathcal{G}^{\min}/T$ is a locally trivial fibration, where $\mathcal{G}^{\min}/T$ is equipped with the quotient topology (cf. Exercise 11.3.E.3).

For any simple reflection s_i, $1 \leq i \leq \ell$, recall the definitions of the operator $D_{s_i} : H^*(\mathcal{X}, \mathbb{Z}) \to H^{*-2}(\mathcal{X}, \mathbb{Z})$ from 11.3.6 and the operator $A_{s_i} : S^*(\mathfrak{h}_{\mathbb{Z}}^*) \to S^{*-1}(\mathfrak{h}_{\mathbb{Z}}^*)$ from Exercise 11.3.E.2, where S^i is the i-th symmetric power. Since the projection map $\pi : \mathcal{G}^{\min}/T \to \mathcal{X}$ induces the isomorphism $\pi^* : H^*(\mathcal{X}, \mathbb{Z}) \xrightarrow{\sim} H^*(\mathcal{G}^{\min}/T, \mathbb{Z})$ (cf. 11.3.4), we can interchangeably view D_{s_i}'s as operators on $H^*(\mathcal{G}^{\min}/T, \mathbb{Z})$ (under the identification π^*). Also recall the definition of the Borel homomorphism

$$\beta : S^*(\mathfrak{h}_{\mathbb{Z}}^*) \to H^{2*}(\mathcal{X}, \mathbb{Z}) \simeq H^{2*}(\mathcal{G}^{\min}/T, \mathbb{Z})$$

from 11.3.5.

For any field k (in fact, any $\mathbb{Z}$-algebra), we can consider the operators $D_{s_i}^k$, $A_{s_i}^k$ obtained by extending the scalars:

$$\begin{aligned} D_{s_i}^k &: H^*(\mathcal{X}, k) \simeq H^*(\mathcal{X}, \mathbb{Z}) \otimes_{\mathbb{Z}} k \to H^{*-2}(\mathcal{X}, k), \text{ and} \\ A_{s_i}^k &: S^*(\mathfrak{h}_k^*) \to S^{*-1}(\mathfrak{h}_k^*), \end{aligned}$$

where $\mathfrak{h}_k^* := \mathfrak{h}_{\mathbb{Z}}^* \otimes_{\mathbb{Z}} k$.

Similarly, we have the Borel homomorphism

$$\beta^k : S^*(\mathfrak{h}_k^*) \to H^{2*}(\mathcal{X}, k).$$

The following is the main result of this section. This was proved by Kac–Peterson. Observe that, by Exercise 11.3.E.3, $\mathcal{G}^{\min}/T$ is connected and simply-connected.

11.5.1 Theorem. *Let k be any field. Then the Leray–Serre spectral sequence (cf. Theorem E.11) corresponding to the fibration $f : \mathcal{G}^{\min} \to \mathcal{G}^{\min}/T$, for the singular cohomology with coeffients in k, collapses at the E_3-term, i.e.,*

$$E_3^{p,q} = E_\infty^{p,q}, \quad \text{for all } p, q \geq 0.$$

Before we come to the proof of the theorem, we need the following preparatory work 11.5.2–6, wherein k denotes any field.

11.5.2 Definition. Define the homogeneous ideal $I_k \subset S^*(\mathfrak{h}_k^*)$ by

$$I_k = \big\{P \in S^+(\mathfrak{h}_k^*) : A_{s_{i_1}}^k \circ \cdots \circ A_{s_{i_n}}^k P \in S^+(\mathfrak{h}_k^*) \\ \text{for all the sequences } i_1, \dots, i_n \ (1 \leq i_j \leq \ell)\big\},$$

where $S^+(\mathfrak{h}_k^*)$ is the standard augmentation ideal of $S(\mathfrak{h}_k^*)$. By the twisted derivation property of A_{s_i}'s (cf. 11.3.E.2), I_k is indeed an ideal. This is called the *ideal of generalized invariants*. Clearly I_k contains $S^+(\mathfrak{h}_k^*)^W$. Moreover, I_k is stable under the action of A_{s_i} for any $1 \leq i \leq \ell$, and also under the canonical action of the Weyl group W on $S^*(\mathfrak{h}_k^*)$. To prove the latter, write $s_i^k P = P - \alpha_i\big(A_{s_i}^k P\big)$ for any $P \in S(\mathfrak{h}_k^*)$.

11.5.3 Lemma. *With the notation as above,*

$$I_k = \operatorname{Ker} \beta^k.$$

Proof. Since $\beta \circ A_{s_i} = D_{s_i} \circ \beta$ (cf. 11.3.E.2), and $\beta_0^k : S^0(\mathfrak{h}_k^*) \xrightarrow{\sim} H^0(\mathcal{X}, k)$, we get

(1) $$\operatorname{Ker} \beta^k \subset I_k.$$

Conversely, take a homogeneous polynomial $P \in I_k$ of degree d. Assume, if possible, that $\beta^k(P) \neq 0$. Write (cf. Proposition 11.3.2)

$$\beta^k(P) = \sum_{\ell(w)=d} c_w \varepsilon^w, \quad \text{for } c_w \in k,$$

and some $c_{w_o} \neq 0$.

By (11.3.17.4), for any $w, w_o \in W$ of the same length,

(2) $$D_{w_o} \varepsilon^w = 0, \quad \text{if } w \neq w_o \\ = 1, \quad \text{if } w = w_o.$$

Hence, taking a reduced decomposition $w_o = s_{i_1} \cdots s_{i_d}$,

$$\beta^k\left(A_{s_{i_1}}^k \circ \cdots \circ A_{s_{i_d}}^k P\right) = D_{s_{i_1}}^k \circ \cdots \circ D_{s_{i_d}}^k (\beta^k(P)) = c_{w_o} \neq 0.$$

In particular, $A_{s_{i_1}}^k \circ \cdots \circ A_{s_{i_d}}^k P \neq 0 \in S^0(\mathfrak{h}_k^*)$. A contradiction to the assumption that $P \in I_k$. This proves the inclusion $I_k \subset \operatorname{Ker} \beta^k$ and hence the lemma. □

11.5.4 Lemma. *Let $R = R_k$ be the k-algebra $\operatorname{Im} \beta^k \subset H(\mathcal{X}, k)$. Then $H(\mathcal{X}, k)$ is a free R-module.*

In fact, as a graded R-module,

$$H(\mathcal{X}, k) \simeq R \otimes_k \big(H(\mathcal{X}, k)/J_k\big),$$

where R acts on the right side by multiplication on the first factor and $J_k \subset H(\mathcal{X}, k)$ is the ideal generated by $\beta^k(S^+(\mathfrak{h}_k^*))$.

Proof. Choose homogeneous elements $\{e_\theta\}$, $e_\theta \in H^{d_\theta}(\mathcal{X}, k)$, such that $\{\bar{e}_\theta\}$ is a k-vector space basis of $H(\mathcal{X}, k)/J_k$, where $\bar{e}_\theta := e_\theta \mod J_k$. We claim that $\{e_\theta\}$ is an R-basis of $H(\mathcal{X}, k)$: By induction on the degree, it is easy to see that $H(\mathcal{X}, k)$ is spanned by $\{e_\theta\}$ over R. So, it suffices to prove that $\{e_\theta\}$ are R-linearly independent. Fix d and, if possible, consider a nontrivial R-relation in $H^d(\mathcal{X}, k)$:

$$\sum_\theta \gamma_\theta e_\theta = 0, \tag{1}$$

where γ_θ are homogeneous elements of R of degree $d - d_\theta$; in particular, all but finitely many γ_θ are zero. Choose $\theta = \theta_o$ such that $\gamma_{\theta_o} \neq 0$ and d_{θ_o} is maximum with this property.

Since $\{\bar{e}_\theta\}$ are k-linearly independent, $\deg \gamma_\theta > 0$ for all those θ such that $\gamma_\theta \neq 0$. By the proof of Lemma 11.5.3, we can choose $1 \leq j \leq \ell$ such that

$$D^k_{s_j} \gamma_{\theta_o} \neq 0. \tag{2}$$

Applying $D^k_{s_j}$ to the relation (1), and using the twisted derivation property of $D^k_{s_j}$ (cf. Corollary 11.3.12), we get

$$\sum_\theta \left(D^k_{s_j} \gamma_\theta\right) e_\theta + \sum_\theta (s_j^k \gamma_\theta)\left(D^k_{s_j} e_\theta\right) = 0. \tag{3}$$

Express

$$D^k_{s_j} e_\theta = \sum_{\theta'} p^\theta_{\theta'} e_{\theta'}, \tag{4}$$

where $p^\theta_{\theta'} \in R$ is homogeneous of degree $d_\theta - 2 - d_{\theta'}$. In particular, $p^\theta_{\theta'} = 0$, unless $d_\theta - d_{\theta'} \geq 2$.

Substituting (4) in (3), we get

$$\sum_\theta \left(D^k_{s_j} \gamma_\theta\right) e_\theta + \sum_\theta \sum_{\theta' : d_{\theta'} < d_\theta} (s_j^k \gamma_\theta)\, p^\theta_{\theta'}\, e_{\theta'} = 0.$$

By (2), the coefficient of e_{θ_o} in the left side is nonzero. Thus, we get a nontrivial R-relation between $\{e_\theta\}$ in $H^{d-2}(\mathcal{X}, k)$. Observe that, by Exercise 11.3.E.2, R is W-stable. Hence, by induction on d, $\{e_\theta\}$ is R-linearly independent. □

11.5.5 Lemma. *The ideal $I = I_k$ of generalized invariants (cf. 11.5.2) is generated by a regular sequence of certain homogeneous elements $\{P_1, \dots, P_m\}$, where*

$$m = \dim \mathfrak{h} - \dim R, \tag{1}$$

$R := S(\mathfrak{h}_k^*)/I$ and $\dim R$ is the Krull dimension of the ring R.

Let $Q_1, \dots, Q_m \in I$ be any homogeneous polynomials. Then $Q_1, \dots, Q_m$ is a regular sequence generating the ideal I iff $\bar{Q}_1, \dots, \bar{Q}_m$ is a k-basis of $I/(S^+(\mathfrak{h}_k). I)$, where $\bar{Q}_i := Q_i \mod (S^+(\mathfrak{h}_k) \cdot I)$. Hence, the unordered collection $\{\deg P_i\}$ does not depend on the choice of the homogeneous regular sequence $\{P_i\}$ generating the ideal I.

If char. $k > 0$ or else k is arbitrary but $\mathcal{G}$ is of finite type,

$$m = \dim \mathfrak{h}\,. \tag{2}$$

(There are examples when $m < \dim \mathfrak{h}$; cf. Exercises 11.5.E.4(a) and 11.5.E.6.)

Proof. By a result of Vasconcelos [Vasconcelos–67] (for a more precise reference suitable for our purposes see [Kac–Peterson-85b, Lemma 2.3]), to prove that I is generated by a regular sequence consisting of homogeneous elements, it suffices to show that $M := I/I^2$ is a free R-module. Choose a set of homogeneous generators $\{e_1, \dots, e_m\}$ of the ideal I such that no proper subset generates I, and let $d_i := \deg e_i$. We prove that $\{\bar{e}_1, \dots, \bar{e}_m\}$ is an R-basis of M, where $\bar{e}_i := e_i \mod I^2$. Clearly, the elements $\{\bar{e}_1, \dots, \bar{e}_m\}$ span M as an R-module. So we need to prove that $\{\bar{e}_1, \dots, \bar{e}_m\}$ are linearly independent over R.

Take a homogeneous R-relation of degree d in M:

$$\sum_j \bar{p}_j \bar{e}_j = 0, \tag{3}$$

where p_j is a homogeneous polynomial of degree $d - d_j$ and $\bar{p}_j := p_j \mod I$. Since no proper subset of $\{e_1, \dots, e_m\}$ generates I, $\deg p_j > 0$ for all those j such that $p_j \neq 0$. Clearly (3) is equivalent to

$$\sum p_j e_j \in I^2. \tag{4}$$

Assume, if possible, that some $\bar{p}_j \neq 0$, i.e., $p_j \notin I$. Among such j's, pick j_o such that d_{j_o} is maximum. By the definition of I, there exists $1 \leq i \leq \ell$ such that $A_{s_i} p_{j_o} \notin I$. Applying $A_{s_i} = A^k_{s_i}$ to (4), we get

$$(A_{s_i} p_{j_o}) e_{j_o} + \sum_{j \neq j_o} (A_{s_i} p_j) e_j + \sum_j (s_i p_j)(A_{s_i} e_j) \in I^2, \tag{5}$$

since $A_{s_i}(I^2) \subset I^2$ (by 11.5.2). Since $A_{s_i}(I) \subset I$, we can write

(6) $$A_{s_i} e_j = \sum_{d_{j'} < d_j} q^j_{j'} e_{j'},$$

for some $q^j_{j'} \in S^{d_j - 1 - d_{j'}}(\mathfrak{h}^*_k)$.

Substituting (6) in (5), we get

(7) $$(A_{s_i} p_{j_o}) e_{j_o} + \sum_{j \neq j_o} (A_{s_i} p_j) e_j + \sum_{d_j \leq d_{j_o}} \sum_{d_{j'} < d_j} \left((s_i p_j) q^j_{j'}\right) e_{j'} + \sum_{d_j > d_{j_o}} \sum_{d_{j'} < d_j} \left((s_i p_j) q^j_{j'}\right) e_{j'} \in I^2.$$

But, by the choice of j_o, for $d_j > d_{j_o}$, p_j and hence $s_i p_j \in I$. So (7) gives a nontrivial homogeneous R-relation of degree $d-1$ in M: $\sum \bar{t}_j \bar{e}_j = 0$, with each t_j homogeneous and $\bar{t}_{j_o} \neq 0$. So, by induction on d, we get that $\{\bar{e}_1, \dots, \bar{e}_m\}$ are R-linearly independent. This proves the first part of the lemma.

To prove (1), use [Serre–89, Chap. IV, B]. The statement in the second paragraph of the lemma follows from the general properties of regular sequences (cf., e.g., [Kac–Peterson–85b, §2]). In the finite case, (2) follows from (1) since R is finite-dimensional in this case. Finally, we prove (2) in the case when $p = $ char. $k > 0$. Let W_p be the image of W under

$$W \to \mathrm{Aut}(\mathfrak{h}^*_{\mathbb{Z}}) \to \mathrm{Aut}(\mathfrak{h}^*_{\mathbb{F}_p}) \hookrightarrow \mathrm{Aut}(\mathfrak{h}^*_k),$$

where the first map is given by the representation of W in $\mathfrak{h}^*_{\mathbb{Z}}$ and the other maps are the canonical maps. Then, of course, W_p is a finite group. So $S(\mathfrak{h}^*_k)$ is integral over the subring of W_p-invariants $S(\mathfrak{h}^*_k)^{W_p}$ (cf. [Serre–88, Chap. III, §3, Proof of Proposition 18]). But, by 11.5.2,

$$I \supset \langle S^+(\mathfrak{h}^*_k)^{W_p} \rangle,$$

where $\langle S^+(\mathfrak{h}^*_k)^{W_p} \rangle$ denotes the ideal generated by $S^+(\mathfrak{h}^*_k)^{W_p}$. Hence, any homogeneous element of positive degree of R is nilpotent, giving that dim $R = 0$. Hence (2) follows in this case from (1). This completes the proof of the lemma. □

We also recall the following well-known *Koszul resolution* (cf. [Serre–89, Chap. IV, Proposition 2]).

11.5.6 Lemma. *Let A be a commutative ring with identity and $\{x_1, \dots, x_m\}$ a regular sequence of elements of A. Let $A^m := A \oplus \cdots \oplus A$ be the free A-module of rank m with standard basis $\{e_1, \dots, e_m\}$. Then the following complex is exact:*

$$0 \to \Lambda^m(A^m) \otimes_A A \to \cdots \to \Lambda^p(A^m) \otimes_A A \xrightarrow{\delta_p} \Lambda^{p-1}(A^m) \otimes_A A \to \cdots$$
$$\to \Lambda^0(A^m) \otimes_A A \xrightarrow{\varepsilon} A/\langle x_1, \cdots, x_m \rangle \to 0,$$

where ε is the canonical quotient map, and δ_p is the A-linear map defined by

$$(1) \quad \delta_p(e_{i_1} \wedge \cdots \wedge e_{i_p} \otimes a) = \sum_{j=1}^{p} (-1)^j e_{i_1} \wedge \cdots \wedge \widehat{e_{i_j}} \wedge \cdots \wedge e_{i_p} \otimes (x_{i_j} a),$$

for $1 \leq i_1 < \cdots < i_p \leq m$ and $a \in A$.

In particular, for any A-module M such that $x_i M = 0$ for all $1 \leq i \leq m$,

$$(2) \quad \mathrm{Tor}_j^A(M, A/\langle x_1, \ldots, x_m \rangle) \simeq \Lambda^j(A^m) \otimes_A M \quad (\textit{for all } j \geq 0). \quad \square$$

With these preparations, we come to the proof of the theorem.

11.5.7 *Proof of Theorem 11.5.1.* The Leray–Serre cohomology spectral sequence (with coefficients in k) for the fibration $f : \mathcal{G}^{\min} \to \mathcal{G}^{\min}/T$ has

$$\begin{aligned} E_2^{p,q} &= H^q(T, k) \otimes_k H^p(\mathcal{X}, k) \\ &\cong \Lambda^q(\mathfrak{h}_k^*) \otimes_k H^p(\mathcal{X}, k), \qquad \text{by Exercise 11.5.E.1.} \end{aligned}$$

Moreover, $d_2 : E_2^{p,q} \to E_2^{p+2,q-1}$, being a bigraded derivation, is completely determined by $d_2|_{E_2^{p,0}}$ and $d_2|_{E_2^{0,1}}$. By the degree consideration, $d_2|_{E_2^{p,0}} \equiv 0$. Moreover, $d_2 : E_2^{0,1} = \mathfrak{h}_k^* \to E_2^{2,0} = H^2(\mathcal{X}, k)$ is nothing but the negative of the Borel homomorphism $\beta^k_{|\mathfrak{h}_k^*}$ (cf. Exercise 11.5.E.2). Hence the bigraded derivation

$$(1) \qquad d_2 : \Lambda^q(\mathfrak{h}_k^*) \otimes_k H^p(\mathcal{X}, k) \to \Lambda^{q-1}(\mathfrak{h}_k^*) \otimes_k H^{p+2}(\mathcal{X}, k)$$

is given by

$$(2) \quad d_2(x_1 \wedge \cdots \wedge x_q \otimes a) = \sum_{i=1}^{q} (-1)^i x_1 \wedge \cdots \wedge \widehat{x_i} \wedge \cdots \wedge x_q \otimes \beta^k(x_i) a\,,$$

for $x_i \in \mathfrak{h}_k^*$ and $a \in H(\mathcal{X}, k)$. Thus, by Lemma 11.5.6,

$$\begin{aligned} (3) \qquad E_3^{*,q} &= \mathrm{Tor}_q^S(k, H^*(\mathcal{X}, k)) \\ &\simeq \mathrm{Tor}_q^S(k, R \otimes_R H^*(\mathcal{X}, k)) \\ &\simeq \mathrm{Tor}_q^S(k, R) \otimes_R H(\mathcal{X}, k), \quad \text{by Lemma 11.5.4,} \end{aligned}$$

where $R := \mathrm{Im}\, \beta^k$ (as in Lemma 11.5.4), $S := S(\mathfrak{h}_k^*)$, k is S-module under the standard augmentation map and $H(\mathcal{X}, k)$ (and R) is an S-module under β^k.

Now we calculate $\mathrm{Tor}^S_*(k, R)$: It is the homology of the complex

$$\text{(4)} \qquad \cdots \to \Lambda^q(\mathfrak{h}_k^*) \otimes_k R \xrightarrow{d} \cdots \xrightarrow{d} \Lambda^0(\mathfrak{h}_k^*) \otimes_k R,$$

where d is given by the same formula (2) for d_2 with $H(\mathcal{X}, k)$ replaced by R. Since d is a derivation, $\mathrm{Tor}^S_*(k, R)$ has a natural bigraded algebra structure. On the other hand, by Lemmas 11.5.3, 11.5.5 and (11.5.6.2),

$$\text{(5)} \qquad \mathrm{Tor}^S_q(k, R) \simeq \Lambda^q(k^m),$$

where m is as in Lemma 11.5.5.

The complex (4) is obtained from the S-free Koszul resolution

$$\Lambda^\bullet(\mathfrak{h}_k^*) \otimes_k S \to k$$

and (5) is obtained from the S-free resolution (cf. Lemma 11.5.6)

$$\Lambda^\bullet(S^m) \otimes_S S \to R,$$

and differentials of both of the complexes $\Lambda^\bullet(\mathfrak{h}_k^*) \otimes_k S$ and $\Lambda^\bullet(S^m) \otimes_S S$ are, in fact, $\mathbb{Z}_+$-graded derivations (with grading, in both the cases, coming from the exterior degree). Taking their tensor product complex we get that the $\mathbb{Z}_+$-graded algebra $\mathrm{Tor}^S_\bullet(k, R)$ is isomorphic with the $\mathbb{Z}_+$-graded algebra $\Lambda^\bullet(k^m)$. In particular, the $\mathbb{Z}_+$-graded algebra $\mathrm{Tor}^S_\bullet(k, R)$ is generated by $\mathrm{Tor}^S_1(k, R)$. Further, by (5),

$$\dim_k \mathrm{Tor}^S_1(k, R) = m.$$

By Exercise 11.5.E.3, $\mathrm{Tor}^S_1(k, R)$ has a k-basis $\{y_1, \dots, y_m\}$ such that, under the identification (3), $y_i \otimes 1 \in E_3^{2(d_i-1),1}$, where $\{d_1, \dots, d_m\}$ are the degrees of $\{P_1, \dots, P_m\}$ respectively — $\{P_1, \dots, P_m\}$ being a regular sequence consisting of homogeneous polynomials generating the ideal I.

In particular, E_3 is generated as an algebra by $E_3^{*,0}$ and $E_3^{*,1}$. Hence, the differential $d_3 : E_3 \to E_3$ is zero and, similarly, all the higher differentials d_r $(r \geq 4)$ are zero. This proves that $E_3 = E_\infty$. □

11.5.8 Corollary (of the proof of Theorem 11.5.1). *Let $\{P_1, \dots, P_m\}$ be a regular sequence consisting of homogeneous polynomials generating the ideal $I_k = \operatorname{Ker} \beta^k$ (cf. Lemmas 11.5.3 and 11.5.5). Let $d_i := \deg P_i$, i.e., $P_i \in S^{d_i}(\mathfrak{h}_k^*)$. Then there exist elements $\{y_1, \dots, y_m\}$ with $y_i \in E_\infty^{2(d_i-1),1}$ such that (as a bigraded algebra)*

$$E_\infty \simeq \Lambda(y_1, \dots, y_m) \otimes_k (H^*(\mathcal{X}, k)/J_k),$$

where $J_k \subset H(\mathcal{X}, k)$ is the ideal as in Lemma 11.5.4, $\Lambda(y_1, \dots, y_m)$ is the exterior algebra generated by $\{y_1, \dots, y_m\}$ and $H^*(\mathcal{X}, k)$ has bidegree $(*, 0)$.

Note that $\{d_1, \dots, d_m\}$ does, in general, depend on char. k, but of course not on k itself.

Proof. Use $E_3 = E_\infty$ (by Theorem 11.5.1) together with the proof of Theorem 11.5.1 as in 11.5.7, specifically (11.5.7.3) and (11.5.7.5). Observe that the standard augmentation ideal R^+ of R acts trivially on $\mathrm{Tor}_q^S(k, R)$. □

11.5.9 Definition. Let $V = \oplus_{i \geq 0} V_i$ be a $\mathbb{Z}_+$-graded vector space. Then V is said to be of *finite type* if $\dim V_i < \infty$ for each i. For such a V, the *Poincaré series* is, by definition, the formal series

$$P(V) = \sum_{i \geq 0} (\dim V_i)\, t^i .$$

For any field k, $H^*(\mathcal{G}^{\min}, k)$ is of finite type. (Use the Leray–Serre spectral sequence for the fibration $f : \mathcal{G}^{\min} \to \mathcal{G}^{\min}/T$.) Since $\mathcal{G}^{\min}$ is a topological group, $H^*(\mathcal{G}^{\min}, k)$ is a Hopf algebra over k.

11.5.10 Proposition. *The image $A_k := f^*(H^*(\mathcal{G}^{\min}/T, k)) \subset H^*(\mathcal{G}^{\min}, k)$ is stable under the coproduct on $H^*(\mathcal{G}^{\min}, k)$.*

Proof. Consider the commutative diagram

$$\text{(D)} \qquad \begin{array}{ccc} \mathcal{G}^{\min} \times \mathcal{G}^{\min} & \xrightarrow{\mu} & \mathcal{G}^{\min} \\ \downarrow{\scriptstyle \bar{f}} & & \downarrow{\scriptstyle f} \\ \mathcal{G}^{\min} \times_T (\mathcal{G}^{\min}/T) & \xrightarrow[\bar{\mu}]{} & \mathcal{G}^{\min}/T , \end{array}$$

where μ and $\bar{\mu}$ are the multiplication maps and $\bar{f}$ is induced from $\mathrm{I} \times f$. Also the principal T-bundle f induces a bundle map $\bar{\theta}$:

$$\begin{array}{ccc} \mathcal{G}^{\min} & \xrightarrow{\bar{\theta}} & E(T) \\ \downarrow & & \downarrow \\ \mathcal{G}^{\min}/T & \xrightarrow[\theta]{} & B(T), \end{array}$$

and hence we obtain the continuous map

$$\hat{\theta} : \mathcal{G}^{\min} \times_T \left(\mathcal{G}^{\min}/T\right) \longrightarrow E(T) \times_T \left(\mathcal{G}^{\min}/T\right)$$

(induced from the T-equivariant map $\bar{\theta}$). For any $w \in W$, recall the Schubert cohomology class $\hat{\varepsilon}_w \in H_T^{2\ell(w)}(\mathcal{X}, \mathbb{Z}) = H_T^{2\ell(w)}(\mathcal{G}^{\min}/T, \mathbb{Z})$ from Proposition 11.3.10. We denote by the same symbol the corresponding class in $H_T^{2\ell(w)}(\mathcal{X}, k)$. Then the pullback $\hat{\theta}^*(\hat{\varepsilon}_w)$ satisfies

(1) $$(\hat{\theta}^*\hat{\varepsilon}_w)_{|_{1\times\mathcal{G}^{\min}/T}} = \varepsilon_w, \quad \text{by using (11.3.10.2), and}$$

(2) $$\bar{f}^*\,\hat{\theta}^*(\hat{\varepsilon}_w) = 1 \boxtimes f^*(\varepsilon_w).$$

To prove (2), use the commutative diagram:

$$\begin{array}{ccc} \mathcal{G}^{\min} \times \mathcal{G}^{\min} & \xrightarrow{\bar{\theta}\times f} & E(T) \times (\mathcal{G}^{\min}/T) \\ \big\downarrow{\scriptstyle \bar{f}} & & \big\downarrow \\ \mathcal{G}^{\min} \times_T (\mathcal{G}^{\min}/T) & \xrightarrow{\hat{\theta}} & E(T) \times_T (\mathcal{G}^{\min}/T). \end{array}$$

By the Leray–Hirsch Theorem C.5, and (1), any cohomology class $c \in H^*(\mathcal{G}^{\min} \times_T (\mathcal{G}^{\min}/T), k)$ can be (uniquely) written as

(3) $$c = \sum_{w\in W} \pi^*(c_w) \cup \hat{\theta}^*(\hat{\varepsilon}_w), \qquad \text{for some } c_w \in H^*(\mathcal{G}^{\min}/T, k),$$

where $\pi : \mathcal{G}^{\min} \times_T (\mathcal{G}^{\min}/T) \to \mathcal{G}^{\min}/T$ is the projection on the first factor.

For any $x \in H^*(\mathcal{G}^{\min}/T, k)$,

$$\begin{aligned} \mu^* f^*(x) &= \bar{f}^*\bar{\mu}^*(x), && \text{from the diagram (D)} \\ &= \bar{f}^*\Big(\sum_{w\in W} \pi^*\big((\bar{\mu}^*x)_w\big) \cup \hat{\theta}^*(\hat{\varepsilon}_w)\Big), && \text{by (3)} \\ &= \sum_{w\in W} (f^*((\bar{\mu}^*x)_w)) \boxtimes f^*(\varepsilon_w), && \text{by (2).} \end{aligned}$$

This proves the proposition. □

11.5.11 Remark. There is a stronger result due to D. Peterson (cf. [Kitchloo–98, Theorem 1.26]) asserting that, for any $w \in W$,

$$\mu^* f^*(\varepsilon_w) = \sum_{uv=w,\ \ell(u)+\ell(v)=\ell(w)} f^*(\varepsilon_u) \boxtimes f^*(\varepsilon_v)\,.$$

11.5.12 Theorem. *Let $\mathcal{G}^{\min}$ be any Kac–Moody group (with the strong topology), and let k be a field of an arbitrary char. p. Then, as graded algebras,*

(1) $$H^*(\mathcal{G}^{\min}, k) \simeq \Lambda(\bar{y}_1, \cdots, \bar{y}_m) \otimes_k A_k, \text{ provided } p \neq 2,$$

where $A_k := \operatorname{Im} f^*$ is as in Proposition 11.5.10, and $\bar{y}_i \in H^{2d_i-1}(\mathcal{G}^{\min}, k)$ ($\{d_1, \ldots, d_m\}$ being as in Corollary 11.5.8).

Moreover,

(2) $$A_k \simeq H^*(\mathcal{X}, k)/J_k, \text{ as graded algebras},$$

where J_k is as in Lemma 11.5.4. Further, A_k is a graded Hopf subalgebra of $H^*(\mathcal{G}^{\min}, k)$ concentrated only in the even degrees. In particular, if k is a perfect field, there exist even integers (possibly finitely many, including none) $2 \leq \ell_1 \leq \ell_2 \leq \cdots$ and positive integers (including ∞) $t_1, t_2, \ldots$, such that, as a graded algebra,

(3) $$A_k \simeq k[x_1, x_2, \cdots]/\langle x_1^{p^{t_1}}, x_2^{p^{t_2}}, \cdots \rangle,$$

where $k[x_1, x_2, \cdots]$ is the polynomial algebra over k in the variables $x_1, x_2, \ldots$ with $\deg x_i = \ell_i$ and x_j, t_j and ℓ_j are all indexed by the same set. (If $t_j = \infty$ or $p = 0$, $x_j^{p^{t_j}}$ is interpreted as 0.)

(The integers ℓ_j, t_j do depend on p in general, but not of course on the perfect field k itself. Further, we impose the restriction $p \neq 2$ only for (1).)

Proof. For the Leray–Serre spectral sequence for any orientable fibration $\pi : E \to B$, we have

$$H^*(B, k) = E_2^{*,0} \twoheadrightarrow E_\infty^{*,0} \hookrightarrow H^*(E, k),$$

and the composite map is nothing but π^*. Applying this to the fibration $f : \mathcal{G}^{\min} \to \mathcal{G}^{\min}/T$ and using Corollary 11.5.8, we get (2).

By Proposition 11.5.10, A_k is a Hopf subalgebra of $H^*(\mathcal{G}^{\min}, k)$ and, of course, it is evenly graded. Now, by a theorem of Hopf–Borel [Borel–67, Theorems 3.1 and 3.2], (3) follows.

We now prove (1): With the notation as in Corollary 11.5.8, lift the class $y_i \in E_\infty^{2(d_i-1),1}$ to a class $\bar{y}_i \in H^{2d_i-1}(\mathcal{G}^{\min}, k)$ and define the graded algebra homomorphism

$$\psi : \Lambda(\bar{y}_1, \ldots, \bar{y}_m) \otimes_k A_k \to H^*(\mathcal{G}^{\min}, k)$$

by $\bar{y}_i \mapsto \bar{y}_i$ and $x \mapsto x$ for $x \in A_k$. (Since $p \neq 2$, the odd degree classes of $H^*(\mathcal{G}^{\min}, k)$ anticommute, thus giving rise to the algebra homomorphism ψ.) The surjectivity of ψ follows from the corresponding result at the associated 'gr' level (cf. Corollary 11.5.8) and (2). In view of Corollary 11.5.8, the injectivity of ψ follows from the dimensional consideration applied to each homogeneous component. □

11.5.13 Corollary. *Let the notation and assumptions be as in Theorem 11.5.12 (with no restriction on p). To highlight the dependence on p, denote d_i by $d_i(p)$, ℓ_j by $\ell_j(p)$ and t_j by $t_j(p)$. Then*

(1) $$P(A_k) = \prod_j \left(\frac{1 - t^{\ell_j(p)p^{t_j(p)}}}{1 - t^{\ell_j(p)}} \right),$$

where $t^{\ell_j(p)p^{t_j(p)}}$ is interpreted as 0 if $t_j(p) = \infty$ or $p = 0$. Further, we have the following identities (2)–(5):

(2) $$P\big(H^*(\mathcal{G}^{\min}, k)\big) = P(A_k) \cdot \prod_{i=1}^{m} \Big(1 + t^{2d_i(p)-1}\Big).$$

(3) $$P(H^*(\mathcal{X}, \mathbb{Q})) = P(H^*(\mathcal{X}, k)) = P(R_k) \cdot P(A_k),$$

where R_k is the image of the Borel homomorphism β^k as in Lemma 11.5.4.

(4) $$P(H^*(\mathcal{X}, k)) = \sum_{w \in W} t^{2\ell(w)}.$$

(5) $$P(R_k) = \frac{\prod_{i=1}^{m} \big(1 - t^{2d_i(p)}\big)}{(1 - t^2)^{\dim \mathfrak{h}}}.$$

In particular, in the case when $\mathcal{G}^{\min} = \mathcal{G}$ is of finite type, the Borel homomorphism $\beta^{\mathbb{Q}}$ is surjective. This gives rise to the well-known identity in this case:

(6) $$\sum_{w \in W} t^{2\ell(w)} = \prod_{i=1}^{\ell} \left(\frac{1 - t^{2d_i(0)}}{1 - t^2} \right),$$

where $\ell = $ rank $\mathcal{G}$.

Proof. (1) follows from (11.5.12.3) and (2) follows from (11.5.12.2) and Corollary 11.5.8. Since $H^*(\mathcal{X}, \mathbb{Z})$ is torsion free (cf. Proposition 11.3.2), the first identity in (3) follows, whereas the second identity of (3) follows from (11.5.12.2) and Lemma 11.5.4. The description of the cohomology of $\mathcal{X}$ (cf. Proposition 11.3.2) clearly gives (4). By the Koszul resolution (Lemma 11.5.6) and Lemmas 11.5.3 and 11.5.5, (5) follows.

Since a finite-dimensional evenly graded Hopf algebra over a field k of char. 0 is one-dimensional, A_k in the finite case reduces to constants. Hence the surjectivity of $\beta^{\mathbb{Q}}$ (in the finite case) follows from (3). Combining (4), (5) and (11.5.5.2) for $k = \mathbb{Q}$, and using the surjectivity of $\beta^{\mathbb{Q}}$, we get (6). This finishes the proof of the corollary. □

11.5.14 Definition. Let $\beta_T^{\mathbb{Q}} : S^*(\mathfrak{h}_{\mathbb{Q}}^*) \to H_T^{2*}(\mathcal{X}, \mathbb{Q})$ be the Borel homomorphism as in 11.3.5. Define the S-linear map

$$\hat{\beta}_T^{\mathbb{Q}} : S \otimes_{S^W} S \to H_T(\mathcal{X}, \mathbb{Q})$$

by $\hat{\beta}_T^{\mathbb{Q}}(P \otimes Q) = (-1)^d P\, \beta_T^{\mathbb{Q}}(Q)$, for $P, Q \in S := S^*(\mathfrak{h}_{\mathbb{Q}}^*)$ and Q homogeneous of degree d, where both the copies of S are S^W-modules under the multiplication and $S \otimes_{S^W} S$ is thought of as an S-module via the multiplication on the first factor. (The map $\hat{\beta}_T^{\mathbb{Q}}$ is well defined, i.e., $\hat{\beta}_T^{\mathbb{Q}}(PR \otimes Q) = \hat{\beta}_T^{\mathbb{Q}}(P \otimes RQ)$ for $R \in S^W$, as can be easily seen from (11.3.9.2) and the definition of $\bar{\beta}_T$.)

The standard Weyl group action on the second copy S (which is clearly S^W-linear) gives rise to a Weyl group action on $S \otimes_{S^W} S$ by extending the scalars. Similarly, the BGG–Demazure operators A_{s_i}, for any simple reflection s_i, defined on S (cf. Exercise 11.3.E.2) give rise to the corresponding operators on $S \otimes_{S^W} S$, denoted by the same symbol.

As an easy consequence of the surjectivity of $\beta^{\mathbb{Q}} : S(\mathfrak{h}_{\mathbb{Q}}^*) \to H^{2*}(\mathcal{X}, \mathbb{Q})$ in the finite case (cf. Corollary 11.5.13), one obtains the following.

11.5.15 Theorem. *Let $\mathcal{G}$ be a Kac–Moody group of finite type. Then the above map*

$$\hat{\beta}_T^{\mathbb{Q}} : S \otimes_{S^W} S \to H_T(\mathcal{X}, \mathbb{Q})$$

is an isomorphism of graded S-algebras, where $S^p \otimes S^q$ is assigned the degree $2(p+q)$. Moreover, for any $w \in W$, $\gamma \in S \otimes_{S^W} S$ and simple reflection s_i,

(1) $$\hat{\beta}_T^{\mathbb{Q}}(w \cdot \gamma) = w \cdot \hat{\beta}_T^{\mathbb{Q}}(\gamma), \qquad \textit{and}$$

(2) $$\hat{\beta}_T^{\mathbb{Q}}(A_{s_i}\gamma) = -\hat{D}_{s_i}\big(\hat{\beta}_T^{\mathbb{Q}}(\gamma)\big).$$

Proof. Let C be the image of $\hat{\beta}_T^{\mathbb{Q}}$. Then C is a graded S-submodule of $H_T(\mathcal{X}, \mathbb{Q})$. By the surjectivity of $\beta^{\mathbb{Q}}$ and Proposition 11.3.7, we get the following:

(3) $$C + (S^+ \cdot H_T(\mathcal{X}, \mathbb{Q})) = H_T(\mathcal{X}, \mathbb{Q}).$$

Now the surjectivity of $\hat{\beta}_T^{\mathbb{Q}}$ follows from the graded version of Nakayama's Lemma (which is trivial to prove).

By a result of Chevalley, S is a free S^W-module of rank $|W|$ and hence $S \otimes_{S^W} S$ is a free S-module of rank $|W|$. Further, $H_T(\mathcal{X}, \mathbb{Q})$ is a free S-module of rank $|W|$ (by Proposition 11.3.7). Hence $\hat{\beta}_T^{\mathbb{Q}}$ is an isomorphism by the following lemma.

The assertions (1) and (2) follow immediately from the corresponding facts for $\beta_T^{\mathbb{Q}}$ (cf. Exercise 11.3.E.2). □

The following lemma is well known.

11.5.16 Lemma. *Let R be a commutative ring with identity. Then a surjective R-module map between two free R-modules of the same finite rank is an isomorphism.* □

11.5.17 Remark. Let $\{\hat{\varepsilon}_w\}_{w \in W}$ be the Schubert basis of $H_T^*(\mathcal{X})$ (cf. Theorem 11.3.9). Then, in the finite case, Graham has given an expression for $(\hat{\beta}_T^{\mathbb{Q}})^{-1}(\hat{\varepsilon}_{w_o})$, where w_o is the longest element of W (cf. [Graham–97, Proposition 4.2]). In particular, applying the BGG–Demazure operators, we get an expression for $(\hat{\beta}_T^{\mathbb{Q}})^{-1}(\hat{\varepsilon}_w)$, for any $w \in W$.

The (a) part of the following conjecture is due to [Kac–85b] and (b) is due to Kumar (cf. [Lyons–96]).

11.5.18 Conjectures. *(a) The Leray–Serre spectral sequence corresponding to the fibration $f : \mathcal{G}^{\min} \to \mathcal{G}^{\min}/T$, for the singular cohomology with integral coeffients, collapses at the E_3-term, i.e.,*

$$E_3^{p,q}(\mathbb{Z}) = E_\infty^{p,q}(\mathbb{Z}). \tag{1}$$

(b) For any word $\mathfrak{w}$ (not necessarily reduced), let $Z_{\mathfrak{w}}$ be the Bott–Samelson–Demazure–Hansen variety (cf. 7.1.3). Consider the morphism $m_{\mathfrak{w}} : Z_{\mathfrak{w}} \to \mathcal{X}$ (cf. 7.1.11 and Proposition 7.1.12); and choose a continuous section σ of the fiber bundle $\mathcal{G}^{\min}/T \to \mathcal{X}$. This exists since its fiber is contractible (cf. 11.3.4). Thus, we get the continuous map $\sigma \circ m_{\mathfrak{w}} : Z_{\mathfrak{w}} \to \mathcal{G}^{\min}/T$. In particular, we can pullback the principal T-bundle $f : \mathcal{G}^{\min} \to \mathcal{G}^{\min}/T$ to get a principal T-bundle $\hat{f}_{\mathfrak{w}}$ over $Z_{\mathfrak{w}}$. Let $\hat{E}_r(\mathbb{Z})$ be the corresponding Leray–Serre cohomology spectral sequence with coefficients in $\mathbb{Z}$. Then it is conjectured that

$$\hat{E}_3(\mathbb{Z}) = \hat{E}_\infty(\mathbb{Z}). \tag{2}$$

The canonical map $E_3(\mathbb{Z}) \to \hat{E}_3(\mathbb{Z})$ is injective (cf. [Lyons–96]). In particular, the validity of (2) would give the validity of (1).

11.5.E EXERCISES

(1) For the torus $T := \mathrm{Hom}_{\mathbb{Z}}(\mathfrak{h}_{\mathbb{Z}}^*, \mathbb{C}^*)$, prove that (as graded $\mathbb{Z}$-algebras)

$$H^*(T, \mathbb{Z}) \simeq \Lambda^*(\mathfrak{h}_{\mathbb{Z}}^*),$$

where Λ^* is the exterior algebra over $\mathbb{Z}$.

In particular, for any field k,

$$H^*(T, k) \simeq \Lambda^*(\mathfrak{h}_k^*).$$

Hint. Use the covering projection $\mathrm{Hom}_{\mathbb{Z}}(\mathfrak{h}^*_{\mathbb{Z}}, \mathbb{C}) \to T$ with fiber $\mathfrak{h}_{\mathbb{Z}}$ induced from the exponential map $\mathbb{C} \to \mathbb{C}^*$, $z \mapsto e^{2\pi i z}$.

(2) With the notation as in 11.5.7, prove that

$$d_2 : E_2^{0,1} = \mathfrak{h}_k^* \to E_2^{2,0} = H^2(\mathcal{X}, k)$$

is the same as the negative of the Borel homomorphism $\beta^k_{|_{\mathfrak{h}_k^*}}$.

(3) With notation as is 11.5.7, prove that $\mathrm{Tor}_1^S(k, R)$ has a k-basis $\{y_1, \dots, y_m\}$ such that, under the identification (11.5.7.3), $y_i \otimes 1 \in E_3^{2(d_i-1),1}$.

(4) Let $\mathcal{G}$ be the (affine) Kac–Moody group associated to the affine Kac–Moody algebra $\hat{\mathcal{L}}(\mathring{\mathfrak{g}})$ corresponding to a finite-dimensional simple Lie algebra $\mathring{\mathfrak{g}}$ of rank ℓ (cf. 13.1.1). Then prove the following:

(a) The cardinality m of any homogeneous regular sequence generating the ideal of generalized invariants $I_{\mathbb{Q}}$ is equal to $\ell + 1 = \dim \mathfrak{h} - 1$.

(b) $H^*(\mathcal{G}^{\min}, \mathbb{Q}) \simeq \Lambda(c, y_1, \dots, y_\ell) \otimes \mathbb{Q}[x_2, \dots, x_\ell]$, as graded algebras, where $c \in H^1(\mathcal{G}^{\min}, \mathbb{Q})$, $y_i \in H^{2e_i-1}(\mathcal{G}^{\min}, \mathbb{Q})$, $x_i \in H^{2e_i-2}(\mathcal{G}^{\min}, \mathbb{Q})$ ($\{2 = e_1 < e_2 \le \cdots \le e_\ell\}$ being the exponents of $\mathring{\mathfrak{g}}$).

(c) $A_{\mathbb{Q}} \simeq \mathbb{Q}[x_2, \dots, x_\ell]$, where A_k is as in Proposition 11.5.10.

(5) For a field k of char. 0 and an arbitrary generalized flag variety $\mathcal{X}$, prove that

$$H^*(\mathcal{X}, k) \simeq R_k \otimes_k \left(H(\mathcal{X}, k)/J_k\right),$$

as graded algebras.

(Compare this with Lemma 11.5.4.)

(6) Let $\mathcal{G} = \mathcal{G}(A)$ be the Kac–Moody group associated to an indecomposable GCM A which is neither of finite nor of affine type. Then show that $I_{\mathbb{Q}} = 0$ if A is nonsymmetrizable and $I_{\mathbb{Q}}$ is generated by a single degree 2 polynomial if A is symmetrizable.

(7)[1] Prove the surjectivity of the Borel homomorphism $\beta^{\mathbb{Q}} : S(\mathfrak{h}^*_{\mathbb{Q}}) \to H^{2*}(\mathcal{X}, \mathbb{Q})$ in the finite case by considering the Leray–Serre spectral sequence of the fibration

$$\mathcal{G}/T \to B(T) \to B(\mathcal{G}).$$

Hint. Prove that $H^*(B(\mathcal{G}), \mathbb{Q})$ is concentrated in even degrees by considering the fibration

$$\mathcal{G}/N \to B(N) \to B(\mathcal{G}),$$

[1] I learned this proof from H. Pittie. The surjectivity of $\beta^{\mathbb{Q}}$ is originally due to Borel.

where N is the normalizer of T in $\mathcal{G}$.

(8)[2] Show that $H^*(\mathcal{X}, \mathbb{C})$ in the finite case is the regular representation of the associated Weyl group W.
Hint. Use the Lefschetz formula to show that, for any $w \neq 1$, the action of w on the total cohomology $H^*(\mathcal{X}, \mathbb{C})$ has trace 0. This gives that W-character of $H^*(\mathcal{X}, \mathbb{C})$ is the same as that of the regular representation of W and hence they are isomorphic W-modules.

(9) Let $\mathfrak{g} = \mathfrak{g}(A)$ be of finite type. Fix a regular $h \in \mathfrak{h}$, i.e., W-isotropy of h is trivial. Show that, for any $w \in W$, there exists a (in general nonhomogeneous) function $f_w \in S(\mathfrak{h}^*)$ satisfying the following three conditions:
(a) $f_w(vh) = 0$, for $\ell(v) \leq \ell(w)$, $v \neq w$,
(b) $f_w(wh) = 1$, and
(c) $\deg f_w \leq \ell(w)$.
Moreover, if f'_w is another such function, then $f'_w - f_w \in I(W \cdot h) := \{f \in S(\mathfrak{h}^*) : f_{|W \cdot h} \equiv 0\}$ (the ideal of $W \cdot h$). In fact, $\deg f_w = \ell(w)$.

Hint. By a result of Chevalley, $S(\mathfrak{h}^*)$ is free over the invariants $S(\mathfrak{h}^*)^W$. Take a basis $\{P_w\}_{w \in W}$ of $S(\mathfrak{h}^*)$ over $S(\mathfrak{h}^*)^W$ consisting of homogeneous polynomials. Then each P_w is automatically of degree $\leq \ell(w_o)$, w_o being the longest element of W. Prove the existence of f_w by the downward induction on $\ell(w)$. For $w = w_o$, take any $g \in S(\mathfrak{h}^*)$ such that $g(wh) = 0$ for all $w \neq w_o$ and $g(w_o h) = 1$. Write $g = \sum_{w \in W} g_w P_w$ with $g_w \in S(\mathfrak{h}^*)^W$. Set $f_{w_o} = \sum_{w \in W} g_w(h) P_w$. Now take $w < w_o$ and choose a simple reflection s_i such that $s_i w > w$. Set $f_w = -\alpha_i(wh) A_{s_i}(f_{s_i w})$, where A_{s_i} is the BGG–Demazure operator (cf. Exercise 11.3.E.2).
To prove the uniqueness, observe first that $I(W \cdot h)$ is stable under the action of A_{s_i}'s. Write $f'_w - f_w = g + \sum_{\ell(v) > \ell(w)} \lambda_v f_v$, for $g \in I(W \cdot h)$ and $\lambda_v \in \mathbb{C}$. Choose v_o of maximum length such that $\lambda_{v_o} \neq 0$. Applying $A_{v_o^{-1}}$ to the above, get a contradiction.

(10) Let the notation and assumptions be as in the above exercise and let $\beta = \beta^{\mathbb{C}} : S(\mathfrak{h}^*) \to H^*(\mathcal{X}, \mathbb{C})$ be the Borel homomorphism (cf. 11.3.5). Then show that, for any $w \in W$,

$$\beta(f_w^o) = \Big(\prod_{\alpha \in \Delta^- \cap w^{-1}\Delta^+} \alpha(h)^{-1} \Big) \varepsilon_{w^{-1}} ,$$

where f_w^o is the homogeneous component of f_w of top degree $\ell(w)$ and $\{\varepsilon_w\}_w$ is the Schubert basis of $H^*(\mathcal{X}, \mathbb{C})$ (cf. 11.3.2).

Hint. Use the BGG–Demazure operators and (11.3.17.4).

---

[2] I learned this proof from H. Pittie. Of course, this result is well known.

(11) Let $\mathfrak{g} = \mathfrak{g}(A)$ be of finite type and let $\mathcal{N} \subset \mathfrak{g}$ be the nilpotent cone (consisting of all the ad nilpotent elements). Then, prove that the affine coordinate ring $\mathbb{C}[\mathcal{N} \cap \mathfrak{h}]$ of the scheme theoretic intersection of $\mathcal{N}$ with $\mathfrak{h}$ is isomorphic, as a graded algebra, with the singular cohomology $H^*(\mathcal{X}, \mathbb{C})$. (The grading on $\mathbb{C}[\mathcal{N} \cap \mathfrak{h}]$ comes from the grading on $S(\mathfrak{h}^*)$, where we assign degree 2 to the elements of $\mathfrak{h}^*$.)

Hint. Under the Borel homomorphism $\beta : S(\mathfrak{h}^*) \to H^*(\mathcal{X}, \mathbb{C})$, $S(\mathfrak{h}^*)/\langle S^+(\mathfrak{h}^*)^W \rangle \simeq H^*(\mathcal{X}, \mathbb{C})$, where $\langle S^+(\mathfrak{h}^*)^W \rangle$ denotes the ideal generated by the positive degree W-invariants in $S(\mathfrak{h}^*)$. By the definition,

$$\mathbb{C}[\mathcal{N} \cap \mathfrak{h}] \simeq S(\mathfrak{g}^*)/(I(\mathfrak{h}) + I(\mathcal{N})) \simeq S(\mathfrak{h}^*)/I(\mathcal{N})_{|\mathfrak{h}},$$

where $I(\mathfrak{h})$, resp. $I(\mathcal{N})$, is the ideal of $\mathfrak{h}$, resp. $\mathcal{N}$, in $S(\mathfrak{g}^*)$ and $I(\mathcal{N})_{|\mathfrak{h}} := \{f \in S(\mathfrak{h}^*) : f = g_{|\mathfrak{h}} \text{ for some } g \in I(\mathcal{N})\}$. Now use a theorem of Kostant [Kostant–63a] asserting that $I(\mathcal{N}) = \langle S^+(\mathfrak{g}^*)^{\mathfrak{g}} \rangle$ and a theorem of Chevalley asserting that $S(\mathfrak{g}^*)^{\mathfrak{g}} \simeq S(\mathfrak{h}^*)^W$.)

(12) Let $\mathfrak{g}$ be an arbitrary Kac–Moody algebra. Define the S-algebra homomorphism $\tilde{\beta}_T : S \otimes_{S^W} S \to \Lambda$ by $\tilde{\beta}_T := \nu \circ \hat{\beta}_T$, where $S := S(\mathfrak{h}^*)$, $S^W := S(\mathfrak{h}^*)^W$, ν is the isomorphism of Theorem 11.3.9 and $\hat{\beta}_T : S \otimes_{S^W} S \to H_T(\mathcal{X}, \mathbb{C})$ is as in 11.5.14 (over the field $\mathbb{C}$). Then show that

$$(\tilde{\beta}_T(P \otimes Q))(w) = P(wQ), \quad \text{for } P, Q \in S, \text{ and } w \in W.$$

Now assume that $\mathfrak{g}$ is of finite type. In this case, by Theorems 11.3.9 and 11.5.15, $\tilde{\beta}_T$ is an isomorphism. Let $\gamma_o \in S \otimes_{S^W} S$ be the (unique) element such that $\tilde{\beta}_T(\gamma_o) = \xi^{w_o}$, where w_o is the longest element of the Weyl group W and ξ^{w_o} is as in (11.1.5.1). View γ_o as a regular function on Spec $(S \otimes_{S^W} S) \subset \mathfrak{h} \times \mathfrak{h}$. It is easy to see that Spec $(S \otimes_{S^W} S)$ is the closed (reduced) subvariety $\cup_{w \in W} \{(h, wh) : h \in \mathfrak{h}\}$ of $\mathfrak{h} \times \mathfrak{h}$. Then, show that γ_o is characterized by the following two properties:

(a) $\gamma_o(h, wh) = 0$ for any $h \in \mathfrak{h}$ and any $w \neq w_o$.
(b) $\gamma_o(h, w_o h) = \prod_{\alpha \in \Delta^+} \alpha(h)$, for any $h \in \mathfrak{h}$.

Applying the BGG–Demazure operators A_{s_i}'s (cf. 11.5.14) to γ_o we get the preimage of all the basis elements ξ^w.

(13) Use the above to show that, when $\mathfrak{g} = sl_n$,

$$\gamma_o = \prod_{i+j \le n} (X_i - Y_j),$$

where X_i, resp. Y_j, is the i-th, resp. j-th, coordinate function on the first copy, resp. second copy, of $\mathfrak{h}$. (Of course, we have the relation $\sum_i X_i = \sum_j Y_j = 0$.)

11.C Comments. Proposition 11.1.11 and its Corollary 11.1.12 in the finite case are due to [Billey–99] which was extended by Kumar to the general Kac–Moody case (cf. [Billey–99, Appendix]). (See also [Andersen–Jantzen–Soergel–94, Remark D.3.].) Proposition 11.1.19 and its Corollary 11.1.20 are due to [Dyer–93]. As mentioned earlier, Corollary 11.1.20 in the case $u = v$ was conjectured by [Deodhar–85]. It was also proved geometrically by Carrell–Peterson and Polo (cf. 12.C). I learned of Proposition 11.1.15 from D. Peterson. All the other results in Sections 11.1 and 11.2 (including Theorem 11.1.2, Lemma 11.1.5, Theorem 11.1.7, Lemma 11.1.8, Theorem 11.1.13, Lemma 11.1.17, and Theorem 11.2.1) are due to [Kostant–Kumar–86b]. Lemma 11.1.22 appears in [Kostant–Kumar–90].

Proposition 11.3.2 is due to [Gutkin–Slodowy–83]. The operators D_{s_i} acting on $H^*(\mathcal{X})$, as in 11.3.6, were defined by Kac–Peterson and Corollary 11.3.12, (11.3.13.2), and (11.3.17.4) were obtained by them (cf. [Kac–85b]). Theorem 11.3.9 for the T-equivariant cohomology is due to [Arabia–89] who generalized the corresponding nonequivariant result of [Kostant–Kumar–86b]. The proof of Theorem 11.3.9 given here is adapted from the corresponding result in T-equivariant K-theory as in [Kostant–Kumar–90]. The operators A_{s_i} were introduced by [Bernstein–Gelfand–Gelfand–73] and [Demazure–74].

We also refer to the paper [Akyildiz–Carrell–Lieberman–86] for a different approach to study the cohomology of $\mathcal{X}^Y$ in the finite case using the holomorphic vector fields with isolated zeroes. (See [Carrell–91] for a survey of this approach.)

Schubert Calculus, i.e., study of the cohomology algebra of the flag varieties in the finite case, has of course been very extensively studied, particularly for the classical groups. We refer to a forthcoming survey article by Sottile on this subject.

[Kumar–Nori–98] proved the nonequivariant analogue of Theorem 11.4.11, i.e., when B is unipotent so that T is trivial, as well as the nonequivariant analogue of its Corollary 11.4.12. The nonequivariant analogue of Theorem 11.4.2 was obtained independently by [Moser–Jauslin–92, Lemma 6.1], [Brion–93, Theorem 1.3] (Brion assumed X to be normal and projective), and [Kumar–Nori–98, Proposition 5]. See also [Fulton–MacPherson–Sottile–Sturmfels–95, Theorem 1] and [Hirschowitz–84, Theorem 1] for a related result. The equivariant extension of all these resuls is due to [Graham–2001]. Graham's proof of these results is adapted from that of [Kumar–Nori–98]. However, the proofs given here are slightly different from his. We give a new characterization of the cohomology basis $\{\varepsilon_i^E(Y)\}$ in Proposition 11.4.7 and subsequently make crucial use of it. Lemma 11.4.10 appears in [Graham–2001] and is attributed there to Brion, but the proof given here is different. Corollary 11.4.12 was conjectured (unpublished) by Dale Peterson (and proved, as mentioned above, by [Graham–2001]). The nonequivariant part of Corollary 11.4.12, in the case when $\mathcal{G}$ is a finite-dimensional algebraic group, is well known and follows easily from the fact that the algebraic group $\mathcal{G}$ acts transitively on the flag variety $\mathcal{X}$. However, even in this case, no combinatorial proofs are known (to my knowledge).

Theorem 11.5.1, in the finite case, for $k = \mathbb{Q}$ is due to [Leray–50]. In general, Theorem 11.5.1 (including Lemmas 11.5.3–11.5.5, Corollary 11.5.8 and Theorem 11.5.12) is due to Kac–Peterson [Kac–85a]. The proof given here is an elaboration of the sketch of the proof as in [Kac–85a]. Proposition 11.5.10 is due to D. Peterson. In fact, he proved a sharper result (cf. Remark 11.5.11), which appeared in [Kitchloo–98]. The proof (of Proposition 11.5.10) given here is much shorter (though of a weaker result). Theorem 11.5.15 is well known. Exercise 11.5.E.5 is taken from [Kumar–85, Lemma 3.5]. Exercise 11.5.E.6 is taken from [Kac–85b, pages 203–204]. Exercises 11.5.E.9 and 11.5.E.10 are unpublished results of Kostant (cf. [Bernstein–Gelfand–Gelfand–73, Theorem 5.9]), and Exercise 11.5.E.11 is due to [Kostant–63a]. Exercises 11.5.E.12 and 11.5.E.13 are due to [Graham–97] although his approach is somewhat different. The results of Section 11.5 are used to unify several important (though scattered) results on $H^*(K, \mathbb{F}_p)$ (for any compact Lie group K) in [Kac–85a], where $\mathbb{F}_p$ is the finite field with p elements. (However, the proofs of some of the results in this paper are not clear to me.) We refer to the two survey articles [Samelson–52] and [Borel–55] on the topology of Lie groups.

Bott pioneered the study of the topology of loop spaces of compact symmetric spaces via Morse Theory [Bott–56,58,59] (see also [Bott–Samelson–58]), thus obtaining the celebrated (Bott) Periodicity for the classical groups. [Garland–Raghunathan–75] and Quillen (unpublished) proved that for a compact simply-connected Lie group K, the based (continuous) loop group $\Omega_e(K)$ is homotopically equivalent to the 'algebraic' loop group $\Omega_e^{\mathrm{alg}}(K)$. They used this homotopy equivalence and the identification $\Omega_e^{\mathrm{alg}}(K) \simeq \mathcal{G}/\mathcal{P}_{\{1,\dots,\ell\}}$ (together with the combinatorics of the affine Weyl group) to give another proof of the Bott Periodicity for the unitary group. Quillen's proof, in fact, works for any compact symmetric space and appeared in [Mitchell–88], also see [Pressley–Segal–86, §8.6]). There are subsequent generalizations of this approximation result due to [Hammack–96], [Teleman–98], and [Raghunathan–2000].

An explicit formula for the Poincaré polynomial, over any field, of an affine Kac–Moody group is given in [Kac–85b]. [Kitchloo–98] has determined the cohomology $H^*(\mathcal{G}^{\min}, \mathbb{F}_p)$ explicitly for any Kac–Moody group $\mathcal{G}^{\min}$ of rank 2 (in the symmetric rank 2 case the result was stated in [Kac–85b]). There is an inductive formula for the total Steenrod p-power operation on $H^*(\mathcal{X}, \mathbb{F}_p)$ in general in [Kac–85b]. (Also see [Kitchloo–98] in the rank 2 case.)

It is known that $\mathcal{X}^Y$ (for any subset Y) is a formal space in the sense of rational homotopy theory (cf. [Kac–85b] and [Kumar–85]). In fact, as observed by Deligne (cf. [Kac–85b]), any affinely paved projective variety satisfying the condition (∗) of Proposition 11.4.7 is a formal space. Using this, the minimal model of $\mathcal{X}$ is determined in [Kac–85b] and [Kumar–85].

There is another approach to study the cohomology of $\mathcal{X}^Y$ (and also that of $\mathcal{G}^{\min}$) due to [Kumar–84,85] in the symmetrizable case, via differential forms. Following [Kostant–63b] in the finite case, Kumar develops the "$d - \delta$" Hodge theory for the invariant forms on $\mathcal{X}^Y$ and shows that there exists a unique $d - \delta$-harmonic representative in each cohomology class. Moreover, again following [Kostant–63b] in the finite case, an explicit expression for this representative is given in [Kumar–84]. Further, it is shown that $H^*(\mathcal{X}^Y, \mathbb{C})$ is isomorphic as a graded algebra with the Lie algebra cohomology $H^*(\mathfrak{g}, \mathfrak{g}_Y)$ under an "integration" map, and also $H^*(\mathcal{G}^{\min}, \mathbb{C})$ is isomorphic with the Lie algebra cohomology $H^*(\mathfrak{g})$ (cf. [Kumar–85]). Both of these isomorphisms are conjectured to hold in the nonsymmetrizable case as well (cf. [Kumar–90, Conjecture 4.4]). However, in contrast to the finite case, "very few" classes in $H^*(\mathcal{G}^{\min}, \mathbb{C})$ are represented by biinvariant forms in the infinite-dimensional case (cf. [Kumar–86b]). There is a parallel approach to study the cohomology of $\mathcal{X}^Y$ in the finite case by using Poisson geometry (cf. [Evens–Lu–99] and [Lu–99]).

XII

Smoothness and Rational Smoothness of Schubert Varieties

We discuss various criteria to determine which points $\dot{v} := v\mathcal{B} \in X_w \subset \mathcal{G}/\mathcal{B}$ are smooth or rationally smooth, where $\mathcal{G}$ is any Kac–Moody group and $v \leq w \in W$.

The first general result in this direction was due to Chevalley, who proved that (in the finite case) the singular locus $\Sigma(X_w)$ of any X_w is of codimension at least two in X_w. Of course, this result follows immediately from the normality of X_w (although Chevalley's proof is different and more direct, as normality was not known at that time; a different proof of Chevalley's result is outlined in Exercise 12.1.E.1). Curiously, he went on to conjecture that any X_w is smooth.

Since $\mathcal{B}$ acts on X_w by algebraic automorphisms, and each $\mathcal{B}$-orbit has a unique T-fixed point $\dot{v}$, for some $v \leq w$, the study of the singular locus of X_w reduces to that of the points $\dot{v}$. In Section 12.1, we give a general smoothness criterion due to Kumar which asserts that, for any $v \leq w$, the point $\dot{v} \in X_w$ is smooth iff $c_{w,v} = (-1)^{\ell(w)-\ell(v)} \prod_{\beta \in S(w,v)} \beta^{-1}$, where $c_{w,v}$ are the entries of the C-matrix defined in the last chapter and $S(w, v) := \{\beta \in \Delta_{\mathrm{re}}^{+} : s_\beta v \leq w\}$.

The proof relies on the formula for the T-character of the ring of functions $\operatorname{gr} \mathcal{O}_{\dot{v}, X_w}$ on the tangent cone of X_w at $\dot{v}$ (cf. Theorem 12.1.3; the proof of which requires the Demazure character formula), determination of the set $E(X_w, \dot{v})$ of all the one-dimensional T-orbit closures in X_w passing through $\dot{v}$ (cf. Proposition 12.1.7) and the result proved in Section 8.2 that X_w is Cohen–Macaulay. Also, as an immediate consequence of the latter (i.e., Proposition 12.1.7), we obtain another proof of Deodhar's conjecture (proved earlier in the last chapter using the nil-Hecke ring).

The smoothness criterion given above is applied to codimension-one Schubert varieties of $\mathcal{G}/\mathcal{B}$ in the finite case in Exercise 12.1.E.7, thereby providing a complete list of codimension-one smooth Schubert varieties. The above smoothness criterion is also used to determine the singular locus of any X_w in the case of rank 2 groups of finite type (in Exercise 12.2.E.4). Also, an expression for the Zariski tangent space $T_{\dot{v}}(X_w)$ in terms of representation theory is given in Exercise 12.1.E.9.

Section 12.2 is devoted to the discussion of various criteria to determine the rational smoothness of points $\dot{v} \in X_w$. The first and most basic criterion is in terms of the fundamental Kazhdan–Lusztig (for short KL) polynomials. We give a quick (but self-contained) review of the definition and elementary properties of the KL polynomials $P_{v,w} \in \mathbb{Z}[q]$ $(v \leq w \in \mathcal{W})$ for any Coxeter group $\mathcal{W}$. Then, as proved by Kazhdan–Lusztig, a point $\dot{v} \in X_w$ is rationally smooth iff $P_{v,w} = 1$. In fact, they proved that if $\mathcal{W} = W$ is the Weyl group of a Kac–Moody group, then the coefficients of $P_{v,w}$, in some sense, measure the deviation from rational smoothness of $\dot{v} \in X_w$. More precisely, they proved that the coefficient of q^i in $P_{v,w}$ is the dimension of the stalk at $\dot{v}$ of the middle intersection cohomology sheaf of X_w in degree $2i$. As a corollary of the above criterion of rational smoothness, it is shown that the subvariety of points in X_w, which are not rationally smooth, is of codimension at least three in X_w.

Using the KL criterion, i.e., the description of rational smoothness in terms of the KL-polynomials given above, criteria for rational smoothness in terms of the number of T-stable curves and of the elements $c_{w,v}$ are given, respectively, in Theorems 12.2.14 and 12.2.16. Specifically, the first theorem (due to Carrell–Peterson) asserts that, for any $v \leq w$, $\dot{v} \in X_w$ is rationally smooth iff $\# E(X_w, \dot{y}) = \ell(w)$, for all $v \leq y \leq w$. The proof of this theorem relies on Deodhar's conjecture mentioned earlier and the following identity (due to Deodhar) which is valid in any Coxeter group $(\mathcal{W}, S)$:

$$\frac{d}{dq}\left(q^{\ell(w)-\ell(v)} P_{v,w}(q^{-2})\right)_{q=1} = \sum_{\substack{t\in T \\ v<tv\leq w}} P_{tv,w}(1), \quad \text{for any } v \leq w \in \mathcal{W},$$

where T is the set of reflections $\cup_{w\in\mathcal{W}} wSw^{-1}$. The second theorem (due to Kumar) asserts that, for any $v \leq w$, $\dot{v} \in X_w$ is rationally smooth $\Leftrightarrow$ for all $v \leq y \leq w$, $c_{w,y} = k_{w,y} \prod_{\beta\in S(w,y)} \beta^{-1}$, for some $k_{w,y}$ such that $(-1)^{\ell(w)-\ell(y)} k_{w,y} \in \mathbb{Z}_+$. In this case, the number $(-1)^{\ell(w)-\ell(v)} k_{w,v}$ is the multiplicity of the point $\dot{v} \in X_w$, when $\mathcal{G}$ is of finite type with no G_2-factors.

Finally, some other criteria to check the rational smoothness of X_w as a whole are given in terms of the Poincaré polynomial of X_w (cf. Exercise 12.2.E.2). This exercise can be used to show that any X_w is rationally smooth for the Kac–Moody group associated to a 2×2 GCM (cf. Exercise 12.2.E.3). Rational smoothness of the codimension-one Schubert varieties (in the finite case) is discussed in Exercises 12.2.E.5 and 12.2.E.7. In particular, a complete list of codimension-one rationally smooth Schubert varieties is obtained.

A different realization of the Hecke algebra for the Weyl group associated to a Kac–Moody algebra is outlined in Exercise 12.2.E.9.

More recently, a direct connection between equivariant cohomology, T-stable curves and rational smoothness has been established in the papers [Arabia–98] and [Brion–98, 99].

12.1. Singular Locus of Schubert Varieties

In this section, $\mathcal{G}$ is any Kac–Moody group, $\mathcal{B}$, T, and W are its standard Borel subgroup, standard maximal torus, and Weyl group respectively. Let $\mathfrak{h}$ be the standard Cartan subalgebra of the associated Kac–Moody Lie algebra $\mathfrak{g}$. For any $w \in W$, let $X_w \subset \mathcal{X} = \mathcal{G}/\mathcal{B}$ be the Schubert variety equipped with the stable variety structure (cf. 7.1.19). Under the left multiplication of $\mathcal{B}$ on $\mathcal{X}$, X_w is $\mathcal{B}$-stable and, moreover, the action is regular. Hence a point $b\dot{v} \in X_w$, for $v \leq w$ and $b \in \mathcal{B}$, is smooth iff $\dot{v} \in X_w$ is smooth, where $\dot{v} := v\mathcal{B} \in \mathcal{X}$. In particular, the singular locus $\Sigma(X_w)$, consisting of all the nonsmooth points of X_w, is a $\mathcal{B}$-stable closed subvariety of X_w. Hence $\Sigma(X_w)$ is the union of certain Schubert subvarieties X_v (for some $v < w$). So, if $\Sigma(X_w) \neq \emptyset$, then $\dot{e} \in \Sigma(X_w)$.

We recall the following result due to Chevalley. Of course, this follows from the normality of X_w (cf. Theorem 8.2.2(b)). But it admits a direct proof (cf. Exercise 12.1.E.1).

12.1.1 Proposition. *For any X_w, any irreducible component of $\Sigma(X_w)$ is of codimension at least two in X_w.* □

From the above discussion, to understand the singular locus $\Sigma(X_w)$, it suffices to determine which of the points $\dot{v} \in X_w$ are singular. To this end, we first study the local ring $\mathcal{O}_{\dot{v},X_w}$ of X_w at $\dot{v}$. Recall the definition of $\operatorname{gr}\mathcal{O}_{x,X}$ (for a closed point x of a variety X) from A.4 and also recall the definition of an admissible T-module and its formal character from 9.3.3.

12.1.2 Definition. (a) As in 8.2.7, let $A(T) := \mathbb{Z}[X(T)]$ be the group algebra of the character group $X(T)$. Define its certain "completion" $\tilde{A}(T)$ as the set of all the formal sums $\sum_{e^\lambda \in X(T)} n_\lambda e^\lambda$, with $n_\lambda \in \mathbb{Z}$ (where we allow infinitely many of the n_λ's to be nonzero). Then $\tilde{A}(T)$ is an $A(T)$-module under multiplication. Observe that $\tilde{A}(T)$ is not a ring. Let $Q(T)$ be the quotient field of $A(T)$ and define the $Q(T)$-module $\tilde{Q}(T) := Q(T) \otimes_{A(T)} \tilde{A}(T)$, by extending the scalars.

Since $Q(T)$ is a flat $A(T)$-module, $Q(T)$ canonically embeds in $\tilde{Q}(T)$.

For an admissible T-module M, its formal character

$$\operatorname{ch} M := \sum_{e^\lambda \in X(T)} (\dim M_\lambda) e^\lambda$$

can be viewed as an element of $\tilde{A}(T)$ and, hence, also as an element of $\tilde{Q}(T)$ by taking $1 \otimes \operatorname{ch} M$.

For any directed set Λ and any "sequence" $\theta : \Lambda \to \tilde{A}(T)$ given by $\theta(\alpha) = \sum_{e^\lambda \in X(T)} n_\lambda(\alpha) e^\lambda$ (with $n_\lambda(\alpha) \in \mathbb{Z}$), we say that

$$\operatorname*{limit}_{\alpha \in \Lambda} \theta(\alpha) = \sum n_\lambda e^\lambda ,$$

if, for any $\lambda \in X(T)$, there exists $\alpha_\lambda \in \Lambda$ such that $n_\lambda(\alpha) = n_\lambda$ for all $\alpha \geq \alpha_\lambda$. Of course, $\operatorname*{limit}_{\alpha \in \Lambda} \theta(\alpha)$ may not exist in general. Observe that if $\operatorname*{limit}_{\alpha \in \Lambda} \theta(\alpha)$ exists, then so does $\operatorname*{limit}_{\alpha \in \Lambda} (p\,\theta(\alpha))$, for any fixed $p \in A(T)$. Moreover,

(1) $$\operatorname*{limit}_{\alpha \in \Lambda} (p\,\theta(\alpha)) = p \operatorname*{limit}_{\alpha \in \Lambda} \theta(\alpha).$$

We need a certain "K-theoretic" analogue of the ring Q_W defined in 11.1.1.

(b) Let $Q(T)_W$ be the $Q(T)$-vector space with basis $\{\delta_w\}_{w \in W}$. Analogous to (11.1.1.1), define the product in $Q(T)_W$ by

(2) $$\Big(\sum_v q_v \delta_v\Big) \cdot \sum_w q'_w \delta_w = \sum_{v,w} q_v (v q'_w) \delta_{vw}, \quad \text{for } q_v, q'_w \in Q(T).$$

This product makes $Q(T)_W$ into an associative ring with multiplicative identity $1 = \delta_e$.

For any simple reflection s_i, $1 \leq i \leq \ell$, define the element

(3) $$y_{s_i} := \frac{1}{1 - e^{\alpha_i}} \delta_{s_i} + \frac{1}{1 - e^{-\alpha_i}} \delta_e \in Q(T)_W.$$

Now, for any $w \in W$, define

(4) $$y_w := y_{s_{i_1}} \cdots y_{s_{i_n}} \in Q(T)_W,$$

where $w = s_{i_1} \cdots s_{i_n}$ is a reduced decomposition. Analogous to Theorem 11.1.2(b), y_w does not depend on the choice of the reduced decomposition of w (cf. Exercise 12.1.E.2). Write

(5) $$y_w = \sum_{v \in W} b_{w,v} \delta_v, \qquad \text{for } b_{w,v} \in Q(T).$$

Then it is easy to see that $b_{w,v} = 0$ unless $v \leq w$ (cf. Theorem 11.1.2(e_1)). An explicit expression for $b_{w,v}$ is given in Exercise 12.1.E.4.

The ring $Q(T)_W$ has a canonical left representation in $Q(T)$ defined by

(6) $$(q\delta_w) \bullet q' = q(wq'), \qquad \text{for } q, q' \in Q(T) \text{ and } w \in W.$$

Then $y_{s_i} \bullet A(T) \subset A(T)$ and hence

$$(7) \qquad y_w \bullet A(T) \subset A(T), \quad \text{for any } w \in W.$$

The field $Q(T)$ admits an involution, i.e., a field automorphism of order 2, defined by $\overline{e^\lambda} = e^{-\lambda}$.

For any word $\mathfrak{w} = (s_{i_1}, \dots, s_{i_n})$, recall the definition of the Demazure operator $D_{\mathfrak{w}} : A(T) \to A(T)$ from 8.2.7. Then, if $\mathfrak{w}$ is reduced,

$$(8) \qquad D_{\mathfrak{w}}\, p = y_w \bullet p, \text{ for any } p \in A(T),$$

where $w := s_{i_1} \cdots s_{i_n}$ (cf. Exercise 12.1.E.3).

(c) For any $n \in \mathbb{Z}_+$ and $a = \sum_{\lambda \in \mathfrak{h}^*_{\mathbb{Z}}} a_\lambda e^\lambda \in A(T)$, define

$$(a)_n := \sum_\lambda a_\lambda \frac{\lambda^n}{n!} \in S^n(\mathfrak{h}^*).$$

Denote $[a] = (a)_{n_o}$, whose n_o is the smallest nonnegative integer such that $(a)_{n_o} \neq 0$. By deg $[a]$ we mean n_o. (If $a = 0$, set $[a] = 0$ and deg $[a]$ is not defined in this case.)

Now, for $q = \frac{a}{b} \in Q(T)$ with $a, b \in A(T)$ and b nonzero, define $[q] := \frac{[a]}{[b]} \in Q(\mathfrak{h}^*)$ where, as in 11.1.1, $Q(\mathfrak{h}^*)$ is the quotient field of the symmetric algebra $S(\mathfrak{h}^*)$. Clearly $[q]$ is well defined. When $q \neq 0$ and $\deg[a] \leq \deg[b]$, we say that q has *a pole of order* $\deg[b] - \deg[a]$ (at the identity $1 \in T$).

The following result determines $\operatorname{ch}(\operatorname{gr} \mathcal{O}_{\dot{v}, X_w})$. The proof uses a mixture of geometric information (in the form of the coordinate ring of some affine open neighborhood of $\dot{v}$ in X_w) and representation theory — more specifically, the Demazure character formula.

12.1.3 Theorem. *For any $v \leq w$, $\operatorname{gr} \mathcal{O}_{\dot{v}, X_w}$ is an admissible T-module and, moreover,*

$$(1) \qquad \operatorname{ch}(\operatorname{gr} \mathcal{O}_{\dot{v}, X_w}) = \bar{b}_{w,v},$$

as elements of $\tilde{Q}(T)$. In particular, $\operatorname{ch}(\operatorname{gr} \mathcal{O}_{\dot{v}, X_w}) \in Q(T) \hookrightarrow \tilde{Q}(T)$.

Proof. Fix a (dominant regular) weight $\lambda \in D^o$ (cf. 7.1.1) and choose a nonzero weight vector $e_{v\lambda} \in L^{\max}(\lambda)$ of weight $v\lambda$. This defines a linear form $\hat{\sigma}_v \in L^{\max}(\lambda)^*$ by $\hat{\sigma}_v(e_{v\lambda}) = 1$ and $\hat{\sigma}_v(x) = 0$, for any weight vector $x \in L^{\max}(\lambda)$ of weight $\neq v\lambda$. This gives rise to the section $\sigma_v := \beta(\hat{\sigma}_v) \in H^0(\mathcal{X}, \mathcal{L}(\lambda))$, where β is as in 8.1.21. It is easy to see (by using the Birkhoff decomposition Theorem

6.2.8) that σ_v vanishes precisely on $\mathcal{X}\backslash v\mathcal{B}^e$ and σ_v is a weight vector of weight $-v\lambda$ (cf. Proof of Lemma 9.3.12), where, as in Section 7.1, $\mathcal{B}^e := \mathcal{U}^-\mathcal{B}/\mathcal{B} \subset \mathcal{G}/\mathcal{B}$.

Consider the T-module embedding, for any $n \geq 1$,

$$\varphi_n : H^0(X_w, \mathcal{L}_w(n\lambda)) \otimes \mathbb{C}_{vn\lambda} \hookrightarrow H^0(X_w, \mathcal{L}_w((n+1)\lambda)) \otimes \mathbb{C}_{v(n+1)\lambda},$$

defined by $\varphi_n(s \otimes 1_n) = s \cdot (\sigma_{v|X_w}) \otimes 1_{n+1}$ for $s \in H^0(X_w, \mathcal{L}_w(n\lambda))$, where $1_n \in \mathbb{C}_{vn\lambda}$ is a fixed nonzero vector. Also, define the T-module embedding

$$\delta_n : H^0(X_w, \mathcal{L}_w(n\lambda)) \otimes \mathbb{C}_{vn\lambda} \hookrightarrow \mathbb{C}[X_w \cap v\mathcal{B}^e]$$

by $\delta_n(s \otimes 1) = (s\sigma_v^{-n})_{|(X_w \cap v\mathcal{B}^e)}$. Then we have the commutative diagram

$$\begin{array}{ccc} H^0(X_w, \mathcal{L}_w(n\lambda)) \otimes \mathbb{C}_{vn\lambda} & \xhookrightarrow{\varphi_n} & H^0(X_w, \mathcal{L}_w((n+1)\lambda)) \otimes \mathbb{C}_{v(n+1)\lambda} \\ & {\searrow}^{\delta_n} \quad {\swarrow}^{\delta_{n+1}} & \\ & \mathbb{C}[X_w \cap v\mathcal{B}^e]. & \end{array}$$

By Lemma A.20, the induced map

$$\delta : \lim_{n\to\infty} \left(H^0(X_w, \mathcal{L}_w(n\lambda)) \otimes \mathbb{C}_{vn\lambda}\right) \xrightarrow{\sim} \mathbb{C}[X_w \cap v\mathcal{B}^e] \tag{2}$$

is a T-module isomorphism.

Now, by the Demazure character formula (8.2.9.3), (12.1.2.5) and (12.1.2.8),

$$\begin{aligned} \operatorname{ch}\left(H^0(X_w, \mathcal{L}_w(n\lambda)) \otimes \mathbb{C}_{vn\lambda}\right) &= e^{vn\lambda} \bar{D}_{\mathfrak{w}}(e^{n\lambda}) \\ &= e^{vn\lambda} \sum_{u \leq w} (\bar{b}_{w,u}) e^{-un\lambda}, \end{aligned} \tag{3}$$

where $\mathfrak{w}$ is a reduced word with $\pi(\mathfrak{w}) = w$. By Proposition 7.3.7, $\mathbb{C}[X_w \cap v\mathcal{B}^e]$ is an admissible T-module. Moreover, from the definition of $b_{w,u}$, it is easy to see that there exist positive real roots (possibly with repetitions) $\{\beta_1, \ldots, \beta_n\}$ depending on w such that

$$P\bar{b}_{w,u} \in A(T), \quad \text{for all } u \leq w, \quad \text{where } P := \prod_{k=1}^{n} \left(1 - e^{-\beta_k}\right).$$

By (2)-(3) and (12.1.2.1), we get in $\tilde{A}(T)$,

$$\begin{aligned} P \operatorname{ch} \mathbb{C}[X_w \cap v\mathcal{B}^e] &= \lim_{n\to\infty} \left(e^{vn\lambda} \sum_{u \leq w} P\, \bar{b}_{w,u} e^{-un\lambda}\right) \\ &= P\, \bar{b}_{w,v} + \lim_{n\to\infty} \left(\sum_{\substack{u \leq w \\ u \neq v}} P\, \bar{b}_{w,u} e^{n(v\lambda - u\lambda)}\right) \\ &= P\, \bar{b}_{w,v}, \end{aligned} \tag{4}$$

since, λ being regular, $v\lambda - u\lambda \neq 0$ for $u \neq v$. Dividing (4) by P, we get

(5) $$\operatorname{ch} \mathbb{C}[X_w \cap v\mathcal{B}^e] = \bar{b}_{w,v}, \quad \text{as elements of } \tilde{Q}(T).$$

Since $X_w \cap v\mathcal{B}^e$ is an affine open subset of X_w containing $\dot{v}$ (cf. Lemma 7.3.5), by (A.4.3), $\operatorname{gr} \mathcal{O}_{\dot{v},X_w}$ is an admissible T-module and

(6) $$\operatorname{ch}(\operatorname{gr} \mathcal{O}_{\dot{v},X_w}) = \operatorname{ch} \mathbb{C}[X_w \cap v\mathcal{B}^e].$$

Using (5), this proves (1) and hence the theorem is proved. □

12.1.4 Lemma. *For any $v, w \in W$,*

(1) $$[\bar{b}_{w,v}] = c_{w,v},$$

where $c_{w,v}$ is as defined in (11.1.2.2). Hence $b_{w,v} \neq 0$ iff $v \leq w$, and in this case it has a pole of order exactly equal to $\ell(w)$. In particular, for any $v \leq w$,

(2) $$[\operatorname{ch}(\operatorname{gr} \mathcal{O}_{\dot{v},X_w})] = c_{w,v}, \quad \textit{as elements of } Q(\mathfrak{h}^*).$$

Proof. For any $w = w's_i$ such that $w' < w$, from 12.1.2, we get $y_w = y_{w'}y_{s_i}$. This translates into

$$y_w = \sum_v \frac{b_{w',v} + b_{w',vs_i}}{1 - e^{-v\alpha_i}} \delta_v, \text{ i.e.,}$$

(3) $$b_{w,v} = \frac{b_{w',v} + b_{w',vs_i}}{1 - e^{-v\alpha_i}}, \quad \text{for any } v \in W.$$

Similarly, we obtain

(4) $$c_{w,v} = \frac{c_{w',v} + c_{w',vs_i}}{(-v\alpha_i)}.$$

Clearly (1) is true for $w = e$. Now, by induction on $\ell(w)$, assume the validity of (1) for w' and any $v \in W$. This gives

(5) $$[\bar{b}_{w',v}] = c_{w',v} \quad \text{and} \quad [\bar{b}_{w',vs_i}] = c_{w',vs_i}.$$

For any $0 \neq f \in Q(\mathfrak{h}^*)$, which can be written as $f = f'/f''$ such that $f', f'' \in S(\mathfrak{h}^*)$ are both homogeneous, define $\deg f = \deg f' - \deg f''$. Then, from Theorem 11.1.2(f),

(6) $$\deg c_{w',v} = -\ell(w'), \text{ for any } v \text{ such that } c_{w',v} \neq 0.$$

By Theorem 11.1.2(e$_1$), $c_{w,v} = 0$ unless $v \leq w$ and similarly $b_{w,v} = 0$ unless $v \leq w$. So, to prove (1), we can assume that $v \leq w$. In this case, by Proposition 11.1.19, $c_{w,v} \neq 0$. Now, combining (3)–(6), we get

$$[\bar{b}_{w,v}] = c_{w,v}\,; \text{ in particular, } b_{w,v} \neq 0.$$

This completes the induction and proves (1). Combining (1) with Theorem 12.1.3, we get (2). □

12.1.5 Definition. Let Y be a variety over $\mathbb{C}$ on which a complex torus T acts algebraically. Let $E(Y)$ be the set of all the one-dimensional T-orbit closures in Y. For $y \in Y^T$ (where Y^T is the set of T-fixed points of Y), let $E(Y, y)$ be the set of all those $C \in E(Y)$ such that $y \in C$.

For any representation V of T, it is easy to see that

(1) $$\mathbb{P}(V)^T = \{\text{lines } [v] : v \text{ is a nonzero weight vector under } T\}.$$

12.1.6 Proposition. *Let Y be an irreducible projective variety on which a torus T acts such that Y^T is finite. Assume further that there exists a finite-dimensional (algebraic) representation V of T such that there is a T-equivariant embedding $Y \hookrightarrow \mathbb{P}(V)$. Then, for any $y \in Y^T$,*

(1) $$\# E(Y, y) \geq \dim Y.$$

Proof. We prove (1) by induction on $\dim V$. If $\dim V = 1$, there is nothing to prove. So assume that $\dim V > 1$. If $\dim Y = 1$, again (1) is trivially true. So assume that $\dim Y > 1$. Choose a codimension-one T-stable subspace $L \subset V$ such that $y \in \mathbb{P}(L)$. This is possible by (12.1.5.1). Choose an irreducible component Y_L of $Y \cap \mathbb{P}(L)$ through y. Then, by induction,

(2) $$\# E(Y, y) \geq \# E(Y_L, y) \geq \dim Y_L \geq \dim Y - 1.$$

The last inequality follows from Theorem A.1. Since $\dim Y > 1$ (by assumption), by (2) there exists a $C \in E(Y_L, y)$. By Exercise 12.1.E.5, C^T consists of exactly 2 points $\{y, z\}$, $z \neq y$. Let L' be a T-stable codimension-one subspace of V containing the line y but complementary to the line z. Then by (2), for L replaced by L', $\# E(Y_{L'}, y) \geq \dim Y - 1$. Moreover, $C \notin E(Y_{L'}, y)$. Hence $\# E(Y, y) \geq \dim Y$, proving the proposition. □

Recall from 11.1.18 that, for any $v \leq w \in W$, $S(w, v) := \{\beta \in \Delta^+_{\text{re}} : s_\beta v \leq w\}$. Observe that, by Lemma 7.1.22(a), X_w^T is finite and, moreover, by

Corollary 7.1.18, there exists a large enough $\lambda \in D^o$ (depending on w) and a finite-dimensional $\mathcal{B}$-submodule V_w of $L^{\max}(\lambda)$ with a T-equivariant embedding $X_w \hookrightarrow \mathbb{P}(V_w)$ induced from i_λ. Thus, X_w satisfies all the assumptions of Proposition 12.1.6.

The following result is crucial to many of the criteria determining the smooth and rationally smooth locus of the Schubert varieties X_w. This result is behind the notion of the oriented graph $\Gamma(v, w)$ defined in Exercise 12.2.E.1.

12.1.7 Proposition. *For any $v \leq w$, for the T-variety X_w,*

$$E(X_w, \dot{v}) = \left\{\overline{\mathcal{U}_\beta \dot{v}} : \beta \in \Delta^+_{\mathrm{re}} \text{ with } s_\beta v < v\right\} \bigsqcup \left\{\overline{\mathcal{U}_{-\beta} \dot{v}} : \beta \in S(w, v) \text{ and } s_\beta v > v\right\}$$

$$(1) \qquad = \left\{\overline{v\mathcal{U}_{-\alpha} \dot{e}} : \alpha \in \Delta^+_{\mathrm{re}} \text{ with } v s_\alpha \leq w\right\},$$

where $\mathcal{U}_\beta$ is the one-parameter subgroup corresponding to the real root β as in (6.2.7.1). Moreover, the above orbit closures are all distinct.

In particular,

$$(2) \qquad \# E(X_w, \dot{v}) = \# S(w, v).$$

Proof. Since $X_w := \bigsqcup_{\theta \leq w} \mathcal{B}\theta\mathcal{B}/\mathcal{B}$ (cf. 7.1.13), one obtains from Exercise 6.2.E.1 that one-dimensional T-orbits contained in X_w are precisely of the form $I_{\theta,\beta} := (\mathcal{U}_\beta \backslash e)\dot{\theta}$, where $\theta \leq w$ and $\beta \in \Phi_\theta := \Delta^+ \cap \theta\Delta^-$. Recall from (11.1.15.4) that $\Phi_\theta = \left\{\beta \in \Delta^+_{\mathrm{re}} : s_\beta\theta < \theta\right\}$. Further, for any $\beta \in \Phi_\theta$, by Lemma 1.3.13,

$$(3) \qquad \mathcal{U}_{-\beta}\dot{\theta} = \theta\,\mathcal{U}_{-\theta^{-1}\beta}\dot{e} = \dot{\theta}.$$

So, we can write,

$$(4) \qquad I_{\theta,\beta} = (\mathcal{U}_\beta \backslash e)\mathcal{U}_{-\beta}\dot{\theta}.$$

From SL_2-theory, it is easy to see (cf. Exercise 7.1.E.2) that

$$s_\beta\dot{\theta} \in \bar{I}_{\theta,\beta}.$$

In particular, by Exercise 12.1.E.5,

$$(5) \qquad \bar{I}_{\theta,\beta} \backslash I_{\theta,\beta} = \{\dot{\theta}, s_\beta\dot{\theta}\}.$$

Hence, using (11.1.15.4), we get

$$(6)\qquad \{(\theta,\beta)\in W\times\Delta^+_{\text{re}}:\theta\le w,\ \beta\in\Phi_\theta \text{ and } \dot v\in\bar I_{\theta,\beta}\}$$
$$=\{(v,\beta):\beta\in\Delta^+_{\text{re}} \text{ with } s_\beta v<v\}$$
$$\bigsqcup\{(s_\beta v,\beta):\beta\in\Delta^+_{\text{re}},\ s_\beta v\le w \text{ and } s_\beta v>v\}.$$

From the above, the first equality of (1) follows by observing that $\bar I_{\theta,\beta}$ is stable (by (4)) under the (left) action of s_β. The "Moreover" part of the proposition follows by left translating the orbit closures by v^{-1}, since the map $g\mapsto g\dot e$, for $g\in\mathcal{U}^-$, is injective by Theorem 5.2.3(e). The second equality of (1) follows from the first by using (11.1.15.4), since $v\mathcal{U}_{-\alpha}\dot e=\mathcal{U}_{-v\alpha}\dot v$ and, for $\alpha\in\Delta^+_{\text{re}}$, $v\alpha\in\Delta^-_{\text{re}}$ iff $vs_\alpha<v$ (by Lemma 1.3.13). Finally (2) follows from the second equality of (1) and Exercise 12.1.E.6. □

From the above proposition, we get another proof of Deodhar's conjecture. Recall that a more general result was obtained in Corollary 11.1.20 by using the nil-Hecke ring.

12.1.8 Corollary. *For any* $v\le w$,

$$\# S(w,v)\ge \ell(w).$$

Proof. Combine Proposition 12.1.6 with (12.1.7.2). □

12.1.9 Definition. For any $v\le w$, the Zariski tangent space $T_{\dot v}(X_w)$ (cf. A.3) is clearly vector space isomorphic with $T_{\dot e}(v^{-1}X_w)$. Consider the affine open subset $v^{-1}X_w\cap\mathcal{B}^e$ of $v^{-1}X_w$ (cf. Lemma 7.3.5). By Proposition 7.3.7, for all large enough k, there is a T-equivariant closed embedding

$$i(k)=i_{v,w}(k):v^{-1}X_w\cap\mathcal{B}^e\hookrightarrow \hat{\mathcal{U}}^{-(k)}.$$

In particular, the derivative of $i(k)$ at the T-fixed point $\dot e$ is a T-module injective map

$$di(k):T_{\dot e}(v^{-1}X_w)\hookrightarrow T_e(\hat{\mathcal{U}}^{-(k)})\simeq\mathfrak{n}^-/\mathfrak{n}^-(k),$$

where $\mathfrak{n}^-(k):=\bigoplus_{\substack{\alpha\in\Delta^-\\ \text{ht}\,\alpha\ge k}}\mathfrak{g}_\alpha$. Since $di(k)$ is an isomorphism for all $k\gg 0$, it is easy to see that there exists a unique T-submodule $V(v,w)\subset\mathfrak{n}^-$, independent of k, such that $\pi_k(V(v,w))=\operatorname{Im} di(k)$ and $\pi_{k|V(v,w)}:V(v,w)\to\operatorname{Im} di(k)$ is an isomorphism for all $k\gg 0$, where π_k is the quotient map $\mathfrak{n}^-\to\mathfrak{n}^-/\mathfrak{n}^-(k)$.

From now on we identify

$$T_{\dot e}(v^{-1}X_w)\simeq V(v,w)\subset\mathfrak{n}^-.$$

As another corollary of Proposition 12.1.7, we obtain the following:

12.1.10 Corollary. *For any $v \leq w$,*

$$(1) \qquad T_{\dot{e}}(v^{-1}X_w) \supset \bigoplus_{\substack{\alpha \in \Delta^+_{\text{re}} \text{ with} \\ vs_\alpha \leq w}} \mathfrak{g}_{-\alpha}.$$

In particular,

$$(2) \qquad \dim T_{\dot{e}}(v^{-1}X_w) \geq \# S(w, v).$$

In general, we do not have equality in (2). In fact, by Theorem 12.2.14 and Exercise 12.1.E.6, for any rationally smooth point $\dot{v} \in X_w$ which is not smooth, the inclusion in (1) is proper.

Proof. By Proposition 12.1.7, for any $\alpha \in \Delta^+_{\text{re}}$ such that $vs_\alpha \leq w$, one has

$$\mathcal{U}_{-\alpha}\dot{e} \subset v^{-1}X_w \cap \mathcal{B}^e.$$

In particular,

$$T_{\dot{e}}(\mathcal{U}_{-\alpha}\dot{e}) \subset T_{\dot{e}}(v^{-1}X_w),$$

which gives

$$\mathfrak{g}_{-\alpha} \subset T_{\dot{e}}(v^{-1}X_w),$$

proving (1). Now (2) follows from (1) and Exercise 12.1.E.6. □

Now we come to the main theorem of this section which gives geometric information (more specifically, the singular locus) of the Schubert varieties in terms of combinatorics of the Weyl group.

12.1.11 Theorem. *Let $\mathcal{G}$ be any Kac–Moody group with the associated Weyl group W. Then, for any $v \leq w \in W$, $\dot{v} \in X_w$ is a smooth point iff*

$$(1) \qquad c_{w,v} = (-1)^{\ell(w)-\ell(v)} \prod_{\beta \in S(w,v)} \beta^{-1}.$$

As a preparation, we first prove the following proposition. This proposition (except the positivity of d) also follows from Proposition 11.1.19 together with (12.1.4.2). However, the geometric proof given below is crucially used in the proof of the above theorem.

12.1.12 Proposition. *Let $v \leq w \in W$. Assume that $\# S(w, v) = \ell(w)$. Then*

$$[\mathrm{ch}(\mathrm{gr}\, \mathcal{O}_{\dot{v}, X_w})] = (-1)^{\ell(w)-\ell(v)} d \prod_{\beta \in S(w,v)} \beta^{-1},$$

where $d > 0$ is an integer (cf. Exercise 12.1.E.10 for an interpretation of d in the finite case).

Proof. Fix any (dominant regular) $\lambda \in D^o$ and a nonzero highest weight vector $v_\lambda \in L^{\max}(\lambda)$, and let $v_\lambda^* \in L^{\max}(\lambda)^*$ be the element such that $v_\lambda^*(v_\lambda) = 1$ and $v_\lambda^*(x) = 0$ for any weight vector $x \in L^{\max}(\lambda)$ of weight $< \lambda$. For any root $\alpha \in \Delta_{\mathrm{re}}^+$, choose a nonzero root vector $X_\alpha \in \mathfrak{g}_\alpha$ and let $X_{-\alpha} \in \mathfrak{g}_{-\alpha}$ be the element such that $[X_\alpha, X_{-\alpha}] = \alpha^\vee$ (cf. 1.3.8). Choose a representative $\bar{v} \in N$ of v and, for any $\alpha \in \Delta_{\mathrm{re}}^+$, define the map $\theta_\alpha : v\mathcal{B}^e \to \mathbb{C}$ by $\theta_\alpha(\bar{v}g\mathcal{B}) = v_\lambda^*(X_\alpha g v_\lambda)$, for $g \in \mathcal{U}^-$. By Lemma 7.3.9, θ_α is a morphism (under the open ind-subvariety structure on $v\mathcal{B}^e \subset \mathcal{G}/\mathcal{B}$). We claim that

(1) $$\theta_\alpha(\bar{v}g\mathcal{B}) \neq 0, \quad \text{for any } g \neq e \in \mathcal{U}_{-\alpha} :$$

Write $g = \mathrm{Exp}(zX_{-\alpha})$, for $z \neq 0 \in \mathbb{C}$. Then

$$\begin{aligned} \theta_\alpha(\bar{v}g\mathcal{B}) &= v_\lambda^*(X_\alpha \,\mathrm{Exp}(zX_{-\alpha})v_\lambda) \\ &= v_\lambda^*(zX_\alpha X_{-\alpha} v_\lambda), && \text{by Proposition 6.2.11} \\ &= v_\lambda^*(z\alpha^\vee v_\lambda) \\ &= z\lambda(\alpha^\vee) \\ &\neq 0, && \text{since } \lambda \text{ is regular.} \end{aligned}$$

Consider the affine open subset $A = A(v, w) := X_w \cap v\mathcal{B}^e$ of X_w. Restricting θ_α to A, we get the regular function $\theta_\alpha^A : A \to \mathbb{C}$. Now define the T-stable closed subvariety (with the reduced structure):

$$Z = \left\{ x \in A : \theta_\alpha^A(x) = 0, \quad \text{for all } \alpha \in \Delta_{\mathrm{re}}^+ \text{ such that } vs_\alpha \leq w \right\}.$$

We claim that

(2) $$Z = \{\dot{v}\} :$$

To prove this, observe first that any T-orbit in A has $\dot{v}$ in its closure, which follows from Proposition 7.3.7 (since, in the notation of loc. cit., all the roots of $\hat{\mathcal{U}}^{-(k)}$ lie in the cone $\bigoplus_{i=1}^{\ell} \mathbb{Z}_+(-\alpha_i)$). In particular, any irreducible component of Z (being T-stable and closed in A) passes through $\dot{v}$. Now let $Z' \subset Z$ be an irreducible component (necessarily passing through $\dot{v}$), and let $\overline{Z'}$ be its closure in X_w. By Proposition 12.1.7, any $C \in E(\overline{Z'}, \dot{v})$ is of the form $\overline{v\mathcal{U}_{-\alpha}\dot{e}}$, for some $\alpha \in \Delta^+_{\mathrm{re}}$ with $vs_\alpha \leq w$. But, by the definition of Z and (1), $v(\mathcal{U}_{-\alpha}\backslash 1)\dot{e} \cap Z = \emptyset$, and hence $v(\mathcal{U}_{-\alpha}\backslash 1)\dot{e} \subset \overline{Z'}\backslash Z'$. In particular, $v(\mathcal{U}_{-\alpha}\backslash 1)\dot{e} \subset X_w \backslash v\mathcal{B}^e$. But this is clearly a contradiction since $v\mathcal{U}_{-\alpha}\dot{e} \subset v\mathcal{B}^e$. This shows that $E(\overline{Z'}, \dot{v}) = \emptyset$; in particular, by Proposition 12.1.6, $\overline{Z'} = \{\dot{v}\}$. This proves the assertion (2). Observe that we have, so far, not used the assumption $\# S(w, v) = \ell(w)$. But now we make use of it.

Order the elements $\{\alpha \in \Delta^+_{\mathrm{re}} : vs_\alpha \leq w\}$ as $\{\beta_1, \dots, \beta_{\ell(w)}\}$ (cf. Exercise 12.1.E.6). Let $\mathbb{C}[A]_{\mathfrak{m}}$ denote the ring $\mathbb{C}[A]$ localized at the maximal ideal $\mathfrak{m}$ of $\mathbb{C}[A]$ corresponding to the point $\dot{v}$, and let $\bar{\theta}^A_{\beta_i}$ be the same element as $\theta^A_{\beta_i}$ but considered as an element of $\mathbb{C}[A]_{\mathfrak{m}}$. Since Z is 0-dimensional and A is of dimension $\ell(w)$, we conclude from Theorems A.35 and 8.2.2(c) that $\{\bar{\theta}^A_{\beta_i}\}_{1\leq i\leq \ell(w)}$ is a regular sequence in the local ring $\mathcal{O}_{\dot{v}, X_w} \simeq \mathbb{C}[A]_{\mathfrak{m}}$ and, moreover, the $\mathbb{C}[A]_{\mathfrak{m}}/I_{\mathfrak{m}}$-algebra homomorphism

(3) $$\frac{\mathbb{C}[A]_{\mathfrak{m}}}{I_{\mathfrak{m}}}[X_1, \dots, X_{\ell(w)}] \xrightarrow{\sim} \bigoplus_{m\geq 0} \frac{(I_{\mathfrak{m}})^m}{(I_{\mathfrak{m}})^{m+1}}, \quad X_i \mapsto \bar{\theta}^A_{\beta_i} \bmod I^2_{\mathfrak{m}},$$

is an isomorphism, where $I_{\mathfrak{m}}$ is the ideal $\langle \bar{\theta}^A_{\beta_1}, \dots, \bar{\theta}^A_{\beta_{\ell(w)}} \rangle \subset \mathbb{C}[A]_{\mathfrak{m}}$ and the left side is the polynomial ring in the variables $\{X_1, \dots, X_{\ell(w)}\}$ over the ring $\frac{\mathbb{C}[A]_{\mathfrak{m}}}{I_{\mathfrak{m}}}$. By (2) we get

(4) $$\mathfrak{m}^d \subset I \subset \mathfrak{m}, \qquad \text{for some } d > 0,$$

where $I := \langle \theta^A_{\beta_1}, \dots, \theta^A_{\beta_{\ell(w)}} \rangle \subset \mathbb{C}[A]$. From this it is easy to see that, for any $m \geq 0$, the following canonical map is an isomorphism:

(5) $$\frac{I^m}{I^{m+1}} \xrightarrow{\sim} \frac{(I_{\mathfrak{m}})^m}{(I_{\mathfrak{m}})^{m+1}}.$$

So (3) takes the form

(6) $$\frac{\mathbb{C}[A]}{I}[X_1, \dots, X_{\ell(w)}] \xrightarrow{\sim} \bigoplus_{m\geq 0} \frac{I^m}{I^{m+1}}.$$

Further, $\theta^A_{\beta_i}$ is a T-weight vector with

(7) $$\text{weight } \theta^A_{\beta_i} = e^{v\beta_i}.$$

From (6) we get (in the $Q(T)$-module $\tilde{Q}(T)$):

(8) $$\mathrm{ch}\Big(\frac{\mathbb{C}[A]}{I}\Big) \cdot \prod_{i=1}^{\ell(w)} \big(1 - e^{v\beta_i}\big)^{-1} = \mathrm{ch}\, \mathbb{C}[A] = \mathrm{ch}\big(\mathrm{gr}\, \mathcal{O}_{\dot{v}, X_w}\big),$$

where the last equality follows from (12.1.3.6).

From (8), taking $[\,\cdot\,]$ (cf. Definition 12.1.2(c)), we get

$$(9)\qquad \dim_{\mathbb{C}}\Big(\frac{\mathbb{C}[A]}{I}\Big)\prod_{\substack{\alpha\in\Delta^+_{\mathrm{re}}:\\ vs_\alpha\le w}}(-v\alpha)^{-1}=\big[\mathrm{ch}\big(\mathrm{gr}\,\mathcal{O}_{\dot v,X_w}\big)\big],$$

which is, in view of (11.1.15.4) (by decomposing the set $\{\alpha\in\Delta^+_{\mathrm{re}}: vs_\alpha\le w\}=\{\alpha\in\Delta^+_{\mathrm{re}}: v<vs_\alpha\le w\}\sqcup\{\alpha\in\Delta^+_{\mathrm{re}}: vs_\alpha<v\}$), equivalent to

$$(10)\qquad \dim_{\mathbb{C}}\Big(\frac{\mathbb{C}[A]}{I}\Big)(-1)^{\ell(w)-\ell(v)}\prod_{\beta\in S(w,v)}\beta^{-1}=\big[\mathrm{ch}\big(\mathrm{gr}\,\mathcal{O}_{\dot v,X_w}\big)\big].$$

This proves the proposition by taking $d=\dim\frac{\mathbb{C}[A]}{I}$. $\square$

Now we are ready to prove Theorem 12.1.11.

12.1.13 *Proof of Theorem 12.1.11.* We first show the implication "$\Leftarrow$", i.e., we show that if $c_{w,v}$ satisfies (1), then $\dot v\in X_w$ is smooth. By (12.1.4.6), for any nonzero $c_{w',v'}$, we have $\deg c_{w',v'}=-\ell(w')$. In particular, by assumption, $\#S(w,v)=\ell(w)$. So, Proposition 12.1.12 is applicable, giving (in view of (12.1.4.2)) $\dim\frac{\mathbb{C}[A]}{I}=1$ (in the notation of the proof of Proposition 12.1.12). In particular, by (12.1.12.4), $I=\mathfrak{m}$.

Thus $I_{\mathfrak{m}}$ is the maximal ideal of $\mathbb{C}[A]_{\mathfrak{m}}$ and hence, by (12.1.12.3), we get the graded algebra isomorphism

$$(1)\qquad \mathbb{C}[X_1,\dots,X_{\ell(w)}]\simeq \mathrm{gr}(\mathcal{O}_{\dot v,X_w}),$$

where each X_i is assigned degree 1. This proves that $\dot v\in X_w$ is smooth (cf. A.4).

We now prove the implication "$\Rightarrow$": Since $\dot v\in X_w$ is smooth (by assumption), $T_{\dot v}(X_w)\simeq T_{\dot e}(v^{-1}X_w)$ is of dim $\ell(w)$ (cf. (A.3.4)). In particular, by Corollaries 12.1.8 and 12.1.10 (and Exercise 12.1.E.6), we have

$$(2)\qquad T_{\dot e}(v^{-1}X_w)=\bigoplus_{\substack{\alpha\in\Delta^+_{\mathrm{re}}\text{ with}\\ vs_\alpha\le w}}\mathfrak{g}_{-\alpha}.$$

Moreover, as graded T-algebras,

$$(3)\qquad S[T_{\dot v}(X_w)^*]\simeq \mathrm{gr}\,\mathcal{O}_{\dot v,X_w},$$

where $S[T_{\dot v}(X_w)^*]$ is the symmetric algebra of the dual vector space $T_{\dot v}(X_w)^*$. Hence, by (2) and (3),

$$\mathrm{ch}\big(\mathrm{gr}\,\mathcal{O}_{\dot v,X_w}\big)=\prod_{\substack{\alpha\in\Delta^+_{\mathrm{re}}\text{ with}\\ vs_\alpha\le w}}(1-e^{v\alpha})^{-1}.$$

Taking $[\,\cdot\,]$ and using (12.1.4.2), we get

$$c_{w,v} = \prod_{\substack{\alpha\in\Delta^+_{\text{re}} \text{ with}\\ vs_\alpha \le w}} (-v\alpha)^{-1} = (-1)^{\ell(w)-\ell(v)} \prod_{\beta\in S(w,v)} \beta^{-1},$$

where the last equality follows by comparing (12.1.12.9) and (12.1.12.10). This finishes the proof of the theorem. □

The following example can be worked out by an easy calculation.

12.1.14 Example. For any Kac–Moody Lie algebra $\mathfrak{g}$ and any simple reflections $s_1, s_2 \in W$, we have the following (as elements of the ring Q_W defined in 11.1.1):

(a) $$x_{s_1}x_{s_2} = \frac{1}{\alpha_1}\left(\frac{1}{\alpha_2}(\delta_e - \delta_{s_2}) - \frac{1}{s_1\alpha_2}(\delta_{s_1} - \delta_{s_1s_2})\right).$$

(b) $$x_{s_1}x_{s_2}x_{s_1} = \frac{1}{\alpha_1}\left(\frac{\alpha_2(\alpha_1^\vee)}{\alpha_2(s_1\alpha_2)}(\delta_e - \delta_{s_1}) + \frac{1}{\alpha_2(s_2\alpha_1)}(\delta_{s_2} - \delta_{s_2s_1}) - \frac{1}{(s_1\alpha_2)(s_1s_2\alpha_1)}(\delta_{s_1s_2} - \delta_{s_1s_2s_1})\right)$$

(c) $$x_{s_1}x_{s_2}x_{s_1}x_{s_2} = \frac{1}{\alpha_1}\left(\frac{(m-1)}{\alpha_2(s_1\alpha_2)(s_2\alpha_1)}(\delta_e - \delta_{s_2}) - \frac{(m-1)}{\alpha_2(s_1\alpha_2)(s_1s_2\alpha_1)}(\delta_{s_1} - \delta_{s_1s_2}) + \frac{1}{\alpha_2(s_2\alpha_1)(s_2s_1\alpha_2)}(\delta_{s_2s_1} - \delta_{s_2s_1s_2}) - \frac{1}{(s_1\alpha_2)(s_1s_2\alpha_1)(s_1s_2s_1\alpha_2)}(\delta_{s_1s_2s_1} - \delta_{s_1s_2s_1s_2})\right)$$

(d) $$\begin{aligned} x_{s_1}x_{s_2}x_{s_1}x_{s_2}x_{s_1} = \frac{1}{\alpha_1}\Bigg(&\frac{(m-1)(2-m)}{\alpha_2(s_1\alpha_2)(s_2\alpha_1)(s_1s_2\alpha_1)}(\delta_e - \delta_{s_1}) \\ &+ \frac{(2-m)\alpha_2(\alpha_1^\vee)}{\alpha_2(s_1\alpha_2)(s_2\alpha_1)(s_2s_1\alpha_2)}(\delta_{s_2} - \delta_{s_2s_1}) \\ &+ \frac{(m-2)\alpha_2(\alpha_1^\vee)}{\alpha_2(s_1\alpha_2)(s_1s_2\alpha_1)(s_1s_2s_1\alpha_2)}(\delta_{s_1s_2} - \delta_{s_1s_2s_1}) \\ &+ \frac{1}{\alpha_2(s_2\alpha_1)(s_2s_1\alpha_2)(s_2s_1s_2\alpha_1)}(\delta_{s_2s_1s_2} - \delta_{s_2s_1s_2s_1}) \\ &- \frac{1}{(s_1\alpha_2)(s_1s_2\alpha_1)(s_1s_2s_1\alpha_2)(s_1s_2s_1s_2\alpha_1)}(\delta_{s_1s_2s_1s_2} - \delta_{s_1s_2s_1s_2s_1})\Bigg), \end{aligned}$$

where $m := \alpha_1(\alpha_2^\vee)\alpha_2(\alpha_1^\vee)$.

12.1.E EXERCISES

(1) Give a direct proof of Proposition 12.1.1.

Hint. It suffices to show that, for any $v \to w$, $\dot{v} \in X_w$ is smooth. Now use Lemma 7.3.10 to identify a neighborhood of $\dot{v} \in X_w$ with $\mathcal{U}_v \times (X_w \cap \mathcal{B}^v)$. Further, identify the one-parameter subgroup $\mathcal{U}_{-\alpha} \xrightarrow{\sim} X_w \cap \mathcal{B}^v$ under $g \mapsto g\dot{v}$, where $\alpha \in \Delta^+_{\text{re}}$ is such that $s_\alpha v = w$.

(2) Prove that $y_w \in Q(T)_W$ defined by (12.1.2.4) does not depend on the choice of the reduced decomposition of w.

Hint. For simple reflections $s_i \neq s_j$, let $s_i s_j$ be of order $m_{ij} \in \{2, 3, 4, 6, \infty\}$ (cf. Proposition 1.3.21). Then it suffices to show that, if $m_{ij} < \infty$, $y_{s_i} y_{s_j} \cdots = y_{s_j} y_{s_i} \cdots$, where both sides have m_{ij} terms.

(3) Recall the definition of the Demazure operator $D_{s_i} : A(T) \to A(T)$ (for any $1 \le i \le \ell$) from 8.2.7. Show that $D_{s_i} e^\lambda = y_{s_i} \bullet e^\lambda$. Hence, for any reduced word $\mathfrak{w} = (s_{i_1}, \dots, s_{i_n})$ with $w := s_{i_1} \cdots s_{i_n}$,

$$D_{\mathfrak{w}} e^\lambda = (y_{s_{i_1}} \cdots y_{s_{i_n}}) \bullet e^\lambda = y_w \bullet e^\lambda.$$

This, together with Corollary 8.2.10, provides another solution to the previous exercise (2).

(4) Take any $v \le w \in W$ and fix a reduced decomposition $w = s_{i_1} \cdots s_{i_n}$. Then prove that

$$b_{w,v} = \sum \left(\left(1 - e^{-s_{i_1}^{\varepsilon_1} \alpha_{i_1}}\right)\left(1 - e^{-s_{i_1}^{\varepsilon_1} s_{i_2}^{\varepsilon_2} \alpha_{i_2}}\right) \cdots \left(1 - e^{-s_{i_1}^{\varepsilon_1} \cdots s_{i_n}^{\varepsilon_n} \alpha_{i_n}}\right)\right)^{-1},$$

where the summation runs over all those $(\varepsilon_1, \dots, \varepsilon_n) \in \{0, 1\}^n$ satisfying $s_{i_1}^{\varepsilon_1} \cdots s_{i_n}^{\varepsilon_n} = v$, and the notation s_i^0 means the identity element.

(5) Let C be an irreducible projective curve on which a torus T acts algebraically. Assume that C^T is finite and there exists a finite-dimensional representation V of T with a T-equivariant embedding $C \hookrightarrow \mathbb{P}(V)$. Then show that C^T has exactly two points.

Hint. Show that any one-dimensional T-orbit O in $\mathbb{P}(V)$ has exactly two points in $\bar{O} \backslash O$.

(6) Show that, for any $v \le w \in W$,

$$\# S(w, v) = \# \{\alpha \in \Delta^+_{\text{re}} : v s_\alpha \le w\}.$$

(7) Assume for this exercise that $\mathcal{G}$ is of finite type so that its associated Weyl group W is finite. Let $w_o \in W$ be the longest element. Prove the following:

(a) The codimension-one Schubert varieties in $\mathcal{G}/\mathcal{B}$ are precisely of the form $\{X_i := X_{w_o s_i}\}_{1 \leq i \leq \ell}$.

(b) For any $v \leq w_o s_i$,

$$(*) \qquad c_{w_o s_i, v} = (-1)^{|\Delta^+| - \ell(v)} \frac{1}{\prod_{\beta \in \Delta^+} \beta} (w_o \omega_i - v\omega_i),$$

where ω_i is the i-th fundamental weight defined by $\omega_i(\alpha_j^\vee) = \delta_{i,j}$.

In particular, by Corollary 12.1.4.2,

$$\left[\mathrm{ch}\left(\mathrm{gr}\, \mathcal{O}_{\dot{v}, X_i}\right)\right] = (-1)^{|\Delta^+| - \ell(v)} \frac{1}{\prod_{\beta \in \Delta^+} \beta} (w_o \omega_i - v\omega_i).$$

Give a direct geometric proof of the above.

Hint. To show $(*)$, prove the recurrence formula

$$\begin{aligned} s_i c_{w, s_i v} - c_{w,v} &= \alpha_i c_{s_i w, v}, && \text{if } s_i w > w \\ &= 0, && \text{otherwise.} \end{aligned}$$

(c) For any simple reflection s_i and any $v \in W$,

$$v \leq w_o s_i \quad \text{iff} \quad \omega_i \neq v^{-1} w_o \omega_i.$$

Hint. Note that $v \leq w_o s_i$ iff $s_i \leq v^{-1} w_o$.

(d) Assume $v \leq w_o s_i$. Then $\omega_i - v^{-1} w_o \omega_i$ is a multiple of a root β iff $\pm v\beta \notin S(w_o s_i, v)$. In particular, $\omega_i - v^{-1} w_o \omega_i$ is a multiple of a root iff $\# S(w_o s_i, v) = |\Delta^+| - 1$.

Hint. Use (c) and Deodhar's conjecture Corollary 12.1.8.

(e) For any $v \leq w_o s_i$,

$$\dot{v} \in X_i \text{ is smooth iff } \omega_i - v^{-1} w_o \omega_i \text{ is a root.}$$

In particular, X_i itself is smooth iff $\omega_i - w_o \omega_i$ is a root.

(f) Among the simple groups of finite type, show that following is the complete list of codimension-one smooth Schubert varieties (following the Bourbaki notation [Bourbaki–81, Planches I–IX]):

f$_1$) $\quad A_n \ (n \geq 1) : X_1, X_n$;
f$_2$) $\quad C_n \ (n \geq 2) : X_1$.

(8) For any $v \leq w$, prove that

$$\left(\prod_{\beta \in S(w,v)} (1 - e^{\beta})\right) b_{w,v} \in A(T).$$

Hint. Use ideas similar to the proof of Proposition 11.1.19.

(9) For any $v \leq w$, prove that

$$(**) \qquad T_{\dot e}(v^{-1}X_w) = \{X \in \mathfrak{n}^- : Xv_\lambda \in v^{-1}L_w^{\max}(\lambda), \text{ for all } \lambda \in D_{\mathbb{Z}}\},$$

where v_λ is a nonzero highest weight vector of $L^{\max}(\lambda)$ and the Demazure module $L_w^{\max}(\lambda) \subset L^{\max}(\lambda)$ is as defined in 8.1.22.

Hint. First consider the finite case and prove the following steps (a)–(b).

For $\lambda \in D$ and $\theta \in L(\lambda)^*$, consider the function $\varphi_\lambda^\theta : \mathfrak{n}^- \to \mathbb{C}$ defined by

$$\varphi_\lambda^\theta(X) = \theta((\operatorname{Exp} X)v_\lambda) = \theta(v_\lambda) + \theta(Xv_\lambda) + \text{order two and higher terms}.$$

So, the linear part $L(\varphi_\lambda^\theta)$ of φ_λ^θ is given by $L(\varphi_\lambda^\theta)\, X = \theta(Xv_\lambda)$ for $X \in \mathfrak{n}^-$.

Let $\mathcal{I}$ be the ideal of the closed subvariety $v^{-1}X_w \cap \mathcal{B}^e$ of $\mathcal{B}^e \simeq \mathcal{U}^- \simeq \mathfrak{n}^-$. Then,

$$T_{\dot e}(v^{-1}X_w) = \{X \in \mathfrak{n}^- : L(f)\, X = 0, \text{ for all } f \in \mathcal{I}\}.$$

(a) Show that any $f \in \mathbb{C}[\mathfrak{n}^-]$ is of the form φ_λ^θ for some $\lambda \in D$ and $\theta \in L(\lambda)^*$ (cf. the proof of Theorem 12.1.3). (In fact $\lambda \in \{n\lambda_o\}_{n \gg 0}$ suffice for a fixed $\lambda_o \in D^o$.)

(b) $\varphi_\lambda^\theta \in \mathcal{I}$ iff $\theta \in \left(\frac{L(\lambda)}{v^{-1}L_w(\lambda)}\right)^*$ (cf. proof of Lemma 8.1.23).

(c) Adapt the above argument to cover the general Kac–Moody case.

Remark. In the finite case, for any $\lambda_o \in D$, it is known that the kernel of the graded algebra homomorphism obtained by the restriction:

$$\bigoplus_{n \geq 0} H^0(\mathcal{G}/\mathcal{B}, \mathcal{L}(n\lambda_o)) \to \bigoplus_{n \geq 0} H^0(v^{-1}X_w, \mathcal{L}(n\lambda_o)_{|v^{-1}X_w})$$

is generated as an ideal by the kernel of

$$H^0(\mathcal{G}/\mathcal{B}, \mathcal{L}(\lambda_o)) \to H^0(v^{-1}X_w, \mathcal{L}(\lambda_o)_{|v^{-1}X_w})$$

(cf. [Ramanathan–87, Theorem 3.11(ii)]). From this, it is easy to prove the following refinement of (**): Fix any $\lambda \in D^o$. Then

$$T_{\dot e}(v^{-1}X_w) = \{X \in \mathfrak{n}^- : Xv_{\lambda_o} \in v^{-1}L_w(\lambda_o)\}.$$

In particular, for any $X \in \mathfrak{n}^-$,

$$Xv_{\lambda_o} \in v^{-1}L_w(\lambda_o) \Leftrightarrow Xv_\lambda \in v^{-1}L_w(\lambda) \text{ for all } \lambda \in D.$$

(10) Let $\mathcal{G}$ be a Kac–Moody group of finite type which does not contain any factors of type G_2. Then show that, under the assumptions of Proposition 12.1.12, d is the multiplicity of the point $\dot{v} \in X_w$.

Hint. Use [Whitney–72, Chapter 7, Theorems 4D and 7P] and [Carrell–97, Theorem 1].

12.2. Rational Smoothness of Schubert Varieties

The aim of this section is to give various criteria for rational smoothness.

We begin by recalling the definition of the Hecke algebra.

12.2.1 Definition. Let $(\mathcal{W}, S)$ be any Coxeter group (cf. 1.3.12) and let $\mathcal{H} = \mathcal{H}(\mathcal{W}, S)$ be the associative algebra with identity over the ring $A := \mathbb{Z}[q^{\frac{1}{2}}, q^{-\frac{1}{2}}]$ of Laurent polynomials with generators $\{T_s\}_{s\in S}$ and the following relations:

(R_1) $(T_s + 1)(T_s - q) = 0$, for $s \in S$.

(R_2) braid property: For $s \neq t \in S$ such that st is of order $m_{s,t} < \infty$,

$$T_sT_t \cdots = T_tT_s \cdots, \text{ where both the sides have } m_{s,t} \text{ terms.}$$

Then $\mathcal{H}$ is called the *Hecke algebra* associated to $(\mathcal{W}, S)$.

For any $w \in \mathcal{W}$, define

$$T_w := T_{s_1} \cdots T_{s_n} \in \mathcal{H}, \tag{1}$$

where $w = s_1 \cdots s_n$ is a reduced decomposition with $s_i \in S$. We set $T_e = 1$. By the braid property (R_2), (1) uniquely defines T_w, i.e., T_w does not depend on the choice of the reduced decomposition of w. Now, it is easy to see (cf. [Bourbaki–81, Chap. IV, §2, Exercise 23]) that $\mathcal{H}$ is free as an A-module with basis $\{T_w\}_{w\in\mathcal{W}}$. By the definition,

$$T_vT_w = T_{vw}, \quad \text{for } v, w \in \mathcal{W} \text{ such that } \ell(vw) = \ell(v) + \ell(w). \tag{2}$$

Moreover, by (R_1), T_s is an invertible element of $\mathcal{H}$ with inverse

$$T_s^{-1} = q^{-1}T_s + (q^{-1} - 1). \tag{3}$$

Hence T_w (for any $w \in \mathcal{W}$) is invertible.

Define an involution, i.e., an algebra automorphism of order 2, of the $\mathbb{Z}$-algebra $\mathcal{H}$ by

(4) $$\overline{q^{\frac{1}{2}}} = q^{-\frac{1}{2}}, \; \bar{T}_s = T_s^{-1} \quad \text{for } s \in S.$$

(It is easy to see that the relations (R_1)–(R_2) are stable under - , and hence it indeed goes down to an involution of $\mathcal{H}$.) An element $a \in \mathcal{H}$ is called *self-dual* if $\bar{a} = a$. From (4), we get

(5) $$\bar{T}_w = T_{w^{-1}}^{-1}, \quad \text{for any } w \in \mathcal{W}.$$

Set

(6) $$T'_w = q^{-\frac{1}{2}\ell(w)} T_w.$$

12.2.2 Theorem. *Let $\mathcal{W}$ be any Coxeter group. For any $w \in \mathcal{W}$, there exists a unique self-dual element C'_w satisfying*

(1) $$C'_w \in T'_w + \sum_{x<w} v\mathbb{Z}[v]\, T'_x, \quad \textit{where } v := q^{-\frac{1}{2}}.$$

In fact, for any $w \in \mathcal{W}$, there exists a unique self-dual element in $T'_w + \sum_x v\mathbb{Z}[v]\, T'_x$.

Proof. We first prove the existence by induction on $\ell(w)$. For $\ell(w) = 0$, i.e., $w = e$, take $C'_e = 1$. Next, for $w = s \in S$, define

(2) $$C'_s = T'_s + v.$$

By (12.2.1.3), C'_s is self-dual. Moreover, C'_s satisfies

(3) $$\begin{aligned} T'_w C'_s &= T'_{ws} + vT'_w, && \text{if } ws > w \\ &= T'_{ws} + v^{-1}T'_w, && \text{if } ws < w. \end{aligned}$$

Now, we come to the existence of C'_w for $\ell(w) > 1$. Choose $s \in S$ such that $ws < w$. By induction, self-dual C'_x's exist satisfying (1) for all $x < w$. By (3) and Corollary 1.3.19, it is easy to see that

$$C'_{ws} \cdot C'_s = T'_w + \sum_{x<w} h_x T'_x, \text{ for some } h_x \in \mathbb{Z}[v].$$

Define

(4) $$C'_w = C'_{ws} C'_s - \sum_{x<w} h_x(0)\, C'_x.$$

Then C'_w is self-dual and satisfies (1). This completes the proof of the existence.

To prove the stronger uniqueness, it suffices to show that any self-dual element $a \in \sum_x v\mathbb{Z}[v]\, T'_x$ satisfies

$$(5) \qquad a = 0.$$

By the property (1) of C'_w,

$$(6) \qquad T'_w \in C'_w + \sum_{x<w} v\mathbb{Z}[v]\; C'_x\,;$$

in particular,

$$(7) \qquad \overline{T'_w} \in T'_w + \sum_{x<w} \mathbb{Z}[v, v^{-1}]\, T'_x.$$

Write $a = \sum_x a_x T'_x$ for $a_x \in v\mathbb{Z}[v]$ and, if $a \neq 0$, choose x_o of maximal length such that $a_{x_o} \neq 0$. Then $a = \bar{a}$ implies $a_{x_o} = \bar{a}_{x_o}$, a contradiction, since $a_{x_o} \in v\mathbb{Z}[v]$. This proves (5), thereby establishing the uniqueness. □

12.2.3 Definition. By virtue of Theorem 12.2.2 write, for any $w \in \mathcal{W}$,

$$(1) \qquad C'_w = \sum_{x \leq w} P'_{x,w}\, T'_x\,,$$

for some unique $P'_{x,w} \in A$. In fact, $P'_{w,w} = 1$ and, for $x < w$, $P'_{x,w} \in q^{-\frac{1}{2}}\mathbb{Z}[q^{-\frac{1}{2}}]$.

The *Kazhdan–Lusztig polynomials* $P_{x,w}$, for $x \leq w$, are defined by

$$(2) \qquad P_{x,w} := q^{\frac{1}{2}\,\ell(x,w)}\, P'_{x,w}\,,$$

where $\ell(x, w) := \ell(w) - \ell(x)$. Then

$$(3) \qquad P_{w,w} = 1.$$

Also write, for any $w \in \mathcal{W}$,

$$(4) \qquad \overline{T'_w} = \sum_{x \in \mathcal{W}} q^{\frac{1}{2}\ell(x,w)}\, \bar{R}_{x,w}\, T'_x\,, \qquad \text{for } R_{x,w} \in A.$$

The following lemma is easy to prove from (12.2.1.2) (using Corollary 1.3.19). It provides an inductive procedure to calculate $R_{x,w}$.

12.2.4 Lemma. *Let s be a simple reflection such that $ws < w$. Then*

$$\text{(1)} \qquad \begin{aligned} R_{x,w} &= R_{xs,ws}\,, && \text{if } xs < x \\ &= (q-1)R_{x,ws} + qR_{xs,ws}\,, && \text{if } xs > x. \end{aligned}$$

Thus, by induction on $\ell(w)$, $R_{x,w} \in \mathbb{Z}[q]$. Moreover, again by induction on $\ell(w)$, $R_{x,w} \neq 0$ iff $x \leq w$ and, in this case, $R_{x,w}$ is of degree exactly equal to $\ell(x, w)$. Further,

$$\text{(2)} \qquad R_{w,w} = 1, \qquad \text{for any } w.$$

Similarly,

$$\text{(3)} \quad \begin{aligned} R_{x,w} &= R_{sx,sw}\,, && \text{if } sx < x \text{ and } sw < w \\ &= (q-1)R_{x,sw} + qR_{sx,sw}\,, && \text{if } sx > x \text{ and } sw < w. \end{aligned}$$ □

12.2.5 Proposition. *For any $y \leq w$, $P_{y,w}$ belongs to $\mathbb{Z}[q]$ and has constant term 1. Moreover,*

$$\text{(1)} \qquad \deg P_{y,w} \leq \frac{1}{2}(\ell(y,w) - 1), \ \text{if } y < w.$$

Proof. We first prove, by induction on $\ell(w)$, that $P_{y,w} \in \mathbb{Z}[q^{\frac{1}{2}}]$ with constant term 1 (for any $y \leq w$): This is clearly true for $\ell(w) = 0$, so assume that $\ell(w) > 0$ and choose $s \in S$ such that $ws < w$. By (12.2.2.2) and (12.2.2.4), if we set $P'_{x,y} = 0$ if $x \not\leq y$,

$$\begin{aligned} C'_w = \sum_{ys>y} q^{-\frac{1}{2}} P'_{y,ws} T'_y + \sum_{ys<y} q^{\frac{1}{2}} P'_{y,ws} T'_y \\ + \sum_y P'_{ys,ws} T'_y - \sum_{x<w} \sum_y h_x(0)\, P'_{y,x} T'_y. \end{aligned}$$

Hence, for any $y \in \mathcal{W}$,

$$P'_{y,w} = q^{\frac{1}{2}c} P'_{y,ws} + P'_{ys,ws} - \sum_{x<w} h_x(0)\, P'_{y,x},$$

where $c = 1$, resp. -1, if $ys < y$, resp. $ys > y$. This gives the inductive formula, setting $P_{x,y} = 0$ if $x \not\leq y$,

$$\text{(2)} \qquad P_{y,w} = q^{\frac{1+c}{2}} P_{y,ws} + q^{\frac{1-c}{2}} P_{ys,ws} - \sum_{x<w} h_x(0)\, q^{\frac{1}{2}\ell(x,w)} P_{y,x}.$$

By the induction hypothesis and the above expression, we see that, for any $y \leq w$, $P_{y,w} \in \mathbb{Z}[q^{\frac{1}{2}}]$ with constant term 1.

We next show that, for any $y \leq w$, $P_{y,w} \in \mathbb{Z}[q]$: From (12.2.3.1) and (12.2.3.4),

$$\begin{aligned} C'_w &= \sum_{x \leq w} \bar{P}'_{x,w} \bar{T}'_x \\ &= \sum_{x \leq w} \bar{P}'_{x,w} \sum_y q^{\frac{1}{2}\ell(y,x)} \bar{R}_{y,x} T'_y. \end{aligned}$$

This gives, for any $y \leq w$,

$$P'_{y,w} = \sum_{x \leq w} q^{\frac{1}{2}\ell(y,x)} \bar{P}'_{x,w} \bar{R}_{y,x},$$

i.e., by Lemma 12.2.4,

(3) $$P_{y,w} - q^{\ell(y,w)} \bar{P}_{y,w} = q^{\ell(y,w)} \sum_{y < x \leq w} \bar{R}_{y,x} \bar{P}_{x,w}.$$

But $P_{y,w} \in \mathbb{Z}[q^{\frac{1}{2}}]$ (as proved above) and, moreover, by Definition (12.2.3.2), for $y < w$,

(4) $$\deg_q P_{y,w} \leq \frac{1}{2}(\ell(y, w) - 1).$$

For a fixed w, assume by induction that $P_{x,w} \in \mathbb{Z}[q]$ for all $y < x \leq w$. Then, from (3)–(4) and Lemma 12.2.4, $P_{y,w} \in \mathbb{Z}[q]$ (since, by (4), there are no cancellations in the left side of (3)). This, together with (4), completes the proof of the proposition. □

For any $y \leq w \in \mathcal{W}$, define $N_{y,w} = q^{\ell(y)} \sum_{y \leq x \leq w} R_{y,x}$.

12.2.6 Lemma. *Let $y \leq w \in \mathcal{W}$. Then,*

(a) $P_{x,w} = 1$, *for all $y \leq x \leq w$ iff $N_{x,w} = q^{\ell(w)}$, for all $y \leq x \leq w$.*

(b) *For $\ell(w) \leq \ell(y) + 2$,*

(b$_1$) $R_{y,w} = (q-1)^{\ell(y,w)}$, *and*
(b$_2$) $P_{y,w} = 1$.

Proof. (a): By induction we can assume that

(1) $P_{x,w} = 1$, for all $y < x \leq w$ iff $N_{x,w} = q^{\ell(w)}$, for all $y < x \leq w$.

Now assume that the equivalent conditions of (1) are satisfied. Then, by the formula (12.2.5.3),

(2) $$P_{y,w} - q^{\ell(y,w)}\,\bar{P}_{y,w} = q^{\ell(w)}\,\bar{N}_{y,w} - q^{\ell(y,w)}.$$

So $P_{y,w} = 1 \Rightarrow N_{y,w} = q^{\ell(w)}$. Conversely, $N_{y,w} = q^{\ell(w)} \Rightarrow$

(3) $$P_{y,w} - q^{\ell(y,w)}\,\bar{P}_{y,w} = 1 - q^{\ell(y,w)}.$$

But, by Proposition 12.2.5, deg $P_{y,w} \leq \frac{1}{2}(\ell(y,w) - 1)$, for $y < w$. Thus, (3) gives $P_{y,w} = 1$. This proves (a).

(b_1): If $\ell(y) = \ell(w)$, (b_1) is true by (12.2.4.2) and (b_2) is true by (12.2.3.3). Next, assume that $\ell(w) = \ell(y) + 1$ and write $y = s_1 \cdots \hat{s}_i \cdots s_n$, where $w = s_1 \cdots s_n$ is a reduced decomposition (cf. Lemma 1.3.16). Then, by Lemma 12.2.4,

$$\begin{aligned} R_{y,w} &= R_{s_1\cdots\hat{s}_i,s_1\cdots s_i} \\ &= (q-1)R_{s_1\cdots\hat{s}_i,s_1\cdots\hat{s}_i} \\ &= q - 1. \end{aligned}$$

Finally, consider the case $\ell(w) = \ell(y) + 2$ and write $y = s_1 \cdots \hat{s}_i \cdots \hat{s}_j \cdots s_n$, $i < j$, where $w = s_1 \cdots s_n$ is a reduced decomposition. Then, by Lemma 12.2.4,

$$\begin{aligned} R_{y,w} &= R_{s_1\cdots\hat{s}_i\cdots\hat{s}_j,s_1\cdots s_j} \\ &= R_{\hat{s}_i\cdots\hat{s}_j,s_i\cdots s_j} \\ &= (q-1)R_{\hat{s}_i\cdots\hat{s}_j,s_i\cdots\hat{s}_j}, && \text{since } \hat{s}_i \cdots s_j \not\leq s_i \cdots s_{j-1} \\ &= (q-1)^2, && \text{from the previous case.} \end{aligned}$$

This proves (b_1).

To prove (b_2), in view of the (a) part, it suffices to show that

(4) $$N_{y,w} = q^{\ell(w)}, \qquad \text{for } \ell(w) \leq \ell(y) + 2.$$

But (4) follows from (b_1) and Lemma 9.2.2(a). □

12.2.7 Definition. Let X be an algebraic variety over $\mathbb{C}$ of pure dimension d, i.e., each irreducible component of X is of dimension d. Then a point $x \in X$ is said to be *rationally smooth* if there exists an open set (in the analytic topology) $U_x \subset X$ containing x such that, for all $y \in U_x$, the singular cohomology

(1) $$\begin{aligned} H^j(X, X\backslash\{y\}, \mathbb{Q}) &= 0, && \text{if } j \neq 2d, \\ &\simeq \mathbb{Q}, && \text{if } j = 2d. \end{aligned}$$

The variety X itself is said to be rationally smooth if each point of X is rationally smooth. Clearly, each smooth point of X is rationally smooth and the set of rationally smooth points of X is open in the analytic topology. In fact, the set of rationally smooth points is open even in the Zariski topology, as can be seen by considering the Whitney stratification of X. Let $\Sigma_r(X)$ denote the Zariski closed subset consisting of those point $x \in X$ which are *not* rationally smooth. Then $\Sigma_r(X) \subset \Sigma(X)$.

Following the discussion as in the beginning of Section 12.1, for a Schubert variety $X_w \subset \mathcal{G}/\mathcal{B}$, where $\mathcal{G}$ is any Kac–Moody group, the point $\dot{v} \in X_w$ is rationally smooth iff $b\dot{v}$ is rationally smooth for any $b \in \mathcal{B}$. Thus $\Sigma_r(X_w)$ is a closed $\mathcal{B}$-stable subset of X_w. Hence, to determine $\Sigma_r(X_w)$, it suffices to check the rational smoothness of the points $\{\dot{v}\}_{v \leq w}$.

We recall the following most basic criterion of rational smoothness due to Kazhdan–Lusztig:

12.2.8 Theorem. *For any $v \leq w \in W$, the point $\dot{v} \in X_w$ is rationally smooth iff $P_{y,w} = 1$ for all $v \leq y \leq w$, where W is the Weyl group associated to the Kac–Moody group $\mathcal{G}$.*

We also recall the following fundamental result, again due to Kazhdan–Lusztig:

12.2.9 Theorem. *For any $v \leq w \in W$,*

$$(1) \qquad P_{v,w} = \sum_i \dim \mathcal{H}^{2i}_{\dot{v}}(X_w)\, q^i,$$

where $\mathcal{H}^{2i}_{\dot{v}}(X_w)$ denotes the stalk at $\dot{v}$ of the middle intersection cohomology sheaf of X_w in degree $2i$.

In particular, the coefficients of the Kazhdan–Lusztig polynomial $P_{v,w}$ are nonnegative (integers) for any crystallographic Coxeter group W.

Proofs of both of the above theorems involve considering the flag variety $\mathcal{G}/\mathcal{B}$ over finite fields, the ℓ-adic cohomology and the Weil conjecture (proved by P. Deligne). The proof of Theorem 12.2.8 is given in [Kazhdan–Lusztig–79, Appendix] in the finite case and the proof of Theorem 12.2.9 is given in [Kazhdan–Lusztig–80] in the finite as well as the affine case. The proof of Theorem 12.2.9 in the general Kac–Moody case is similar and is given in [Haddad–84]. A different proof of Theorem 12.2.9 over $\mathbb{C}$ in the finite case, due to MacPherson, is sketched in [Springer–82].

12.2.10 Remark. It is still an open conjecture [Kazhdan–Lusztig–79, §1] that, for any Coxeter group $\mathcal{W}$ and $v \leq w \in \mathcal{W}$, the polynomial $P_{v,w}$ has nonnegative coefficients.

Combining Theorem 12.2.8 with Lemma 12.2.6(b), we immediately get the following:

12.2.11 Corollary. *With the notation as in 12.2.7, any irreducible component of $\Sigma_r(X_w)$ is of codimension ≥ 3.* □

Contrast the above corollary with the singular locus $\Sigma(X_w)$ which is, in general, only of codimension ≥ 2.

As a consequence of Lemma 12.2.4, we prove the following, which will crucially be used in the proof of Proposition 12.2.13.

12.2.12 Lemma. *Let $\mathcal{W}$ be any Coxeter group. Then, for any $v < w \in \mathcal{W}$,*

(a) $R_{v,w}$ is divisible by $q-1$. In fact,

(b$_1$) $R_{v,w}$ is divisible by $(q-1)^2$, if $v^{-1}w \notin T$, and

(b$_2$) $R_{v,w} - (q-1)\,q^{\frac{1}{2}(\ell(v,w)-1)}$ is divisible by $(q-1)^2$, if $v^{-1}w \in T$, where, as in 1.3.15, $T := \bigcup_{w\in\mathcal{W}} wSw^{-1}$ is the set of "reflections."

Observe that if $v^{-1}w \in T$, $\ell(v,w)$ is odd.

Proof. Take a simple reflection $s \in S$ such that $ws < w$. If $vs < v$ then, by Lemma 12.2.4, $R_{v,w} = R_{vs,ws}$. Further, in this case, $vs < ws$ by Corollary 1.3.19. So, by induction on $\ell(w)$, the lemma follows in this case. Observe that $v^{-1}w \in T$ iff $(vs)^{-1}ws \in T$.

Assume now that $vs > v$. Then, by (12.2.4.1),

$$(1) \qquad R_{v,w} = (q-1)\,R_{v,ws} + q\,R_{vs,ws}.$$

So, again by induction on $\ell(w)$, (a) and (b$_1$) follow. Observe that, in the case (b$_1$), $v < ws$ and if $vs \nleq ws$ then $R_{vs,ws} = 0$ by Lemma 12.2.4.

Finally, we come to the case (b$_2$) (under the assumption $vs > v$). By (1),

(2)
$$R_{v,w} - (q-1)\,q^{\frac{1}{2}(\ell(v,w)-1)} = (q-1)\,R_{v,ws} + q\Big[R_{vs,ws} - (q-1)\,q^{\frac{1}{2}(\ell(vs,ws)-1)}\Big].$$

Now, by Corollary 1.3.19, $v \leq ws$. Further, by the definition of the Bruhat–Chevalley order (cf. Definition 1.3.15), either $v = ws$ or $vs < ws$. (If $v \neq ws$ and $vs \nleq ws$, then $vs > ws$ since $wv^{-1} \in T$ by assumption, which contradicts $v < ws$.) If $v = ws$, by Lemma 12.2.6(b), $R_{v,w} = q-1$. Thus $R_{v,w}$ satisfies (b$_2$) in this case. So assume that $vs < ws$. In this case, by induction on $\ell(w)$, (a) and (2), $R_{v,w}$ again satisfies (b$_2$). This proves (b$_2$) and hence the lemma. □

12.2.13 Proposition. *For any $v \leq w \in \mathcal{W}$,*

$$(1) \qquad \frac{d}{dq}\Big(q^{\ell(v,w)}P_{v,w}(q^{-2})\Big)_{q=1} = \sum_{\substack{t\in T\\ v<tv\leq w}} P_{tv,w}(1).$$

Proof. Applying the involution to the two sides of (12.2.5.3), we get

$$\begin{aligned} q^{\ell(v,w)}\bar{P}_{v,w} - P_{v,w} &= \sum_{v<x\leq w} R_{v,x}\, P_{x,w} \\ &= \sum_{\substack{v<x\leq w\\ v^{-1}x\in T}} R_{v,x}\, P_{x,w} + \sum_{\substack{v<x\leq w\\ v^{-1}x\notin T}} R_{v,x}\, P_{x,w} \\ &= \sum_{\substack{v<x\leq w\\ v^{-1}x\in T}} \left[R_{v,x} - (q-1)\, q^{\frac{1}{2}(\ell(v,x)-1)}\right] P_{x,w} \end{aligned}$$

$$(2) \qquad + \sum_{\substack{v<x\leq w\\ v^{-1}x\in T}} (q-1)\, q^{\frac{1}{2}(\ell(v,x)-1)}\, P_{x,w} + \sum_{\substack{v<x\leq w\\ v^{-1}x\notin T}} R_{v,x}\, P_{x,w}.$$

Now take $\frac{d}{dq}$ of (2) and evaluate at $q=1$ to get (using Lemma 12.2.12):

$$\frac{d}{dq}\left(q^{\ell(v,w)}\,\bar{P}_{v,w} - P_{v,w}\right)_{q=1} = \sum_{\substack{v<x\leq w\\ v^{-1}x\in T}} P_{x,w}(1).$$

But it is easy to see that the left side of the above equation is equal to $\frac{d}{dq}\left(q^{\ell(v,w)}\, P_{v,w}(q^{-2})\right)_{q=1}$. This proves the proposition. □

As an easy consequence of the above proposition and Deodhar's conjecture Corollary 12.1.8, we get the following interesting characterization of rationally smooth points due to Carrell–Peterson.

12.2.14 Theorem. *Let W be the Weyl group associated to a Kac–Moody group $\mathcal{G}$. Then, for any $v \leq w \in W$, the following are equivalent:*

(a) $P_{v,w} = 1$,

(b) $P_{y,w} = 1$, *for all* $v \leq y \leq w$,

(c) $\# S(w, y) = \ell(w)$, *for all* $v \leq y \leq w$,

(d) $\# E(X_w, \dot{y}) = \ell(w)$, *for all* $v \leq y \leq w$,

where $S(w, y) := \{\beta \in \Delta^+_{\mathrm{re}} : s_\beta y \leq w\}$ is as in 11.1.18 and $E(X_w, \dot{y})$ is as defined in 12.1.5.

Hence, by Theorem 12.2.8, $\dot{v} \in X_w \subset \mathcal{G}/\mathcal{B}$ is rationally smooth iff any of the above equivalent conditions is satisfied.

Proof. (a) ⇒ (b): By Proposition 12.2.13, applied to $P_{v,w} = 1$, we get

$$(1) \qquad \ell(v, w) = \sum_{\substack{\beta\in\Delta^+_{\mathrm{re}}\\ v<s_\beta v\leq w}} P_{s_\beta v, w}(1).$$

But, by Corollary 12.1.8,

$$(2) \qquad \# \{\beta \in \Delta^+_{\text{re}} : v < s_\beta v \leq w\} \geq \ell(v, w),$$

since

$$(3) \quad \# \{\beta \in \Delta^+_{\text{re}} : s_\beta v < v\} = \ell(v), \qquad \text{by (11.1.15.4) and Lemma 1.3.14.}$$

Moreover, for any $x \leq w$, $P_{x,w}$ has nonnegative integral coefficients (cf. Theorem 12.2.9) and the constant term of $P_{x,w}$ equals 1 (cf. Proposition 12.2.5). So (1) and (2) together force

$$(4) \qquad \# \{\beta \in \Delta^+_{\text{re}} : v < s_\beta v \leq w\} = \ell(v, w),$$

as well as

$$(5) \qquad P_{s_\beta v, w} = 1, \qquad \text{for any } \beta \in \Delta^+_{\text{re}} \text{ satisfying } v < s_\beta v \leq w.$$

Iterating (5), we get that $P_{y,w} = 1$, for any $v \leq y \leq w$. This proves (b).

The identities (3) and (4) show that, for any $y \leq w$ with $P_{y,w} = 1$, we have $\# S(w, y) = \ell(w)$. This proves (b) $\Rightarrow$ (c).

(c) $\Rightarrow$ (a): By induction on $\ell(v, w)$, we can assume that for any $\beta \in \Delta^+_{\text{re}}$ such that $v < s_\beta v \leq w$, we have

$$(6) \qquad P_{s_\beta v, w} = 1.$$

Substituting this in Proposition 12.2.13, we get

$$(7) \qquad \begin{aligned} \frac{d}{dq}\left(q^{\ell(v,w)}\, P_{v,w}(q^{-2})\right)_{q=1} &= \# \{\beta \in \Delta^+_{\text{re}} : v < s_\beta v \leq w\} \\ &= \ell(v, w) \qquad \text{by (c) and (3).} \end{aligned}$$

But since the constant term of $P_{v,w}$ is 1, $\deg P_{v,w}(q) \leq \frac{1}{2}(\ell(v, w) - 1)$ and all the coefficients of $P_{v,w}$ are non-negative, (7) forces $P_{v,w} = 1$. This proves (a). The equivalence of (c) and (d) follows from (12.1.7.2). This completes the proof of the theorem. □

12.2.15 Remark. The equivalence of (a), (b), and (c) in Theorem 12.2.14 is valid (by the same proof) for any Coxeter group $\mathcal{W}$ such that all the $P_{y,w}$ have nonnegative coefficients.

The following theorem provides yet another criterion for rational smoothness of points $\dot{v}$ in Schubert varieties X_w. Compare this with the smoothness criterion of Theorem 12.1.11.

12.2.16 Theorem. *Let $\mathcal{G}$ be a Kac–Moody group with the associated Weyl group W. Then, for any $v \leq w \in W$, $\dot{v} \in X_w$ is rationally smooth iff for all $v \leq y \leq w$,*

$$(1) \qquad c_{w,y} = k_{w,y} \prod_{\beta \in S(w,y)} \beta^{-1}, \textit{ for some (nonzero) } k_{w,y} \in \mathbb{Q}.$$

In fact, if $\dot{v} \in X_w$ is rationally smooth,

$$(2) \qquad (-1)^{\ell(w,y)} k_{w,y} \textit{ is a positive integer for all } v \leq y \leq w.$$

Proof. Assume first that $\dot{v} \in X_w$ is rationally smooth. Then, by Theorem 12.2.14, for all $v \leq y \leq w$, $\# S(w, y) = \ell(w)$. Now, applying Proposition 12.1.12 and (12.1.4.2), we get (1) and (2). (Observe that (1) also follows from Proposition 11.1.19.)

Conversely, assuming (1), from degree considerations (cf. Theorem 11.1.2(f)) we get $\# S(w, y) = \ell(w)$, for all $v \leq y \leq w$. This proves that $\dot{v} \in X_w$ is rationally smooth by Theorem 12.2.14. □

12.2.17 Remark. There are examples of $v \leq w \in W$ such that $c_{w,v}$ satisfies the property (1) of the above theorem but $c_{w,y}$ fails to satisfy (1) for some $v < y \leq w$. In particular, to check the rational smoothness of $\dot{v} \in X_w$ it is *not* sufficient to check (1) only for $c_{w,v}$ (in contrast to the corresponding condition (12.1.11.1) for smoothness).

For example, in all the cases covered by Exercise 12.2.E.6 which do *not* appear in Exercise 12.2.E.7, the property (1) is satisfied for $c_{w_o s_i, e}$ but is violated for some $c_{w_o s_i, y}$ $(e < y \leq w_o s_i)$.

12.2.E EXERCISES

(1) Let $(\mathcal{W}, S)$ be any Coxeter group and $v \leq w \in \mathcal{W}$. Define the oriented graph $\Gamma = \Gamma(v, w)$ called the *Bruhat graph* as follows: The set of vertices of Γ consists of all $v \leq x \leq w$. Further, there is an oriented edge from x to y if $x < y$ and $xy^{-1} \in T := \bigcup_{w \in \mathcal{W}} wSw^{-1}$.

Counting the edges of Γ in two different ways, show that

$$\sum_{v \leq x \leq w} \#\{t \in T : x < tx \leq w\} = \sum_{v \leq y \leq w} \#\{t \in T : v \leq ty < y\}.$$

(2) For any $w \in \mathcal{W}$, define the *Poincaré polynomial* $\chi_w(q) := \sum_{v \leq w} q^{\ell(v)}$, and set $a_w = \frac{\frac{d}{dq}(\chi_w)_{q=1}}{\chi_w(1)}$. Then a_w can be thought of as the average of $\ell(v)$ for $v \leq w$. Show that $a_w \geq \frac{1}{2}\ell(w)$.

Now let W be the Weyl group associated to a Kac–Moody group. Then prove that the following four properties are equivalent for any $w \in W$:

(a) χ_w is palindromic, i.e., $q^{\ell(w)}\chi_w(q^{-1}) = \chi_w(q)$.

(b) $a_w = \frac{1}{2}\,\ell(w)$.

(c) $\#\, S(w, v) = \ell(w)$, for all $v \leq w$.

(d) The Schubert variety X_w is rationally smooth.

Hint. The assertion "(a) $\Rightarrow$ (b)" follows easily from the definition of a_w; to prove "(b) $\Rightarrow$ (c)" use Exercise (1) and Deodhar's conjecture Corollary 12.1.8; "(c) $\Rightarrow$ (d)" is given by Theorem 12.2.14; and "(d) $\Rightarrow$ (a)" follows from the validity of the Poincaré duality for X_w over $\mathbb{Q}$.

(3) Let $\mathcal{G}$ be the Kac–Moody group associated to a 2×2 GCM. Then show that, for any $w \in W$, X_w is rationally smooth.

Hint. Use the equivalence of (a) and (d) in the previous exercise.

(4) Prove the following complete description of the singular locus of Schubert varieties in the case of rank two simple groups of finite type. We follow the convention as in [Bourbaki–81].

(a) $\mathcal{G}$ of type A_2: In this case all the six Schubert varieties are smooth.

(b) $\mathcal{G}$ of type C_2: There are, in all, eight Schubert varieties. Out of these only $X_{s_1s_2s_1}$ is singular and $\Sigma(X_{s_1s_2s_1}) = X_{s_1}$.

(c) $\mathcal{G}$ of type G_2: There are, in all, twelve Schubert varieties. Following is the complete list of singular ones and their singular loci:

	$\Sigma(X_w)$
(1) $X_{s_1s_2s_1}$	$-\ X_{s_1}$
(2) $X_{s_1s_2s_1s_2}$	$-\ X_{s_1s_2}$
(3) $X_{s_2s_1s_2s_1}$	$-\ X_{s_2s_1}$
(4) $X_{s_1s_2s_1s_2s_1}$	$-\ X_{s_1s_2s_1}$
(5) $X_{s_2s_1s_2s_1s_2}$	$-\ X_{s_2}$.

Hint. Use Exercise (3), Example 12.1.14 and Theorems 12.1.11 and 12.2.16.

(5) Let $\mathcal{G}$ be of finite type. For any $1 \leq i \leq \ell$, let X_i be the codimension-one Schubert variety as in Exercise 12.1.E.7. Show that, for any $v \leq w_o s_i$, $\dot{v} \in X_i$ is rationally smooth iff $\omega_i - y^{-1}w_o\omega_i$ is a multiple of a root β_y (depending on y) for all $v \leq y \leq w_o s_i$.

Hint. Use Theorem 12.2.14 together with Exercise 12.1.E.7(d).

(6) Let $\mathfrak{g}$ be a finite-dimensional simple Lie algebra of rank ℓ. Then, for any fundamental weight ω_i, $1 \leq i \leq \ell$, prove that $\omega_i - w_o\omega_i$ is a multiple of a root but

not a root itself precisely in the following cases (again following the convention from [Bourbaki–81]):

	i
(a_1) B_n ($n \geq 3$)	: 1, 2
(a_2) C_n ($n \geq 2$)	: 2
(a_3) D_n ($n \geq 4$)	: 2
(a_4) E_6	: 2
(a_5) E_7	: 1
(a_6) E_8	: 8
(a_7) F_4	: 1, 4
(a_8) G_2	: 1, 2.

(7) Let $\mathcal{G}$ be a simple group of finite type. Then show that following is the complete list of codimension-one Schubert varieties X_i which are rationally smooth but *not* smooth:

(a) C_2 : X_2
(b) G_2 : X_1, X_2
(c) B_n ($n \geq 3$) : X_1.

(8) Let $\mathcal{G}$ be a Kac–Moody group with the associated Weyl group W and let $v \leq w \in W$. Assume that the reduced tangent cone (cf. A.4) $C^{\text{red}}_{\dot{y}}(X_w)$ is an affine space for all $v \leq y \leq w$. Then show that $\dot{v} \in X_w$ is rationally smooth.

Assume now that $\mathcal{G}$ is simply-laced of finite type. Then, show that the converse is true, i.e., if $\dot{v} \in X_w$ is rationally smooth then $C^{\text{red}}_{\dot{y}}(X_w)$ is an affine space for all $v \leq y \leq w$.

The converse is, in general, false for non-simply-laced $\mathcal{G}$. Take, e.g., $\mathcal{G}$ of type G_2 or C_2 and $w = s_1s_2s_1$, $v = e$. Then show that $C^{\text{red}}_{\dot{e}}(X_w)$ is *not* an affine space.

Hint. For the first part, use Proposition 12.1.7 and Deodhar's conjecture together with the general fact that $\dim C^{\text{red}}_y(S) = \dim S$ for any irreducible variety S and any $y \in S$ (cf. (A.4.2)).

(9) Recall the definition of the ring $Q(T)$ from 12.1.2 and define the ring (obtained by extending the scalars from $\mathbb{Z}$ to $\mathbb{Z}[q^{\frac{1}{2}}, q^{-\frac{1}{2}}]$):

$$Q_q(T) = \mathbb{Z}[q^{\frac{1}{2}}, q^{-\frac{1}{2}}] \underset{\mathbb{Z}}{\otimes} Q(T).$$

Exactly similar to the definition of the ring $Q(T)_W$ as in 12.1.2(b), define the ring $Q_q(T)_W$ with $q^{\pm\frac{1}{2}}$ being central. We define a certain q-deformation $y^q_{s_i}$ of the element y_{s_i} (for any simple reflection s_i) as follows:

$$y^q_{s_i} := \frac{qe^{-\alpha_i} - 1}{1 - e^{\alpha_i}}\delta_{s_i} + \frac{q - 1}{1 - e^{-\alpha_i}}\delta_e \in Q_q(T)_W.$$

Then prove the following:

(a) $(y^q_{s_i} + 1)(y^q_{s_i} - q) = 0$.

(b) $y^q_{s_i}$'s satisfy the braid property.

Thus the $\mathbb{Z}[q^{\frac{1}{2}}, q^{-\frac{1}{2}}]$-subalgebra $\mathcal{H}'$ of $Q_q(T)_W$ generated by $\{y^q_{s_i}\}_{1\le i\le \ell}$ is isomorphic with the Hecke algebra $\mathcal{H}$ associated to (W, S) defined in 12.2.1 (for the Weyl group W of any Kac–Moody algebra) under the map $y^q_{s_i} \mapsto T_{s_i}$.

Further, define an involution τ of $Q_q(T)_W$ by $\tau(q^{\frac{1}{2}}) = q^{-\frac{1}{2}}$, $\tau(e^\lambda) = e^{-\lambda}$ for any $e^\lambda \in X(T)$, and $\tau(\delta_w) = e^{-2\rho}\delta_w e^{2\rho}$. Then show that

(c) τ keeps $\mathcal{H}'$ stable, and

(d) τ corresponds to the involution as defined in (12.2.1.4) under the above identification.

12.C Comments. Definition 12.1.2(b) is taken from [Kostant–Kumar–90]. The identity (12.1.4.2) is due to [Rossmann–89]. Propositions 12.1.6 and 12.1.7 are due to Carrell–Peterson (cf. [Carrell–94]). Corollary 12.1.8 was obtained by [Polo–94] in the finite case and by Carrell–Peterson [Carrell–94] in the general Kac–Moody case. Recall that, in Corollary 11.1.20, we gave another proof due to [Dyer–93] of a more general result for an arbitrary Coxeter group, using the nil-Hecke ring. The description of the Zariski tangent space given in Exercise 12.1.E.9 is due to [Polo–94] in the finite case (also see [Kumar–96] for a different proof). [Kumar–96] attributes Exercise 12.1.E.8 to the referee of his paper (although a different geometric proof is given in [Kumar–96]), and the proof of (12.1.4.1) given here is somewhat simpler than the original proof due to [Kumar–96] and is attributed to the referee of this paper. All the other main results of Section 12.1, including Theorems 12.1.3, 12.1.11, Lemma 12.1.4, Proposition 12.1.12, and Exercises 12.1.E.4, 7, 10 are taken from [Kumar–96]. Exercise 12.1.E.7(e) and (f) parts are also due to [Carrell–95].

The results in 12.2.2–12.2.11 are taken from the classic papers [Kazhdan–Lusztig–79, 80]. (We have given a slightly different formulation of Theorem 12.2.2, taken from [Soergel–97].) For a survey of KL-theory, see [Deodhar–94]. Some explicit calculations of KL polynomials and also the R-polynomials can be found on various websites, for example:

http://www.math.ias.edu/~goresky/tables.html ,

http://www.desargues.univ-lyons.fr/home/ducloux/coxeter.html.

Lemma 12.2.12 and Proposition 12.2.13 are due to [Deodhar–82b], Theorem 12.2.14 is due to Carrell–Peterson [Carrell–94] and Theorem 12.2.16 is due to [Kumar–96]. Exercises 12.2.E.1, 2, 8 are due to Carrell–Peterson [Carrell–94]. (A different proof of Exercise 12.2.E.8 is given in [Kumar–96].) Exercise 12.2.E.3 is well known. Exercises 12.2.E.4, 5, 6, 7 are taken from [Kumar–96]. Exercise

12.2.E.7(c) appears also in [Boe–88].

We recall only a few special results in the finite case. For a more detailed treatment in this case, we refer to the book [Billey–Lakshmibai–00]. D. Peterson proved that, in the simply-laced groups of finite type, a Schubert variety is smooth iff it is rationally smooth (cf. [Carrell–Kuttler–99]) (This generalizes the corresponding result for SL_n due to [Deodhar–85].) In the case of classical groups of finite type, [Billey–98b], [Lakshmibai–Sandhya–90], [Lakshmibai–Song–97] explicitly determined which of the Schubert varieties are smooth/rationally smooth in terms of "pattern avoidance." An explicit description of the irreducible components of the singular locus of any Schubert variety in the case of SL_n is obtained by [Billey–Warrington–2001], [Manivel–2001] and [Kassel–Lascoux–Reutenauer–2001] by proving a conjecture of [Lakshmibai–Sandhya–90].

Finally, we refer to the articles [Brion–98, 99] and [Arabia–98] for different and more general viewpoints to smoothness/rational smoothness questions.

XIII

An Introduction to Affine Kac–Moody Lie Algebras and Groups

The collection of indecomposable Kac–Moody algebras is divided into three mutually exclusive types: finite, affine, and indefinite. The finite type indecomposable Kac–Moody algebras are precisely the finite-dimensional simple Lie algebras. The aim of this chapter is to explicitly realize the Kac–Moody algebras of affine type (also called the affine Kac–Moody algebras) and the associated groups. Most of the important applications of Kac–Moody theory so far center around this type. Actually, we only consider the Kac–Moody algebras $\mathfrak{g}$ and the associated groups $\mathcal{G}$ of "untwisted" affine type. The "twisted" ones are obtained from the untwisted $\mathfrak{g}$ as the subalgebra consisting of the elements fixed under an automorphism of $\mathfrak{g}$ of finite order. The twisted affine Kac–Moody groups are obtained similarly.

In Section 13.1, we realize any (untwisted) affine Kac–Moody algebra $\mathfrak{g} = \mathfrak{g}(A)$ as a specific (in fact, universal) central extension $\hat{\mathcal{L}}(\mathring{\mathfrak{g}})$ of the loop algebra $\mathring{\mathfrak{g}} \otimes \mathbb{C}[t, t^{-1}]$ of a finite-dimensional simple Lie algebra $\mathring{\mathfrak{g}}$ adjoined with the degree derivation d. (The GCM $A = (a_{i,j})_{0 \le i,j \le \ell}$ is obtained from the Cartan matrix of $\mathring{\mathfrak{g}}$ by adding the 0-th row and the 0-th column as defined by (13.1.1.7).)

The realization of $\mathfrak{g}$ as $\hat{\mathcal{L}}(\mathring{\mathfrak{g}})$ allows us to explicitly determine the set of roots and the root space decomposition of $\mathfrak{g}$. We identify the Weyl group W of $\mathfrak{g}$ with the affine Weyl group $\mathrm{Aff}(\mathring{W})$, where $\mathring{W}$ is the (finite) Weyl group of $\mathring{\mathfrak{g}}$. In addition, we explicitly determine an invariant bilinear form on $\mathfrak{g}$ in terms of that of $\mathring{\mathfrak{g}}$ ($\mathfrak{g}(A)$ is indeed symmetrizable by Exercise 13.1.E.5). The set of imaginary roots, Cartan involution, Tits cone, $K^{\text{w.g.}}$ and the length function on $\mathrm{Aff}(\mathring{W})$ are explicitly determined in Exercises 13.1.E. As mentioned earlier in Chapter 1, this section can be read without a knowledge of Chapters 2–12.

In Section 13.2, we study the associated groups and the generalized flag varieties. Analogously the Kac–Moody group $\mathcal{G}$ associated to the affine Kac–Moody algebra $\mathfrak{g} \simeq \hat{\mathcal{L}}(\overset{\circ}{\mathfrak{g}})$ is explicitly realized as a central extension by $\mathbb{C}^*$ of the loop group consisting of the K-rational points $\overset{\circ}{G}(K)$ of $\overset{\circ}{G}$ extended by $\operatorname{Exp} d$, where $K := \mathbb{C}[[t]][t^{-1}]$ and $\overset{\circ}{G}$ is the simply-connected algebraic group with $\overset{\circ}{\mathfrak{g}}$ as its Lie algebra.

Various parabolic subgroups $\mathcal{P}_Y$ of $\mathcal{G}$ take a concrete form under this realization, e.g., for any $Y \subset \{1, \dots, \ell\}$, $\mathcal{P}_Y = \bar{e}^{-1}(P_Y)$ (up to the center $\mathbb{C}^*$ and $\operatorname{Exp} d$), where P_Y is the corresponding parabolic subgroup of $\overset{\circ}{G}$ and $\bar{e} : \overset{\circ}{G}(\mathbb{C}[[t]]) \to \overset{\circ}{G}$ is the homomorphism obtained by the evaluation at $t = 0$. The Bruhat decomposition takes the form $\mathcal{G} = \sqcup_{\bar{n} \in \operatorname{Mor}(\mathbb{C}^*, N(\overset{\circ}{T}))/\overset{\circ}{T}} (\mathcal{B}\bar{n}\mathcal{B})$, where $\overset{\circ}{T}$ is the maximal torus of $\overset{\circ}{G}$, $N(\overset{\circ}{T})$ is its normalizer in $\overset{\circ}{G}$, $\operatorname{Mor}(\mathbb{C}^*, N(\overset{\circ}{T}))$ denotes the group of all the regular maps from $\mathbb{C}^*$ to $N(\overset{\circ}{T})$ under the pointwise multiplication, and $\overset{\circ}{T}$ sits inside $\operatorname{Mor}(\mathbb{C}^*, N(\overset{\circ}{T}))$ as the subgroup of constant maps with values in $\overset{\circ}{T}$. The subgroup $\mathcal{G}^{\min}$ is realized as $\overset{\circ}{G}(\mathbb{C}[t, t^{-1}])$ (up to the center and $\operatorname{Exp} d$). Also an explicit realization of the adjoint representation of $\mathcal{G}$ is given in 13.2.3.

The generalized flag variety $\mathcal{X}^Y = \mathcal{G}/\mathcal{P}_Y$, for the subset $Y = \{1, \dots, \ell\}$ of $\{0, 1, \dots, \ell\}$, plays a particularly important role in several applications. In the case $\overset{\circ}{G} = \mathrm{SL}_N$, we realize $\mathcal{X}^Y$ as the set of $\mathcal{R}$-lattices L in the K-vector space $K \otimes_{\mathbb{C}} \mathbb{C}^N$ such that $t^n L_o \subset L \subset t^{-n} L_o$ and $\dim_{\mathbb{C}}(L/t^n L_o) = nN$ (for some $n \geq 0$), where $\mathcal{R} := \mathbb{C}[[t]]$, K is as above and $L_o := \mathcal{R} \otimes_{\mathbb{C}} \mathbb{C}^N$.

This identification gives rise to a natural projective ind-variety structure on $\mathcal{X}^Y$. It is further shown (by using the Bott–Samelson–Demazure–Hansen varieties) that this ind-variety structure on $\mathcal{X}^Y$ coincides with the projective ind-variety structure defined in Section 7.1. (The case of general $\overset{\circ}{G}$ is handled by using an embedding $\overset{\circ}{G} \hookrightarrow \mathrm{SL}_N$.) Finally, we determine the Picard group of $\mathcal{X}^Y$ and show that it is isomorphic with $\mathbb{Z}$ generated by the homogeneous line bundle corresponding to the 0-th fundamental weight.

13.1. Affine Kac–Moody Lie Algebras

13.1.1 Definition. Let $\overset{\circ}{\mathfrak{g}} := \mathfrak{g}(\overset{\circ}{A})$ be the (finite-dimensional) simple Lie algebra associated to an indecomposable GCM of finite type $\overset{\circ}{A} = (a_{i,j})_{1 \leq i,j \leq \ell}$. Let $\mathcal{A} := \mathbb{C}[t, t^{-1}]$ be the algebra of Laurent polynomials. Define the *(untwisted) affine Kac–Moody Lie algebra*

$$\hat{\mathcal{L}}(\overset{\circ}{\mathfrak{g}}) := (\mathcal{A} \otimes_{\mathbb{C}} \overset{\circ}{\mathfrak{g}}) \oplus \mathbb{C}c \oplus \mathbb{C}d, \tag{1}$$

under the Lie bracket

$$\text{(2)}\quad \left[t^m \otimes x + \lambda c + \mu d, t^{m'} \otimes x' + \lambda' c + \mu' d\right] = t^{m+m'} \otimes [x, x'] + \mu m' t^{m'} \otimes x' - \mu' m t^m \otimes x + m\delta_{m,-m'} \langle x, x' \rangle c,$$

for $\lambda, \mu, \lambda', \mu' \in \mathbb{C}$, $m, m' \in \mathbb{Z}$ and $x, x' \in \overset{\circ}{\mathfrak{g}}$,

where $\langle \cdot, \cdot \rangle$ is the invariant (symmetric nondegenerate) bilinear form on $\overset{\circ}{\mathfrak{g}}$ (cf. Theorem 1.5.4) normalized so that the induced form on the dual space $\overset{\circ}{\mathfrak{h}}{}^*$ satisfies

$$\text{(3)}\qquad \langle \theta, \theta \rangle = 2, \ \text{ for the highest root } \theta \text{ of } \overset{\circ}{\mathfrak{g}},$$

where $\overset{\circ}{\mathfrak{h}}$ is the Cartan subalgebra of $\overset{\circ}{\mathfrak{g}}$.

It is easy to see that (2) indeed defines a Lie algebra structure and, clearly, c is a central element of $\hat{\mathcal{L}}(\overset{\circ}{\mathfrak{g}})$. Define the Lie subalgebra

$$\text{(4)}\qquad \hat{\mathcal{L}}(\overset{\circ}{\mathfrak{g}})' := (\mathcal{A} \otimes \overset{\circ}{\mathfrak{g}}) \oplus \mathbb{C}c.$$

By the *loop algebra* we mean

$$\mathcal{L}(\overset{\circ}{\mathfrak{g}}) := \mathcal{A} \otimes_{\mathbb{C}} \overset{\circ}{\mathfrak{g}},$$

under the bracket

$$[t^m \otimes x, t^{m'} \otimes x'] = t^{m+m'}[x, x'].$$

Then $\mathcal{L}(\overset{\circ}{\mathfrak{g}})$ can be thought of as the quotient Lie algebra $\hat{\mathcal{L}}(\overset{\circ}{\mathfrak{g}})'/\mathbb{C}c$ of $\hat{\mathcal{L}}(\overset{\circ}{\mathfrak{g}})'$.

Thus, we get a central extension of Lie algebras:

$$\text{(5)}\qquad 0 \to \mathbb{C}c \to \hat{\mathcal{L}}(\overset{\circ}{\mathfrak{g}})' \to \mathcal{L}(\overset{\circ}{\mathfrak{g}}) \to 0.$$

As proved by [Garland–80, Theorem (3.14)] and also independently by V. Chari (unpublished), the above is a universal central extension of $\mathcal{L}(\overset{\circ}{\mathfrak{g}})$ (see also [Kac–78b] and [Kac–90, Exercises 3.14 and 7.8]). Its twisted analogue is due to [Wilson–82]. (For a geometric proof, see [Kumar–85, Corollary 1.9(c)].)

Similarly, define the Lie algebra $\bar{\mathcal{L}}(\overset{\circ}{\mathfrak{g}}) = \mathcal{L}(\overset{\circ}{\mathfrak{g}}) \oplus \mathbb{C}\, d$ with the bracket:

$$\left[t^m \otimes x + \mu d, t^{m'} \otimes x' + \mu' d\right] = t^{m+m'} \otimes [x, x'] + \mu m' t^{m'} \otimes x' - \mu' m t^m \otimes x.$$

Then we have a central extension of Lie algebras:

$$0 \to \mathbb{C}c \to \hat{\mathcal{L}}(\overset{\circ}{\mathfrak{g}}) \to \bar{\mathcal{L}}(\overset{\circ}{\mathfrak{g}}) \to 0.$$

The map $x \mapsto 1 \otimes x$ is a Lie algebra embedding $\overset{\circ}{\mathfrak{g}} \hookrightarrow \hat{\mathcal{L}}(\overset{\circ}{\mathfrak{g}})$. Henceforth we consider $\overset{\circ}{\mathfrak{g}}$ as a subalgebra of $\hat{\mathcal{L}}(\overset{\circ}{\mathfrak{g}})$ (under this embedding). Set

(6) $$\mathfrak{h} = \overset{\circ}{\mathfrak{h}} \oplus \mathbb{C}c \oplus \mathbb{C}d.$$

We will see that the abelian Lie algebra $\mathfrak{h}$ serves as a Cartan subalgebra of $\hat{\mathcal{L}}(\overset{\circ}{\mathfrak{g}})$.

Let $A = (a_{i,j})_{0 \le i,j \le \ell}$ be the GCM obtained by adding the 0-th row and the 0-th column to $\overset{\circ}{A}$ as follows:

(7) $$a_{0,0} = 2, \quad a_{0,j} = -\alpha_j(\theta^\vee), \quad a_{j,0} = -\theta(\alpha_j^\vee), \qquad \text{for } 1 \le j \le \ell,$$

where $\{\alpha_1, \dots, \alpha_\ell\}$ are the simple roots of $\overset{\circ}{\mathfrak{g}}$, θ is the highest root, and $\{\alpha_1^\vee, \dots, \alpha_\ell^\vee, \theta^\vee\} \in \overset{\circ}{\mathfrak{h}}$ are the corresponding coroots.

Let $\delta \in \mathfrak{h}^*$ be defined by

(8) $$\delta_{|\overset{\circ}{\mathfrak{h}} \oplus \mathbb{C}c} \equiv 0, \quad \delta(d) = 1.$$

We view $\overset{\circ}{\mathfrak{h}}{}^* \hookrightarrow \mathfrak{h}^*$ by demanding

(9) $$\lambda_{|\mathbb{C}c \oplus \mathbb{C}d} \equiv 0 \text{ for } \lambda \in \overset{\circ}{\mathfrak{h}}{}^*.$$

Set
$$\pi = \{\alpha_0 := \delta - \theta, \alpha_1, \dots, \alpha_\ell\} \subset \mathfrak{h}^*, \qquad \text{and}$$
$$\pi^\vee = \{\alpha_0^\vee := c - \theta^\vee, \alpha_1^\vee, \cdots, \alpha_\ell^\vee\} \subset \mathfrak{h}.$$

Then $(\mathfrak{h}, \pi, \pi^\vee)$ is a realization of the GCM A (cf. 1.1.2). (It is easy to see that corank $A = 1$.)

The following lemma describes the weight space decomposition of $\hat{\mathcal{L}}(\overset{\circ}{\mathfrak{g}})$ with respect to the adjoint action of $\mathfrak{h}$. Its proof is straightforward and hence left to the reader.

13.1.2 Lemma. *As an $\mathfrak{h}$-module, $\hat{\mathcal{L}}(\overset{\circ}{\mathfrak{g}})$ decomposes as follows:*

$$\hat{\mathcal{L}}(\overset{\circ}{\mathfrak{g}}) = \mathfrak{h} \oplus \Big(\bigoplus_{j \in \mathbb{Z}\backslash 0} t^j \otimes \overset{\circ}{\mathfrak{h}} \Big) \oplus \Big(\bigoplus_{j \in \mathbb{Z}, \beta \in \overset{\circ}{\Delta}} t^j \otimes \overset{\circ}{\mathfrak{g}}_\beta \Big),$$

where $\overset{\circ}{\Delta}$ is the set of (nonzero) roots of $\overset{\circ}{\mathfrak{g}}$. Moreover, $\mathfrak{h}$ acts on $t^j \otimes \overset{\circ}{\mathfrak{h}}$ (resp. $t^j \otimes \overset{\circ}{\mathfrak{g}}_\beta$) via the weight $j\delta$ (resp. $j\delta + \beta$). In particular, the set of $\mathfrak{h}$-weights of $\hat{\mathcal{L}}(\overset{\circ}{\mathfrak{g}})$ is given by

$$\{0\} \cup \{j\delta;\ j \in \mathbb{Z}\backslash 0\} \cup \{j\delta + \beta;\ j \in \mathbb{Z}, \beta \in \overset{\circ}{\Delta}\}. \qquad \square$$

Choose $x_0 \in \overset{\circ}{\mathfrak{g}}_\theta$, a root vector corresponding to the highest root θ, such that $\langle x_0, \overset{\circ}{\omega} x_0 \rangle = -1$, where $\overset{\circ}{\omega}$ is the Cartan involution of $\overset{\circ}{\mathfrak{g}}$ as in 1.1.2. Set

$$E_0 = -t \otimes \overset{\circ}{\omega}(x_0), \quad F_0 = t^{-1} \otimes x_0 \in \hat{\mathcal{L}}(\overset{\circ}{\mathfrak{g}}).$$

13.1.3 Theorem. *Let $\mathfrak{g} = \mathfrak{g}(A)$ be the Kac–Moody algebra with generators $\{\mathfrak{h}, e_i, f_i; 0 \le i \le \ell\}$ associated to the* GCM *A defined in 13.1.1. Then, there exists a unique Lie algebra isomorphism $\phi : \mathfrak{g} \xrightarrow{\sim} \hat{\mathcal{L}}(\mathring{\mathfrak{g}})$ satisfying:*

$$e_0 \mapsto E_0, \;\; e_i \mapsto \mathring{e}_i \; (1 \le i \le \ell),$$

$$(1) \qquad f_0 \mapsto F_0, \;\; f_i \mapsto \mathring{f}_i \; (1 \le i \le \ell), \;\; \phi_{|\mathfrak{h}} = \mathrm{I},$$

where $\{\mathring{e}_i, \mathring{f}_i; 1 \le i \le \ell\}$ are the Chevalley generators of $\mathring{\mathfrak{g}} := \mathfrak{g}(\mathring{A})$ such that $\mathring{e}_i$, resp. $\mathring{f}_i$, belongs to the root space $\mathring{\mathfrak{g}}_{\alpha_i}$, resp. $\mathring{\mathfrak{g}}_{-\alpha_i}$.

Proof. To show the existence of ϕ as a Lie algebra homomorphism, it suffices to show that $\{\mathfrak{h}, \phi(e_i), \phi(f_i); 0 \le i \le \ell\}$ satisfy the relations (R_1)–(R_5) of 1.1.2, where $\phi(e_i), \phi(f_i)$ are given by (1).

The relations (R_1)–(R_3) are easy to verify using Theorem 1.5.4(b_2) for $\mathring{\mathfrak{g}}$ and the choice (13.1.1.3). Thus, we get a Lie algebra homomorphism $\tilde{\phi} : \tilde{\mathfrak{g}} \to \hat{\mathcal{L}}(\mathring{\mathfrak{g}})$ (cf. 1.1.2), satisfying $\tilde{\phi}(e_i) = \phi(e_i)$ and $\tilde{\phi}(f_i) = \phi(f_i)$ for all $0 \le i \le \ell$. We next show that $\tilde{\phi}$ is surjective:

Since $\{\mathring{e}_i, \mathring{f}_i; 1 \le i \le \ell\}$ generate $\mathring{\mathfrak{g}}$ (as a Lie algebra), we see that $\operatorname{Im}\tilde{\phi} \supset 1 \otimes \mathring{\mathfrak{g}}$. Moreover, $\mathring{\mathfrak{g}}$ being simple, its adjoint representation is irreducible giving $t \otimes \mathring{\mathfrak{g}} \subset \operatorname{Im}\tilde{\phi}$ (since $t \otimes \mathring{\omega}(x_0) \in \operatorname{Im}\tilde{\phi}$). Since $[\mathring{\mathfrak{g}}, \mathring{\mathfrak{g}}] = \mathring{\mathfrak{g}}$, we get $t^k \otimes \mathring{\mathfrak{g}} \subset \operatorname{Im}\tilde{\phi}$ for any $k > 0$. Similarly, $t^{-k} \otimes \mathring{\mathfrak{g}} \subset \operatorname{Im}\tilde{\phi}$ for any $k > 0$. This gives the surjectivity of $\tilde{\phi}$.

We now show that for any ideal I of $\hat{\mathcal{L}}(\mathring{\mathfrak{g}})$ such that $I \cap \mathfrak{h} = (0)$, I itself is 0: Assuming $I \neq (0)$, from the weight space decomposition (Lemma 13.1.2), either there exist $j \in \mathbb{Z}$ and $\beta \in \mathring{\Delta}$ such that $t^j \otimes \mathring{\mathfrak{g}}_\beta \subset I$ or there exists $j \in \mathbb{Z}\backslash 0$ such that $t^j \otimes x \in I$ for some nonzero $x \in \mathring{\mathfrak{h}}$. In the first case $[t^{-j} \otimes \mathring{\mathfrak{g}}_{-\beta}, t^j \otimes \mathring{\mathfrak{g}}_\beta] \cap \mathfrak{h} \neq 0$, and in the second case $[t^{-j} \otimes \mathring{\mathfrak{h}}, t^j \otimes x] \cap \mathfrak{h} \neq 0$. This is a contradiction to the assumption that $I \cap \mathfrak{h} = 0$, proving that $I = (0)$.

Let $\tilde{\mathfrak{r}} := \operatorname{Ker}\gamma$, where $\gamma : \tilde{\mathfrak{g}} \to \mathfrak{g}$ is the canonical surjective homomorphism (cf. 1.1.2). Then $\tilde{\phi}(\tilde{\mathfrak{r}})$ is an ideal of $\hat{\mathcal{L}}(\mathring{\mathfrak{g}})$ (since $\tilde{\phi}$ is surjective) and, moreover, $\tilde{\phi}(\tilde{\mathfrak{r}}) \cap \mathfrak{h} = 0$ (by (1.2.1.4)). In particular, from the above, $\tilde{\phi}(\tilde{\mathfrak{r}}) = 0$, showing that $\tilde{\phi}$ descends to a Lie algebra homomorphism $\phi : \mathfrak{g} \to \hat{\mathcal{L}}(\mathring{\mathfrak{g}})$. Since $\{\mathfrak{h}, e_i, f_i; 0 \le i \le \ell\}$ generate the Lie algebra $\mathfrak{g}$, there is a unique Lie algebra homomorphism satisfying (1).

The injectivity of ϕ follows from Corollary 3.2.10. (By Exercise 13.1.E.5, the GCM A is symmetrizable.) This completes the proof of the theorem. □

As an immediate consequence of Lemma 13.1.2 and the above theorem, we get the following:

13.1.4 Corollary. *The set of the roots of the Kac–Moody algebra* $\mathfrak{g} = \mathfrak{g}(A)$, *with* A *as in 13.1.1, is given by*

$$\Delta = \{j\delta;\ j \in \mathbb{Z}\backslash 0\} \cup \{j\delta + \beta;\ j \in \mathbb{Z}, \beta \in \overset{\circ}{\Delta}\}.$$

In fact (recalling from 13.1.1 that $\delta = \alpha_0 + \theta$*),*

$$\Delta^+ = \{j\delta;\ j > 0\} \cup \{j\delta + \beta;\ j > 0, \beta \in \overset{\circ}{\Delta}\} \cup \{\beta;\ \beta \in \overset{\circ}{\Delta}{}^+\}.$$

Moreover, the root multiplicities are given by: $\operatorname{mult}(j\delta) = \ell := \operatorname{rank} \overset{\circ}{\mathfrak{g}}$ *for any* $j \in \mathbb{Z}\backslash 0$, *and* $\operatorname{mult}(j\delta+\beta) = 1$ *for any* $j \in \mathbb{Z}$ *and* $\beta \in \overset{\circ}{\Delta}$. *Hence, the triangular decomposition (cf. Theorem 1.2.1(a)) of* $\mathfrak{g}$ *is given by*

$$\mathfrak{g} \simeq \left((t^{-1}\mathbb{C}[t^{-1}] \otimes \overset{\circ}{\mathfrak{g}}) \oplus \overset{\circ}{\mathfrak{n}}{}^-\right) \oplus \mathfrak{h} \oplus \left((t\mathbb{C}[t] \otimes \overset{\circ}{\mathfrak{g}}) \oplus \overset{\circ}{\mathfrak{n}}{}^+\right),$$

where $\overset{\circ}{\mathfrak{n}}{}^+$, *resp.* $\overset{\circ}{\mathfrak{n}}{}^-$ *is the positive, resp. negative, part of* $\overset{\circ}{\mathfrak{g}}$. □

13.1.5 Remark. Henceforth we identify the Kac–Moody algebra $\mathfrak{g}(A)$ with $\hat{\mathcal{L}}(\overset{\circ}{\mathfrak{g}})$ as given by Theorem 13.1.3 and thus also refer to $\mathfrak{g}(A)$ as an *(untwisted) affine Kac–Moody algebra.* This is symmetrizable by Exercise 13.1.E.5. There are also the "twisted" analogues of the affine Kac–Moody algebras, which we do not deal with here. They are obtained from $\hat{\mathcal{L}}(\overset{\circ}{\mathfrak{g}})$ as the subalgebra consisting of those elements fixed under a certain automorphism of $\hat{\mathcal{L}}(\overset{\circ}{\mathfrak{g}})$ of finite order induced from an automorphism of the Dynkin diagram of $\overset{\circ}{\mathfrak{g}}$ (cf. [Kac–90, Chapter 8]).

13.1.6 Definition. Let $\overset{\circ}{W}$ be the (finite) Weyl group of $\overset{\circ}{\mathfrak{g}}$ and let $\overset{\circ}{Q}{}^\vee := \oplus_{i=1}^{\ell}\mathbb{Z}\alpha_i^\vee \subset \overset{\circ}{\mathfrak{h}}$ be the coroot lattice of $\overset{\circ}{\mathfrak{g}}$. Then $\overset{\circ}{W}$ has a canonical action on $\overset{\circ}{Q}{}^\vee$ denoted by $\bullet$, obtained from restricting its action on $\overset{\circ}{\mathfrak{h}}$.

Define the *affine Weyl group* $\operatorname{Aff}\overset{\circ}{W}$ as the semidirect product

$$\text{(1)} \qquad \operatorname{Aff}\overset{\circ}{W} := \overset{\circ}{W} \ltimes \overset{\circ}{Q}{}^\vee.$$

For $q \in \overset{\circ}{Q}{}^\vee$, we denote the corresponding element of $\operatorname{Aff}\overset{\circ}{W}$ by τ_q.

13.1.7 Proposition. *Let* W *be the Weyl group associated to the affine Kac–Moody algebra* $\hat{\mathcal{L}}(\overset{\circ}{\mathfrak{g}})$. *Then there exists a (unique) isomorphism of groups*

$$\beta : W \to \operatorname{Aff}\overset{\circ}{W},$$

taking $s_0 \mapsto \tau_{\theta^\vee}\gamma_\theta$ *and* $s_i \mapsto \overset{\circ}{s}_i$, *for* $1 \le i \le \ell$, *where* $\{s_0, \ldots, s_\ell\}$ *are the simple reflections of* W, $\{\overset{\circ}{s}_1, \ldots, \overset{\circ}{s}_\ell\}$ *are the simple reflections of* $\overset{\circ}{W}$, *and* $\gamma_\theta \in \overset{\circ}{W}$ *is the reflection corresponding to the highest root* θ, *i.e.*, $\gamma_\theta\lambda := \lambda - \lambda(\theta^\vee)\theta$, *for* $\lambda \in \overset{\circ}{\mathfrak{h}}^*$.

Proof. For $h \in \overset{\circ}{\mathfrak{h}}$, define $T_h \in \mathrm{Aut}(\mathfrak{h}^*)$ by

(1) $$T_h\lambda = \lambda + \lambda(c)\,\nu(h) - \big[\lambda(h) + \frac{1}{2}\langle h, h\rangle\,\lambda(c)\big]\delta, \quad \text{for } \lambda \in \mathfrak{h}^*,$$

where $\nu : \overset{\circ}{\mathfrak{h}} \to \overset{\circ}{\mathfrak{h}}^*$ is the isomorphism induced by the nondegenerate $\overset{\circ}{W}$-invariant form $\langle \cdot, \cdot \rangle$ on $\overset{\circ}{\mathfrak{h}}^*$ normalized by (13.1.1.3). (It is easy to see that T_h indeed is an automorphism.)

Consider the hyperplane $H_c := \{\lambda \in \mathfrak{h}^* : \lambda(c) = 0\}$. Then T_h keeps H_c stable. Moreover, for $\lambda \in H_c$,

(2) $$T_h\lambda = \lambda - \lambda(h)\,\delta.$$

We next have the following additivity property for any $h_1, h_2 \in \overset{\circ}{\mathfrak{h}}$:

(3) $$T_{h_1+h_2} = T_{h_1} \circ T_{h_2} \qquad \text{(cf. Exercise 13.1.E.2).}$$

Further, for any $w \in \overset{\circ}{W}$ and $h \in \overset{\circ}{\mathfrak{h}}$,

(4) $$T_{w \bullet h} = w T_h w^{-1}, \quad \text{acting on } \mathfrak{h}^*,$$

where we think of $\overset{\circ}{W}$ canonically embedded in W under $\overset{\circ}{s}_i \mapsto s_i$ $(1 \le i \le \ell)$ and W is canonically embedded in $\mathrm{Aut}(\mathfrak{h}^*)$ as in 1.3.1.

From (2) and (3), we see that the map $h \mapsto T_h$ gives a group embedding

(5) $$T : \overset{\circ}{\mathfrak{h}} \hookrightarrow \mathrm{Aut}(\mathfrak{h}^*).$$

We identify W and $\overset{\circ}{\mathfrak{h}}$ with their images in $\mathrm{Aut}(\mathfrak{h}^*)$. Then we claim that

(6) $$\overset{\circ}{Q}{}^\vee \subset W.$$

Observe first by an explicit calculation that

(7) $$T_{\theta^\vee} \circ \gamma_\theta = s_0,$$

where $\gamma_\theta \in \overset{\circ}{W}$ is thought of as an element of W. In particular, $\theta^\vee \in W$ and hence, by (4),

$$\overset{\circ}{W} \bullet \theta^\vee \subset W.$$

But since $\overset{\circ}{Q}{}^\vee$ is spanned by $\overset{\circ}{W} \bullet \theta^\vee$ (cf. Exercise 13.1.E.9), we obtain (6).

By (4), the subgroup $\overset{\circ}{Q}{}^\vee \subset W$ is normalized by $\overset{\circ}{W}$ and, moreover, $\overset{\circ}{W}$ being finite, $\overset{\circ}{W} \cap \overset{\circ}{Q}{}^\vee = \{1\}$. Finally, by (7), the group W is generated by its subgroups $\overset{\circ}{W}$ and $\overset{\circ}{Q}{}^\vee$. This proves the proposition. □

Let $\langle \cdot, \cdot \rangle$ be the invariant (symmetric nondegenerate) bilinear form on $\overset{\circ}{\mathfrak{g}}$ normalized by (13.1.1.3).

13.1.8 Lemma. *Define the symmetric bilinear form on $\hat{\mathcal{L}}(\overset{\circ}{\mathfrak{g}})$ as follows:*

$$(1) \qquad \langle p \otimes x, q \otimes y \rangle = \operatorname{res}(t^{-1} pq) \langle x, y \rangle, \quad \textit{for } x, y \in \overset{\circ}{\mathfrak{g}},\ p, q \in \mathcal{A},$$
$$\langle \mathbb{C}c + \mathbb{C}d, \mathcal{A} \otimes \overset{\circ}{\mathfrak{g}} \rangle = \langle c, c \rangle = \langle d, d \rangle = 0, \quad \langle c, d \rangle = 1,$$

where 'res' denotes the residue, i.e., the coefficient of t^{-1}. Then this form satisfies the properties of Theorem 1.5.4.

Observe that the above form restricted to $\overset{\circ}{\mathfrak{g}}$ coincides with the original form on $\overset{\circ}{\mathfrak{g}}$ and hence we keep the same notation for both of them.

The induced form on $\mathfrak{h}^*$, still denoted by $\langle \cdot, \cdot \rangle$, has the orthogonal decomposition:

$$(2) \qquad \mathfrak{h}^* = \overset{\circ}{\mathfrak{h}}{}^* \perp (\mathbb{C}\delta \oplus \mathbb{C}\omega_0),$$

where $\omega_0 \in \mathfrak{h}^*$ is defined by $\omega_0{}_{|\overset{\circ}{\mathfrak{h}} \oplus \mathbb{C}d} \equiv 0$ and $\omega_0(c) = 1$. Moreover, $\langle \cdot, \cdot \rangle_{|\overset{\circ}{\mathfrak{h}}{}^*}$ coincides with the form induced from the original form on $\overset{\circ}{\mathfrak{g}}$. Further,

$$(3) \qquad \langle \delta, \delta \rangle = \langle \omega_0, \omega_0 \rangle = 0, \quad \langle \delta, \omega_0 \rangle = 1.$$

Proof. The lemma follows easily from Exercises 13.1.E.4–5. □

13.1.E EXERCISES

(1) Let $\overset{\circ}{\mathfrak{g}}$ and the normalized invariant bilinear form $\langle \cdot , \cdot \rangle$ on $\overset{\circ}{\mathfrak{g}}$ be as in 13.1.1. Choose a basis $\{e_\alpha\}_{\alpha \in \overset{\circ}{\Delta}}$ of root vectors such that $\langle e_\alpha, e_{-\alpha} \rangle = 1$ for all $\alpha \in \overset{\circ}{\Delta}$.

Also choose a basis $\{u_i\}_i$ and the dual basis $\{v_i\}_i$ of $\mathring{\mathfrak{h}}$. Then show that the Casimir–Kac element Ω (cf. 1.5.8) for $\hat{\mathcal{L}}(\mathring{\mathfrak{g}})$ is given as follows:

$$\Omega = 2d(c+h^\vee)+\mathring{\Omega}+2\sum_{n\geq 1}\Big(\sum_{\alpha\in\mathring{\Delta}}(t^{-n}\otimes e_{-\alpha})(t^n\otimes e_\alpha)+\sum_i(t^{-n}\otimes u_i)(t^n\otimes v_i)\Big),$$

where $h^\vee$ is the dual Coxeter number $1+\langle\mathring{\rho},\theta^\vee\rangle$, $\mathring{\rho}$ is the half sum of positive roots of $\mathring{\mathfrak{g}}$, and $\mathring{\Omega}$ is the Casimir element for $\mathring{\mathfrak{g}}$ (cf. Exercise 1.5.E.3 for an explicit expression for $\mathring{\Omega}$).

(By the following exercise (5), $\hat{\mathcal{L}}(\mathring{\mathfrak{g}})$ is indeed symmetrizable.)

(2) Prove the formula (13.1.7.3).

Hint. Show that the two sides of (13.1.7.3) are the same acting separately on H_c and on $\omega_0 \in \mathfrak{h}^*$ defined in 13.1.8.

(3) By virtue of Proposition 13.1.7, the elements of Aff $\mathring{W}$ acquire a length. Show that, for any $q \in \mathring{Q}^\vee$,

$$\ell(\tau_q) = \sum_{\alpha\in\mathring{\Delta}^+}|\alpha(q)|.$$

Hint. Use Lemma 1.3.14, Corollary 13.1.4 and (13.1.7.2).

(4) Show that the invariant bilinear form on $\mathring{\mathfrak{g}}$ normalized by (13.1.1.3) restricts to the standard invariant form on $\mathring{\mathfrak{h}}$ (i.e., satisfies (1.5.2.1)).

(5) Let $\mathring{A}$ and A be the GCM's as in 13.1.1 and let $\mathring{D} = \operatorname{diag}(\epsilon_1, \dots, \epsilon_\ell)$ be the minimal diagonal matrix associated to $\mathring{A}$ (cf. 1.5.1). Define the diagonal matrix $D := \operatorname{diag}(1, \epsilon_1, \dots, \epsilon_\ell)$. Then show that D^{-1} is symmetric, thus A is symmetrizable and, moreover, D is the minimal diagonal matrix associated to A.

Hint. Use the above Exercise 4.

(6) Let $\mathring{\omega}$ be the Cartan involution of $\mathring{\mathfrak{g}}$. Then show that the Cartan involution ω of $\hat{\mathcal{L}}(\mathring{\mathfrak{g}})$ is given by

$$\omega(t^n\otimes x+\lambda c+\mu d) = t^{-n}\otimes\mathring{\omega}(x)-\lambda c-\mu d,$$

for $x \in \mathring{\mathfrak{g}}$, $\lambda, \mu \in \mathbb{C}$ and $n \in \mathbb{Z}$.

(7) With the notation and assumptions as in Corollary 13.1.4, show that

$$\Delta_{\text{im}} = \{j\delta;\ j \in \mathbb{Z}\backslash 0\}.$$

(8) Fix a real form $\mathfrak{h}_{\mathbb{R}}$ of the Cartan subalgebra $\mathfrak{h}$ of $\hat{\mathcal{L}}(\overset{\circ}{\mathfrak{g}})$ as follows:

$$\mathfrak{h}_{\mathbb{R}} := \oplus_{i=0}^{\ell} \mathbb{R}\alpha_i^\vee \oplus \mathbb{R}d.$$

Then $\mathfrak{h}_{\mathbb{R}}$ satisfies (a)–(b) of 1.4.1. Set $\mathfrak{h}_{\mathbb{R}}^* = \mathrm{Hom}_{\mathbb{R}}(\mathfrak{h}_{\mathbb{R}}, \mathbb{R}) \subset \mathfrak{h}^*$. Prove the following:

(a) The Tits cone C (cf. 1.4.1) for $\hat{\mathcal{L}}(\overset{\circ}{\mathfrak{g}})$ is given by

$$C = \{\lambda \in \mathfrak{h}_{\mathbb{R}}^* : \lambda(c) > 0\} \cup \mathbb{R}\delta.$$

(b) $K^{\text{w.g.}} = \{\lambda \in \mathfrak{h}^* : (\lambda + \rho)(c) \neq 0\}$, where $K^{\text{w.g.}}$ is as defined in 2.3.7 and $\rho \in \mathfrak{h}_{\mathbb{R}}^*$ is any element such that $\rho(\alpha_i^\vee) = 1$ for all $0 \leq i \leq \ell$.

(9) With the notation as in 13.1.6, show that $\overset{\circ}{Q}{}^\vee$ is spanned by $\overset{\circ}{W} \bullet \theta^\vee$, where θ is the highest root of $\overset{\circ}{\mathfrak{g}}$.

Hint. The $\mathbb{Z}$-span $L = \mathbb{Z}\overset{\circ}{W} \bullet \theta^\vee$ is, of course, $\overset{\circ}{W}$-invariant. If there are unequal root lengths, choose a short simple root α and a long simple root β such that $< \alpha, \beta^\vee > \neq 0$. Then $s_\alpha \beta^\vee = \beta^\vee + \alpha^\vee \in L$. Thus $\alpha^\vee \in L$.

13.2. Affine Kac–Moody Groups

Let $\overset{\circ}{\mathfrak{g}} := \mathfrak{g}(\overset{\circ}{A})$ be the (finite-dimensional) simple Lie algebra associated to an indecomposable GCM of finite type $\overset{\circ}{A} = (a_{i,j})_{1 \leq i,j \leq \ell}$ and let $\mathfrak{g} = \hat{\mathcal{L}}(\overset{\circ}{\mathfrak{g}})$ be the corresponding affine Kac–Moody Lie algebra as in 13.1.1. Let $\overset{\circ}{G}$ be the connected, simply-connected complex algebraic group with Lie algebra $\overset{\circ}{\mathfrak{g}}$ and let $\mathcal{G}$ be the Kac–Moody group associated to the (Kac–Moody) Lie algebra $\mathfrak{g}$ as in Theorem 6.1.17, with the choice $\mathfrak{h}_{\mathbb{Z}} := \bigoplus_{i=1}^{\ell} \mathbb{Z}\alpha_i^\vee \oplus \mathbb{Z}c \oplus \mathbb{Z}d$.

13.2.1 Definition. For any commutative associative $\mathbb{C}$-algebra R with identity, by $\overset{\circ}{G}(R)$ we mean the set of R-rational points of the algebraic group $\overset{\circ}{G}$. Recall that this is, by definition, the set of all $\mathbb{C}$-algebra homomorphisms from the affine coordinate ring $\mathbb{C}[\overset{\circ}{G}] \to R$ (cf. [Mumford–88, Definition on page 158]). Then $\overset{\circ}{G}(R)$ has a canonical group structure (cf. [Springer–98, §2.1.6]). For example, if $\overset{\circ}{G} = \mathrm{SL}_N(\mathbb{C})$, then $\overset{\circ}{G}(R)$ can canonically be identified with the group $\mathrm{SL}_N(R)$ of $N \times N$ matrices of determinant 1 with entries in R. In general, embed $\overset{\circ}{G} \hookrightarrow \mathrm{SL}_N(\mathbb{C})$ as an algebraic subgroup (for some $N > 0$) and let $I \subset \mathbb{C}[\mathrm{SL}_N(\mathbb{C})]$ be the full ideal defining $\overset{\circ}{G}$. Then it is easy to see that, under a canonical identification,

(1) $$\overset{\circ}{G}(R) = \{g = (g_{ij}) \in \mathrm{SL}_N(R) : P(g_{ij}) = 0, \text{ for all } P \in I\},$$

and the group structure on $\overset{\circ}{G}(R)$ makes it a subgroup of $\mathrm{SL}_N(R)$.

In particular, consider the $\mathbb{C}$-algebra $K := \mathbb{C}[[t]][t^{-1}]$ of Laurent power series and define the *loop group*

$$\mathcal{L}(\overset{\circ}{G}) := \overset{\circ}{G}(K). \tag{2}$$

The algebra homomorphism $\mathbb{C} \hookrightarrow R$, $z \mapsto z \cdot 1$, gives rise to an embedding of groups $\overset{\circ}{G} \hookrightarrow \overset{\circ}{G}(R)$.

We extend $\mathcal{L}(\overset{\circ}{G})$ by adding "Exp d" as follows. Consider the group homomorphism (obtained by "rotation") $\gamma : \mathbb{C}^* \to \mathrm{Aut}(K)$, $\gamma(z)(P(t)) = P(zt)$ for $z \in \mathbb{C}^*$ and $P \in K$, where $\mathrm{Aut}(K)$ denotes the group of $\mathbb{C}$-algebra automorphisms of K. This canonically induces a group homomorphism

$$\gamma_{\overset{\circ}{G}} : \mathbb{C}^* \to \mathrm{Aut}\,\mathcal{L}(\overset{\circ}{G}),$$

where $\mathrm{Aut}\,\mathcal{L}(\overset{\circ}{G})$ is the group of all the automorphisms of the group $\mathcal{L}(\overset{\circ}{G})$. Now define the semidirect product group

$$\bar{\mathcal{L}}(\overset{\circ}{G}) := \mathbb{C}^* \ltimes \mathcal{L}(\overset{\circ}{G}).$$

For any $z \in \mathbb{C}^*$, we denote by d_z the corresponding element $(z, 1)$ in $\bar{\mathcal{L}}(\overset{\circ}{G})$. Observe that the semidirect product $\mathbb{C}^* \ltimes \mathcal{L}(\overset{\circ}{G})$ restricted to $\mathbb{C}^* \ltimes \overset{\circ}{G}$, in fact, is a direct product.

The question of how is $\bar{\mathcal{L}}(\overset{\circ}{G})$ related to $\mathcal{G}$ will be solved later in Theorem 13.2.8.

13.2.2 Certain subgroups of $\bar{\mathcal{L}}(\overset{\circ}{G})$. Let $\mathcal{R} := \mathbb{C}[[t]]$ be the power series ring. The group homomorphism $\gamma : \mathbb{C}^* \to \mathrm{Aut}(K)$ restricts to (again denoted by) $\gamma : \mathbb{C}^* \to \mathrm{Aut}(\mathcal{R})$ and hence we have the subgroup $\bar{\overset{\circ}{G}}(\mathcal{R}) := \mathbb{C}^* \ltimes \overset{\circ}{G}(\mathcal{R}) \subset \bar{\mathcal{L}}(\overset{\circ}{G})$. The evaluation map $e : \mathcal{R} \to \mathbb{C}$, $P \mapsto P(0)$, induces the group homomorphism

$$\bar{e} : \mathbb{C}^* \ltimes \overset{\circ}{G}(\mathcal{R}) \to \mathbb{C}^* \times \overset{\circ}{G}, \quad (z, g) \mapsto (z, g(0)).$$

For any subset $Y \subset \{1, \cdots, \ell\}$, let $\overset{\circ}{P}_Y \subset \overset{\circ}{G}$ be the corresponding parabolic subgroup containing the standard Borel subgroup $\overset{\circ}{B}$ of $\overset{\circ}{G}$ (with the Lie algebra $\overset{\circ}{\mathfrak{b}} := \overset{\circ}{\mathfrak{h}} \oplus (\oplus_{\alpha \in \overset{\circ}{\Delta}^+} \overset{\circ}{\mathfrak{g}}_\alpha)$). Define the corresponding *standard parahoric subgroup of* $\bar{\mathcal{L}}(\overset{\circ}{G})$ by

$$\bar{\mathcal{P}}_Y := \bar{e}^{-1}(\mathbb{C}^* \times \overset{\circ}{P}_Y). \tag{1}$$

For $Y = \emptyset$, we denote $\bar{\mathcal{P}}_Y$ by $\bar{\mathcal{B}}$ and it is called the *standard Iwahori subgroup* of $\bar{\mathcal{L}}(\overset{\circ}{G})$. (The terms Iwahori and parahoric come from analogous subgroups of p-adic groups.)

Let $\overset{\circ}{T} := \operatorname{Hom}_{\mathbb{Z}}(\mathfrak{h}^*_{\mathbb{Z}}, \mathbb{C}^*) \subset \overset{\circ}{G}$ be the standard maximal torus, where $\overset{\circ}{\mathfrak{h}}{}^*_{\mathbb{Z}} := \operatorname{Hom}_{\mathbb{Z}}(\overset{\circ}{Q}{}^\vee, \mathbb{Z})$. Then the subgroup

$$\bar{T} := \mathbb{C}^* \times \overset{\circ}{T} \subset \mathbb{C}^* \times \overset{\circ}{G} \subset \bar{\mathcal{L}}(\overset{\circ}{G})$$

is called the *standard maximal torus* of $\bar{\mathcal{L}}(\overset{\circ}{G})$. Let $\bar{\mathfrak{h}}_{\mathbb{Z}} := \mathbb{Z}d \oplus \overset{\circ}{Q}{}^\vee$. Then

(2) $$\bar{T} \simeq \operatorname{Hom}_{\mathbb{Z}}(\bar{\mathfrak{h}}^*_{\mathbb{Z}}, \mathbb{C}^*).$$

For any $h \in \overset{\circ}{Q}{}^\vee$, define the algebraic group morphism $t^h : \mathbb{C}^* \to \overset{\circ}{T} \subset \overset{\circ}{G}$, i.e., a cocharacter of $\overset{\circ}{T}$, by $t^h(z)\lambda = z^{\lambda(h)}$, for $z \in \mathbb{C}^*$ and $\lambda \in \overset{\circ}{\mathfrak{h}}{}^*_{\mathbb{Z}}$. This gives rise to a $\mathbb{C}$-algebra homomorphism $\mathbb{C}[\overset{\circ}{G}] \to \mathbb{C}[t, t^{-1}] \subset K$, and hence an element (still denoted by) $t^h \in \mathcal{L}(\overset{\circ}{G})$. Clearly,

(3) $$t^h = 1 \quad \text{iff} \quad h = 0.$$

Moreover, it is easy to see that (under the product in $\mathcal{L}(\overset{\circ}{G})$)

(4) $$t^{h_1} t^{h_2} = t^{h_1+h_2}, \qquad \text{for } h_1, h_2 \in \overset{\circ}{Q}{}^\vee.$$

Further, for any $z \in \mathbb{C}^*, n \in N_{\overset{\circ}{G}}(\overset{\circ}{T})$ and $h \in \overset{\circ}{Q}{}^\vee$ (where $N_{\overset{\circ}{G}}(\overset{\circ}{T})$ is the normalizer of $\overset{\circ}{T}$ in $\overset{\circ}{G}$),

(5) $$nt^h n^{-1} = t^{\operatorname{Ad}(n)h},$$

(6) $$d_z t^h d_z^{-1} = z^h t^h, \qquad \text{where } z^h := t^h(z) \in \overset{\circ}{T}.$$

Let $\bar{N} \subset \bar{\mathcal{L}}(\overset{\circ}{G})$ be the subgroup generated by $d_{\mathbb{C}^*} := \{d_z : z \in \mathbb{C}^*\}$, $N_{\overset{\circ}{G}}(\overset{\circ}{T})$ and $t^{\overset{\circ}{Q}{}^\vee} := \{t^h; h \in \overset{\circ}{Q}{}^\vee\}$. By (5), $N_{\overset{\circ}{G}}(\overset{\circ}{T})$ normalizes $t^{\overset{\circ}{Q}{}^\vee}$ and, by (6), $d_{\mathbb{C}^*}$ normalizes $\overset{\circ}{T} \cdot t^{\overset{\circ}{Q}{}^\vee}$. By the definition, $d_{\mathbb{C}^*}$ commutes with $\overset{\circ}{G}$. Further, it is easy to see that $\big(d_{\mathbb{C}^*} \times N_{\overset{\circ}{G}}(\overset{\circ}{T})\big) \cap t^{\overset{\circ}{Q}{}^\vee} = \{1\}$. Hence $\bar{N}$ is the double semidirect product

(7) $$\bar{N} = d_{\mathbb{C}^*} \ltimes \big(N_{\overset{\circ}{G}}(\overset{\circ}{T}) \ltimes t^{\overset{\circ}{Q}{}^\vee}\big).$$

From (5)–(6), we see that $\bar{N}$ normalizes $\bar{T}$. Moreover, by (3) and (7), as groups,

$$(7) \qquad \bar{N}/\bar{T} \simeq \operatorname{Aff} \overset{\circ}{W} := \overset{\circ}{W} \ltimes \overset{\circ}{Q}{}^{\vee},$$

where $\overset{\circ}{W}$ is the (finite) Weyl group $N_{\overset{\circ}{G}}(\overset{\circ}{T})/\overset{\circ}{T}$ of $\overset{\circ}{G}$ and $\overset{\circ}{W}$ acts canonically on $\overset{\circ}{Q}{}^{\vee}$ (cf. 13.1.6).

13.2.3 Adjoint representation. Recall the definition of the affine Kac–Moody Lie algebra $\hat{\mathcal{L}}(\overset{\circ}{\mathfrak{g}})$ from 13.1.1. We need to consider a certain completion $\hat{\mathcal{L}}(\overset{\circ}{\mathfrak{g}})_{\text{comp}}$ of $\hat{\mathcal{L}}(\overset{\circ}{\mathfrak{g}})$ in the "positive direction":

$$(1) \qquad \hat{\mathcal{L}}(\overset{\circ}{\mathfrak{g}})_{\text{comp}} := (K \otimes_{\mathbb{C}} \overset{\circ}{\mathfrak{g}}) \oplus \mathbb{C}c \oplus \mathbb{C}d,$$

with the bracket defined by the same formula as (13.1.1.2). (Observe that since only finitely many negative powers of t are allowed, the formula (13.1.1.2) extends.)

Now we define the *adjoint representation* Ad of $\bar{\mathcal{L}}(\overset{\circ}{G})$ in $\hat{\mathcal{L}}(\overset{\circ}{\mathfrak{g}})_{\text{comp}}$ as follows. The adjoint representation of $\overset{\circ}{G}$ in $\overset{\circ}{\mathfrak{g}}$ (cf. [Borel–91, Chap. I, §3.13]) over $\mathbb{C}$ extends to give a representation Ad_R of $\overset{\circ}{G}(R)$ in $R \otimes_{\mathbb{C}} \overset{\circ}{\mathfrak{g}}$ for any commutative associative $\mathbb{C}$-algebra R with identity. If we choose an algebraic group embedding $\overset{\circ}{G} \hookrightarrow \mathrm{SL}_N(\mathbb{C})$, this is given by

$$(2) \qquad \big(\operatorname{Ad}_R\, g\big)\, x = g x g^{-1},$$

for $g \in \overset{\circ}{G}(R) \subset \mathrm{SL}_N(R)$ and $x \in R \otimes_{\mathbb{C}} \overset{\circ}{\mathfrak{g}} \subset M_N(R)$ (where $M_N(R)$ is the space of all the $N \times N$ matrices over R). In particular, we get a representation Ad_K of$\overset{\circ}{G}(K)$ in $K \otimes_{\mathbb{C}} \overset{\circ}{\mathfrak{g}}$.

For $g \in \mathcal{L}(\overset{\circ}{G})$, $x \in K \otimes_{\mathbb{C}} \overset{\circ}{\mathfrak{g}}$, $\lambda, \mu \in \mathbb{C}$, $z \in \mathbb{C}^*$, define

(3)

$$\begin{aligned} \operatorname{Ad}(g)(x + \lambda c + \mu d) &:= (\operatorname{Ad}_K g)(x) - \mu t \big(\frac{dg}{dt}\big) g^{-1} \\ &\quad + \big(\lambda - \operatorname{res} \langle g^{-1} \frac{dg}{dt}, x - \frac{1}{2} \mu t g^{-1} \frac{dg}{dt} \rangle_t\big) c + \mu d, \\ \operatorname{Ad}(d_z)(x + \lambda c + \mu d) &:= \gamma_{\overset{\circ}{\mathfrak{g}}}(z)(x) + \lambda c + \mu d, \end{aligned}$$

where $\langle \cdot, \cdot \rangle_t$ is the K-bilinear form on $K \otimes_{\mathbb{C}} \overset{\circ}{\mathfrak{g}}$ extending the normalized invariant bilinear form $\langle \cdot, \cdot \rangle$ on $\overset{\circ}{\mathfrak{g}}$ given in 13.1.1, res denotes the coefficient of t^{-1} as in

13.1.8, $\gamma_{\overset{\circ}{\mathfrak{g}}} : \mathbb{C}^* \to \operatorname{Aut}(K \otimes_{\mathbb{C}} \overset{\circ}{\mathfrak{g}})$ is the map induced from γ (similar to the map $\gamma_{\overset{\circ}{G}}$) and, representing $g = (g_{ij})_{1\le i,j\le N} \in \mathrm{SL}_N(K)$, $\frac{dg}{dt}$ is defined to be $\left(\frac{dg_{ij}}{dt}\right)_{i,j} \in M_N(K)$. (Observe that, by Exercise 13.2.E.2, $(\frac{dg}{dt})g^{-1} = g^{-1}\frac{dg}{dt}$ and it belongs to $K \otimes \overset{\circ}{\mathfrak{g}}$.) By Exercise 13.2.E.1, (3) indeed defines a representation of $\bar{\mathcal{L}}(\overset{\circ}{G})$.

One deduces from Theorem 13.1.3 and Corollary 13.1.4 that there is a canonical identification $\hat{\mathfrak{g}} \xrightarrow{\sim} \hat{\mathcal{L}}(\overset{\circ}{\mathfrak{g}})_{\mathrm{comp}}$, where $\hat{\mathfrak{g}}$ is the completion of $\mathfrak{g} = \mathfrak{g}(A)$ defined in 6.1.1. (Use the fact that, for any $j > 0$, $\overset{\circ}{\mathfrak{g}} \otimes t^j$ is a finite sum of positive root spaces and conversely positive root spaces are contained in $\overset{\circ}{\mathfrak{g}} \otimes t^j$, $j \ge 0$.) As observed in 6.1.1, any $\mathfrak{g}$-module $M \in \mathcal{O}$ canonically becomes a $\hat{\mathfrak{g}}$-module, extending the $\mathfrak{g}$-module structure.

For any vector space V, let $\mathrm{GL}(V) = \operatorname{Aut} V$ be the group of all the linear automorphisms of V and let $\mathrm{PGL}(V) := \mathrm{GL}(V)/\mathbb{C}^*$ be the quotient of $\mathrm{GL}(V)$ by the central subgroup $\mathbb{C}^*$ consisting of the scalar automorphisms. Then $\mathrm{PGL}(V)$ acts on End V via the adjoint action "Ad", where End V is the vector space of all the linear endomorphisms of V.

13.2.4 Proposition. *Let $\pi : \hat{\mathcal{L}}(\overset{\circ}{\mathfrak{g}}) \to \operatorname{End} V$ be the integrable highest weight representation of $\hat{\mathcal{L}}(\overset{\circ}{\mathfrak{g}})$ with highest weight $\lambda \in D_{\mathbb{Z}}$ (so that $V = L(\lambda)$), where $D_{\mathbb{Z}} := \mathfrak{h}_{\mathbb{Z}}^* \cap D$ is as in 6.1.6 with the choice of $\mathfrak{h}_{\mathbb{Z}}$ as in the beginning of this section. Then there exists a unique group homomorphism $\hat{\pi} : \bar{\mathcal{L}}(\overset{\circ}{G}) \to \mathrm{PGL}(V)$ such that the following holds for any $g \in \bar{\mathcal{L}}(\overset{\circ}{G})$ and $X \in \hat{\mathcal{L}}(\overset{\circ}{\mathfrak{g}})_{\mathrm{comp}}$:*

$$(1) \qquad (\operatorname{Ad} \hat{\pi}(g))(\bar{\pi}(X)) = \bar{\pi}(\operatorname{Ad} g\, X),$$

where $\bar{\pi} : \hat{\mathcal{L}}(\overset{\circ}{\mathfrak{g}})_{\mathrm{comp}} \to \operatorname{End} V$ is the extension of π.

Proof. Fix $g \in \bar{\mathcal{L}}(\overset{\circ}{G})$. We first prove that if there exists an element $\phi \in \mathrm{PGL}(V)$ such that $(\operatorname{Ad}\phi)(\bar{\pi}(X)) = \bar{\pi}(\operatorname{Ad} g\, X)$ for all $X \in \hat{\mathcal{L}}(\overset{\circ}{\mathfrak{g}})_{\mathrm{comp}}$, then ϕ is unique:

If possible, let ϕ_1 be another such element. Then

$$(\operatorname{Ad}(\phi_1^{-1}\phi))(\bar{\pi}(X)) = \bar{\pi}(X), \qquad \text{for all } X \in \hat{\mathcal{L}}(\overset{\circ}{\mathfrak{g}})_{\mathrm{comp}}.$$

But, V being a highest weight module, $\phi_1^{-1}\phi = 1$ in $\mathrm{PGL}(V)$. (To show this observe that, for any highest weight $\mathfrak{g}$-module M, any $\delta \in \operatorname{End} M$ commuting with the action of $\mathfrak{g}$ is a scalar, since the one-dimensional highest weight space M_o is stable under δ and, moreover, M is generated by M_o as a $\mathfrak{g}$-module.) This proves the uniqueness assertion.

Define the set

$$S := \{g \in \bar{\mathcal{L}}(\overset{\circ}{G}) : \hat{\pi}(g) \text{ is defined satisfying (1) for all } X \in \hat{\mathcal{L}}(\overset{\circ}{\mathfrak{g}})_{\text{comp}}\}\,.$$

By the uniqueness, it is clear that S is a subgroup of $\bar{\mathcal{L}}(\overset{\circ}{G})$ and, moreover, the map $\hat{\pi} : S \to \mathrm{PGL}(V)$ is a group homomorphism. We next prove that $S = \bar{\mathcal{L}}(\overset{\circ}{G})$.

For any root vector $x \in \overset{\circ}{\mathfrak{g}}$ and $p \in K$, define

(2) $$\hat{\pi}(\mathrm{Exp}(p \otimes x)) = \exp(\bar{\pi}(p \otimes x)) \qquad \text{projected to } \mathrm{PGL}(V),$$

where $\mathrm{Exp} : K \otimes_{\mathbb{C}} \overset{\circ}{\mathfrak{n}}{}^{+} \to \overset{\circ}{U}(K)$ is theexponential map (cf. [Demazure–Gabriel–70, Corollaire 3.5, Chap. II, §6]) and similarly for $\overset{\circ}{\mathfrak{n}}{}^{-}$ ($\overset{\circ}{\mathfrak{n}}{}^{+} \subset \overset{\circ}{\mathfrak{g}}$ being the positive part and $\overset{\circ}{U}$ being the unipotent subgroup of $\overset{\circ}{G}$ corresponding to the Lie algebra $\overset{\circ}{\mathfrak{n}}{}^{+}$). (Observe that, since x is a root vector and V is integrable, $\bar{\pi}(p\otimes x) \in \mathrm{End}\,V$ is locally nilpotent; in particular, $\exp(\bar{\pi}(p \otimes x))$ is well defined. It is easy to see that $\hat{\pi}(\mathrm{Exp}(p \otimes x))$, as defined by (2), satisfies (1) for every $x \in \hat{\mathcal{L}}(\overset{\circ}{\mathfrak{g}})_{\text{comp}}$. Further, by [Steinberg–67, Corollary 3, page 115], the group generated by the elements $\{\mathrm{Exp}(p \otimes x);\ p \in K \text{ and root vectors } x \in \overset{\circ}{\mathfrak{g}}\}$ is the whole group $\overset{\circ}{G}(K)$. Finally, define $\hat{\pi}(d_z) = z^{\pi(d)}$, for $z \in \mathbb{C}^*$, where $z^{\pi(d)}$ acts via the multiplication by z^n on a vector $v \in V$ such that $\pi(d)v = nv$. (Observe that since $\lambda \in \mathfrak{h}^*_{\mathbb{Z}}$, $n \in \mathbb{Z}$.) This proves the proposition. □

We follow the notation from 13.1.1 freely; in particular, let $\mathfrak{h}$ be the Cartan subalgebra of the affine Kac–Moody algebra $\mathfrak{g} = \hat{\mathcal{L}}(\overset{\circ}{\mathfrak{g}})$.

13.2.5 Proposition. *For any $\lambda \in D$ such that $\lambda(c) \neq 0$, the map*

$$\pi_\lambda : \mathfrak{g} \to \mathrm{End}(L(\lambda)) \textit{ is injective,}$$

where $L(\lambda)$ is the integral highest weight $\mathfrak{g}$-module with highest weight λ.

Proof. Let I be the kernel of π_λ. By Corollary 3.2.10, it suffices to show that $I_o := I \cap \mathfrak{h} = \{0\}$.

Since $L(\lambda)$ is integrable, by Lemma 1.3.5(a), I_o is W-stable. As a $\overset{\circ}{W}$-module (considering $\overset{\circ}{W}$ as a subgroup of W as in the proof of Proposition 13.1.7), $\mathfrak{h}$ decomposes as: $\mathfrak{h} = \overset{\circ}{\mathfrak{h}} \oplus (\mathbb{C}c \oplus \mathbb{C}d)$, where $\mathbb{C}c \oplus \mathbb{C}d$ is the trivial $\overset{\circ}{W}$-module and $\overset{\circ}{\mathfrak{h}}$ is the standard representation of $\overset{\circ}{W}$. Since $\lambda(c) \neq 0$ (by assumption), from the complete reducibility of $\overset{\circ}{W}$-modules, the only possibilities for I_o are

$$I_o = \overset{\circ}{\mathfrak{h}};\ \mathbb{C}(\lambda_o c + \mu_o d);\ \overset{\circ}{\mathfrak{h}} \oplus \mathbb{C}(\lambda_o c + \mu_o d); \quad \text{or } \{0\}, \quad \text{for some } \mu_o \neq 0.$$

Since $s_0(\lambda_o c + \mu_o d) = \lambda_o c + \mu_o(d - c + \theta^\vee)$, the second possibility is not possible. Since $s_0(\theta^\vee) = 2c - \theta^\vee$, the first and the third possibilities are also ruled out. This proves the lemma. □

13.2.6 Corollary. *For any $\lambda \in D_{\mathbb{Z}}$ such that $\lambda(c) \neq 0$, the map*

$$\hat{\pi} : \bar{\mathcal{L}}(\overset{\circ}{G}) \to \mathrm{PGL}(L(\lambda))$$

of Proposition 13.2.4 has kernel precisely equal to C, where $C \subset \overset{\circ}{G} \subset \bar{\mathcal{L}}(\overset{\circ}{G})$ is the center of $\overset{\circ}{G}$.

Proof. Let $g \in \operatorname{Ker} \hat{\pi}$. Then, by (13.2.4.1),

$$(1) \qquad \bar{\pi}(X) = \bar{\pi}(\operatorname{Ad} g X), \quad \text{for all } X \in \hat{\mathcal{L}}(\overset{\circ}{\mathfrak{g}})_{\text{comp}},$$

where $\bar{\pi} : \hat{\mathcal{L}}(\overset{\circ}{\mathfrak{g}})_{\text{comp}} \to \operatorname{End}(L(\lambda))$ is the extension of π. Hence, by Proposition 13.2.5,

$$(2) \qquad \operatorname{Ad} g\, X = X, \quad \text{for all } X \in \hat{\mathcal{L}}(\overset{\circ}{\mathfrak{g}})_{\text{comp}}.$$

(Even though Proposition 13.2.5 is proved for $\mathfrak{g} = \hat{\mathcal{L}}(\overset{\circ}{\mathfrak{g}})$, the same result remains true for $\mathfrak{g}$ replaced by $\hat{\mathcal{L}}(\overset{\circ}{\mathfrak{g}})_{\text{comp}}$, since, if $x \in \operatorname{Ker} \bar{\pi}$, the β-th component $x_\beta \in \operatorname{Ker} \bar{\pi}$ for all $\beta \in \Delta$.) Write $g = g_1 d_z$, for some $z \in \mathbb{C}^*$ and $g_1 \in \mathcal{L}(\overset{\circ}{G})$. For $n \in \mathbb{Z}$ and $x \in \overset{\circ}{\mathfrak{g}}$, we get

$$(3) \quad \operatorname{Ad}(g_1 d_z)(t^n \otimes x) = g_1 z^n (t^n \otimes x)\, g_1^{-1} - z^n \operatorname{res} \langle g_1^{-1} \frac{dg_1}{dt}, t^n \otimes x \rangle_t\, c.$$

Hence, by (2),

$$(4) \qquad g_1 z^n (t^n \otimes x)\, g_1^{-1} = t^n \otimes x, \quad \text{for all } n \in \mathbb{Z} \text{ and } x \in \overset{\circ}{\mathfrak{g}}.$$

But since conjugation by g_1 is K-linear, (4) forces $z = 1$. Hence $g = g_1 \in \mathcal{L}(\overset{\circ}{G})$ and further, by (5) below, $g \in C$.

We next show that

$$(5) \qquad \operatorname{Ker} \operatorname{Ad}_K = C,$$

where $\operatorname{Ker} \operatorname{Ad}_K := \{h \in \overset{\circ}{G}(K) : (\operatorname{Ad}_K h)x = x, \text{ for all } x \in K \otimes_{\mathbb{C}} \overset{\circ}{\mathfrak{g}}\}$.

To prove (5), by [Borel–91, Chap. I, §3.15], $\operatorname{Ker} \operatorname{Ad}_K = Z(K)$, where Z is the center of the algebraic group $\overset{\circ}{G}$, and $Z(K)$ denotes its K-rational points. Since $Z(\mathbb{C})$ is discrete ($\overset{\circ}{G}$ being a simple algebraic group), $Z(\mathbb{C}) = Z(K)$. This proves (5).

Conversely, for $g \in C$, (1) is satisfied for all $X \in \hat{\mathcal{L}}(\overset{\circ}{\mathfrak{g}})_{\text{comp}}$ by (5) and (3). So $g \in \operatorname{Ker}\hat{\pi}$, by the proof of Proposition 13.2.4. This proves the corollary. □

13.2.7 Pro-group structure on certain subgroups of $\bar{\mathcal{L}}(\overset{\circ}{G})$. Consider the inverse system of algebraic groups:

$$(*) \qquad \cdots \to \overset{\circ}{G}(\mathcal{R}_3) \xrightarrow{\rho_2} \overset{\circ}{G}(\mathcal{R}_2) \xrightarrow{\rho_1} \overset{\circ}{G}(\mathcal{R}_1),$$

where $\mathcal{R}$ (as in 13.2.2) is the power series ring $\mathbb{C}[[t]]$, $\mathcal{R}_n := \mathcal{R}/\langle t^n \rangle$, and $\rho_n : \overset{\circ}{G}(\mathcal{R}_{n+1}) \to \overset{\circ}{G}(\mathcal{R}_n)$ is induced by the canonical $\mathbb{C}$-algebra homomorphism $\mathcal{R}_{n+1} \to \mathcal{R}_n$.

By Exercise 13.2.E.3, the canonical map $\xi_n : \overset{\circ}{G}(\mathcal{R}) \to \overset{\circ}{G}(\mathcal{R}_n)$ is surjective and hence so is ρ_n for all $n \geq 1$. Thus, by Example 4.4.3(3), the inverse limit $\mathcal{H}$ of $(*)$ is a pro-group over $\mathbb{C}$. The group homomorphisms ξ_n induce a group homomorphism

$$\xi : \overset{\circ}{G}(\mathcal{R}) \to \mathcal{H}.$$

It is easy to see that ξ is injective (since, for $0 \neq P \in \mathcal{R}$, there exists $n \geq 1$ such that $P \bmod \langle t^n \rangle$ is nonzero in $\mathcal{R}_n$). We now prove that ξ is surjective:

Fix an embedding of algebraic groups $\overset{\circ}{G} \hookrightarrow \mathrm{SL}_N(\mathbb{C}) \subset M_N(\mathbb{C})$. Take any $x = (x_n)_n \in \mathcal{H}$, where $x_n \in \overset{\circ}{G}(\mathcal{R}_n)$. Since $\operatorname{Inv.lt.}_n \mathcal{R}_n \simeq \mathcal{R}$, there exists $\hat{x} \in M_N(\mathcal{R})$ such that its image $\hat{x}_n$ in $M_N(\mathcal{R}_n)$ coincides with x_n for all $n \geq 1$. But then $P(\hat{x}_n) = 0$, for all $P \in I$ and $n \geq 1$, where $I \subset \mathbb{C}[X_{ij}]$ is the full ideal of $\overset{\circ}{G}$ in $M_N(\mathbb{C})$. From this it is easy to see that $P(\hat{x}) = 0$ for all $P \in I$, and hence $\hat{x} \in \overset{\circ}{G}(\mathcal{R})$. Clearly $\xi(\hat{x}) = x$. This proves the surjectivity of ξ.

Thus, transporting the pro-group structure from $\mathcal{H}$ via ξ, we realize $\overset{\circ}{G}(\mathcal{R})$ as a pro-group. By the definition of the pro-group structure on $\mathcal{H}$, the map $\xi_1 : \overset{\circ}{G}(\mathcal{R}) \to \overset{\circ}{G}(\mathcal{R}_1) = \overset{\circ}{G}$ is a pro-group morphism. Hence, for any algebraic subgroup $\overset{\circ}{H} \subset \overset{\circ}{G}$, $\xi_1^{-1}(\overset{\circ}{H})$ is a pro-subgroup of $\overset{\circ}{G}(\mathcal{R})$. In particular, taking $\overset{\circ}{H}$ to be any parabolic subgroup $\overset{\circ}{P}_Y \subset \overset{\circ}{G}$ (cf. 13.2.2), we get the pro-subgroup $\xi_1^{-1}(\overset{\circ}{P}_Y)$. Similarly, taking $\overset{\circ}{U} \subset \overset{\circ}{G}$ to be the unipotent radical of $\overset{\circ}{B}$, we get the pro-subgroup $\bar{\mathcal{U}} := \xi_1^{-1}(\overset{\circ}{U})$ of $\overset{\circ}{G}(\mathcal{R})$. It is easy to see that there is a canonical isomorphism of pro-Lie algebras

$$(1) \qquad \hat{\mathfrak{n}} \simeq \operatorname{Lie}\bar{\mathcal{U}},$$

where $\hat{\mathfrak{n}}$ is the completion of $\mathfrak{n} \subset \mathfrak{g}$ as in 6.1.1 ($\mathfrak{g}$ being the affine Kac–Moody Lie algebra $\hat{\mathcal{L}}(\overset{\circ}{\mathfrak{g}})$). Moreover, $\bar{\mathcal{U}}$ is pro-unipotent. Hence, by Theorem 4.4.19, the pro-Lie algebra isomorphism (1) induces a pro-group isomorphism

$$i : \mathcal{U} \xrightarrow{\sim} \bar{\mathcal{U}}, \tag{2}$$

where $\mathcal{U}$ is the pro-unipotent pro-group with Lie $\mathcal{U} \simeq \hat{\mathfrak{n}}$ (cf. 6.1.1). Recall that, as in Section 6.1, $\mathcal{U}$ is a subgroup of the Kac–Moody group $\mathcal{G}$ associated to the Kac–Moody algebra $\hat{\mathcal{L}}(\mathring{\mathfrak{g}})$.

Similarly, we can realize $\bar{\mathring{G}}(\mathcal{R}) = \mathbb{C}^* \ltimes \mathring{G}(\mathcal{R})$ as a pro-group and the subgroups $\bar{\mathcal{P}}_Y \subset \bar{\mathring{G}}(\mathcal{R})$ as pro-subgroups (cf. 13.2.2 for the definitions of $\bar{\mathring{G}}(\mathcal{R})$ and $\bar{\mathcal{P}}_Y$).

As above, let $\mathcal{G}$ be the Kac–Moody group associated to the affine Kac–Moody algebra $\mathfrak{g} = \hat{\mathcal{L}}(\mathring{\mathfrak{g}})$.

13.2.8 Theorem. *There exists a unique group homomorphism* $\psi : \mathcal{G} \to \bar{\mathcal{L}}(\mathring{G})/C$ *making the following diagram commutative for all the integrable highest weight* $\mathfrak{g}$*-modules* $L(\lambda)$ *with* $\lambda \in D_{\mathbb{Z}}$*:*

$$(\mathcal{D}_\lambda) \qquad \begin{array}{ccc} \mathcal{G} & \xrightarrow{\tilde{\pi}} & \mathrm{GL}(L(\lambda)) \\ \downarrow{\scriptstyle\psi} & & \downarrow \\ \bar{\mathcal{L}}(\mathring{G})/C & \xrightarrow{\hat{\pi}} & \mathrm{PGL}(L(\lambda)), \end{array}$$

where $\hat{\pi} = \hat{\pi}_\lambda$ *is the map guaranteed by Proposition 13.2.4 and the proof of Corollary 13.2.6 and* $\tilde{\pi} = \tilde{\pi}_\lambda$ *is the map induced from* π *via Theorem 6.2.3.*

The map ψ*, in addition, satisfies:*

(a) $\psi(g) = i(g)C$*, for* $g \in \mathcal{U}$*, where* i *is the isomorphism as in (13.2.7.2).*

(b) $\psi(T) = \bar{T}/C$ *and* $\psi_{|T}$ *is the map*

$$T = \mathrm{Hom}_{\mathbb{Z}}(\mathfrak{h}^*_{\mathbb{Z}}, \mathbb{C}^*) \to \mathrm{Hom}_{\mathbb{Z}}(\bar{\mathfrak{h}}^*_{\mathbb{Z}}, \mathbb{C}^*)/C \quad \textit{(cf. (13.2.2.2))}$$

induced from the $\mathbb{Z}$*-module map* $\mathfrak{h}_{\mathbb{Z}} \to \bar{\mathfrak{h}}_{\mathbb{Z}}$ *taking*

$$\alpha_i^\vee \mapsto \alpha_i^\vee \ (1 \le i \le \ell), \ d \mapsto d \text{ and } c \mapsto 0.$$

(c) *For* $\alpha \in \mathring{\Delta}$ *and* $n \in \mathbb{Z}$*,*

$$\psi(\mathrm{Exp}(t^n \otimes X_\alpha)) = \mathrm{Exp}(t^n \otimes X_\alpha) \mod C,$$

where $X_\alpha \in \mathring{\mathfrak{g}}$ *is any root vector corresponding to the root* α*.*

(d) ψ *is surjective.*

(e) $\mathrm{Ker}\,\psi =$ *center of* $\mathcal{G}$*.*

Proof. The uniqueness of ψ follows from the commutativity of the diagrams $(\mathcal{D}_\lambda)$ and Corollary 13.2.6. (In fact, the uniqueness of ψ follows from the commutativity of $(\mathcal{D}_\lambda)$ for a single $\lambda \in D_{\mathbb{Z}}$ such that $\lambda(c) \neq 0$.) Let

$$\mathcal{E} := \left\{g \in \mathcal{G} : \text{there exists } \psi(g) \text{ making } (\mathcal{D}_\lambda) \text{ commutative for all } \lambda \in D_{\mathbb{Z}}\right\}.$$

Then, it is easy to see that $\mathcal{E}$ is a subgroup of $\mathcal{G}$.

For $g \in \mathcal{U}$, setting $\psi(g) = i(g)C$, we see that $\mathcal{U} \subset \mathcal{E}$ (by using Proposition 6.2.11 and (6.2.10.3)). Defining $\psi(t)$ for $t \in T$ by (b), we similarly see that $T \subset \mathcal{E}$. (We have used Exercise 13.2.E.4 here for $g \in \mathcal{U} \cup T$.) For $\alpha \in \overset{\circ}{\Delta}$ and $n \in \mathbb{Z}$, $t^n \otimes X_\alpha \in \mathfrak{g}$ is a real root vector (cf. Corollary 13.1.4 and Exercise 13.1.E.7) and hence, by Proposition 6.2.11, $\operatorname{Exp}(t^n \otimes X_\alpha) \in \mathcal{G}$ makes sense. Moreover, defining $\psi(\operatorname{Exp}(t^n \otimes X_\alpha))$ by (c), we get (from (13.2.4.2) and Proposition 6.2.11) that $\operatorname{Exp}(t^n \otimes X_\alpha) \in \mathcal{E}$. Since $\mathcal{U}$, T and the elements $\{\operatorname{Exp}(t^n \otimes X_\alpha); \alpha \in \overset{\circ}{\Delta}, n \in \mathbb{Z}\}$ generate $\mathcal{G}$ as a group (cf. 6.1.16), we see that $\mathcal{E} = \mathcal{G}$. This shows that there exists a (unique) map ψ making the diagrams $(\mathcal{D}_\lambda)$ commutative for all $\lambda \in D_{\mathbb{Z}}$. (Observe that if $\lambda(c) = 0$, then the representation $L(\lambda)$ is one-dimensional and hence $(\mathcal{D}_\lambda)$ is automatically commutative.) The map ψ is a group homomorphism since all the other maps of the diagram $(\mathcal{D}_\lambda)$ are group homomorphisms and $\hat{\pi}$ is injective for some $L(\lambda)$, in fact, for any $\lambda \in D_{\mathbb{Z}}$ such that $\lambda(c) > 0$.

Since the elements $\{\operatorname{Exp}(p \otimes X_\alpha); \alpha \in \overset{\circ}{\Delta}, p \in K\}$ generate $\overset{\circ}{G}(K)$ (cf. the proof of Proposition 13.2.4), the elements $\bar{\mathcal{U}} \cup \bar{T} \cup \{\operatorname{Exp}(t^n \otimes X_\alpha); \alpha \in \overset{\circ}{\Delta}, n \in \mathbb{Z}\}$ generate $\bar{\mathcal{L}}(\overset{\circ}{G})$ as a group. (To prove this, write any $p \in K$ as $p = p_1 + \sum_{-d \le n \le 0} a_n t^n$, for some $p_1 \in t\mathbb{C}[[t]]$, $d \geq 0$ and $a_n \in \mathbb{C}$. Then $\operatorname{Exp}(p \otimes X_\alpha) = \operatorname{Exp}(p_1 \otimes X_\alpha) \cdot \prod_{-d \le n \le 0} \operatorname{Exp}(t^n \otimes a_n X_\alpha)$.) Hence the map ψ is surjective.

To prove (e), from the commutativity of $(\mathcal{D}_\lambda)$, we see that

$$\text{(1)} \qquad \operatorname{Ker} \psi = \left\{g \in \mathcal{G} : \tilde{\pi}_\lambda(g) = z_\lambda(g)\,\mathrm{I}, \text{ for all } \lambda \in D_{\mathbb{Z}}\right\},$$

for some $z_\lambda(g) \in \mathbb{C}^*$ depending only on λ and g.

By the Bruhat decomposition of $\mathcal{G}$, we get $\operatorname{Ker} \psi \subset \mathcal{B}$. Take $g = tu \in \operatorname{Ker} \psi$, for some $t \in T$ and$u \in \mathcal{U}$.For any $\lambda \in D_{\mathbb{Z}}$, from (1), we get

$$\text{(2)} \qquad \tilde{\pi}_\lambda(u) = z_\lambda(g)\,\tilde{\pi}_\lambda(t)^{-1}.$$

Take any finite-dimensional $\mathcal{B}$-stable subspace $V \subset L(\lambda)$. Then $\tilde{\pi}_\lambda(u)_{|V}$ is unipotent whereas $(\tilde{\pi}_\lambda(t)^{-1})_{|V}$ is semisimple, giving

$$\text{(3)} \qquad \tilde{\pi}_\lambda(u) = \mathrm{I} = z_\lambda(g)\,\tilde{\pi}_\lambda(t)^{-1}.$$

If $u \neq 1$, it is easy to see that there exists $\lambda \in D_{\mathbb{Z}}$ (depending upon u) such that $\bar{\pi}_\lambda(u) \neq \mathrm{I}$ (cf. proof of Corollary 6.2.6). This contradicts (3) and hence $u = 1$, so that $\operatorname{Ker} \psi \subset T$. Thus we get

$$\begin{aligned}\operatorname{Ker} \psi &= \big\{t \in T : t(\alpha_i) = 1, \text{ for all the simple roots } \alpha_i, 0 \leq i \leq \ell\big\} \\ &= \text{center of } \mathcal{G}, \text{ from Lemma 6.2.9(c).}\end{aligned}$$

This proves (e) and hence the proof of the theorem is completed. □

There is a canonical embedding $\overset{\circ}{G} \hookrightarrow \mathcal{G}$ corresponding to the Lie subalgebra $\overset{\circ}{\mathfrak{g}} \subset \hat{\mathcal{L}}(\overset{\circ}{\mathfrak{g}}_{\text{comp}})$ (cf. 6.1.13) and also $\overset{\circ}{G} \subset \bar{\mathcal{L}}(\overset{\circ}{G})$. It is easy to see, from the (c) part of the above theorem, that $\psi_{|\overset{\circ}{G}}$ is the quotient map $\overset{\circ}{G} \to \overset{\circ}{G}/C$. Further, for any subset $Y \subset \{1, \dots, \ell\}$, the standard parabolic subgroup $\mathcal{P}_Y$ of $\mathcal{G}$, defined in 6.1.18, is generated (as a group) by $\mathcal{U}$, T and $\overset{\circ}{P}_Y$ and similarly $\bar{\mathcal{P}}_Y \subset \bar{\mathcal{L}}(\overset{\circ}{G})$ is generated by $\bar{\mathcal{U}}$, $\bar{T}$ and $\overset{\circ}{P}_Y$. Thus, from the (a) and (b) parts of the above theorem, $\psi(\mathcal{P}_Y) = \bar{\mathcal{P}}_Y/C$.

13.2.9 Corollary. *For any subset $Y \subset \{1, \dots, \ell\}$, the group homomorphism ψ induces a bijection*

$$\psi_Y : \mathcal{G}/\mathcal{P}_Y \xrightarrow{\sim} \bar{\mathcal{L}}(\overset{\circ}{G})/\bar{\mathcal{P}}_Y.$$

Moreover, $\psi(\mathcal{G}^{\min}) = (\mathbb{C}^* \ltimes \overset{\circ}{G}(\mathcal{A}))/C$, where (as in 13.1.1) $\mathcal{A} = \mathbb{C}[t, t^{-1}]$ and $\mathcal{G}^{\min} \subset \mathcal{G}$ is as in 7.4.1.

Proof. The surjectivity of ψ_Y follows immediately from Theorem 13.2.8(d). Since $\psi(\mathcal{P}_Y) = \bar{\mathcal{P}}_Y/C$, $\operatorname{Ker} \psi =$ center of $\mathcal{G}$ (by Theorem 13.2.8(e)) and the center of $\mathcal{G} \subset T$ (by Lemma 6.2.9(c)), the injectivity of ψ_Y follows.

Recall that $\mathcal{G}^{\min}$, by the definition, is generated by $\{\mathcal{U}_\alpha\}_{\alpha \in \Delta_{\text{re}}}$ and T. Hence, by Theorem 13.2.8-(b)-(c),

$$(1) \qquad \psi(\mathcal{G}^{\min}) \subset (\mathbb{C}^* \ltimes \overset{\circ}{G}(\mathcal{A}))/C \qquad \text{and } \psi(\mathcal{G}^{\min}) \supset \bar{T}/C.$$

Conversely, since $\overset{\circ}{G}(\mathcal{A})$ is generated by $\big\{\operatorname{Exp}(t^n \otimes X_\alpha); \alpha \in \overset{\circ}{\Delta} \text{ and } n \in \mathbb{Z}\big\}$ (cf. [Steinberg–67, Corollary 3, p. 115]), we get that $\psi(\mathcal{G}^{\min}) \supset \overset{\circ}{G}(\mathcal{A})/C$. Combining this with (1), we get $\psi(\mathcal{G}^{\min}) = (\mathbb{C}^* \ltimes \overset{\circ}{G}(\mathcal{A}))/C$. □

As another corollary of Theorem 13.2.8, the Bruhat decomposition of $\mathcal{G}$ (cf. Corollary 6.1.20) readily gives the Bruhat decomposition of $\bar{\mathcal{L}}(\overset{\circ}{G})$.

13.2.10 Corollary. *We have the disjoint union*

$$\bar{\mathcal{L}}(\overset{\circ}{G}) = \bigsqcup_{\bar{n}\in\bar{N}/\bar{T}} \bar{\mathcal{B}}\,\bar{n}\,\bar{\mathcal{B}},$$

where $\bar{N}, \bar{\mathcal{B}}$ are as in 13.2.2. (Observe that, by (13.2.2.8), $\bar{N}/\bar{T}$ is isomorphic with the affine Weyl group $\overset{\circ}{W} \ltimes \overset{\circ}{Q}{}^\vee$.)

Proof. Combine Theorem 13.2.8 with Exercise 13.2.E.5. □

13.2.11 Remark. The group homomorphism

$$\psi : \mathcal{G} \to \bar{\mathcal{L}}(\overset{\circ}{G})/C$$

can be lifted to a surjective homomorphism $\hat{\psi} : \mathcal{G} \to \bar{\mathcal{L}}(\overset{\circ}{G})$ with kernel precisely equal to $\mathbb{C}^* = \mathrm{Exp}(\mathbb{C}c)$, thus providing a central extension

$$1 \to \mathbb{C}^* \to \mathcal{G} \to \bar{\mathcal{L}}(\overset{\circ}{G}) \to 1.$$

We do not give the details of the proof here, however, it is given in [Garland–80]. The proof involves lifting $\psi_{|\mathcal{P}_i}$, for any minimal parabolic subgroup $\mathcal{P}_i \subset \mathcal{G}$ $(0 \le i \le \ell)$, and also lifting $\psi_{|N}$ and showing that these lifts can be so chosen that they coincide on the various intersections.

13.2.12. The generalized flag variety $\mathcal{X}^Y := \mathcal{G}/\mathcal{P}_Y$, for $Y = \{1, \dots, \ell\}$, is particularly important. We shall freely identify $\mathcal{X}^Y$ with $\overset{\circ}{G}(K)/\overset{\circ}{G}(\mathcal{R})$ set theoretically via Corollary 13.2.9. Let W, resp. $\overset{\circ}{W}$, be the Weyl group of the affine Kac–Moody Lie algebra $\mathfrak{g} = \hat{\mathcal{L}}(\overset{\circ}{\mathfrak{g}})$, resp. of $\overset{\circ}{\mathfrak{g}}$. By Proposition 13.1.7, $W \simeq \mathrm{Aff}\,\overset{\circ}{W} := \overset{\circ}{W} \ltimes \overset{\circ}{Q}{}^\vee$, where $\overset{\circ}{Q}{}^\vee \subset \overset{\circ}{\mathfrak{h}}$ is the coroot lattice of $\overset{\circ}{\mathfrak{g}}$.

From now on we shall abbreviate $\mathcal{X}^Y$ by $\mathcal{Y}$ and $\mathcal{P}_Y$ by $\mathcal{P}$ for the above choice of Y. Recall from 7.1.19 that $\mathcal{Y}$ is a projective ind-variety under the filtration

$$X_0^Y \subset X_1^Y \subset X_2^Y \subset \cdots,$$

where

(1)
$$X_n^Y := \bigcup_{\substack{w\in W'_Y \\ \ell(w)\le n}} \mathcal{B}\,w\,\mathcal{P}/\mathcal{P}.$$

The ind-variety $\mathcal{Y}$ is often called the *infinite Grassmannian* (associated to $\overset{\circ}{G}$) in the literature.

13.2.13 Realizing $\mathcal{Y}$ as an ind-variety via lattices (SL_N case). We first consider the case of $\overset{\circ}{G} = \mathrm{SL}_N$. Denote $V = \mathbb{C}^N$, and (as earlier) $\mathcal{R} = \mathbb{C}[[t]]$. For any $n \geq 0$, consider the set $\mathcal{F}_n$ of $\mathcal{R}$-submodules $L \subset K \otimes_{\mathbb{C}} V$ such that (denoting $\mathcal{R} \otimes_{\mathbb{C}} V$ by L_o)

$$t^n L_o \subset L \subset t^{-n} L_o \,, \text{ and } \dim_{\mathbb{C}}(L/t^n L_o) = nN \,.$$

Consider the complex vector space $V_n := t^{-n} L_o / t^n L_o$ of dimension $2nN$. Then the multiplication by t induces a nilpotent endomorphism $\bar{t}_n$ of V_n and, hence, $1 + \bar{t}_n$ is a (unipotent) automorphism of V_n. In particular, $1 + \bar{t}_n$ induces a biregular isomorphism (denoted by the same symbol) of the Grassmannian $\mathrm{Gr}(nN, 2nN)$ of nN-dimensional subspaces of the $2nN$-dimensional space V_n. Let $\mathrm{Gr}(nN, 2nN)^{1+\bar{t}_n}$ denote its fixed point set. Then, clearly, the map $j_n : \mathcal{F}_n \to \mathrm{Gr}(nN, 2nN)^{1+\bar{t}_n}$ given by $L \mapsto L/t^n L_o$ is a bijection. We pull the (reduced) subvariety structure of $\mathrm{Gr}(nN, 2nN)^{1+\bar{t}_n}$ via j_n to equip $\mathcal{F}_n$ with a projective variety structure. We next claim that the canonical inclusion $\mathcal{F}_n \hookrightarrow \mathcal{F}_{n+1}$ is a closed embedding.

Consider the commutative diagram:

$$\begin{array}{ccc} \mathcal{F}_n & \xrightarrow{j_n} & \mathrm{Gr}(nN, 2nN)^{1+\bar{t}_n} \\ \downarrow & & \downarrow \theta_n \\ \mathcal{F}_{n+1} & \xrightarrow{j_{n+1}} & \mathrm{Gr}((n+1)N, 2(n+1)N)^{1+\bar{t}_{n+1}} \,, \end{array}$$

where the map θ_n takes $V' \subset t^{-n} L_o / t^n L_o \approx t^{n-1} V \oplus t^{n-2} V \oplus \cdots \oplus t^{-n} V$ to $t^n V \oplus V'$ ($t^k V := t^k \otimes V$). It is easy to see that θ_n is a closed embedding and hence so is $\mathcal{F}_n \hookrightarrow \mathcal{F}_{n+1}$. This equips $\mathcal{F} = \cup_{n \geq 0} \mathcal{F}_n$ with a projective ind-variety structure.

Let $\mathcal{G}^o$ be the Kac–Moody group corresponding to the affine Kac–Moody Lie algebra $\hat{\mathcal{L}}(sl_N)$ and let $\mathcal{P}^o$ be the standard parabolic subgroup of $\mathcal{G}^o$ corresponding to the subset $Y = \{1, \dots, N-1\}$, where sl_N is the Lie algebra of $N \times N$ matrices of trace 0.

Set $\mathcal{Y}^o := \mathcal{G}^o / \mathcal{P}^o = \mathrm{SL}_N(K)/\mathrm{SL}_N(\mathcal{R})$. By virtue of the following lemma, the map $\beta : \mathcal{Y}^o \to \mathcal{F}$ (defined below) is a bijection. By transporting the projective ind-variety structure from $\mathcal{F}$ to $\mathcal{Y}^o$ (via β), we equip $\mathcal{Y}^o$ with a projective ind-variety structure. With this structure we denote $\mathcal{Y}^o$ by $\mathcal{Y}^o_{\mathrm{lat}}$. We also define the filtration $\hat{X}^o_n$ of $\mathcal{Y}^o$ by

$$\hat{X}^o_n := \beta^{-1}(\mathcal{F}_n).$$

The group $\mathrm{SL}_N(K)$ acts canonically on $K \otimes V$.

13.2.14 Lemma. *The map* $g\mathrm{SL}_N(\mathcal{R}) \mapsto gL_o$ *(for* $g \in \mathrm{SL}_N(K)$*) induces a bijection* $\beta : \mathcal{Y}^o \to \mathcal{F}$.

Proof. Let $g \in \mathrm{SL}_N(K)$. It is easy to see that there exists some n (depending upon g) such that

$$(1) \qquad t^n L_o \subset gL_o \subset t^{-n} L_o.$$

Of course, gL_o is t-stable. We next calculate the dimension of $gL_o/t^n L_o$:

By the Bruhat decomposition (Corollary 13.2.10), we may assume that g is an algebraic group morphism $\mathbb{C}^* \to D$, where D is the diagonal subgroup of SL_N. Write

$$g(t) = \begin{pmatrix} t^{n_1} & & 0 \\ & \ddots & \\ 0 & & t^{n_N} \end{pmatrix}, \qquad \text{for } t \in \mathbb{C}^* \text{ and } n_i \in \mathbb{Z}.$$

Then, since $\mathrm{Im}\, g \subset \mathrm{SL}_N$, $\sum n_i = 0$. Now

$$\dim(gL_o/t^n L_o) = (n - n_1) + \cdots + (n - n_N) = Nn - \sum n_i = Nn.$$

This proves that $gL_o \in \mathcal{F}_n$.

Conversely, take $L \in \mathcal{F}_n$. Since $\mathcal{R}$ is a PID and $t^k L_o$ is $\mathcal{R}$-free of rank N (for any $k \in \mathbb{Z}$), we get that L is $\mathcal{R}$-free of rank N. Further, $K \otimes_{\mathcal{R}} L \to V((t))$ is an isomorphism, where $V((t)) := K \otimes_{\mathbb{C}} V$. Let $\{e_1, \ldots, e_N\}$ be the standard $\mathbb{C}$-basis of V and take an $\mathcal{R}$-basis $\{v_1, \ldots, v_N\}$ of L. Now define the K-linear automorphism g of $V((t))$ by $ge_i = v_i$ $(1 \leq i \leq N)$. We prove that $\det g$ is a unit of $\mathcal{R}$: Write $\det g = t^k u$, where $k \in \mathbb{Z}$ and u is a unit of $\mathcal{R}$. Consider the K-linear automorphism α of $V((t))$ defined by

$$\begin{aligned} \alpha\, e_i &= e_i\,, \quad \text{for } 1 \leq i < N, \\ &= t^{-k} u^{-1} e_N\,, \quad \text{for } i = N. \end{aligned}$$

Then $\det(g\alpha) = 1$, and $t^{n+|k|} L_o \subset (g\alpha) L_o \subset t^{-n-|k|} L_o$. Hence, by the earlier part of the proof, we get

$$(2) \qquad \dim\left(\frac{g\alpha(L_o)}{t^{n+|k|} L_o}\right) = (n + |k|)N.$$

On the other hand,

$$(3) \qquad \begin{aligned} \dim\left(\frac{g\alpha(L_o)}{t^{n+|k|} L_o}\right) &= \dim \frac{gL_o}{t^n L_o} + |k|N + k \\ &= Nn + |k|N + k\,, \qquad \text{since } L \in \mathcal{F}_n. \end{aligned}$$

Now, combining (2) and (3), we get $k = 0$, hence $(g\alpha)L_o = gL_o = L$. This proves the surjectivity of β. The injectivity of β is clear. This proves the lemma. □

13.2.15 Realizing $\mathcal{Y}$ as an ind-variety via lattices (general case). We now come to the case of general (connected, simply-connected, simple) $\overset{\circ}{G}$. Fix an embedding $\overset{\circ}{G} \hookrightarrow \mathrm{SL}_N$ such that the Borel subgroup $\overset{\circ}{B}$ of $\overset{\circ}{G}$ is the inverse image of the standard Borel subgroup of SL_N (consisting of the upper triangular matrices of determinant 1) and, in addition, the maximal torus $\overset{\circ}{T}$ goes inside the diagonal subgroup D of SL_N. This gives rise to the injective map

$$i : \mathcal{Y} = \mathcal{G}/\mathcal{P} \hookrightarrow \mathcal{Y}^o = \mathcal{G}^o/\mathcal{P}^o,$$

where $\mathcal{G}$, resp. $\mathcal{G}^o$, corresponds to $\overset{\circ}{G}$, resp. SL_N. Observe that the injectivity of i follows from Corollary 13.2.9, since $\overset{\circ}{G}(K) \cap \mathrm{SL}_N(\mathcal{R}) = \overset{\circ}{G}(\mathcal{R})$.

The filtration $\hat{X}_n^o$ of $\mathcal{Y}^o$ (given in 13.2.13), on restriction, gives the filtration $\hat{X}_n$ of $\mathcal{Y}$, i.e.,

$$\hat{X}_n := \hat{X}_n^o \cap \mathcal{Y}.$$

By (a subsequent) Lemma 13.2.17, for any $n \geq 0$, $\hat{X}_n$ is a Zariski closed subset of $\hat{X}_n^o$, and thus $\mathcal{Y}$ is closed in $\mathcal{Y}^o$. This allows us to put the closed ind-subvariety structure on $\mathcal{Y}$ (cf. 4.1.1). Equipped with this ind-variety structure, we denote $\mathcal{Y}$ by $\mathcal{Y}_{\mathrm{lat}}$.

Recall the filtration $\{X_n = X_n^Y\}_{n\geq 0}$ of $\mathcal{Y}$ given by (13.2.12.1).

13.2.16 Lemma. *The two filtrations $\{X_n\}_{n\geq 0}$ and $\{\hat{X}_n\}_{n\geq 0}$ of $\mathcal{Y}$ are compatible, i.e., for any n there exists $k(n)$ such that*

$$X_n \subseteq \hat{X}_{k(n)} \text{ and } \hat{X}_n \subseteq X_{k(n)}.$$

Proof. Fix an embedding $\overset{\circ}{G} \hookrightarrow \mathrm{SL}_N$ as above. There is a bijection $W/\overset{\circ}{W} \simeq \mathrm{Mor}(\mathbb{C}^*, \overset{\circ}{T})$, where $\mathrm{Mor}(\mathbb{C}^*, \overset{\circ}{T})$ denotes the set of all the algebraic group morphisms $\mathbb{C}^* \to \overset{\circ}{T}$ (cf. Exercise 13.2.E.5 and 13.2.2). Since the set $\{\bar{w} := w\overset{\circ}{W} \in W/\overset{\circ}{W} : \ell(w) \leq n\}$ is finite, it is easy to see that $X_n \subset \hat{X}_{k(n)}$ (for some large enough $k(n)$).

Conversely, for a fixed n, we show that, for all but finitely many $\bar{w} \in W/\overset{\circ}{W}$, $(\mathcal{B}\bar{w}\mathcal{P}/\mathcal{P}) \cap \hat{X}_n = \emptyset$. Represent $\bar{w}$ as the morphism $\mathbb{C}^* \to \overset{\circ}{T} \hookrightarrow D$,

$$t \mapsto \begin{pmatrix} t^{n_1(\bar{w})} & & \\ & \ddots & \\ & & t^{n_N(\bar{w})} \end{pmatrix},$$

for some $n_i(\bar{w}) \in \mathbb{Z}$. We first claim that any $\bar{w}$ such that $n_i(\bar{w}) < -n$, for some i, satisfies $(\mathcal{B}\bar{w}\mathcal{P}/\mathcal{P}) \cap \hat{X}_n = \emptyset$: Otherwise, if for some $b \in \mathcal{B}$, $b\bar{w}L_o \in \mathcal{F}_n$, then $\bar{w}L_o \in b^{-1}\mathcal{F}_n = \mathcal{F}_n$, a contradiction to the choice of $\bar{w}$! Now, observe that the set $\{\bar{w} \in W/\overset{\circ}{W} : n_i(\bar{w}) \geq -n \text{ for all } i\}$ is finite since $\sum n_i(\bar{w}) = 0$. From this, it follows that $\hat{X}_n \subset X_{k(n)}$, for some large enough $k(n)$. This proves the lemma. □

13.2.17 Lemma. *With the notation as in 13.2.15, $\hat{X}_n$ is a Zariski closed subset of $\hat{X}^o_n$ for all $n \geq 0$.*

Proof. Fix $\bar{w} \in W/\overset{\circ}{W}$ and take a coset representative w of $\bar{w}$ of minimal length. Choose any reduced decomposition $w = s_{i_1} \dots s_{i_p}$, where the s_j's are the simple reflections in W, and consider the Bott–Samelson–Demazure–Hansen variety $Z_{\mathfrak{w}}$ defined in 7.1.3 for the reduced word $\mathfrak{w} := (s_{i_1}, \dots , s_{i_p})$. Let $\mathcal{P}_j$ be the minimal parabolic subgroup of $\mathcal{G}$ corresponding to the simple reflection s_j. Recall that, set theoretically, $Z_{\mathfrak{w}} := \mathcal{P}_{i_1} \times \cdots \times \mathcal{P}_{i_p}/\mathcal{B}^p$, where $\mathcal{B}^p$ acts on $\mathcal{P}_{i_1} \times \mathcal{P}_{i_2} \times \cdots \times \mathcal{P}_{i_p}$ from the right via

$$(x_1, \dots, x_p)(b_1, \dots, b_p) = (x_1 b_1, b_1^{-1} x_2 b_2, \dots, b_{p-1}^{-1} x_p b_p),$$

for $x_j \in \mathcal{P}_{i_j}$ and $b_j \in \mathcal{B}$.

As in 7.1.11, define the map $m_{\mathfrak{w}} = m^Y_{\mathfrak{w}} : Z_{\mathfrak{w}} \to \mathcal{Y}$ by

$$m_{\mathfrak{w}}((x_1, \dots, x_p) \bmod \mathcal{B}^p) = x_1 \dots x_p \mathcal{P}.$$

Since $\operatorname{Im} m_{\mathfrak{w}} = X_{\bar{w}}$ (cf. (7.1.11.2)), by Lemma 13.2.16, $\operatorname{Im}(i \circ m_{\mathfrak{w}}) \subset \hat{X}^o_m$ for some m, where $i : \mathcal{Y} \hookrightarrow \mathcal{Y}^o$ is the inclusion as in 13.2.15 and $X_{\bar{w}} = X^Y_{\bar{w}}$ is the Schubert variety defined by (7.1.13.1). It can easily be seen that $i \circ m_{\mathfrak{w}} : Z_{\mathfrak{w}} \to \mathcal{Y}^o_{\text{lat}}$ is a morphism by showing that $i \circ m_{\mathfrak{w}|Z_{\mathfrak{w}}(\mathfrak{a})}$ is a morphism, where $Z_{\mathfrak{w}}(\mathfrak{a})$ is defined by (7.1.3.5) and it is open in $Z_{\mathfrak{w}}$ by the proof of Proposition 7.1.8. In particular, $Z_{\mathfrak{w}}$ being projective, $i(X_{\bar{w}})$ is closed in $\hat{X}^o_m$ (cf. Theorem A.2). We now prove that $i(\hat{X}_n)$ is closed in $\hat{X}^o_n$:

Observe first that $\hat{X}_n$ is left $\mathcal{B}$-stable. Take any $\bar{w} \in W/\overset{\circ}{W}$ such that $\mathcal{B}\bar{w}\mathcal{P}/\mathcal{P} \subset \hat{X}_n$. Then we claim that $X_{\bar{w}} \subset \hat{X}_n$: There is an open (dense) subset $Z^o_{\mathfrak{w}} \subset Z_{\mathfrak{w}}$ such that $m_{\mathfrak{w}}(Z^o_{\mathfrak{w}}) = \mathcal{B}\bar{w}\mathcal{P}/\mathcal{P}$ (cf. Proposition 7.1.15(a)). Hence, considering the morphism $i \circ m_{\mathfrak{w}} : Z_{\mathfrak{w}} \to \mathcal{Y}^o_{\text{lat}}$, we see that $i \circ m_{\mathfrak{w}}(Z_{\mathfrak{w}}) \subset \hat{X}^o_n$ (since $\hat{X}^o_n$ is projective). In particular, $X_{\bar{w}} \subset \hat{X}_n$ and thus $\hat{X}_n$ is a finite union (by Lemma 13.2.16) of Schubert varieties $X_{\bar{w}}$. Now, since $i(X_{\bar{w}})$ is closed in $\hat{X}^o_n$, so is $i(\hat{X}_n)$. This proves the lemma. □

13.2.18 Proposition. *The identity map $\mathcal{Y}_{\mathrm{rep}} \to \mathcal{Y}_{\mathrm{lat}}$ is an isomorphism of ind-varieties, where we denote $\mathcal{Y}$ equipped with the projective ind-variety structure as in 7.1.19 by $\mathcal{Y}_{\mathrm{rep}}$.*

Proof. Embed $\overset{\circ}{G} \hookrightarrow \mathrm{SL}_N$ as in 13.2.15 and follow the same notation as in 13.2.13 and 13.2.15. By the definition, $\mathcal{Y}_{\mathrm{lat}} \hookrightarrow \mathcal{Y}^o_{\mathrm{lat}}$ is a closed embedding. Similarly, we claim that $\mathcal{Y}_{\mathrm{rep}} \hookrightarrow \mathcal{Y}^o_{\mathrm{rep}}$ is a closed embedding:

Take the integrable highest weight $\mathfrak{g}_o$-module $L = L(\omega_0)$, where $\mathfrak{g}_o := \hat{\mathcal{L}}(s\ell_N)$ and ω_0 is the zeroth fundamental weight for $\mathfrak{g}_o$, and let $L' \subseteq L$ be the (integrable highest weight) $\mathfrak{g}$-submodule generated by the highest weight vector of L. Then we have the commutative diagram:

$$\begin{array}{ccc} \mathcal{Y}_{\mathrm{rep}} & \hookrightarrow & \mathcal{Y}^o_{\mathrm{rep}} \\ \downarrow & & \downarrow \\ \mathbb{P}(L') & \hookrightarrow & \mathbb{P}(L), \end{array}$$

where both the vertical maps (obtained by taking the orbits through the highest weight vector) are closed embeddings by Proposition 7.1.15(c) and, moreover, $\mathbb{P}(L') \hookrightarrow \mathbb{P}(L)$ is, of course, a closed embedding. This proves that $\mathcal{Y}_{\mathrm{rep}} \hookrightarrow \mathcal{Y}^o_{\mathrm{rep}}$ is a closed embedding. So, to prove the lemma, we can take $\overset{\circ}{G} = \mathrm{SL}_N$.

Fix $\bar{w} \in W/\overset{\circ}{W}$ (where W is the affine Weyl group corresponding to $\overset{\circ}{G} = \mathrm{SL}_N$). By the proof of Lemma 13.2.17 (following the same notation), the map $m_{\mathfrak{w}} : Z_{\mathfrak{w}} \to \mathcal{Y}^o_{\mathrm{lat}}$ is a morphism with its image precisely equal to the Schubert variety $X^o_{\bar{w}}$. We denote $X^o_{\bar{w}}$ endowed with the reduced closed subvariety structure induced from $\mathcal{Y}^o_{\mathrm{rep}}$ by $X^o_{\bar{w},\mathrm{rep}}$ (and a similar meaning for $X^o_{\bar{w},\mathrm{lat}}$). Then the map $\bar{m}_{\mathfrak{w}} : Z_{\mathfrak{w}} \to X^o_{\bar{w},\mathrm{rep}}$ is a surjective morphism (cf. Proposition 7.1.15) and, moreover, $X^o_{\bar{w},\mathrm{rep}}$ is an irreducible normal variety (cf. Theorem 8.2.2(b)), where the map $\bar{m}_{\mathfrak{w}}$ at the level of sets is nothing but $m_{\mathfrak{w}}$. We claim that the inclusion map $I_{\bar{w}} : X^o_{\bar{w},\mathrm{rep}} \to \mathcal{Y}^o_{\mathrm{lat}}$ is a morphism (and thus the identity map $\mathcal{Y}^o_{\mathrm{rep}} \to \mathcal{Y}^o_{\mathrm{lat}}$ is a morphism). To prove the assertion in parenthesis, use the fact that, for any $n \geq 0$, there exists $\bar{w} = \bar{w}_n \in W/\overset{\circ}{W}$ such that $X^o_{\bar{w}} \supset X^o_n$ by Lemma 1.3.20.

First, by Lemma 13.2.16, Im $I_{\bar{w}} \subset \hat{X}^o_n$ for some n. Now, the map $\bar{m}_{\mathfrak{w}} : Z_{\mathfrak{w}} \to X^o_{\bar{w},\mathrm{rep}}$ being a proper surjective morphism, the (Zariski) topology on $X^o_{\bar{w},\mathrm{rep}}$ is the quotient topology. Let $V \subset \hat{X}^o_n$ be an open subset. Then $m_{\mathfrak{w}}^{-1}(V) = (\bar{m}_{\mathfrak{w}})^{-1} I_{\bar{w}}^{-1}(V)$ is open in $Z_{\mathfrak{w}}$ and hence $I_{\bar{w}}^{-1}(V)$ is open in $X^o_{\bar{w},\mathrm{rep}}$. To prove that $I_{\bar{w}}$ is a morphism, it suffices to show that, for any affine open $V \subset \hat{X}^o_n$, the map $I_{\bar{w}|I_{\bar{w}}^{-1}(V)} : I_{\bar{w}}^{-1}(V) \to V$ is a morphism. But this follows from Proposition A.12 by taking a closed embedding $V \subset \mathbb{A}^N$, since the map $I_{\bar{w}} \circ \bar{m}_{\mathfrak{w}} = m_{\mathfrak{w}}$ is a morphism (and $X^o_{\bar{w},\mathrm{rep}}$ is normal).

Conversely, we show that the identity map $\mathcal{Y}^o_{\mathrm{lat}} \to \mathcal{Y}^o_{\mathrm{rep}}$ is a morphism: Fix $n \geq 0$ and consider $\hat{X}^o_n$. Then for the identity map $I : \mathcal{Y}^o_{\mathrm{rep}} \to \mathcal{Y}^o_{\mathrm{lat}}$, which is a morphism as proved above, $I^{-1}(\hat{X}_n^o) \subset \mathcal{Y}_{\text{rep}}^o$ is a Zariski closed subset and, moreover, by Lemma 13.2.16, $I^{-1}(\hat{X}_n^o) \subset X_m^o$ (for some m); in particular, $I^{-1}(\hat{X}_n^o)$ acquires the structure of a closed subvariety of $\mathcal{Y}_{\text{rep}}^o$. Further, the bijective map $I_n : I^{-1}(\hat{X}_n^o) \to \hat{X}_n^o$ is a morphism since I is a morphism (where $I_n := I_{|I^{-1}(\hat{X}_n^o)}$), and the variety $\hat{X}_n^o$ is isomorphic with the variety $\mathrm{Gr}(nN, 2nN)^{1+\bar{t}_n}$ (cf. 13.2.13). But $\mathrm{Gr}(nN, 2nN)^{1+\bar{t}_n}$ is known to be irreducible and normal by using a result of Kostant (cf. Exercise 13.2.E.6). Moreover, I_n being a proper bijective morphism, it is a homeomorphism; in particular, $I^{-1}(\hat{X}_n^o)$ is irreducible as well. Hence, by Theorem A.11, I_n is a biregular isomorphism. This shows that the identity map $\mathcal{Y}_{\text{lat}}^o \to \mathcal{Y}_{\text{rep}}^o$ also is a morphism, proving the proposition. □

So we identify $\mathcal{Y}_{\text{lat}}$ with $\mathcal{Y}_{\text{rep}}$ and simply denote them by $\mathcal{Y}$. For any $\lambda \in \mathfrak{h}_{\mathbb{Z},Y}^*$, recall the definition of the line bundle $\mathcal{L}(\lambda) = \mathcal{L}^Y(\lambda)$ on $\mathcal{G}/\mathcal{P}_Y$ from 7.2.1. We have the following proposition determining $\mathrm{Pic}(\mathcal{Y})$.

13.2.19 Proposition. *The map $\mathbb{Z} \to \mathrm{Pic}(\mathcal{Y})$ given by $d \mapsto \mathcal{L}(d\omega_0)$ is an isomorphism, where $\omega_0 \in \mathfrak{h}_{\mathbb{Z}}^*$ is the zeroth fundamental weight defined by $\omega_0(\alpha_i^\vee) = \delta_{0,i}$, for any $0 \le i \le \ell$, and $\omega_0(d) = 0$ (cf. 13.1.1).*

Proof. For any $\bar{w} \in W/\overset{\circ}{W}$, since $X_{\bar{w}}$ is a projective variety, by GAGA [Serre–56], the natural map

$$\text{(1)} \qquad \mathrm{Pic}(X_{\bar{w}}) \xrightarrow{\sim} \mathrm{Pic}_{\text{an}}(X_{\bar{w}})$$

is an isomorphism, where $\mathrm{Pic}_{\text{an}}(X_{\bar{w}})$ is the set of isomorphism classes of analytic line bundles on $X_{\bar{w}}$.

We have the sheaf exact sequence

$$\text{(2)} \qquad 0 \to \mathbb{Z} \to \mathcal{O}_{\text{an}} \to \mathcal{O}_{\text{an}}^* \to 0,$$

where $\mathcal{O}_{\text{an}}$, resp. $\mathcal{O}_{\text{an}}^*$, denotes the sheaf of analytic functions, resp. the sheaf of invertible analytic functions, on $X_{\bar{w}}$ (under the analytic topology). Taking the associated long exact cohomology sequence, we get

$$\text{(3)} \qquad \cdots \to H^1(X_{\bar{w}}, \mathcal{O}_{\text{an}}) \to H^1(X_{\bar{w}}, \mathcal{O}_{\text{an}}^*) \xrightarrow{c_1} H^2(X_{\bar{w}}, \mathbb{Z}) \to H^2(X_{\bar{w}}, \mathcal{O}_{\text{an}}) \to \dots,$$

where the map c_1 associates to any line bundle its first Chern class. Now,

$$\text{(4)} \qquad H^i(X_{\bar{w}}, \mathcal{O}) = 0, \qquad \text{for all } i > 0,$$

by Theorem 8.2.2(c), and by GAGA

$$\text{(5)} \qquad H^i(X_{\bar{w}}, \mathcal{O}) \approx H^i(X_{\bar{w}}, \mathcal{O}_{\text{an}}),$$

and hence the map c_1 is an isomorphism. But

(6) $$\mathrm{Pic}_{\mathrm{an}}(X_{\bar{w}}) \approx H^1(X_{\bar{w}}, \mathcal{O}^*_{\mathrm{an}}) .$$

Hence, by combining (1) and (3)–(6), we get the following isomorphism (again denoted by)

(7) $$c_1 : \mathrm{Pic}(X_{\bar{w}}) \overset{\approx}{\to} H^2(X_{\bar{w}}, \mathbb{Z}) .$$

Further, the following diagram is commutative whenever $\bar{w} \leq \bar{v}$.

($\mathcal{D}$) $$\begin{array}{ccc} \mathrm{Pic}(X_{\bar{v}}) & \xrightarrow[\sim]{c_1} & H^2(X_{\bar{v}}, \mathbb{Z}) \\ \downarrow & & \downarrow \\ \mathrm{Pic}(X_{\bar{w}}) & \xrightarrow[\sim]{c_1} & H^2(X_{\bar{w}}, \mathbb{Z}) , \end{array}$$

where the vertical maps are the canonical restriction maps, and we define $\bar{w} \leq \bar{v}$ iff $X_{\bar{w}} \subseteq X_{\bar{v}}$. But, from the Bruhat decomposition and Proposition 11.3.2, for any $\bar{w} \geq \bar{s}_0$, the restriction map

(8) $$H^2(X_{\bar{w}}, \mathbb{Z}) \to H^2(X_{\bar{s}_0}, \mathbb{Z})$$

is an isomorphism, where $s_0 \in W$ is the (simple) reflection corresponding to the simple coroot $\alpha_0^\vee$, and $\bar{s}_0 := s_0 \bmod \overset{\circ}{W}$. Moreover, $X_{\bar{s}_0}$ being isomorphic with the complex projective line $\mathbb{P}^1$, $H^2(X_{\bar{s}_0}, \mathbb{Z})$ is a free $\mathbb{Z}$-module of rank 1, which is generated by the first Chern class of the line bundle $\mathcal{L}(\omega_0)_{|X_{\bar{s}_0}}$. Thus, for any $\bar{w} \geq \bar{s}_0$, $\mathrm{Pic}(X_{\bar{w}})$ is freely generated by $\mathcal{L}(\omega_0)_{|X_{\bar{w}}}$.

We next prove that the canonical map

$$\alpha : \mathrm{Pic}\ (\mathcal{Y}) \to \underset{\bar{w} \in W/\overset{\circ}{W}}{\mathrm{Inv.\ lt.}}\ \mathrm{Pic}\ (X_{\bar{w}})$$

is an isomorphism: Since $\mathcal{L}(\omega_0)$ is an algebraic line bundles on $\mathcal{Y}$, the surjectivity of the map α follows. Now we come to the injectivity of α:

Let $\mathcal{L} \in \mathrm{Ker}\ \alpha$. Fix a nonzero vector v_o in the fiber of $\mathcal{L}$ over the base point $\mathfrak{e} := e\mathcal{P} \in \mathcal{Y}$. Then $\mathcal{L}_{|X_{\bar{w}}}$ being a trivial line bundle over any $X_{\bar{w}}$, we can choose a nowhere-vanishing section $s_{\bar{w}}$ of $\mathcal{L}_{|X_{\bar{w}}}$ such that $s_{\bar{w}}(\mathfrak{e}) = v_o$. We next show that, for any $\bar{v} \geq \bar{w}$, $s_{\bar{v}|X_{\bar{w}}} = s_{\bar{w}}$: Clearly $s_{\bar{v}|X_{\bar{w}}} = f s_{\bar{w}}$, for a regular function $f : X_{\bar{w}} \to \mathbb{C}^*$. But, $X_{\bar{w}}$ being projective and irreducible, f is constant and, in fact, $f \equiv 1$ since $s_{\bar{v}}(\mathfrak{e}) = s_{\bar{w}}(\mathfrak{e})$. So the sections $s_{\bar{w}}$ patch together to give rise to a nowhere-vanishing (regular) section s of $\mathcal{L}$ on the whole of $\mathcal{Y}$ such that $s_{|X_{\bar{w}}} = s_{\bar{w}}$ for any $\bar{w} \in W/\overset{\circ}{W}$. From this it is easy to see that $\mathcal{L}$ is isomorphic with the trivial line bundle on $\mathcal{Y}$. This proves that α is injective, thereby completing the proof of the proposition. □

13.2.E EXERCISES

(1) Prove that (13.2.3.3) indeed defines a representation of $\bar{\mathcal{L}}(\overset{\circ}{G})$ in $\hat{\mathcal{L}}(\overset{\circ}{\mathfrak{g}})_{\text{comp}}$.

(2) Show that for any $g \in \overset{\circ}{G}(K)$, $g^{-1}\frac{dg}{dt} = (\frac{dg}{dt})g^{-1}$ and, moreover, it belongs to $K \otimes \overset{\circ}{\mathfrak{g}}$ (cf. 13.2.3).

Hint. For $g, h \in \overset{\circ}{G}(K)$, $(gh)^{-1}\frac{d(gh)}{dt} = h^{-1}(g^{-1}\frac{dg}{dt})h + h^{-1}\frac{dh}{dt}$. Thus the elements $g \in \overset{\circ}{G}(K)$ such that $g^{-1}\frac{dg}{dt} \in K \otimes \overset{\circ}{\mathfrak{g}}$ form a subgroup. Further, if Log g exists in $K \otimes \overset{\circ}{\mathfrak{g}}$, $\frac{d}{dt}(\text{Log}\, g) = g^{-1}\frac{dg}{dt} \in K \otimes \overset{\circ}{\mathfrak{g}}$. Now use the result of Steinberg as in the proof of Proposition 13.2.4.

(3) With the notation as in 13.2.7, show that the canonical map $\overset{\circ}{G}(\mathcal{R}) \to \overset{\circ}{G}(\mathcal{R}_n)$ is surjective for all $n \geq 1$.

Hint. Show first that, for any unipotent subgroup $H \subset \overset{\circ}{G}$, $H(\mathcal{R}) \to H(\mathcal{R}_n)$ is surjective. Now apply Steinberg's result as in the proof of Proposition 13.2.4.

(4) With the notation as in Theorem 13.2.8, show that for any $g \in \mathcal{G}$, $X \in \hat{\mathcal{L}}(\overset{\circ}{\mathfrak{g}})_{\text{comp}}$,

$$\text{Ad}\, g(X) = (\text{Ad}\, \psi(g))(X),$$

where Ad g refers to the adjoint representation defined in 6.2.10 and Ad $\psi(g)$ is defined by (13.2.3.3).

(5) Show that the group homomorphism

$$\psi : \mathcal{G} \to \bar{\mathcal{L}}(\overset{\circ}{G})/C \text{ of Theorem 13.2.8}$$

induces an isomorphism

$$N_{\mathcal{G}}(T)/T \simeq \bar{N}/\bar{T},$$

where $N_{\mathcal{G}}(T)$ is the normalizer of T in $\mathcal{G}$ and $\bar{N}$, $\bar{T}$ are as in 13.2.2. (By Lemma 6.2.9(b), $N_{\mathcal{G}}(T)/T \simeq W$, the Weyl group of $\mathfrak{g}$.)

(6) Under the notation of 13.2.13, show that $\text{Gr}(nN, 2nN)^{1+\bar{t}_n}$ is irreducible and normal.

Hint. There exists an open dense subspace $Z \subset \text{Gr}(nN, 2nN)^{1+\bar{t}_n}$ containing the "worst" point z_o of $\text{Gr}(nN, 2nN)^{1+\bar{t}_n}$ such that $V \times Z$ is biregular isomorphic with an open subset of the nilpotent cone $\mathcal{N} := \{X \in sl_d : X \text{ is a nilpotent matrix}\}$ in sl_d for some d, where V is a smooth variety. Now use a result of Kostant that $\mathcal{N}$ is a normal and irreducible variety.

(7) Consider the GCM $A = \begin{pmatrix} 2 & -2 \\ -2 & 2 \end{pmatrix}$, so that $\mathcal{G}(A)$ is the Kac–Moody group corresponding to the affine Kac–Moody algebra $\hat{\mathcal{L}}(sl_2)$. Let $\mathcal{Y}$ be the

associated infinite Grassmannian (as in 13.2.12). Then, for any $d \geq 0$, show that there is a unique Schubert variety $Y_d \subset \mathcal{Y}$ of dimension d. It is explicitly given by $Y_d = \mathcal{B}\begin{pmatrix} t^d & 0 \\ 0 & t^{-d} \end{pmatrix}\mathcal{P}/\mathcal{P} \subset \mathcal{Y}$, where $\mathcal{P} := \mathcal{P}_{\{1\}}$.

Show further that the singular locus of Y_d is precisely equal to Y_{d-2} for any $d \geq 2$. (Of course, Y_0, Y_1 are smooth.)

13.C Comments. The material of Section 13.1 is fairly standard. For a more detailed treatment, including of the twisted affine Kac–Moody algebras, we refer to [Kac–90, Chap. 6–8].

A detailed study of the "loop group" $\overset{\circ}{G}(K)$ (including its Bruhat decomposition), for K a field with nontrivial discrete valuation, was done by [Iwahori–Matsumoto–65]. Subsequently [Garland–80] studied the affine Kac–Moody groups $\mathcal{G}(k)$ and their representations over an arbitrary field k. In particular, he constructed the "simply-connected" $\mathcal{G}(k)$ for an arbitrary k. He showed the existence of a Chevalley lattice in any integrable highest weight module of $\mathcal{G}(k)$ for k a field of char. 0 (cf. [Garland–78, §11], [Garland–80, §6]) and used it to develop a theory of arithmetic subgroups $\Gamma \subset \mathcal{G}(k)$ ($k = \mathbb{R}$ or $\mathbb{C}$).

The explicit expression of the adjoint representation as in 13.2.3 is taken from [Pressley–Segal–86, Proposition 4.9.4], [Kac–90, Exercise 7.20] (also available in the 1983 edition of Kac's book). Proposition 13.2.4 is standard. Theorem 13.2.8 is essentially available in [Garland–80], although not exactly in the form stated here. In fact, there is a stronger result (cf. Remark 13.2.11) in loc. cit. As mentioned in 7.C, the realization of the generalized flag varieties $\mathcal{X}^Y$ in the affine case, as the set of certain lattices, is due to [Kazhdan–Lusztig–80] and [Lusztig–83]. This way they obtained a projective ind-variety structure on $\mathcal{X}^Y$. The identification of the two projective ind-variety structures on $\mathcal{Y}$ as in Proposition 13.2.18 is taken from [Kumar–97a, Appendix C], and so is the proof of Proposition 13.2.19.

For applications of the infinite Grassmannian $\mathcal{Y}$ to certain questions in the moduli space of $\overset{\circ}{G}$-bundles on a projective curve, the reader is referred to the articles [Beauville–Laszlo–94], [Faltings–94], [Kumar–Narasimhan–Ramanathan–94], [Kumar–Narasimhan–97], [Laszlo–Sorger–97], among others.

For the origin of the Bruhat, resp. Birkhoff, decomposition in the affine case, the reader is referred to [Grothendieck–57a], resp. [Birkhoff–13]. Finally, we refer to the book [Pressley–Segal–86] for a detailed study of various aspects of the loop groups (although we have followed a more algebro-geometric approach in contrast to loc. cit.).

Appendix A

Results from Algebraic Geometry

We collect some results from algebraic geometry that are frequently used in the book. The standard references are [Hartshorne–77], [Mumford–88], [Šafarevič–94]. We do not state the results in their full generality, but we state whatever is required for their applications in the book.

In this appendix k is an arbitrary algebraically closed field (of any characteristic). By a variety X, we always mean a quasiprojective (reduced) variety over k as defined in [Šafarevič–94, Chapter I, §4.1], i.e., an open subset of a closed, not necessarily irreducible, subset of a projective space over k. Its structure sheaf is denoted by $\mathcal{O}_X$ and its coordinate ring, i.e., the ring of global regular functions on X is denoted by $k[X]$. By the dimension dim X of a variety X, we mean the maximum of the dimensions of its irreducible components. By a point of a variety, we mean a closed point.

A.1 Theorem. ([Šafarevič–94, Chap. I, §6, Theorem 5].) *Let X be an irreducible projective variety and let F be a nonzero form on X, i.e., F is the restriction of a homogeneous polynomial on an ambient projective space $\mathbb{P}^n$ for some (closed) embedding of X in $\mathbb{P}^n$. Then each irreducible component of the zero set X_F has dimension equal to* $\dim X - 1$.

A.2 Theorem. ([Šafarevič–94, Chap. I, §5, Theorem 2].) *The image of a projective variety under a morphism is closed.*

A.3 Definition. Let X be a variety and $x \in X$. The *Zariski tangent space* $T_x(X)$ of X at x is, by definition, the vector space over k :

$$(1) \qquad T_x(X) := \mathrm{Hom}_k(\mathfrak{m}_x/\mathfrak{m}_x^2, k),$$

where $\mathfrak{m}_x$ is the maximal ideal of the local ring $\mathcal{O}_{x,X}$.

Any morphism $f : X \to Y$ of varieties induces a canonical map (for any $x \in X$)

$$(2) \qquad (df)_x : T_x(X) \to T_{f(x)}(Y),$$

called the *derivative* of f at x.

By [Mumford–88, Chap. III, §4, Proposition 2], for an irreducible variety X of dimension ℓ and any point $x \in X$,

(3) $$\dim T_x(X) \geq \ell.$$

Moreover,

(4) $$x \in X \text{ is a smooth point} \Leftrightarrow \dim T_x(X) = \ell.$$

A.4 Definition. For any local ring R with maximal ideal $\mathfrak{m}$, define the graded $R/\mathfrak{m}$-algebra

(1) $$\operatorname{gr} R := \sum_{n \geq 0} \mathfrak{m}^n/\mathfrak{m}^{n+1}.$$

Let X be a variety over k and let $x \in X$ be a point. Then the *tangent cone* $C_x(X)$ of X at x is, by definition, $\operatorname{Spec}(\operatorname{gr} \mathcal{O}_{x,X})$, where (as above) $\mathcal{O}_{x,X}$ is the local ring of X at x. By [Mumford–88, Corollary on pg. 226], if X is an irreducible variety of dimension ℓ, then

(2) $$\dim C_x(X) = \ell.$$

In fact, each irreducible component of the scheme $C_x(X)$ is of dimension ℓ. By [Mumford–88, Chapter 3, §4], $x \in X$ is a smooth point iff the tangent cone is linear, i.e., $\operatorname{gr} \mathcal{O}_{x,X}$ is graded isomorphic with the polynomial ring $k[t_1, \dots, t_\ell]$ with $\deg t_i = 1$ for all i.

Let $x \in U \subset X$ be an affine neighborhood of x in X. Then, there is a canonical graded algebra isomorphism (cf. [Mumford–88, Chapter 3, §3]):

(3) $$\sum_{n \geq 0} \mathfrak{m}_x(U)^n/\mathfrak{m}_x(U)^{n+1} \simeq \operatorname{gr} \mathcal{O}_{x,X},$$

where $\mathfrak{m}_x(U)$ is the maximal ideal of the coordinate ring $k[U]$ consisting of the functions vanishing at x.

Observe that the k-algebra $\operatorname{gr} \mathcal{O}_{x,X}$ has nilpotents in general. Let $\mathcal{N}$ be the ideal of $\operatorname{gr} \mathcal{O}_{x,X}$ consisting of all the nilpotent elements. Then the k-algebra $(\operatorname{gr} \mathcal{O}_{x,X})/\mathcal{N}$ is of course reduced. Define the *reduced tangent cone* $C_x^{\text{red}}(X)$ as $\operatorname{Spec}(\operatorname{gr} \mathcal{O}_{x,X}/\mathcal{N})$.

A.5 Definition. Let X be an irreducible variety. Then a point $x \in X$ is said to be a *normal point of* X or X is said to be *normal at* x, if the local ring $\mathcal{O}_{x,X}$ is integrally closed in its quotient field. The variety X itself is called *normal* if it is normal at every point.

For example, a smooth irreducible variety is normal. It is well known (cf. [Mumford-88, III.8, Proposition 1]) that the codimension of the (closed) set of singular points $\Sigma(X)$ of a normal variety X is at least 2, i.e., each irreducible component of $\Sigma(X)$ is of codimension at least 2 in X.

A *normalization* of an irreducible variety X is a normal (irreducible) variety $\tilde{X}$ together with a finite morphism $\pi : \tilde{X} \to X$ which is a birational isomorphism.

The following result can be found in [Mumford–88, III.8, Theorems 3,4] and [Šafarevič–94, Chap. II, §5, Theorem 4].

A.6 Proposition. *An irreducible variety X admits a normalization $\pi : \tilde{X} \to X$. Moreover, it is unique, in the sense that if $\pi' : \tilde{X}' \to X$ is another normalization, then there exists an isomorphism $f : \tilde{X} \to \tilde{X}'$ making the following diagram commutative:*

$$\begin{array}{ccc} \tilde{X} & \xrightarrow{f} & \tilde{X}' \\ & {}_{\pi}\searrow \quad \swarrow_{\pi'} & \\ & X & . \end{array}$$

The normalization of an (irreducible) affine, resp. projective, variety is affine, resp. projective.

The normalization satisfies the following universal property (cf. [Hartshorne–77, Chap. II, Exercise 3.8]):

A.7 Proposition. *With the notation as is the above proposition, let $f : Y \to X$ be a dominant morphism, i.e., $f(Y)$ is dense in X, such that Y is a normal (irreducible) variety. Then there exists a unique lift $\tilde{f} : Y \to \tilde{X}$ such that $\pi \circ \tilde{f} = f$.*

Let $f : X \to Y$ be a morphism of varieties. For any $\mathcal{O}_X$-module $\mathcal{S}$, resp. $\mathcal{O}_Y$-module $\mathcal{T}$, recall the definition of the *direct image* $\mathcal{O}_Y$-module $f_*\mathcal{S}$, resp. the *inverse image* $\mathcal{O}_X$-module $f^*\mathcal{T}$, from [Hartshorne–77, Chap. II, §5]. Then, by loc. cit., f_* and f^* are adjoint functors. More specifically,

A.8 Lemma. $\mathrm{Hom}_{\mathcal{O}_X}(f^*\mathcal{T}, \mathcal{S}) \simeq \mathrm{Hom}_{\mathcal{O}_Y}(\mathcal{T}, f_*\mathcal{S})$.

A proof of the following Zariski's connectedness theorem (also known as the Zariski's main theorem) can be found, e.g., in [Hartshorne–77, Chap. III, Corollary 11.4 and its proof].

A.9 Theorem. *Let $f : X \to Y$ be a birational projective morphism between irreducible varieties. Assume further that Y is normal. Then, for any $y \in Y$, $f^{-1}(y)$ is connected. Moreover,*

(1) $$f_*\mathcal{O}_X = \mathcal{O}_Y .$$

(Observe that, for projective varieties X, Y, any morphism $f : X \to Y$ is a projective morphism.)

A subset Y of a topological space X is called *locally closed* if Y is open in the closure $\bar{Y}$ of Y, or equivalently, if Y is the intersection of an open subset with a closed subset of X. A *constructible subset* of X is, by definition, a finite union of locally closed subsets of X.

We recall the following result due to Chevalley (cf. [Borel–91, Chap. AG, Corollary 10.2]).

A.10 Theorem. *Let $f : X \to Y$ be a morphism of varieties. Then the image of any constructible subset of X is constructible in Y. In particular, by Exercise A.E.3, $f(X)$ contains a dense open subset of $\overline{f(X)}$.*

A.11 Theorem. *Assume char. $k = 0$. Let $f : X \to Y$ be a bijective morphism between irreducible varieties. Assume further that Y is normal. Then f is a biregular isomorphism.*

Proof. Use [Springer–98, Theorems 5.1.6(iii) and 5.2.8] together with the fact that any field extension in char. 0 is separable. □

The following proposition is taken from [Kumar–Narasimhan–Ramanathan–94].

A.12 Proposition. *Assume char. $k = 0$. Let $f : X \to Y$ be a surjective morphism between irreducible varieties over k. Assume that Y is normal and let $\mathcal{E} \to Y$ be an algebraic vector bundle over Y.*

Then any set theoretic section σ of the vector bundle $\mathcal{E}$ is regular if and only if the induced section $f^*(\sigma)$ of the induced bundle $f^*(\mathcal{E})$ is regular. In particular, a set map $\sigma : Y \to k$ is regular iff $\sigma \circ f : X \to k$ is regular.

Proof. The "only if" part is of course trivially true. So we come to the "if" part. Since the question is local (in Y), we can assume that Y is affine and, moreover, the vector bundle $\mathcal{E}$ is trivial, i.e., it suffices to show that any (set theoretic) map $\sigma : Y \to k$ is regular, provided $\bar{\sigma} := \sigma \circ f : X \to k$ is regular (under the assumption that $Y = \operatorname{Spec} R$ is irreducible normal and affine):

Since the map f is surjective (in particular, dominant), the ring R is canonically embedded in $\Gamma(X) := H^0(X, \mathcal{O}_X)$. Let $R[\bar{\sigma}]$ denote the subring of $\Gamma(X)$ generated by R and $\bar{\sigma} \in \Gamma(X)$. Then $R[\bar{\sigma}]$ is a (finitely generated) domain (as X is irreducible by assumption), and we get a dominant morphism $\hat{f} : Z \to \operatorname{Spec} R$, where $Z := \operatorname{Spec}(R[\bar{\sigma}])$. Consider the commutative diagram:

$$\begin{array}{ccc} & X & \\ {\scriptstyle\theta}\swarrow & & \searrow{\scriptstyle f} \\ Z & \xrightarrow[\hat{f}]{} & Y, \end{array}$$

where θ is the dominant morphism induced from the inclusion $R[\bar{\sigma}] \hookrightarrow \Gamma(X)$ (cf. [Hartshorne–77, Chap. I, Proposition 3.5]). In particular, Im θ contains a nonempty Zariski open subset U of Z (cf. Theorem A.10). Let $x_1, x_2 \in X$ be (closed) points such that $f(x_1) = f(x_2)$. Then $r(x_1) = r(x_2)$ for all $r \in R$, and also $\bar{\sigma}(x_1) = \bar{\sigma}(x_2)$. This forces $\theta(x_1) = \theta(x_2)$; in particular, $\hat{f}_{|U}$ is injective on the closed points of U.

Since $\hat{f}$ is dominant, by cutting down U if necessary, we can assume that $\hat{f}_{|U} : U \to V$ is a bijection, for some open subset $V \subset Y$. Now, since Y is (by assumption) normal and Z is irreducible, by Theorem A.11, $\hat{f}_{|U} : U \to V$ is an isomorphism, and hence σ is regular on V.

Assume, if possible, that $\sigma_{|_V}$ does not extend to a regular function on the whole of Y. Then, by [Borel–91, Lemma 18.3, Chapter AG], there exists a point $y_o \in Y$ and a regular function h on a Zariski neighborhood W of y_o in Y such that $h(y_o) = 0$ and $h\sigma \equiv 1$ on $W \cap V$. But then $\bar{h}\bar{\sigma} \equiv 1$ on $f^{-1}(W \cap V)$, where $\bar{h} := h \circ f$, and hence, $\bar{\sigma}$ being regular on the whole of X, $\bar{h}\bar{\sigma} \equiv 1$ on $f^{-1}(W)$. Taking $\bar{y}_o \in f^{-1}(y_o)$ (f is, by assumption, surjective), we get $\bar{h}(\bar{y}_o)\bar{\sigma}(\bar{y}_o) = 0$. This contradiction shows that $\sigma_{|_V}$ does extend to a regular function (say σ') on the whole of Y. Hence $\bar{\sigma} = \bar{\sigma}'$ on the whole of X; in particular, by the surjectivity of f, $\sigma = \bar{\sigma}'$. This proves the proposition. □

A.13 Definition. A morphism $f : X \to Y$ of varieties is called *affine* if there is an open affine cover $\{V_i\}$ of Y such that $f^{-1}(V_i)$ is affine for each i.

By [Hartshorne–77, Chap. II, Exercise 5.17(a)], for an affine morphism f and any affine open subset $V \subset Y$, $f^{-1}(V)$ is affine.

A.14 Definition. Recall [Hartshorne–77, Chap. III, §10] that a morphism $f : X \to Y$ of varieties is called *smooth of relative dimension n* if the following three conditions are satisfied:

(1) f is flat,

(2) if $X' \subset X$ and $Y' \subset Y$ are irreducible components such that $f(X') \subset Y'$, then dim $X' =$ dim $Y' + n$, and

(3) the sheaf of relative differentials $\Omega^1_{X/Y}$ is a locally free sheaf of rank n.

Clearly, an open embedding is smooth of relative dimension 0.

Smooth morphisms have the base change property, i.e., if $f : X \to Y$ is a smooth morphism of relative dimension n and $g : Y' \to Y$ is a morphism, then the morphism $f' : X' \to Y'$ obtained by base change is also smooth of relative dimension n.

Moreover, the composition of two smooth morphisms is smooth. More specifically, if $f : X \to Y$ is smooth of relative dimension m and $g : Y \to Z$ is smooth of relative dimension n , then $g \circ f : X \to Z$ is smooth of relative dimension $m+n$. By [Hartshorne–77,Chap. III, Exercise 9.1], a smooth morphism is open, i.e., sends open sets to open sets.

A locally iso-trivial fibration of varieties with smooth fiber of dimension n is a smooth morphism of relative dimension n (cf. [Altman–Kleiman–70] or [Hartshorne–77, Chapter III, Theorem 10.2]).

A.15 Definition. Let X be an irreducible variety over k. Let $\mathcal{H} = \mathcal{H}_X$ be the collection of all the codimension-one closed irreducible subvarieties of X. By div X, we meanthe free $\mathbb{Z}$-module generated by the elements of $\mathcal{H}$. Any element D of div X is called a *divisor*. So D can be written as $D = \sum_{H\in\mathcal{H}} k_H H$, where $k_H \in \mathbb{Z}$ and all but finitely many k_H are 0. Often we omit in the above sum those H for which $k_H = 0$.

If each $k_H = 0$, we write $D = 0$. If each $k_H \geq 0$ and some $k_H > 0$, we write $D > 0$. If each $k_H \geq 0$, D is said to be *effective*. If $k_H = 0$ for all but one H_o and $k_{H_o} = 1$, then D is called a *prime divisor.* The *support* of D, denoted supp D, is defined to be the subvariety of X:

$$\text{supp } D := \bigcup_{k_H \neq 0} H.$$

For any closed (reduced) subvariety Y of pure codimension-one of X (i.e., each irreducible component of Y is of codimension-one) with irreducible decomposition $Y = \cup_i Y_i$, we define the *divisor associated to* Y, denoted $[Y]$, by

$$[Y] := \sum_i Y_i.$$

If X is smooth (and irreducible), then any divisor D gives rise to a line bundle denoted $\mathcal{O}_X(D)$ on X (cf. [Šafarevič-94, Chap. VI, §1.4; and Chap. III, §1.1–1.2]). (In loc. cit. $\mathcal{O}_X(D)$ is denoted by E_D.) For any closed subvariety $Y \subset X$ of pure codimension-one, we often abbreviate $\mathcal{O}_X([Y])$ by $\mathcal{O}_X(Y)$. Then, there is a sheaf exact sequence of $\mathcal{O}_X$-modules (cf. [Hartshorne–77, Chap. II, Proposition 6.18]):

$$0 \to \mathcal{O}_X([-Y]) \to \mathcal{O}_X \to i_*(\mathcal{O}_Y) \to 0, \tag{1}$$

where $i : Y \subset X$ denotes the inclusion.

For $D_1, D_2 \in$ div X, we have

$$\mathcal{O}_X(D_1 + D_2) \simeq \mathcal{O}_X(D_1) \otimes \mathcal{O}_X(D_2) \tag{2}$$

$$\mathcal{O}_X(-D_1) \simeq \mathcal{O}_X(D_1)^* \tag{3}$$

$$\mathcal{O}_X(0) \simeq \varepsilon^1_X, \tag{4}$$

where ε^1_X is the trivial line bundle on X. Also, recall from [Šafarevič–94, Chap. III, §1.2] that, for any morphism of smooth irreducible varieties $f : X \to Y$ and a divisor D in Y such that

$$f(X) \not\subset \text{supp } D, \tag{5}$$

one defines the *pullback* (or the *inverse image*) $f^*D \in \operatorname{div} X$ of the divisor D.

Further, for a morphism $f : X \to Y$ of smooth irreducible varieties and a divisor D on Y satisfying (5), by [Šafarevič–94, Chap. VI, §1.4],

$$f^*(\mathcal{O}_Y(D)) \simeq \mathcal{O}_X(f^*D). \tag{6}$$

A.16 Lemma. *Let $f : X \to Y$ be a surjective smooth morphism of smooth irreducible varieties, and let $D \in \operatorname{div} Y$ be a prime divisor. Then $f^*(D) = \sum_{H \in \mathcal{H}_X} k_H H$, where $k_H = 1$ if H is an irreducible component of the closed subvariety $f^{-1}(\text{supp } D)$, and $k_H = 0$ otherwise.*

Proof. By the base change property (cf. A.14), the scheme theoretic inverse image $f^{-1}(\text{supp } D)$ is smooth over supp D. But then supp D being reduced, so is $f^{-1}(\text{supp } D)$. Thus the lemma follows from the definition of $f^*(D)$. □

A.17 Lemma. *Let Z be a smooth irreducible variety and let Y, H be smooth irreducible closed subvarieties of Z such that H is of codimension-one. Assume further that Y intersects H transversally (cf. [Šafarevič–94, Chap. II, §2.1]). Then*

$$\mathcal{O}_Y \otimes_{\mathcal{O}_Z} \mathcal{O}_Z(H) \simeq \mathcal{O}_Y(Y \cap H),$$

as line bundles on Y, where $Y \cap H$ is thought of as a closed (reduced) subvariety of Y. (Observe that, because of the transversality assumption, $Y \cap H$ is reduced and is of pure codimension-one in Y.)

Proof. By the transversality assumption, for any $p \in Y \cap H$, there exist local equations f_1 for H in Z and $\{f_2, \dots, f_k\}$ for Y in Z at p such that $\{f_1, \dots, f_k, f_{k+1}, \dots, f_n\}$ is a local parameter for Z at p, for some regular functions $f_{k+1}, \dots, f_n$ defined on a neighborhood of p in Z. Thus $f_{1|Y}$ provides a local equation for $Y \cap H$ in Y at p. This proves the lemma. □

Recall that the *degree* of a line bundle $\mathcal{L}$ on $\mathbb{P}^1$ is the number which gives the first Chern class $c_1(\mathcal{L})$ under the canonical identification $H^2(\mathbb{P}^1, \mathbb{Z}) \simeq \mathbb{Z}$.

A proof of the following lemma can be found in [Ramanathan–85, Lemma 3].

A.18 Lemma. *Let $f : X \to Y$ be a Zariski locally trivial $\mathbb{P}^1$-fibration between smooth irreducible varieties, and let $\sigma : Y \to X$ be an algebraic section of f. Let $D := \sigma(Y)$ be the codimension-one closed subvariety of X. (By Exercise A.E.1, D is indeed closed in X.) Then*

$$K_X \simeq f^*K_Y \otimes \mathcal{O}_X(D)^{-2} \otimes (\sigma \circ f)^*(\mathcal{O}_X(D)), \tag{1}$$

where K_X is the canonical bundle of X.

Moreover, if $\mathcal{L}$ is any line bundle on X whose degree along the fibers of f is 1, then the relative canonical bundle $K_{X/Y} := K_X \otimes f^*(K_Y^{-1})$ is given by

$$K_{X/Y} = \mathcal{O}_X(D)^{-1} \otimes \mathcal{L}^{-1} \otimes (\sigma \circ f)^*\mathcal{L}. \tag{2}$$

A.19 Definition. Recall that a line bundle $\mathcal{L}$ on a variety X is called *very ample* if there exists an embedding $\phi : X \to \mathbb{P}^n$ (i.e., ϕ is an isomorphism onto a locally closed subset of some $\mathbb{P}^n$) such that $\phi^*(\mathcal{O}(1)) \simeq \mathcal{L}$, where $\mathcal{O}(1)$ is the dual of the tautological line bundle on $\mathbb{P}^n$ (cf. Example 4.2.7(c)). A line bundle $\mathcal{L}$ on X is called *ample* if some positive power $\mathcal{L}^m$ is very ample.

A.20 Lemma. *Let X be a projective variety with an ample line bundle $\mathcal{L}$. Then, for any non-zero $\sigma \in H^0(X, \mathcal{L})$, the open subvariety $X^o := X \backslash Z(\sigma)$ is affine, where $Z(\sigma) := \{x \in X : \sigma(x) = 0\}$ is the zero set of σ. Moreover, for any $f \in k[X^o]$, there exists some $n > 0$ (depending upon f) such that the section $f \cdot \sigma^n$ of $\mathcal{L}^n_{|X^o}$ extends to an element of $H^0(X, \mathcal{L}^n)$.*

Proof. The first part follows by taking sufficiently high power $\mathcal{L}^m$, embedding X in a projective space $\mathbb{P}^N$ via $H^0(X, \mathcal{L}^m)$, and observing that

$$X^o = X \backslash Z(\sigma) = X \backslash Z(\sigma^m) = X \cap (\mathbb{P}^N \backslash H),$$

where H is a hyperplane in $\mathbb{P}^N$.

A more general statement than the second part of the lemma is proved in [Hartshorne–77, Chap. II, Lemma 5.14(b)]. ☐

A.21 Lemma. *([Hartshorne–77, Chap. III, Lemma 2.10].) If $f : Y \hookrightarrow X$ is a closed embedding of varieties, then $H^p(Y, f_*\mathcal{S}) \simeq H^p(X, \mathcal{S})$, for any $p \geq 0$.*

The following result is known as the *projection formula* (cf. [Hartshorne–77, Chap. III, Exercise 8.3]).

A.22 Theorem. *Let $f : X \to Y$ be a morphism of varieties, let $\mathcal{F}$ be an $\mathcal{O}_X$-module, and let $\mathcal{E}$ be a locally free $\mathcal{O}_Y$-module of finite rank. Then*

$$R^p f_*(\mathcal{F} \otimes f^*\mathcal{E}) \simeq R^p f_*(\mathcal{F}) \otimes \mathcal{E}, \text{for any } p \geq 0. \tag{1}$$

In particular, taking $p = 0$ in (1), we get

$$f_*(\mathcal{F} \otimes f^*\mathcal{E}) \simeq f_*(\mathcal{F}) \otimes \mathcal{E}\,. \tag{2}$$

A proof of the following *Leray spectral sequence* (1) can be found, e.g., in [Godement–58, Chap. II, Theorem 4.17.1]. For (2), see [Hartshorne–77, Chap. III, Exercise 8.2].

A.23 Theorem. *Let $f : X \to Y$ be a continuous map of topological spaces. Let $\mathcal{F}$ be a sheaf of abelian groups on X. Then there is a convergent cohomology spectral sequence with*

$$E_2^{p,q} = H^p(Y, R^q f_*(\mathcal{F})) \tag{1}$$

which converges to the sheaf cohomology $H^{p+q}(X, \mathcal{F})$.

In particular, let $f : X \to Y$ be an affine morphism of varieties (cf. A.13). Then, for any quasi-coherent sheaf $\mathcal{F}$ of $\mathcal{O}_X$-modules on X,

$$H^p(X, \mathcal{F}) \simeq H^p(Y, f_*\mathcal{F}), \text{ for any } p \geq 0. \tag{2}$$

A.24 Definition. Let $f : X \to Y$ be a morphism of varieties. Following Kempf [Kempf–76, page 567], f is called *trivial* if the induced map $\mathcal{O}_Y \to f_*\mathcal{O}_X$ is surjective and the direct images $R^i f_*\mathcal{O}_X$ vanish for $i > 0$.

Assume char. $k = 0$. A trivial morphism $f : X \to Y$ is called a *rational resolution* if X is smooth, X and Y are both irreducible projective varieties and f is birational.

(In char. $p > 0$, one also adds the assumption that $R^i f_* K_X = 0$, for all $i > 0$. By a result of [Grauert–Riemenschneider–70b], this is automatically satisfied in char. 0.)

It is known that, for a given irreducible projective variety Y, if there exists one rational resolution then any other smooth resolution is automatically trivial (cf. [Kempf–Knudsen–Mumford–Saint-Donat–73, pp.50–51]).

A.25 Lemma. *Let $f : X \to Y$ be a trivial morphism between varieties such that $\mathcal{O}_Y = f_*\mathcal{O}_X$. Then, for any locally free sheaf $\mathcal{S}$ on Y,*

$$H^i(Y, \mathcal{S}) \xrightarrow{\sim} H^i(X, f^*\mathcal{S}), \qquad \text{for all } i \geq 0.$$

Proof. For any (quasi-coherent) sheaf $\mathcal{F}$ on X, the Leray spectral sequence Theorem A.23 has its E_2-terms:

$$E_2^{p,q} = H^p(Y, R^q f_*\mathcal{F}),$$

and it converges to $H^{p+q}(X, \mathcal{F})$. Further, when $\mathcal{F} = f^*(\mathcal{S})$ for some locally free sheaf $\mathcal{S}$ on Y, then, by the projection formula (A.22.1),

$$R^q f_*(f^*\mathcal{S}) \simeq (R^q f_*\mathcal{O}_X) \otimes \mathcal{S}.$$

Thus the lemma follows from the assumption that f is trivial with $f_*\mathcal{O}_X = \mathcal{O}_Y$. □

A proof of the following result of Serre can be found in [Hartshorne–77, Chap. III, Theorem 5.2 and Proposition 5.3; and Chap. II, Theorem 7.6 and Definition on page 153].

A.26 Theorem. *Let X be a projective variety over k and let $\mathcal{F}$ be a coherent sheaf on X. Then,*

(a) for any $p \geq 0$,

$$H^p(X, \mathcal{F}) \text{ is a finite-dimensional vector space over } k.$$

(b) Let $\mathcal{L}$ be an ample line bundle on X. Then there exists a positive integer $n_{\mathcal{F}}$ (depending on $\mathcal{F}$) such that, for any $p > 0$,

$$H^p(X, \mathcal{F} \otimes \mathcal{L}^n) = 0, \quad \text{for } n \geq n_{\mathcal{F}}.$$

Moreover, for all $n \geq n_{\mathcal{F}}$, the sheaf $\mathcal{F} \otimes \mathcal{L}^n$ is generated as an $\mathcal{O}_X$-module by (a finite number of) its global sections.

A proof of the following (a) and (b) parts can be found in [Hartshorne–77, Chap. III, Corollary 8.6 and Theorem 8.8(b)], and for the (c) part, see, e.g., [Mathieu–88, Lemme 19]. For generalities on H-equivariant sheaves, see, e.g., [Thomason–87].

A.27 Theorem. *(a) Let $f : X \to Y$ be a morphism between varieties. Then, for any quasi-coherent sheaf $\mathcal{F}$ on X, the sheaves $R^p f_*(\mathcal{F})$ are quasi-coherent on Y for any $p \geq 0$.*

(b) If in (a) we assume, in addition, that f is a projective morphism and $\mathcal{F}$ is a coherent sheaf on X, then, for any $p \geq 0$, $R^p f_*(\mathcal{F})$ is a coherent sheaf on Y.

(c) Let H be an algebraic group which acts on the varieties X and Y and let $f : X \to Y$ be an H-equivariant separated morphism. Then, for any H-equivariant quasi-coherent sheaf $\mathcal{F}$ on X, $R^p f_*(\mathcal{F})$ are naturally H-equivariant (quasi-coherent) sheaves on Y (for any $p \geq 0$).

A proof of the following *Serre vanishing* can be found in [Hartshorne–77, Chap. III, Theorem 3.7].

A.28 Theorem. *Let X be an affine variety and let $\mathcal{F}$ be a quasi-coherent sheaf on X. Then, for any $p > 0$,*

$$H^p(X, \mathcal{F}) = 0\,.$$

A proof of the following *semicontinuity theorem* due to Grauert and Grothendieck can be found in [Hartshorne–77, Chap. III, Corollary 12.9].

A.29 Theorem. *Let $f : X \to Y$ be a projective morphism between varieties, where Y is assumed to be irreducible. Let $\mathcal{F}$ be a coherent sheaf on X, flat over Y. Fix any $p \geq 0$ and assume that $\dim_k H^p(X_y, \mathcal{F}_y)$ is constant for $y \in Y$, where X_y is the scheme theoretic fiber of f over y. Then, $R^p f_*(\mathcal{F})$ is a locally free sheaf on Y and, for every $y \in Y$, the natural map*

$$R^p f_*(\mathcal{F}) \otimes k(y) \to H^p(X_y, \mathcal{F}_y)$$

is an isomorphism, where $k(y)$ is the residue field of $y \in Y$.

A.30 Proposition. ([Hartshorne–77, Chap. III, Exercise 5.7(d)].) *Let $f : X \to Y$ be a finite and surjective morphism between projective varieties and let $\mathcal{L}$ be a line bundle on Y. Then $\mathcal{L}$ is an ample line bundle on Y iff $f^*\mathcal{L}$ is an ample line bundle on X.*

We recall the following lemma due to Kempf (cf. [Demazure–74, §5, Proposition 2]).

A.31 Lemma. *Let $f : X \to Y$ be a morphism between projective varieties. Assume that $f_*\mathcal{O}_X = \mathcal{O}_Y$ and, moreover, there exists an ample line bundle $\mathcal{L}$ on Y such that $H^i(X, f^*(\mathcal{L}^n)) = 0$, for all $i > 0$ and all sufficiently large n. Then*

$$R^i f_*(\mathcal{O}_X) = 0, \qquad \textit{for all } i > 0.$$

A.32 Lemma. *Let $f : X \to Y$ be a surjective morphism between projective varieties. Assume that there is an ample line bundle $\mathcal{L}$ on Y such that the canonical map $H^0(Y, \mathcal{L}^n) \to H^0(X, f^*\mathcal{L}^n)$ is an isomorphism for all $n \geq n_o$, where n_o is some fixed positive integer. Then $f_*\mathcal{O}_X = \mathcal{O}_Y$.*

Proof. Consider the sheaf exact sequence on Y:

$$0 \to \mathcal{O}_Y \to f_*\mathcal{O}_X \to \mathcal{Q} \to 0,$$

where $\mathcal{Q}$, by definition, is the quotient sheaf $f_*\mathcal{O}_X/\mathcal{O}_Y$. Tensoring this sequence over $\mathcal{O}_Y$ with the locally free sheaf $\mathcal{L}^n$ and taking cohomology (and using the projection formula (A.22.2)), we get

$$0 \to H^0(Y, \mathcal{L}^n) \to H^0(X, f^*\mathcal{L}^n) \to H^0(Y, \mathcal{Q} \otimes \mathcal{L}^n) \to H^1(Y, \mathcal{L}^n) \to \cdots .$$

But $\mathcal{L}$ being ample, by Theorem A.26(b), there exists $\bar{n}_o > 0$ such that $H^1(Y, \mathcal{L}^n) = 0$, for all $n \geq \bar{n}_o$. In particular, by the assumption, $H^0(Y, \mathcal{Q} \otimes \mathcal{L}^n) = 0$, for all $n \geq \max(n_o, \bar{n}_o)$. Now by Theorem A.27(b), $f_*\mathcal{O}_X$, and hence $\mathcal{Q}$, is a coherent sheaf on Y. But then, $\mathcal{L}$ being ample, we conclude that $\mathcal{Q}$ itself is 0 by Theorem A.26(b), i.e., $\mathcal{O}_Y \approx f_*\mathcal{O}_X$, proving the lemma. □

The following result is due to [Grauert–Riemenschneider-70a].

A.33 Theorem. *Assume char. $k = 0$. Let X be a smooth irreducible projective variety over k, and $\mathcal{L}$ a line bundle on X such that there is an integer $N > 0$ and a birational morphism $\phi : X \to Y \subset \mathbb{P}^{N_o}$ onto a variety Y such that $\phi^*(\mathcal{O}(1)) \simeq \mathcal{L}^N$. Then*

(1) $$H^p(X, \mathcal{L}^{-1}) = 0, \qquad \textit{for any } 0 \leq p < \dim X.$$

A.34 Definition. A local noetherian ring A is said to be *Cohen–Macaulay* if depth $A = \dim A$. A variety itself is said to be Cohen–Macaulay if all of its local rings are Cohen–Macaulay.

A projective variety $X \subset \mathbb{P}^n$ is said to be *projectively Cohen–Macaulay*, resp. *projectively normal*, also called *arithmetically Cohen–Macaulay*, resp. *arithmetically normal*, with respect to the given embedding inside $\mathbb{P}^n$ if the cone over X (in $\mathbb{A}^{n+1}$) is Cohen–Macaulay, resp. normal.

We remark that both of these properties depend upon the choice of the embedding of X in $\mathbb{P}^n$; in particular, these are *not* intrinsic properties (cf. [Hartshorne–77, Chap. I, Exercise 3.18(c)]).

Recall the following from [Hartshorne–77, Chap. II, Theorem 8.21 A].

A.35 Theorem. *Let A be a (local noetherian) Cohen–Macaulay ring with maximal ideal $\mathfrak{m}$. Then, we have the following:*

(a) A set of elements $x_1, \dots, x_r \in \mathfrak{m}$ forms a regular sequence for A iff $\dim A/\langle x_1, \dots, x_r\rangle = \dim A - r$.

(b) Assume that $x_1, \dots, x_r \in \mathfrak{m}$ is a regular sequence for A. Then the map from the polynomial ring

$$\frac{A}{I}[t_1, \dots, t_r] \to \mathrm{gr}_I\, A := \oplus_{n\geq 0} I^n/I^{n+1}, \quad \textit{taking } t_i \mapsto x_i \bmod I^2 \in I/I^2,$$

is an isomorphism, where $I := \langle x_1, \dots, x_r\rangle$.

A.36 Theorem. ([Hartshorne–77, Chap. III, Theorem 7.6 and its proof].) *Let Y be an equidimensional projective variety of dimension m (i.e., all the irreducible components of Y have the same dimension m). Then, Y is Cohen–Macaulay iff*

$$(1) \qquad H^p(Y, \mathcal{L}_o^{-n}) = 0, \qquad \textit{for all } p < m \textit{ and } n \gg 0,$$

where $\mathcal{L}_o$ is any (fixed) very ample line bundle on Y.

A proof of the following *Serre duality* can be found, e.g., in [Hartshorne–77, Chap. III, Corollary 7.7].

A.37 Theorem. *Let X be a projective Cohen–Macaulay equidimensional variety of dimension n over k. Then, for any locally free $\mathcal{O}_X$-module $\mathcal{F}$, there is a natural isomorphism for any $p \geq 0$:*

$$H^p(X, \mathcal{F}) \simeq H^{n-p}(X, \mathcal{F}^\vee \otimes \omega_X)^*,$$

where $\mathcal{F}^\vee$ denotes the dual sheaf $Hom_{\mathcal{O}_X}(\mathcal{F}, \mathcal{O}_X)$, ω_X denotes the dualizing sheaf of X (cf. [Hartshorne–77, Chap. III, §7]) and, for any k-vector space V, V^* denotes its dual.

Recall that, in the case of smooth X, ω_X is the canonical bundle of X (cf. [Hartshorne–77, Chap. III, Corollary 7.12]).

The following lemma is taken from [Ramanathan–85, Proposition 4].

A.38 Lemma. *Assume char. $k = 0$. Let $f : X \to Y$ be a rational resolution of an irreducible projective variety Y (cf. A.24). Then Y is Cohen–Macaulay. In fact, in this case, for any ample line bundle $\mathcal{L}$ on Y,*

$$(1) \qquad H^p(Y, \mathcal{L}^{-n}) = 0, \qquad \textit{for all } p < \dim Y \textit{ and } n > 0.$$

Proof. Since f is a trivial morphism, by Lemma A.25, for all $p \geq 0$ and $n \in \mathbb{Z}$,

$$(2) \qquad H^p(Y, \mathcal{L}^{-n}) \simeq H^p(X, f^*(\mathcal{L})^{-n}).$$

Observe that $f_* \mathcal{O}_X = \mathcal{O}_Y$ since f is a trivial morphism and it is surjective (being a proper birational map). Since $\mathcal{L}$ is ample, there exists $N > 0$ and an embedding $\phi : Y \to \mathbb{P}^{N_o}$ (for some N_o) such that $\mathcal{L}^N = \phi^*(\mathcal{O}(1))$. Thus, by Theorem A.33,

$$(3) \qquad H^p(X, f^*(\mathcal{L})^{-1}) = 0, \qquad \text{for all } p < \dim X \text{ and } n > 0.$$

Combining (2)–(3), we get (1) for $n = 1$. But a line bundle $\mathcal{L}$ on Y is ample iff $\mathcal{L}^n$ is ample for any $n > 0$ (cf. [Hartshorne–77, Chap. II, Proposition 7.5]). Hence replacing $\mathcal{L}$ by $\mathcal{L}^n$, (1) follows. Thus Y is Cohen–Macaulay by Theorem A.36. This proves the lemma. □

A.39 Proposition. ([Hartshorne–77, Chap. II, Exercise 5.14 (d)].) *Let $X \subset \mathbb{P}^n$ be a closed irreducible subvariety. Then X is projectively normal iff it is normal and, for each $k \geq 0$, the restriction map*

$$H^0(\mathbb{P}^n, \mathcal{O}(k)) \to H^0(X, \mathcal{O}(k)_{|X})$$

is surjective, where $\mathcal{O}(k) := \mathcal{O}(1)^{\otimes k}$ and (as earlier) $\mathcal{O}(1)$ is the dual of the tautological line bundle on $\mathbb{P}^n$.

Recall the following from [Borel–91, Chap. I, Corollary 1.4(a)].

A.40 Theorem. *Let $f : G \to H$ be an algebraic group morphism between algebraic groups over k. Then $f(G)$ is a closed subgroup of H.*

For the following see [Serre–58].

A.41 Definition. Let H be an algebraic group. By a *principal H-bundle* on a variety Y, we mean a variety X on which H acts algebraically from the right and an H-equivariant morphism $\pi : X \to Y$ (where$H$ acts trivially on Y), such that π is *locally isotrivial*, i.e., there exists an open cover $\{U_i\}_i$ of Y and an étale cover $f_i : V_i \to U_i$ such that the pullback bundle $f_i^*(X)$ is a trivial H-bundle for each i.

Let H act algebraically on a variety Z from the left. We can then form the *associated bundle with fiber* Z, denoted $X \times_H Z$, such that $\hat{\pi} : X \times Z \to X \times_H Z$ is a principal H-bundle under the right action of H on $X \times Z$ via

$$(e, f) \cdot g = (eg, g^{-1} f), \quad \text{for } g \in H, e \in X, \quad \text{and } f \in Z.$$

Then the projection $\pi_1 : X \times Z \to X$ descends to give a locally isotrivial fibration $\pi_Z : X \times_H Z \to Y$ with fiber Z.

If Z is a finite-dimensional H-module, then the associated bundle $X \times_H Z$ with fiber Z acquires a canonical (algebraic) vector bundle structure denoted $\mathcal{L}_\pi(Z)$.

With this notation, we have the following:

A.42 Lemma. *Let $\mathcal{M}$ be a H-equivariant vector bundle on a left H-variety Z (cf. Definition 4.2.6) and let $\pi : X \to Y$ be a principal H-bundle as above. Then the H- equivariant vector bundle $\varepsilon_X^1 \boxtimes \mathcal{M}$ on $X \times Z$ descends uniquely to a vector bundle $\mathcal{M}_\pi$ on $X \underset{H}{\times} Z$, i.e., there exists a unique vector bundle $\mathcal{M}_\pi$ on $X \times_H Z$ such that*

$$\text{(1)} \qquad \hat{\pi}^*(\mathcal{M}_\pi) \simeq \varepsilon_X^1 \boxtimes \mathcal{M},$$

as H-equivariant vector bundles, under the canonical H-equivariant vector bundle structure on $\hat{\pi}^*(\mathcal{M}_\pi)$. Recall that ε_X^1 is the trivial line bundle $X \times k \to X$ as in Example 4.2.7(a), $\varepsilon_X^1 \boxtimes \mathcal{M}$ is the external tensor product as in Example 4.2.7(e) and, moreover, H acts on $X \times k$ via

$$(x, z) \cdot g = (xg, z), \quad \textit{for } g \in H, x \in X, \quad \textit{and } z \in k.$$

Assume, in addition, that Z is projective. Then, there is a canonical isomorphism of $\mathcal{O}_Y$-modules:

$$\text{(2)} \qquad R^i \pi_{Z*}(\mathcal{M}_\pi) \simeq \mathcal{L}_\pi(H^i(Z, \mathcal{M})), \quad \textit{for all } i \geq 0.$$

Proof. Recall that the map $\pi^* : \operatorname{Vect} Y \to \operatorname{Vect}_H X$, $\mathcal{V} \mapsto \pi^*\mathcal{V}$, is a bijection, where $\operatorname{Vect} Y$, resp. $\operatorname{Vect}_H X$, denotes the set of isomorphism classes of vector bundles on Y, resp. H-equivariant vector bundles on X (cf. [Kraft–91, Proposition 6.4]). Applying this to the principal H-bundle $\hat{\pi} : X \times Z \to X \times_H Z$, we get the bijection $\hat{\pi}^* : \operatorname{Vect}(X \times_H Z) \xrightarrow{\sim} \operatorname{Vect}_H (X \times Z)$. Thus the first part of the lemma follows by taking $\mathcal{M}_\pi := (\hat{\pi}^*)^{-1}(\varepsilon_X^1 \boxtimes \mathcal{M})$.

We now prove (2). By the semicontinuity Theorem A.29, $R^i \pi_{Z*}(\mathcal{M}_\pi)$ is a locally free sheaf on Y. Further, since cohomology commutes with flat base extension (cf. [Hartshorne–77, Chapter III, Proposition 9.3]), there is a canonical (H-equivariant) isomorphism

$$\pi^*(R^i \pi_{Z*}(\mathcal{M}_\pi)) \simeq R^i \pi_{1*}(\varepsilon_X^1 \boxtimes \mathcal{M}),$$

where $\pi_1 : X \times Z \to X$ is the projection on the first factor. It is easy to see that $R^i\pi_{1*}(\varepsilon^1_X \boxtimes \mathcal{M})$ is the H-equivariant vector bundle $X \times H^i(Z, \mathcal{M}) \to X$. Thus (2) follows by using the isomorphism π^*. □

A.E EXERCISES

(1) Let $f : X \to Y$ be a surjective morphism of varieties and let $\sigma : Y \to X$ be an algebraic section of f, i.e., σ is a morphism and $f \circ \sigma$ is the identity map of Y. Then prove that Im σ is a closed subset of X.
Hint. Show that $(\sigma \circ f)(\overline{\sigma(Y)}) = \overline{\sigma(Y)}$ and also clearly $(\sigma \circ f)(\overline{\sigma(Y)}) = \sigma(Y)$.

(2) Let $\phi : G \to G'$ be a bijective morphism of algebraic groups over an algebraically closed field of char. 0. Then prove that ϕ is an isomorphism.
(This exercise is taken from [Springer–98, Exercise 5.3.5.1].)

(3) Show that a dense constructible subset of a variety X contains a dense open subset of X.

Appendix B

Local Cohomology

We recall the definition of local cohomology and some of its basic properties used in the book. For more detailed treatment, see [Grothendieck–67], [Hartshorne–66], [Kempf–78].

B.1 Definition. Let X be a topological space together with closed subspaces $Z \subset Y$, and let $\mathcal{S}$ be a sheaf of abelian groups (for short, an abelian sheaf) on X. For any open subset $U \subset X$, the space of global sections of $\mathcal{S}_{|U}$ is denoted by $\Gamma(U, \mathcal{S})$. A global section $\gamma \in \Gamma(X, \mathcal{S})$ is said to have *support in* Y if $\gamma_{|X \backslash Y} = 0$. The space of sections $\gamma \in \Gamma(X, \mathcal{S})$ with support in Y is denoted by $\Gamma_Y(X, \mathcal{S})$. Thus, by definition, there is a short exact sequence:

$$(1) \qquad 0 \to \Gamma_Y(X, \mathcal{S}) \to \Gamma(X, \mathcal{S}) \to \Gamma(X \backslash Y, \mathcal{S}).$$

Clearly,

$$(2) \qquad \Gamma_X(X, \mathcal{S}) = \Gamma(X, \mathcal{S}), \qquad \text{and} \qquad \Gamma_\emptyset(X, \mathcal{S}) = 0.$$

Furthermore, define

$$(3) \qquad \Gamma_{Y/Z}(X, \mathcal{S}) := \Gamma_Y(X, \mathcal{S}) / \Gamma_Z(X, \mathcal{S}),$$

giving rise to the exact sequence

$$(4) \qquad 0 \to \Gamma_Z(X, \mathcal{S}) \to \Gamma_Y(X, \mathcal{S}) \to \Gamma_{Y/Z}(X, \mathcal{S}) \to 0.$$

Clearly,

$$(5) \qquad \Gamma_{Y/\emptyset}(X, \mathcal{S}) = \Gamma_Y(X, \mathcal{S}), \qquad \text{and}$$

$$(6) \qquad \Gamma_{Y/Y}(X, \mathcal{S}) = 0.$$

For closed subspaces $Y' \supset Z'$ of X such that $Y \subset Y'$, $Z \subset Z'$, there exists a canonical homomorphism

$$\Gamma_{Y/Z}(X, \mathcal{S}) \to \Gamma_{Y'/Z'}(X, \mathcal{S}). \tag{7}$$

Noether's isomorphism gives rise to the following short exact sequence for any sequence of closed subspaces $X_3 \subset X_2 \subset X_1 \subset X$:

$$0 \to \Gamma_{X_2/X_3}(X, \mathcal{S}) \to \Gamma_{X_1/X_3}(X, \mathcal{S}) \to \Gamma_{X_1/X_2}(X, \mathcal{S}) \to 0. \tag{8}$$

Let $Z \subset Y$ be closed subspaces of X. Then $\Gamma_{Y/Z}(X, -)$ is an additive covariant functor from the (abelian) category $\mathfrak{Ab}(X)$ to the category $\mathfrak{Ab}$, where $\mathfrak{Ab}(X)$ is the category of sheaves of abelian groups on X and $\mathfrak{Ab}$ is the category of abelian groups. By [Hartshorne–77, Chap. III, Corollary 2.3], $\mathfrak{Ab}(X)$ has enough injectives. Let $H^i_{Y/Z}(X, \mathcal{S})$ be the i-th right derived functor of $\Gamma_{Y/Z}(X, \mathcal{S})$, for any $i \geq 0$ (cf. [Hartshorne–66, Chap. I, Corollary 5.3], see also [Hartshorne–77, Chap. III, Theorem 1.1.A]). The groups $H^i_{Y/Z}(X, \mathcal{S})$ are called the *local cohomology groups*, also called the *cohomology with supports*.

Take an injective resolution of $\mathcal{S}$ in the category $\mathfrak{Ab}(X)$:

$$0 \to \mathcal{S} \to \mathcal{I}_0 \to \mathcal{I}_1 \to \cdots. \tag{9}$$

Applying the functor $\Gamma_{Y/Z}(X, -)$ to the sequence (9), we get the complex

$$\Gamma_{Y/Z}(X, \mathcal{I}_0) \to \Gamma_{Y/Z}(X, \mathcal{I}_1) \to \Gamma_{Y/Z}(X, \mathcal{I}_2) \to \cdots. \tag{10}$$

Recall that $H^i_{Y/Z}(X, \mathcal{S})$ is, by definition, the i-th cohomology of the above complex (10). By a basic fact from homological algebra, the group $H^i_{Y/Z}(X, \mathcal{S})$ does not depend (up to a canonical isomorphism) on the choice of the injective resolution of $\mathcal{S}$ in the category $\mathfrak{Ab}(X)$ (cf. loc. cit.).

From general properties of derived functors (cf. loc. cit.), for any short exact sequence of abelian sheaves on X:

$$0 \to \mathcal{S}_1 \to \mathcal{S} \to \mathcal{S}_2 \to 0,$$

there is a natural long exact cohomology sequence:

$$0 \to H^0_{Y/Z}(X, \mathcal{S}_1) \to H^0_{Y/Z}(X, \mathcal{S}) \to H^0_{Y/Z}(X, \mathcal{S}_2) \to H^1_{Y/Z}(X, \mathcal{S}_1) \to \cdots. \tag{11}$$

Observe that, in general, $\Gamma_{Y/Z}(X, -)$ is not a left exact functor, and hence $H^0_{Y/Z}(X, \mathcal{S})$ is not always isomorphic with $\Gamma_{Y/Z}(X, \mathcal{S})$. However, $\Gamma_{Y/\emptyset}(X, -) = \Gamma_Y(X, -)$ is indeed a left exact functor (as is easy to see); hence, by loc. cit.,

$$H^0_{Y/\emptyset}(X, \mathcal{S}) \simeq \Gamma_Y(X, \mathcal{S}). \tag{12}$$

Abbreviate $H^i_{Y/\emptyset}(X, \mathcal{S})$ by $H^i_Y(X, \mathcal{S})$. By (2),

$$H^i_X(X, \mathcal{S}) \cong H^i(X, \mathcal{S}) \text{ and } H^i_\emptyset(X, \mathcal{S}) = 0. \tag{13}$$

The canonical homomorphism of (7) (for closed subspaces $Z \subset Y$ and $Z' \subset Y'$ of X such that $Y \subset Y'$ and $Z \subset Z'$) gives rise to the canonical homomorphism

$$H^i_{Y/Z}(X, \mathcal{S}) \to H^i_{Y'/Z'}(X, \mathcal{S}). \tag{14}$$

Further, for any sequence of closed subspaces $X_3 \subset X_2 \subset X_1 \subset X$, the short exact sequence (8) gives rise to the long exact sequence

$$\begin{aligned} 0 \to H^0_{X_2/X_3}(X, \mathcal{S}) \to H^0_{X_1/X_3}(X, \mathcal{S}) \to H^0_{X_1/X_2}(X, \mathcal{S}) \xrightarrow{\delta} \\ H^1_{X_2/X_3}(X, \mathcal{S}) \to \dots, \end{aligned} \tag{15}$$

where the map $\delta : H^i_{X_1/X_2}(X, \mathcal{S}) \to H^{i+1}_{X_2/X_3}(X, \mathcal{S})$ is referred to as the *connecting homomorphism* for the triple $X_3 \subset X_2 \subset X_1$.

Moreover, we have the following lemma showing the functoriality of (15). (A proof of the lemma can be found in [Kempf–78, Lemma 11.3].)

B.2 Lemma. *Let X and Y be topological spaces with a sequence of closed subspaces $X_3 \subset X_2 \subset X_1 \subset X$ and $Y_3 \subset Y_2 \subset Y_1 \subset Y$, and let $f : X \to Y$ be a continuous map such that $X_p \supset f^{-1}(Y_p)$ for $p = 1, 2, 3$. Then, for any abelian sheaves $\mathcal{S}$ on X and $\mathcal{T}$ on Y together with a sheaf morphism $\phi : \mathcal{T} \to f_*\mathcal{S}$, there exists a natural homomorphism*

$$H^i_{Y_1/Y_2}(Y, \mathcal{T}) \to H^i_{X_1/X_2}(X, \mathcal{S}), \quad \textit{for all } i \geq 0. \tag{1}$$

Further, these homomorphisms give a homomorphism of the (exact) cochain complex (B.1.15) for the triple $Y_3 \subset Y_2 \subset Y_1$ to the cochain complex for the triple $X_3 \subset X_2 \subset X_1$.

B.3 Lemma (Excision). *For closed subsets $Z \subset Y$ of X and open subset U of X containing Y, we have a natural isomorphism*

$$H^i_{Y/Z}(X, \mathcal{S}) \simeq H^i_{Y/Z}(U, \mathcal{S}_{|U}), \tag{1}$$

for any abelian sheaf $\mathcal{S}$ on X. The isomorphism (1) is induced by the canonical restriction map

$$\Gamma_{Y/Z}(X, -) \to \Gamma_{Y/Z}(U, -_{|U}).$$

Proof. For any abelian sheaf $\mathcal{T}$ on X, the canonical restriction map

$$\gamma : \Gamma_Y(X, \mathcal{T}) \to \Gamma_Y(U, \mathcal{T}_{|U})$$

is an isomorphism:

The injectivity of γ is clear. To prove the surjectivity of γ, take $\sigma \in \Gamma_Y(U, \mathcal{T}_{|U})$ and let $\tilde{\sigma} \in \Gamma(X, \mathcal{T})$ be the element such that $\tilde{\sigma}_{|U} = \sigma$ and $\tilde{\sigma}_{|X\backslash Y} = 0$.

The isomorphism γ gives rise to the isomorphism (again denoted by)

$$\gamma : \Gamma_{Y/Z}(X, \mathcal{T}) \tilde{\to} \Gamma_{Y/Z}(U, \mathcal{T}_{|U}).$$

This gives rise to the isomorphism (1) since, for an injective sheaf $\mathcal{T}$, $\mathcal{T}_{|U}$ is again injective (cf. [Bredon–97, Chap. II, Proposition 3.4]). □

B.4 Lemma. *For closed subspaces $Z \subset Y$ of X, there is a canonical isomorphism*

$$H^i_{Y/Z}(X, \mathcal{S}) \to H^i_{Y\backslash Z}(X\backslash Z, \mathcal{S}).$$

Proof. By [Hartshorne–77, Chap. III, Lemma 2.4], any injective sheaf $\mathcal{I} \in \mathfrak{Ab}(X)$ is flasque. (Recall from [Hartshorne–77, Chap. II, Exercise 1.16] that a sheaf $\mathcal{I}$ on X is called *flasque* if for any open subsets $U \subset V$ of X, the restriction map $\Gamma(V, \mathcal{I}) \to \Gamma(U, \mathcal{I})$ is surjective.) Next, observe that for any flasque sheaf $\mathcal{I}$, the canonical restriction map $\Gamma_{Y/Z}(X, \mathcal{I}) \to \Gamma_{Y\backslash Z}(X\backslash Z, \mathcal{I})$ is an isomorphism. From this the lemma follows since $\mathcal{I}_{|X\backslash Z}$ is again an injective sheaf. □

B.5 Corollary. *For any closed subset $Y \subset X$, there is a natural exact sequence (for $U := X\backslash Y$)*

(1) $$0 \to H^0_Y(X, \mathcal{S}) \to H^0(X, \mathcal{S}) \to H^0(U, \mathcal{S}_{|U}) \to H^1_Y(X, \mathcal{S}) \to \cdots.$$

Proof. The exact sequence (B.1.15) reduces to (1) if we take $X_1 = X$, $X_2 = Y$, $X_3 = \emptyset$ and use Lemma B.4 and (B.1.13). □

B.6 Lemma. *Let Y, Z be two closed subsets of X. Then there is a natural long exact Mayer–Vietoris sequence*

(1) $$0 \to H^0_{Y\cap Z}(X, \mathcal{S}) \to H^0_Y(X, \mathcal{S}) \oplus H^0_Z(X, \mathcal{S}) \to H^0_{Y\cup Z}(X, \mathcal{S}) \to H^1_{Y\cap Z}(X, \mathcal{S}) \to \cdots.$$

Proof. For any abelian sheaf $\mathcal{I}$, we have the exact sequence:

(2) $$0 \to \Gamma(X, \mathcal{I}) \xrightarrow{i} \Gamma(X, \mathcal{I}) \oplus \Gamma(X, \mathcal{I}) \xrightarrow{\pi} \Gamma(X, \mathcal{I}) \to 0,$$

where $i(\sigma) = \sigma \oplus \sigma$ and $\pi(\sigma \oplus \sigma') = \sigma - \sigma'$. If, moreover, $\mathcal{I}$ is flasque (in particular, for injective $\mathcal{I}$), (2) gives rise to the exact sequence on restriction:

(3) $$0 \to \Gamma_{Y\cap Z}(X, \mathcal{I}) \to \Gamma_Y(X, \mathcal{I}) \oplus \Gamma_Z(X, \mathcal{I}) \xrightarrow{\pi'} \Gamma_{Y\cup Z}(X, \mathcal{I}) \to 0;$$

to prove the surjectivity of π', take $\sigma \in \Gamma_{Y\cup Z}(X, \mathcal{I})$ and define

$$\sigma' \in \Gamma(X\backslash(Y \cap Z), \mathcal{I}) \text{ by } \sigma'_{|X\backslash Z} = 0, \quad \sigma'_{|X\backslash Y} = \sigma_{|X\backslash Y}.$$

Now, since $\mathcal{I}$ is flasque, we can extend σ' to $\tilde{\sigma}' \in \Gamma(X, \mathcal{I})$. Then

$$s := (\sigma - \tilde{\sigma}', -\tilde{\sigma}') \in \Gamma_Y(X, \mathcal{I}) \oplus \Gamma_Z(X, \mathcal{I}) \text{ and } \pi'(s) = \sigma.$$

Taking an injective resolution $0 \to \mathcal{S} \to \mathcal{I}_0 \to \mathcal{I}_1 \to \cdots$, we get the following short exact sequence of cochain complexes by virtue of (3):

$$0 \to \Gamma_{Y\cap Z}(X, \mathcal{I}_\bullet) \to \Gamma_Y(X, \mathcal{I}_\bullet) \oplus \Gamma_Z(X, \mathcal{I}_\bullet) \to \Gamma_{Y\cup Z}(X, \mathcal{I}_\bullet) \to 0.$$

The corresponding cohomology long exact sequence gives (1). □

B.7 Grothendieck–Cousin Complex. Let X be a topological space with a filtration by closed subspaces $X = X_0 \supset X_1 \supset X_2 \supset \cdots$, and let $\mathcal{S}$ be an abelian sheaf on X. Consider the sequence

(1)

$$0 \to H^0(X, \mathcal{S}) \xrightarrow{\varepsilon} H^0_{X_0/X_1}(X, \mathcal{S}) \xrightarrow{d^0} H^1_{X_1/X_2}(X, \mathcal{S}) \xrightarrow{d^1} H^2_{X_2/X_3}(X, \mathcal{S}) \to \cdots,$$

where ε is the restriction map

$$H^0(X, \mathcal{S}) \to H^0(X\backslash X_1, \mathcal{S}) \approx H^0_{X_0/X_1}(X, \mathcal{S}),$$

and $d^i : H^i_{X_i/X_{i+1}}(X, \mathcal{S}) \to H^{i+1}_{X_{i+1}/X_{i+2}}(X, \mathcal{S})$ is the connecting homomorphism for the triple $X_{i+2} \subset X_{i+1} \subset X_i$ (cf. B.1).

B.8 Proposition. [Kempf–78, Lemma 7.8]. *The above sequence (1) is a complex, i.e., composite of any two successive maps is zero.*

This complex is known as the *global Cousin complex* of $\mathcal{S}$ with respect to the decreasing filtration $(X_i)_{i\geq 0}$ of X.

We recall the following result due to Kempf [Kempf–78, Theorem 10.9]. In fact, we only state a weaker version of his theorem, which is sufficient for our purposes.

B.9 Theorem. *Let X be a Cohen–Macaulay irreducible variety (over an algebraically closed field) together with a filtration by closed subvarieties $X = X_0 \supset X_1 \supset X_2 \supset \cdots$, and let $\mathcal{S}$ be a locally free sheaf of $\mathcal{O}_X$-modules on X. Assume further that*

(a) $X_i\backslash X_{i+1}$ are affine varieties (under the locally closed subvariety structure) and $X_i\backslash X_{i+1} \hookrightarrow X$ are affine morphisms for all $i \geq 0$, and

(b) the codimension of X_i in X is at least i for all $i \geq 1$.

Then, the global Cousin complex of $\mathcal{S}$ with respect to the filtration (X_i) of X is exact if and only if $H^n(X, \mathcal{S}) = 0$, for all $n \geq 1$.

B.10 Lemma. *(a) Let K be a (finite-dimensional) affine algebraic group over $\mathbb{C}$ with Lie algebra $\mathfrak{k}$, let X be a K-variety over $\mathbb{C}$, and let $\mathcal{S}$ be a K-equivariant vector bundle on X. Then, for closed subspaces $Y \supset Z$ of X, the local cohomology $H^p_{Y/Z}(X, \mathcal{S})$ for any $p \geq 0$ admits a natural structure of a $\mathfrak{k}$-module such that, for any closed subspace $W \subset Z$, the connecting homomorphism $H^p_{Y/Z}(X, \mathcal{S}) \to H^{p+1}_{Z/W}(X, \mathcal{S})$ is a $\mathfrak{k}$-module map. Further, it is functorial in the following sense:*

Let X' be another K-variety over $\mathbb{C}$ with closed subspaces $Y' \supset Z'$, and a K-morphism $f : X' \to X$ such that $Y' \supset f^{-1}(Y)$ and $Z' \supset f^{-1}(Z)$. Then, the induced map $H^p_{Y/Z}(X, \mathcal{S}) \to H^p_{Y'/Z'}(X', f^*(\mathcal{S}))$ (cf. (B.2.1)) is a $\mathfrak{k}$-module map.

Observe that, by Lemma A.8, there is a canonical sheaf morphism $\mathcal{S} \to f_* f^*(\mathcal{S})$.

(b) If we assume in addition (in the first paragraph of (a)) that Y and Z are both K-stable, then the $\mathfrak{k}$-module structure on $H^p_{Y/Z}(X, \mathcal{S})$ integrates to give a locally finite algebraic K-module structure. In particular, in this case, the $\mathfrak{k}$-module structure on $H^p_{Y/Z}(X, \mathcal{S})$ is locally finite as well.

Even though not stated exactly in this form, a proof of the above lemma can be found in [Kempf–78, Sect. 11]. (Actually [Kempf–78, Sect. 11] contains more general results.)

B.11 Lemma. *Let $\mathbb{A}^d$ be the affine space of* dim d *over a field k. Then*

(a) $H^p_{\{0\}}(\mathbb{A}^d, \mathcal{O}_{\mathbb{A}^d}) = 0$, *for $p \neq d$, and*

(b) $H^d_{\{0\}}(\mathbb{A}^d, \mathcal{O}_{\mathbb{A}^d})$ *is "canonically" isomorphic with $\sum\limits_{n_1, \dots, n_d < 0} k x_1^{n_1} \cdots x_d^{n_d}$ as k-vector spaces, where 0 is the origin of $\mathbb{A}^d$ and $(x_1, \dots, x_d)$ are the coordinate functions on $\mathbb{A}^d$. Moreover, if $\mathbb{A}^d$ is a T-module (for a torus T) such that $x_1, \dots, x_d$ are T-eigenfunctions, then the isomorphism (b) is T-equivariant.*

For a proof of the above see, e.g., [Kempf–78, Proposition 11.9].

B.12 Lemma. *Let $f : X \to Y$ be a continuous map of topological spaces and let Y' be a closed subspace of Y. Then, for any abelian sheaf $\mathcal{S}$ on X, there is a spectral sequence with*

$$E_2^{p,q} = H^p_{Y'}(Y, R^q f_*(\mathcal{S})) \Longrightarrow H^n_{X'}(X, \mathcal{S}),$$

where $X' := f^{-1}(Y')$.

In particular, if $R^q f_*(\mathcal{S}) = 0$ for all $q \geq 1$, then the above spectral sequence degenerates at E^2 and

$$H^n_{X'}(X, \mathcal{S}) \simeq H^n_{Y'}(Y, f_*\mathcal{S}).$$

For a proof of the above see [Grothendieck–67, Proposition 5.5].

Appendix C

Results from Topology

We recall the following Universal Coefficient Theorem in singular homology (cf. [Spanier–66, Chap. 5, §5, Theorem 3 and Corollary 4]).

C.1 Theorem. *For a topological pair (X, A), i.e., X is a topological space and A is any subspace, there is a split short exact sequence*

$$0 \to \mathrm{Ext}^1_{\mathbb{Z}}(H_{q-1}(X, A), \mathbb{Z}) \to H^q(X, A) \to \mathrm{Hom}_{\mathbb{Z}}(H_q(X, A), \mathbb{Z}) \to 0,$$

where $H_q(X, A)$, resp. $H^q(X, A)$, denotes the q-th singular homology, resp. singular cohomology, of the pair (X, A) with integral coefficients.

C.2 Corollary. *If (X, A) is a topological pair such that $H_q(X, A)$ is a finitely generated $\mathbb{Z}$-module for each q. Then $H^q(X, A)$'s are finitely generated. Moreover, for each q, the ranks of $H^q(X, A)$ and $H_q(X, A)$ are the same and the torsion submodules of $H^q(X, A)$ and $H_{q-1}(X, A)$ are isomorphic, where the rank means the rank of any maximal free $\mathbb{Z}$-submodule.*

Let X be a topological space with subspaces A_1, A_2 such that A_1 and A_2 are both open in $A_1 \cup A_2$. Recall the definition of the *cup product* from [Spanier–66, Chap. 5, §6]:

$$H^p(X, A_1) \otimes H^q(X, A_2) \to H^{p+q}(X, A_1 \cup A_2), \quad u \otimes v \mapsto u \cup v\,,$$

and also the *cap product* from loc. cit.:

$$H^p(X, A_1) \otimes H_n(X, A_1 \cup A_2) \to H_{n-p}(X, A_2), \quad u \otimes a \mapsto u \cap a.$$

We have the following result from [Spanier–66, Chap. 5, § 6, 16,18].

C.3 Lemma. *(a) Let $f : X \to Y$ be a continuous map between topological spaces. Let A_1, A_2, resp. B_1, B_2, be subsets of X, resp. Y, such that A_1, A_2 are both open in $A_1 \cup A_2$, resp. B_1, B_2 are both open in $B_1 \cup B_2$. Assume that $f(A_i) \subset B_i, 1 \leq i \leq 2$. Let $f_1 : (X, A_1) \to (Y, B_1)$, $f_2 : (X, A_2) \to (Y, B_2)$ and $\overline{f} : (X, A_1 \cup A_2) \to (Y, B_1 \cup B_2)$ be maps induced by f. Then, for $u \in H^p(Y, B_1)$ and $z \in H_n(X, A_1 \cup A_2)$, we have*

$$(1) \qquad f_{2*}(f_1^* u \cap z) = u \cap \overline{f}_* z, \quad \text{as elements of } H_{n-p}(Y, B_2).$$

(b) Let X be a topological space with three subsets A_1, A_2, A_3 such that A_1, A_2, A_3 are all open in $A_1 \cup A_2 \cup A_3$. Then, for $u \in H^p(X, A_1)$, $v \in H^q(X, A_2)$ and $z \in H_n(X, A_1 \cup A_2 \cup A_3)$, we have

$$(2) \qquad u \cap (v \cap z) = (u \cup v) \cap z, \quad \text{as elements of } H_{n-p-q}(X, A_3).$$

C.4 Definition. A *fiber bundle pair* with base space B consists of a *total pair* $(E, \dot{E})$, a *fiber pair* $(F, \dot{F})$, and a projection $p : E \to B$ such that there exists an open cover $\{V_\alpha\}_\alpha$ of B and, for each V_α, a homeomorphism φ_α making the following diagram commutative:

$$\begin{array}{ccc} V_\alpha \times (F, \dot{F}) & \xrightarrow[\sim]{\varphi_\alpha} & (p^{-1}(V_\alpha), p^{-1}(V_\alpha) \cap \dot{E}) \\ & {\scriptstyle \pi_1} \searrow \quad \swarrow {\scriptstyle p} & \\ & V_\alpha , & \end{array}$$

where π_1 is the projection on the first factor.

By a *cohomology extension of fiber* (of a fiber bundle pair), we mean a graded $\mathbb{Z}$-module homomorphism $\theta : H^*(F, \dot{F}) \to H^*(E, \dot{E})$ of degree 0 such that, for each $b \in B$, the composite

$$H^*(F, \dot{F}) \xrightarrow{\theta} H^*(E, \dot{E}) \xrightarrow{i_b^*} H^*(E_b, \dot{E}_b)$$

is an isomorphism, where $(E_b, \dot{E}_b) := (p^{-1}(b), p^{-1}(b) \cap \dot{E})$ and i_b^* is induced by the inclusion $i_b : (E_b, \dot{E}_b) \hookrightarrow (E, \dot{E})$.

Clearly, a necessary condition for the existence of a cohomology extension of fiber θ is that $H^*(F, \dot{F})$ is a graded $\mathbb{Z}$-submodule of $H^*(E, \dot{E})$. In particular, θ does not exist in general.

Assume that B is path-connected. Then, by [Spanier–66, Chap. 5, Exercise $E(2)$], a graded $\mathbb{Z}$-module map $\theta : H^*(F, \dot{F}) \to H^*(E, \dot{E})$ is a cohomology extension of fiber iff $i_b^* \circ \theta$ is an isomorphism for some $b \in B$.

We recall the following important result due to Leray–Hirsch (cf. [Spanier–66, Chap. 5, §7, Theorem 9]).

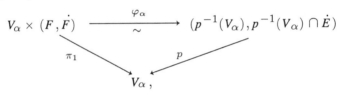

C.5 Theorem. *Let $(E, \dot{E})$ be the total pair of a fiber bundle pair with base B, fiber pair $(F, \dot{F})$ and the projection $p : E \to B$. Assume that the total homology $H_*(F, \dot{F})$ is free and finitely generated as a $\mathbb{Z}$-module, and that there is a cohomology extension of fiber θ. Then the maps*

$$(1)\quad \Phi : H_*(E, \dot{E}) \to H_*(B) \otimes H_*(F, \dot{F}),\ \ \Phi(c) = \sum_i p_*(\theta(m_i^*) \cap c) \otimes m_i,$$

and

$$(2)\qquad \Phi' : H^*(B) \otimes H^*(F, \dot{F}) \to H^*(E, \dot{E}),\ \ \phi'(u \otimes v) = p^*(u) \cup \theta(v),$$

are both graded $\mathbb{Z}$-module isomorphisms, where $\{m_i\}_i$ is any graded $\mathbb{Z}$-basis of $H_*(F, \dot{F})$ and $\{m_i^*\}$ is the corresponding dual basis of $H^*(F, \dot{F})$.

We recall the basic definitions and properties of equivariant oohomology (we need). We refer to [Borel–60], [Allday-Puppe–93] for details.

C.6 Definition and Elementary Properties. Let K be a real Lie group and let $\pi : E(K) \to B(K)$ be the universal principal K-bundle (cf. [Husemoller–94, Chap. 4]).

For a topological space X with a continuous action of K (we call such a space a *K-space*), consider the associated bundle with fiber X:

$$\pi_X : E(K) \times_K X \to B(K).$$

We abbreviate $E(K) \times_K X$ by X_K. Following Borel, the *K-equivariant cohomology of X with integer coefficients $H_K^*(X)$* is defined to be the singular cohomology:

$$(1)\qquad H_K^*(X) := H^*(X_K, \mathbb{Z}).$$

Clearly, $H_K^*(X)$ is a graded $\mathbb{Z}$-algebra. Moreover, the $\mathbb{Z}$-algebra homomorphism

$$(2)\qquad \pi_X^* : H^*(B(K)) \to H^*(X_K)$$

induces a graded $H^*(B(K))$-algebra structure on $H_K^*(X)$.

A continuous K-equivariant map $\varphi : X \to Y$ between K-spaces canonically induces a (continuous) map $\varphi_K : X_K \to Y_K$, making the following diagram commutative:

$$\begin{array}{ccc} X_K & \xrightarrow{\varphi_K} & Y_K \\ & {\scriptstyle \pi_X}\searrow \quad \swarrow{\scriptstyle \pi_Y} & \\ & B(K)\,. & \end{array}$$

In particular, φ_K induces a graded $H^*(B(K))$-algebra homomorphism

$$\varphi_K^* : H_K^*(Y) \to H_K^*(X). \tag{3}$$

When the reference to K is clear from the context, we sometimes write φ_K^* just as φ^*.

Fixing any base point $e \in E(K)$, we get the inclusion $i_e : X \hookrightarrow X_K$, $x \mapsto [e, x]$, where $[e, x]$ is the K-orbit of (e, x). The inclusion i_e induces the graded $\mathbb{Z}$-algebra homomorphism (called the *evaluation* map)

$$\eta = i_e^* : H_K^*(X) \to H^*(X). \tag{4}$$

For a different $e' \in E(K)$, $i_{e'}$ is homotopic to i_e; in particular, η does not depend on the choice of the base point in $E(K)$.

For a K-space X and a K-stable subspace Y, we get the following long exact cohomology sequence from the pair (X_K, Y_K):

$$0 \to H_K^0(X, Y) \to H_K^0(X) \to H_K^0(Y) \to H_K^1(X, Y) \to \cdots, \tag{5}$$

where $H_K^i(X, Y) := H^i(X_K, Y_K)$.

Assume that K is a connected Lie group, and X a K-space. Then the Leray-Serre spectral sequence corresponding to the fibration π_X (cf. Theorem E.11) has

$$E_2^{p,q} = H^p(B(K), H^q(X)) \Rightarrow H_K^*(X). \tag{6}$$

(Observe that, since K is connected by assumption, $B(K)$ is simply-connected.)

C.7 Lemma. *For a topological space X with the trivial action of K, X_K is homeomorphic with $B(K) \times X$. In particular, if $H^*(B(K))$ is torsion free (e.g., if K is a compact or a complex torus), by the Kunneth Theorem (cf. [Spanier–66, Theorem 1, §6, Chap. 5]),*

$$H_K^*(X) \simeq H^*(B(K)) \otimes_{\mathbb{Z}} H^*(X)$$

as graded $H^*(B(K))$-algebras, where $H^*(B(K))$ acts on the right side via multiplication on the first factor.

We recall a special case of the Borel–Atiyah–Segal Localization Theorem (cf. [Allday-Puppe–93, Theorem 3.2.6]) which is sufficient for our purposes:

C.8 Theorem. *Let T be a compact (connected) torus and let X be a compact Hausdorff T-space. Then the restriction map induces an isomorphism*

$$Q \otimes_{H^*(B(T))} \check{H}_T^*(X) \xrightarrow{\sim} Q \otimes_{H^*(B(T))} \check{H}_T^*(X^T) \text{ of } Q\text{-algebras},$$

where Q is the quotient field of the integral domain $H^*(B(T))$ and $\check{H}_T^*(X)$ is the Čech cohomology of the space X_T with integral coefficients (cf. [Spanier–66, §7, Chap. 6]).

Observe that, for any paracompact Hausdorff space Y which is locally contractible,

$$\check{H}^*(Y) \simeq H^*(Y) \tag{1}$$

(cf. [Spanier–66, Corollary 5, §9, Chap. 6]). In particular, for a T-space Y with Y a CW−complex,

$$\check{H}_T^*(Y) \simeq H_T^*(Y). \tag{2}$$

Appendix D

Relative Homological Algebra

Basic references for this appendix are [Cartan–Eilenberg–56] and [Hochschild–56].

In this appendix we take R to be a (not necessarily commutative) ring with identity element 1 and S a subring containing 1. All the R-modules M are assumed to be unitary in the sense that 1 acts as the identity operator on M. Of course, an R-module M is an S-module under restriction. We will have occasion to consider both the left and right R-modules. When we just say R-module, we will mean a left R-module.

The aim of this appendix is to define the relative Tor and Ext functors and establish their basic properties. We also define the Koszul resolution.

D.1 Definition. An exact sequence of R-modules and R-module homomorphisms

$$(1) \qquad \cdots \to M_i \xrightarrow{t_i} M_{i-1} \to \cdots$$

is called (R, S)-*exact* if, for all i, Ker t_i is a direct S-module summand of M_i. (The sequence is allowed to terminate in either of the directions.)

There is a similar notion of (R, S)-exact sequence

$$(2) \qquad \cdots \to M_i \xrightarrow{t^{i+1}} M_{i+1} \to \cdots .$$

An R-module M is called (R, S)-*injective*, if for every (R, S)-exact sequence

$$(3) \qquad 0 \to M_1 \xrightarrow{f_1} M_2 \xrightarrow{f_2} M_3 \to 0$$

and every R-module map $f : M_1 \to M$, there is an R-module map $\tilde{f} : M_2 \to M$ extending f.

Dually, an R-module M is called (R, S)-*projective* if for any (R, S)-exact sequence (3) and R-module map $f : M \to M_3$, there is an R-module lift of f to M_2, i.e., an R-module map $\tilde{f} : M \to M_2$ such that $f_2 \circ \tilde{f} = f$.

Clearly an R-injective, resp. R-projective, module is (R, S)-injective, resp. (R, S)-projective.

It is easy to see that for an (R, S)-exact sequence (1) and any (R, S)-injective module M, the induced sequence

$$\cdots \leftarrow \mathrm{Hom}_R(M_i, M) \leftarrow \mathrm{Hom}_R(M_{i-1}, M) \leftarrow \cdots$$

is exact, where $\mathrm{Hom}_R(M_i, M)$ is the abelian group of all the R-module homomorpisms from M_i to M.

Dually, for any (R, S)-exact sequence (1) and any (R, S)-projective module M, the induced sequence

$$\cdots \to \mathrm{Hom}_R(M, M_i) \to \mathrm{Hom}_R(M, M_{i-1}) \to \cdots$$

is exact.

Also, for any (R, S)-exact sequence (1) and any right (R, S)-projective module M, the induced sequence

$$\cdots \to M \otimes_R M_i \to M \otimes_R M_{i-1} \to \cdots$$

is exact.

Let M be an S-module; then the abelian group $\mathrm{Hom}_S(R, M)$, where R is an S-module under the left multiplication, is made into an R-module under

$$(r \cdot f)(r') = f(r'r), \text{ for } r, r' \in R \text{ and } f \in \mathrm{Hom}_S(R, M).$$

D.2 Lemma. *For any S-module M, the R-module $\mathrm{Hom}_S(R, M)$ is (R, S)-injective.*

Dually, the R-module $R \otimes_S M$ is (R, S)-projective, where S acts on R via the right mltiplication and R acts on $R \otimes_S M$ via its left multiplication on the first factor.

Proof. We have shown in Lemma 3.1.7 that $R \otimes_S M$ is (R, S)-projective. The (R, S)-injectivity of $\mathrm{Hom}_S(R, M)$ is proved similarly by using the following isomorphism instead of the isomorphism (3.1.7.1):

For any S-module M and R-module N,

$$\varphi : \mathrm{Hom}_R(N, \mathrm{Hom}_S(R, M)) \simeq \mathrm{Hom}_S(N, M), \quad \varphi(f)(n) = f(n)(1),$$

for $f \in \mathrm{Hom}_R(N, \mathrm{Hom}_S(R, M))$ and $n \in N$. □

D.3 Definition. By an (R, S)-*projective resolution* of an R-module M, we mean an (R, S)-exact sequence

$$\cdots \xrightarrow{\delta_2} C_1 \xrightarrow{\delta_1} C_0 \xrightarrow{\delta_0} M \to 0,$$

in which each C_j is an (R, S)-projective module.

Similarly, an (R, S)-exact sequence

$$0 \to M \xrightarrow{d^0} B_0 \xrightarrow{d^1} B_1 \xrightarrow{d^2} \cdots$$

is called an (R, S)-*injective resolution* if each B_j is an (R, S)-injective module.

D.4 Lemma. *Any R-module M admits an (R, S)-projective (as well as (R, S)-injective) resolution.*

Proof. We prove the existence of an (R, S)-projective resolution; the proof of the existence of an (R, S)-injective resolution is similar. Consider the surjective R-module map

$$\epsilon_M : R \otimes_S M \to M, r \otimes m \mapsto rm, \text{ for } r \in R \text{ and } m \in M.$$

Set $P_0 = R \otimes_S M$, $\delta_0 = \epsilon_M$, and let $P_1 = R \otimes_S \operatorname{Ker} \epsilon_M$ with the map $\delta_1 : P_1 \to P_0$ defined as $\epsilon_{\operatorname{Ker} \epsilon_M}$. Now define $P_2 = R \otimes_S \operatorname{Ker} \delta_1$ and continue in this manner. Thus, we get an exact sequence of R-modules (and R-module maps between them):

$$\cdots \xrightarrow{\delta_2} P_1 \xrightarrow{\delta_1} P_0 \xrightarrow{\delta_0} M \to 0.$$

To prove that the above sequence is (R, S)-exact, use the S-module map $\theta_N : N \to R \otimes_S N, n \mapsto 1 \otimes n$ (for any S-module N). By Lemma D.2, each $R \otimes_S M$ is (R, S)-projective, thus the lemma follows. □

D.5 Definition. The (R, S)-projective resolution of M constructed in the above proof is called the *standard* (R, S)*-projective resolution* of M. Similarly, for any R-module M, the R-module injective map

$$M \overset{i}{\hookrightarrow} \operatorname{Hom}_S(R, M), i(m)r = r.m,$$

gives rise to an (R, S)-injective resolution (called the *standard* (R, S)*-injective resolution*) of M:

$$0 \to M \to I_0 \to I_1 \to \cdots,$$

where $I_0 := \operatorname{Hom}_S(R, M)$, $I_1 := \operatorname{Hom}_S(R, I_0/M)$, and so on.

Now we come to the following basic result of relative homological algebra.

D.6 Theorem. *Let*

$$\cdots \xrightarrow{s_2} C_1 \xrightarrow{s_1} C_0 \xrightarrow{s_0} M \to 0$$

be a chain complex (i.e., the composition of any two successive maps is zero) of R-module maps, where each C_i is (R, S)-projective, and let

$$\cdots \xrightarrow{t_2} D_1 \xrightarrow{t_1} D_0 \xrightarrow{t_0} N \to 0$$

be a (R, S)-resolution of an R-module N. (We do not assume that D_i's are (R, S)-projective.) Assume further that there is given an R-module map $f : M \to N$. Then, there exists an R-module map of chain complexes $f_\bullet : C_\bullet \to D_\bullet$ covering

f, i.e., we have R-module maps $f_i : C_i \to D_i$ (for each $i \geq 0$) making the following diagram commutative:

(1)
$$\begin{array}{ccccccccc} \cdots & \longrightarrow & C_1 & \xrightarrow{s_1} & C_0 & \xrightarrow{s_0} & M & \longrightarrow & 0 \\ & & \downarrow{\scriptstyle f_1} & & \downarrow{\scriptstyle f_0} & & \downarrow{\scriptstyle f} & & \\ \cdots & \longrightarrow & D_1 & \xrightarrow{t_1} & D_0 & \xrightarrow{t_0} & N & \longrightarrow & 0\,. \end{array}$$

Moreover, if $g_\bullet : C_\bullet \to D_\bullet$ is another R-module map of chain complexes covering f, then there exists an R-module homotopy connecting them, i.e., there exists a sequence of R-module maps $h_i : C_i \to D_{i+1}$ such that

(2_i) $\quad f_i - g_i = t_{i+1} \circ h_i + h_{i-1} \circ s_i$, *for all* $i \geq 0$ (*where* $h_{-1} := 0$).

Proof. The existence of f_0 making the rightmost rectangle commutative follows since C_0 is (R, S)-projective. Assume, by induction, that R-module maps $f_0, \dots, f_i$ have been constructed so that all the rectangles to the right of (and including) the arrow f_i in diagram (1) are commutative. We now define f_{i+1} making the next rectangle commutative: Consider the (R, S)-exact sequence

$$0 \to \operatorname{Ker} t_{i+1} \to D_{i+1} \xrightarrow{t_{i+1}} \operatorname{Im} t_{i+1} = \operatorname{Ker} t_i \to 0.$$

By the induction hypothesis, the map $f_i \circ s_{i+1}$ has image contained in $\operatorname{Ker} t_i$. Thus, the existence of f_{i+1} making the corresponding rectangle commutative follows since C_{i+1} is (R, S)-projective. This completes the induction, thereby proving the existence of $f_\bullet$.

For two chain maps $f_\bullet$ and $g_\bullet$, the existence of homotopy $h_\bullet$ follows by a similar argument. Since $t_0 \circ (f_0 - g_0) = 0$ and C_0 is (R, S)-projective, the existence of $h_0 : C_0 \to D_1$ satisfying (2_0) follows. Assume now, by induction, that R-module maps $h_0, \cdots, h_i$ satisfying $(2_0), \dots, (2_i)$ have been constructed. From the commutativity of diagram (1) and identity (2_i), we get

$$t_{i+1}(f_{i+1} - g_{i+1} - h_i \circ s_{i+1}) = (f_i - g_i)s_{i+1} - (f_i - g_i)s_{i+1} = 0.$$

From the above identity, we see that $\operatorname{Im}(f_{i+1} - g_{i+1} - h_i \circ s_{i+1}) \subset \operatorname{Ker} t_{i+1} = \operatorname{Im} t_{i+2}$. Thus, C_{i+1} being (R, S)-projective, we can construct $h_{i+1} : C_{i+1} \to D_{i+2}$ satisfying (2_{i+1}). This completes the induction and hence the theorem is proved. □

D.7 Definition. For any right R-module M and (left) R-module N, define the abelian group $\operatorname{Tor}_n^{(R,S)}(M, N)$ (for any $n \geq 0$), called the *relative Tor functor*, as follows:

Take the standard (R, S)-projective resolution of the left R-module N:

$$\cdots \to P_1 \to P_0 \to N \to 0,$$

and consider the chain complex of abelian groups:

$$\cdots \to M \otimes_R P_1 \to M \otimes_R P_0 \to 0.$$

The n-th homology of this chain complex is denoted by $\mathrm{Tor}_n^{(R,S)}(M, N)$ or simply by $\mathrm{Tor}_n(M, N)$ when the reference to (R, S) is clear.

Similarly, for two (left) R-modules N and Q, we define the abelian group $\mathrm{Ext}^n_{(R,S)}(N, Q)$ (called the *relative Ext functor*) as the n-th cohomology of the cochain complex

$$0 \to \mathrm{Hom}_R(N, I_0) \to \mathrm{Hom}_R(N, I_1) \to \cdots,$$

where

$$0 \to Q \to I_0 \to I_1 \to \cdots$$

is the standard (R, S)-injective resolution of Q. Again, we abbreviate $\mathrm{Ext}^n_{(R,S)}(N, Q)$ by $\mathrm{Ext}^n(N, Q)$ when the reference to (R, S) is clear. Then $\mathrm{Ext}^1(N, Q)$ is isomorphic with the group of equivalence classes of the S-trivial extensions of Q by N (cf. [Hochschild–56, §2]).

D.8 Proposition. *Let M be a right R-module and let N be a (left) R-module. Take any right (R, S)-projective resolution of M:*

$$\cdots \to C_1 \xrightarrow{s_1} C_0 \xrightarrow{s_0} M \to 0,$$

and a (left) (R, S)-projective resolution of N:

$$\cdots \to D_1 \xrightarrow{t_1} D_0 \xrightarrow{t_0} N \to 0.$$

Then there are canonical isomorphisms of abelian groups:

$$(1) \qquad H_n(C_\bullet \otimes_R N) \simeq H_n(M \otimes_R D_\bullet) \simeq \mathrm{Tor}_n(M, N),$$

where $C_\bullet \otimes_R N$ denotes the chain complex

$$\cdots \to C_1 \otimes_R N \to C_0 \otimes_R N \to 0.$$

Similarly, let Q be a (left) R-module and let

$$0 \to Q \to B_0 \to B_1 \to \cdots$$

be an (R, S)*-injective resolution of* Q*. Then there is a canonical isomorphism of abelian groups:*

$$(2) \qquad H^n(\mathrm{Hom}_R(D_\bullet, Q)) \simeq H^n(\mathrm{Hom}_R(N, B_\bullet)) \simeq \mathrm{Ext}^n(N, Q),$$

where $\mathrm{Hom}_R(D_\bullet, Q)$ *is the cochain complex*

$$0 \to \mathrm{Hom}_R(D_0, Q) \to \mathrm{Hom}_R(D_1, Q) \to \cdots.$$

Proof. We first prove that $H_n(M \otimes_R D_\bullet) \simeq \mathrm{Tor}_n(M, N)$: By Theorem D.6, there exists a chain map $f_\bullet : D_\bullet \to P_\bullet$ covering I_N, where $P_\bullet$ is the standard (R, S)-projective resolution of N. This gives rise to the chain map $I_M \otimes f_\bullet : M \otimes_R D_\bullet \to M \otimes_R P_\bullet$. Moreover, for a different choice of the chain map $g_\bullet : D_\bullet \to P_\bullet$ covering I_N, the chain map $I_M \otimes g_\bullet$ induces the same map in homology $H_i(M \otimes_R D_\bullet) \to H_i(M \otimes_R P_\bullet)$ (by virtue of (2_i) of (D.6)). Similarly, we get a unique map $H_*(M \otimes_R P_\bullet) \to H_*(M \otimes_R D_\bullet)$. From this the canonical isomorphism

$$(3) \qquad H_n(M \otimes_R D_\bullet) \simeq \mathrm{Tor}_n(M, N)$$

follows.

To prove the other part of the isomorphism (1), consider the double complex $(C_\bullet \otimes D_\bullet)_{p,q} := C_p \otimes_R D_q$ $(p, q \geq 0)$ with $s_\bullet \otimes I_{D_\bullet}$ and $I_{C_\bullet} \otimes t_\bullet$ as the two differentials. Of course, this gives rise to the associated single complex (which we denote by $[C_\bullet \otimes D_\bullet]$):$[C_\bullet \otimes D_\bullet]_n := \bigoplus_{p+q=n} C_p \otimes_R D_q$ with the differential

$$\partial(x \otimes y) = s_p x \otimes y + (-1)^p x \otimes t_q y, \quad \text{for } x \in C_p \text{ and } y \in D_q.$$

Let $N_\bullet$ be the chain complex with $N_0 = N$ and $N_i = 0$ for all $i \neq 0$. Consider the chain map $D_\bullet \to N_\bullet$, where $D_0 \to N$ is the original map t_0. This gives rise to a surjective chain map $\theta : C_\bullet \otimes D_\bullet \to C_\bullet \otimes N_\bullet$ of double complexes and thus a chain map $\overline{\theta} : [C_\bullet \otimes D_\bullet] \to C_\bullet \otimes_R N$ of single chain complexes. We next show that the chain complex $\mathrm{Ker}\,\overline{\theta}$ has all its homologies zero (thus $\overline{\theta}$ induces an isomorphism in homology). Let $D'_\bullet \subset D_\bullet$ be the chain subcomplex defined by $D'_n := D_n$ if $n > 0$ and $D'_0 := \mathrm{Ker}\, t_0$. Since any C_n is (R, S)-projective, tensoring with $C_\bullet$ is an exact functor on the category of (R, S)-exact sequences (cf. D.1). Thus, $\mathrm{Ker}\,\theta$ is the double complex $C_\bullet \otimes D'_\bullet$. Consider the increasing filtration $\mathcal{F}(p)$ (by chain subcomplexes) of the single chain complex $[C_\bullet \otimes D'_\bullet]$ defined by

$$\mathcal{F}(p) := \bigoplus_{r \leq p} C_r \otimes_R D'_\bullet.$$

Then the associated spectral sequence (cf. Appendix E) has

$$E^1_{p,q} = H_{p+q}(\mathcal{F}(p)/\mathcal{F}(p-1)) \simeq H_{p+q}(C_p \otimes_R D'_\bullet). \tag{4}$$

But since $D'_\bullet$ is an (R, S)-exact complex (and C_p is (R, S)-projective), $C_p \otimes_R D'_\bullet$ remains exact. Thus $H_*(C_p \otimes_R D'_\bullet) = 0$. Combining this with (4), we get that $E^1_{p,q} = 0$ for all $p, q \geq 0$, and thus the chain complex $\operatorname{Ker}\bar{\theta} = [\operatorname{Ker}\theta]$ is exact, thereby $\overline{\theta}$ induces an isomorphism in homology.

A similar argument shows that the chain map $[C_\bullet \otimes D_\bullet] \to M \otimes_R D_\bullet$ induces an isomorphism in homology. Combining these two homology isomorphisms, we get that there is a canonical isomorphism

$$H_*(M \otimes_R D_\bullet) \simeq H_*(C_\bullet \otimes_R N). \tag{5}$$

This completes the proof of (1). The proof of (2) follows by a similar argument. □

D.9 Remark. Observe that the isomorphism (3) (in the above proof) also follows from the isomorphism (5). But we have retained a more direct proof of (3).

As an immediate consequence of the above proposition, we get the following:

D.10 Corollary. *Let M be a right R-module and N a (left) R-module. Assume that at least one of M or N is a (R, S)-projective module. Then*

$$\operatorname{Tor}_n(M, N) = 0, \quad \textit{for all } n > 0. \tag{1}$$

Similarly, for any (left) R-modules M and N,

$$\operatorname{Ext}^n(M, N) = 0, \quad \textit{for all } n > 0, \tag{2}$$

provided either M is a projective (R, S)-module or N is an injective (R, S)-module. □

D.11 Lemma. *For any right, resp. left, R-module M and any left R-module homomorphism $f : N_1 \to N_2$, there exists a functorial homomorphism of abelian groups (defined in the proof below)*

$$f_* : \operatorname{Tor}_*(M, N_1) \to \operatorname{Tor}_*(M, N_2)$$

and

$$f^* : \operatorname{Ext}^*(M, N_1) \to \operatorname{Ext}^*(M, N_2)$$

respectively.

Similarly, for any left R-module N and right, resp. left, R-module homomorphism $g : M_1 \to M_2$, *there exists a functorial homomorphism of abelian groups*

$$g_* : \mathrm{Tor}_*(M_1, N) \to \mathrm{Tor}_*(M_2, N)$$

and

$$g^* : \mathrm{Ext}^*(M_2, N) \to \mathrm{Ext}^*(M_1, N)$$

respectively.

Thus, for any $n \geq 0$ *and any left R-module N, resp. right R-module M,* $\mathrm{Tor}_n(-, N)$, *resp.* $\mathrm{Tor}_n(M, -)$, *is a covariant functor from the category of right R-modules, resp. left R-modules, to the category of abelian groups.*

Similarly, for a R-module Q, $\mathrm{Ext}^n(-, Q)$, *resp.* $\mathrm{Ext}^n(Q, -)$, *is a contravariant, resp. covariant, functor from the category of R-modules to the category of abelian groups.*

Proof. Take a right (R, S)-projective resolution of M:

$$\cdots \to C_1 \to C_0 \to M \to 0.$$

Then the map f_* is induced from the chain map

$$C_\bullet \otimes_R N_1 \xrightarrow{I_{C_\bullet} \otimes f} C_\bullet \otimes_R N_2.$$

The definition of g_* is very similar.

The map f^* is induced from the cochain map

$$\mathrm{Hom}_R(C_\bullet, N_1) \to \mathrm{Hom}_R(C_\bullet, N_2),\ \chi \mapsto f \circ \chi.$$

Take an (R, S)-injective resolution of N:

$$0 \to N \to B_0 \to B_1 \to \cdots .$$

Then the map g^* is induced from the cochain map

$$\mathrm{Hom}_R(M_2, B_\bullet) \to \mathrm{Hom}_R(M_1, B_\bullet),\ \chi \mapsto \chi \circ g. \qquad \square$$

D.12 Lemma. *Let*

$$0 \to M_1 \xrightarrow{f_1} M_2 \xrightarrow{f_2} M_3 \to 0$$

be an (R, S)-*exact sequence of right R-modules. Then, for any R-module N, there is a functorial long exact sequence of abelian groups:*

$$\begin{aligned} \cdots \to \mathrm{Tor}_n(M_1, N) \xrightarrow{f_{1*}} \mathrm{Tor}_n(M_2, N) \xrightarrow{f_{2*}} \mathrm{Tor}_n(M_3, N) \to \\ \mathrm{Tor}_{n-1}(M_1, N) \to \cdots \to \mathrm{Tor}_0(M_3, N) \to 0 . \end{aligned} \tag{1}$$

Similarly, for an (R, S)-exact sequence

$$0 \to N_1 \xrightarrow{g_1} N_2 \xrightarrow{g_2} N_3 \to 0$$

of left R-modules, right R-module M, and (left) R-module Q, there are functorial long exact sequences of abelian groups:

$$\tag{2} \begin{aligned} \cdots \to \operatorname{Tor}_n(M, N_1) \xrightarrow{g_{1*}} \operatorname{Tor}_n(M, N_2) \xrightarrow{g_{2*}} \operatorname{Tor}_n(M, N_3) \to \\ \operatorname{Tor}_{n-1}(M, N_1) \to \cdots \to \operatorname{Tor}_0(M, N_3) \to 0, \end{aligned}$$

$$\tag{3} \begin{aligned} 0 \to \operatorname{Ext}^0(Q, N_1) \to \cdots \to \operatorname{Ext}^n(Q, N_1) \xrightarrow{g_1^*} \operatorname{Ext}^n(Q, N_2) \xrightarrow{g_2^*} \\ \operatorname{Ext}^n(Q, N_3) \to \operatorname{Ext}^{n+1}(Q, N_1) \to \cdots, \end{aligned}$$

and

$$\tag{4} \begin{aligned} 0 \to \operatorname{Ext}^0(N_3, Q) \to \cdots \to \operatorname{Ext}^n(N_3, Q) \xrightarrow{g_2^*} \operatorname{Ext}^n(N_2, Q) \xrightarrow{g_1^*} \\ \operatorname{Ext}^n(N_1, Q) \to \operatorname{Ext}^{n+1}(N_3, Q) \to \cdots. \end{aligned}$$

Proof. Take an (R, S)-projective resolution of N:

$$\cdots \to C_1 \to C_0 \to N.$$

Since each C_i is (R, S)-projective, this gives rise to a short exact sequence of chain complexes:

$$0 \to M_1 \otimes_R C_\bullet \xrightarrow{f_1 \otimes I_{C_\bullet}} M_2 \otimes_R C_\bullet \xrightarrow{f_2 \otimes I_{C_\bullet}} M_3 \otimes_R C_\bullet \to 0.$$

Then (1) is the associated long exact homology sequence (cf. [Spanier–66, Chap. 4, §5, Theorem 4]). The derivation of (2) is exactly similar.

Take an (R, S)-injective resolution of Q:

$$0 \to Q \to B_0 \to B_1 \to \cdots.$$

This gives rise to a short exact sequence of cochain complexes (cf. D.1):

$$0 \to \operatorname{Hom}_R(N_3, B_\bullet) \to \operatorname{Hom}_R(N_2, B_\bullet) \to \operatorname{Hom}_R(N_1, B_\bullet) \to 0.$$

Then (4) is the associated long exact cohomology sequence. The derivation of (3) is similar. □

D.13 Definition. (Koszul resolution) Let

$$0 \to V' \xrightarrow{p_1} V \xrightarrow{p_2} V'' \to 0$$

be a short exact sequence of vector spaces (over any field k). For any $n > 0$, consider the sequence

$$\begin{aligned} 0 \to \wedge^n(V') \to \cdots \to S^{n-i}(V) \otimes \wedge^i(V') \xrightarrow{\delta_{i-1}} S^{n-i+1}(V) \otimes \wedge^{i-1}(V') \\ \to \cdots \to S^{n-1}(V) \otimes V' \xrightarrow{\delta_0} S^n(V) \xrightarrow{\hat{p}_2} S^n(V'') \to 0, \end{aligned} \tag{1}$$

where $\hat{p}_2$ is induced by the map p_2 and

$$\delta_{i-1} : S^{n-i}(V) \otimes \wedge^i(V') \to S^{n-i+1}(V) \otimes \wedge^{i-1}(V')$$

is defined by

$$\text{(2)}\quad \delta_{i-1}(P \otimes v_1 \wedge \cdots \wedge v_i) = \sum_{j=1}^{i} (-1)^{j-1}(p_1 v_j) P \otimes v_1 \wedge \cdots \wedge \hat{v}_j \wedge \cdots \wedge v_i.$$

Then, as is well known, the above sequence (1) is an exact complex, called the *Koszul complex* (cf. [Serre–89, Chap. IV.A]).

D.E EXERCISE. For any right R-module M and (left) R-module N, show that

$$\operatorname{Tor}_0^{(R,S)}(M, N) \simeq M \otimes_R N.$$

Similarly, for R-modules M, N, show that

$$\operatorname{Ext}^0_{(R,S)}(M, N) \simeq \operatorname{Hom}_R(M, N).$$

Appendix E

An Introduction to Spectral Sequences

Basic references for this appendix are [Cartan–Eilenberg–56, Chap. XV], [Godement–58, Chap. I, §4] and [Spanier–66, Chap. 9].

The aim of this appendix is to define the homology and cohomology spectral sequences and associate a homology, resp. cohomology, spectral sequence to anincreasing, resp. decreasing, filtration of a chain, resp. cochain, complex. We give examples of two spectral sequences associated to a double complex. Further, we recall the Leray–Serre spectral sequence associated to a fibration and the Hochschild–Serre spectral sequence associated to a Lie algebra pair.

E.1 Definition. Let R be a commutative ring with identity. A *bigraded module* E over R is an indexed collection of R-modules $\{E_{s,t}\}_{s,t\in\mathbb{Z}}$. A *differential* $d : E \to E$ *of bidegree* $(-r, r-1)$ is a collection of R-module maps $d : E_{s,t} \to E_{s-r,t+r-1}$, for all s and t, such that $d^2 = 0$. The *homology module* $H(E)$ is the bigraded module defined by

$$H_{s,t}(E) = \mathrm{Ker}(d : E_{s,t} \to E_{s-r,t+r-1})/d(E_{s+r,t-r+1}).$$

Note that if $[E]_q$ is defined to equal $\oplus_{s+t=q} E_{s,t}$, the differential d defines $\hat{d}$: $[E]_q \to [E]_{q-1}$ such that $\{[E], \hat{d}\}$ is a chain complex. Furthermore, the q-th homology module of this chain complex equals $\oplus_{s+t=q} H_{s,t}(E)$.

A *homology spectral sequence* E is a sequence $\{E^r, d^r\}_{r\geq 0}$ such that

(a) E^r is a bigraded module over R and d^r is a differential of bidegree $(-r, r-1)$ on E^r.

(b) For $r \geq 0$, there is given a bigraded isomorphism $H(E^r) \simeq E^{r+1}$.

The above spectral sequence is said to *collapse* (or *degenerate*) at E^{r_o} (for some $r_o \geq 0$) if $d^r = 0$ for all $r \geq r_o$. Thus, in this case, $E^{r_o} \simeq E^{r_o+1} \simeq \cdots$. A *homomorphism* $\varphi : E \to E'$ between two spectral sequences is a collection of R-module maps $\varphi^r : E^r_{s,t} \to E'^r_{s,t}$ for $r \geq 0$ (and all s and t) commuting with the differentials and such that the induced map $\varphi^r_* : H(E^r) \to H(E'^r)$ corresponds to the map $\varphi^{r+1} : E^{r+1} \to E'^{r+1}$ under the isomorphisms of the spectral sequences. The composite of homomorphisms is a homomorphism, and so there is a category of spectral sequences and homomorphisms.

To define the limit term of a spectral sequence, we regard E^{r+1} as identified with $H(E^r)$ by the isomorphism of the spectral sequence. Let Z^0 be the bigraded module $Z^0_{s,t} := \text{Ker}\,(d^0 : E^0_{s,t} \to E^0_{s,t-1})$ and let B^0 be the bigraded module $B^0_{s,t} := d^0(E^0_{s,t+1})$. Then $B^0 \subset Z^0$ and $E^1 \simeq Z^0/B^0$. Let $Z(E^1)$ be the bigraded module $Z(E^1)_{s,t} := \text{Ker}\,(d^1 : E^1_{s,t} \to E^1_{s-1,t})$ and let $B(E^1)$ be the bigraded module $B(E^1)_{s,t} := d^1(E^1_{s+1,t})$. By the Noether isomorphism, there exist unique bigraded submodules Z^1 and B^1 of Z^0 both containing B^0 such that $Z(E^1)_{s,t} = Z^1_{s,t}/B^0_{s,t}$ and $B(E^1)_{s,t} = B^1_{s,t}/B^0_{s,t}$ for all s and t. It follows that $B^1 \subset Z^1$, and we have

$$B^0 \subset B^1 \subset Z^1 \subset Z^0.$$

Continuing by induction, we obtain graded submodules for any $r \geq 0$:

$$B^0 \subset B^1 \subset \cdots \subset B^r \subset \cdots \subset Z^r \subset \cdots \subset Z^1 \subset Z^0 \subset E^0,$$

such that $E^{r+1} \simeq Z^r/B^r$. We define bigraded modules $Z^\infty := \bigcap_r Z^r$, $B^\infty := \bigcup_r B^r$, and $E^\infty := Z^\infty/B^\infty$. The bigraded module E^∞ is called the *limit of the spectral sequence* E. Thus the terms E^r of the spectral sequence can be considered as successive approximations to E^∞.

A homomorphism $\varphi : E \to E'$ between spectral sequences induces a bigraded R-module map $\varphi^\infty : E^\infty \to E'^\infty$ between their limit terms. Therefore, there is a covariant functor from the category of spectral sequences to the category of bigraded modules which assigns to every spectral sequence its limit.

The spectral sequence E is said to be *convergent* if, for every s and t, there exists a nonnegative integer $r(s,t)$ such that, for $r \geq r(s,t)$, $d^r : E^r_{s,t} \to E^r_{s-r,t+r-1}$ is trivial. In this case $E^{r+1}_{s,t}$ is isomorphic to a quotient of $E^r_{s,t}$ and $E^\infty_{s,t}$ is isomorphic to the direct limit of the sequence

$$E^{r(s,t)}_{s,t} \twoheadrightarrow E^{r(s,t)+1}_{s,t} \twoheadrightarrow \cdots.$$

Observe that a *first quadrant spectral sequence*, i.e., a spectral sequence E such that $E^r_{s,t} = 0$ if $s < 0$ or $t < 0$ is convergent.

The following very useful result is trivial to prove.

E.2 Proposition. *Let $\varphi : E \to E'$ be a homomorphism of spectral sequences, which is an isomorphism for some $r \geq 0$. Then φ is an isomorphism for all $r' \geq r$. Furthermore, if E and E' are convergent, then φ^∞ is an isomorphism of their limits.*

E.3 Definition. An *increasing filtration* F of an R-module A is a sequence of submodules F_sA for all integers s such that $F_sA \subset F_{s+1}A$. Given a filtration F

of A, the *associated graded module* $\mathrm{Gr}(A)$ is defined by $\mathrm{Gr}(A)_s := F_sA/F_{s-1}A$. A filtration F of A is said to be *convergent* if $\cap_s F_sA = 0$ and $\cup_s F_sA = A$.

If A is a graded module and the filtration F is compatible with the gradation (i.e., F_sA is graded by $\{F_sA_t\}$), the associated graded module $\mathrm{Gr}(A)$ is brigraded by the modules $\mathrm{Gr}(A)_{s,t} = F_sA_{s+t}/F_{s-1}A_{s+t}$. In this case, s is called the *filtered degree*, t the *complementary degree*, and $s+t$ the *total degree* of an element of $\mathrm{Gr}(A)_{s,t}$.

A *chain filtration* F of a chain complex C is a filtration of C compatible with its gradation as well as with the differential of C (i.e., each F_sC is a chain subcomplex of C consisting of $\{F_sC_t\}$). The filtration F of C induces a filtration of $H_*(C)$ defined by

$$F_sH_*(C) := \mathrm{Im}\,(H_*(F_sC) \to H_*(C)).$$

Because the homology functor commutes with direct limits (cf. [Spanier–66, Chap. 4, §1, Theorem 7]), if F is a convergent filtration of C, it follows that $\cup_s F_sH_*(C) = H_*(C)$; however, it is not true in general that $\cap_s F_sH_*(C) = 0$. Thus, to ensure that F induces a convergent filtration of $H_*(C)$, we need a stronger assumption on the filtration F. A filtration F of a graded module A compatible with the gradation is said to be *bounded below* if for any t there is $s(t)$ such that $F_{s(t)}A_t = 0$. It is clear that if F is a chain filtration bounded below of a chain complex C, then the induced filtration of $H_*(C)$ is also bounded below. Thus, if a chain filtration F of C is convergent and bounded below, the same is true for the induced filtration of $H_*(C)$.

The following theorem associates a spectral sequence to a chain filtration of a chain complex. This is one of the most important ways in which a spectral sequence arises naturally.

E.4 Theorem. *Let F be a convergent chain filtration bounded below of a chain complex C. Then there is a convergent homology spectral sequence with*

$$E^1_{s,t} \simeq H_{s+t}(F_sC/F_{s-1}C),$$

and $d^1 : E^1_{s,t} \to E^1_{s-1,t}$ corresponds to the boundary operator in the long exact homology sequence associated to the short exact sequence of chain complexes:

$$(1) \qquad 0 \to F_{s-1}C/F_{s-2}C \to F_sC/F_{s-2}C \to F_sC/F_{s-1}C \to 0$$

(cf. [Spanier–66, Chap. 4, §5, Lemma 3]).

Moreover, E^∞ is isomorphic to the bigraded module $\mathrm{Gr}\ H_*(C)$ *associated to the filtration* $F_sH_*(C) := \mathrm{Im}\,(H_*(F_sC) \to H_*(C))$.

Proof. For an arbitrary $r \geq 0$, define

$$Z^r_s := \{c \in F_sC : \partial c \in F_{s-r}C\}, \text{ and } Z^\infty_s := \{c \in F_sC : \partial c = 0\}.$$

These are graded modules with $Z^r_{s,t} = \{c \in F_s C_{s+t} : \partial c \in F_{s-r} C\}$ and $Z^\infty_{s,t} = \{c \in F_s C_{s+t} : \partial c = 0\}$. We then have a sequence of graded modules

$$\partial Z^0_s \subset \partial Z^1_{s+1} \subset \cdots \subset \partial C \cap F_s C \subset Z^\infty_s \subset \cdots\cdots \subset Z^1_s \subset Z^0_s = F_s C.$$

We first define

$$E^0_{s,t} := F_s C_{s+t} / F_{s-1} C_{s+t} = \mathrm{Gr}(C)_{s,t} \,,$$

and $d^0 : F_s C_{s+t}/F_{s-1}C_{s+t} \to F_s C_{s+t-1}/F_{s-1}C_{s+t-1}$ as the boundary operator of the quotient complex $F_s C/F_{s-1}C$.

Now, we define (for any $r \geq 1$)

$$E^r_s := Z^r_s/(Z^{r-1}_{s-1} + \partial Z^{r-1}_{s+r-1}), \text{ and } E^\infty_s := Z^\infty_s/(Z^\infty_{s-1} + (\partial C \cap F_s C)).$$

The map ∂ sends Z^r_s to Z^r_{s-r} and $Z^{r-1}_{s-1} + \partial Z^{r-1}_{s+r-1}$ to ∂Z^{r-1}_{s-1}. Therefore, it induces a homomorphism (for any $r \geq 1$)

$$d^r : E^r_s \to E^r_{s-r} \,.$$

Then E^r is a bigraded module and d^r is a differential of bidegree $(-r, r-1)$ on it.

It is easy to see that $E^1_{s,t} \simeq H_{s+t}(F_s C/F_{s-1}C)$ by the Noether isomorphism. The fact that, under this isomorphism, d^1 corresponds to the boundary operator in the long exact homology sequence, associated to the short exact sequence (1) of chain complexes, is proved by a direct verification using the definitions.

We prove that $E = \{E^r\}_{r\geq 0}$ is a spectral sequence by computing the homology of E^r with respect to d^r. We have

$$\begin{aligned}
&\{c \in Z^r_s : \partial c \in Z^{r-1}_{s-r-1} + \partial Z^{r-1}_{s-1}\} \\
&\qquad = \{c \in Z^r_s : \partial c \in F_{s-r-1}C\} + \{c \in Z^r_s : \partial c \in \partial Z^{r-1}_{s-1}\} \\
&\qquad = Z^{r+1}_s + (Z^{r-1}_{s-1} + Z^\infty_s) = Z^{r+1}_s + Z^{r-1}_{s-1} \,.
\end{aligned}$$

Therefore, Ker $(d^r : E^r_s \to E^r_{s-r}) = (Z^{r+1}_s + Z^{r-1}_{s-1})/(Z^{r-1}_{s-1} + \partial Z^{r-1}_{s+r-1})$. By definition,

$$\mathrm{Im}\,(d^r : E^r_{s+r} \to E^r_s) = (\partial Z^r_{s+r} + Z^{r-1}_{s-1})/(Z^{r-1}_{s-1} + \partial Z^{r-1}_{s+r-1}).$$

Hence, by the Noether isomorphism, in E^r_s we have

$$\begin{aligned}
\mathrm{Ker}\, d^r/\mathrm{Im}\, d^r &\simeq (Z^{r+1}_s + Z^{r-1}_{s-1})/(\partial Z^r_{s+r} + Z^{r-1}_{s-1}) \\
&\simeq Z^{r+1}_s/(Z^{r+1}_s \cap (\partial Z^r_{s+r} + Z^{r-1}_{s-1})) \\
&= Z^{r+1}_s/(\partial Z^r_{s+r} + Z^r_{s-1}) = E^{r+1}_s \,.
\end{aligned}$$

Therefore, we have obtained a canonical isomorphism $H_*(E^r) \simeq E^{r+1}$, and thus E is a spectral sequence.

We now compute the limit of this spectral sequence. By definition and the Noether isomorphism,

$$E_s^r = Z_s^r/(Z_{s-1}^{r-1} + \partial Z_{s+r-1}^{r-1}) \simeq (Z_s^r + F_{s-1}C)/(F_{s-1}C + \partial Z_{s+r-1}^{r-1}).$$

In the last expression, the numerators descrease as r increases and the denominators increase as r increases. Since the filtration F is bounded below, for a fixed pair s, t, $Z_{s,t}^r = Z_{s,t}^\infty$ for all large enough r. Thus, by definition, the limit equals

$$\begin{aligned}(\cap_r(Z_s^r + F_{s-1}C))/(\cup_r(F_{s-1}C + \partial Z_{s+r-1}^{r-1}))\\ = ((\cap_r Z_s^r) + F_{s-1}C)/(F_{s-1}C + \cup_r \partial Z_{s+r-1}^{r-1}).\end{aligned}$$

Since $\cup_s F_s C = C$, we have $\cup_r \partial Z_{s+r-1}^{r-1} = \partial C \cap F_s C$. Therefore, the limit term equals

$$(Z_s^\infty + F_{s-1}C)/(F_{s-1}C + (\partial C \cap F_s C)) \simeq Z_s^\infty/(Z_{s-1}^\infty + (\partial C \cap F_s C)) = E_s^\infty .$$

To show that the spectral sequence is convergent, note that, because the filtration F is bounded below, for fixed $s+t$, $E_{s,t}^r = 0$ for s small enough. Therefore, for fixed s and t, there exists r such that for $r' \geq r$, $E_{s,t}^{r'+1}$ is a quotient of $E_{s,t}^{r'}$, thus the spectral sequence is convergent.

To complete the proof, we interpret the limit E^∞ as Gr $H_*(C)$: By definition, Gr $H_*(C)_{s,t} = F_s H_{s+t}(C)/F_{s-1}H_{s+t}(C)$. Clearly, the graded module $F_s H_*(C) = Z_s^\infty/\partial C \cap F_s C$, and thus

$$\begin{aligned}F_s H_*(C)/F_{s-1}H_*(C) &= (Z_s^\infty/\partial C \cap F_s C)/(Z_{s-1}^\infty/\partial C \cap F_{s-1}C)\\ &\simeq Z_s^\infty/(Z_{s-1}^\infty + (\partial C \cap F_s C))\\ &= E_s^\infty . \qquad \square\end{aligned}$$

In Theorem E.4 note that, even in the most favorable circumstances, E^∞ does not determine $H_*(C)$ completely, but only up to module extensions.

It should be observed that the spectral sequence of the above theorem is functorial on the category of chain complexes with convergent chain filtrations which are bounded below. Combining this with Proposition E.2, we obtain the following result.

E.5 Corollary. *Let C and C' be chain complexes having convergent chain filtrations bounded below and let $\tau : C \to C'$ be a chain map preserving the filtrations. If, for some $r \geq 1$, the induced map $\tau^r : E^r \to E'^r$ is an isomorphism, then τ induces an isomorphism*

$$\tau_* : H_*(C) \simeq H_*(C').$$

Proof. By Proposition E.2, τ^∞ is an isomorphism. We have the following commutative diagram with exact rows:

$$\begin{array}{ccccccccc} 0 & \longrightarrow & F_{s-1}H_n(C) & \longrightarrow & F_sH_n(C) & \longrightarrow & E^{\infty}_{s,n-s} & \longrightarrow & 0 \\ & & \downarrow{\scriptstyle \tau_*} & & \downarrow{\scriptstyle \tau_*} & & \downarrow{\scriptstyle \tau^\infty} & & \\ 0 & \longrightarrow & F_{s-1}H_n(C') & \longrightarrow & F_sH_n(C') & \longrightarrow & E'^{\infty}_{s,n-s} & \longrightarrow & 0\,. \end{array}$$

For fixed n, $F_{s-1}H_n(C)$ and $F_{s-1}H_n(C')$ are both 0 for s small enough (because the filtrations are bounded below). It follows by induction on s, using the five lemma and the fact that τ^∞ is an isomorphism, that $\tau_* : F_sH_n(C) \simeq F_sH_n(C')$ for all s. Because the filtrations are convergent, $H_n(C) = \cup_s F_sH_n(C)$ and $H_n(C') = \cup_s F_sH_n(C')$, and so $\tau_* : H_n(C) \simeq H_n(C')$. $\square$

E.6 Example. Let C' and C'' be nonnegative chain complexes consisting of free R-modules with boundary operators ∂' and ∂'', respectively, and let $C = C' \otimes_R C''$ be their tensor product with the boundary operator ∂. Recall that

$$\partial(x \otimes y) = \partial' x \otimes y + (-1)^p x \otimes \partial'' y\,, \quad \text{for } x \in C'_p \text{ and } y \in C''.$$

There is a convergent filtration bounded below of C defined by

$$F_sC = \oplus_{q \leq s} C'_q \otimes_R C''.$$

For the corresponding spectral sequence,

$$E^1_{s,t} \simeq C'_s \otimes_R H_t(C''),$$

and $E^2_{s,t} \simeq H_s(C' \otimes_R H_t(C''))$, where $C' \otimes_R H_t(C'')$ is the chain complex under the differential

$$\partial(x \otimes y) = \partial' x \otimes y, \quad \text{for } x \in C' \text{ and } y \in H_t(C'').$$

A similar result is obtained by filtering the tensor product by the gradation of the second factor.

There is a completely parallel theory of cohomology spectral sequences.

E.7 Definition. A *cohomology spectral sequence* E is a sequence $\{E_r, d_r\}_{r\geq 0}$ such that

(a) E_r is a bigraded module over R and d_r is a differential of bidegree $(r, 1-r)$ on E_r.

(b) For $r \geq 0$, there is given a bigraded isomorphism $H(E_r) \simeq E_{r+1}$.

A homology spectral sequence is distinguished from a cohomology spectral sequence by using a different indexing convention. A homology, resp. cohomology, spectral sequence is denoted by E^r, resp. E_r.

The notion of a homomorphism $\varphi : E \to E'$ between cohomology spectral sequences is exactly parallel. Also, the same way as in E.1, we can define the limit E_∞ of a cohomology spectral sequence, which is a bigraded module. The spectral sequence E is said to be *convergent* if, for every s and t, there exists a nonnegative integer $r(s,t)$ such that for $r \geq r(s,t)$, $d_r : E_r^{s,t} \to E_r^{s+r,t-r+1}$ is trivial. In this case $E_{r+1}^{s,t}$ is isomorphic to a quotient of $E_r^{s,t}$ and $E_\infty^{s,t}$ is isomorphic to the direct limit of the sequence

$$E_{r(s,t)}^{s,t} \twoheadrightarrow E_{r(s,t)+1}^{s,t} \twoheadrightarrow \cdots .$$

The analogue of Proposition E.2 is true for cohomology spectral sequences.

Let M be a graded module with a filtration compatible with its gradation. Then we (loosely) say that a spectral sequence *abuts* (or *converges*) to M if $E_\infty \simeq \mathrm{Gr}\, M$ (as bigraded R-modules). This is denoted as $E_r \Rightarrow M$.

E.8 Definition. A *cochain filtration* F of a cochain complex $C = \{C^n\}_n$ is a decreasing filtration

$$\cdots \supset F^sC \supset F^{s+1}C \supset \cdots$$

of C compatible with the gradation of C as well as with the differential of C, i.e., F^sC is a cochain subcomplex of C consisting of $\{F^sC^t\}$. The filtration F of C induces a filtration of $H^*(C)$ defined by

$$F^sH^*(C) := \mathrm{Im}\,(H^*(F^sC) \to H^*(C)).$$

The filtration F of C is defined to be *convergent* if $\cup_s F^sC = C$ and $\cap_s F^sC = 0$. It is said to be *bounded above* if for each n there is $s(n)$ such that $F^{s(n)}C^n = 0$.

We have the following cohomological analogue of Theorem E.4.

E.9 Theorem. *Let F be a convergent cochain filtration bounded above of a cochain complex C. Then there is a convergent cohomology spectral sequence with*

$$E_1^{s,t} \simeq H^{s+t}(F^sC/F^{s+1}C),$$

and $d_1 : E_1^{s,t} \to E_1^{s+1,t}$ *corresponds to the coboundary operator in the long exact cohomology sequence associated to the short exact sequence of cochain complexes:*

$$(1) \qquad 0 \to F^{s+1}C/F^{s+2}C \to F^sC/F^{s+2}C \to F^sC/F^{s+1}C \to 0\,.$$

Moreover, E_∞ *is isomorphic to the bigraded module* Gr $H^*(C)$ *associated to the filtration* $F^sH^*(C) := \mathrm{Im}\,(H^*(F^sC) \to H^*(C))$.

Proof. For an arbitrary $r \geq 0$, define

$$Z_r^s := \{c \in F^sC : \partial c \in F^{s+r}C\}, \text{ and } Z_\infty^s := \{c \in F^sC : \partial c = 0\}.$$

These are graded modules with $Z_r^{s,t} = \{c \in F^sC^{s+t} : \partial c \in F^{s+r}C\}$ and $Z_\infty^{s,t} = \{c \in F^sC^{s+t} : \partial c = 0\}$. We define (for any $r \geq 1$):

$$E_r^s = Z_r^s/(Z_{r-1}^{s+1} + \partial Z_{r-1}^{s-r+1}), \text{ and } E_\infty^s = Z_\infty^s/(Z_\infty^{s+1} + (\partial C \cap F^sC)).$$

We also define $E_0^s = F^sC/F^{s+1}C$ and the map $d_0 : E_0^s \to E_0^s$ as the coboundary map of the quotient complex E_0^s.

The map ∂ sends Z_r^s to Z_r^{s+r} and $Z_{r-1}^{s+1} + \partial Z_{r-1}^{s-r+1}$ to ∂Z_{r-1}^{s+1}. Therefore, it induces a homomorphism (for any $r \geq 1$)

$$d_r : E_r^s \to E_r^{s+r}\,.$$

Then E_r is a bigraded module and d_r is a differential of bidegree $(r, 1-r)$ on it.

It is easy to see that $E_1^{s,t} \simeq H^{s+t}(F^sC/F^{s+1}C)$ by the Noether isomorphism. The assertion that $H^*(E_r) \simeq E_{r+1}$ is obtained similar to the corresponding fact in homology and so are the remaining assertions of the theorem. □

A particular example of the above theorem given by a double complex is often used. This generalizes Example E.6.

E.10 Example. A *double cochain complex* is a bigraded group

$$K^{*,*} = \oplus_{p,q\geq 0} K^{p,q}$$

together with differentials

$$d : K^{p,q} \to K^{p+1,q},\, \delta : K^{p,q} \to K^{p,q+1},$$

satisfying

$$d^2 = \delta^2 = d\delta + \delta d = 0.$$

The double complex is denoted by $(K^{*,*}; d, \delta)$. The *associated single cochain complex* $([K]^*, D)$ is defined by

$$[K]^n = \oplus_{p+q=n} K^{p,q}\ , \text{ and } D := d + \delta.$$

There are two cochain filtrations of $([K]^*, D)$ given by

$$'F^p[K]^n = \oplus_{p' \geq p} K^{p',n-p'}, \text{ and}$$

$$''F^p[K]^n = \oplus_{p' \geq p} K^{n-p',p'}.$$

Observe that both of these filtrations are convergent and bounded above. Thus, by Theorem E.9, there are two convergent spectral sequences $('E_r)$ and $(''E_r)$ both abutting to $H^*([K])$. Let us consider the first one. (The second one is similar by symmetry.) We have

$$'E_0^{p,q} = \frac{K^{p,q} + K^{p+1,q-1} + \cdots}{K^{p+1,q-1} + \cdots} \simeq K^{p,q},$$

and the differential d_0 is induced from D by passing to the quotient. Thus, under the above isomorphism, $d_0 = \delta$ and

$$'E_1^{p,q} \simeq H_\delta^q(K^{p,*}),$$

where the right side denotes the q-th cohomology of the complex:

$$\cdots \to K^{p,q-1} \xrightarrow{\delta} K^{p,q} \xrightarrow{\delta} K^{p,q+1} \to \cdots .$$

The differential d_1 is computed from $D = d + \delta$ on $'E_1$. Since $\delta = 0$ on $'E_1$ we see that $d_1 = d$ and thus

$$'E_2^{p,q} \simeq H^p('E_1^{*,q}, d_1) \simeq H_d^p(H_\delta^q(K^{*,*})).$$

The last expression denotes the p-th cohomology of

$$\cdots \to H_\delta^q(K^{p-1,*}) \xrightarrow{\bar{d}} H_\delta^q(K^{p,*}) \xrightarrow{\bar{d}} H_\delta^q(K^{p+1,*}) \to \cdots ,$$

where $\bar{d}$ is induced from d, which is possible since $\delta d + d\delta = 0$. Summarizing:

Associated to a bigraded cochain complex $(K^{*,*}; d, \delta)$ are two spectral sequences both abutting to the cohomology of the total complex $[K]$ and where

$$'E_2^{p,q} \simeq H_d^p(H_\delta^q(K^{*,*})), \text{ and}$$

$$''E_2^{p,q} \simeq H_\delta^p(H_d^q(K^{*,*})).$$

We recall the following fundamental Leray–Serre spectral sequence for a fibration. For a proof see, e.g., [Spanier–66, Chap. 9, §4].

Let M be a $\mathbb{Z}$-module. For any topological pair (B, A), let $H^*(B, A, M)$ denote the singular cohomology of the pair with coefficients in M.

E.11 Theorem. *Let $\pi : E \to B$ be a fibration over a connected simply-connected base B and let $F := \pi^{-1}(b_o)$ be a fiber. Given any subspace $A \subset B$, there is a convergent cohomology spectral sequence with*

$$E_2^{s,t} \simeq H^s(B, A, H^t(F, M))$$

and abutting to $H^*(E, \pi^{-1}A, M)$.

In fact, the theorem is true more generally for any "orientable" fibration (with no simply-connectedness assumption on the base).

We recall the following Hochschild–Serre spectral sequence (cf. [Hochschild–Serre–53] for the cohomology spectral sequence and [Cartan–Eilenberg–56, Chap. XVI, §6] for both).

E.12 Theorem. *Let $\mathfrak{s}$ be a (not necessarily finite-dimensional) Lie algebra, $\mathfrak{t}$ be an ideal and let M be a $\mathfrak{s}$-module. There exists a convergent homology spectral sequence with*

$$E^2_{p,q} \simeq H_p(\mathfrak{s}/\mathfrak{t}, H_t(\mathfrak{t}, M))$$

and abutting to $H_*(\mathfrak{s}, M)$.

Similarly, there is a convergent cohomology spectral sequence with

$$E_2^{p,q} \simeq H^p(\mathfrak{s}/\mathfrak{t}, H^t(\mathfrak{t}, M))$$

and abutting to $H^*(\mathfrak{s}, M)$.

(Observe that , by Subsections 3.1.1–3.1.2, $\mathfrak{t}$ acts trivially on $H_*(\mathfrak{t}, M)$ and also on $H^*(\mathfrak{t}, M)$. Thus, these are modules for the quotient Lie algebra $\mathfrak{s}/\mathfrak{t}$.)

We also recall the following from [Hochschild–Serre–53]. Though the following theorem is proved in loc. cit. under the additional assumption that $\mathfrak{s}$ is finite-dimensional, and only for the Lie algebra cohomology, the same proof applies.

E.13 Theorem. *Let $\mathfrak{s}$ be a Lie algebra, $\mathfrak{t}$ be a finite-dimensional subalgebra and let M be a $\mathfrak{s}$-module which is finitely semisimple as a $\mathfrak{t}$-module (cf. 3.1.6). Assume further that the adjoint action of $\mathfrak{t}$ on $\mathfrak{s}$ is finitely semisimple. Then, there exists a convergent homology spectral sequence with*

$$E^2_{p,q} \simeq H_p(\mathfrak{s}, \mathfrak{t}, M) \otimes_{\mathbb{C}} H_q(\mathfrak{t}, \mathbb{C})$$

and abutting to $H_*(\mathfrak{s}, M)$, where $H_*(\mathfrak{s}, \mathfrak{t}, M)$ denotes the (Chevalley–Eilenberg) Lie algebra homology of the pair $(\mathfrak{s}, \mathfrak{t})$ (cf. 3.1.3).

Similarly, there is a convergent cohomology spectral sequence with

$$E_2^{p,q} \simeq H^p(\mathfrak{s}, \mathfrak{t}, M) \otimes_{\mathbb{C}} H^q(\mathfrak{t}, \mathbb{C})$$

and abutting to $H^*(\mathfrak{s}, M)$.

Bibliography

Abe, E., Morita, J.

[88] Some Tits systems with affine Weyl groups in Chevalley groups over Dedekind domains, *J. Algebra* **115** (1988), 450–465.

Abe, E., Takeuchi, M.

[92] Groups associated with some types of infinite dimensional Lie algebras, *J. Algebra* **146** (1992), 385–404.

Adams, J.F.

[69] *Lectures on Lie Groups*, W.A. Benjamin, Inc., 1969.

Adler, M., van Moerbeke, P.

[80] Completely integrable systems, Euclidean Lie algebras, and curves, *Advances in Math.* **38** (1980), 267–317.

[91] The Toda lattice, Dynkin diagrams, singularities and abelian varieties, *Invent. Math.* **103** (1991), 223–278.

Akyildiz, E., Carrell, J.B., Lieberman, D.I.

[86] Zeroes of holomorphic vector fields on singular spaces and intersection rings of Schubert varieties, *Compositio Math.* **57** (1986), 237–248.

Allday, C., Puppe, V.

[93] *Cohomological Methods in Transformation Groups*, Cambridge Studies in Advanced Mathematics 32, Cambridge University Press, 1993.

Altman, A., Kleiman, S.

[70] *Introduction to Grothendieck Duality Theory*, Lecture Notes in Math. 146, Springer-Verlag, 1970.

Andersen, H.H.

[80] Vanishing theorems and induced representations, *J. Algebra* **62** (1980), 86–100.

[85] Schubert varieties and Demazure's character formula, *Invent. Math.* **79** (1985), 611–618.

Andersen, H.H., Jantzen, J.C., Soergel, W.

[94] Representations of quantum groups at a p-th root of unity and of semisimple groups in characteristic p: independence of p, *Astérisque* **220** (1994), 1–321.

Andrews, G.E.

[84] Hecke modular forms and the Kac–Peterson identities, *Trans. Amer. Math. Soc.* **283** (1984), 451–458.

Arabia, A.

[89] Cohomologie T-équivariante de la variété de drapeaux d'un groupe de Kač–Moody, *Bull. Soc. Math. France* **117** (1989), 129–165.

[98] Classes d'Euler équivariantes et points rationnellement lisses, *Ann. Inst. Fourier, Grenoble* **48** (1998), 861–912.

Arbarello, E., De Concini, C., Kac, V.G.

[89] The infinite wedge representation and the reciprocity law for algebraic curves, *Proc. Symp. Pure Math.* **49**, Part 1 (1989), AMS, pp. 171–190.

Arbarello, E., De Concini, C., Kac, V.G., Procesi, C.

[88] Moduli spaces of curves and representation theory, *Comm. Math. Phys.* **117** (1988), 1–36.

Atiyah, M.F., Hirzebruch, F.

[61] Vector bundles and homogeneous spaces, in: *Proc. Symp. Pure Math.* **3**, AMS, 1961, pp. 7–38.

Atiyah, M.F., Macdonald, I.G.

[69] *Introduction to Commutative Algebra*, Addison-Wesley, 1969.

Atiyah, M.F., Pressley, A.N.

[83] Convexity and loop groups, in: *Arithmetic and Geometry*, Vol. II, Progress in Math. 36, Birkhäuser, 1983, pp. 33–64.

Baum, P., Fulton, W., MacPherson, R.

[75] Riemann–Roch for singular varieties, *Publ. Math. IHES* **45** (1975), 101–145.

Bausch, J., Rousseau, G.

[89] *Algèbres de Kac–Moody Affines (Automorphismes et Formes Réelles)*, Institut E. Cartan 11, Paris, 1989.

Beauville, A., Laszlo, Y.

[94] Conformal blocks and generalized theta functions, *Comm. Math. Phys.* **164** (1994), 385–419.

Beilinson, A., Bernstein, J.

[81] Localisation de $\mathfrak{g}$-modules, *C.R. Acad. Sci. Paris* **292** (1981), 15–18.

Berman, S.

[76] On derivations of Lie algebras, *Canad. J. Math.* **28** (1976), 174–180.

[89] On generators and relations for certain involutory subalgebras of Kac–Moody Lie algebras, *Comm. in Alg.* **17** (1989), 3165–3185.

Berman, S., Lee, Y.S., Moody, R.V.

[89] The spectrum of a Coxeter transformation, affine Coxeter transformations, and the defect map, *J. Algebra* **121** (1989), 339–357.

Berman, S., Moody, R.V.

[79] Lie algebra multiplicities, *Proc. Amer. Math. Soc.* **76** (1979), 223–228.

Berman, S., Pianzola, A.

[87] Generators and relations for real forms of some Kac–Moody Lie algebras, *Comm. in Alg.* **15** (1987), 935–959.

Bernard, D., Felder, G.

[90] Fock representations and BRST cohomology in *SL(2)* current algebra, *Comm. Math. Phys.* **127** (1990), 145–168.

Bernstein, I.N., Gelfand, I.M., Gelfand, S.I.

[71] Structure of representations generated by vectors of highest weight, *Funct. Anal. Appl.* **5** (1971), 1–8.

[73] Schubert cells and cohomology of the spaces G/P, *Russ. Math. Surv.* **28** (1973), 1–26.

[75] Differential operators on the base affine space and a study of $\mathfrak{g}$-modules, in: *Lie groups and Their Representations* (I.M. Gelfand, ed.), Summer school of the Bolyai János Math. Soc., Halsted Press, 1975, pp. 21–64.

[76] Category of $\mathfrak{g}$-modules, *Funct. Anal. Appl.* **10** (1976), 87–92.

Billey, S.C.

[98] Pattern avoidance and rational smoothness of Schubert varieties, *Advances in Math.* **139** (1998), 141–156.

[99] Kostant polynomials and the cohomology ring for G/B, *Duke Math. J.* **96** (1999), 205–224.

Billey, S., Lakshmibai, V.

[2000] *Singular Loci of Schubert Varieties*, Prog. Math. 182, Birkhäuser, 2000.

Billey, S.C., Warrington, G.S.

[2001] Maximal singular loci of Schubert varieties in $\mathrm{SL}(n)/B$, preprint (2001).

Birkhoff, G.D.

[13] A theorem on matrices of analytic functions, *Math. Annalen* **74** (1913), 122–133.

Björner, A., Wachs, M.

[82] Bruhat order of Coxeter groups and shellability, *Advances in Math.* **43** (1982), 87–100.

Boe, B.D.

[88] Kazhdan–Lusztig polynomials for Hermitian symmetric spaces, *Trans. Amer. Math. Soc.* **309** (1988), 279–294.

Borcherds, R.

[86] Vertex algebras, Kac–Moody algebras, and the Monster, *Proc. Natl. Acad. Sci. USA* **83** (1986), 3068–3071.

[88] Generalized Kac–Moody algebras, *J. Algebra* **115** (1988), 501–512.

Borel, A.

[54] Représentations linéaires et espaces homogènes kähleriens des groupes simples compacts (1954); in Borel's *Collected Papers I*, Springer-Verlag, 1983, pp. 392–396.

[55] Topology of Lie groups and characteristic classes, *Bull. Amer. Math. Soc.* **61** (1955), 397–432.

[60] *Seminar on Transformation Groups*, Annals of Math. Studies 46, Princeton University Press, Princeton, NJ, 1960.

[67] *Topics in the Homology Theory of Fibre Bundles*, Lecture Notes in Math. 36, Springer-Verlag, 1967.

[91] *Linear Algebraic Groups*, GTM 126, Springer-Verlag, 1991.

Borel, A., Tits, J.

[65] Groupes réductifs, *Publ. Math. IHES* **27** (1965), 55–152.

Borho, W., Jantzen, J.C.

[77] Über primitive ideale in der einhüllenden einer halbeinfachen Lie-algebra, *Invent. Math.* **39** (1977), 1–53.

Bott, R.

[56] An application of the Morse theory to the topology of Lie groups, *Bull. Soc. Math. France* **84** (1956), 251–282.

[57] Homogeneous vector bundles, *Annals Math.* **66** (1957), 203–248.

[58] The space of loops on a Lie group, *Michigan Math. J.* **5** (1958), 35–61.

[59] The stable homotopy of the classical groups, *Annals Math.* **70** (1959), 313–337.

Bott, R., Samelson, H.

[58] Applications of the theory of Morse to symmetric spaces, *Amer. J. Math.* **80** (1958), 964–1029.

Bourbaki, N.

[61] *Algèbre Commutative, Ch. 1–2*, Hermann, Paris, 1961.

[62] *Algèbre Linéaire, Ch. 2*, 3rd ed., Hermann, Paris, 1962.

[71] *Éléments de Mathématique, Topologie Générale, Chapitres 1 à 4*, Hermann, Paris, 1971.

[75a] *Groupes et Algèbres de Lie, Ch. 7–8*, Hermann, Paris, 1975.

[75b] *Lie Groups and Lie Algebras, Part I, Chaps. 1–3*, Hermann, Paris, 1975.

[81] *Groupes et Algèbres de Lie, Ch. 4–6*, Masson, Paris, 1981.

Boutot, J.-F.

[87] Singularités rationnelles et quotients par les groupes réductifs, *Invent. Math.* **88** (1987), 65–68.

Bredon, G.E.

[97] *Sheaf Theory*, GTM 170, Springer-Verlag, 1997.

Brion, M.

[93] Variétés sphériques et théorie de Mori, *Duke Math. J.* **72** (1993), 369–404.

[98] Equivariant cohomology and equivariant intersection theory, in: *Representation Theories and Algebraic Geometry* (A. Broer et al., eds.), Kluwer Academic Publishers, 1998, pp. 1–37.

[99] Rational smoothness and fixed points of torus actions, *Transformation Groups* **4** (1999), 127–156.

Bröcker, T., tom Dieck, T.

[85] *Representations of Compact Lie Groups*, GTM 98, Springer-Verlag, 1985.

Brown, K.S.

[89] *Buildings*, Springer-Verlag, 1989.

Brylinski, J.-L.

[81] Differential operators on the flag varieties, *Astérisque* **87–88** (1981), 43–60.

[90] Representations of loop groups, Dirac operators on loop space, and modular forms, *Topology* **29** (1990), 461–480.

Brylinski, J.-L., Kashiwara, M.

[81] Kazhdan–Lusztig conjecture and holonomic systems, *Invent. Math.* **64** (1981), 387–410.

Buchsbaum, D.A.

[59] A note on homology in categories, *Annals Math.* **69** (1959), 66–74.

Carrell, J.B.

[91] Vector fields, flag varieties and Schubert calculus, in: *Proceedings of the Hyderabad Conference on Algebraic Groups* (S. Ramanan, ed.), Manoj Prakashan, 1991, pp. 23–57.

[94] The Bruhat graph of a Coxeter group, a conjecture of Deodhar, and rational smoothness of Schubert varieites, in: *Proc. Symp. Pure Math.* **56**, Part 1 (W.J. Haboush and B.J. Parshall, eds.), AMS, 1994, pp. 53–61.

[95] On the smooth points of a Schubert variety, in: *Representations of Groups*, CMS Conf. Proc. **16**, 1995, pp. 15–33.

[97] The span of the tangent cone of a Schubert variety, in: *Algebraic Groups and Lie Groups*, Australian Math. Soc. Lect. Ser. **9**, Cambridge University Press, Cambridge, 1997, pp. 51–59.

Carrell, J.B., Kuttler, J.

[99] On the smooth points of a T-stable subvariety of G/B and the Peterson map, preprint (1999).

Carrell, J.B., Lieberman, D.I.

[77] Vector fields and Chern numbers, *Math. Annalen* **225** (1977), 263–273.

Cartan, H., Eilenberg, S.

[56] *Homological Algebra*, Princeton Univ. Press, Princeton, N.J., 1956.

Casian, L.

[96] Proof of the Kazhdan–Lusztig conjecture for Kac–Moody algebras (The characters $\mathrm{ch}L_{w\rho-\rho}$), *Advances in Math.* **119** (1996), 207–281.

Chari, V.

[85] Annihilators of Verma modules of Kac–Moody Lie algebras, *Invent. Math.* **81** (1985), 47–58.

[86] Integrable representations of affine Lie-algebras, *Invent. Math.* **85** (1986), 317–335.

Chari, V., Ilangovan, S.

[84] On the Harishchandra homomorphism for infinite-dimensional Lie algebras, *J. Algebra* **90** (1984), 476–490.

Chari, V., Pressley, A.

[86] New unitary representations of loop groups, *Math. Annalen* **275** (1986), 87–104.

[87] A new family of irreducible, integrable modules for affine Lie algebras, *Math. Annalen* **277** (1987), 543–562.

[88] Integrable representations of twisted affine Lie algebras, *J. Algebra* **113** (1988), 438–464.

[89] Integrable representations of Kac–Moody algebras: Results and open problems, in: *Infinite Dimensional Lie Algebras and Groups*, Adv. Ser. in Math. Phys. **7**, World Sci., 1989, pp. 3–24.

Chevalley, C.

[48] Sur la classification des algèbres de Lie simples et de leurs représentations, *C.R. Acad. Sci. Paris* **227** (1948), 1136–1138.

Chevalley, C., Eilenberg, S.

[48] Cohomology theory of Lie groups and Lie algebras, *Trans. Amer. Math. Soc.* **63** (1948), 85–124.

Cline, E., Parshall, B., Scott, L.

[78] Induced modules and extensions of representations, *Invent. Math.* **47** (1978), 41–51.

[80] Cohomology, hyperalgebras, and representations, *J. Algebra* **63** (1980), 98–123.

Coleman, A.J.

[58] The Betti numbers of the simple Lie groups, *Canad. J. Math.* **10** (1958), 349–356.

[89] Killing and the Coxeter transformation of Kac–Moody algebras, *Invent. Math.* **95** (1989), 447–477.

Coleman, A.J., Howard, M.

[89] Root multiplicities for general Kac–Moody algebras, *C.R. Math. Rep. Acad. Sci. Canada* **11** (1989), 15–18.

Conway, J.H., Norton, S.P.

[79] Monstrous moonshine, *Bull. London Math. Soc.* **11** (1979), 308–339.

Dadok, J., Kac, V.

[85] Polar representations, *J. Algebra* **92** (1985), 504–524.

Date, E., Jimbo, M., Kashiwara, M., Miwa, T.

[81] Transformation groups for soliton equations III. Operator approach to the Kadomtsev–Petviashvili equation, *J. Phys. Soc. Japan* **50** (1981), 3806–3812.

[82a] Transformation groups for soliton equations IV. A new hierarchy of soliton equations of KP-type, *Physica* **D 4** (1982), 343–365.

[82b] Transformation groups for soliton equations. Euclidean Lie algebras and reduction of the KP hierarchy, *Publ. RIMS* **18** (1982), 1077–1110.

[83] Transformation groups for soliton equations, in: *Nonlinear Integrable Systems–Classical Theory and Quantum Theory*, World Sci., 1983, pp. 39–119.

De Concini, C., Kac, V.G., Kazhdan, D.A.

[89] Boson-fermion correspondence over $\mathbb{Z}$, in: *Infinite Dimensional Lie Algebras and Groups*, Adv. Ser. in Math. Phys. 7, World Sci., 1989, pp. 124–137.

Demazure, M.

[68] Une démonstration algébrique d'un théorème de Bott, *Invent. Math.* **5** (1968), 349–356.

[74] Désingularisation des variétés de Schubert généralisées, *Ann. Sci. Éc. Norm. Sup.* **7** (1974), 53–88.

[76] A very simple proof of Bott's theorem, *Invent. Math.* **33** (1976), 271–272.

Demazure, M., Gabriel, P.

[70] *Groupes Algébriques Tome I*, Masson & Cie, Paris, 1970.

Deodhar, V.V.

[77] Some characterizations of Bruhat ordering on a Coxeter group and determination of the relative Möbius function, *Invent. Math.* **39** (1977), 187–198.

[82a] On the root system of a Coxeter group, *Comm. in Alg.* **10** (1982), 611–630.

[82b] On the Kazhdan–Lusztig conjectures, *Indag. Math.* **44** (1982), 1–17.

[85] Local Poincaré duality and non-singularity of Schubert varieties, *Comm. in Alg.* **13** (1985), 1379–1388.

[86] Some characterizations of Coxeter groups, *L'Enseign. Math.* **32** (1986), 111–120.

[94] A brief survey of Kazhdan–Lusztig theory and related topics, in: *Proc. Symp. Pure Math.* **56**, Part 1 (W.J. Haboush and B.J. Parshall, eds.), AMS, 1994, pp. 105–124.

Deodhar, V.V., Gabber, O., Kac, V.

[82] Structure of some categories of representations of infinite-dimensional Lie algebras, *Advances in Math.* **45** (1982), 92–116.

Deodhar, V.V., Kumaresan, S.

[86] A finiteness theorem for affine Lie algebras, *J. Algebra* **103** (1986), 403–426.

Dixmier, J.

[74] *Algèbres Enveloppantes*, Gauthier-Villars, Paris, 1974. English translation: *Enveloping Algebras*, Minerva Translations, North Holland, 1977.

Dyer, M.J.

[93] The nil Hecke ring and Deodhar's conjecture on Bruhat intervals, *Invent. Math.* **111** (1993), 571–574.

Dynkin, E.B.

[57] Semisimple subalgebras of semisimple Lie algebras, *Amer. Math. Soc. Transl. Ser. II* **6** (1957), 111–244.

Dyson, F.J.

[72] Missed opportunities, *Bull. Amer. Math. Soc.* **78** (1972), 635–652.

Eisenbud, D.

[95] *Commutative Algebra with a View Toward Algebraic Geometry*, GTM 150, Springer-Verlag, 1995.

Enright, T.J.

[79] Fundamental series of a semisimple Lie algebra, *Annals Math.* **110** (1979), 1–82.

Evens, S., Bressler, P.

[87] On certain Hecke rings, *Proc. Natl. Acad. Sci. USA* **84** (1987), 624–625.

Evens, S., Lu, J-H.

[99] Poisson harmonic forms, Kostant harmonic forms, and the S^1-equivariant cohomology of K/T, *Advances in Math.* **142** (1999), 171–220.

Faltings, G.

[94] A proof for the Verlinde formula, *J. Alg. Geom.* **3** (1994), 347–374.

Feigin, B.L., Frenkel, E.V.

[90a] Representations of affine Kac–Moody algebras and bosonization, in: *Physics and Mathematics of Strings*, World Sci., 1990, pp. 271–316.

[90b] Representations of affine Kac–Moody algebras, bosonization and resolutions, *Lett. Math. Phys.* **19** (1990), 307–317.

[90c] Affine Kac–Moody algebras and semi-infinite flag manifolds, *Comm. Math. Phys.* **128** (1990), 161–189.

Feigin, B.L., Fuchs, D.B.

[88] Cohomology of some nilpotent subalgebras of the Virasoro and Kac–Moody Lie algebras, *J. Geom. Phys.* **5** (1988), 209–235.

Feingold, A.J.

[80] A hyperbolic GCM Lie algebra and the Fibonacci numbers, *Proc. Amer. Math. Soc.* **80** (1980), 379–385.

Feingold, A.J., Frenkel, I.B.

[83] A hyperbolic Kac–Moody algebra and the theory of Siegel modular forms of genus 2, *Math. Annalen* **263** (1983), 87–144.

[85] Classical affine algebras, *Advances in Math.* **56** (1985), 117–172.

Fialowski, A.

[84] Deformations of nilpotent Kac–Moody algebras, *Studia Sci. Math. Hung.* **19** (1984), 465–483.

Finkelberg, M.

[96] An equivalence of fusion categories, *Geom. Funct. Anal.* **6** (1996), 249–267.

Foda, O., Misra, K.C., Okado, M.

[98] Demazure modules and vertex models: The $\widehat{s\ell}(2)$ case, *J. Math. Phys.* **39** (1998), 1601–1622.

Freed, D.S.

[85] Flag manifolds and infinite dimensional Kähler geometry, in: *Infinite Dimensional groups with applications*, MSRI publ. **4**, Springer-Verlag, 1985, pp. 83–124.

Frenkel, I.B.

[81] Two constructions of affine Lie algebra representations and boson-fermion correspondence in quantum field theory, *J. Funct. Anal.* **44** (1981), 259–327.

[82] 'Representations of affine Lie algebras, Hecke modular forms and Korteweg-de Vries type equations,' in: Lecture Notes in Math. 933, Springer-Verlag, 1982, pp. 71–110.

[84] Orbital theory for affine Lie algebras, *Invent. Math.* **77** (1984), 301–352.

[85] 'Representations of Kac–Moody algebras and dual resonance models,' in: Lectures in Applied Math. 21, 1985, pp. 325–353.

Frenkel, I.B., Garland, H., Zuckerman, G.J.

[86] Semi-infinite cohomology and string theory, *Proc. Natl. Acad. Sci. USA* **83** (1986), 8442–8446.

Frenkel, I.B., Kac, V.G.

[80] Basic representations of affine Lie algebras and dual resonance models, *Invent. Math.* **62** (1980), 23–66.

Frenkel, I., Lepowsky, J., Meurman, A.

[88] *Vertex Operator Algebras and the Monster*, Pure and Applied Math. 134, Academic Press, 1988.

Fulton, W.

[97] *Young Tableaux*, Cambridge University Press, 1997.

[98] *Intersection Theory*, Second edition, Springer-Verlag, 1998.

Fulton, W., MacPherson, R., Sottile, F., Sturmfels, B.

[95] Intersection theory on spherical varieties, *J. Algebraic Geom.* **4** (1995), 181–193.

Gabber, O., Kac, V.G.

[81] On defining relations of certain infinite-dimensional Lie algebras, *Bull. Amer. Math. Soc.* **5** (1981), 185–189.

Garfinkle, D.

[82] *A New Construction of the Joseph Ideal*, Ph.D. Thesis, M.I.T., 1982.

Garland, H.

[75] Dedekind's η-function and the cohomology of infinite dimensional Lie algebras, *Proc. Natl. Acad. Sci. USA* **72** (1975), 2493–2495.

[78] The arithmetic theory of loop algebras, *J. Algebra* **53** (1978), 480–551.

[80] The arithmetic theory of loop groups, *Publ. Math. IHES* **52** (1980), 5–136.

Garland, H., Lepowsky, J.

[76] Lie algebra homology and the Macdonald–Kac formulas, *Invent. Math.* **34** (1976), 37–76.

Garland, H., Raghunathan, M.S.

[75] A Bruhat decomposition for the loop space of a compact group: A new approach to results of Bott, *Proc. Natl. Acad. Sci. USA* **72** (1975), 4716–4717.

Goddard, P., Kent, A., Olive, D.

[86] Unitary representations of the Virasoro and super-Virasoro algebras, *Comm. Math. Phys.* **103** (1986), 105–119.

Godement, R.

[58] *Topologie Algébrique et Théorie des Faisceaux*, Hermann, Paris, 1958.

Goodman, R., Wallach, N.R.

[84] Structure and unitary cocycle representations of loop groups and the group of diffeomorphisms of the circle, *J. für die Reine und Angewandte Math.* **347** (1984), 69–133.

Graham, W.

[97] The class of the diagonal in flag bundles, *J. Diff. Geometry* **45** (1997), 471–487.

[2001] Positivity in equivariant Schubert calculus, *Duke Math. J.* **109** (2001), 599–614.

Grauert, H., Riemenschneider, O.

[70a] 'Verschwindungssätze für analytische kohomologiegruppen auf komplexen räumen,' in: Lecture Notes in Math. 155, Springer-Verlag, 1970, pp. 97–109.

[70b] Verschwindungssätze für analytische kohomologiegruppen auf komplexen räumen, *Invent. Math.* **11** (1970), 263–292.

Griffiths, P., Harris, J.

[78] *Principles of Algebraic Geometry*, John Wiley & Sons, Inc., New York, 1978.

Grothendieck, A.

[57] Sur quelques points d'algèbre homologique, *Tôhoku Math. J.* **9** (1957), 119–221.

[57a] Sur la classification des fibres holomorphes sur la sphere de Riemann, *Amer. J. Math.* **79** (1957), 121–138.

[58] Sur quelques propriétés fondamentales en théorie des intersections, in: *Anneaux de Chow et Applications*, Séminaire C. Chevalley 2e année, 1958, pp. 4-01–4-36.

[61] Éléments de géométrie algébrique III, *Publ. Math. IHES* **11** (1961).

[65] Éléments de géométrie algébrique IV (Seconde Partie), *Publ. Math. IHES* **24** (1965).

[67] *Local Cohomology*, Lecture Notes in Math. 41, Springer-Verlag, 1967.

Gutkin, E., Slodowy, P.

[83] Cohomologie des variétés de drapeaux infinies, *C. R. Acad. Sci. Paris* **296** (1983), 625–627.

Haddad, Z.S.

[84] *Infinite-dimensional Flag Varieties*, Ph. D. Thesis, MIT, 1984.

Hammack, R.

[96] *A Homotopy Equivalence Between Spaces of Algebraic and Holomorphic Maps from an Affine Curve to a Flag Manifold*, Ph. D. Thesis, University of North Carolina, Chapel Hill, 1996.

Hansen, H.C.

[73] On cycles in flag manifolds, *Math. Scand.* **33** (1973), 269–274.

Harish-Chandra

[51] On some applications of the universal enveloping algebra of a semisimple Lie algebra, *Trans. Amer. Math. Soc.* **70** (1951), 28–96.

Hartshorne, R.

[66] *Residues and Duality*, Lecture Notes in Math. 20, Springer-Verlag, 1966.

[70] *Ample Subvarieties of Algebraic Varieties*, Lecture Notes in Math. 156, Springer-Verlag, 1970.

[77] *Algebraic Geometry*, GTM 52, Springer-Verlag, 1977.

Hayashi, T.

[88] Sugawara operators and Kac–Kazhdan conjecture, *Invent. Math.* **94** (1988), 13–52.

Helgason, S.

[78] *Differential Geometry, Lie Groups, and Symmetric Spaces*, Academic Press, 1978.

Hiller, H.

[82] *Geometry of Coxeter Groups*, Pitman, Boston, 1982.

Hilton, P.J., Stammbach, U.

[97] *A Course in Homological Algebra*, Second edition, GTM 4, Springer-Verlag, 1997.

Hirschowitz, A.

[84] Le groupe de Chow équivariant, *C. R. Acad. Sci. Paris* **298** (1984), 87–89.

Hochschild, G.

[56] Relative homological algebra, *Trans. Amer. Math. Soc.* **82** (1956), 246–269.

[81] *Basic Theory of Algebraic Groups and Lie Algebras*, GTM 75, Springer-Verlag, 1981.

Hochschild, G., Serre, J-P.

[53] Cohomology of Lie algebras, *Annals Math.* **57** (1953), 591–603.

Hsiang, W.Y.

[75] *Cohomology Theory of Topological Transformation Groups*, Springer-Verlag, 1975.

Humphreys, J.E.

[72] *Introduction to Lie Algebras and Representation Theory*, GTM 9, Springer-Verlag, 1972.

[90] *Reflection Groups and Coxeter Groups*, Cambridge Studies in Advanced Mathematics 29, Cambridge University Press, 1990.

[95] *Linear Algebraic Groups*, GTM 21, Springer-Verlag, 1995.

Husemoller, D.

[94] *Fibre Bundles*, Third edition, GTM 20, Springer-Verlag, 1994.

Iwahori, N., Matsumoto, H.

[65] On some Bruhat decomposition and the structure of the Hecke rings of $\mathfrak{p}$-adic Chevalley groups, *Publ. Math. IHES* **25** (1965), 5–48.

Jacobson, N.

[62] *Lie Algebras*, Interscience Publishers, New York, 1962.

Jantzen, J.C.

[79] *Moduln mit Einem Höchsten Gewicht*, Lecture Notes in Math. 750, Springer-Verlag, 1979.

[87] *Representations of Algebraic Groups*, Pure and Applied Mathematics 131, Academic Press, Inc., 1987.

Jimbo, M., Miwa, T.

[84] Irreducible decomposition of fundamental modules for $A_\ell^{(1)}$ and $C_\ell^{(1)}$, and Hecke modular forms, *Adv. Stud. Pure Math.* **4** (1984), 97–119.

[85] On a duality of branching rules for affine Lie algebras, *Adv. Stud. Pure Math.* **6** (1985), 17–65.

Joseph, A.

[84] On the variety of a highest weight module, *J. Algebra* **88** (1984), 238–278.

[85] On the Demazure character formula, *Ann. Sci. Éc. Norm. Sup.* **18** (1985), 389–419.

[86] On the Demazure character formula II–generic homologies, *Compositio Math.* **58** (1986), 259–278.

Kac, V.G.

[67] Simple graded Lie algebras of finite growth, *Funct. Anal. Appl.* **1** (1967), 328–329.

[68a] Graded Lie algebras and symmetric spaces, *Funct. Anal. Appl.* **2** (1968), 183–184.

[68b] Simple irreducible graded Lie algebras of finite growth, *Math. USSR Izvestija* **2** (1968), 1271–1311.

[69a] Automorphisms of finite order of semisimple Lie algebras, *Funct. Anal. Appl.* **3** (1969), 252–254.

[69b] An algebraic definition of compact Lie groups, *Trudy MIEM* **5** (1969), 36–47.

[74] Infinite-dimensioned Lie algebras and Dedekind's η-function, *Funct. Anal. Appl.* **8** (1974), 68–70.

[77] Lie superalgebras, *Advances in Math.* **26** (1977), 8–96.

[78a] Infinite-dimensional algebras, Dedekind's η-function, classical Möbius function and the very strange formula, *Advances in Math.* **30** (1978), 85–136.

[78b] Highest weight representations of infinite-dimensional Lie algebras, in: *Proceeding of ICM Helsinki*, 1978, pp. 299–304.

[80a] An elucidation of "infinite-dimensional algebras ... and the very strange formula." $E_8^{(1)}$ and the cube root of the modular invariant j, *Advances in Math.* **35** (1980), 264–273.

[80b] On simplicity of certain infinite dimensional Lie algebras, *Bull. Amer. Math. Soc.* **2** (1980), 311–314.

[80c] Infinite root systems, representations of graphs and invariant theory, *Invent. Math.* **56** (1980), 57–92.

[80d] Some remarks on nilpotent orbits, *J. Algebra* **64** (1980), 190–213.

[82] 'Some problems on infinite dimensional Lie algebras and their representations,' in: Lecture Notes in Mathematics 933, Springer-Verlag, 1982, pp. 117–126.

[84] Laplace operators of infinite-dimensional Lie algebras and theta functions, *Proc. Natl. Acad. Sci. USA* **81** (1984), 645–647.

[85a] Torsion in cohomology of compact Lie groups and Chow rings of reductive algebraic groups, *Invent. Math.* **80** (1985), 69–79.

[85b] Constructing groups associated to infinite dimensional Lie algebras, in: *Infinite Dimensional Groups with Applications*, MSRI publ. 4, Springer-Verlag, 1985, pp. 167–216.

[88] Modular invariance in mathematics and physics, *Address at the Centennial of the AMS*, 1988.

[90] *Infinite Dimensional Lie Algebras*, Third edition, Cambridge University Press, 1990.

[97] *Vertex Algebras for Beginners*, University Lecture Series 10, AMS, 1997.

Kac, V.G., Kazhdan, D.A.

[79] Structure of representations with highest weight of infinite-dimensional Lie algebras, *Advances in Math.* **34** (1979), 97–108.

Kac, V.G., Kazhdan, D.A., Lepowsky, J., Wilson, R.L.

[81] Realization of the basic representation of the Euclidean Lie algebras, *Advances in Math.* **42** (1981), 83–112.

Kac, V.G., Moody, R.V., Wakimoto, M.

[88] On E_{10}, in: *Proceedings of the 1987 Conference on Differential-geometrical Methods in Physics*, Kluwer, 1988, pp. 109–128.

Kac, V.G., Peterson, D.H.

[80] Affine Lie algebras and Hecke modular forms, *Bull. Amer. Math. Soc.* **3** (1980), 1057–1061.

[81] Spin and wedge representations of infinite-dimensional Lie algebras and groups, *Proc. Natl. Acad. Sci. USA* **78** (1981), 3308–3312.

[83] Regular functions on certain infinite-dimensional groups, in: *Arithmetic and Geometry II* (M. Artin and J. Tate, eds.), Progress in Math. 36, Birkhäuser, 1983, pp. 141–166.

[84a] Infinite-dimensional Lie algebras, theta functions and modular forms, *Advances in Math.* **53** (1984), 125–264.

[84b] Unitary structure in representations of infinite-dimensional groups and a convexity theorem, *Invent. Math.* **76** (1984), 1–14.

[85a] Defining relations of certain infinite dimensional groups, in: *The Mathematical Heritage of Elie Cartan*, Lyon 1984, Astérisque, Numéro hors série, 1985, pp. 165–208.

[85b] Generalized invariants of groups generated by reflections, in: *Proceedings of the Conference Giornate di Geometria*, Rome 1984, Progress in Math. 60, Birkhäuser, 1985, pp. 231–249.

[85c] 112 constructions of the basic representation of the loop group of E_8, in: *Proceedings of the Conference "Anomalies, Geometry, Topology,"* Argonne 1985, World Sci., 1985, pp. 276–298.

[86] Lectures on the infinite wedge representation and the MKP hierarchy, in: *Séminaire de Math. Supérieures*, Les Presses de L' Université de Montréal 102, 1986, pp. 141–186.

[87] 'On geometric invariant theory for infinite-dimensional groups,' in: Lecture Notes in Mathematics 1271, Springer-Verlag, 1987, 109–142.

Kac, V.G., Popov, V.L., Vinberg, E.B.

[76] Sur les groupes linéaires algébriques dont l'algèbre des invariants est libre, *C. R. Acad. Sci. Paris* **283** (1976), 875–878.

Kac, V.G., Raina, A.K.

[87] *Highest Weight Representations of Infinite Dimensional Lie Algebras*, World Sci., Singapore, 1987.

Kac, V.G., Wakimoto, M.

[88a] Modular and conformal invariance constraints in representation theory of affine algebras, *Advances in Math.* **70** (1988), 156–236.

[88b] Modular invariant representations of infinite-dimensional Lie algebras and superalgebras, *Proc. Natl. Acad. Sci. USA* **85** (1988), 4956–4960.

[89a] Classification of modular invariant representations of affine algebras, in: *Infinite Dimensional Lie Algebras and Groups*, Adv. Ser. in Math. Phys. **7**, World Sci., 1989, pp. 138–177.

[89b] Exceptional hierarchies of soliton equations, *Proc. Symp. Pure Math.* **49**, Part 1, AMS, 1989, pp. 191–237.

[90] Branching functions for winding subalgebras and tensor products, *Acta Applicandae Math.* **21** (1990), 3–39.

[94] Integrable highest weight modules over affine superalgebras and number theory, in: *Lie Theory and Geometry (in honor of Bertram Kostant)*, Progress in Math. 123, Birkhäuser, 1994, pp. 415–456.

[2001] Integrable highest weight modules over affine superalgebras and Appell's function, *Comm. Math. Phys.* **215** (2001), 631–682.

Kac, V.G., Wang, S.P.

[92] On automorphisms of Kac–Moody algebras and groups, *Advances in Math.* **92** (1992), 129–195.

Kambayashi, T.

[96] Pro-affine algebras, ind-affine groups and the Jacobian problem, *J. Algebra* **185** (1996), 481–501.

Kashiwara, M.

[89] The flag manifold of Kac–Moody Lie algebra, in: *Algebraic Analysis, Geometry, and Number Theory* (J-I. Igusa, ed.), The Johns Hopkins University Press, Baltimore, 1989, pp. 161–190.

[90] Kazhdan–Lusztig conjecture for a symmetrizable Kac–Moody Lie algebra, in: *The Grothendieck Festschrift, Vol. II*, Progress in Math. 87, Birkhäuser, Boston, 1990, pp. 407–433.

[93] The crystal base and Littelmann's refined Demazure character formula, *Duke Math. J.* **71** (1993), 839–858.

Kashiwara, M., Tanisaki, T.

[95] Kazhdan–Lusztig conjecture for affine Lie algebras with negative level, *Duke Math. J.* **77** (1995), 21–62.

[96] Kazhdan–Lusztig conjecture for affine Lie algebras with negative level II: Nonintegral case, *Duke Math. J.* **84** (1996), 771–813.

Kassel, C.

[84] Kähler differentials and coverings of complex simple Lie algebras extended over a commutative algebra, *J. Pure Applied Algebra* **34** (1984), 265–275.

Kassel, C., Lascoux, A., Reutenauer, C.

[2001] The singular locus of a Schubert variety, preprint (2001).

Kazhdan, D., Lusztig, G.

[79] Representations of Coxeter groups and Hecke algebras, *Invent. Math.* **53** (1979), 165–184.

[80] Schubert varieties and Poincaré duality, in: *Proc. Symp. Pure Math.* **36**, AMS, 1980, pp. 185–203.

[88] Fixed point varieties on affine flag manifolds, *Israel J. Math.* **62** (1988), 129–168.

[93a] Tensor structures arising from affine Lie algebras. I, *J. of Amer. Math. Soc.* **6** (1993), 905–947.

[93b] Tensor structures arising from affine Lie algebras. II, *J. of Amer. Math. Soc.* **6** (1993), 949–1011.

[94a] Tensor structures arising from affine Lie algebras. III, *J. of Amer. Math. Soc.* **7** (1994), 335–381.

[94b] Tensor structures arising from affine Lie algebras. IV, *J. of Amer. Math. Soc.* **7** (1994), 383–453.

Kempf, G.

[76] Linear systems on homogeneous spaces, *Annals Math.* **103** (1976), 557–591.

[78] The Grothendieck–Cousin complex of an induced representation, *Advances in Math.* **29** (1978), 310–396.

Kempf, G., Knudsen, F., Mumford, D., Saint-Donat, B.

[73] *Toroidal Embeddings I*, Lecture Notes in Math. 339, Springer-Verlag, 1973.

Kitchloo, N.

[98] Topology of Kac–Moody groups, preprint (1998).

Koch, P.O.

[73] On the product of Schubert classes, *J. Diff. Geometry* **8** (1973), 349–358.

Kostant, B.

[59a] A formula for the multiplicity of a weight, *Trans. Amer. Math. Soc.* **93** (1959), 53–73.

[59b] The principal three-dimensional subgroup and the Betti numbers of a complex simple Lie group, *Amer. J. Math.* **81** (1959), 973–1032.

[61] Lie algebra cohomology and the generalized Borel–Weil theorem, *Annals Math.* **74** (1961), 329–387.

[63a] Lie group representations on polynomial rings, *Amer. J. Math.* **85** (1963), 327–404.

[63b] Lie algebra cohomology and generalized Schubert cells, *Annals Math.* **77** (1963), 72–144.

[66] Groups over $\mathbb{Z}$, in: *Proc. Symp. Pure Math.* **9**, AMS, 1966, pp. 90–98.

[70] 'Quantization and unitary representations I. Prequantization,' in: Lecture Notes in Mathematics 170, Springer-Verlag, 1970, pp. 87–207.

[76] On Macdonald's η-function formula, the Laplacian and generalized exponents, *Advances in Math.* **20** (1976), 179–212.

[99] A cubic Dirac operator and the emergence of Euler number multiplets of representations for equal rank subgroups, *Duke Math. J.* **100** (1999), 447–501.

Kostant, B., Kumar, S.

[86a] The nil Hecke ring and cohomology of G/P for a Kac–Moody group G, *Proc. Natl. Acad. Sci. USA* **83** (1986), 1543–1545.

[86b] The nil Hecke ring and cohomology of G/P for a Kac–Moody group G, *Adv. in Math.* **62** (1986), 187–237.

[87] T-equivariant K-theory of generalized flag varieties, *Proc. Natl. Acad. Sci. USA* **84** (1987), 4351–4354.

[90] T-equivariant K-theory of generalized flag varieties, *J. Diff. Geometry* **32** (1990), 549–603.

Kovacic, J.

[73] Pro-algebraic groups and the Galois theory of differential fields, *Amer. J. Math.* **95** (1973), 507–536.

Ku, J-M.

[87a] The Jantzen filtration of a certain class of Verma modules, *Proc. Amer. Math. Soc.* **99** (1987), 35–40.

[87b] Local submodules and the multiplicity of irreducible subquotients in category $\mathcal{O}$, *J. Algebra* **106** (1987), 403–412.

[88] On the uniqueness of embeddings of Verma modules defined by the Shapovalov elements, *J. Algebra* **118** (1988), 85–101.

[89] Structure of the Verma module $M(-\rho)$ over Euclidean Lie algebras, *J. Algebra* **124** (1989), 367–387.

[90] Relative version of Weyl–Kac character formula, *J. Algebra* **130** (1990), 191–197.

Kumar, S.

[84] Geometry of Schubert cells and cohomology of Kac–Moody Lie–algebras, *J. Diff. Geometry* **20** (1984), 389–431.

[85] Rational homotopy theory of flag varieties associated to Kac–Moody groups, in: *Infinite Dimensional Groups with Applications*, MSRI publ. 4, Springer-Verlag, 1985, pp. 233–273.

[86a] A homology vanishing theorem for Kac–Moody algebras with coefficients in the category $\mathcal{O}$, *J. Algebra* **102** (1986), 444–462.

[86b] Non-representability of cohomology classes by bi-invariant forms (gauge and Kac–Moody groups), *Comm. Math. Phys.* **106** (1986), 177–181.

[87a] Demazure character formula in arbitrary Kac–Moody setting, *Invent. Math.* **89** (1987), 395–423.

[87b] Extension of the category $\mathcal{O}^g$ and a vanishing theorem for the Ext functor for Kac–Moody algebras, *J. Algebra* **108** (1987), 472–491.

[88] Proof of the Parthasarathy–Ranga Rao–Varadarajan conjecture, *Invent. Math.* **93** (1988), 117–130.

[89a] Existence of certain components in the tensor product of two integrable highest weight modules for Kac–Moody algebras, in: *Infinite Dimensional Lie Algebras and Groups*, Adv. Ser. in Math. Phys. **7**, World Sci., 1989, pp. 25–38.

[89b] A refinement of the PRV conjecture, *Invent. Math.* **97** (1989), 305–311.

[90] Bernstein–Gelfand–Gelfand resolution for arbitrary Kac–Moody algebras, *Math. Annalen* **286** (1990), 709–729.

[92] Proof of Wahl's conjecture on surjectivity of the Gaussian map for flag varieties, *Amer. J. Math.* **114** (1992), 1201–1220.

[94a] The nil Hecke ring and singularity of Schubert varieties, in: *Lie Theory and Geometry (in honor of Bertram Kostant)*, Progress in Math. 123, Birkhäuser, 1994, pp. 497–507.

[94b] Symmetric and exterior powers of homogeneous vector bundles, *Math. Annalen* **299** (1994), 293–298.

[94c] Toward proof of Lusztig's conjecture concerning negative level representations of affine Lie algebras, *J. Algebra* **164** (1994), 515–527.

[96] The nil Hecke ring and singularity of Schubert varieties, *Invent. Math.* **123** (1996), 471–506.

[97a] Infinite Grassmannians and moduli spaces of G-bundles, in: *Vector Bundles on Curves – New Directions*, Lecture Notes in Math. 1649, Springer-Verlag, 1997, pp. 1–49.

[97b] Fusion product of positive level representations and Lie algebra homology, in: *Geometry and Physics* (J.E. Andersen et al., eds.), Lecture Notes in Pure and Applied Mathematics 184, Marcel Dekker, Inc., 1997, pp. 253–259.

[99] Homology of certain truncated Lie algebras, in: *Recent Developments in Quantum Affine Algebras and Related Topics* (N. Jing, K.C. Misra, eds.), Contemporary Mathematics 248, 1999, pp. 309–325.

Kumar, S., Littelmann, P.

[2002] Algebraization of Frobenius splitting via quantum groups, *Annals Math.* **155** (2002), 491–551.

Kumar, S., Narasimhan, M.S.

[97] Picard group of the moduli spaces of G-bundles, *Math. Annalen* **308** (1997), 155–173.

Kumar, S., Narasimhan, M.S., Ramanathan, A.

[94] Infinite Grassmannians and moduli spaces of G-bundles, *Math. Annalen* **300** (1994), 41–75.

Kumar, S., Nori, M.V.

[98] Positivity of the cup product in cohomology of flag varieties associated to Kac–Moody groups, *IMRN* **14** (1998), 757–763.

Kuniba, A., Misra, K.C., Okado, M., Takagi, T., Uchiyama, J.

[98] Characters of Demazure modules and solvable lattice models, *Nucl. Phys.* **B 510** [PM] (1998), 555–576.

Lakshmibai, V.

[87] Singular loci of Schubert varieties for classical groups, *Bull. Amer. Math. Soc.* **16** (1987), 83–90.

[95] Tangent spaces to Schubert varieties, *Math. Res. Letters* **2** (1995), 473–477.

Lakshimibai, V., Sandhya, B.

[90] Criterion for smoothness of Schubert varieties in $Sl(n)/B$, *Proc. Indian Acad. Sci. (Math. Sci.)* **100** (1990), 45–52.

Lakshmibai, V., Seshadri, C.S.

[84] Singular locus of a Schubert variety, *Bull. Amer. Math. Soc.* **11** (1984), 363–366.

[89] Standard monomial theory for $\widetilde{SL}_2$, in: *Infinite Dimensional Lie Algebras and Groups*, Adv. Ser. in Math. Phys. **7**, World Sci., 1989, pp. 178–234.

[91] Standard monomial theory, in: *Proceedings of the Hyderabad Conference on Algebraic Groups* (S. Ramanan, ed.), Manoj Prakashan, Madras, 1991, pp. 279–322.

Lakshmibai, V., Song, M.

[97] A criterion for smoothness of Schubert varieties in $Sp(2n)/B$, *J. Algebra* **189** (1997), 332–352.

Lakshmibai, V., Weyman, J.

[90] Multiplicities of points on a Schubert variety in a minuscule G/P, *Advances in Math.* **84** (1990), 179–208.

Lancaster, G., Towber, J.

[79] Representation-functors and flag-algebras for the classical groups. I, *J. Algebra* **59** (1979), 16–38.

Landweber, G.D.

[2001] Multiplets of representations and Kostant's Dirac operator for equal rank loop groups, *Duke Math. J.* **110** (2001), 121–160.

Lang, S.

[65] *Algebra*, Addison-Wesley Publishing Co., Inc., 1965.

Laszlo, Y., Sorger, C.

[97] The line bundles on the moduli of parabolic G-bundles over curves and their sections, *Ann. Sci. Éc. Norm. Sup. 30* (1997), 499–525.

Lepowsky, J.

[77] A generalization of the Bernstein–Gelfand–Gelfand resolution, *J. Algebra* **49** (1977), 496–511.

[78] Lie algebras and combinatorics, in: *Proceedings of ICM*, Helsinki, 1978, pp. 579–584.

[79] Generalized Verma modules, loop space cohomology and Macdonald-type identities, *Ann. Sci. Éc. Norm. Sup. 12* (1979), 169–234.

[82] Affine Lie algebras and combinatorial identities, in: *Proc. 1981 Rutgers Conference on Lie Algebras and Related Topics*, Lecture Notes in Math. 933, Springer-Verlag, 1982, pp. 130–156.

Lepowsky, J., Moody, R.V.

[79] Hyperbolic Lie algebras and quasi-regular cusps on Hilbert modular surfaces, *Math. Annalen* **245** (1979), 63–88.

Lepowsky, J., Wilson, R.L.

[78] Construction of the affine Lie algebra $A_1^{(1)}$, *Comm. Math. Phys.* **62** (1978), 43–53.

[85] The structure of standard modules, II. The case $A_1^{(1)}$, principal gradation, *Invent. Math.* **79** (1985), 417–442.

Leray, J.

[50] Sur l'homologie des groupes de Lie, des espaces homogènes et des espaces fibrés principaux, in: *Colloque de Topologie (Espace Fibrés)*, Bruxelles, 1950, pp. 101–115.

Levstein, F.

[88] A classification of involutive automorphisms of an affine Kac–Moody Lie algebra, *J. Algebra* **114** (1988), 489–518.

Littelmann, P.

[94] A Littlewood–Richardson rule for symmetrizable Kac–Moody algebras, *Invent. Math.* **116** (1994), 329–346.

[95] Paths and root operators in representation theory, *Annals Math.* **142** (1995), 499–525.

[98] Contracting modules and standard monomial theory for symmetrizable Kac–Moody algebras, *J. of Amer. Math. Soc.* **11** (1998), 551–567.

Liu, L.

[89] Kostant's formula for Kac–Moody Lie algebras, preprint (1989).

Looijenga, E.

[76] Root systems and elliptic curves, *Invent. Math.* **38** (1976), 17–32.

[80] Invariant theory for generalized root systems, *Invent. Math.* **61** (1980), 1–32.

Lu, J-H.

[99] Coordinates on Schubert cells, Kostant's harmonic forms, and the Bruhat–Poisson structure on G/B, *Transformation Groups* **4** (1999), 355–374.

Lusztig, G.

[83] Singularities, character formulas, and a q-analog of weight multiplicities, *Astérisque* **101–102** (1983), 208–229.

Lyons, D.

[96] *Some Results on Integral Weyl Invariants and Leray–Serre Spectral Sequence for Compact Lie Groups*, Ph. D. Thesis, University of North Carolina, Chapel Hill, 1996.

Macdonald, I.G.

[72] Affine root systems and Dedekind's η-function, *Invent. Math.* **15** (1972), 91–143.

[86] Kac–Moody–algebras, in: *Lie Algebras and Related Topics*, Canadian Math. Soc. Conference Proc. **5**, 1986, pp. 69–109.

[95] *Symmetric Functions and Hall Polynomials*, Second edition, Oxford Science Publications, The Clarendon Press, 1995.

Manivel, L.

[2001] Le lieu singulier des variétés de Schubert, *IMRN* **16** (2001), 849–871.

Marcuson, R.

[75] Tits' systems in generalized nonadjoint Chevalley groups, *J. Algebra* **34** (1975), 84–96.

Mathieu, O.

[86] Formules de Demazure–Weyl, et généralisation du théorème de Borel–Weil–Bott, *C. R. Acad. Sci. Paris* **303** (1986), 391–394.

[87] Classes canoniques des variétés de Schubert et algèbres affines, *C. R. Acad. Sci. Paris* **305** (1987), 105–107.

[88a] Formules de caractères pour les algèbres de Kac–Moody générales, *Astérisque* **159–160** (1988), 1–267.

[88b] Construction du groupe de Kac–Moody et applications, *C. R. Acad. Sci. Paris* **306** (1988), 227–230.

[89] Construction d'un groupe de Kac–Moody et applications, *Compositio Math.* **69** (1989), 37–60.

Matsumura, H.

[89] *Commutative Ring Theory*, Cambridge University Press, 1989.

Mehta, V.B., Ramanathan, A.

[85] Frobenius splitting and cohomology vanishing for Schubert varieties, *Annals Math.* **122** (1985), 27–40.

Mickelsson, J.

[87] Kac–Moody groups, topology of the Dirac determinant bundle, and fermionization, *Comm. Math. Phys.* **110** (1987), 173–183.

Milnor, J.

[63] *Morse Theory*, Annals of Mathematics Studies 51, Princeton Univ. Press, 1963.

[84] Remarks on infinite-dimensional Lie groups, in: *Relativity, Groups and Topology II* (B.S. DeWitt and R. Stora, eds.), North–Holland Physics Publishing, 1984, pp. 1007-1057.

Milnor, J.W., Stasheff, J.D.

[74] *Characteristic Classes*, Annals of Mathematics Studies 76, Princeton Univ. Press, 1974.

Mimura, H., Toda, H.

[91] *Topology of Lie Groups, I and II*, Translations of Mathematical Monographs 91, AMS, 1991.

Misra, K.C.

[84] Structure of certain standard modules for $A_n^{(1)}$ and the Rogers–Ramanujan identities, *J. Algebra* **88** (1984), 196–227.

[84a] Structure of some standard modules for $C_n^{(1)}$, *J. Algebra* **90** (1984), 385–409.

[88] Specialized characters for affine Lie algebras and the Rogers–Ramanujan identities, in: *Proc. of the Centenary Conference "Ramanujan Revisited,"* Academic Press, 1988, pp. 85–109.

Mitchell, S.A.

[88] Quillen's theorem on buildings and the loops on a symmetric space, *L'Enseignement Math.* **34** (1988), 123–166.

Mitzman, D.

[85] *Integral Bases for Affine Lie Algebras and Their Universal Enveloping Algebras*, Contemporary Math. 40, AMS, 1985.

Moody, R.V.

[67] Lie algebras associated with generalized Cartan matrices, *Bull. Amer. Math. Soc.* **73** (1967), 217–221.

[68] A new class of Lie algebras, *J. Algebra* **10** (1968), 211-230.

[69] Euclidean Lie algebras, *Canad. J. Math.* **21** (1969), 1432–1454.

[75] Macdonald identities and Euclidean Lie algebras, *Proc. Amer. Math. Soc.* **48** (1975), 43–52.

[79] Root systems of hyperbolic type, *Advances in Math.* **33** (1979), 144-160.

Moody, R.V., Pianzola, A.

[89] On infinite root systems, *Trans. Amer. Math. Soc.* **315** (1989), 661–696.

[95] *Lie Algebras with Triangular Decompositions*, Canadian Math. Soc. Series of Monographs and Advanced Texts, John Wiley & Sons, 1995.

Moody, R.V., Teo, K.L.

[72] Tits' systems with crystallographic Weyl groups, *J. Algebra* **21** (1972), 178–190.

Moody, R.V., Yokonuma, T.

[82] Root systems and Cartan matrices, *Canad. J. Math.* **34** (1982), 63–79.

Moreno, C.J., Rocha-Caridi, A.

[87] Rademacher-type formulas for the multiplicities of irreducible highest–weight representations of affine Lie algebras, *Bull. Amer. Math. Soc.* **16** (1987), 292–296.

Morita, J.

[79] Tits' systems in Chevalley groups over Laurent polynomial rings, *Tsukuba J. Math.* **3** (1979), 41–51.

[87] Commutator relations in Kac–Moody groups, *Proc. Japan Acad.* **63** (1987), 21–22.

Morita, J., Rehmann, U.

[90] A Matsumoto-type theorem for Kac–Moody groups, *Tôhoku Math. J.* **42** (1990), 537–560.

Moser-Jauslin, L.

[92] The Chow rings of smooth complete $SL(2)$-embeddings, *Compositio Math.* **82** (1992), 67–106.

Mumford, D.

[88] *The Red Book of Varieties and Schemes*, Lecture Notes in Math. 1358, Springer-Verlag, 1988.

Neidhardt, W.

[84] Irreducible subquotients and imbeddings of Verma modules, *Algebras, Groups and Geometries* **1** (1984), 127–136.

[86] The BGG resolution, character and denominator formulas, and related results for Kac–Moody algebras, *Trans. Amer. Math. Soc.* **297** (1986), 487–504.

[87] Verma module imbeddings and the Bruhat order for Kac–Moody algebras, *J. Algebra* **109** (1987), 430–438.

Neumann, F., Neusel, M.D., Smith, L.

[96] Rings of generalized and stable invariants of pseudoreflections and pseudoreflection groups, *J. Algebra* **182** (1996), 85–122.

Peterson, D.H.

[82] 'Affine Lie algebras and theta-functions,' in: Lecture Notes in Math. 933, Springer-Verlag, 1982, pp. 166-175.

[83] Freudenthal-type formulas for root and weight multiplicities, preprint (1983).

Peterson, D.H., Kac, V.G.

[83] Infinite flag varieties and conjugacy theorems, *Proc. Natl. Acad. Sci. USA* **80** (1983), 1778–1782.

Pittie, H.V.

[91] The integral homology and cohomology rings of SO(n) and Spin(n), *J. Pure Applied Algebra* **73** (1991), 105–153.

Polo, P.

[94] On Zariski tangent spaces of Schubert varieties, and a proof of a conjecture of Deodhar, *Indag. Math.* **5** (1994), 483–493.

Pressley, A.N.

[80] Decompositions of the space of loops on a Lie group, *Topology* **19** (1980), 65–79.

Pressley, A., Segal, G.

[86] *Loop Groups*, Clarendon Press, Oxford, 1986.

Raghunathan, M.S.

[2000] On spaces of morphisms of curves in algebraic homogeneous spaces, preprint (2000).

Ramanan, S., Ramanathan, A.

[85] Projective normality of flag varieties and Schubert varieties, *Invent. Math.* **79** (1985), 217–224.

Ramanathan, A.

[85] Schubert varieties are arithmetically Cohen–Macaulay, *Invent. Math.* **80** (1985), 283–294.

[87] Equations defining Schubert varieties and Frobenius splitting of diagonals, *Publ. Math. IHES* **65** (1987), 61–90.

Rao, S.E.

[89] A new class of unitary representations for affine Lie algebras, *J. Algebra* **120** (1989), 54–73.

Roberts, A.W., Varberg, D.E.

[73] *Convex Functions*, Pure and Applied math. 57, Academic Press, New York, 1973.

Rocha-Caridi, A.

[80] Splitting criteria for $\mathfrak{g}$-modules induced from a parabolic and the Bernstein-Gelfand-Gelfand resolution of a finite dimensional, irreducible $\mathfrak{g}$-module, *Trans. Amer. Math. Soc.* **262** (1980), 335–366.

Rocha-Caridi, A., Wallach, N.R.

[82] Projective modules over graded Lie algebras. I, *Math. Z.* **180** (1982), 151–177.

[83] Highest weight modules over graded Lie algebras: Resolutions, filtrations and character formulas, *Trans. Amer. Math. Soc.* **277** (1983), 133–162.

Ronan, M.

[89] *Lectures on Buildings*, Perspectives in Math. 7, Academic Press, Inc., 1989.

Rosenlicht, M.

[61] On quotient varieties and the affine embedding of certain homogeneous spaces, *Trans. Amer. Math. Soc.* **101** (1961), 211–223.

Rossmann, W.

[89] Equivariant multiplicities on complex varieties, *Astérisque* **173–174** (1989), 313–330.

Rousseau, G.

[89] Almost split K-forms of Kac–Moody algebras, in: *Infinite Dimensional Lie Algebras and Groups*, Adv. Ser. in Math. Phys. **7**, World Sci., 1989, pp. 70–85.

Rudin, W.

[73] *Functional Analysis*, McGraw-Hill Book Company, New York, 1973.

Šafarevič, I.R.

[82] On some infinite–dimensional groups. II, *Math. USSR Izvestija* **18** (1982), 185–194.

[94] *Basic Algebraic Geometry, I and II*, Second revised and expanded edition, Springer-Verlag, 1994.

[95] Letter to the editors: "On some infinite-dimensional groups. II," *Izv. Ross. Akad. Nauk. Ser. Mat.* **59** (1995), 224–224.

Samelson, H.

[52] Topology of Lie groups, *Bull. Amer. Math. Soc.* **58** (1952), 2–37.

Sanderson, Y.B.

[2000] On the connection between Macdonald polynomials and Demazure characters, *J. Alg. Comb.* **11** (2000), 269–275.

Segal, G., Wilson, G.

[85] Loop groups and equations of KdV type, *Publ. Math. IHES* **61** (1985), 5–65.

Seligman, G.B.

[89] 'Kac–Moody modules and generalized Clifford algebras,' in: Lecture Notes in Math. 1373, Springer-Verlag, 1989, pp. 124–143.

Sen, C.

[84] The homology of Kac–Moody Lie algebras with coefficients in a generalized Verma module, *J. Algebra* **90** (1984), 10–17.

Serre, J-P.

[56] Géométrie algébrique et géométrie analytique, *Ann. Inst. Fourier, Grenoble* **6** (1956), 1–42.

[58] Espaces fibrés algébriques, in: *Anneaux de Chow et Applications*, Séminaire C. Chevalley 2e année, 1958, pp. 1-01–1-37.

[60] Groupes proalgébriques, *Publ. Math. IHES* **7** (1960), 341–403.

[64] *Cohomologie Galoisienne*, Lecture Notes in Math. 5, Springer-Verlag, 1964.

[80] *Trees*, Springer-Verlag, 1980.

[87] *Complex Semisimple Lie Algebras*, Springer-Verlag, 1987.

[88] *Algebraic Groups and Class Fields*, GTM 117, Springer-Verlag, 1988.

[89] *Algèbre Locale–Multiplicités*, Lecture Notes in Math. 11, Springer-Verlag, 1989.

Seshadri, C.S.

[87] Line bundles on Schubert varieties, in: *Vector Bundles on Algebraic Varieties*, Bombay (1984), Oxford University Press, 1987, pp. 499–528.

Shapovalov, N.N.

[72] On a bilinear form on the universal enveloping algebra of a complex semisimple Lie algebra, *Funct. Anal. Appl.* **6** (1972), 307–312.

Slodowy, P.

[82] 'Chevalley groups over $\mathbb{C}((t))$ and deformations of simply elliptic singularities,' in: Lecture Notes in Math. 961, Springer-Verlag, 1982, pp. 285–301.

[84a] *Singularitäten, Kac–Moody–Liealgebren, Assoziierte Gruppen und Verallgemeinerungen*, Habilitationsschrift, Universität zu Bonn, 1984.

[84b] On the geometry of Schubert varieties attached to Kac–Moody Lie algebras, in: *Proceedings of the 1984 Vancouver Conference in Algebraic Geometry*, Canadian Math. Soc. Conf. Proc. **6**, pp. 405–442.

[85a] A character approach to Looijenga's invariant theory for generalized root systems, *Compositio Math.* **55** (1985), 3–32.

[85b] An adjoint quotient for certain groups attached to Kac–Moody algebras, in: *Infinite Dimensional Groups with Applications*, MSRI publ. **4**, Springer-Verlag, 1985, pp. 307–334.

[86] Beyond Kac–Moody algebras, and inside, in: *Lie Algebras and Related Topics*, Canadian Math. Soc. Conf. Proc. **5**, 1986, pp. 361–371.

Soergel, W.

[97] Kazhdan–Lusztig polynomials and a combinatoric for tilting modules, *Representation Theory* **1** (1997), 83–114.

Spanier, E.H.

[66] *Algebraic Topology*, McGraw-Hill Book Company, Inc., New York, 1966.

Springer, T.A.

[82] Quelques applications de la cohomologie d'intersection, *Astérisque* **92–93** (1982) (Séminaire Bourbaki, Exposé 589), 249–273.

[98] *Linear Algebraic Groups*, Second edition, Progress in Math. 9, Birkhäuser, 1998.

Steenrod, N.

[51] *The Topology of Fibre Bundles*, Princeton University Press, Princeton, NJ, 1951.

Steinberg, R.

[62] Générateurs, relations et revêtements de groupes algébriques, in: *Colloque Théorie des Groupes Algébriques*, Bruxelles, 1962, pp. 113-127.

[67] *Lectures on Chevalley Groups*, Mimeographed Notes, Yale University, New Haven, CT, 1967.

Teleman, C.

[98] Borel–Weil–Bott theory on the moduli stack of G-bundles over a curve, *Invent. Math.* **134** (1998), 1–57.

Thomason, R.W.

[87] Equivariant resolution, linearization, and Hilbert's fourteenth problem over arbitrary base schemes, *Advances in Math.* **65** (1987), 16–34.

Tits, J.

[65] Structures et groupes de Weyl, *Séminaire Bourbaki 17e année* (1964–65), no 288, pp. 288-01–288-15.

[66] Normalisateurs de tores I. Groupes de Coxeter étendus, *J. Algebra* **4** (1966), 96–116.

[74] *Buildings of Spherical Type and Finite BN-Pairs*, Lecture Notes in Math. 386, Springer-Verlag, 1974.

[81a] Définition par générateurs et relations de groupes avec BN-paires, *C. R. Acad. Sci. Paris* **293** (1981), 317–322.

[81b] Résumé de cours, *Annuaire Collège de France* **81**, 1980-81, pp. 75–87.

[82] Résumé de cours, *Annuaire Collège de France* **82**, 1981-82, pp. 91–106.

[85] 'Groups and group functors attached to Kac–Moody data,' in: Lecture Notes in Math. 1111, Springer-Verlag, 1985, pp. 193–223.

[87] Uniqueness and presentation of Kac–Moody groups over fields, *J. Algebra* **105** (1987), 542–573.

[89] Groupes associés aux algèbres de Kac–Moody, *Astérisque* **177-178** (1989) (Séminaire Bourbaki, Exposé 700), 7–31.

Tsuchiya, A., Ueno, K., Yamada, Y.

[89] Conformal field theory on universal family of stable curves with gauge symmetries, *Adv. Stud. Pure Math.* **19** (1989), 459–566.

Vasconcelos, W.V.

[67] Ideals generated by R-sequences, *J. Algebra* **6** (1967), 309–316.

Verlinde, E.

[88] Fusion rules and modular transformations in 2D conformal field theory, *Nucl. Phys.* **B 300** (1988), 360–376.

Verma, D.-N.

[68] Structure of certain induced representations of complex semisimple Lie algebras, *Bull. Amer. Math. Soc.* **74** (1968), 160–166.

[71] Möbius inversion for the Bruhat ordering on a Weyl group, *Ann. Sci. Éc. Norm. Sup.* **4** (1971), 393–398.

[75a] The role of affine Weyl groups in the representation theory of algebraic Chevalley groups and their Lie algebras, in: *Lie groups and their representations* (I.M. Gelfand, ed.), Summer School of the Bolyai János Math. Soc., Halsted Press, 1975, pp. 653–705.

[75b] Review of I.G. Macdonald's paper "Affine root systems and Dedekind's η-function," *Math Reviews* **50** (1975) (MR 50-9996).

Vershik, A.M., Gelfand, I.M., Graev, M.I.

[80] Representations of the group of functions taking values in a compact Lie group, *Compositio Math.* **42** (1980), 217–243.

Vinberg, E.B.

[71] Discrete linear groups generated by reflections, *Math. USSR Izvestija* **5** (1971), 1083–1119.

[76] The Weyl group of a graded Lie algebra, *Math. USSR Izvestija* **10** (1976), 463–495.

Vinberg, E.B., Kac, V.G.

[67] Quasi-homogeneous cones, *Mat. Zametki* **1** (1967), 347–354.

Wahl, J.

[91] Gaussian maps and tensor products of irreducible representations, *Manuscripta Math.* **73** (1991), 229–259.

Wallach, N.

[88] *Real Reductive Groups I*, Academic Press, Inc., 1988.

Warner, F.W.

[71] *Foundations of Differentiable Manifolds and Lie Groups*, Scott, Foresman and Co., 1971.

Whitney, H.

[72] *Complex Analytic Varieties*, Addison-Wesley Publishing Company, 1972.

Williams, F.L.

[78] The cohomology of semisimple Lie algebras with coefficients in a Verma module, *Trans. Amer. Math. Soc.* **240** (1978), 115–127.

Wilson, G.

[85] Infinite-dimensional Lie groups and algebraic geometry in soliton theory, *Phil. Trans. R. Soc. Lond.* **A 315** (1985), 393–404.

Wilson, R.L.

[82] 'Euclidean Lie algebras are universal central extensions,' in: Lecture Notes in Math. 933, Springer-Verlag, 1982, pp. 210–213.

Witten, E.

[88] Quantum field theory, Grassmannians, and algebraic curves, *Comm. Math. Phys.* **113** (1988), 529–600.

Wolper, J.S.

[89] A combinatorial approach to the singularities of Schubert varieties, *Advances in Math.* **76** (1989), 184–193.

[95] The Riccati flow and singularities of Schubert varieties, *Proc. Amer. Math. Soc.* **123** (1995), 703–709.

Index of Notation

Subsections

Index

Progress in Mathematics

Edited by:

Hyman Bass
Dept. of Mathematics
Columbia University
New York, NY 10027
USA

Joseph Oesterlé
Institut Henri Poincaré
11, rue Pierre et Marie Curie
75231 Paris Cedex 05
FRANCE

Alan Weinstein
Dept. of Mathematics
University of California
Berkeley, CA 94720
USA

Progress in Mathematics is a series of books intended for professional mathematicians and scientists, encompassing all areas of pure mathematics. This distinguished series, which began in 1979, includes authored monographs and edited collections of papers on important research developments as well as expositions of particular subject areas.

We encourage preparation of manuscripts in some form of $\mathrm{T_EX}$ for delivery in camera-ready copy which leads to rapid publication, or in electronic form for interfacing with laser printers or typesetters.

Proposals should be sent directly to the editors or to: Birkhäuser Boston, 675 Massachusetts Avenue, Cambridge, MA 02139, USA
or
Birkhäuser Verlag, 40–44 Viadukstrasse, Basel Ch-4051, Switzerland

124 AEBISCHER/BORER/KALIN/LEUENBERGER/REIMANN (eds). Symplectic Geometry
125 LUBOTZKY. Discrete Groups, Expanding Graphs and Invariant Measures
126 RIESEL. Prime Numbers and Computer Methods for Factorization
127 HÖRMANDER. Notions of Convexity
128 SCHMIDT. Dynamical Systems of Algebraic Origin
129 DIJKGRAAF/FABER/VAN DER GEER (eds). The Moduli Space of Curves
130 DUISTERMAAT. Fourier Integral Operators
131 GINDIKIN/LEPOWSKY/WILSON (eds). Functional Analysis on the Eve of the 21st Century. In Honor of the Eightieth Birthday of I. M. Gelfand, Vol. 1
132 GINDIKIN/LEPOWSKY/WILSON (eds.) Functional Analysis on the Eve of the 21st Century. In Honor of the Eightieth Birthday of I. M. Gelfand, Vol. 2
133 HOFER/TAUBES/WEINSTEIN/ZEHNDER (eds). The Floer Memorial Volume
134 CAMPILLO LOPEZ/NARVAEZ MACARRO (eds). Algebraic Geometry and Singularities
135 AMREIN/BOUTET DE MONVEL/GEORGESCU. C_0-Groups, Commutator Methods and Spectral Theory of N-Body Hamiltonians
136 BROTO/CASACUBERTA/MISLIN (eds). Algebraic Topology: New Trends in Localization and Periodicity
137 VIGNERAS. Représentations l-modulaires d'un groupe réductif p-adique avec $l = p$
138 BERNDT/DIAMOND/HILDEBRAND (eds). Analytic Number Theory, Vol. 1 In Honor of Heini Halberstam
139 BERNDT/DIAMOND/HILDEBRAND (eds). Analytic Number Theory, Vol. 2 In Honor of Heini Halberstam
140 KNAPP. Lie Groups Beyond an Introduction
141 CABANES (eds). Finite Reductive Groups: Related Structures and Representations
142 MONK. Cardinal Invariants on Boolean Algebras
143 GONZALEZ-VEGA/RECIO (eds). Algorithms in Algebraic Geometry and Applications
144 BELLAÏCHE/RISLER (eds). Sub-Riemannian Geometry

145 ALBERT/BROUZET/DUFOUR (eds). Integrable Systems and Foliations Feuilletages et Systèmes Intégrables
146 JARDINE. Generalized Etale Cohomology
147 DIBIASE. Fatou TypeTheorems. Maximal Functions and Approach Regions
148 HUANG. Two-Dimensional Conformal Geometry and Vertex Operator Algebras
149 SOURIAU. Structure of Dynamical Systems. A Symplectic View of Physics
150 SHIOTA. Geometry of Subanalytic and Semialgebraic Sets
151 HUMMEL. Gromov's Compactness Theorem for Pseudo-holomorphic Curves
152 GROMOV. Metric Structures for Riemannian and Non-Riemannian Spaces
153 BUESCU. Exotic Attractors: From Liapunov Stability to Riddled Basins
154 BÖTTCHER/KARLOVICH. Carleson Curves, Muckenhoupt Weights, and Toeplitz Operators
155 DRAGOMIR/ORNEA. Locally Conformal Kähler Geometry
156 GUIVARC'H/JI/TAYLOR. Compactifications of Symmetric Spaces
157 MURTY/MURTY. Non-vanishing of L-functions and Applications
158 TIRAO/VOGAN/WOLF (eds). Geometry and Representation Theory of Real and p-adic Groups
159 THANGAVELU. Harmonic Analysis on the Heisenberg Group
160 KASHIWARA/MATSUO/SAITO/SATAKE (eds). Topological Field Theory, Primitive Forms and Related Topics
161 SAGAN/STANLEY (eds). Mathematical Essays in Honor of Gian-Carlo Rota
162 ARNOLD/GREUEL/STEENBRINK. Singularities. The Brieskorn Anniversary Volume
163 BERNDT/SCHMIDT. Elements of the Representaiton Theory of the Jacobi Group
164 ROUSSARIE. Bifurcations of Planar Vector Fields and Hilbert's Sixteenth Problem
165 MIGLIORE. Introduction to Liaison Theory and Deficiency Modules
166 ELIAS/GIRAL/MIRO-ROIG/ZARZUELA (eds). Six Lectures on Commutative Algebra
167 FACCHINI. Module Theory
168 BALOG/KATONA/SZA'SZ/RECSKI (eds). European Congress of Mathematics, Budapest, July 22-26, 1996. Vol. I
169 BALOG/KATONA/SZA'SZ/RECSKI (eds). European Congress of Mathematics, Budapest, July 22-26, 1996. Vol. II
168/169 Sets Vols I, II
170 PATERSON. Groupoids, Inverse Semigroups, and their Operator Algebras
171 REZNIKOV/SCHAPPACHER (eds). Regulators in Analysis, Geometry and Number Theory
172 BRYLINSKI/BRYLINSKI/NISTOR/TSYGAN/XU (eds). Advances in Geometry
173 DRÄXLER/MICHLER/RINGEL (eds). Computational Methods for Representation of Groups and Algebras: Euroconference in Essen
174 GOERSS/JARDINE. Simplicial Homotopy Theory
175 BAÑUELOS/MOORE. Probabilistic Behavior of Harmonic Functions
176 BASS/LUBOTZKY. Tree Lattices
177 BIRKENHAKE/LANGE. Complex Tori
178 PUIG. On the Local Structure of Morita and Rickard Equivalence Between Brauer Blocks
179 RUIZ/JOSE. Differential Galois Theory and Non-integrability of Hamiltonian Systems
180 PATERNAIN. Geodesic Flows
181 HAUSER/LIPMAN/OORT/QUIRÓS. Resolution of Singularities: In Tribute to Oscar Zariski
182 BILLEY/LAKSHMIBAI. Singular Loci of Schubert Varieties
183 KAPOVICH. Hyperbolic Manifolds and Discrete Groups
84 du SAUTOY/SEGAL/SHALEV (eds). New Horizons in pro-p Groups
185 FARAUT/KANEYUKI/KORÁNYI/LU/ROOS. Analysis and Geometry on Complex Homogeneous Domains
186 KLEINERT. Units in Skew Fields
187 BOST/LOESER/RAYNAUD. Courbes Semi-stables et Groupe Fondamental en Géométrie Algébrique
188 DOLBEAULT/IORDAN/HENKIN/SKODA/TRÉPREAU (eds). Complex Analysis and Geometry: International Conference in Honor of Pierre Lelong
189 ANDRÉ/BALDASSARI. DeRham Cohomology of Differential Modules on Algebraic Varieties